MATHPOWER™ *Nine*

Authors and Reviewers

George Knill, B.Sc., M.S.Ed.
Hamilton, Ontario

Dino Dottori, B.Sc., M.S.Ed.
North Bay, Ontario

Rene Baxter, B.A., B.Ed., M.Ed.
Saskatoon, Saskatchewan

George Fawcett, B.A.
Hamilton, Ontario

Mary Lou Forest, B.Ed.
Edmonton, Alberta

David Kennedy, B.Sc., Ph.D.
Langley, British Columbia

Stan Pasko, B.A., B.Ed., M.Ed., Ph.D.
North Bay, Ontario

Harry Traini, B.Sc.
Hamilton, Ontario

McGraw-Hill Ryerson Limited

Toronto Montreal New York Auckland Bogotá Caracas
Lisbon London Madrid Mexico Milan New Delhi Paris
San Juan Singapore Sydney Tokyo

ISBN 0-07-549890-1

1 2 3 4 5 6 7 8 9 10 TRI 3 2 1 0 9 8 7 6 5 4

Printed and bound in Canada

Canadian Cataloguing in Publication Data

Main entry under title:

Mathpower™ 9

Includes index.

ISBN 0-07-549890-1

1. Mathematics. 2. Mathematics — Problems, exercises, etc. I. Knill, George., date

QA107.M37 1993 510 C93-094594-8

Executive Editor: Michael Webb
Editors: Jean Ford, Bonnie Di Malta, and Susan Marshall
Senior Supervising Editor: Carol Altilia
Permissions Editor: Jacqueline Donovan
Cover and Interior Design: Pronk&Associates
Electronic Assembly: Pronk&Associates
Cover Illustration: Doug Martin
Photo Researcher: Lois Browne / In a Word Communications Services

This book was manufactured in Canada using acid-free and recycled paper.

CONTENTS

CANADA

CHAPTER 9
Ratio, Rate, and Percent 305

CHAPTER 10
Statistics and Probability 357

CHAPTER 11
Perimeter, Area, and Volume 399

CHAPTER 12
Geometry 441

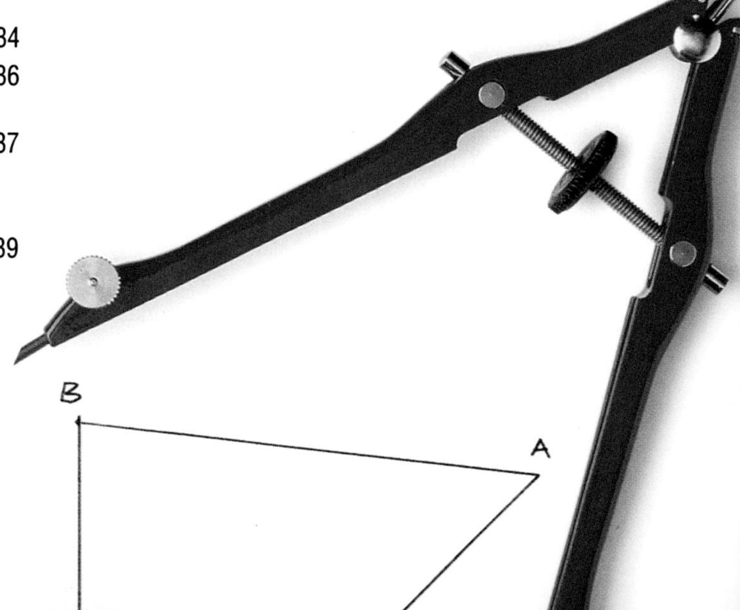

USING MATHPOWER™ 9

Each chapter contains a number of sections.
In a typical section, you find the following features.

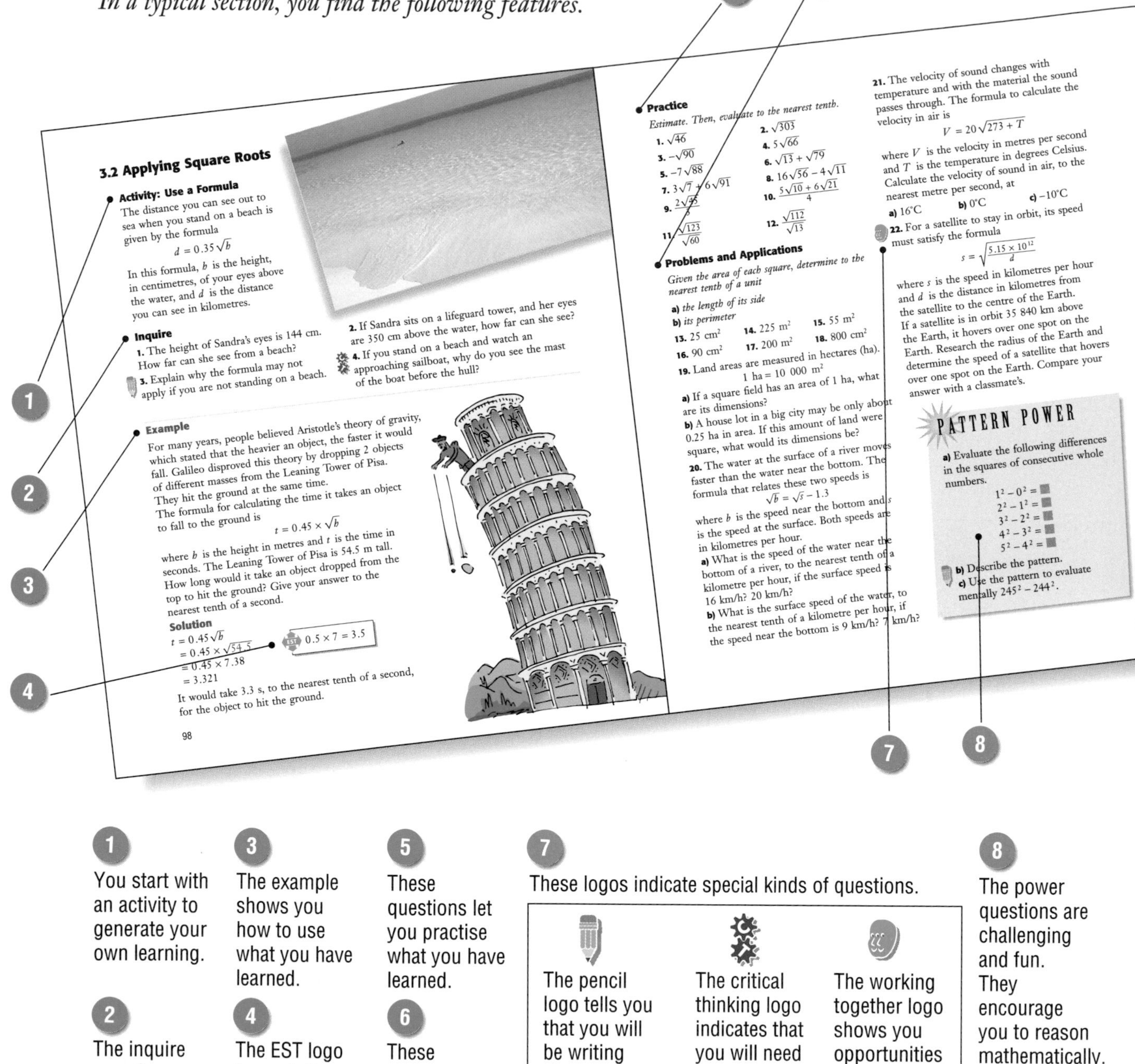

3.2 Applying Square Roots

Activity: Use a Formula

The distance you can see out to sea when you stand on a beach is given by the formula

$$d = 0.35\sqrt{h}$$

In this formula, h is the height, in centimetres, of your eyes above the water, and d is the distance you can see in kilometres.

Inquire

1. The height of Sandra's eyes is 144 cm. How far can she see from a beach?
2. If Sandra sits on a lifeguard tower, and her eyes are 350 cm above the water, how far can she see?
3. Explain why the formula may not apply if you are not standing on a beach.
4. If you stand on a beach and watch an approaching sailboat, why do you see the mast of the boat before the hull?

Example

For many years, people believed Aristotle's theory of gravity, which stated that the heavier an object, the faster it would fall. Galileo disproved this theory by dropping 2 objects of different masses from the Leaning Tower of Pisa. They hit the ground at the same time. The formula for calculating the time it takes an object to fall to the ground is

$$t = 0.45 \times \sqrt{h}$$

where h is the height in metres and t is the time in seconds. The Leaning Tower of Pisa is 54.5 m tall. How long would it take an object dropped from the top to hit the ground? Give your answer to the nearest tenth of a second.

Solution

$t = 0.45\sqrt{h}$
$= 0.45 \times \sqrt{54.5}$
$\doteq 0.45 \times 7.38$
$= 3.321$

EST $0.5 \times 7 = 3.5$

It would take 3.3 s, to the nearest tenth of a second, for the object to hit the ground.

98

Practice

Estimate. Then, evaluate to the nearest tenth.

1. $\sqrt{46}$
2. $\sqrt{303}$
3. $-\sqrt{90}$
4. $5\sqrt{66}$
5. $-7\sqrt{88}$
6. $\sqrt{13} + \sqrt{79}$
7. $3\sqrt{7} + 6\sqrt{91}$
8. $16\sqrt{56} - 4\sqrt{11}$
9. $\frac{2\sqrt{45}}{5}$
10. $\frac{5\sqrt{10} + 6\sqrt{21}}{4}$
11. $\frac{\sqrt{123}}{\sqrt{60}}$
12. $\frac{\sqrt{112}}{\sqrt{13}}$

Problems and Applications

Given the area of each square, determine to the nearest tenth of a unit

a) *the length of its side*
b) *its perimeter*

13. 25 cm^2
14. 225 m^2
15. 55 m^2
16. 90 cm^2
17. 200 m^2
18. 800 cm^2

19. Land areas are measured in hectares (ha).
$$1 \text{ ha} = 10\ 000 \text{ m}^2$$
a) If a square field has an area of 1 ha, what are its dimensions?
b) A house lot in a big city may be only about 0.25 ha in area. If this amount of land were square, what would its dimensions be?

20. The water at the surface of a river moves faster than the water near the bottom. The formula that relates these two speeds is
$$\sqrt{b} = \sqrt{s} - 1.3$$
where b is the speed near the bottom and s is the speed at the surface. Both speeds are in kilometres per hour.
a) What is the speed of the water near the bottom of a river, to the nearest tenth of a kilometre per hour, if the surface speed is 16 km/h? 20 km/h?
b) What is the surface speed of the water, to the nearest tenth of a kilometre per hour, if the speed near the bottom is 9 km/h? 7 km/h?

21. The velocity of sound changes with temperature and with the material the sound passes through. The formula to calculate the velocity in air is
$$V = 20\sqrt{273 + T}$$
where V is the velocity in metres per second and T is the temperature in degrees Celsius. Calculate the velocity of sound in air, to the nearest metre per second, at
a) 16°C **b)** 0°C **c)** −10°C

22. For a satellite to stay in orbit, its speed must satisfy the formula
$$s = \sqrt{\frac{5.15 \times 10^{12}}{d}}$$
where s is the speed in kilometres per hour and d is the distance in kilometres from the satellite to the centre of the Earth. If a satellite is in orbit 35 840 km above the Earth, it hovers over one spot on the Earth. Research the radius of the Earth and determine the speed of a satellite that hovers over one spot on the Earth. Compare your answer with a classmate's.

PATTERN POWER

a) Evaluate the following differences in the squares of consecutive whole numbers.

$1^2 - 0^2 = \blacksquare$
$2^2 - 1^2 = \blacksquare$
$3^2 - 2^2 = \blacksquare$
$4^2 - 3^2 = \blacksquare$
$5^2 - 4^2 = \blacksquare$

b) Describe the pattern.
c) Use the pattern to evaluate mentally $245^2 - 244^2$.

99

1 You start with an activity to generate your own learning.

2 The inquire questions help you learn from the activity.

3 The example shows you how to use what you have learned.

4 The EST logo in a solution indicates an estimate.

5 These questions let you practise what you have learned.

6 These questions let you apply and extend what you have learned.

7 These logos indicate special kinds of questions.

The pencil logo tells you that you will be writing about math.	The critical thinking logo indicates that you will need to think carefully before you answer a question.	The working together logo shows you opportunities for working with a classmate or in a larger group.

8 The power questions are challenging and fun. They encourage you to reason mathematically.

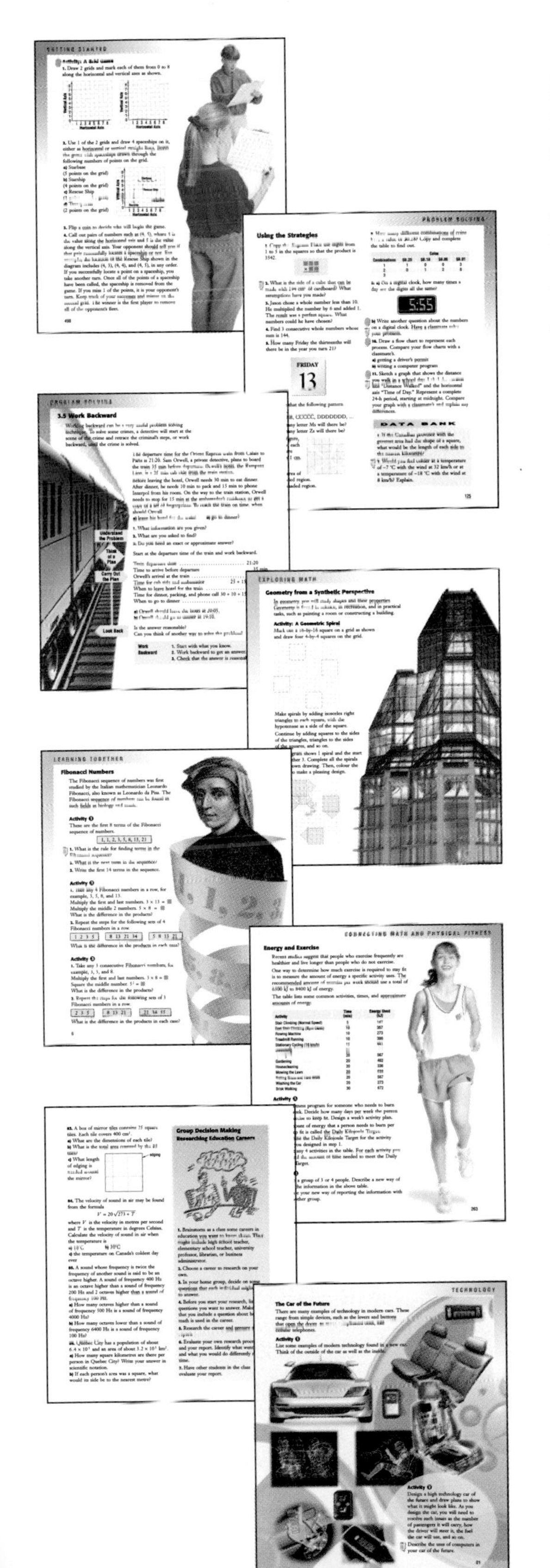

In the first 3 chapters of the book, there are 12 PROBLEM SOLVING sections that help you to use different problem solving strategies.

At or near the end of each chapter is a page headed PROBLEM SOLVING: Using the Strategies. To solve the problems on this page, you use the strategies you have studied. In all chapters except chapter 1, this page ends with DATA BANK questions. To solve them, you need to look up information in the Data Bank on pages 564 to 569.

There are 14 EXPLORING MATH pages before chapter 1. The activities on these pages let you explore 14 mathematical standards that will be essential for citizens of the twenty-first century.

A GETTING STARTED section begins each chapter. This section reviews, in a fun way, what you should know before you work on the chapter. Each GETTING STARTED section includes a set of Mental Math questions.

In LEARNING TOGETHER sections, you learn by completing activities with your classmates.

The TECHNOLOGY sections show you some uses of technology and how you can apply technology to solve problems.

Some sections are headed CONNECTING MATH AND.... In these sections, you apply math to other subject areas, such as art, language, science, and the environment.

Each chapter includes sets of questions called REVIEW and CHAPTER CHECK, so that you can test your progress.

At the end of each review is a column headed GROUP DECISION MAKING. Here, you work with your classmates to research careers and do other projects.

Chapters 4, 8, and 12 end with sets of questions headed CUMULATIVE REVIEW. These three reviews cover the work you did in chapters 1-4, 5-8, and 9-12.

The GLOSSARY on pages 598 to 602 helps you to understand mathematical terms.

On pages 570 to 597, there are ANSWERS to most of the problems in this book.

A Problem Solving Model

The world is full of mathematical problems. A problem exists when you are presented with a situation and are unable to make sense of it. To solve any problem, you must make decisions.

MATHPOWER™ 9 will help you to become actively involved in problem solving by providing the experiences and strategies you need.

George Polya was one of the world's best problem solvers. The problem solving model used in this book has been adapted from a model developed by George Polya.

The problem solving model includes the following 4 stages.

Understand the Problem

First, read the problem and make sure that you understand it. Ask yourself these questions.

- Do I understand all the words?
- What information am I given?
- What am I asked to find?
- Can I state the problem in my own words?
- Am I given enough information?
- Am I given too much information?
- Have I solved a similar problem?

Think of a Plan

Organize the information you need. Decide whether you need an exact or approximate answer. Plan how to use the information. The following list includes some of the problem solving strategies that may help.

- Act out the problem.
- Manipulate materials.
- Work backward.
- Account for all possibilities.
- Change your point of view.
- Draw a diagram.
- Look for a pattern.
- Make a table.
- Use a formula.
- Guess and check.
- Solve a simpler problem.
- Use logical reasoning.

Carry Out the Plan

Choose the calculation method you will use to carry out your plan.
Estimate the answer to the problem.
Carry out your plan, using paper and pencil, a calculator, a computer, or manipulatives.
Write a final statement that gives the solution to the problem.

Look Back

Check all your calculations.
Check your answer against the original problem. Is your answer reasonable? Does it agree with your estimate?
Look for an easier way to solve the problem.

EXPLORING MATH

The following 14 pages of activities will let you explore 14 mathematical standards that will be essential for citizens of the twenty-first century.

Problem Solving

Being a good problem solver is an important life skill. *MATHPOWER™ 9* will help you develop this skill.

Activity

Solve these problems.
In each case, describe the method you used.

1. How many different, even 4-digit house numbers can you make with the digits 2, 3, 4, and 5?

2. A basketball team is standing in a circle. The team members are evenly spaced and numbered in order, clockwise, from 1 up. Player number 5 is directly opposite player number 12. How many players are in the circle?

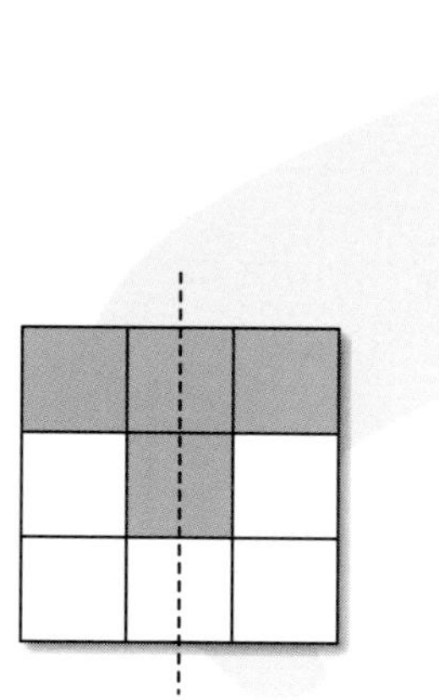

3. The diagram shows how to shade 4 of the 9 squares to give just 1 line of symmetry. In how many other ways can you shade 4 of the 9 squares to give 1 line of symmetry?

4. The number that represents the area of a square is twice the number that represents the sum of its sides. What is the side length of the square?

5. Take 6 dominoes, the 0/0, 0/1, 0/2, 1/1, 1/2, and 2/2. Arrange them in a square so that each side of the square contains the same number of dots.

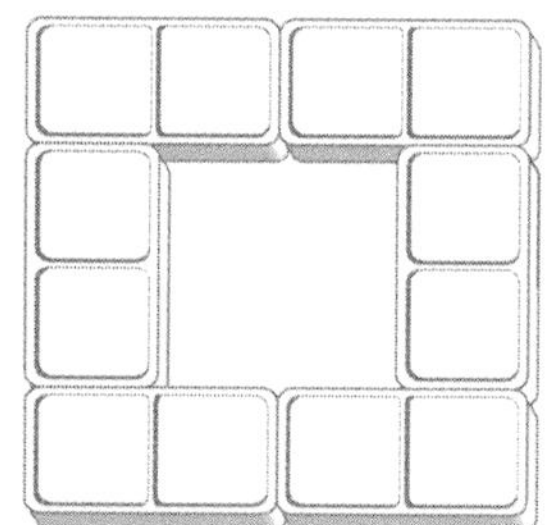

6. The diagram shows how to make 2 non-overlapping triangles with 4 straight lines. How many non-overlapping triangles can you make with 5 straight lines?

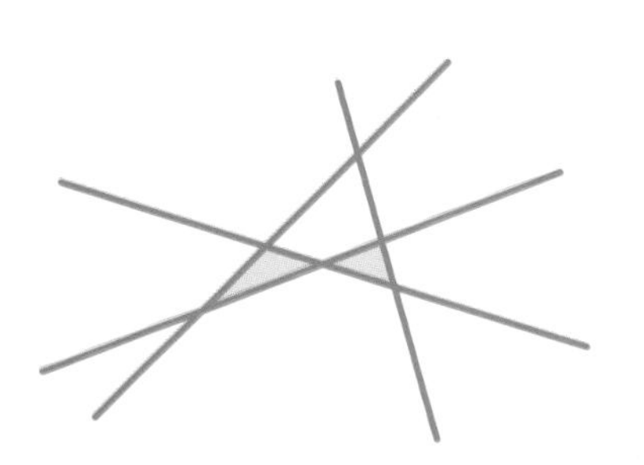

Mathematics as Communication

As you develop new ideas in any field, it is important to be able to communicate these ideas to others.

Activity ❶

There are 2 research camps located in the desert at A and B. Each camp has a jeep. A water station is to be located so that the total distance the 2 jeeps must travel to the water and back is as short as possible.

1. What is the total distance the jeeps travel if the water station is placed at position 3? at position 4? at position 1? at A? at B?

1 km 1 km 1 km 1 km 1 km 1 km
A 1 2 3 4 5 B

2. Where should the water station be placed so that the total distance travelled by the jeeps is the shortest?

Activity ❷

There are now 3 research camps, A, B, and C, in the desert. Each camp has a jeep.

1. What is the total distance travelled by the jeeps to get water if the water station is placed at position 3? at position 2? at A? at B? at C?

1 km 1 km 1 km 1 km 1 km 1 km
A 1 B 2 3 4 C

2. Where should the water station be placed so that the total distance travelled by the jeeps is the shortest?

Activity ❸

Where should the water station be placed so that the total distance travelled by the jeeps is the shortest in each diagram?

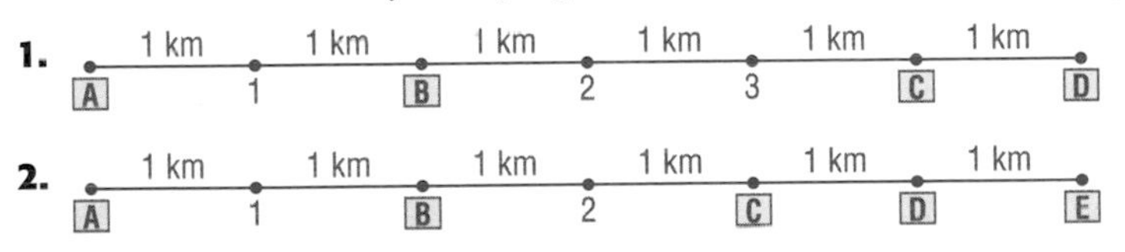

Activity ❹

1. Write a rule for finding where the station should be placed if there is an even number of camps.

2. Write a rule for finding where the station should be placed if there is an odd number of camps.

3. Test your rules for 6 camps and 7 camps.

Mathematics as Reasoning

The ability to reason logically, to think your way through a problem, is a skill you can develop.

Activity ❶

For each card, use all the numbers only once. Use as many of the operations +, −, ×, ÷, and brackets ( ) as you wish. Write a number sentence with the value 24 for each card.

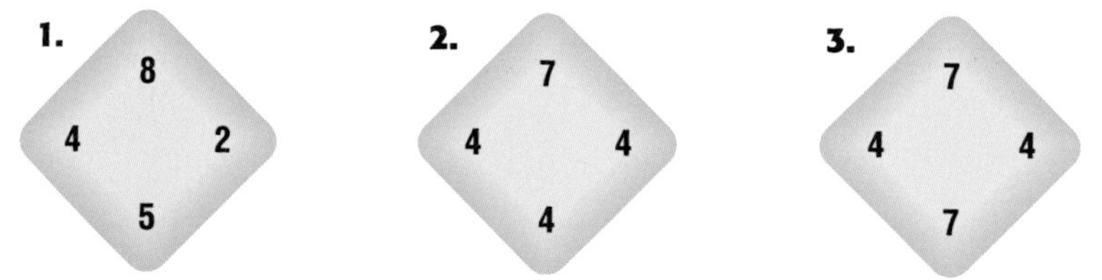

Activity ❷

There are three books on the counter. The books are blue, green, and black. The titles of the books are *MATHPOWER*™, *SCIENCEPOWER*, and *COMPUTERPOWER*. The books belong to Sari, Terri, and Dmitri. Use the clues to reason who owns which book.

1. Sari's book is not about computers.

2. Dmitri's book is not blue or green.

3. *SCIENCEPOWER* is not green.

4. *MATHPOWER*™ is a blue book.

Copy and complete these tables to sort out the owners, colours, and book titles.

	Colour of Book		
	Blue	Green	Black
Sari			
Terri			
Dmitri			

	Colour of Book		
	Blue	Green	Black
MATHPOWER™			
SCIENCEPOWER			
COMPUTERPOWER			

	Colour of Book	Title of Book
Sari		
Terri		
Dmitri		

Activity ❸

Use the table to make up the clues to a problem that can be solved by reasoning.

	Bicycles	
Jim	green	3–speed
Ali	blue	5–speed
Sue	red	10–speed

Mathematical Connections

Mathematics can be found in many places. The following activities explore some of them.

Activity ❶

Work with a classmate to plan a 2-week vacation.

You will be travelling by car. Decide where you want to go and the route you are going to take to get there. You will need some maps. Assume that you can travel for a maximum of 8 h/day and that you will sleep in motels or hotels. You will want to spend at least 7 days at your vacation spot.

Keep a list of what you will spend every day for gas, food, lodging, and entertainment. Determine the total cost of the vacation.

Activity ❷

1. Select a sport and list 5 ways in which mathematics is used in this sport.
2. Describe how the sport would be different if it did not include any mathematics.

Activity ❸

The people responsible for playing the music you listen to on the radio use mathematics every day. List some of the ways in which mathematics is used at a radio station.

Activity ❹

There are many geometric shapes that you see and use every day.

1. List 5 geometric shapes found in your classroom.
2. List 5 geometric shapes you see on your way to school.

Activity ❺

There are many examples of mathematics in nature.

The chambers in a nautilus shell form a spiral.

Animals and plants have lines of symmetry.

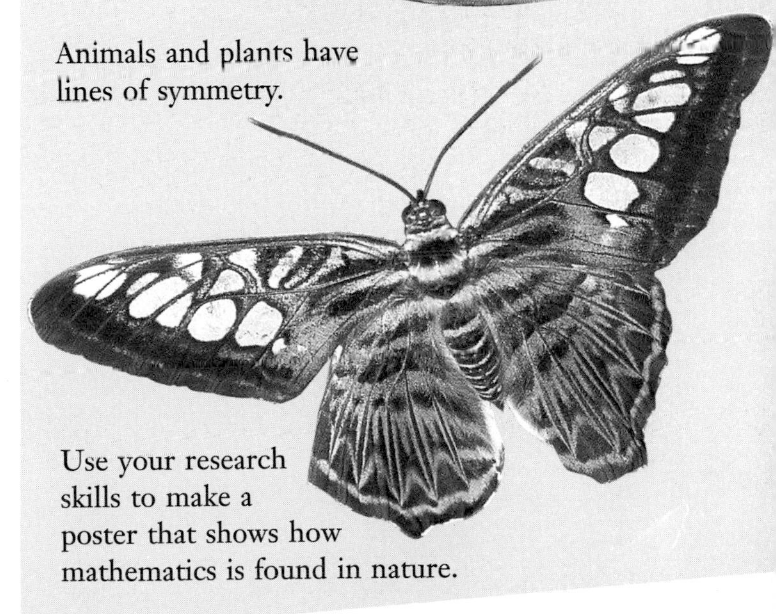

Use your research skills to make a poster that shows how mathematics is found in nature.

Algebra

Algebra is the language of mathematics.

Activity ❶

The concept of a variable is one of the most important in algebra. A variable can represent different numbers at different times.

In the following equation there are two variables, a triangle and a square.

▲ + ▲ + ■ = ■ + ■

If you replace the ▲ with 1 and the ■ with 2, you get

$$1 + 1 + 2 = 2 + 2$$

This is a true statement.

If you replace the ▲ with 2 and the ■ with 3, you get

$$2 + 2 + 3 = 3 + 3$$

This is not a true statement.

1. a) Find 3 other pairs of whole numbers that can replace the ▲ and the ■ to make a true statement. Put the pairs in a table.

b) Describe the pattern.

2. Repeat the activity for the following equations. If you cannot find 3 pairs of whole numbers that make true statements, explain why.

a) ▲ + ▲ + ▲ + ■ = ■ + ■

b) ▲ + ▲ = 2 × ■

c) ▲ + 2 = ■ + ■

d) ▲ + ▲ = ■ + ■ + ■

e) ▲ + ■ = ▲ × ■

f) ▲ × ▲ × ■ = ▲ + ▲ + ■

Activity ❷

Find the value of the letter to make each equation true.

1. $m + 8 = 15$

2. $x - 5 = 7$

3. $6 - t = 4$

4. $x + x + 5 = 17$

5. $n \div 2 = 10$

6. $2 \times t + 7 = 11$

Activity ❸

1. $s + u = 15$
$r + s + t = 15$
If $r = 6$ and $t = 4$, find u.

2. $y + z = 5$
$w + x = 7$
$x + y = 6$
If $w = 3$, find z.

3. $a = b$
$x = a + b + c + d$
$c = d$
If $x = 10$ and $b = 2$, find d.

4. $w = y$
$x = w + z$
$z = y + m$
If $w = 3$ and $x = 10$, find m.

5. $y + w = u$
$x = y - z$
If $x = 8$, $z = 3$, and $u = 11$, find w.

Functions

In much of the mathematics you do, you study patterns. There are many patterns in nature.

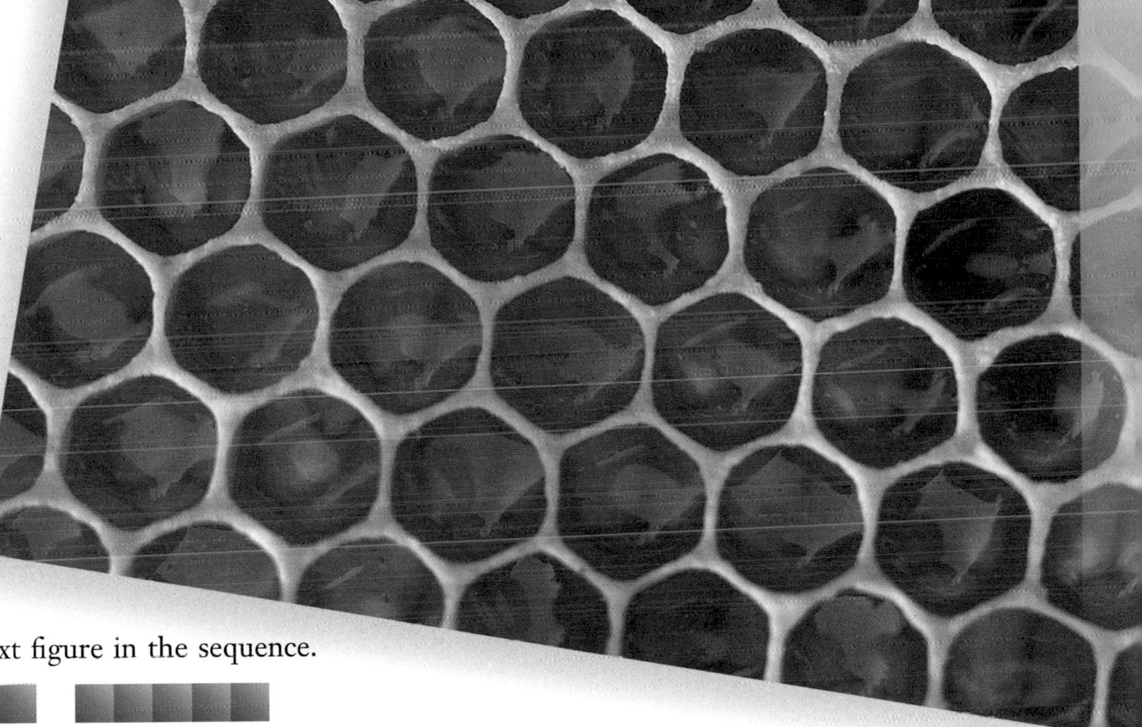

Activity ❶

1. a) Draw the next figure in the sequence.

b) Copy and complete the table for the sequence.

Figure	Perimeter
1	4
2	8
3	12
4	
5	
10	
100	

c) Describe the pattern in words.

2. Repeat question 1 for these sequences.

a)

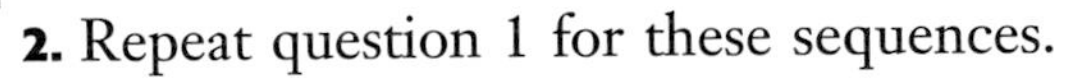

b)

c)

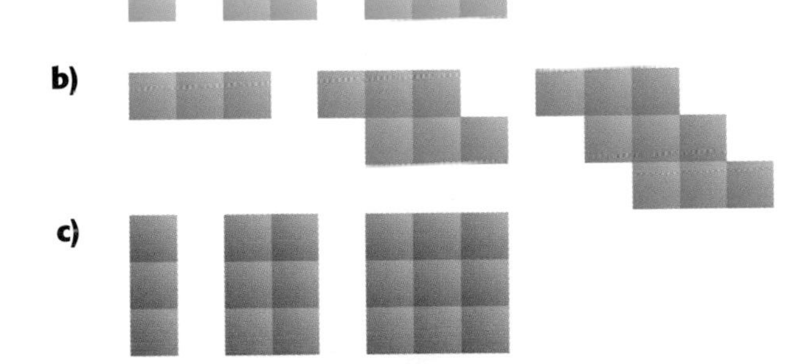

Activity ❷

Copy each table.

Describe the rule that lets you find y if you know x or find x if you know y.

Use the rule to complete each table.

1.

x	y
9	1
15	7
22	
8.1	
	11
	3.3

2.

x	y
8	4
36	18
40	
6.4	
	8.1
	9.7

3.

x	y
3	17
9	11
13	7
4	
18	
	6.2
	15.5

4.

x	y
1	3
5	11
9	19
6	
12	
2.5	
	9

Geometry from a Synthetic Perspective

In geometry, you will study shapes and their properties. Geometry is found in science, in recreation, and in practical tasks, such as painting a room or constructing a building.

Activity: A Geometric Spiral

Mark out a 16-by-16 square on a grid as shown and draw four 4-by-4 squares on the grid.

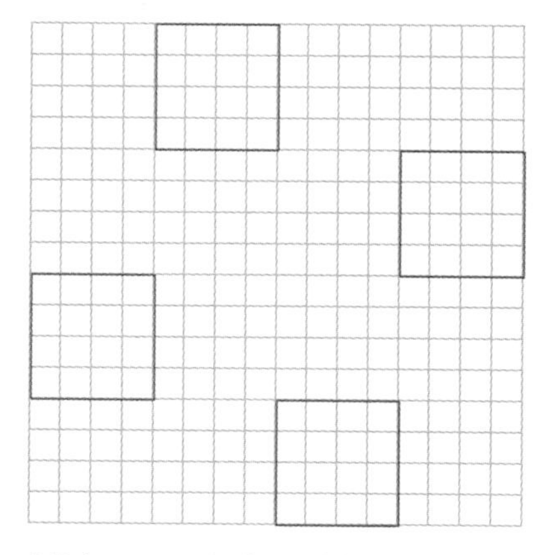

Make spirals by adding isosceles right triangles to each square, with the hypotenuse as a side of the square.

Continue by adding squares to the sides of the triangles, triangles to the sides of the squares, and so on.

The diagram shows 1 spiral and the start of the other 3. Complete all the spirals in your own drawing. Then, colour the figures to make a pleasing design.

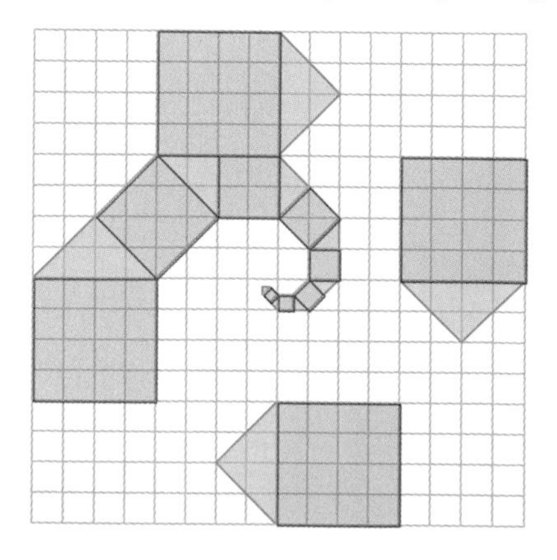

Geometry from an Algebraic Perspective

The study of geometry through the use of transformations has changed geometry from static to dynamic. For example, geometry can now be used to make moving images on film or videotape.

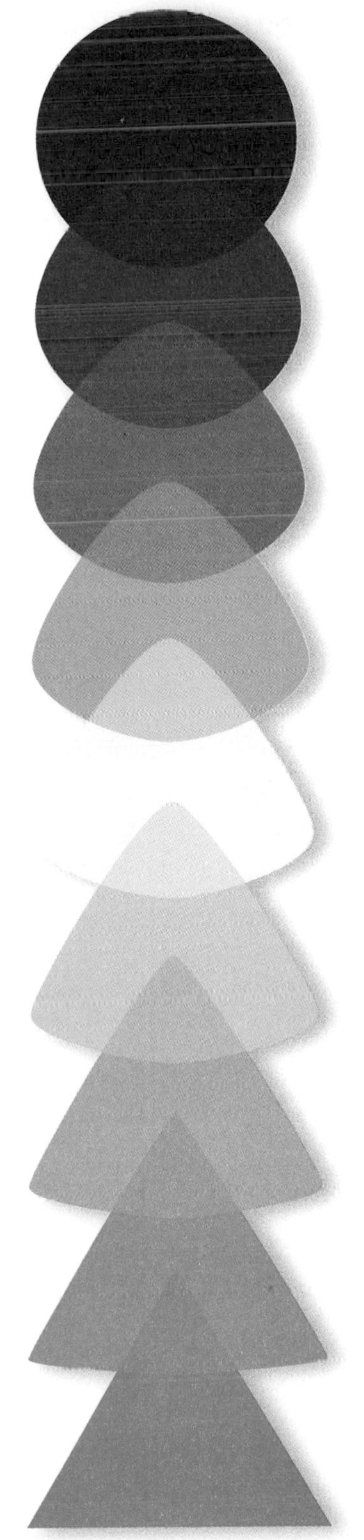

Activity ❶ Patterns

Copy each pattern onto a grid and continue it.

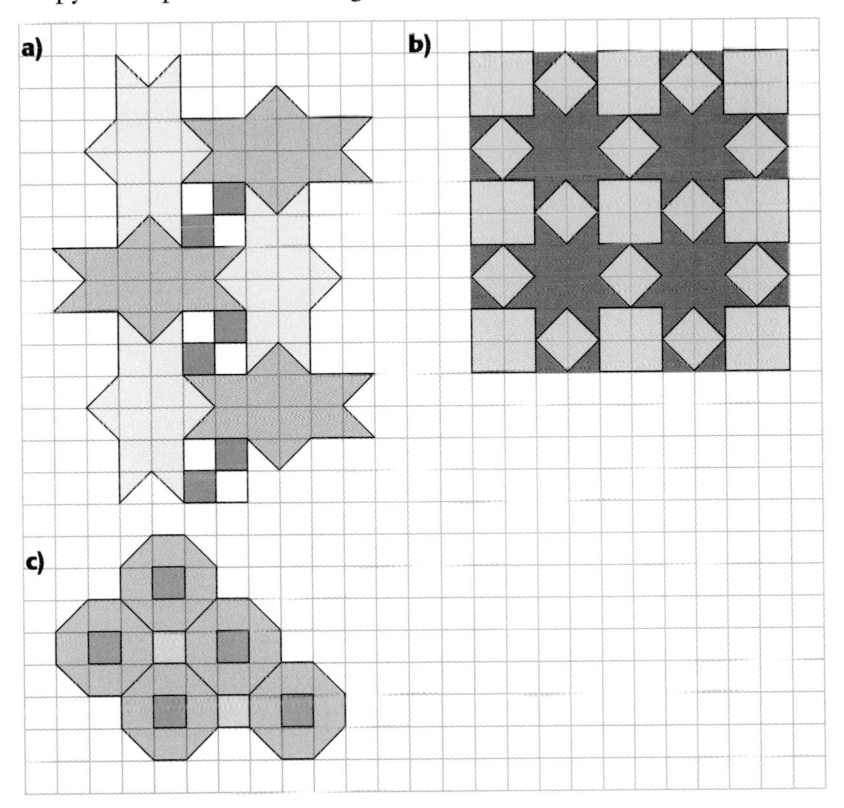

Activity ❷ Reflections

Copy each diagram onto grid paper and draw its reflection in the reflection line.

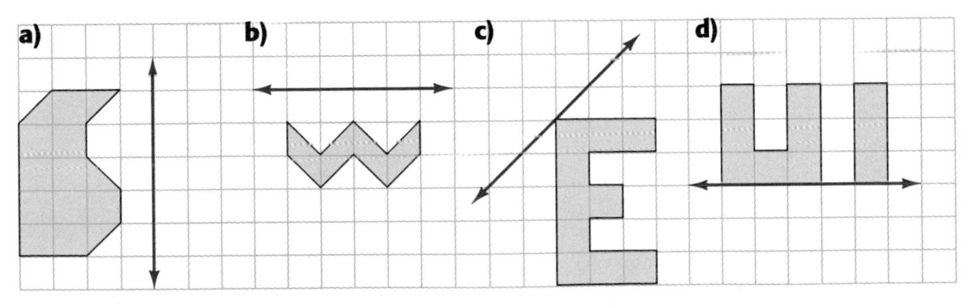

Trigonometry

Trigonometry is the study of the measures of triangles.
Trigonometry has many uses, including surveying and navigation.

Activity

1. In each of these right triangles, ∠B is 27° and ∠A is 63°.
Measure sides BC and AC of each triangle.

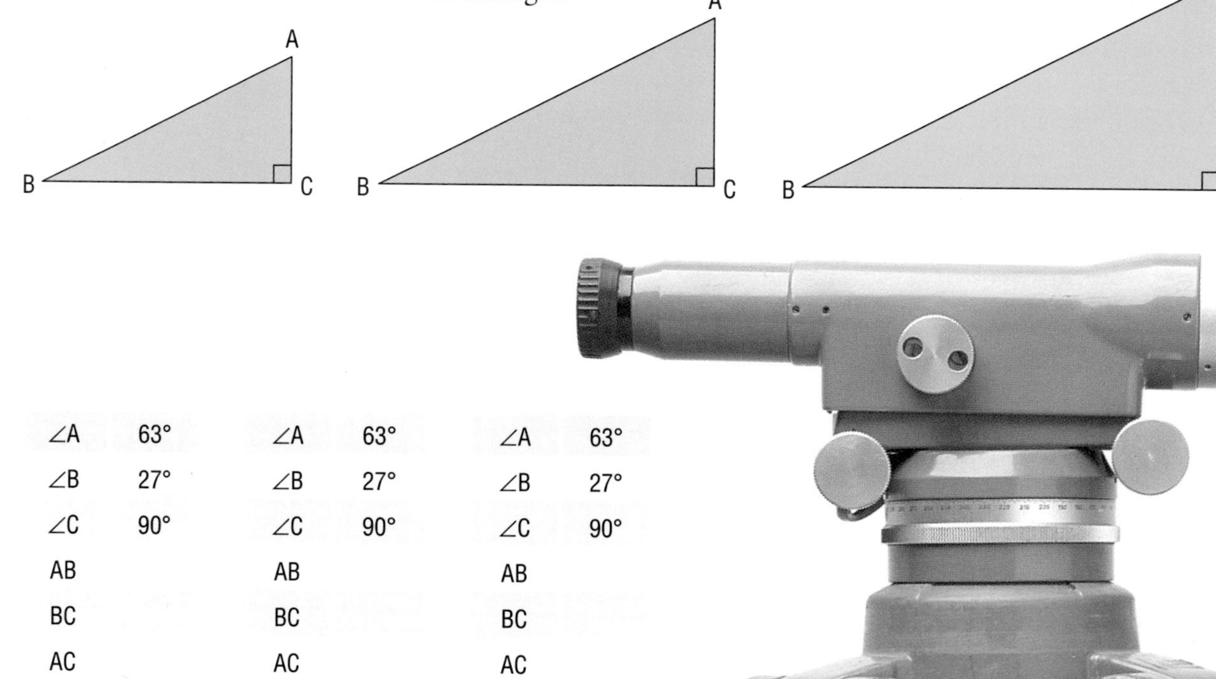

∠A	63°
∠B	27°
∠C	90°
AB	
BC	
AC	

∠A	63°
∠B	27°
∠C	90°
AB	
BC	
AC	

∠A	63°
∠B	27°
∠C	90°
AB	
BC	
AC	

2. a) Calculate the quotient $\frac{\text{BC}}{\text{AC}}$ to the nearest hundredth for each triangle.
b) What do you notice about the values of the quotient?

3. Use your result from question 2 and the information in the diagram to find the height of the flagpole.

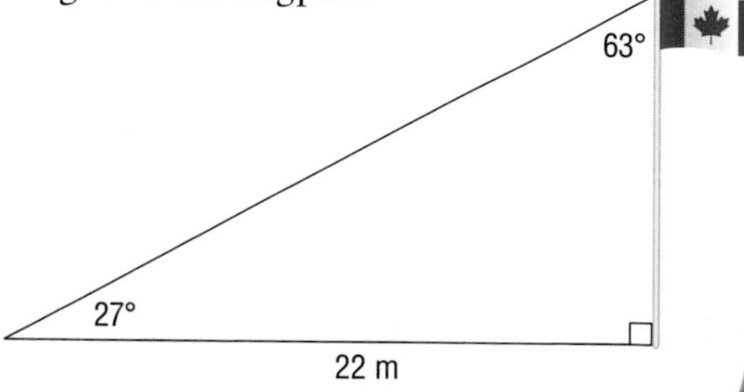

Statistics

Statistics play a very important part in our lives. They are found everywhere, even in the machines we use and the games we play.

Activity ❶ The QWERTY Keyboard

The picture shows a manual typewriter first manufactured in 1872. The keys were arranged in this way to avoid the locking of the type bars of the most frequently used letters. This keyboard is often called the QWERTY keyboard.

The table gives the average number of times a letter appears in 100 letters of written material.

A	8.2	H	5.3	O	8	V	0.9
B	1.4	I	6.5	P	2	W	1.5
C	2.8	J	0.1	Q	0.1	X	0.2
D	3.8	K	0.4	R	6.8	Y	2
E	13	L	3.4	S	6	Z	0.05
F	3	M	2.5	T	10.5		
G	2	N	7	U	2.5		

1. Draw a QWERTY keyboard.

2. On each key, write the number of times that letter will appear in 100 letters of written material.

3. Explain the arrangement of the letters on the keyboard.

4. How might you have placed the letters differently?

5. Why is the QWERTY keyboard still used today?

Activity ❷ Breaking Codes

Codes or puzzles are constructed by letting one letter stand for another.

1. How could the table in Activity 1 be used to help decipher codes or puzzles?

2. Make up a coded message and have a classmate decipher it.

Activity ❸ SCRABBLE®

1. List the letters of the alphabet. Beside each letter, write how many tiles of that letter are found in a SCRABBLE® game and the point value of that letter.

2. Explain why you think the makers of the game used these numbers of tiles and this point system.

3. Choose any 7 tiles to make the word that will give you the most points. Compare your word with your classmates'.

Probability

Probability has been called the mathematics of chance. Knowing about probability will help you make informed decisions about the likelihood of events.

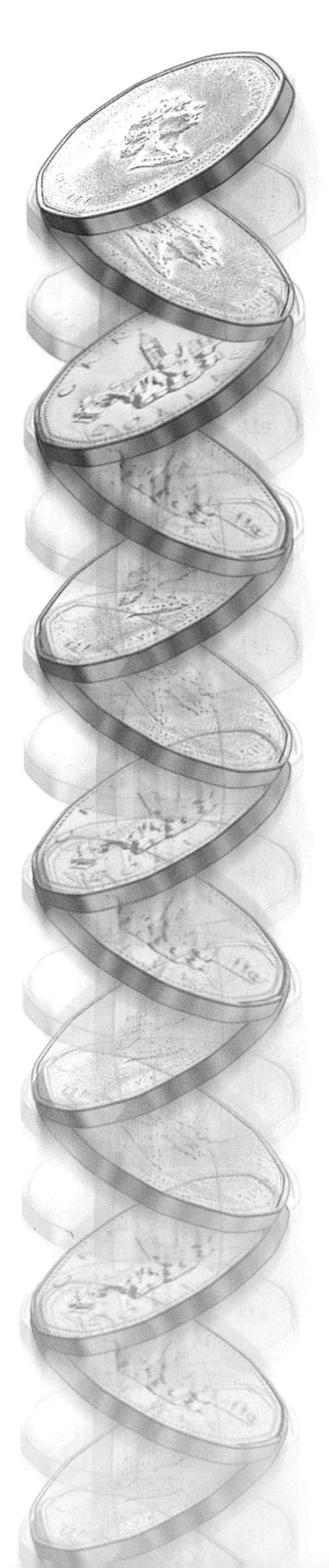

Activity ❶

Suppose there are 2 blue marbles, 3 red marbles, and 5 white marbles in a bag. You select 1 marble, look at the colour, and return it to the bag.

1. a) The probability of picking a red marble is $\frac{3}{10}$. Why?
b) What is the probability of picking a blue marble? a white marble?

2. What percent of the time should you pick a white marble?

3. If you selected a marble 100 times, how many times should you pick a white marble?

Activity ❷

1. Estimate the number of times a baseball cap will land right side up if you toss it 10 times.

2. Toss a baseball cap 10 times and compare the result with your estimate.

3. What is the probability that a baseball cap will land right side up?

Activity ❸

Some homes have automatic garage doors that are operated by a garage-door opener. An opener has 8 switches. Each switch can be set in 2 positions, on and off.

What is the chance that 2 people have the same code for their garage-door openers? To find out, start with simple cases. What if there is only 1 switch that can be set to on or off? In this case, there are only 2 possibilities. This means that there is 1 chance out of 2 that both people will have the same code.

What if there are 2 switches?

When switch 1 is on, switch 2 can be on or off. When switch 1 is off, switch 2 can be on or off. This means that there are 4 possibilities in total, so the chance is now 1 in 4. What if there are 3 switches? 4 switches? 8 switches?

SWITCH 1 SWITCH 2

on → on
on → off
off → on
off → off

Mathematics and Counting (Discrete Mathematics)

Discrete mathematics is the study of things that can be counted.

Activity ❶

1. If you follow the arrows, how many different paths spell COUGAR?

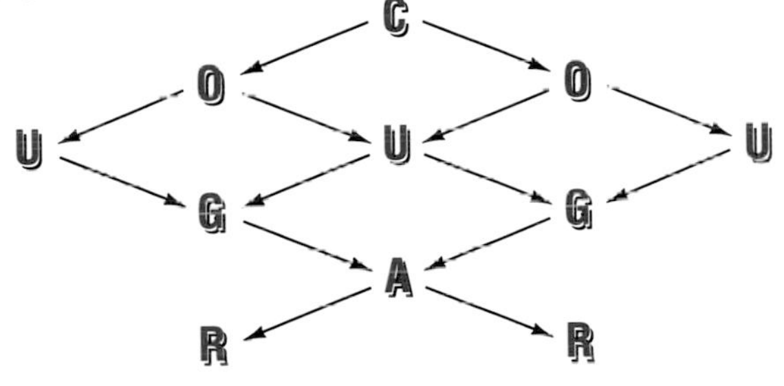

2. Make up your own path puzzle using the name of a rock group. Ask a classmate to solve your puzzle.

Activity ❷

There are 2 roads from Hart to Adams.
There are 4 roads from Adams to Young.
How many different routes are there from Hart to Young?

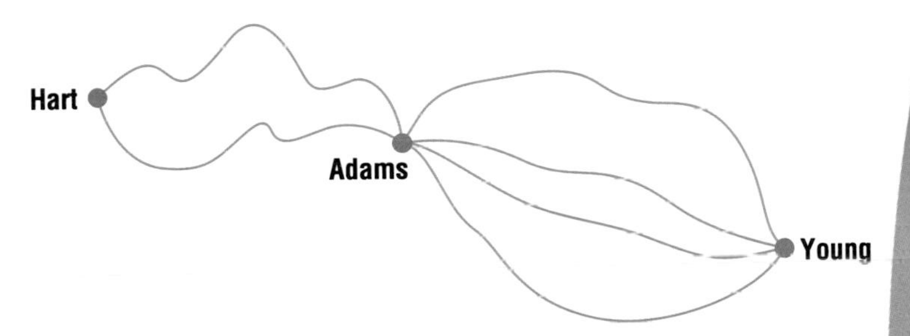

Activity ❸

Four people, Lisa, Ben, Ona, and Dino, are going on a trip in a car. The car will seat 2 in the front and 2 in the back. Only Lisa and Dino can drive. In how many different ways can the 4 people sit in the car?

Activity ❹

Copy each sequence and predict the next 3 terms.

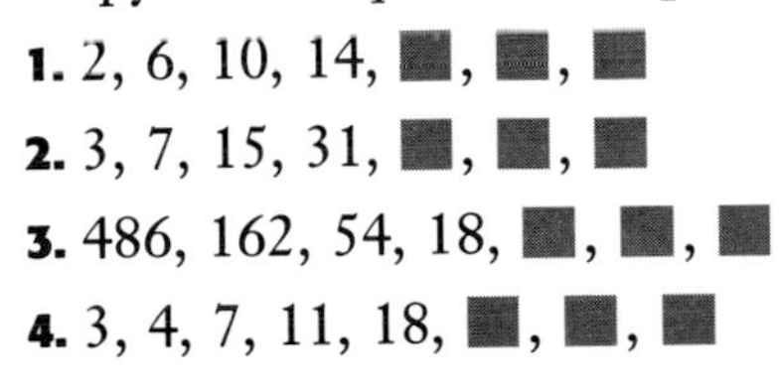

1. 2, 6, 10, 14, ■, ■, ■

2. 3, 7, 15, 31, ■, ■, ■

3. 486, 162, 54, 18, ■, ■, ■

4. 3, 4, 7, 11, 18, ■, ■, ■

Investigating Limits

A very important branch of mathematics is known as calculus. It is applied in many fields, including the sciences, the social sciences, and business. The study of limits plays an essential part in understanding calculus.

Activity 1

Consider this series.

$\frac{1}{2} + \frac{1}{4} + \frac{1}{8} + \frac{1}{16} + \frac{1}{32} + \cdots$

The series can be summed in parts as shown.

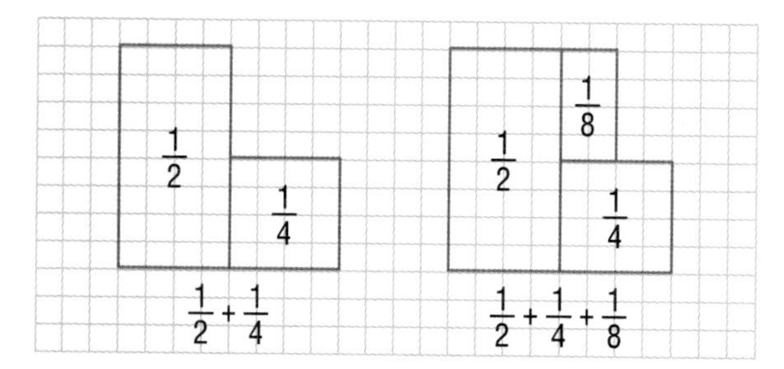

1. Draw a 16-by-16 grid and continue the pattern.

2. What number does the sum of the series approach? Check your answer with a calculator.

Activity 2

1. Draw a circle with radius 5 cm.

2. Divide the circle into 10 rectangles, each 1 cm wide.

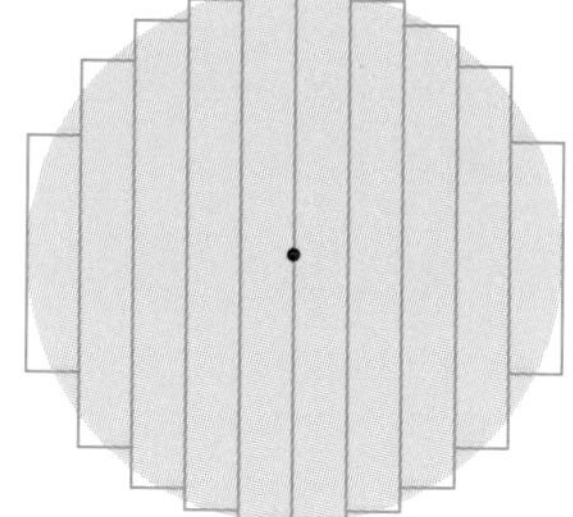

3. Measure the length of each rectangle and find the area of each rectangle.

4. Find the sum of the areas of the rectangles.

5. Find the area of the circle using the formula $A = \pi r^2$. You can assume that π is about 3.14.

6. Compare your answers from steps 4 and 5.

7. If you made the rectangles narrower, would the sum of their areas be closer to the area found using the formula? Explain.

Mathematical Structure

When is $10 + 4 = 2$?
When is $10 + 10 = 8$?
When you are telling time.
If it is now 10 o'clock, it will be
2 o'clock in 4 h, and 8 o'clock in 10 h.
Clock arithmetic is an example
of modular arithmetic.

Activity ❶

A scientist is doing an experiment that takes 41 h to complete. The experiment was started at 12 noon.

1. At what time will the experiment be completed?

2. How long had the experiment been running at 8 a.m. on the second day?

3. Complete the following calculations using the arithmetic of the 12-h clock.

a) $3 + 5$ **b)** $9 + 9$
c) $13 + 15$ **d)** $15 + 22$
e) 3×6 **f)** 4×7
g) $5 - 7$ **h)** $3 - 8$

Activity ❷

On the 4-h clock shown, the hours are 1, 2, 3, and 0.
Use the 4-h clock for the following.

1. State whether the expressions in each pair are equal.
a) $2 + 3$ and $3 + 2$
b) $3 - 2$ and $2 - 3$
c) $2 + 2$ and $2 - 2$
d) $3 + 3$ and $3 - 3$

2. Simplify.
a) $2 \times (3 + 2)$
b) $(2 + 3) \times 3$
c) $(2 \times 3) \times 2$
d) $(2 \times 3) \times 3$
e) $3 \times 3 - 2 + 1$

3. State whether the expressions in each pair are equal.
a) $3 \times (3 + 2)$ and $3 \times 3 + 3 \times 2$
b) $(3 + 1) \times (2 + 3)$ and
$3 \times 1 + 3 \times 2 + 1 \times 2 + 1 \times 3$

Solar Café
NEXT EXIT

Connecting Numbers and Variables

Astronomers define an astronomical unit (AU) as the average distance of the Earth from the sun. This distance is about 150 000 000 km.

Neptune is about 30 AU from the sun. How far is that in kilometres?

Jupiter is about 5.2 AU from the sun. How far is that in kilometres?

The circumference of the Earth at the Equator is about 40 000 km. How many times greater is 1 AU?

Explain why we do not use the astronomical unit to measure distances on Earth.

Why do we not use the astronomical unit to measure distances to the stars? What unit do we use?

Math Magic

Activity 1 The Stubborn Coin Trick

1. a) Place a coin on any red card.
b) Move the coin horizontally, to the left or right, onto the nearest black card.
c) Move the coin vertically, up or down, onto the nearest red card.
d) Move the coin diagonally onto the nearest black card.
e) Move the coin down or to the right onto the nearest red card.

2. Try the trick a few times to find out how it works. You might try working backward.

3. Explain how the trick works.

Activity 2 Repeat, Repeat, Repeat

1. a) Pick a number from 2 to 9.
b) Multiply it by 41.
c) Multiply your answer by 271.

2. Pick another number from 2 to 9 and repeat the process.

3. Explain how the trick works.

Activity 3 Lucky Number

1. a) Mark 20 lines in a row.
b) Cross out any number of lines, up to 9.
c) Count the number of lines you have left. Find the sum of the two digits of this number.
d) Cross out a number of lines equal to the sum you found in step c).
e) Cross out 2 more lines. How many are left?

2. Repeat the process several times, but vary the number you choose in part b).

3. Explain how the trick works.

Activity 4 The Stubborn Card Trick

1. a) Pick a number larger than 4.
b) Starting from the card marked A, count your chosen number of places along the card trail.

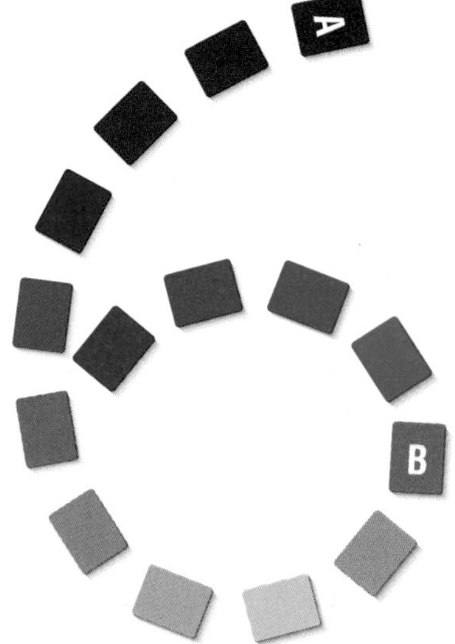

c) Count the same number of places backward, but go around the circle instead of returning to A.

2. Repeat the process several times, starting with different numbers larger than 4. Where do you always land?

3. Explain how the trick works.

Activity 5 Predicting the Hour

1. a) Ask a classmate to think of a favourite hour of the day on a 12-h clock.
b) Ask your classmate to add 1 to the hour every time you point at the clock, until the total reaches 20.
c) Point at any 7 numbers on the clock. Then, point at the 12, the 11, the 10, the 9, and so on, until your classmate tells you that the total is 20.
d) State that the last number you pointed at is your classmate's favourite hour.

2. Repeat the trick several times with different classmates.

3. Explain how the trick works.

Activity 6 Predicting the Number

1. a) Ask a classmate to write down and keep hidden a number that consists of 3 consecutive ascending digits. Examples are 345 and 567.
b) Ask your classmate to write below the number the 5 other 3-digit numbers that include the same digits.
c) Have your classmate add the 6 numbers, divide the sum by 6, and tell you the answer. The 3 digits in the answer should all be the same.
d) Subtract 1 from the first digit of the answer. Add 1 to the last digit of the answer. State that the resulting number was your classmate's original choice.

2. Try the trick several times, beginning with different 3-digit numbers.

3. Explain how the trick works.

Mental Math

Calculate.

1. $101 + 765$ **2.** $432 + 333$
3. $723 + 145$ **4.** $824 + 134$
5. $666 + 121$ **6.** $545 + 232$
7. $616 + 303$ **8.** $876 + 123$

Calculate.

9. $555 - 121$ **10.** $656 - 234$
11. $875 - 662$ **12.** $285 - 164$
13. $742 - 141$ **14.** $637 - 237$
15. $567 - 456$ **16.** $973 - 663$

Calculate.

17. $7 + 8 - 3$ **18.** $11 - 6 + 7$
19. $13 + 2 - 9$ **20.** $17 + 8 - 20$
21. $21 - 6 + 8$ **22.** $55 - 15 + 11$
23. $44 - 22 - 11$ **24.** $21 + 21 - 20$

Calculate.

25. 102×3 **26.** 111×7
27. 2002×4 **28.** 1010×6
29. 3434×2 **30.** 1212×4
31. 3003×3 **32.** 44×4

Calculate.

33. $5555 \div 5$ **34.** $606 \div 3$
35. $8080 \div 4$ **36.** $963 \div 3$
37. $8642 \div 2$ **38.** $147 \div 7$
39. $515 \div 5$ **40.** $246 \div 6$

Calculate.

41. $5 \times 6 \div 2$ **42.** $3 \times 4 \times 5$
43. $3 \div 3 \times 4$ **44.** $8 \div 2 \times 2$
45. $6 \times 0 \div 2$ **46.** $6 \times 5 \div 2$
47. $4 \div 4 \times 2$ **48.** $8 \times 1 \times 9$

1.1 Choosing a Calculation Method

You use mathematics every day to solve problems. Before you solve a problem, you should decide what kind of answer you need. Sometimes, the answer must be exact. At other times, an approximate answer is enough. Once you decide on the kind of answer you need, you select a calculation method.

Estimation gives an approximate answer. You can often estimate **mentally**. Using a **calculator** or **paper and pencil** can also help you estimate. You can obtain some exact answers mentally, but you more often use a calculator, paper and pencil, or **computer**. A computer program is very efficient when repetitive calculations are required.

Activity: Choose a Calculation Method

Decide if you need an exact answer or an estimate. Then, choose a calculation method.

a) You are playing tennis and keeping score.
b) You are working at a provincial park in the summer and are asked how many deer are in the park.
c) You are responsible for the scheduling and statistics for a 68-team hockey league.
d) You are working at a fast-food restaurant and are asked the cost of a birthday party for 10 five-year-olds.

Inquire

1. List 3 careers in which you would often
a) use estimation skills.
b) use a computer.

2. Are you more likely to obtain exact answers or approximate answers mentally? Explain.

3. If you had a choice, when would you use paper and pencil instead of a calculator to make a calculation? Explain.

4. Estimates that are too high are called **overestimates**. Estimates that are too low are called **underestimates**. Suppose a friend came back from vacation and said: "It rained all week." An estimate of the number of hours of rain might be 20×7 or 140. Would the answer likely be an overestimate or underestimate? Explain.

Practice

State if you need an exact answer or an estimate.

1. How many school jackets are needed for the band?

2. How many hotel rooms will you need to reserve for the school ski trip?

3. How long will the bus take to get to the campground?

4. How much paint will you need to paint the boat?

5. At what time will you land the spacecraft?

6. What is the cost of gasoline needed to drive across Canada?

7. How long will it take you to read a book?

8. How much change should you get when you make a purchase in a store?

State whether you need an exact or approximate answer. Then, choose a calculation method. Give reasons for your answers.

9. How much will it cost you to go to college 4 years from now?

10. You want to know your bowling average after 90 games.

11. You keep score for a game of SCRABBLE® involving 4 players.

12. You record the number of whales seen at a feeding area from 10:00 until 12:00.

13. How much money will you have after you make 7 purchases?

14. You are responsible for recording the sale and distribution of tickets for the home games of a baseball team. The team plays 37 home games in a park with 8766 seats.

15. Do you have enough money to buy lunch?

16. At what time should you leave for the concert?

17. You want to know if the piano fits through the front door.

18. How many pizzas should you buy for your whole class?

Problems and Applications

19. The world's population increases by 3 people each second. Which method would you use to find the population increase in one year? Explain.

20. Which calculation method would you use to decide which shampoo is the best buy? Explain.

21. Should the posted speed limit on a road winding down a mountain be lower or higher than the safe speed? Why?

22. Should a mechanic overestimate or underestimate the cost of repairing a car? Why?

23. State whether each newspaper headline most likely represents an underestimate, overestimate, or exact value. Explain.

a) Flooding May Leave 5000 Homeless
b) Election Just 10 Days Away
c) Airport Construction To Take 3 Years

24. Look in newspapers and magazines to find examples of headlines that you think include exact values, overestimates, and underestimates. Compare your findings with your classmates'.

25. a) List 3 examples of problems for which you would overestimate the answers.
b) List 3 examples of problems for which you would underestimate the answers.
c) Give a reason for each example.
d) Compare your answers with your classmates'.

Fibonacci Numbers

The Fibonacci sequence of numbers was first studied by the Italian mathematician Leonardo Fibonacci, also known as Leonardo da Pisa. The Fibonacci sequence of numbers can be found in such fields as biology and music.

Activity ❶

These are the first 8 terms of the Fibonacci sequence of numbers.

1, 1, 2, 3, 5, 8, 13, 21

1. What is the rule for finding terms in the Fibonacci sequence?

2. What is the next term in the sequence?

3. Write the first 14 terms in the sequence.

Activity ❷

1. Take any 4 Fibonacci numbers in a row, for example, 3, 5, 8, and 13.
Multiply the first and last numbers. $3 \times 13 = ■$
Multiply the middle 2 numbers. $5 \times 8 = ■$
What is the difference in the products?

2. Repeat the steps for the following sets of 4 Fibonacci numbers in a row.

What is the difference in the products in each case?

Activity ❸

1. Take any 3 consecutive Fibonacci numbers, for example, 3, 5, and 8.
Multiply the first and last numbers. $3 \times 8 = ■$
Square the middle number. $5^2 = ■$
What is the difference in the products?

2. Repeat the steps for the following sets of 3 Fibonacci numbers in a row.

What is the difference in the products in each case?

Activity ❹

1. Complete the calculations for the sums of the squares of Fibonacci numbers.

$$1^2 + 1^2 = 2 = 1 \times 2$$
$$1^2 + 1^2 + 2^2 = 6 = 2 \times 3$$
$$1^2 + 1^2 + 2^2 + 3^2 = 15 = \blacksquare$$
$$1^2 + 1^2 + 2^2 + 3^2 + 5^2 = \blacksquare = \blacksquare$$
$$1^2 + 1^2 + 2^2 + 3^2 + 5^2 + 8^2 = \blacksquare = \blacksquare$$

2. Write the next 2 lines of the pattern.

3. What is the sum of the squares of the first 10 Fibonacci numbers?

Activity ❺

The following problem is adapted from one of Fibonacci's books.

A pair of rabbits one month old is too young to produce more rabbits. But at the end of the second month, they produce a pair of rabbits, and a pair of rabbits every month after that. Each new pair of rabbits does the same thing, producing a pair of rabbits every month, starting at the end of the second month.

The diagram shows how the family of rabbits grows. Draw a diagram to show the total number of rabbits after 8 months.

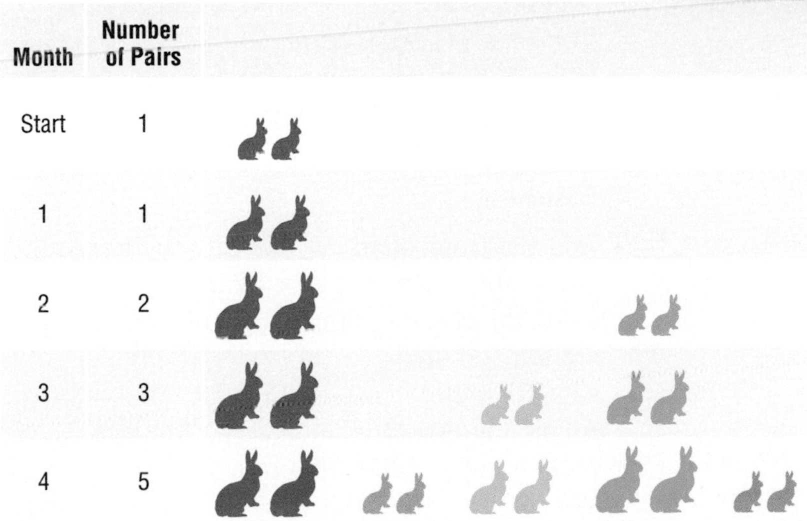

Month	Number of Pairs
Start	1
1	1
2	2
3	3
4	5

Activity ❻

Use your research skills to describe where else the Fibonacci numbers appear in nature.

1.2 Estimating Sums and Differences

Activity: Make a Comparison

The table shows the cost of souvenirs at a Beatles' music and souvenir store. Miguel estimated the cost of a poster, book, and glasses as
$14 + $38 + $14 = $66
Susan estimated the cost as
$10 + $40 + $10 = $60

Item	Price ($)
Cavern poster	14.45
Beatles history book	37.95
Lennon glasses	13.75
T-shirt	28.25
Album covers	10.60
London Palladium picture	11.95

Inquire

1. To what place value were the prices rounded by
a) Miguel? **b)** Susan?

2. What is the advantage of rounding
a) Miguel's way? **b)** Susan's way?

As well as rounding, there are other ways to estimate sums and differences.

Example 1

Use **clustering** to estimate the cost of a poster, glasses, and picture.

Solution

The prices are $14.45, $13.75, and $11.95.
They cluster around $13, so a good estimate is 3 × $13 or $39.

Example 2

Use **compatible numbers** to estimate the cost of a poster, T-shirt, and picture.

Solution

Compatible numbers are numbers we can compute mentally.
The prices are $14.45, $28.25, and $11.95.
$14.45 + $11.95 is about $14 + $11 or $25.
$25 + $28.25 is about $25 + $30 or $55.
The estimate is about $55.

Example 3

Estimate 345 + 513 + 263 + 836 by **front-end estimation.**

Solution

Add the front digits, in this case the hundreds.
300 + 500 + 200 + 800 = 1800
Look for compatible numbers in the parts that are left.
$45 + 63 \doteq 100$ $13 + 36 \doteq 50$
Adjust the estimate.
1800 + 100 + 50 = 1950
So, 345 + 513 + 263 + 836 is about 1950.

$\doteq$ means approximately equals

Practice

In questions 1–14, compare your estimated and calculated answers to check that your answers are reasonable.

Estimate by rounding, then calculate.

1. 346 + 221 + 147

2. 678 – 431

3. $34.79 + $12.56 + $29.89 + $15.21

4. $385.78 – $115.60

Estimate by clustering, then calculate.

5. 56 + 44 + 61 + 51

6. $18.45 + $20.56 + $22.99

7. $387 + $415 + $390 + $410

Estimate by finding compatible numbers, then calculate.

8. 13 + 38 + 61 + 48

9. $23.56 + $78.31 + $101.45

10. $6.35 + $4.12 + $58.23

11. $124.55 – $99.81

Estimate by front-end estimation, then calculate.

12. 623 + 555 + 201 + 777

13. $4.56 + $3.78 + $2.56

14. $23.70 + $11.34 + $23.80 + $56.24

Problems and Applications

In questions 15–19, estimate by the method of your choice, then calculate the answer. Compare the estimate and the calculated answer.

15. Paul bought 3 items for $2.45, $3.56, and $11.18. How much change did he get from a $20 bill?

16. Last month, Carl earned $147.67 working part-time at a restaurant. Jenny earned $217.89. What was the difference in their earnings?

17. Copy and complete the following table to show Leolo's account balance on each date.

Date	Withdrawal	Deposit	Balance
April 27			$567.81
April 28	$23.85		
April 30		$85.00	
May 1	$123.45		
May 6	$67.60		
May 13		$160.00	

18. Samantha received $14.63 change from a $50 bill. Her cash register receipt showed the total of her purchases as $34.37. Samantha claimed that the sales clerk had made an error. Was Samantha right? If so, how great was the sales clerk's error?

19. The table shows the North American sales of some Dr. Seuss books up to 1990. What were the total sales?

Title	Number Sold
The Cat in the Hat	3 693 197
Green Eggs and Ham	3 683 097
One Fish, Two Fish, Red Fish, Blue Fish	2 970 833
Hop on Pop	2 953 324
ABC	2 753 823
The Cat in the Hat Comes Back	2 470 305

20. Use your research skills to estimate the following. Then, compare your estimates with your classmates'.

a) the total number of fans who attended the last 10 Grey Cup finals

b) the total number of fans who attended Toronto Blue Jays home games in the last 10 years

1.3 Estimating Products and Quotients

Activity: Study the Information

Each fall, the blackpoll warbler flies nonstop from Nova Scotia to South America. The warbler flies at 41 km/h, and the trip takes 96 h.

Inquire

1. What operation will you use to estimate the distance the warbler flies?

2. Estimate the distance.

To estimate products or quotients, we can round the numbers, substitute compatible numbers, or combine these methods.

Example 1

The Arctic tern is the long-distance migratory champion of the animal world. It flies from the far north to winter along the shores of Antarctica. The trip is about 17 500 km long and takes about 16 weeks. Estimate how far the tern flies each week.

Solution

Divide to estimate the distance.

Method 1

Round both numbers to the highest place value.

$$\frac{17\ 500}{16} \Rightarrow \frac{20\ 000}{20}$$
$$= 1\ 000$$

The tern flies about 1000 km each week.

Method 2

Round one number to the highest place value. Substitute for the other number to obtain compatible numbers.

$$\frac{17\ 500}{16} \Rightarrow \frac{18\ 000}{20}$$
$$= 900$$

The tern flies about 900 km each week.

Which method in Example 1 gives the better estimate? Can you suggest a third method?

To estimate with decimals less than 1, we substitute 1 or multiples of 0.1 or 0.5.

Example 2

Estimate.

a) 0.85×5.6

b) $0.533 \div 10.2$

c) 0.261×3.1

Solution

a) 0.85×5.6
$\doteq 1 \times 6$
$= 6$

b) $0.533 \div 10.2$
$\doteq 0.5 \div 10$
$= 0.05$

c) 0.261×3.1
$\doteq 0.3 \times 3$
$= 0.9$

Practice

Estimate, then calculate. Compare with your estimates to check that your calculated answer is reasonable.

1. 34.5×21.9

2. $\$23.87 \times 19$

3. $\$130 \div 12$

4. $\$650 \div 20$

5. $\$45.67 \times 12$

6. $\$5613.42 \div 52$

7. 59×0.86

8. 212×0.43

9. $334 \div 0.91$

10. $67 \div 0.52$

11. 139×1.678

12. $47.89 \div 2.22$

Problems and Applications

Estimate, then calculate. Compare the results to check that your calculated answer is reasonable.

13. A monarch butterfly can fly 1000 km without stopping to feed.

a) How many hours does it take the monarch to fly this distance at 16 km/h?

b) How many hours would it take you to walk this distance?

14. It costs $0.37/min to access a database through your computer. What is the cost if you are on-line for 17 min?

15. A long-distance telephone call costs $0.27/min. What is the cost for a 43-min call?

16. Adult tickets to the Movieworld theme park cost $28.50. Children's tickets cost $21.75. What is the cost for 3 adults and 2 children?

17. Colleen worked for 12.5 h and was paid $191.25. How much did she earn per hour?

18. During the car rally, Caroline averaged 78 km/h for 4.5 h. How far did she drive?

19. Carlos can keyboard 43 words/min. How long will it take him to input his 7500-word essay?

20. A Concorde jet cruises at a speed of 2300 km/h. At this speed, how long would it take to fly the following distances?

a) across Canada, a distance of 5514 km

b) to the moon, a distance of 384 365 km

c) to the sun when it is closest to the Earth, 147 000 000 km away

21. The **population density** of a country is the average number of people per square kilometre. Canada has an area of 9 976 000 km^2. Research the current population of Canada, then determine Canada's population density.

22. A blue whale has a mass of about 138 000 kg. If the students in your school had an average mass of 58 kg, would their total mass be greater than a blue whale's?

23. Estimate the following. Explain why you chose your estimation method. Compare your views with your classmates'.

a) $149 \div 23$

b) 12.1×0.43

c) 27.5×0.83

d) $4869 \div 24$

NUMBER POWER

Copy the diagram. Put the numbers from 1 to 9 in the circles so that each line adds to 25.

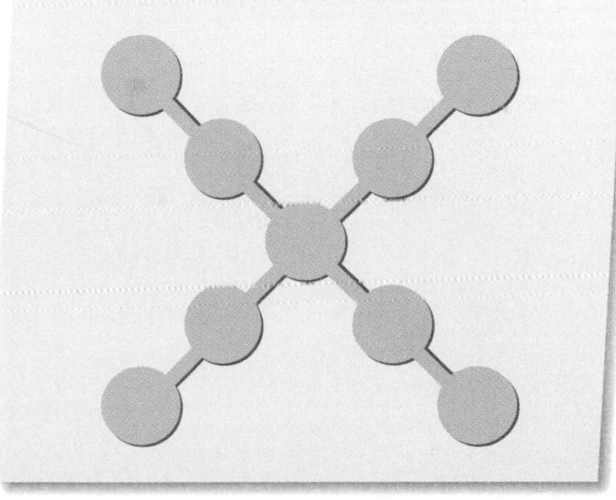

1.4 Sequence the Operations

The solutions to some problems have several steps, which must be performed in the proper sequence.

Bob Robertson is a gold miner in the Klondike. He has 400 claims and pays the Canadian government \$10 per year to operate each one. Bob employs 14 people during the "season," which runs from April to freeze-up in October. Each employee works 12 h a day for \$19.75/h. One year, 9 people worked 143 days, and the other 5 worked 131 days. If gold was worth \$15 000/kg, how many kilograms of gold did Bob's claims have to produce to cover the salaries and government costs? Write your answer to the nearest kilogram.

Understand the Problem

1. What information are you given?

2. What are you asked to find?

3. What is the proper sequence of steps?

Think of a Plan

Calculate the wages of the 14 employees. Add the amount paid to the government. Divide the total expenses by \$15 000 to find out how many kilograms were needed. Use a calculator to find the exact answer.

Carry Out the Plan

Wages of 9 employees
$= 9 \times 143 \times 12 \times \$19.75$
$= \$305\ 019$

EST: $10 \times 140 \times 10 \times 20 = 280\ 000$

Wages of 5 employees
$= 5 \times 131 \times 12 \times \$19.75$
$= \$155\ 235$

EST: $5 \times 100 \times 10 \times 20 = 100\ 000$

Payment to government
$= 400 \times \$10$
$= \$4000$

Total expenses
$= \$305\ 019 + \$155\ 235 + \$4000$
$= \$464\ 254$

EST: $300\ 000 + 160\ 000 + 4000 = 464\ 000$

Kilograms of gold needed
$= \dfrac{\$464\ 254}{\$15\ 000}$
$= 30.950\ 266\ 67$

EST: $450\ 000 \div 15\ 000 = 30$

Bob's claims had to produce 31 kg of gold to cover salaries and government costs.

Look Back

How can you work backward to check your answer?

Sequence the Operations

1. List the given facts.
2. Decide on the proper solution sequence.
3. Complete the calculation.
4. Check that your answer is reasonable.

Problems and Applications

1. At ages 3, 4, and 5, a child learns about 3 new words each day. About how many words does a child learn in these 3 years?

2. The average person normally speaks about 125 words/min. How many words does the average person speak in 2.5 h?

3. The sports arena has 17 624 seats. Tickets for the track meet cost $22.50 each. If 1057 tickets remain unsold, what is the value of the tickets sold?

4. Tamar and Ken left the highway service centre at the same time and travelled in the same direction. Tamar drove at 90 km/h, and Ken drove at 70 km/h.

a) How far did Tamar drive in 3.5 h?
b) How far did Ken drive in 4 h?
c) How far apart were they after 4.5 h?

5. The parking garage charges $5.50 for the first hour or part of an hour. The charge is $2.00 for each additional half hour or part of a half hour. Eileen arrived at 08:17 and left at 11:45. How much was she charged?

6. Mohammed works on a highway survey crew during the summer. He earns $11.70/h for up to 35 h/week. He earns time-and-a-half for hours over 35 h/week. If he works 41 h in one week, how much does he earn?

7. Light travels through space at a speed of 298 000 km/s. A **light year** is the distance light travels through space in one year. How far is a light year?

8. What time will it be 213 000 h from now?

9. Twenty-four students are planning a trip to a theatre festival. The total cost is $10 466. The school will pay $1100 towards the total cost. The students will equally share the rest of the cost. Each student will pay in 5 equal monthly instalments. How much is each instalment?

10. The passenger pigeon was a North American migratory bird. At one time, there were huge numbers of these pigeons, but they became extinct in 1914. A flock of passenger pigeons was once described as follows.

"The column was 500 m wide and flew overhead at 500 m/min. It took three hours to fly by. Each square metre was occupied by ten pigeons."

How many pigeons were in the flock?

LOGIC POWER

Examine the map. Use your knowledge of geography to decide which province this area is in. Then find this area on a complete map.

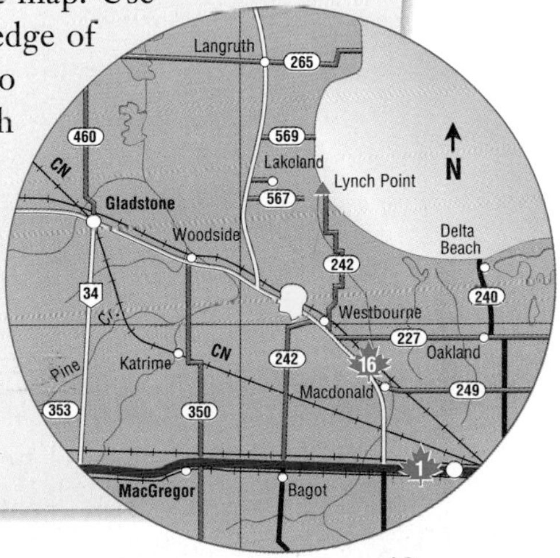

1.5 Variables in Expressions

Activity: Complete the Table

People who write books are paid a set amount for every book sold. This amount is called a **royalty**. Sarah Barker wrote a book on sailboat racing. Copy and complete the table to show Sarah's royalties.

Number of Books Sold	10	20	30	100	352
Sarah's Royalties ($)	30	60			

Inquire

1. How much would Sarah make if the following numbers of books were sold?

a) 1 **b)** 1000 **c)** 5000 **d)** 20 000 **e)** n

2. If the letter n represents the number of books sold, n is called a **variable**. Why?

3. In your own words, define a variable.

Sarah's publisher pays expenses of 35 cents/km when Sarah drives to bookstores to promote her book. We can use the letter d to represent the distance driven in kilometres.

Sarah's expenses, in cents, can be expressed as $35d$. This is an example of an **algebraic expression**. In the expression $35d$, the multiplication symbol is understood, so $35d$ means $35 \times d$. The expression $35d$ represents many different amounts of money, since its value depends on the value of d.

When $d = 80$, $35d$ is 35×80 or 2800, which is \$28.00.
When $d = 200$, $35d$ is 35×200 or 7000, which is \$70.00.

Expressions such as $5t$, $x + 7$, and $2(l + w)$ are also algebraic expressions. The letters t, x, l, and w are variables.

Example

Sarah gives sailboat racing lessons. The cost, in dollars, is given by the expression $30t + 25$, where t represents the time in hours. Calculate the cost of lessons that last

a) 7 h **b)** 20.5 h

Solution

a) When $t = 7$

$$30t + 25 = 30(7) + 25 = 210 + 25 = 235$$

The lessons cost \$235.

multiply first

b) When $t = 20.5$

$$30t + 25 = 30(20.5) + 25 = 615 + 25 = 640$$

The lessons cost \$640.

Practice

Write an expression for each.

1. the product of 7 and 11

2. the product of x and 3

3. the sum of 5 and 8

4. the sum of 9 and y

5. the product of x and y

6. 7 more than m

7. 8 less than y

8. the quotient x divided by y

9. 4 times t

10. Find the value of $4n$ when n has each value.

a) 2 **b)** 3 **c)** 4 **d)** 1
e) 0 **f)** 9 **g)** 10 **h)** 35

11. Find the value of $3t + 1$ when t has each value.

a) 2 **b)** 3 **c)** 0 **d)** 10
e) 3.5 **f)** 1.1 **g)** 6.2 **h)** 8.8

12. Find the value of $2x - 1$ when x has each value.

a) 2 **b)** 1 **c)** 5 **d)** 10
e) 1.5 **f)** 3.1 **g)** 8.2 **h)** 9.5

13. Evaluate for $c = 3$ and $d = 2$.

a) $c + d$ **b)** $3c + 4d$ **c)** $5d + 6c$
d) $3c + d + 7$ **e)** $6c - d$ **f)** $4cd + 8$
g) $2c + 4d$ **h)** $8d - 3c$ **i)** $7 - c - 2d$

14. Evaluate each expression.

a) $2m + 4,\ m = 5.3$ **b)** $6t - 7,\ t = 4.6$
c) $5s + 13,\ s = 9.1$ **d)** $25 - 2x,\ x = 3.5$

15. Evaluate for $r = 4.1$, $s = 3.2$, and $t = 2.3$.

a) $r + s + t$ **b)** $3r + 4s + 5t$
c) $4t - s + 5r$ **d)** $4st + 3rs + rt$
e) $5rst - 9$ **f)** $8.5 + 2rst$

Problems and Applications

16. Sophia earns $11/h installing swimming pools.
a) Write an expression for her earnings if she works n hours.
b) Use the expression to calculate her earnings if she works 41 h; 53.5 h.

17. To put on a banquet, the manager of the Purple Onion restaurant charges $300 for the room plus $35/person.
a) Write an expression for the cost of a graduation banquet for n people.
b) Use the expression to find the cost of a banquet for 250 graduates.

18. a) Write an expression for the distance a car travels in t hours at a speed of s kilometres per hour.
b) Use the expression to calculate the distance travelled at 75 km/h for 3.5 h.

19. Let x represent an even number. Write an expression for each statement.
a) the next even number after x
b) the next odd number after x
c) the next even number before x
d) x plus the next odd number after x
e) x plus the next even number after x

NUMBER POWER

Write the numbers from 1 to 12 in the circles so that each side adds to 25.

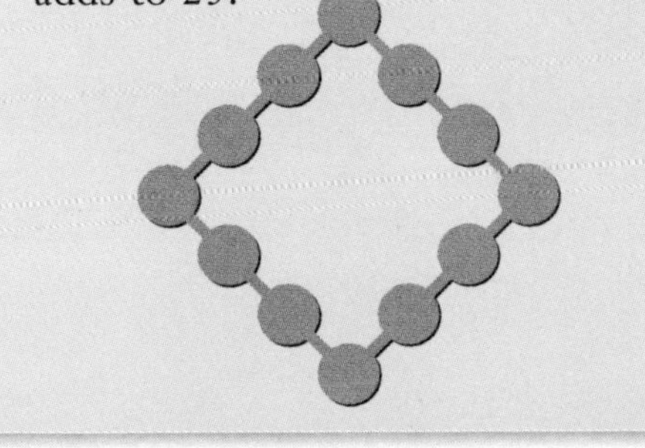

1.6 Make Assumptions

To solve some problems, you must make assumptions.

The Alaska Highway runs from Dawson Creek, B.C., to Fairbanks, Alaska. It is 2400 km long. The speed limit is 80 km/h. The De Marco family is leaving Dawson Creek at 09:00 on a Wednesday to drive to Fairbanks. They plan to stop for 10 h to eat and sleep on Wednesday night and on Thursday night. At what time on what day should they arrive in Fairbanks?

Understand the Problem

1. What information are you given?

2. What are you asked to find?

3. What assumptions should you make?

Think of a Plan

Assume that the De Marco family will drive at the speed limit. Calculate the time they will spend driving. Add 20 h to this time for eating and sleeping.

Carry Out the Plan

If the De Marco family drives at the 80 km/h speed limit, the time needed to drive 2400 km is found by dividing 2400 km by 80 km/h.

$$\frac{2400}{80} = 30$$

Thirty hours from 09:00 on Wednesday is 15:00 on Thursday. Add 20 h for eating and sleeping.

The De Marco family should arrive in Fairbanks at 11:00 on Friday.

Look Back

How could you use subtraction and multiplication to check your answer?

Make Assumptions

1. Decide what assumption(s) to make.
2. Use your assumption(s) to solve the problem.
3. Check that your answer is reasonable.

Problems and Applications

Solve the following problems and state each assumption that you make.

1. Frank earned $215.75 in the first month at his part-time job. How much can he expect to earn in a year?

2. The drama club sells oranges to raise money for charity. Last year, each member of the club sold 20 cases. How many cases can the 43 members of the club expect to sell this year?

3. Clara surveyed 40 students in the school. Ten of them said they would attend the dance. If the school has 800 students, how many can Clara expect to attend the dance?

4. The patrol boat travels at a speed of 15 km/h. How far can the boat travel in 5 h?

5. John trained for 5 weeks and reduced his time in the 100-m dash from 12.0 s to 11.5 s. He reasoned that with 20 more weeks of training he would be able to run 100 m in 9.5 s and break the world record. What assumption did he make? Is he necessarily correct? Explain.

6. Soo Lin surveyed 100 people who bought new bikes. Of those surveyed, 80 said the town needed new bike trails. There are 10 000 people in the town, so Soo Lin reported that 8000 people wanted new bike trails. What assumption did she make? Explain.

7. Six tents are placed 5.2 m apart in a straight line. What is the distance from the first tent to the last tent?

8. How many cuts must you make in a rope to make the following number of pieces?

a) 3 **b)** 4

9. How many seconds are there in February?

10. The distance from Eagle's Nest to Brewsterville is 425 km. The first part of the trip is 150 km of highway, where the speed limit is 100 km/h. The rest of the trip is along a country road, where the speed limit is 50 km/h.

a) How long will it take Paulina to drive from Eagle's Nest to Brewsterville?

b) If she leaves Eagle's Nest at 16:45, at what time will she arrive in Brewsterville?

11. Identify the next 3 terms in each of the following.

a) 200, 100, 50, 25, …

b) 14, 17, 21, 26, …

c) 3, 7, 15, 31, …

12. The world's longest sneezing fit lasted 977 days. In the first year of the fit, Donna Griffiths sneezed about a million times. How many seconds did she average between sneezes?

13. Write a problem in which the solution requires at least one assumption. Have a classmate solve your problem and state the assumption(s).

LOGIC POWER

Move 3 sticks to form 4 squares of the same size.

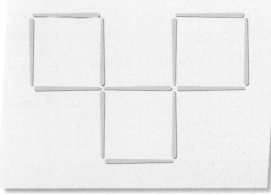

1.7 Exponents and Powers

Exponents are used as a short way to write repeated multiplication.

standard form → $81 = \underbrace{3 \times 3 \times 3 \times 3}_{\text{repeated multiplication}} = 3^4$ ← exponential form

$81 = 3^4$ ← exponent

base, power

Activity: Do an Experiment

When you repeatedly fold a piece of paper in half, the number of layers increases with the number of folds. Fold a standard piece of paper, and copy and complete the table.

Number of Folds	Number of Layers
1	$2 = 2$ or 2^1
2	$2 \times 2 = 4$ or 2^2
3	
4	
5	
6	

Inquire

1. If you were to fold the piece of paper the following numbers of times, how many layers would you have? Express each answer in exponential form.

a) 5 **b)** 7 **c)** 50

2. Explain how you found your answers to question 1.

3. If 10 layers of paper are about 1 mm thick, how thick is a piece of paper after 10 folds?

4. What is the maximum number of times you can fold a piece of paper?

$3 \times 3 = 3^2$ — 3^2 is read as "three to the second" or more commonly "three squared" because it can be pictured as a square.

$2 \times 2 \times 2 = 2^3$ — 2^3 is read as "two to the third" or "two cubed" because it can be pictured as a cube.

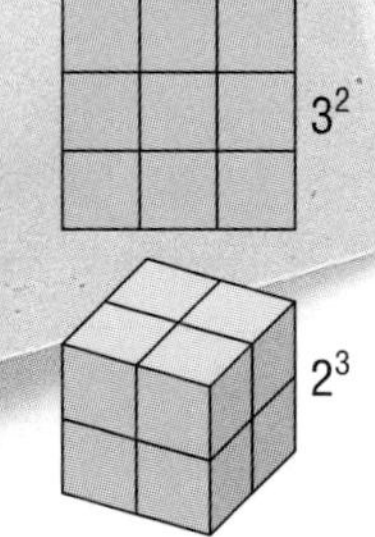

Exponents are also used with variables.

$3y^4$ means $3 \times y \times y \times y \times y$

When an exponent is outside a pair of brackets, the exponent is applied to everything inside the brackets.

$(3y)^4$ means $(3y) \times (3y) \times (3y) \times (3y)$

$= 3 \times 3 \times 3 \times 3 \times y \times y \times y \times y$

$= 81y^4$

Example 1

If $x = 2$ and $y = 3$, evaluate $5x^4 + 6xy$.

Solution

$$\begin{aligned} 5x^4 + 6xy &= 5(2)^4 + 6(2)(3) \\ &= 5(16) + 36 \\ &= 80 + 36 \\ &= 116 \end{aligned}$$

Practice

State the base and the exponent.

1. 5^3 **2.** 10^7 **3.** x^5 **4.** t^2

Write in exponential form.

5. $3 \times 3 \times 3 \times 3 \times 3$

6. $4 \times 4 \times 4 \times 4 \times 4 \times 4$

7. $10 \times 10 \times 10$

8. $6 \times 6 \times 6 \times 6$

9. $m \times m \times m \times m \times m$

10. $r \times r \times r$

Write as a repeated multiplication.

11. 5^2 **12.** 1^6 **13.** 2^5 **14.** 10^4

15. 0^3 **16.** y^4 **17.** $5x^3$ **18.** $(2m)^3$

Evaluate.

19. the third power of 2

20. 3 to the fourth

21. 5 cubed

22. 10 to the fifth

What power does each figure represent?

23.

24.

25.

x

x

x

26.

y

y

Write each number as a power of 10.

27. 100 **28.** 1000

29. 100 000 **30.** 1 000 000

31. 100 000 000 **32.** 10 000 000

Write in standard form.

33. 2^5 **34.** 5^3 **35.** 4^4

36. 7^3 **37.** 10^7 **38.** 3^6

39. 0.5^2 **40.** 1.1^3 **41.** 0.1^4

Write as a power of 2.

42. 16 **43.** 64 **44.** 256

State which is smaller.

45. 5^3 or 3^5 **46.** 2^5 or 5^2

47. 4^2 or 2^4 **48.** 2^3 or 3^2

Evaluate.

49. $7^2 + 3^2$ **50.** $4^3 - 2^4$

51. 3×2^3 **52.** $4^4 \div 2^5$

53. 100×0.1^3 **54.** $0.8^2 \times 0.2^3$

55. 1000×0.2^4 **56.** 0.1×0.1^2

Evaluate.

57. $7^2 + 3^3$ **58.** $4^3 - 2^2$

59. $2^2 \times 2^3$ **60.** $4^2 \div 2^2$

61. Evaluate for $x = 4$.

a) x^3 **b)** $x^2 - 5$

c) $5x^2 - 7$ **d)** $(2x)^2$

62. Evaluate for $t = 3$ and $s = 2$.

a) $t^2 + s^2$ **b)** $(t + s)^3$

c) $t^3 - s^3$ **d)** $2t^2 - s^2$

e) $6s^3 - 2st$ **f)** $(3t)^2 - 4st$

CONTINUED

Problems and Applications

63. A scientist found that the number of bacteria in a culture doubled every hour. If there were 1000 bacteria at 08:00, how many were there at the following times?

a) 09:00 **b)** 11:00 **c)** 14:00

64. The ancestors of Ling Ling, the giant panda, included 2 parents and 4 grandparents. Her grandparents were 2 generations before her.

a) How many ancestors were in the seventh generation before Ling Ling? Express your answer in exponential form.

b) How many ancestors were in the tenth generation before Ling Ling? Express your answer in exponential form.

65. If a ball is thrown straight up at a speed of 30 m/s, its height in metres after t seconds is given by the formula $h = 30t - 5t^2$.

a) What is the height of the ball after 1 s? 3 s? 4 s?

b) After how many seconds will the ball hit the ground? State your assumptions.

66. a) Describe the pattern in this series.

1, 4, 27, 256, …

b) Predict the fifth and tenth numbers in the series.

67. The product of the ages of a set of quadruplets is 16 times the sum of their ages. How old are they?

68. The product of the ages of a set of triplets is 12 times the sum of their ages. How old are they?

69. If $f(x) = 3x^2 + 1$

$$\begin{aligned} \text{then } f(2) &= 3(2)^2 + 1 \\ &= 3 \times 4 + 1 \\ &= 13 \end{aligned}$$

Find the following.

a) $f(4)$ **b)** $f(10)$ **c)** $f(0)$

70. Decide whether each statement is always true, sometimes true, or never true for whole numbers greater than 1. Explain your reasoning.

a) Twice a number is smaller than the number squared.

b) The cube of a number is greater than the square of the number.

c) Powers with the same base but different exponents are equal.

71. A story says that the inventor of chess asked his ruler to give him 1 grain of wheat on the first square of the chessboard, 2 on the second, 4 on the third, 8 on the fourth, and so on for all 64 squares.

Suppose you could stack loonies on a chessboard in the same manner.

a) How many loonies would you stack on the fourth square? the fifth square? the sixth square? the seventh square?

b) About how high would the stack be on the sixty-fourth square? Compare your result with your classmates'.

CALCULATOR POWER

Many calculators have a y^x key to evaluate powers.

To evaluate 2^{12}

Press 2 y^x 12 =

Display 2 2 12 4096

Use a calculator to evaluate the following.

1. 2^6 **2.** 3^7 **3.** 4^8 **4.** 0.3^3

The Car of the Future

There are many examples of technology in modern cars. These range from simple devices, such as the levers and buttons that open the doors, to more complicated ones, like cellular telephones.

Activity ❶

List some examples of modern technology found in a new car. Think of the outside of the car as well as the inside.

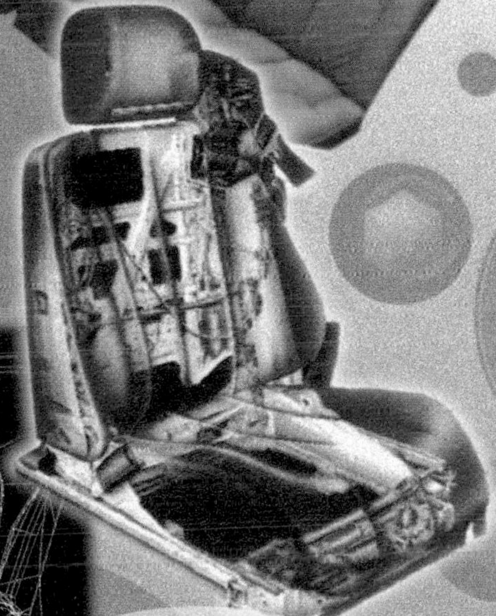

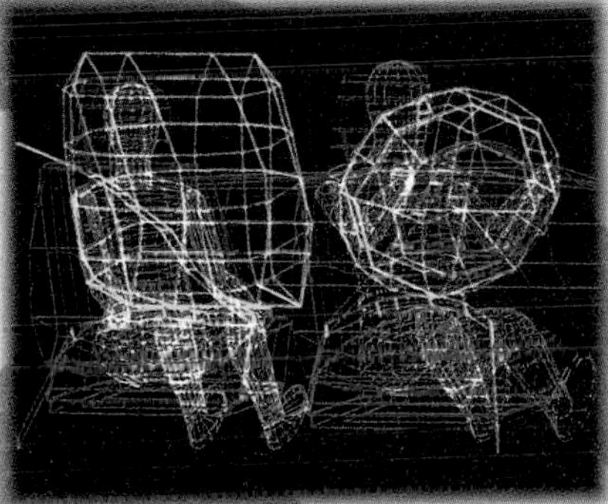

Activity ❷

Design a high technology car of the future and draw plans to show what it might look like. As you design the car, you will need to resolve such issues as the number of passengers it will carry, how the driver will steer it, the fuel the car will use, and so on.

Describe the uses of computers in your car of the future.

1.8 The Exponent Rules

The exponent rules are short cuts for multiplying and dividing powers with the *same base.*

Activity: Discover the Relationship

Copy and complete the table.

Exponential Form	Standard Form	Answer in Standard Form	Answer in Exponential Form
$2^3 \times 2^4$	8×16	128	2^7
$3^2 \times 3^2$			
$2^2 \times 2^3$			
$10^3 \times 10^3$			
$2^5 \div 2^2$			
$3^4 \div 3^2$			
$10^5 \div 10^3$			

$2^3 \times 2^4 = 2^7$

Inquire

1. For the products, how are the exponents in the first column related to the exponents in the last column?
2. Write a rule for multiplying powers with the same base.
3. For the quotients, how are the exponents in the first column related to the exponents in the last column?
4. Write a rule for dividing powers with the same base.

Example 1

Simplify $2^5 \times 2^3$.

Solution

Method 1

$2^5 \times 2^3$
$= 2 \times 2 \times 2 \times 2 \times 2 \times 2 \times 2 \times 2$
$= 2^8$

Method 2

To multiply powers with the same base, add the exponents.

$x^m \times x^n = x^{m+n}$

$2^5 \times 2^3 = 2^{5+3}$
$= 2^8$

Example 2

Simplify $y^7 \div y^2$.

Solution

Method 1

$y^7 \div y^2$
$= \dfrac{y \times y \times y \times y \times y \times y \times y}{y \times y}$
$= y \times y \times y \times y \times y$
$= y^5$

Method 2

To divide powers with the same base, subtract the exponents.

$x^m \div x^n = x^{m-n}$

$y^7 \div y^2 = y^{7-2}$
$= y^5$

Example 3

Simplify.

a) $(2^4)^3$ **b)** $(y^2)^4$

Solution

a) $(2^4)^3 = 2^4 \times 2^4 \times 2^4$
$= 2^{4+4+4}$
$= 2^{12}$

b) $(y^2)^4 = y^2 \times y^2 \times y^2 \times y^2$
$= y^{2+2+2+2}$
$= y^8$

To raise a power to a power, multiply the exponents.
$(x^m)^n = x^{m \times n}$

Practice

Simplify.

1. $5^3 \times 5^4$ **2.** $2^3 \times 2^7$ **3.** $7^5 \times 7^5$
4. $10^6 \times 10$ **5.** $4^2 \times 4^9$ **6.** 3×3^6
7. $y^2 \times y^4$ **8.** $x^3 \times x^6$ **9.** $a \times a^6$

Find the value of x.

10. $5^3 \times 5^x = 5^7$ **11.** $3^4 \times 3^x = 3^6$
12. $8^x \times 8^2 = 8^8$ **13.** $6^x \times 6^5 = 6^6$
14. $4^3 \times 4^x = 4^6$ **15.** $m^7 \times m^x = m^9$
16. $t^x \times t^3 = t^6$ **17.** $y^x \times y^4 = y^5$

Simplify.

18. $4^5 \div 4^3$ **19.** $3^7 \div 3^6$
20. $9^2 \div 9^2$ **21.** $10^6 \div 10^5$
22. $4^7 \div 4$ **23.** $5^8 \div 5^8$
24. $m^5 \div m^4$ **25.** $x^3 \div x$

Find the value of x.

26. $3^6 \div 3^x = 3^2$ **27.** $6^7 \div 6^x = 6^5$
28. $7^x \div 7^4 = 7^3$ **29.** $2^x \div 2^2 = 2^8$
30. $9^5 \div 9^x = 9^4$ **31.** $m^x \div m^2 = m^7$
32. $m^x \div m = m^3$ **33.** $y^6 \div y^x = y$

Simplify.

34. $(2^3)^4$ **35.** $(3^5)^2$ **36.** $(4^2)^7$
37. $(10^5)^3$ **38.** $(5^4)^4$ **39.** $(x^5)^4$
40. $(y^3)^3$ **41.** $(t^6)^7$ **42.** $(m^1)^5$

Find the value of x.

43. $(2^3)^x = 2^6$ **44.** $(3^x)^4 = 3^{12}$
45. $(5^x)^2 = 5^8$ **46.** $(7^5)^x = 7^{10}$
47. $(x^3)^x = x^9$ **48.** $(m^x)^5 = m^{15}$
49. $(t^4)^x = t^{20}$ **50.** $(z^x)^5 = z^5$

Problems and Applications

51. The approximate size of a quantity, expressed as a power of 10, is known as an **order of magnitude**. To the nearest orders of magnitude, the mass of the Earth is 10^{25} kg and the mass of the sun is 10^{30} kg. About how many times greater is the mass of the sun than the mass of the Earth?

52. On a test, a student wrote that $2^3 \times 3^2 = 6^5$.
a) What mistake did the student make?
b) What is the value of $2^3 \times 3^2$?

53. A student said that $6^3 \div 2^2 = 3^1$.
a) What mistake did the student make?
b) What is the value of $6^3 \div 2^2$?

NUMBER POWER

Here is one way to use the digits 1, 2, 3, 5, 7, and 9 to add to 648. Find 3 other ways.

$$\begin{array}{r} 251 \\ +397 \\ \hline 648 \end{array}$$

1.9 Use a Formula

The word "sonar" is an acronym for **so**und **na**vigation and **r**anging. Oceanographers use sonar to explore the floors of oceans and lakes. A sonar instrument on a ship can find the depth of the water or the distance to an object. The instrument works by measuring the time a sound signal takes for the round trip to the ocean or lake floor and back.

The speed of sound in water is 1500 m/s. The formula for the distance to an object is

$$d = \frac{1500 \times t}{2}$$

where d is the distance in metres, and t is the time in seconds for a sonar signal to reach the object and bounce back.

It takes 0.35 s for sound to reach a wreck and bounce back. How deep is the water above the wreck?

Understand the Problem

1. What information are you given?

2. What are you asked to find?

3. What formula should you use?

Think of a Plan

Substitute for t in the formula.

Carry Out the Plan

$$\begin{aligned} d &= \frac{1500 \times t}{2} \\ &= \frac{1500 \times 0.35}{2} \\ &= \frac{525}{2} \\ &= 262.5 \end{aligned}$$

EST
$1500 \times 0.4 = 600$
$600 \div 2 = 300$

The water is 262.5 m deep.

Look Back

To check your answer, how could you use the depth to calculate the time?

Use a Formula

1. Write the formula.
2. Replace variables with known values.
3. Calculate the remaining variable.
4. Check that your answer is reasonable.

Problems and Applications

1. The following formula gives the ideal amount of sleep needed each night by people 19 years old or younger.

$$s = \frac{35 - n}{2}$$

The amount of sleep is s hours, and n is the age in years.

a) How much sleep does a 14-year-old need?

b) Jason is 10. He gets 9 h of sleep each night. Is this enough?

2. You are a sonar operator searching for a wreck that is known to be in 900 m to 1000 m of water. Your sonar has generated the table below, which includes time measurements at 16 locations, A_1, A_2, and so on. Use the formula $d = \frac{1500 \times t}{2}$ to decide where the wreck might be.

	Sonar Measurements (s)			
	1	2	3	4
A	0.9	0.8	1.1	1.0
B	1.0	1.4	1.2	0.7
C	0.6	1.3	1.5	1.6
D	1.1	1.2	0.5	2.1

3. To find the distance to an object in space, a radio telescope bounces radio signals off the object and measures the time it takes for the signals to make a round trip. Radio signals travel at 298 000 km/s in space. The formula for the distance to an object is

$$d = \frac{298\ 000 \times t}{2}$$

where d is the distance in kilometres, and t is the time for the round trip in seconds.

a) When the moon is closest to the Earth, the round-trip time is 2.45 s. How far is the moon from the Earth?

b) When the moon is furthest from the Earth, the round-trip time is 2.71 s. How far is the moon from the Earth?

4. In the following formula, T is the air temperature in degrees Celsius at an altitude of h metres, and t is the ground temperature in degrees Celsius.

$$T = t - \frac{h}{150}$$

If the ground temperature is 25°C, what is the temperature outside an aircraft at the following altitudes?

a) 8000 m **b)** 4500 m

5. The formula relates the time, t seconds, an object takes to fall to the Earth from a height of h metres.

$$h = 4.9t^2$$

The table gives the time it takes an object to fall to the ground from the top of several structures. Calculate the height of each.

Structure	Time (s)
Eiffel Tower	8.08
CN Tower	10.63
Washington Monument	5.88

6. If you stand on the Earth and jump up at a speed of 5 m/s, your approximate height, h metres, above the ground after t seconds is given by this formula.

$$h = 5t - 5t^2$$

If you stand on the moon and jump up at the same speed, the corresponding formula is

$$h = 5t - 0.8t^2$$

a) How far would you be from the Earth after 0.5 s? 1 s?

b) How far would you be from the moon after 0.5 s? 1 s? 3 s? 6 s?

c) About how much longer would you stay off the ground on the moon than on the Earth? Explain.

7. Write a problem that can be solved with one of the formulas on this page. Have a classmate solve your problem.

Powers of Ten: Mental Math

The place value chart shows the powers of 10 as numbers in standard form and exponential form.

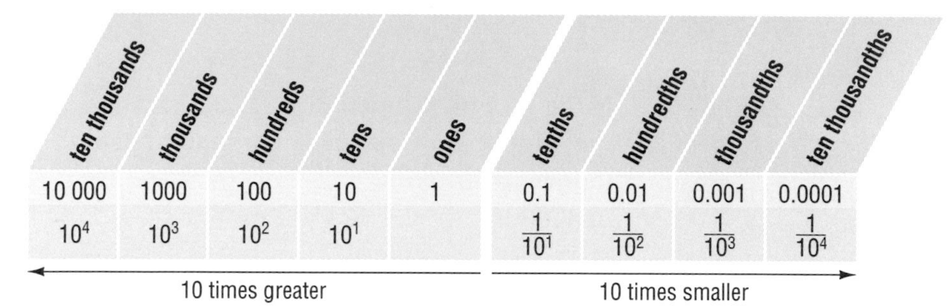

ten thousands	thousands	hundreds	tens	ones	tenths	hundredths	thousandths	ten thousandths
10 000	1000	100	10	1	0.1	0.01	0.001	0.0001
10^4	10^3	10^2	10^1		$\frac{1}{10^1}$	$\frac{1}{10^2}$	$\frac{1}{10^3}$	$\frac{1}{10^4}$

You will complete many calculations using powers of 10. Being able to calculate mentally will simplify the calculations. There will also be times when your skills with powers of 10 will help you to estimate.

Activity ❶ Multiplying: Powers Greater Than 1

1. Copy and complete the charts.

$91.6 \times 10 = \blacksquare$ $\quad$ $0.04 \times 10 = \blacksquare$
$91.6 \times 100 = \blacksquare$ $\quad$ $0.04 \times 100 = \blacksquare$
$91.6 \times 1000 = \blacksquare$ $\quad$ $0.04 \times 1000 = \blacksquare$
$91.6 \times 10\ 000 = \blacksquare$ $\quad$ $0.04 \times 10\ 000 = \blacksquare$

2. Write a rule for multiplying by powers of 10 greater than 1.

3. Calculate mentally.

a) 32×10 **b)** 1.2×100 **c)** 346×1000
d) $2.03 \times 10\ 000$ **e)** 0.6×100 **f)** 0.05×1000
g) 32.8×100 **h)** 5.666×1000 **i)** $7000 \times 10\ 000$

Activity ❷ Multiplying: Powers Less Than 1

1. Copy and complete the charts.

$67.5 \times 0.1 = \blacksquare$ $\quad$ $0.2 \times 0.1 = \blacksquare$
$67.5 \times 0.01 = \blacksquare$ $\quad$ $0.2 \times 0.01 = \blacksquare$
$67.5 \times 0.001 = \blacksquare$ $\quad$ $0.2 \times 0.001 = \blacksquare$

2. Write a rule for multiplying by powers of 10 less than 1.

3. Calculate mentally.

a) 88×0.1 **b)** 5.6×0.01 **c)** 8900×0.001
d) 0.4×0.01 **e)** 0.67×0.1 **f)** 33.3×0.0001
g) 8×0.001 **h)** 1.11×0.1 **i)** 600×0.01

Activity 3 Dividing: Powers Greater Than 1

1. Copy and complete the charts.

$700 \div 10 = ■$	$1.4 \div 10 = ■$
$700 \div 100 = ■$	$1.4 \div 100 = ■$
$700 \div 1000 = ■$	$1.4 \div 1000 = ■$
$700 \div 10\ 000 = ■$	$1.4 \div 10\ 000 = ■$

2. Write a rule for dividing by powers of 10 greater than 1.

3. Calculate mentally.

a) $56 \div 10$ **b)** $6.2 \div 100$ **c)** $970 \div 1000$

d) $2000 \div 10\ 000$ **e)** $6 \div 100$ **f)** $55 \div 100$

g) $32.8 \div 100$ **h)** $66 \div 1000$ **i)** $7000 \div 10$

Activity 4 Dividing: Powers Less Than 1

1. Copy and complete the charts.

$60 \div 0.1 = ■$	$1.2 \div 0.1 = ■$
$60 \div 0.01 = ■$	$1.2 \div 0.01 = ■$
$60 \div 0.001 = ■$	$1.2 \div 0.001 = ■$

2. Write a rule for dividing by powers of 10 less than 1.

3. Calculate mentally.

a) $34 \div 0.1$ **b)** $9.2 \div 0.01$ **c)** $33 \div 0.0001$

d) $234 \div 0.01$ **e)** $9 \div 0.1$ **f)** $0.04 \div 0.01$

g) $0.8 \div 0.001$ **h)** $4500 \div 0.01$ **i)** $9.9 \div 0.001$

Activity 5

1. Have a classmate time you to see how long you take to write the answers to the following questions.

a) 12×100 **b)** $12 \div 100$ **c)** 70×0.01

d) $12 \div 0.01$ **e)** 88×10 **f)** 8.8×0.1

g) 400×0.001 **h)** $3 \div 0.1$ **i)** $3.4 \div 1000$

j) $0.8 \div 10$ **k)** 0.88×10 **l)** $9.6 \div 0.01$

m) 560×0.1 **n)** 0.004×1000 **o)** $9.9 \div 0.001$

2. Check your answers and use the following formula to calculate your score, S. The time you took is t seconds, and W is your number of wrong answers.

$$S = t + 5W$$

Compare your score with your classmates'. The lower your score is the better.

1.10 Large Numbers

A leading Canadian astronomer, Dr. Helen Sawyer Hogg, studied huge balls of stars, known as globular clusters. The stars in these clusters are about 14 000 000 000 years old. The size of a number like 14 000 000 000 depends on the number of zeros. Writing another zero gives 140 000 000 000, which is ten times larger. Removing a zero from 14 000 000 000 gives 1 400 000 000, which is ten times smaller. To avoid mistakes when writing many zeros, we express large numbers in **scientific notation**.

Activity: Look for a Pattern

Copy and complete the table.

Standard Form	Product Form	Scientific Notation
410	4.1×100	4.1×10^2
4 100	$4.1 \times$ ■	■
41 000	$4.1 \times$ ■	■
410 000	$4.1 \times$ ■	■
4 100 000	$4.1 \times$ ■	■
41 000 000	$4.1 \times$ ■	■

Inquire

1. Use the completed table to write the rule for expressing numbers in scientific notation.

2. Why would you not write 2356 in scientific notation?

Example 1

Mercury is about 58 000 000 km from the sun. Write 58 000 000 in scientific notation.

Solution

In scientific notation, a number has the form $x \times 10^n$, where x is greater than or equal to 1 but less than 10, and 10^n is a power of 10.

58 000 000 — The decimal point starts here. Move the decimal point 7 places.

$= 5.8 \times 10\ 000\ 000$

$= 5.8 \times 10^7$

Notice that the number of places you moved the decimal point to the left is the exponent in the power of 10.

Example 2

Calculate $(4.5 \times 10^3) \times (8 \times 10^5)$. Write your answer in scientific notation.

Solution

$$\begin{aligned}(4.5 \times 10^3) \times (8 \times 10^5) &= 4.5 \times 8 \times 10^3 \times 10^5 \\ &= 36 \times 10^{3+5} \\ &= 36 \times 10^8 \\ &= 3.6 \times 10^9\end{aligned}$$

Practice

1 Copy and complete the chart. The first line has been completed for you.

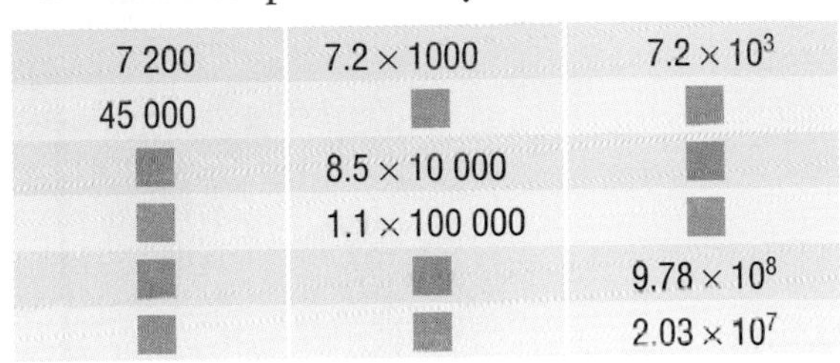

7 200	7.2×1000	7.2×10^3
45 000	■	■
■	$8.5 \times 10\ 000$	■
■	$1.1 \times 100\ 000$	■
■	■	9.78×10^8
■	■	2.03×10^7

State each value of n.

2. $6000 = 6 \times 10^n$

3. $7\ 100\ 000 = 7.1 \times 10^n$

4. $35\ 000 = 3.5 \times 10^n$

5. $54\ 000\ 000 = 5.4 \times 10^n$

6. $145\ 000 = 1.45 \times 10^n$

7. $460\ 000\ 000 = 4.6 \times 10^n$

Write the number that is 10 times as large as each of the following.

8. 77 000

9. 6700

10. 7.6×10^5

11. 9.8×10^7

Write the number that is one-tenth as large as each of the following.

12. 350

13. 2.3×10^5

14. 6.7×10^4

15. 1 300 000

Estimate, then calculate. Write each answer in scientific notation.

16. $(3.4 \times 10^5) \times (5 \times 10^4)$

17. $(4 \times 10^8) \times (1.2 \times 10^7)$

18. $(6.7 \times 10^3) \times (8.9 \times 10^{10})$

Round each number to the highest place value, then calculate using scientific notation.

19. $23\ 000\ 000 \times 341\ 000$

20. $870\ 600\ 000 \times 710\ 000$

21. $92\ 000\ 000 \div 56\ 000$

22. $83\ 000\ 000\ 000 \div 777\ 000\ 000$

Problems and Applications

23. Express each number in scientific notation.

a) Canada is the country with the longest coastline at about 91 000 km.

b) The distance across the universe has been estimated at 800 000 000 000 000 000 000 000 km.

c) Do you want to hunt for sunken treasure? The *HMS Edinburgh* went down with \$95 000 000 in gold on board.

24. How far can you see? The Andromeda Constellation is the furthest object we can see with the unaided eye. It is about 22 000 000 000 000 000 km from Earth.

a) Express this distance in scientific notation.

b) Light travels in space at a speed of about 300 000 km/s. Calculate how many years light takes to reach Earth from the Andromeda Constellation.

25. Why is 56×10^6 not in scientific notation?

26. Work with a classmate.

a) Find the area of Canada in square metres.

b) Find the population of Canada.

c) Calculate how many square metres there are for each Canadian.

CALCULATOR POWER

1. Multiply $1\ 400\ 000 \times 3\ 000\ 000$ using your calculator. How does your calculator display the answer?

2. Scientific calculators display the answer as [4.2 12]. What does it mean?

3. Many calculators have an [EE] key. To input 2.3×10^7, press **2** [·] **3** [EE] **7**. Use the [EE] key to find

a) $(4.5 \times 10^4) \times (5.1 \times 10^7)$

b) $(7.8 \times 10^{11}) \times (6.7 \times 10^9)$

1.11 Using Tables

Activity: Study the Tables

The first table gives the ranges and cruising speeds of several aircraft.

Aircraft	Range (km)	Speed (km/h)
L–1011	8850	880
B–767	9500	855
B–727	4340	850
B–747	6600	890
DC–9	2400	825

The second table gives the flying distances between some Canadian cities. A blank space in the table means that there is no direct flight between the two cities on any of the aircraft listed in the first table. To fly from one city to the other, you would have to land in a city between them.

Flying Distances (km)

	Charlottetown	Edmonton	Fredericton	Halifax	Montreal	Ottawa	Regina	St. John's	Toronto	Vancouver	Winnipeg
Charlottetown						976			1326		
Edmonton						2848	698		2687	826	1187
Fredericton					562				1060		
Halifax					803	958		880	1287		
Montreal			562	803		151		1618	508	3679	1816
Ottawa	976	2848		958	151				363	3350	1687
Regina		698								1330	533
St. John's				880	1618						
Toronto	1326	2687	1060	1287	508	363				3342	1502
Vancouver		826			3679	3350	1330		3342		1862
Winnipeg		1187			1816	1687	533		1502	1862	

Inquire

1. What does the range of an aircraft mean?

2. What is the flying distance between the following cities?

a) Edmonton and Ottawa

b) Regina and Winnipeg

3. You can fly from Fredericton to Vancouver by stopping in Toronto. What is the total flying distance?

4. a) In what other way can you fly from Fredericton to Vancouver?

b) What is the total distance?

c) Is the distance the same as if you travel through Toronto? Explain.

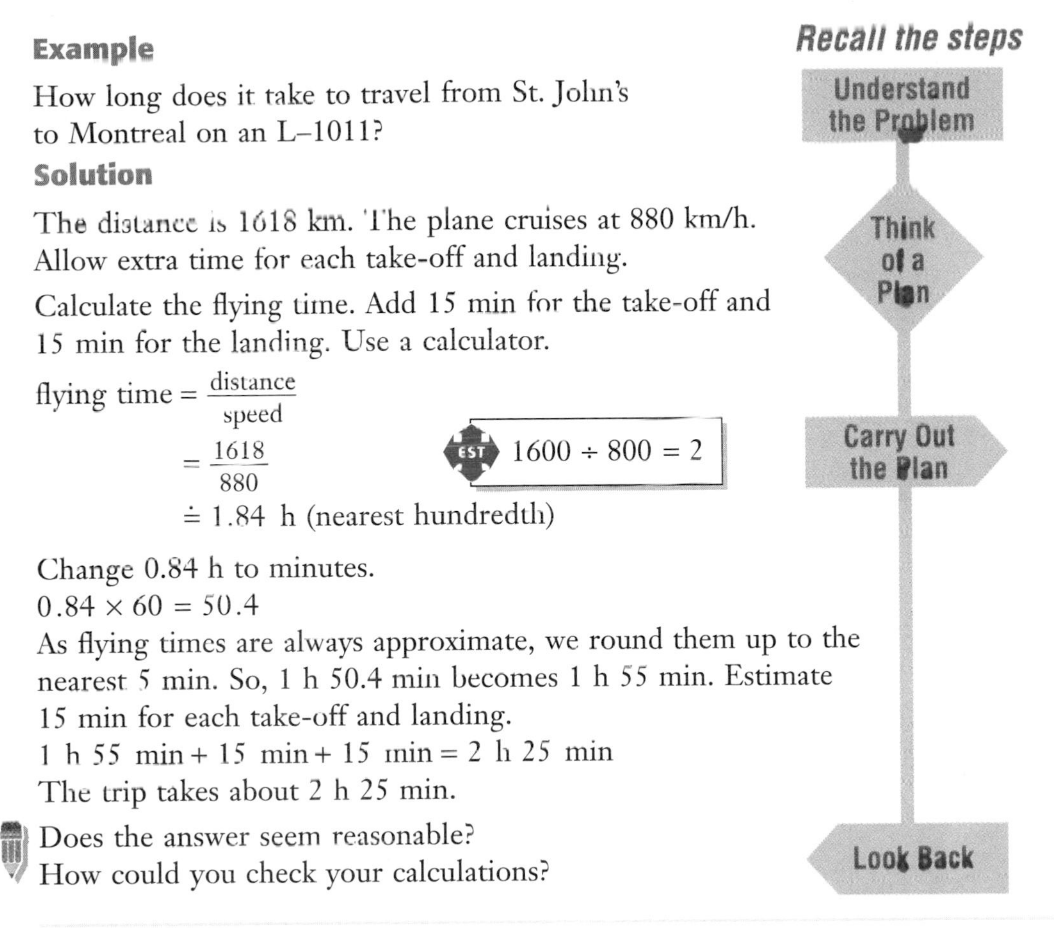

Example

How long does it take to travel from St. John's to Montreal on an L–1011?

Solution

The distance is 1618 km. The plane cruises at 880 km/h. Allow extra time for each take-off and landing.

Calculate the flying time. Add 15 min for the take-off and 15 min for the landing. Use a calculator.

$$\text{flying time} = \frac{\text{distance}}{\text{speed}}$$
$$= \frac{1618}{880}$$
$$\doteq 1.84 \text{ h (nearest hundredth)}$$

EST $1600 \div 800 = 2$

Change 0.84 h to minutes.
$0.84 \times 60 = 50.4$
As flying times are always approximate, we round them up to the nearest 5 min. So, 1 h 50.4 min becomes 1 h 55 min. Estimate 15 min for each take-off and landing.
1 h 55 min + 15 min + 15 min = 2 h 25 min
The trip takes about 2 h 25 min.

Does the answer seem reasonable?
How could you check your calculations?

Problems and Applications

1. From the second table, what is the flying distance between each pair of cities?
a) St. John's and Montreal
b) Halifax and Toronto
c) Regina and Winnipeg
d) Winnipeg and Edmonton
e) Ottawa and Vancouver

2. If you did each of the following journeys on the same type of aircraft, which journey would take the longest? Explain.
a) Charlottetown to Toronto
b) Regina to Vancouver
c) Ottawa to Halifax
d) Winnipeg to Edmonton
e) Fredericton to Montreal

3. How many kilometres will you fly if you travel from Halifax to Ottawa to Edmonton?

4. The distance by train from Regina to Vancouver is 1795 km. How much shorter is the flying distance?

5. Justine is an airline pilot. Last year, she flew from Winnipeg to Ottawa and back a total of 21 times. How far did she fly?

6. Carl planned a trip from Winnipeg to Halifax. There was no direct flight, but he could fly via Ottawa, Toronto, or Montreal. Which trip was the shortest and by how much?

7. How long does a trip from Toronto to Fredericton take on a DC–9, including 15 min for each take-off and landing?

CONTINUED

8. Your B–727 leaves Montreal for Toronto at 11:45. At what time will you arrive in Toronto if you include the time for take-off and landing?

9. Could a DC–9 fly from Vancouver to Toronto without refuelling? Explain.

10. The flying distance from Edmonton to London, England, is 6782 km. Which of the aircraft could fly non-stop from Edmonton to London?

11. What is the total length of time you would spend in the air if you flew from Charlottetown to Toronto in an L–1011?

12. A B–767 left Halifax for St. John's at 08:50. What was the time in St. John's when the plane landed? (Hint: The Data Bank on pages 564 to 569 includes a table of time zones.)

13. A B–727 left Winnipeg for Vancouver at 15:15. What was the time in Vancouver when the plane arrived?

14. A B–747 left Regina for Ottawa at 16:05. The trip included a 25-min stopover in Winnipeg. What was the time in Ottawa when the plane arrived? Allow for all take-offs and landings.

15. The flying distance from Winnipeg to Honolulu, Hawaii, is 6124 km. If your B–767 left Winnipeg on November 12 at 22:30, state the date and time when you arrived in Honolulu.

16. The flying time on a direct flight from Vancouver to Montreal can be as much as 45 min shorter than the return flight on the same plane.

a) Explain why.

b) Determine the flying time for a flight on a B–747 from Vancouver to Montreal. What assumptions have you made?

17. Plan an around-the-world trip that starts at the airport nearest you and finishes there. Start your trip at 09:00 on December 1. Your plane is an L–1011. Travel in an easterly or westerly direction. Stop in major cities on your trip. Stop in at least one city in the southern hemisphere. Allow an extra 2 h at each stop for refuelling, as well as the 15 min for take-off and 15 min for landing. The best trip is the one that gets you back as soon after 09:00 on December 1 as possible. Compare your trip with your classmates'.

PATTERN POWER

Some pairs of 2-digit numbers have the same product when the digits are reversed.

$$12 \times 42 = 21 \times 24$$

$$24 \times 63 = 42 \times 36$$

$$24 \times 84 = 42 \times 48$$

There are 11 other pairs of these numbers. Look for the pattern, then find the pairs of numbers.

1.12 Order of Operations

Many activities require you to perform steps in a certain order. For example, after getting into a car, the driver

- puts on the seat belt
- inserts the key and starts the car
- places the gearshift in "Drive"
- checks for traffic in all directions
- releases the hand brake
- turns on the indicator
- depresses the accelerator and drives

Mathematical operations must also be done in order.

Activity: Study the Problem

The Last Chance Gas Station ran a scratch-and-win contest. For an entry to be valid, a customer had to evaluate this expression correctly.

$$20 \times 3 + 15 \div 5$$

Brad's answer was 15, Cashala's was 63, and Pierre's was 72.

Inquire

1. In what order were the calculations done by
a) Brad? **b)** Cashala? **c)** Pierre?

2. What is the correct value of the expression?

To avoid confusion, mathematicians have agreed on the following **order of operations**. Remember it with the acronym "**BEDMAS.**"

- Do all calculations in **b**rackets first. **B**
- Simplify numbers with **e**xponents. **E**
- **D**ivide and **m**ultiply in order from left to right. **DM**
- **A**dd and **s**ubtract in order from left to right. **AS**

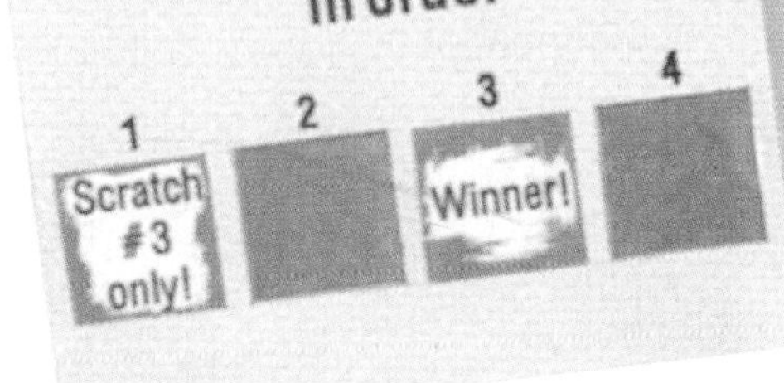

Example 1

Simplify $3(7-2)+10 \div 5$.

Solution

$3(7-2)+10 \div 5$ — **brackets**
$= 7(5)+10 \div 5$ — **divide and multiply**
$= 35+2$ — **add**
$= 37$

CONTINUED

Example 2

Simplify $\frac{9+3}{4}+7$.

Solution

The fraction bar is a division and grouping symbol.

$$\begin{aligned}\frac{9+3}{4}+7 &= \frac{(9+3)}{4}+7 && \textbf{brackets}\\ &= \frac{12}{4}+7 && \textbf{divide}\\ &= 3+7 && \textbf{add}\\ &= 10\end{aligned}$$

Example 3

Simplify $5(9-7)^3+4$.

Solution

$$\begin{aligned}&5(9-7)^3+4 && \textbf{brackets}\\ &= 5(2)^3+4 && \textbf{exponents}\\ &= 5(8)+4 && \textbf{multiply}\\ &= 40+4 && \textbf{add}\\ &= 44\end{aligned}$$

Example 4

Simplify $8.6^2-2(7.2-1.3)$.

Solution

$$\begin{aligned}&8.6^2-2(7.2-1.3) && \textbf{brackets}\\ &= 8.6^2-2(5.9) && \textbf{exponents}\\ &= 73.96-2(5.9) && \textbf{multiply}\\ &= 73.96-11.8 && \textbf{subtract}\\ &= 62.16\end{aligned}$$

C 7 . 2 − 1 . 3 = × 2 = M 8 . 6 x² − MRC = 62.16

or

C 8 . 6 x² − 2 × (7 . 2 − 1 . 3) = 62.16

Practice

List the operations in the order you would perform them.

1. $5 \times 3 + 2$

2. $8 - 3 + 2^3$

3. $7 + 5 - 3$

4. $4 \times 3 \div 2$

5. $36 \div 4 - 2^2$

6. $5(6-2)+3$

7. $(3+4)^3 \times 4$

8. $8 + 6 \div 2$

Find the error in each calculation.

9. $5 \times 3 + 4 \times 2$
$= 5 \times 7 \times 2$
$= 70$

10. $20 - 6 \times 3$
$= 14 \times 3$
$= 42$

11. $18 - 7 + 8$
$= 18 - 15$
$= 3$

12. $9 + 3 \times 2^2$
$= 9 + 6^2$
$= 9 + 36$
$= 45$

13. $36 \div 3 \times 2$
$= 36 \div 6$
$= 6$

14. $3^2 + 2^2 + 1^2$
$= 6^2$
$= 36$

Simplify.

15. $12 + 7 - 6$

16. $12 \times 5 - 6$

17. $14 - 10 \div 5$

18. $20 \div 5 \times 3$

19. $2^3 + 5 \times 6$

20. $3(8-2)+6^2$

21. $(18-6) \div 4$

22. $4^3-(5-3)^3$

23. $\frac{10}{5-3}$

24. $\frac{6 \times 2}{4+2}$

25. $3^2-(4 \times 2)$

26. $2^3+3^3-2^4$

27. $\frac{3+4 \times 3}{5}$

28. $\frac{9 \times 4 - 6}{7-1}$

29. $\frac{20 \div 5}{2 \times 2}$

30. $5^2+\frac{19-11}{18 \div 9}$

Simplify.

31. $3.4 \times 2 - 1.6 \div 2^2$

32. $24.5 + 13 \times 0.6$

33. $\frac{9.8 + 7.8}{4}$

34. $23.5 - 1.2(3.6 + 4.4)$

35. $(5.6 - 4.3)(8.1 + 2.2)$

Simplify.

36. $3.2^3 - (4.6 + 2.1)$

37. $5.2 \times 3 - 2.4^2$

38. $6.3 \div 9 + 11.2$

39. $(14 + 5.6) \div 2$

40. $34.6 - 15.5 \div 5$

41. $1.2 \times 100 + 15$

Problems and Applications

Use brackets to make each statement true.

42. $7 + 1 \div 2 = 4$

43. $5 \times 8 - 6 = 10$

44. $15 \div 3 + 2 + 1 = 4$

45. $6 + 4 \div 2 \times 2 = 2.5$

46. $4.8 + 1.6 - 1.2 \times 5 = 6.8$

47. $3 \times 1.7 + 5.7 \div 3 = 7.4$

48. $5 + 2^2 \div 7 + 12 = 19$

49. When the grade 9 class and 2 teachers took the bus to the Science Museum, they paid 2 adult fares at $1.30 each and 25 student fares at $0.65 each. Admission to the museum was $4.50 for each teacher and $3.00 for each student.

a) Write an expression for the total cost of the trip.

b) Use your expression to find the total cost.

50. A group of 10 skiers ate dinner at the Mountaintop Restaurant. The meal had a set price of $19.95, including the tax and tip. Four of the group had discount coupons and paid only half price. What was the total cost of dinner?

51. Use the following rules to build a sequence of numbers, starting with any number and ending with 1.

- If the previous number in the sequence is odd, find the next number by tripling it and adding 1.
- If the previous number is even, find the next number by taking half of it.

If we start with 5, the sequence is 5, 16, 8, 4, 2, 1.

a) Write the sequences that start with several different even numbers and several different odd numbers.

b) Explain why the sequences always end in 1.

52. Here is a way to make the number 1 using five 2s and the order of operations.

$$2^2 - \frac{2}{2} - 2 = 1$$

Write the numbers from 2 to 9 using five 2s and the order of operations. Compare your expression for each number with a classmate's.

WORD POWER

Lewis Carroll invented a word game called doublets. The object is to go from one word to another by changing one letter at a time. You must form a real word each time you change a letter. Change the word FOOT to the word BALL by changing one letter at a time. Compare your list of words with your classmates'. The best solution is the one with the fewest steps.

1.13 Interpret Graphs

Much of the information you receive from newspapers and magazines is displayed on graphs. Therefore, the ability to interpret graphs correctly is an important life skill. The graph shows the distance from Port Colborne to Brampton and the time taken for a car to travel this distance. What is happening between P and Q; Q and R; R and S; S and T?

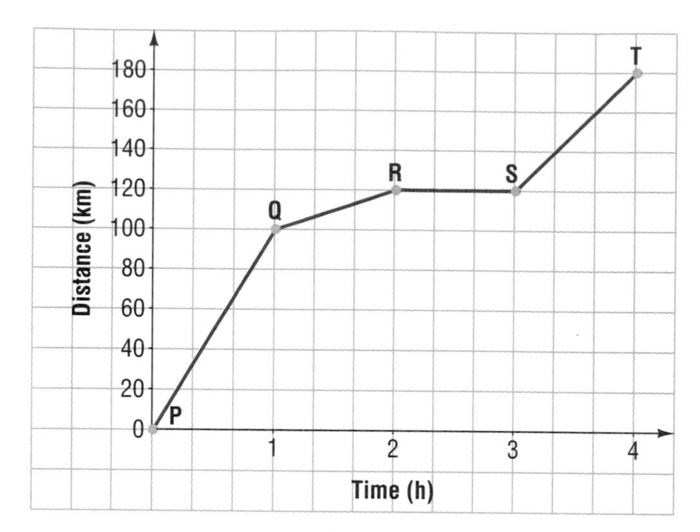

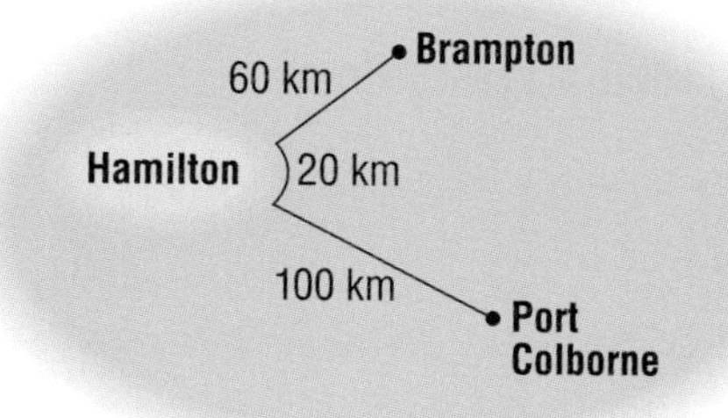

1. What information are you given?

2. What are you asked to find?

3. Do you need an exact or approximate answer?

Read the information from the graph. Use it to solve the problem.

From P to Q The car covered 100 km in 1 h, so it travelled at 100 km/h from Port Colborne to Hamilton.

From Q to R The car took 1 h to drive 20 km around Hamilton. It averaged 20 km/h.

From R to S The car stopped for 1 h on the outskirts of Hamilton on the Brampton side.

From S to T The car travelled the 60 km from Hamilton to Brampton in 1 h. The car travelled at 60 km/h.

Does the answer seem reasonable?

Interpret Graphs
1. Interpret the graph.
2. Use the information to solve the problem.
3. Check that the answer is reasonable.

Problems and Applications

1. The graph shows the amount of gasoline in the tank of a salesperson's car at different times of a work day. Describe how the salesperson might have spent the day.

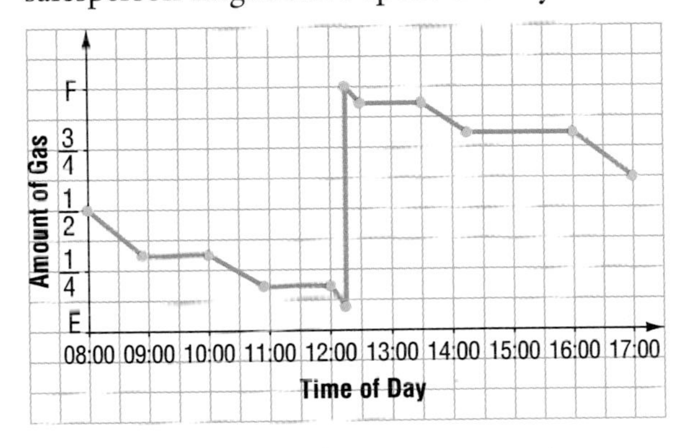

2. The 2 graphs describe 2 pleasure boats. Compare the boats by age, cost, speed, and length.

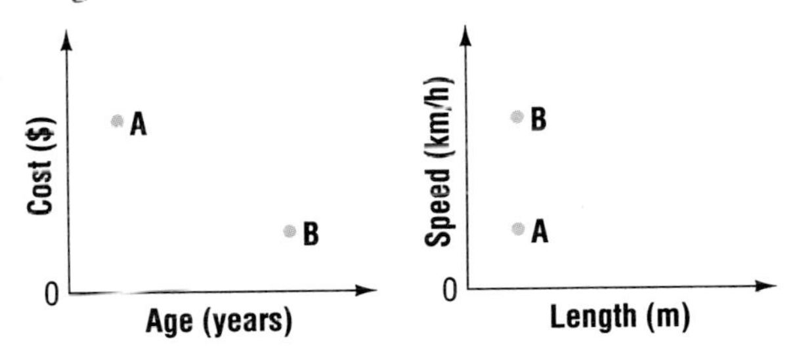

3. a) The graph shows the distance versus time for a car travelling from Fredericton to Moncton. Describe what is happening between A and B; B and C; C and D; D and E.

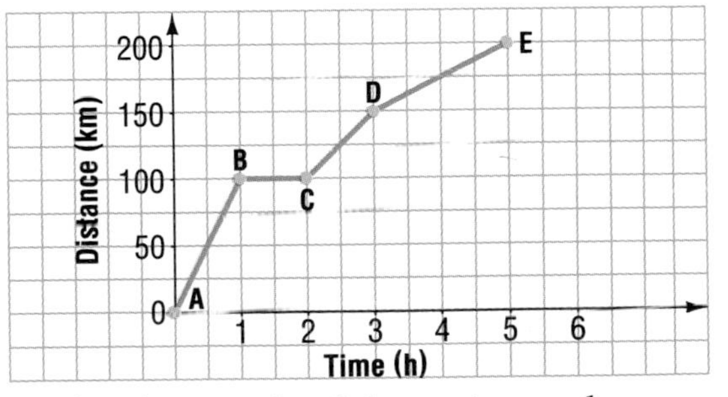

b) Sketch a graph of the car's speed versus time.

4. You are the first person to get on and off a Ferris wheel with 8 cars. Once the cars are loaded, the Ferris wheel makes 3 revolutions before letting people off. Sketch a graph of your height versus time.

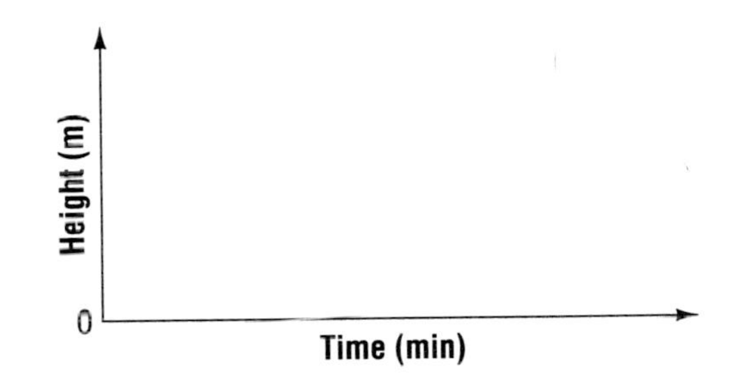

5. Suppose someone turned on the hot-water tap on a kitchen sink and left it running. Sketch a graph of the water temperature versus time as the tap is running.

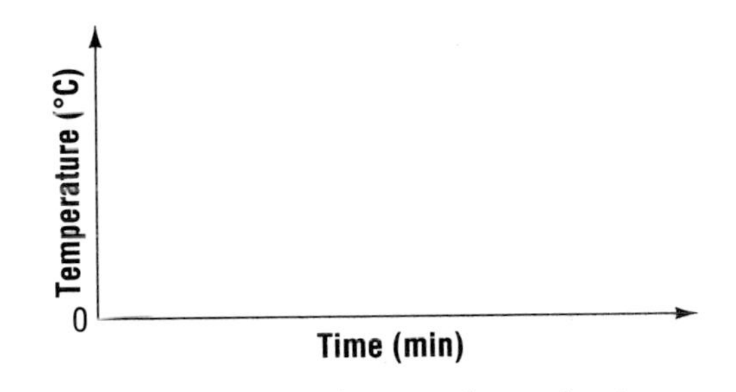

6. Sketch a graph of the number of minutes of daylight versus the month of the year.

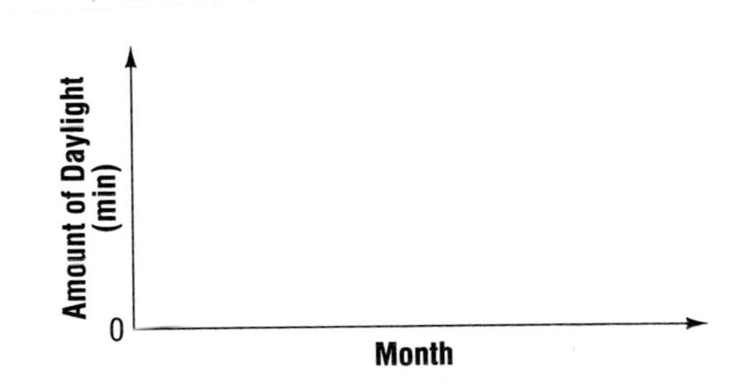

7. Write a graph-sketching problem that involves each of the following. Have a classmate solve each problem.

a) the temperature of water in a kettle versus time

b) the depth of water in a bathtub versus time

Fingerprints

Because your fingerprints never change and are different from everyone else's, they can be used to identify you. There are 3 different patterns of fingerprints, known as arches, whorls, and loops.

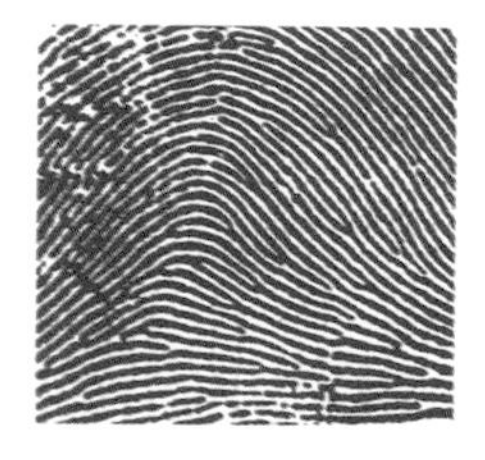

Plain Arch
The ridge lines enter on one side and curve up and exit on the other side.

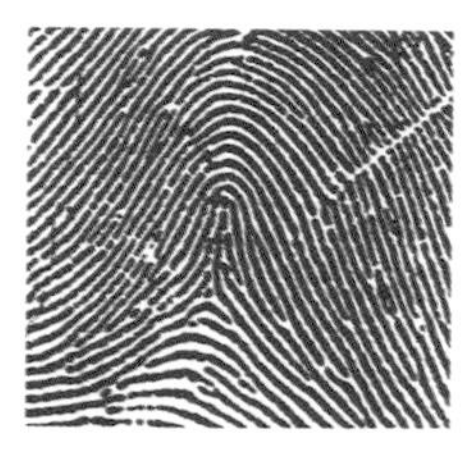

Tented Arch
The ridge lines are the same as a plain arch, except that they make a sharp point or tent in the middle.

Whorls
The ridge lines are circles or ovals.

Loops
The ridge lines enter on one side of the print, curve, and exit on the same side. An **ulnar loop** slants toward the ulna bone in the wrist. A **radial loop** slants toward the radius bone in the wrist.

Right-hand ulnar loop or left-hand radial loop.

Right-hand radial loop or left-hand ulnar loop.

The RCMP receives many sets of prints with requests to identify the people they belong to. It is impossible to check the prints against all the prints on file. The RCMP first classifies the prints.

Activity 1

1. Identify each of your fingerprints as an arch, a loop, or a whorl.

2. According to fingerprint experts, there are 5% arches, 65% loops, and 30% whorls on fingerprints. Are these percents true for your class?

Activity ❷

The primary fingerprint classification is based on the number of whorls. Prints are assigned the following values.

Whorl = 1 Arch = 0 Loop = 0

These values are substituted into the following formulas.

$$(\text{Rt Index}) \times 16 + (\text{Rt Ring}) \times 8 + (\text{Lt Thumb}) \times 4 + (\text{Lt Middle}) \times 2 + (\text{Lt Little}) \times 1 + 1$$

$$(\text{Rt Thumb}) \times 16 + (\text{Rt Middle}) \times 8 + (\text{Rt Little}) \times 4 + (\text{Lt Index}) \times 2 + (\text{Lt Ring}) \times 1 + 1$$

For example, Yoshiko has whorls on her left thumb, left little finger, and right thumb. All her other prints are arches and loops. Her primary classification is as follows.

$$(0) \times 16 + (0) \times 8 + (1) \times 4 + (0) \times 2 + (1) \times 1 + 1$$
$$(1) \times 16 + (0) \times 8 + (0) \times 4 + (0) \times 2 + (0) \times 1 + 1$$

$$0 + 0 + 4 + 0 + 1 + 1 = 6$$
$$16 + 0 + 0 + 0 + 0 + 1 = 17$$

Her primary classification is $\frac{6}{17}$

1. What is your primary classification?
2. How many different primary classifications are there in your class?
3. How many different primary classifications are possible?

Activity ❸

The secondary classification is a fraction made up of letters based on the prints on the index fingers. The fraction is

$$\frac{\textit{right index finger}}{\textit{left index finger}}.$$

The letters are A for plain arch, T for tented arch, W for whorl, U for ulnar loop, and R for radial loop. If the right index finger is a whorl and the left index finger is a radial loop, the secondary classification is $\frac{W}{R}$.

1. What is your secondary classification?
2. How many different secondary classifications are there in your class?
3. How many different secondary classifications are possible?

Activity ❹

The combined classification of a set of prints is written in the form $\frac{6\ W}{17\ R}$.

1. What is your combined classification?
2. How many different combined classifications are there in your class?
3. How many different combined classifications are possible?

Review

1. State if you need an exact answer or an estimate.
a) Your employer asks you if your pay has been calculated correctly.
b) You are the timer for a professional football game. The referee asks you how much time is left in the game.
c) A reporter from Holland asks how long you spend in school each year.
d) The student council wants to know the cost of food for the athletic banquet.

2. Decide if you need an exact answer or an estimate. Then, choose a calculation method.
a) the number of seats in the school auditorium
b) the score in a ball hockey game
c) the batting averages of the Toronto Blue Jays
d) the results of a provincial election

3. Use rounding to estimate.
a) $46.51 + $93.60 + $15.35
b) $513.40 – $480.29
c) $7.60 + $8.35 + $9.74 + $11.33
d) $624.30 – $129.70 – $263.20

4. Use clustering to estimate.
a) $37.40 + $43.80 + $41.40
b) $12.80 + $9.35 + $8.40
c) $15.85 + $17.70 + $12.10 + $14.14

5. Use compatible numbers to estimate.
a) $14.60 + $85.70 + $260.20
b) $3.70 + $8.15 + $73.31
c) $25.10 + $115.20 + $74.18 + $182.20

Estimate, then calculate.

6. $569 - 292$ **7.** $4521 - 2888$

8. $4.56 - 3.87$ **9.** $1.008 - 0.76$

10. 345×54 **11.** 7401×83

12. 4.23×0.72 **13.** 5.67×1.9

14. $966 \div 42$ **15.** $16\ 002 \div 63$

16. $18.72 \div 3.6$ **17.** $4.704 \div 0.84$

18. $4567 + 982 + 3002 + 801$

19. $67.78 + 9.07 + 92.6$

Calculate mentally.

20. 3.21×10 **21.** 0.56×100

22. 72.4×0.1 **23.** 561×0.01

24. 0.009×1000 **25.** 4600×0.001

26. $340 \div 100$ **27.** $18 \div 1000$

28. $42 \div 0.1$ **29.** $2.3 \div 0.01$

30. $9.02 \div 10$ **31.** $76 \div 0.001$

Express in exponential form.

32. $2^3 \times 2^5$ **33.** $5^7 \div 5^3$ **34.** $(2^4)^2$

35. $7^8 \times 7^4$ **36.** $(3^3)^3$ **37.** $6^5 \div 6$

Simplify.

38. $7 \times 5 - (3 \times 2)$ **39.** $8 \times 7 - 3 + 6^2$

40. $(8 - 3) \times (9 - 1)$ **41.** $30 \div 6 - 1 + 11$

42. $6(12.6 - 9.3) + 9.1$

43. $(21.3 - 5.7) \times 6.5$

44. $2.1^2 + 8.9 - 3.1$

Write in scientific notation.

45. 43 000 000 **46.** 710 000

47. 80 000 000 000 **48.** 389 000 000 000

Write in standard form.

49. 3.5×10^3 **50.** 5.6×10^7

51. 9.99×10^1 **52.** 1.25×10^{10}

53. Evaluate for $c = 5$, $d = 6$.
a) $5c + 4d$ **b)** $3d + 4cd$
c) $c^2 + d^2$ **d)** $3cd + c$
e) $14 - c - d$ **f)** $6c - 2d + 8$

54. Find the value of $4t - 2$ when t equals
a) 8.6 **b)** 7.2 **c)** 9.9

Estimate the cost.

55. 15 tickets at \$8.75 each

56. 6 pairs of socks at \$4.20 a pair

57. 21 team jerseys at \$81.50 each

58. 7 notebooks at \$6.85 each

59. Bill earns \$7.80/h for the first 40 h he works in a week and \$11.70/h for each hour over 40 h. Last week, he worked 47 h. How much did he earn?

60. Paula decided she could drive 440 km to the camp at 80 km/h. If she leaves at 08:00, at what time will she arrive at the camp? What assumption did you make?

61. Sandra has two part-time jobs. Last week, she worked 7.5 h at a video store for \$8.25/h. She also worked 4 h at a gas station for \$9.10/h. How much did she earn altogether?

62. A baton is thrown straight up at 25 m/s. The expression $h = 25t - 5t^2$ gives the baton's height, h metres, after t seconds.

a) What is the height of the baton after 1 s? 2 s? 3 s?

b) After how many seconds will the thrower catch the baton? State your assumptions.

63. Use the tables on page 30 to find how long a trip from Edmonton to Toronto takes in a B–747. Include 15 min for take-off and 15 min for landing.

64. What time will it be 2000 h from now?

65. To rent a river boat for a banquet costs \$250, plus an extra \$35 per person.

a) Write an expression for the cost of a banquet for n people.

b) How much would it cost for a banquet for 152 people?

c) If the people shared the cost equally, how much would each have to pay?

66. a) Why is 0.45×10^7 not in scientific notation?

b) Write 0.45×10^7 in scientific notation.

Group Decision Making
Researching Earth Science Careers

To work effectively in groups.

a) Contribute ideas by speaking quietly.

b) Encourage others to participate, and listen to them carefully.

c) Respect the opinions of others.

d) Ask the teacher for help only when your whole group has the same difficulty.

1. Brainstorm as a class 6 earth science careers to investigate. They might include geographer, cartographer, geologist, paleontologist, and seismologist.

2. In home groups of 6, decide which career each person will research.

1 2 3 4 5 6 | 1 2 3 4 5 6

Home Groups

1 2 3 4 5 6 | 1 2 3 4 5 6

3. In your home group, decide what you want to know about each career. Make sure you include the ways in which math is used in each career.

4. Research your career individually.

5. Form an expert group with students who researched the same career as you. Combine your findings.

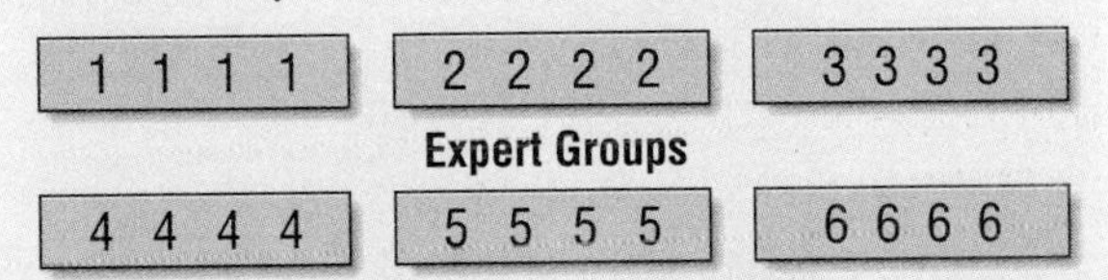

6. In your expert group, prepare a report on the career. The report can take any form the group chooses.

7. In your expert group, evaluate the process and your presentation. What worked well and what would you do differently next time?

Chapter Check

1. **a)** Write a problem for which you would estimate the answer.
   b) Write a problem for which an exact answer is needed.

2. Write a problem for which you would use each calculation method.
   a) mental math
   b) calculator
   c) computer

Estimate, then calculate.

3. $234 + 561 + 88$ **4.** $4.37 + 0.98 + 17.69$

5. $303 - 187$ **6.** $42.03 - 6.75$

7. 45×56 **8.** 3.81×8.6

9. $2025 \div 25$ **10.** $0.364 \div 5.6$

11. 4.8^2

Estimate the cost.

12. 17 posters at $3.85 each

13. 36 stamps at $0.57 each

Evaluate.

14. 3.13×1000 **15.** $23.9 \div 1000$

16. $56.9 \div 0.01$ **17.** 1.56×0.1

Write in scientific notation.

18. 45 000 000 **19.** 213 000

Evaluate.

20. 4^2 **21.** 3^4 **22.** 5^3

Write the answer in exponential form.

23. $3^4 \times 3^5$ **24.** $2^7 \div 2^3$ **25.** $(5^2)^4$

Simplify.

26. $40 \div 5 - 2 \times 3$ **27.** $(7 - 5)^3 + 28 \div 7$

28. $2^3 + 4^2 - 3$ **29.** $5(11 - 8) - 2^2$

30. $10(4.8 - 2.5) + 7.8$ **31.** $0.6^2 + 7.2 \div 0.1$

Evaluate for $r = 4$ and $t = 5$.

32. $5r + 4t$ **33.** $3rt + t^2$

Evaluate for $x = 0.5$ and $y = 1.2$.

34. $3x + 5y$ **35.** $10xy - 2x$

36. At a flea market, Carol bought 4 old records for $4.95 each and 3 books for $7.25 each. How much did she spend?

37. A ball is thrown upward at a speed of 20 m/s. The expression $h = 20t - 5t^2$ gives the ball's height, h metres, after t seconds. Find the height after these lengths of time.
a) 1 s **b)** 2 s **c)** 3 s

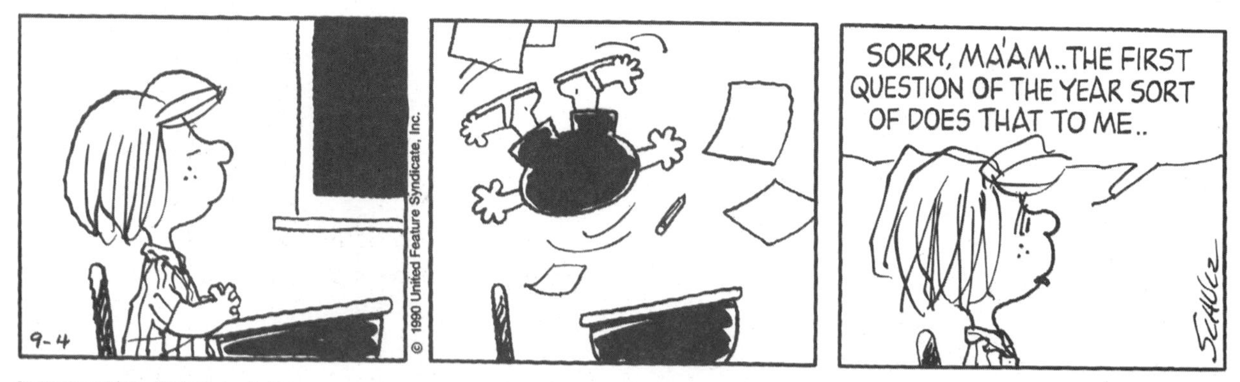

9-4
© 1990 United Feature Syndicate, Inc.
SORRY, MA'AM..THE FIRST QUESTION OF THE YEAR SORT OF DOES THAT TO ME..
SCHULZ

Reprinted by permission of United Feature Syndicate.

Using the Strategies

1. Take 2 sets of the numbers from 1 to 13. You now have 26 numbers. Write these numbers in pairs so that the sum of every pair of numbers is a perfect square.

2. The cost to rent a car is given by the following formula.

$$C = 35n + 0.4d$$

C is the cost in dollars, n is the rental period in days, and d is the distance driven in kilometres. How much would you pay if you rented a car for 5 days and drove 900 km?

3. Students are allowed into the school at 08:00. Classes start at 09:00. There are 15-min breaks that begin at 10:30 and 14:30. Lunch is from 12:00 to 13:00. School is dismissed at 16:00. Students must leave the building by 17:00. Students are allowed to use the pop machine in the school cafeteria before and after school and during their breaks and lunch hour. The machine is filled twice a day, at 07:00 and 14:00. Sketch a graph of the number of cans in the machine versus time on a hot Monday in September.

4. If 4 days before tomorrow is Friday, what day is 2 days after yesterday?

5. The whole number 5 can be written as the difference in the squares of 2 whole numbers.

$$5 = 3^2 - 2^2$$

What other whole numbers between 1 and 10 can be written as the difference in the squares of 2 whole numbers?

6. Suppose you drove from Moosonee, on James Bay, through Canada, the United States, and Mexico, and continued to the tip of South America. The trip started on January 1 and ended on February 27 of the same year. Sketch a graph of the outside temperature versus the distance travelled.

7. If New Year's Day in a leap year is on a Tuesday, on what day of the week will July 1 fall?

8. Two days ago, Robert was 16 years old. Next year, Robert will be 19 years old. What is today's date and when is Robert's birthday?

9. The time is now 07:34. What time was it 15 h and 43 min ago?

10. The front seat of a power boat can hold 3 people. In how many ways can Jose, Alicia, and Jennifer travel in the front seat if only Jose and Alicia can drive the boat?

11. The number 63 can be written as the sum of consecutive whole numbers as follows.

$$63 = 20 + 21 + 22$$

a) Find another way to write 63 as the sum of consecutive whole numbers.
b) Find 4 consecutive whole numbers that add to 138.

12. When you add the digits of the number 45, you create a perfect square.

$$4 + 5 = 9 \text{ or } 3^2$$

How many other numbers between 1 and 100 have digits that add to give perfect squares?

13. The table shows the maximum speeds of some animals.

Animal	Maximum Speed (km/h)
Ostrich	80
Coyote	72
Elephant	40
Porcupine	18

How many minutes less does a coyote take than a porcupine to run 1 km? What assumptions have you made?

CHAPTER 2

Integers

The highest land on Earth is the peak of Mount Everest, 8863 m above sea level. The lowest land on Earth is the shore of the Dead Sea, 400 m below sea level.

What is the difference between the elevations of the peak of Mount Everest and the shore of the Dead Sea?

Research the elevation of your home town. What is the difference between the elevation of your home town and the elevation of the peak of Mount Everest? What is the difference between the elevation of your home town and the elevation of the shore of the Dead Sea?

Activity ❶ Rating Hockey Players

One way to evaluate an NHL player's effectiveness is to use the Plus-Minus System.

Players on the ice when their team scores a goal are each credited with +1. Players on the ice when their opponents score a goal are each credited with −1.

The table gives statistics for 12 players.

Player	GP	G	A	Pts	+/−	PIM
Smith	57	22	50	72	+7	60
Olychuk	56	26	27	53	−17	41
Goldstein	43	21	31	52	+9	44
Lemieux	56	25	25	50	+8	37
Courtner	51	21	23	44	−7	47
Donner	46	20	22	42	−9	34
Delecourt	51	6	18	24	−13	82
Salter	47	1	20	21	+14	50
McDonald	48	7	8	15	+2	38
Sorensen	10	5	6	11	−5	11
Blair	9	1	3	4	−6	6
Jereneskon	10	1	1	2	−1	4

1. List, in order, the 4 players with the best Plus-Minus records.

2. List, in order, the 4 players with the worst Plus-Minus records.

3. The following table shows the numbers of goals scored while certain players were on the ice. Calculate each player's Plus-Minus record.

Player	Own Team's Goals	Opponents' Goals
Carlisle	34	32
Benson	46	50
Korchinski	54	33
Beltz	31	40
Kranich	56	50
Fragnal	26	45
Meechum	41	41
Bolchek	42	57
Mazza	35	33

Activity ❷ Golf Scores

1. If a golf hole is par 4, a good golfer should take 4 strokes to complete the hole. Golfers use the following terms to describe how well they did on a hole.

double bogey: 2 over par
bogey: 1 over par
par
birdie: 1 under par
eagle: 2 under par

Represent each term by a suitable integer.

2. Sportscasters use integers to report golf scores. On 9 holes, Michelle scored as follows.

par, par, birdie, birdie, par, bogey, bogey, double bogey, par

a) Was Michelle over or under par after 9 holes?

b) Express Michelle's score after 9 holes as an integer.

c) If par for the 9 holes is 36, what was Michelle's score?

Activity ❸ Temperature Changes

Copy and complete the table.

Temperature (°C)		
Initial	Change	Final
+21	Down 5	
+3	Down 4	
−7	Up 2	
−3	Up 6	
+3	Up 4	
+4	Down 7	
−5	Down 8	
−11	Up 5	
−13	Down 9	
0	Up 4	
0	Down 8	
−10	Up 10	
−8	Up 7	
−6	Down 6	

Activity ❹ Inequalities

When we say $x > 5$, where x is a whole number, we mean that x is all the whole numbers greater than 5. This means that x has the following values.

6, 7, 8, 9, 10, 11, 12, 13, …

The 3 dots mean that the whole numbers continue above 13.

The graph shows the values of x on a number line.

0 1 2 3 4 5 6 7 8 9 10 11 12 13

The red arrow on the end of the line is like the 3 dots. It means that the values of x continue above 13.

Graph the following whole numbers on a number line.

a) all the whole numbers greater than 3

b) all the whole numbers less than 8

c) all the whole numbers less than 9 and greater than 2

d) all the whole numbers greater than 4 and less than 12

e) x is greater than 7

f) x is less than 5

g) x is less than 10 and greater than 2

h) x is greater than or equal to 3

i) x is less than or equal to 5

j) $x > 3$ **k)** $x < 6$

l) $x \geq 7$ **m)** $x \leq 9$

Mental Math

Calculate.

1. 1000 × 21 **2.** 7 × 100 **3.** 13 × 100 **4.** 44 × 10 000 **5.** 10 × 361

6. 1000 × 234 **7.** 400 × 100 **8.** 60 × 1000 **9.** 1000 × 300 **10.** 10 × 2000

Calculate.

11. 3 × 101 **12.** 5 × 101 **13.** 4 × 202 **14.** 303 × 2 **15.** 505 × 2

16. 606 × 3 **17.** 4 × 707 **18.** 808 × 9 **19.** 404 × 3 **20.** 7 × 909

Calculate.

21. 8 × 1001 **22.** 1001 × 7 **23.** 2002 × 4 **24.** 5 × 3003 **25.** 9 × 2002

26. 6 × 4004 **27.** 7 × 6006 **28.** 4 × 8008 **29.** 7007 × 7 **30.** 5005 × 8

Calculate.

31. 6200 ÷ 100 **32.** 7500 ÷ 10 **33.** 4000 ÷ 1000 **34.** 12 000 ÷ 100 **35.** 7600 ÷ 10

36. 8700 ÷ 100 **37.** 940 ÷ 10 **38.** 46 500 ÷ 100 **39.** 23 000 ÷ 10 **40.** 141 000 ÷ 1000

Calculate.

41. 1000 ÷ 20 **42.** 500 ÷ 20 **43.** 8000 ÷ 40 **44.** 1600 ÷ 40 **45.** 2400 ÷ 60

46. 3300 ÷ 30 **47.** 1800 ÷ 90 **48.** 2800 ÷ 700 **49.** 4800 ÷ 80 **50.** 600 ÷ 50

Calculate.

51. 600 ÷ 25 **52.** 900 ÷ 25 **53.** 300 ÷ 15 **54.** 600 ÷ 15 **55.** 450 ÷ 15

56. 450 ÷ 25 **57.** 325 ÷ 25 **58.** 300 ÷ 75 **59.** 600 ÷ 75 **60.** 725 ÷ 25

2.1 Integers

Activity: Use the Diagram

The diagram shows the temperatures at noon in cities across Canada on one day.

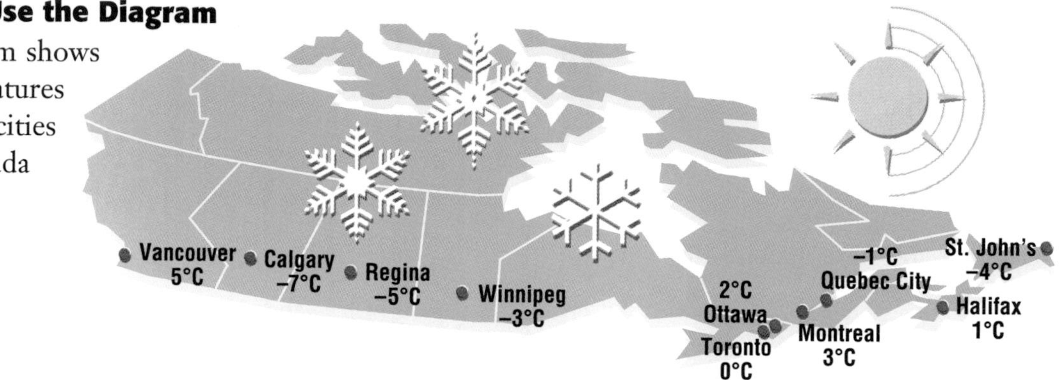

Inquire

1. What was the temperature in Vancouver?

2. What was the temperature in Winnipeg?

3. In what way are the numbers in questions 1 and 2 alike?

4. In what way are the numbers in questions 1 and 2 different?

5. Which temperature is higher, Calgary's or Ottawa's?

6. Which temperature is lower, Regina's or Quebec City's?

7. List all the temperatures from highest to lowest.

Numbers like +5, −7, and −10 are called **integers**. These numbers are used to show direction *above* or *below* some **zero point**. For temperature, we will take the zero point as 0°C.

The integer +5 represents 5°C above zero.
The integer −7 represents 7°C below zero.

The highest point in Canada is the top of Mount Logan in the Yukon Territory. It is 5951 m above sea level. The Marianna Trench is in the Pacific Ocean. One part of the Trench goes to a depth of 11 036 m. These two measurements go in *opposite directions*. One is *above* sea level. The other is *below* sea level. Here, sea level is the zero point.

The integer +5951 represents 5951 m above sea level.
The integer −11 036 represents 11 036 m below sea level.

Integers can be shown on a number line.

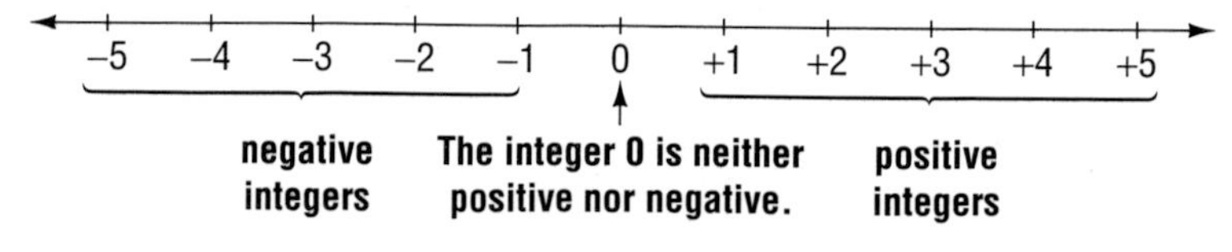

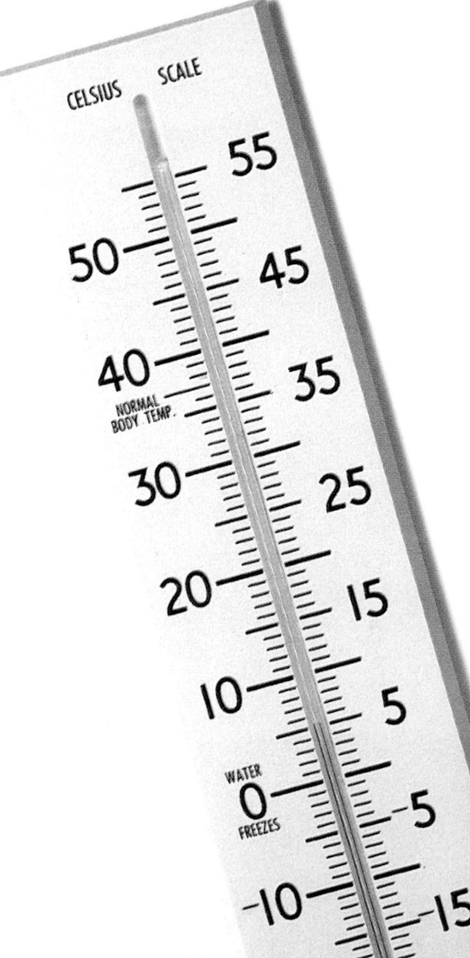

Every integer, except 0, has an **opposite.**

−6 5 −4 −3 −2 −1 0 +1 +2 +3 +4 +5 +6

−4 ← **These are opposite integers.** → +4

When two integers are compared on a number line, the integer to the right is the greater.

−4 −3 −2 −1 0 +1 +2 +3

Since +3 is to the right of −4, we say " +3 is greater than −4," and we write $+3 > -4$. We can also say " −4 is less than +3," and write $-4 < +3$.

A positive integer can be written without the + sign. So, +3 becomes 3, and +15 becomes 15. A negative integer must always be written with the − sign.

We can list the integers, I, as follows.

$$I = \ldots, -3, -2, -1, 0, 1, 2, 3, \ldots$$

Example 1

Graph the following on a number line.

a) the integers less than 4

b) the integers greater than −5

Solution

a) The integers less than 4 are 3, 2, 1, 0, −1, ...

−1 0 +1 +2 +3 +4

The red arrow means that the integers continue without end.

b) The integers greater than −5 are −4, −3, −2, −1, 0, ...

−5 −4 −3 −2 −1 0

Example 2

Graph the following integers on a number line.

a) $x > -3$ **b)** $x \geq -3$

Solution

a) $x > -3$ means the integers greater than −3.

−4 −3 −2 −1 0 +1 +2

b) $x \geq -3$ means the integers greater than or equal to −3.

−4 −3 −2 −1 0 +1 +2

CONTINUED

Practice

Describe each of the following with an integer.

1. 22°C above zero

2. 9°C below zero

3. 567 m above sea level

4. 234 m below sea level

5. a loss of 13 points

6. a profit of $1345

7. 3 over par

8. 7 under par

9. 11 floors up

10. 3 floors down

11. a gain of 56 points

State the opposite of each integer.

12. −7 **13.** 8 **14.** −6

15. −2 **16.** −3 **17.** 13

State the next larger integer.

18. 6 **19.** −4 **20.** 0 **21.** 11

22. −7 **23.** 9 **24.** −11 **25.** −9

State the integer that is 1 less than each of the following.

26. 9 **27.** −3 **28.** 0 **29.** −1

30. 1 **31.** −5 **32.** 16 **33.** −13

Which integer is greater?

34. 7 or 3 **35.** −5 or 2 **36.** −3 or −9

37. 8 or −8 **38.** 6 or 0 **39.** 0 or −1

Which integer is smaller?

40. 6 or 8 **41.** 5 or −9 **42.** −3 or −4

43. 0 or −5 **44.** 4 or 0 **45.** −1 or −2

Describe each of the following in words.

46. 3, 4, 5, 6, …

47. −4, −3, −2, −1, 0, 1, …

48. 9, 8, 7, 6, …

49. −3, −4, −5, −6, …

Describe each of the following in words.

50. 0 +1 +2 +3 +4

51. −4 −3 −2 −1 0

52. −5 −4 −3 −2 −1 0

Replace the ■ with $<$, $=$, or $>$ to make each statement true.

53. 5 ■ 3 **54.** 6 ■ 0

55. −6 ■ 1 **56.** 3 ■ −3

57. −1 ■ −4 **58.** −7 ■ 6

59. +2 ■ 2 **60.** 0 ■ −1

61. −8 ■ −7 **62.** −13 ■ 14

Write each statement in symbols.

63. Positive three is greater than negative two.

64. Negative four is less than positive seven.

65. Negative two is greater than negative nine.

66. Negative five is less than zero.

Write the integers in order from largest to smallest.

67. 0, 3, −2 **68.** −6, −5, −9, −1

69. 4, −3, 0, 5 **70.** 0, −1, −3, 2, 4

Write the integers in order from smallest to largest.

71. −2, 0, −3, 5 **72.** −4, −1, −3, −2

73. −15, 18, −11, 14 **74.** −1, 1, −2, 2

Graph the integers on a number line.

75. integers less than 5

76. integers greater than −2

77. the positive even integers

78. integers less than 0

List and graph the integers.

79. less than 5 and greater than 1

80. less than 3 and greater than -2

81. greater than -1 and less than 4

82. greater than -7 and less than -9

83. less than 0 and greater than -5

List the integers and graph them on a number line.

84. $x > 3$ **85.** $x < 0$ **86.** $x \leq 4$

87. $x \geq -2$ **88.** $x > -6$ **89.** $x \leq -7$

Problems and Applications

90. Write the integer that is

a) 5 less than 7 **b)** 3 more than -6

c) 2 less than -3 **d)** 2 more than 0

e) 4 less than -6 **f)** 5 more than 2

g) 3 less than 1 **h)** 4 more than -1

i) 5 less than 0 **j)** 7 more than -3

k) 8 less than 2 **l)** 6 more than -6

91. Expressions like $-2 < x \leq 3$ are read from the middle in both directions. This expression means that x is less than or equal to 3 and greater than -2. The following integers satisfy these conditions.

$$-1, 0, 1, 2, 3$$

List the integers that satisfy the following.

a) $1 < x < 4$ **b)** $-2 < x \leq 5$

c) $-2 \leq x < 7$ **d)** $-4 \leq x \leq 1$

e) $-6 < x \leq 0$ **f)** $-8 \leq x < -1$

92. Can you have a bank balance of $-\$35$? Explain.

93. The world's time zones were devised by a famous Canadian, Sir Sandford Fleming. The zones can be named with integers. Research the time zones and describe how integers are used.

94. Arrange the temperatures in order from lowest to highest.

The hottest day on Earth	58°C
The hottest place on the moon	134°C
The coldest place on the moon	−170°C
The human body	37°C
A Canada jay's body	43°C
The coldest day on Earth	−88°C
The freezing point of pure water	0°C
The water penguins swim in	−2°C
Canada's coldest day	−63°C

95. Work with a classmate. List occupations that require the use of integers. For each occupation, write an example of a question that might require the use of integers.

PATTERN POWER

Aziza and Efra are twin babies.

1. Determine the sum and product of their ages when they reach 1, 2, 3, 4, and 5 years of age.

2. Determine the quotient of the product over the sum at each of the ages in question 1.

3. Describe the pattern in the quotients.

4. Use the pattern to predict the quotient of the product over the sum when the twins' ages reach

a) 25 **b)** 50 **c)** 70

5. Check your answers to question 4 by calculating the sum and product of their ages and finding the quotient in each case.

6. How old will the twins be when the quotient has these values?

a) 15 **b)** 24.5 **c)** 40

A Model for Integer Operations

Static electricity provides a model for integer operations. To model the operations, we represent a +1 charge with a red disk and a −1 charge with a blue disk. Combining a positive charge and a negative charge results in a charge of zero.

positive charge

negative charge

zero charge

Activity ❶ Modelling Charges

The plates show 3 ways to model a charge of +3.

The plates show 3 ways to model a charge of −2.

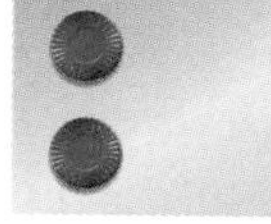
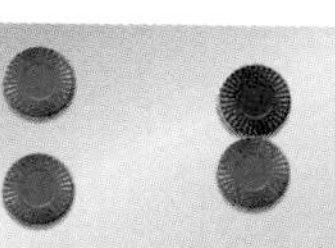

Use integer disks or diagrams to show 3 ways to model each charge.

1. +4 **2.** −1 **3.** −3 **4.** 0 **5.** +1

Activity ❷ Adding Integers

The charge on the plate is +6.

Add 4 negative charges.

After the positive and negative charges combine, we have 2 positive charges left.

This example shows that (+6) + (−4) = +2 or 6 + (−4) = 2.

Copy the table. Use integer disks to help you complete it.

Charge on the Plate	Charge Added	New Charge	Addition Statement
+7	+2		
+5	−3		
0	−6		
−3	+8		
−4	−3		
−9	+7		

Activity ❸ Subtracting Integers

The charge on the plate is +2.

Remove 3 negative charges.

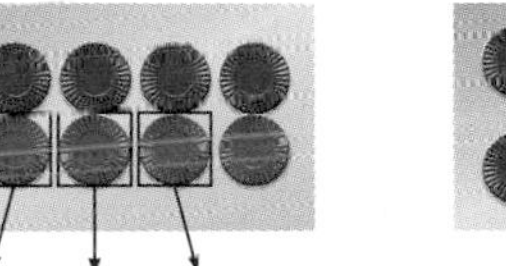

The charge on the plate is +5.

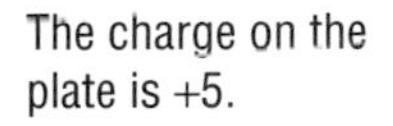

This example shows that $(+2) - (-3) = +5$ or $2 - (-3) = 5$.

Sometimes, there may not be enough charges on the plate of the kind you want to remove. In this case, just add zero charges to the plate until you have enough of the charges you want to remove.

The charge on the plate is –1. You want to remove 3 positive charges.

Add 2 zero charges.

Remove 3 positive charges.

The charge on the plate is –4.

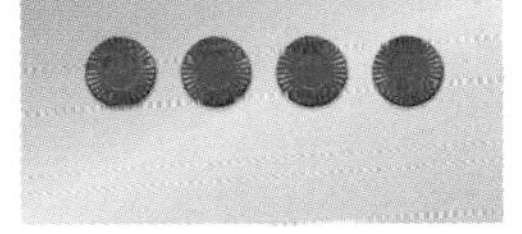

The example shows that $(-1) - (+3) = -4$ or $-1 - (+3) = -4$

Copy the table. Use integer disks to help you complete it.

Charge on the Plate	Charge Removed	New Charge	Subtraction Statement
+5	+2		
+2	–3		
0	–2		
–3	+3		
–2	–3		
–4	+5		

2.2 Look for a Pattern

Patterns are everywhere, and people often use them in their work. Scientists look for patterns in their research findings. Air traffic controllers use traffic patterns.

Using patterns to *predict results* is a very useful problem solving tool. Store managers and hotel managers use patterns to predict and plan for their busy times. Meteorologists use patterns to help predict the weather.

A group of towns has a phone network to call volunteer firefighters in the event of a forest fire. The fire chief phones 2 firefighters and gives them information about the fire. In the second round of calls, these 2 firefighters each phone 2 more. In the third round of calls, each person called in the second round phones 2 more people. How many firefighters are called in the ninth round of calls?

Understand the Problem

1. What information are you given?

2. What are you asked to find?

3. Do you need an exact or approximate answer?

Think of a Plan

Make a table. Write the numbers of people called in the first few rounds of calls and look for a pattern.

Carry Out the Plan

Round of Calls	1	2	3	4	5	6	7	8	9
Number of People Called	2 or 2^1	4 or 2^2	8 or 2^3	16 or 2^4					

The number of calls in a round equals 2 raised to an exponent that equals the round number.

From this pattern, we predict that 2^5 people are called in the fifth round, 2^6 people in the sixth round, and so on.

In the ninth round of calls, 2^9 or 512 people are called.

Look Back

Does the answer seem reasonable?

Can you think of another way to solve the problem?

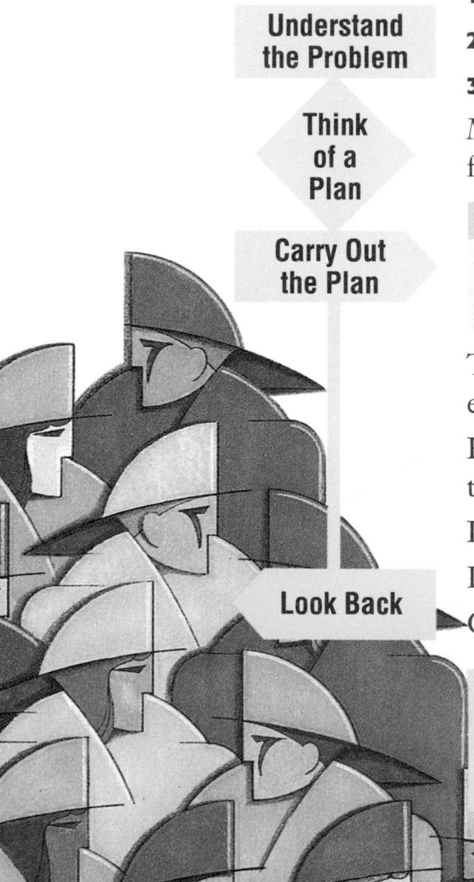

Look for a Pattern

1. Use the given information to find a pattern.
2. Use the pattern to solve the problem.
3. Check that the answer is reasonable.

Problems and Applications

In questions 1–5, determine the pattern in each set of numbers. Then, write the next 2 numbers.

1. 5, 8, 11, 14, ■, ■

2. 14, 12, 10, 8, ■, ■

3. 10, 12, 9, 11, 8, 10, 7, ■, ■

4. 5, 10, 20, 40, ■, ■

5. 256, 128, 64, 32, ■, ■

6. If the pattern continues, how many toothpicks are needed for
a) the 5th diagram?
b) the 6th diagram?
c) the 50th diagram?

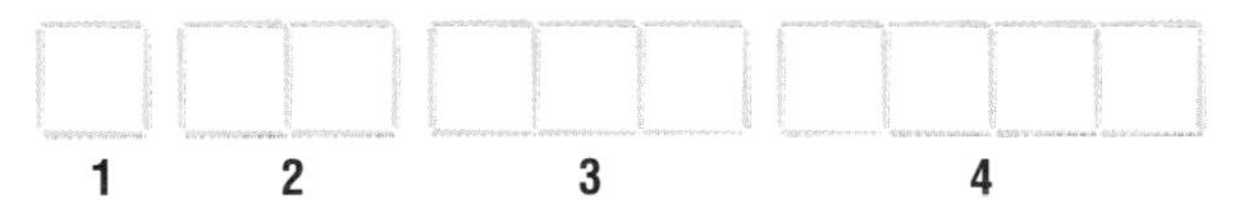

7. The numbers 1, 8, 27, and 64 are the first 4 cubes.
a) Find the sum of the first 2 cubes.
b) Find the sum of the first 3 cubes.
c) Find the sum of the first 4 cubes.
d) Describe the pattern in the sums.
e) Use the pattern to find the sum of the first 9 cubes.

8. Find the pattern and complete the chart.

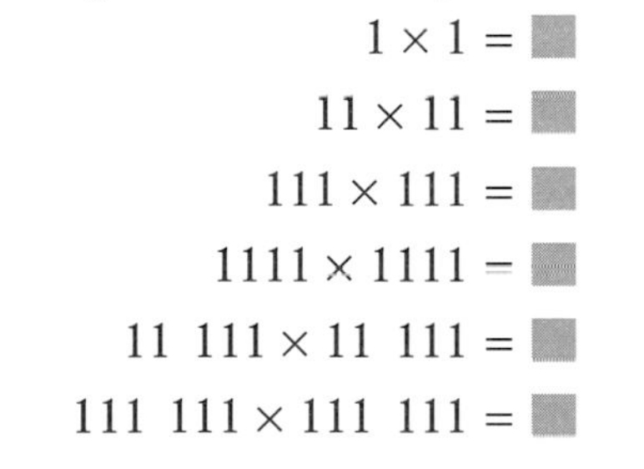

$1 \times 1 =$ ■

$11 \times 11 =$ ■

$111 \times 111 =$ ■

$1111 \times 1111 =$ ■

$11\ 111 \times 11\ 111 =$ ■

$111\ 111 \times 111\ 111 =$ ■

9. A computer company donated money to a university scholarship fund for 10 years. In the first year, the company gave $1 000 000. Every year after that, it gave $100 000 more than in the previous year. How much did the university receive in the tenth year?

10. Six people entered a chess tournament. Each player played until losing a game. If there were no draws, how many games were played?

11. There are 10 rooms on each floor of a hotel. There are 9 floors. The rooms are numbered from 10 to 19 on the first floor, from 20 to 29 on the second floor, and so on. You have been hired to put new brass numbers on each room door. For example, you will put one brass "1" and one brass "0" on the door of the first room. To have enough brass numbers, how many will you need to buy for each digit from 0 to 9?

12. If you place a block on a table, 5 faces are showing. If you add another block as shown, 8 faces are showing. How many faces are showing when there are 21 blocks in a row?

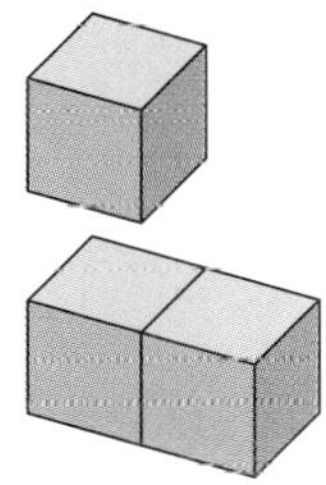

13. How many logs will be in a pile with 9 logs in the bottom row?

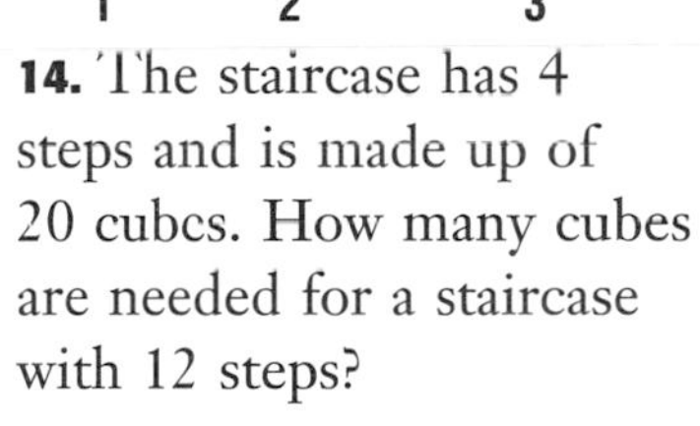

14. The staircase has 4 steps and is made up of 20 cubes. How many cubes are needed for a staircase with 12 steps?

15. Eight people entered a tennis tournament. Each person played everyone else once. How many games were played?

16. A triangle has no diagonals. A quadrilateral has 2 diagonals. A pentagon has 5 diagonals. How many diagonals does a 12-sided figure have?

17. Write a problem you can solve by finding a pattern. Have a classmate solve your problem.

2.3 Adding Integers

One morning in Red Deer, Alberta, the temperature was –2°C. The weather forecast predicted an increase of 3°C by noon. Louiselle wanted to know the expected temperature at noon.

Activity: Use a Model

We can use red and blue disks as **integer disks**.
Each red disk represents the integer +1.
Each blue disk represents the integer –1.
A red and blue disk together represent the integer 0.

The integer is –2 or (–2).

Add the integer (+3) by adding 3 positive disks.

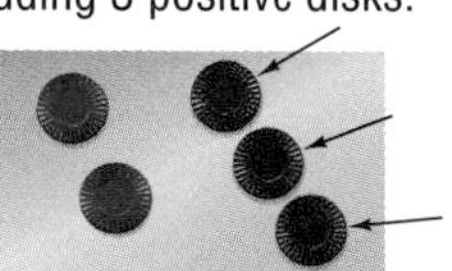

Pairs of red and blue disks are combined. The final integer is +1 or (+1).

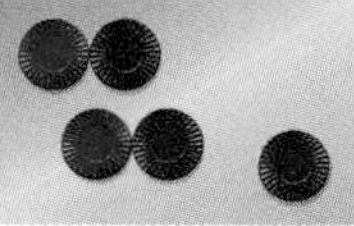

The addition statement is $(-2) + (+3) = +1$ or $-2 + 3 = 1$.

Inquire

1. What is the expected temperature at noon in Red Deer?

2. Use disks or diagrams to model these additions. Write addition statements for each.

a) $(+3) + (-2)$ **b)** $(+2) + (-4)$ **c)** $(-2) + (-4)$
d) $(-5) + (+3)$ **e)** $(-2) + (+5)$ **f)** $(+3) + (-3)$

3. Is the sum of 2 negative integers always a negative integer? Explain.

4. Is the sum of a negative and a positive integer always a positive integer? always a negative integer? Explain.

Integer addition can also be related to a number line. Think of positive integers as trips to the right and negative integers as trips to the left.

To show $(+3) + (-7)$, start at 0 and go 3 units to the right. Then go 7 units to the left. The single trip that gives the same result as $(+3) + (-7)$ is (–4).

$$(+3) + (-7) = (-4) \text{ or } 3 + (-7) = -4$$

Example 1

Use a number line to add $(-6) + (+8)$.

Solution

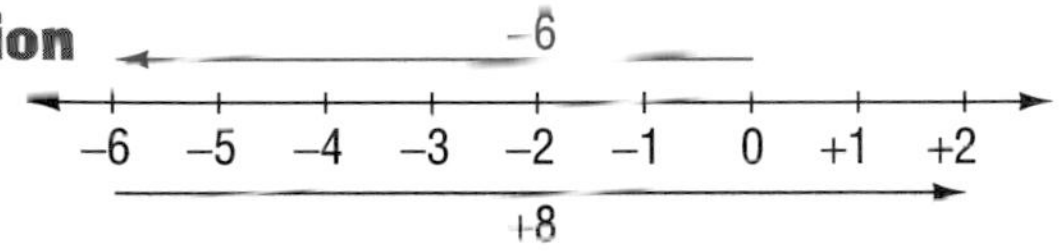

$(-6) + (+8) = +2$ or $-6 + 8 = 2$

Example 2

Use a number line to add $(-3) + (-4)$.

Solution

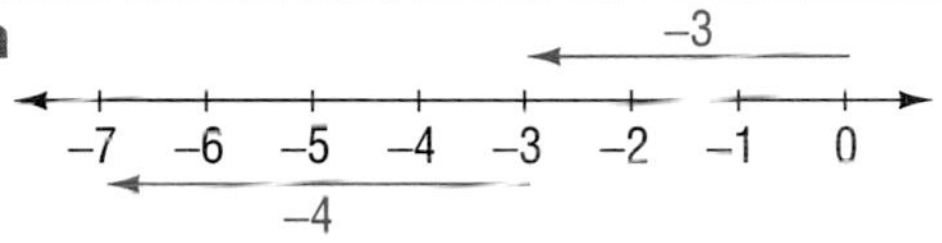

$(-3) + (-4) = -7$ or $-3 + (-4) = -7$

Example 3

Use a number line to add $(+4) + (-3) + (-5)$.

Solution

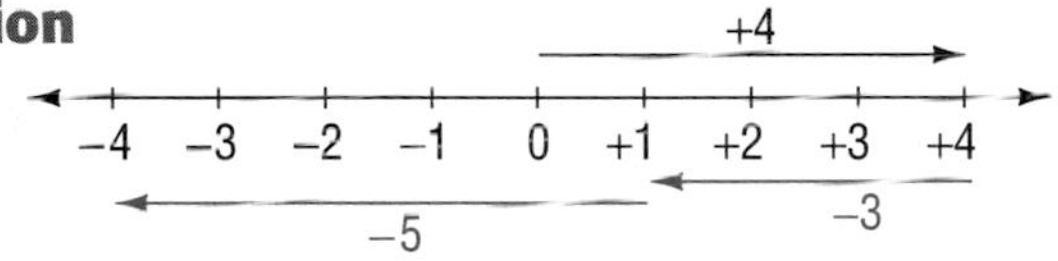

$(+4) + (-3) + (-5) = -4$ or $4 + (-3) + (-5) = -4$

Example 4

Add $7 + (-4) + (-6) + 8 + (-1)$.

Solution

$$\begin{aligned} & \underbrace{7 + (-4)} + (-6) + 8 + (-1) \\ = & \underbrace{3 + (-6)} + 8 + (-1) \\ = & \underbrace{-3 + 8} + (-1) \\ = & \underbrace{5 + (-1)} \\ = & 4 \end{aligned}$$

When adding several integers, it is sometimes easier to add those with the same sign first.

$$\begin{aligned} 7 + (-4) + (-6) + 8 + (-1) &= 7 + 8 + (-4) + (-6) + (-1) \\ &= 15 + (-11) \\ &= 4 \end{aligned}$$

Practice

Add.

1. $(+4) + (+5)$
2. $(+3) + (-2)$
3. $(-3) + (-3)$
4. $(-1) + (+6)$
5. $(-7) + (+2)$
6. $(+3) + (-4)$
7. $(-7) + (-2)$
8. $(-5) + (+5)$
9. $(+7) + (-2)$
10. $(+5) + (-9)$

Add.

11. $(+3) + (+2)$
12. $(+4) + (-8)$
13. $(-3) + (-4)$
14. $(-8) + (+8)$
15. $(-5) + 0$
16. $(+2) + (-2)$
17. $(-4) + (+6)$
18. $(+9) + (-5)$
19. $(-9) + (-8)$
20. $(-2) + (+4)$

CONTINUED

Add.

21. $8 + (-2)$

22. $-5 + (-4)$

23. $5 + (-7)$

24. $-4 + 5$

25. $-3 + 2$

26. $-7 + (-7)$

Add.

27. $-11 + 4$

28. $14 + 3$

29. $8 + (-10)$

30. $-13 + (-6)$

31. $-12 + (-11)$

32. $16 + (-5)$

33. $-17 + 19$

34. $-15 + 15$

35. $23 + (-14)$

36. $34 + (-35)$

Simplify.

37. $(+4) + (+3) + (+5)$

38. $(+7) + (+6) + (-3)$

39. $(-5) + (+4) + (+2)$

40. $(-7) + (-3) + (+1)$

41. $(-6) + (+8) + (-2)$

42. $(-4) + (-3) + (-6)$

43. $(+9) + (-6) + (-8)$

44. $(-10) + (+3) + (-2)$

Simplify.

45. $(+5) + (+2) + (+7) + (+1)$

46. $(+6) + (-3) + (-2) + (+7)$

47. $(-3) + (+2) + (+8) + (-9)$

48. $(-4) + (-7) + (+9) + (-6)$

Simplify.

49. $4 + 7 + (-4)$

50. $-6 + (-2) + 5$

51. $8 + (-3) + (-8)$

52. $9 + (-9) + (-9)$

53. $-7 + 6 + (-8)$

54. $9 + (-3) + (-6)$

55. $7 + (-1) + 0$

56. $-2 + (-2) + (-2)$

Estimate, then add.

57. $123 + (-78)$

58. $-237 + (-324)$

59. $-762 + 456$

60. $678 + (-821)$

61. $-405 + (-306)$

62. $-777 + 888$

Problems and Applications

Find the value of each ■ .

63. $(\blacksquare) + 5 = 4$

64. $-3 + (\blacksquare) = -8$

65. $(\blacksquare) + (-6) = -1$

66. $6 + (\blacksquare) = -2$

67. $(-2) + (\blacksquare) = -9$

68. $(\blacksquare) + (-3) = 4$

Copy and complete the tables.

69.

x	$x + 4$
3	
2	
1	
0	
–1	
–2	
–3	

70.

y	$y + (-2)$
3	
2	
1	
0	
–1	
–2	
–3	

71.

m	$-6 + m$
2	
1	
0	
–1	
–2	

72.

t	$-4 + t$
2	
1	
0	
–1	
–2	

73. The temperature in Red Rock is 4°C. What is the temperature after a rise of 5°C followed by a fall of 10°C?

74. The temperature in Whitehorse at midnight was –4°C. The table shows the hourly temperature changes. Copy and complete the table.

Time	Temperature Change (°C)	New Temperature (°C)
01:00	–5	
02:00	+3	
03:00	–2	
04:00	+2	
05:00	–6	
06:00	–2	

75. Naomi has a $145 overdraft at the bank. She makes a $398 deposit. How much is in her account after she makes the deposit?

76. a) Find the sum of each row, column, and diagonal of this magic square.

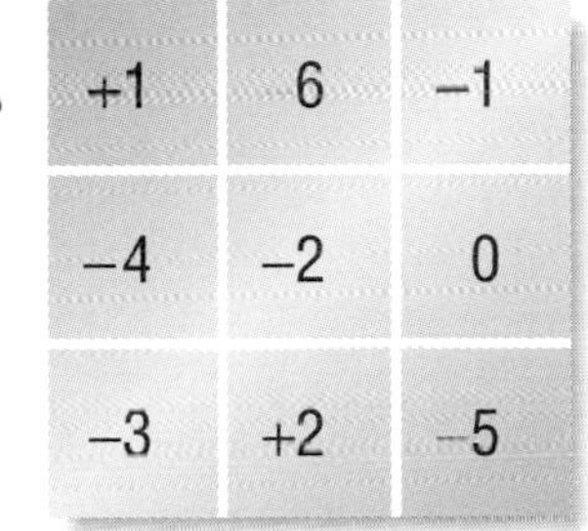

+1	6	−1
−4	−2	0
−3	+2	−5

b) Choose any integer and add it to each integer in the square. Is the result another magic square? Explain.

77. The melting point of oxygen is −218°C. The boiling point of oxygen is 35°C higher. What is the boiling point?

78. Copy and complete each statement.

a) The sum of 2 or more positive integers is always ______ .

b) The sum of 2 or more negative integers is always ______ .

c) The sum of an integer and its opposite is always ______ .

d) The sum of a positive integer and 0 is always ______ .

e) The sum of a negative integer and 0 is always ______ .

79. a) What are the possible totals you can score by rolling 2 six-sided dice?

b) Suppose you have 2 six-sided dice marked differently than usual. One has faces numbered +1 to +6. The other has faces numbered −1 to −6. What possible scores can you roll?

80. Write the following and compare your answers with a classmate's.

a) 2 different integers whose sum is 0

b) 3 different integers whose sum is −1

c) 4 different integers whose sum is +2

81. Here are 3 ways to write the integer −15 as the sum of consecutive integers.

$$(-1) + (-2) + (-3) + (-4) + (-5) = -15$$
$$(-4) + (-5) + (-6) = -15$$
$$(-7) + (-8) = -15$$

Here are 2 ways to write the integer −5 as the sum of consecutive integers.

$$(-2) + (-3) = -5$$
$$1 + 0 + (-1) + (-2) + (-3) = -5$$

a) Write each of the following integers as the sum of consecutive integers. Compare your answers with a classmate's.

−9 26 −22 18

b) Write each of the following integers as the sum of consecutive integers in 3 ways. Compare your answers with a classmate's.

−21 9 −27

c) Write 2 integers that cannot be written as the sum of consecutive integers.

CALCULATOR POWER

Adding Integers

To enter a negative integer on a calculator, press the [+/−] key after you enter the number.

Press 7 [+/−]
Display 7 −7

To calculate −21 + (−12) + 39

Press 21 [+/−] [+] 12 [+/−] [+] 39 [=]
Display 21 −21 −21 12 −12 −33 39 6

EST −20 + (−10) + 40 = 10

Estimate, then calculate the following.

1. −34 + (−76) + (−56) + 98

2. −86 + (−67) + 43 + (−75)

3. 45 + (−47) + (−22) + 88

4. −67 + (−27) + (−90) + (−12)

2.4 Subtracting Integers

The chemical most often used as de-icing salt is called calcium chloride. It works by lowering the freezing point of water from its normal value of 0°C. The freezing point of a solution of calcium chloride in water depends on the percent of calcium chloride in the solution. If the solution contains 6% calcium chloride, the freezing point is −3°C.

Activity: Use a Model

The temperature is +2°C. To find out how much colder it has to get to freeze a 6% solution of calcium chloride in water, subtract −3°C from +2°C.

We can represent the subtraction with integer disks.

The integer is +2.

Subtract the integer (−3) by removing 3 negative disks.

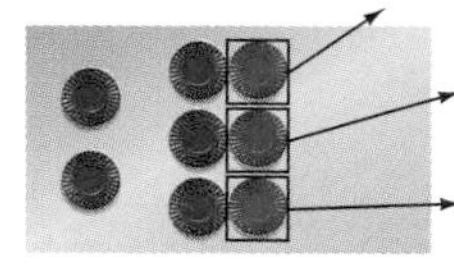

The final integer is +5.

The subtraction statement is $(+2) - (-3) = (+5)$ or $2 - (-3) = 5$.
It must get 5°C colder to freeze the solution.

A 9% solution freezes at a temperature 2°C lower than the 6% solution. To find the freezing point of the 9% solution, subtract 2°C from −3°C.

The integer is −3. Add 2 zero integers. Subtract the integer 2 by removing 2 positive disks. The final integer is −5.

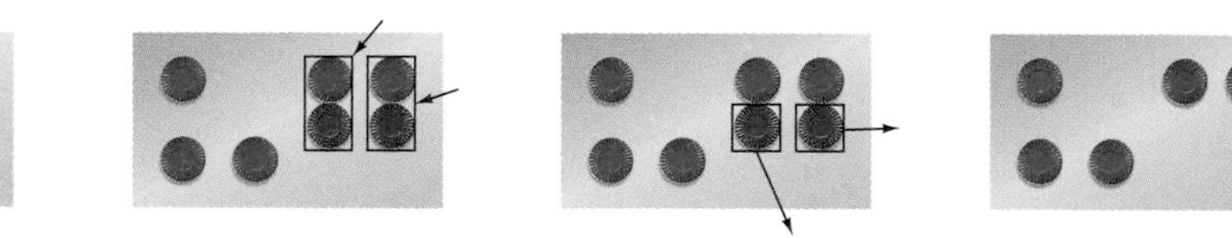

The subtraction statement is $(-3) - (+2) = -5$ or $-3 - 2 = -5$.
The freezing point of a 9% solution is −5°C.

Inquire

1. Use disks or diagrams to model these subtractions. Write a subtraction statement for each.

a) $(+2) - (-1)$ **b)** $(+3) - (+4)$ **c)** $(-3) - (-5)$
d) $(-4) - (+3)$ **e)** $(-2) - (-1)$ **f)** $(+2) - (+2)$

2. Complete each addition sentence.

a) $(+2) + (+1)$ **b)** $(+3) + (-4)$ **c)** $(-3) + (+5)$

d) $(-4) + (-3)$ **e)** $(-2) + (+1)$ **f)** $(+2) + (-2)$

3. Write the number sentences from questions 1 a) and 2 a) next to each other. Repeat for the number sentences from 1 b) and 2 b), and so on, to get 6 pairs of number sentences. Compare the number sentences in each pair and write a rule for using addition to subtract integers. Compare your rule with your classmates'.

Example 1

Subtract.

a) $(+8) - (+4)$

b) $(-6) - (+2)$

c) $(-9) - (-4)$

Solution

To subtract an integer, add its opposite.

a) $(+8) - (+4)$
$= (+8) + (-4)$
$= +4$

b) $(-6) - (+2)$
$= (-6) + (-2)$
$= -8$

c) $(-9) - (-4)$
$= (-9) + (+4)$
$= -5$

Example 2

Simplify.

a) $9 - (-6)$

b) $-5 - (+2)$

c) $14 + (-4) - 2$

Solution

To subtract an integer, add its opposite.

a) $9 - (-6)$
$= 9 + 6$
$= 15$

b) $-5 - (+2)$
$= -5 + (-2)$
$= -7$

c) $14 + (-4) - 2$
$= \underbrace{14 + (-4)} + (-2)$
$= \underbrace{10 + (-2)}$
$= 8$

Example 3

Simplify.

a) $-3 + 4 - 2$

b) $5 - 6 + 7 - 8$

Solution

To subtract an integer, add its opposite.

a) $-3 + 4 - 2$
$= \underbrace{-3 + 4} + (-2)$
$= \underbrace{1 + (-2)}$
$= -1$

b) $5 - 6 + 7 - 8$
$= \underbrace{5 + (-6)} + 7 + (-8)$
$= \underbrace{-1 + 7} + (-8)$
$= \underbrace{6 + (-8)}$
$= -2$

Practice

State the opposite of each integer.

1. 4 **2.** −3 **3.** −11 **4.** 7

5. −1 **6.** −23 **7.** 21 **8.** (-4)

Subtract.

9. $(+4) - (+1)$ **10.** $(+9) - (-4)$

11. $(-2) - (+6)$ **12.** $(-2) - (-3)$

13. $(+7) - (-8)$ **14.** $(+5) - (-6)$

15. $(-7) - (-7)$ **16.** $(-4) - (-3)$

Subtract.

17. $9 - 7$ **18.** $7 - (-3)$

19. $-3 - (-4)$ **20.** $-6 - (-5)$

21. $4 - 9$ **22.** $0 - (-2)$

23. $-3 - 7$ **24.** $-2 - (-2)$

CONTINUED

Subtract.

25. $(+11) - (+6)$
26. $(-9) - (-12)$
27. $(-10) - (+5)$
28. $(+14) - (+4)$
29. $(+9) - (-11)$
30. $(-14) - (-7)$
31. $(-16) - (-14)$
32. $(+18) - (-4)$

Subtract.

33. $-4 - (-11)$
34. $23 - (-6)$
35. $13 - 18$
36. $-17 - (-14)$
37. $28 - 37$
38. $-21 - 37$
39. $-18 - (-16)$
40. $-49 - 21$

Simplify.

41. $(+5) - (+2) + (+6)$
42. $(+8) + (+4) - (-2)$
43. $(-4) - (+1) - (+3)$
44. $(-6) + (-4) - (+1)$
45. $(-3) - (+9) + (-1)$
46. $(-2) - (-3) - (-5)$
47. $(+7) + (-5) - (-6)$
48. $(-11) - (+4) + (-1)$

Simplify.

49. $8 + (-7) - 2$
50. $-9 - (-2) + 4$
51. $7 - (-2) + 5$
52. $-3 - (-4) - (-1)$
53. $-5 + (-2) - 9$
54. $7 - (-6) + (-8)$
55. $2 - 5 - 3$
56. $-4 - 6 + 8$

Simplify.

57. $(+6) - (+3) + (+8) - (+2)$
58. $(+7) + (-3) - (-3) - (+8)$
59. $(-4) + (+3) - (+9) - (-8)$
60. $(-4) - (-8) - (+3) + (-9)$
61. $(-2) - (-6) - (-6) + (-8)$
62. $(+7) - (-4) + (-5) + (+2)$

Simplify.

63. $4 - 2 - 3 + 5$
64. $-3 + 7 + 6 - 7$
65. $8 - 1 - 5 - 6$
66. $4 - 9 + 3 - 1$
67. $-4 + 6 - 1 + 3 - 6$
68. $-2 - 3 - 1 + 5 + 7$
69. $5 - 7 + 2 - 3 + 6$
70. $2 - 3 - 5 + 1 + 3$

Problems and Applications

Find the value of $\blacksquare$.

71. $(+9) - (\blacksquare) = (+6)$
72. $(\blacksquare) - (+3) = (+2)$
73. $(-2) - (\blacksquare) = (-7)$
74. $(\blacksquare) - (-5) = (-2)$
75. $(+7) - (\blacksquare) = (-3)$
76. $(\blacksquare) - (-4) = (+5)$
77. $(\blacksquare) - (+7) = (-8)$

Find the value of $\blacksquare$.

78. $9 - \blacksquare = 2$
79. $9 - \blacksquare = 11$
80. $4 - \blacksquare = -1$
81. $4 - \blacksquare = 9$
82. $2 - \blacksquare = -4$
83. $2 - \blacksquare = 8$

Find the value of $\blacksquare$.

84. $-6 - \blacksquare = 6$
85. $-2 - \blacksquare = -3$
86. $-2 - \blacksquare = 1$
87. $\blacksquare - 3 = 5$
88. $\blacksquare - 3 = -11$
89. $\blacksquare - 4 = 1$

90. Complete the tables.

a)

x	$x - 1$
2	
1	
0	
–1	
–2	

b)

t	$t - 5$
2	
1	
0	
–1	
–2	

91. Some people work at the tops of tall buildings or at the bottoms of mines. The diagram gives some elevations. Calculate the difference in each of the following pairs of elevations.

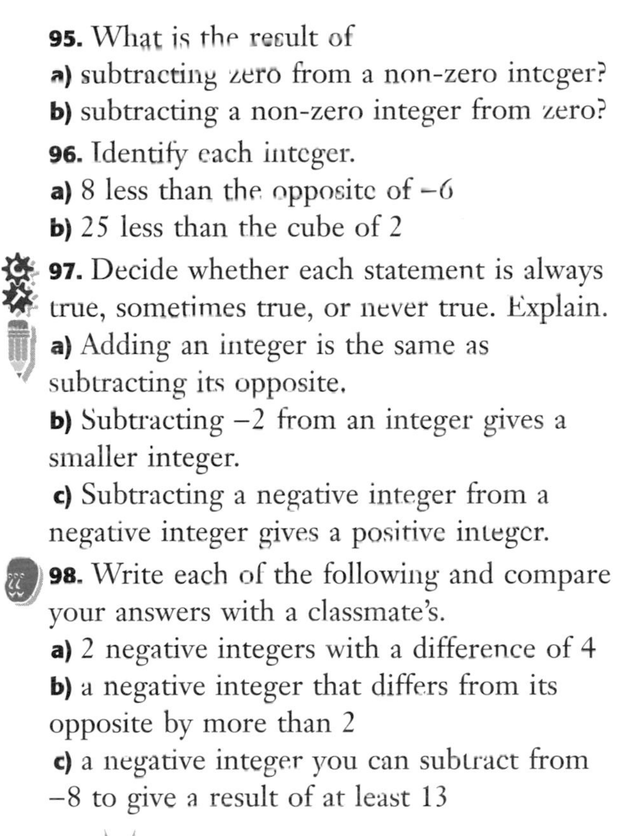

a) the Sears Tower and the Eiffel Tower
b) the CN Tower and the Kolar Mine
c) the Sears Tower and the Nova Lima Mine
d) the Eiffel Tower and the Western Deep Mine
e) the Boksburg Mine and the Kolar Mine
f) the CN Tower and the Western Deep Mine
g) the Nova Lima Mine and the Western Deep Mine

92. a) Mount Everest has an elevation of 8863 m. Death Valley has an elevation of −86 m. What is the difference in these elevations?
b) The surface of the Dead Sea has an elevation of −400 m. Which is higher, the Dead Sea or Death Valley, and by how much?

93. Measurements of elevations are not exact. Two measurements of Mount Everest are 8848 m and 8863 m. What is the difference in these values?

94. The boiling point of liquid nitrogen is −196°C. The boiling point of liquid hydrogen is −253°C. What is the difference in these boiling points?

95. What is the result of
a) subtracting zero from a non-zero integer?
b) subtracting a non-zero integer from zero?

96. Identify each integer.
a) 8 less than the opposite of −6
b) 25 less than the cube of 2

97. Decide whether each statement is always true, sometimes true, or never true. Explain.
a) Adding an integer is the same as subtracting its opposite.
b) Subtracting −2 from an integer gives a smaller integer.
c) Subtracting a negative integer from a negative integer gives a positive integer.

98. Write each of the following and compare your answers with a classmate's.
a) 2 negative integers with a difference of 4
b) a negative integer that differs from its opposite by more than 2
c) a negative integer you can subtract from −8 to give a result of at least 13

CALCULATOR POWER

Subtracting Integers

To simplify 5 − 9,

Press	5	−	9	=
Display	5	5	9	−4

To simplify −34 − (−23),

Press	34	+/−	−	23	+/−	=
Display	34	−34	−34	23	−23	−11

EST −30 − (−20) = −10

Estimate, then calculate the following.

1. 93 − 19
2. 56 − (−14)
3. 40 − (−22)
4. −33 − (−11)
5. −35 − 87
6. 12 − 93
7. −46 − (−21)
8. 40 − (−45)

2.5 Use a Diagram

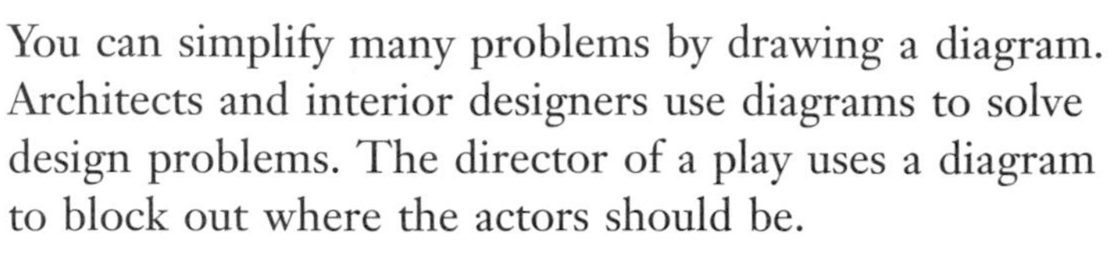

You can simplify many problems by drawing a diagram. Architects and interior designers use diagrams to solve design problems. The director of a play uses a diagram to block out where the actors should be.

Tom and Margarita are park rangers. They patrol the park by driving around the outside in jeeps. They start on opposite sides and drive at different speeds. Margarita takes 15 min to drive halfway around the park. Tom takes 20 min. They start their patrols at 20:00 and travel in opposite directions. If they stay on schedule, what are the approximate times at which they pass each other between 20:00 and 22:00?

Understand the Problem

1. What information are you given?
2. What are you asked to find?
3. Do you need an exact or approximate answer?

Think of a Plan

Draw a diagram to show the time taken to drive around the park. Show Tom in red. Show Margarita in blue. Use the diagram to find when they pass each other.

Carry Out the Plan

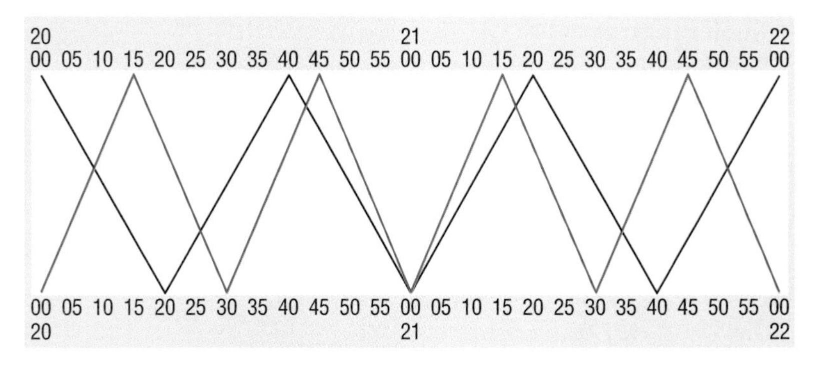

They pass each other 7 times, at about 20:09, 20:26, 20:43, 21:00, 21:17, 21:34, and 21:51.

Look Back

Does the answer seem reasonable?

Use a Diagram	**1.** Draw a diagram to represent the situation.
	2. Use the diagram to solve the problem.
	3. Check that the answer is reasonable.

Problems and Applications

1. How many different-sized rectangles have a perimeter of 16 m and side lengths that are whole numbers?

2. In a scalene triangle, no 2 sides are equal. How many scalene triangles have a perimeter of 12 cm or less, if the side lengths are whole numbers?

3. Roberto bought 4 stamps attached as shown.

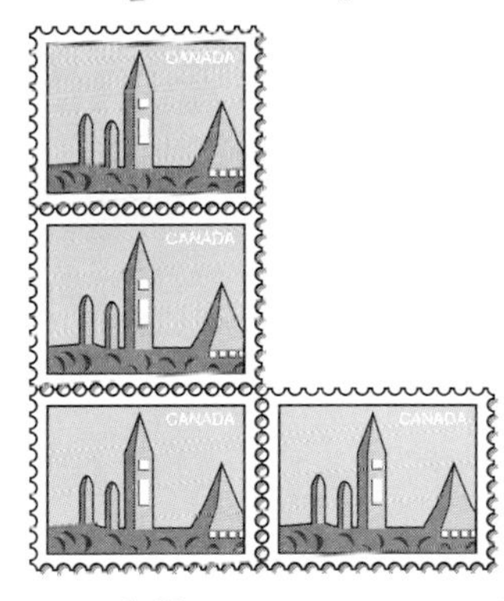

In how many different ways can 4 stamps be attached to each other?

4. Lenore wants to fence a rectangular yard that measures 15 m by 9 m. She wants posts at the corners and every 3 m. How many posts does she need?

5. Yoshiko and Carla are water polo players. Part of their training program is to swim widths of the pool. Carla swims a width in 15 s. Yoshiko swims a width in 12 s. They start swimming widths at the same time from opposite sides of the pool. How many times do they pass each other in 2 min? What assumptions have you made?

6. Terry, Rocky, Indu, and Lennox went out for dinner. They sat at a square table. They decided to use a math problem to see who would leave the tip. The problem was: "In how many different ways can 4 people sit at a square table if there must be 1 person on each side?"

7. Ships at sea use flags to send signals. Different orders of the same flags mean different messages. How many different messages could you send by using 3 flags?

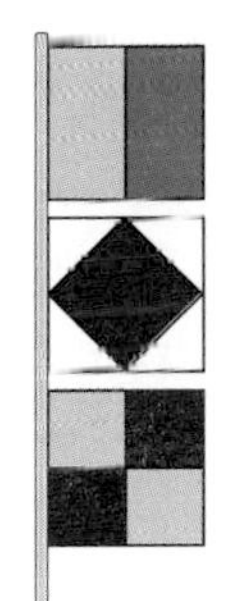

8. You can cut a round pie into 4 pieces with 2 straight cuts.
If you do not care about the sizes of the pieces, you can cut a round pie into 7 pieces using 3 straight cuts.
What is the greatest number of pieces you can get using 5 straight cuts?

9. A farmer must get a wolf, a goat, and some cabbages across a river. The boat will hold the farmer and one of the wolf, the goat, or the cabbages. Only the farmer can row the boat. The farmer cannot leave the goat alone with the cabbages, because the goat will eat them. The farmer cannot leave the wolf alone with the goat, because the wolf will chase the goat away. How does the farmer get the goat, wolf, and cabbages safely across the river?

10. A photography expedition planned an 8-day crossing of a desert. Each person could carry, at most, a 5-day supply of water. What is the smallest number of people needed to start the trip so that one person can cross the desert and the others can return to the starting point?

11. Write a problem that can be solved with a diagram. Have a classmate solve your problem.

2.6 Multiplying Integers

When you climb from one storey to the next in an office tower, you climb about 4 m. The integer +4 or 4 represents your change in height. If you climb 3 storeys, your change in height is represented by $4 + 4 + 4$. You can think of this repeated addition as a multiplication.

$$\begin{aligned} 4 + 4 + 4 &= 3 \times 4 \\ &= 12 \end{aligned}$$

So, you climb 12 m.

You can use the fact that multiplication is repeated addition to develop sign rules for multiplying integers.

Activity: Use a Model

The integer is 0.

Add 3 groups of disks.
Each group represents +2.

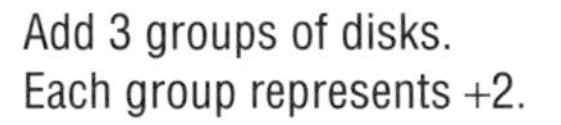

The final integer is (+6) or +6.

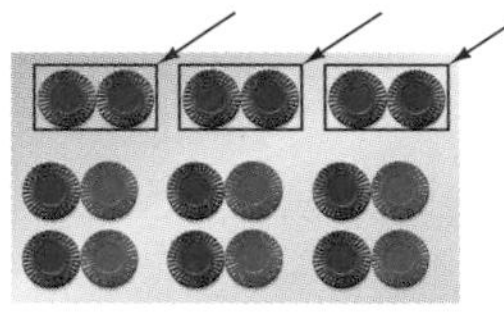

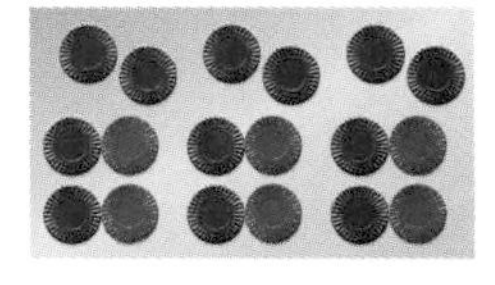

$(+2) + (+2) + (+2) = +6$ or $(+3) \times (+2) = +6$

The integer is 0.

Add 3 groups of disks.
Each group represents –2.

The final integer is (–6) or –6.

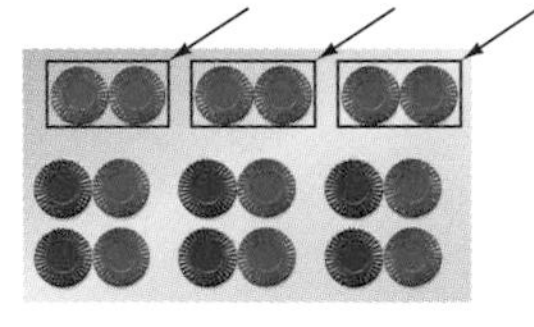

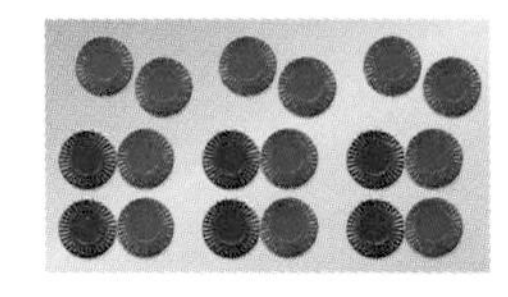

$(-2) + (-2) + (-2) = -6$ or $(+3) \times (-2) = -6$

The integer is 0.

Remove 3 groups of disks.
Each group represents +2.

The final integer is (–6) or –6.

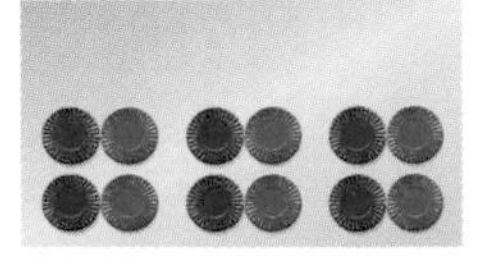

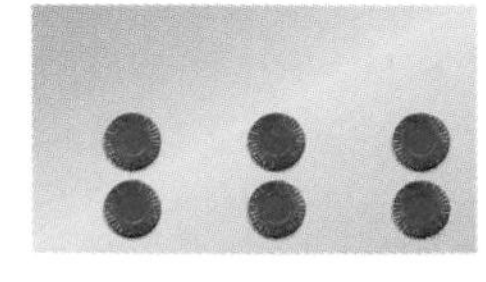

$-(+2) - (+2) - (+2) = -6$ or $(-3) \times (+2) = -6$

The integer is 0.

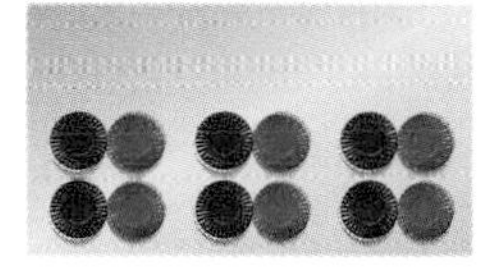

Remove 3 groups of disks.
Each group represents –2.

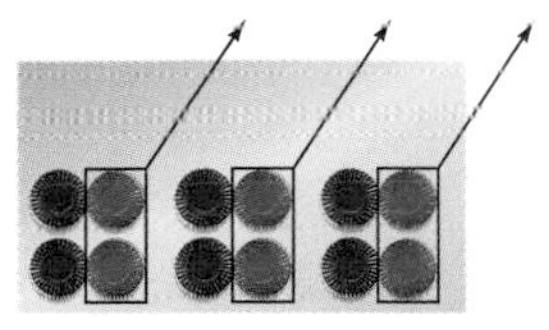

The final integer
is (+6) or +6.

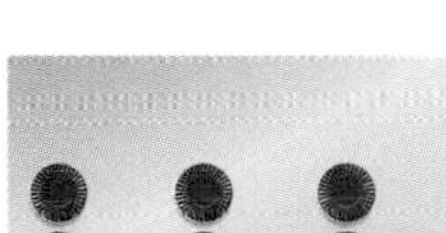

$-(-2) - (-2) - (-2) = +6$ or $(-3) \times (-2) = +6$

Inquire

1. Use disks or diagrams to model these operations, beginning with a diagram that represents the integer 0. Write a multiplication statement for each.

a) Adding 2 groups of 3 positive disks.
b) Adding 2 groups of 3 negative disks.
c) Removing 2 groups of 3 positive disks.
d) Removing 2 groups of 3 negative disks.

2. Repeat question 1 with 2 groups of 2 positive disks and 2 groups of 2 negative disks.

3. Repeat question 1 with 3 groups of 3 positive disks and 3 groups of 3 negative disks.

4. Look for a pattern in the signs in the multiplication statements from questions 1–3. Copy and complete the statements for multiplying integers.

a) $(+) \times (+) = ■$ **b)** $(+) \times (-) = ■$
c) $(-) \times (+) = ■$ **d)** $(-) \times (-) = ■$

Example 1

Multiply.
a) $(-2)(-5)$
b) $4(-3)$
c) $-3(-4)$

Solution

The product of 2 negative integers or 2 positive integers is a positive integer.
The product of a positive integer and a negative integer is a negative integer.

a) $(-2)(-5)$
$= 10$

b) $4(-3)$
$= (+4)(-3)$
$= -12$

c) $-3(-4)$
$= (-3)(-4)$
$= 12$

Example 2

Simplify $(-2)^3$.

Solution

$(-2)^3 = (-2) \times (-2) \times (-2)$
$= -8$

CONTINUED

Practice

1. Identify the missing sign.

a) $(+4)(+3) = \blacksquare\ 12$ **b)** $(-5)(+2) = \blacksquare\ 10$

c) $(+6)(-1) = \blacksquare\ 6$ **d)** $(-5)(-3) = \blacksquare\ 15$

2. Identify the missing sign.

a) $(+6)(\blacksquare\ 7) = 42$ **b)** $(-3)(\blacksquare\ 5) = -15$

c) $(\blacksquare\ 6)(+2) = -12$ **d)** $(-8)(\blacksquare\ 3) = 24$

e) $(\blacksquare\ 5)(+5) = -25$ **f)** $(\blacksquare\ 9)(-5) = 45$

Multiply.

3. $(+4)(+7)$ **4.** $(-4)(+6)$

5. $(-3)(-6)$ **6.** $(+9)(-9)$

Multiply.

7. $(+10)(-3)$ **8.** $(-11)(-3)$

9. $(+12)(+2)$ **10.** $(-3)(+10)$

Multiply.

11. $5 \times (-7)$ **12.** $-6 \times (-5)$

13. $0 \times (-3)$ **14.** $(-7) \times 8$

15. -6×9 **16.** 8×7

17. -6×0 **18.** $11 \times (-2)$

19. $7 \times (-12)$ **20.** $-20 \times (-1)$

Multiply.

21. $9(-5)$ **22.** $-5(-6)$ **23.** $7(-8)$

24. $-1(-7)$ **25.** $12(-6)$ **26.** $-11(-7)$

Simplify.

27. $(-2)(-3)(+4)$ **28.** $(+3)(+2)(-1)$

29. $(+4)(-5)(+2)$ **30.** $(-3)(-2)(-3)$

31. $(+6)(-2)(-1)$ **32.** $(+5)(+5)(+2)$

33. $(0)(-3)(+4)$ **34.** $(-4)(+4)(-2)$

35. $(+5)(+3)(+4)$ **36.** $(+5)(-2)(-7)$

Simplify.

37. $-3(-4)(+5)$ **38.** $-4 \times 2 \times 6$

39. $-2 \times 3 \times (-5)$ **40.** $3 \times 7 \times (-2)$

41. $6(-1) \times 8$ **42.** $4 \times (-3) \times 5$

43. $0(-5)(-7)$ **44.** $-3 \times 2 \times 2$

45. $-1 \times 0 \times 3$ **46.** $8(-2)(-2)$

Simplify.

47. 7^2 **48.** $(-2)^2$ **49.** $(-2)^3$

50. $(+2)^3$ **51.** $(-2)^4$ **52.** $(+2)^4$

53. $(-1)^5$ **54.** $(+1)^9$ **55.** $(-1)^8$

Problems and Applications

Find the value of $\blacksquare$.

56. $(-3)(\blacksquare) = -9$ **57.** $(\blacksquare)(+6) = -12$

58. $(\blacksquare)(+3) = 15$ **59.** $(-5)(\blacksquare) = -20$

60. $(+9)(\blacksquare) = 27$ **61.** $(\blacksquare)(-5) = -15$

62. $(+7)(\blacksquare) = -28$ **63.** $(\blacksquare)(-8) = 32$

64. $(\blacksquare)(+6) = -30$ **65.** $(+7)(\blacksquare) = -35$

Find the value of $\blacksquare$.

66. $(+5)(\blacksquare)(-2) = -20$

67. $(\blacksquare)(-1)(-4) = 36$

68. $(-4)(-3)(\blacksquare) = -24$

69. $(-1)(-5)(\blacksquare) = 0$

70. $(\blacksquare)(+6)(-1) = 12$

71. $(+3)(\blacksquare)(+4) = -36$

72. Copy and complete the tables.

a)

x	$3x$
2	
1	
0	
−1	
−2	

b)

m	$-2m$
2	
1	
0	
−1	
−2	

c)

t	t^2
2	
1	
0	
−1	
−2	

d)

s	$2s-5$
2	
1	
0	
−1	
−2	

73. Evaluate for $x = -3$.

a) $5x$ **b)** $-4x$ **c)** $5x - 4$

d) $x - 2$ **e)** $2x + 1$ **f)** $x + 3$

74. Evaluate for $x = -2$ and $y = -3$.

a) xy **b)** $3xy$ **c)** $-5xy$

d) $(4x)(6y)$ **e)** $-2xy - 3$ **f)** $5xy + 3y$

75. Nicole invested in the Stock Market. Her 25 shares of Arco Gold each lost \$2/day for 4 days. Find the change in value of her shares.

76. The temperature in Quebec City was 3°C at midnight. The average temperature drop was 2°C/h until 05:00.

a) What was the total temperature change?

b) What was the temperature at 05:00?

77. An airplane has an altitude of 5800 m. It then climbs at 400 m/min for 16 min.

a) What is its change in altitude?

b) What is its final altitude?

78. Copy and complete each statement.

a) The product of 3 negative integers is always ▬ .

b) The product of 2 negative integers and 1 positive integer is always ▬ .

c) The product of 2 positive integers and 1 negative integer is always ▬ .

79. Decide whether each statement is always true, sometimes true, or never true. Explain.

a) The product of a non-zero integer and its opposite is the square of the integer.

b) Multiplying a non-zero integer by itself gives a larger integer.

c) If the product of two integers is positive, one of the integers is positive.

d) The fourth power of an integer is not negative.

e) The cube of an integer is larger than the square of the same integer.

80. a) Write (-8) as a power with base (-2).

b) Can you write (-16) as a power with base (-2)? Explain.

81. Write the following integers and compare your answers with a classmate's.

a) 2 different integers with a product of 0

b) 2 different integers with a product of -20

c) 3 different integers with a product of -12

d) 4 different integers with a product of -36

82. Write a problem that involves the multiplication of integers. Have a classmate solve your problem.

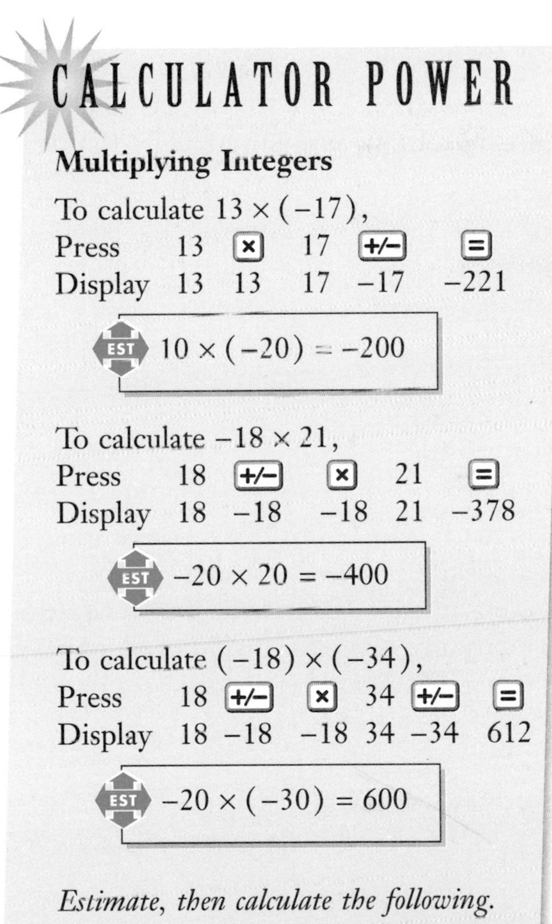

CALCULATOR POWER

Multiplying Integers

To calculate $13 \times (-17)$,

Press	13	×	17	+/–	=
Display	13	13	17	–17	–221

EST $10 \times (-20) = -200$

To calculate -18×21,

Press	18	+/–	×	21	=
Display	18	–18	–18	21	–378

EST $-20 \times 20 = -400$

To calculate $(-18) \times (-34)$,

Press	18	+/–	×	34	+/–	=
Display	18	–18	–18	34	–34	612

EST $-20 \times (-30) = 600$

Estimate, then calculate the following.

1. $(-11)(+15)$ **2.** $(+22)(+5)$

3. $(-72)(-9)$ **4.** $(-16)(+13)$

5. $(+18)(-14)$ **6.** $(-41)(-41)$

7. $(+11)(+17)$ **8.** $(-56)(-21)$

9. $(-71)(+14)$ **10.** $(+80)(-43)$

2.7 Dividing Integers

Activity: Look for a Pattern

An octopus has 8 tentacles. So, 7 octopuses have 8×7 or 56 tentacles. We can also say that if a group of octopuses has 56 tentacles, there are $56 \div 8$ or 7 octopuses in the group. Also, if 7 octopuses have 56 tentacles, each octopus has $56 \div 7$ or 8 tentacles. In mathematics, we say that division is the inverse operation of multiplication.

Since $8 \times 7 = 56$, then $56 \div 7 = 8$

and $56 \div 8 = 7$

It follows that,

since $(-6) \times (-5) = 30$, then $(+30) \div (-5) = -6$

and $(+30) \div (-6) = -5$

Inquire

1. Write two division statements for each multiplication.

a) $(+5) \times (+6) = +30$ **b)** $(+4) \times (-3) = -12$

c) $(-2) \times (+3) = -6$ **d)** $(-4) \times (-2) = +8$

e) $(+5) \times (-3) = -15$ **f)** $(-5) \times (+2) = -10$

g) $(-5) \times (-4) = +20$ **h)** $(+7) \times (+2) = +14$

2. Is the answer positive or negative when

a) a positive integer is divided by a positive integer?

b) a positive integer is divided by a negative integer?

c) a negative integer is divided by a positive integer?

d) a negative integer is divided by a negative integer?

Example 1

Divide.

a) $36 \div 9$ **b)** $(-48) \div (-6)$

Solution

The quotient of 2 integers with the same sign is positive.

a) $36 \div 9$
$= +4$

b) $(-48) \div (-6)$
$= 8$

Example 2

Simplify.

a) $-15 \div 3$ **b)** $\frac{(-8)(-6)}{4(-3)}$

Solution

The quotient of 2 integers with different signs is negative.

a) $-15 \div 3$
$= -5$

b) $\frac{(-8)(-6)}{4(-3)} = \frac{+48}{-12}$
$= -4$

Practice

State the sign of each answer.

1. $(+12) \div (+4)$
2. $(-15) \div (+3)$
3. $(-10) \div (-5)$
4. $(+20) \div (-4)$

Divide.

5. $(+18) \div (-2)$
6. $(-36) \div (-9)$
7. $(+10) \div (+5)$
8. $(-40) \div (+5)$

Divide.

9. $\frac{-27}{-9}$
10. $\frac{-45}{5}$
11. $\frac{0}{-2}$
12. $\frac{-8}{8}$
13. $\frac{6}{-6}$
14. $\frac{72}{9}$

Divide.

15. $49 \div (-7)$
16. $25 \div 5$
17. $-60 \div 12$
18. $-28 \div (-4)$

Simplify.

19. $\frac{(-2)(+8)}{(-4)}$
20. $\frac{-40}{(-5)(-2)}$
21. $\frac{8(-5)}{(-2)(-2)}$
22. $\frac{(-4)(+9)}{(-2)(+3)}$
23. $\frac{-10(-6)}{4(-5)}$
24. $\frac{(-10)(+6)}{(-3)(-2)}$

Problems and Applications

25. The nightly low temperatures in Calgary one week were +1°C, −5°C, −12°C, −3°C, +4°C, +6°C, and −5°C. What was the average low temperature?

26. An aircraft has an altitude of 10 200 m. It descends to 8450 m in 5 min. Calculate its change in altitude in metres per minute.

27. Identify each integer.
a) half the opposite of 2
b) the quotient of the smallest positive integer and the largest negative integer

28. What can you say about 2 integers if their quotient is 1? 0? −1?

29. Decide whether each statement is always true, sometimes true, or never true. Explain.
a) Dividing an integer by −1 changes the sign of the integer.
b) The quotient of 2 positive integers is smaller than either of them.
c) The sign of the quotient of 2 negative integers depends on which integer is the divisor.

30. Write the divisor and the dividend. Compare your answers with a classmate's.
a) 2 different integers with a quotient of −4
b) 2 different integers with a quotient of 3
c) 2 different integers with a quotient of 0

CALCULATOR POWER

Dividing Integers

To calculate $228 \div (-12)$,

Press	228	÷	12	+/−	=
Display	228	228	12	−12	−19

EST $200 \div (-10) = -20$

To calculate $(-315) \div 21$,

Press	315	+/−	÷	21	=
Display	315	−315	−315	21	−15

EST $-300 \div 20 = -15$

To calculate $(-306) \div (-17)$,

Press	306	+/−	÷	17	+/−	=
Display	306	−306	−306	17	−17	18

EST $-300 \div (-20) = 15$

Estimate, then calculate the following.

1. $-406 \div (-29)$
2. $246 \div (-41)$
3. $432 \div 48$
4. $-267 \div 89$

2.8 Solve a Simpler Problem

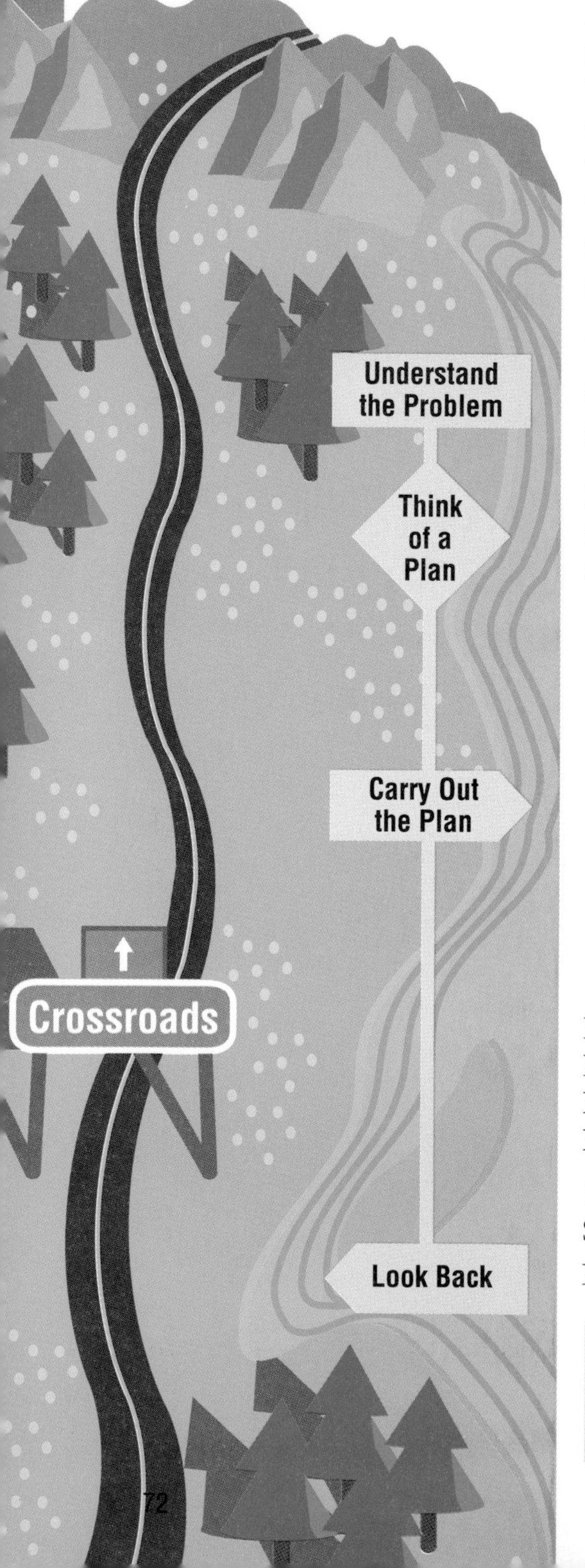

A problem that appears to be difficult because of large numbers can often be simplified if you use smaller numbers. You may need to solve a series of simpler problems, with the numbers getting larger, until you see a pattern. This pattern will help you solve the original problem.

There are 9 towns in the county. Each town is connected directly to each of the others by a road. How many roads are there?

1. What information are you given?

2. What are you asked to find?

3. Do you need an exact or approximate answer?

Find out how many roads connect 2 towns, 3 towns, 4 towns, and so on. Draw diagrams, make a table of the results, and look for a pattern.

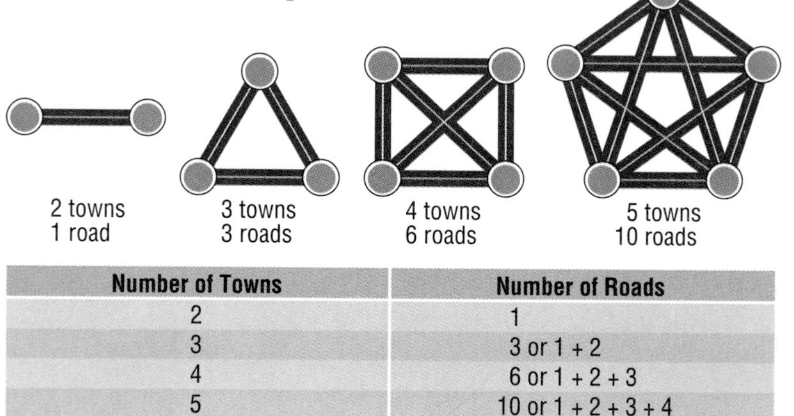

Number of Towns	Number of Roads
2	1
3	3 or 1 + 2
4	6 or 1 + 2 + 3
5	10 or 1 + 2 + 3 + 4

For 3 towns, add the first 2 non-zero whole numbers.
For 4 towns, add the first 3 non-zero whole numbers.
For 5 towns, add the first 4 non-zero whole numbers.
For 9 towns, add the first 8 non-zero whole numbers.

$1 + 2 + 3 + 4 + 5 + 6 + 7 + 8 = 36$
So, there are 36 roads.

How could you check the answer?

Solve a Simpler Problem	1. Break the problem into smaller problems. 2. Solve the problem. 3. Check that your answer is reasonable.

Problems and Applications

1. How would you estimate the number of listings in the white pages of a telephone book?

2. How would you estimate the thickness of 1 page of this textbook?

3. How would you estimate how long it would take you to read a 250-page novel?

4. Caroline owns a rectangular field of length 60 m and width 40 m. She wants to put a fence around the field. The fence posts are to be 2 m apart and there must be a post in each corner. How many posts will she need?

5. Find the sum of the first 100 odd numbers.
$1 + 3 + 5 + 7 + 9 + \cdots$

6. Find the sum of the first 100 even numbers.
$2 + 4 + 6 + 8 + 10 + \cdots$

7. Calculate the following.
$80 - 79 + 78 - 77 \cdots + 2 - 1$

8. Sixteen teams play in a basketball tournament. Each team plays until it loses 1 game. There are no ties. How many games are played?

9. For the final game of an international hockey tournament, each team had 18 players. Before the game, each player shook hands with every player on the opposing team. How many handshakes were exchanged?

10. Each of the 10 provinces sent 1 student to Ottawa for the opening of Parliament. Each student gave each of the other students a provincial pin. How many pins were exchanged?

11. About how many breaths do you take in 1 year?

12. If 1 bacterium divides into 2 bacteria in 1 min, how many bacteria come from 1 bacterium in 12 min? What assumptions have you made?

13. Write a problem that can be solved by means of a simpler problem. Have a classmate solve your problem.

LOGIC POWER

The Riddler has invited you to a party at his apartment. The elevator buttons are shown below. They are lettered from A to I, but the letters are hidden. You must press the buttons in order from A to I to get to the party. Use the clues to find the button letters.

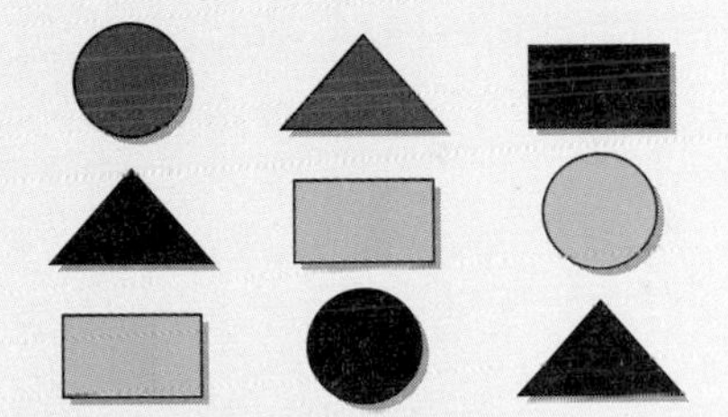

1. The D button is green.

2. The G button is red.

3. The F button is directly below a triangle.

4. The I button is in the left column.

5. The B button is not round.

6. The E button is in the middle row.

7. The H button is in the top row.

8. The F button is lower than the C button.

9. The C button is directly to the left of a round button.

2.9 Order of Operations with Integers

Activity: Examine the Solutions

Maria, Indira, and David gave different solutions to the same problem.

Maria's Solution

$(-24) \div (-4) + (-2) \div 2$
$= +6 + (-2) \div 2$
$= +4 \div 2$
$= 2$

Indira's Solution

$(-24) \div (-4) + (-2) \div 2$
$= +6 + (-1)$
$= +5$

David's Solution

$(-24) \div (-4) + (-2) \div 2$
$= (-24) \div (-6) \div 2$
$= (-24) \div (-3)$
$= 8$

Inquire

1. Is there an error in Maria's solution? If so, what is it?

2. Is there an error in Indira's solution? If so, what is it?

3. Is there an error in David's solution? If so, what is it?

The order of operations agreed upon for whole numbers and decimals is also applied to integers.
Remember the order with the acronym **BEDMAS**.

Example 1

Simplify $4(-3) \div (-2) - (-5)$.

Solution

$4(-3) \div (-2) - (-5)$ — **multiply**
$= (-12) \div (-2) - (-5)$ — **divide**
$= 6 - (-5)$
$= 6 + 5$ — **add**
$= 11$

Example 2

Simplify $2(5 - 8)^2 + 15$.

Solution

$2(5 - 8)^2 + 15$ — **brackets**
$= 2(-3)^2 + 15$ — **exponents**
$= 2(9) + 15$ — **multiply**
$= 18 + 15$ — **add**
$= 33$

Example 3

Simplify $\frac{-3 \times 2 + 16}{20 \div (-4)}$.

Solution

The division bar is a division and grouping symbol.

$\frac{-3 \times 2 + 16}{20 \div (-4)} = \frac{(-6 + 16)}{-5}$ — **brackets**
$= \frac{10}{-5}$ — **divide**
$= -2$

Practice

Simplify.

1. $(+8)+(-2)(-3)$
2. $(-6)(-4)+11$
3. $24 \div (-6) - (-2)$
4. $(-8)(0) \times (-2)$
5. $(-3)(-4) \times (-6)$
6. $(+16) \times (-4)(-5)$
7. $(-18) \div (-9) + (-8) \times 2$

Simplify.

8. $15 \div (-3) + 8$
9. $3 \times 4 - 2$
10. $10 - (-5) + 7$
11. $-6 + (-7) - 3$
12. $-3(-6) + (-5)$
13. $-6 \times 7 - 15$
14. $9 + 4 - 2 - 1$
15. $11 + (-8) \div (-4)$

Simplify.

16. $15 \times 6 \div 3$
17. $-7(-6) \div (-3)$
18. $-2(-9) \div 6$
19. $5(-2)(-3) + 7$
20. $10 - (-4)(+2)$
21. $-3(+6) + 3(-4)$
22. $-20 \div 4 \times 3 + 2$
23. $9 \times (-8) \div (-2)$

Simplify.

24. $5(8-7)+6$
25. $-6+4(3-9)$
26. $-2(9-4)-3$
27. $-(6-3)-(2-6)$
28. $4(2-5)+5(2+1)$

Simplify.

29. $-2+(-3)^2$
30. $(-4)^2+5$
31. -4^2+5
32. $(-2)^3-(-3)$
33. $-4+(-2)^2+5$
34. $(5-2)^2+6$
35. $-3(-7)-(-1)^3$
36. $2(3-7)^2-(-2)(-3)$

Simplify.

37. $\frac{(-3) \times 8}{4}$
38. $\frac{5 \times (-8)}{-10}$
39. $\frac{-8 \div 4}{-2 \div (-1)}$
40. $\frac{4-12}{3-(-5)}$
41. $\frac{-21 \div (-3)}{-7 \times (-1)}$
42. $\frac{-6 \times 8}{-7+(-5)}$

Problems and Applications

43. Add brackets to make each statement true.

a) $-3+4^2 \times 5 = 5$
b) $-1-3-8 \div 4 = -3$
c) $3^2+4 \times 2 - 5 = -3$
d) $6^2 - 20 \div 2 + 6 = 2$

44. For your next birthday, a relative offers you a choice of 2 amounts of money, both in dollars. The amounts are given by

$$(2x-20)^2 \text{ and } \frac{1}{2}(x-9)^3$$

where x is your age on your next birthday. Which amount will you choose and why?

45. There are 2 meanings of "5 less than −3 squared."

a) Use symbols to show the meanings.
b) What value results from each meaning?
c) Rewrite the phrase in words to make the 2 meanings clear. Compare your wording with a classmate's.

CALCULATOR POWER

We can evaluate $-15(18-35)^2-97$ as follows.

15 [+/−] [×] [(] 18 [−] 35 [)] [x²] [−] 97 [=] −4432.

In questions 1–5, estimate, then calculate.

1. $34 \times (-12) \div 51$
2. $(-504) \div (-56) + 77$
3. $75 - 85 \times (-9) + 57$
4. $(56-79)^2 - 64$
5. $-78 + (23+16) \times (32-49)$
6. Questions 1–5 all include brackets. Do you have to use the [(] and [)] keys in each calculation? Explain.

Knowledge-Based and Expert Systems

Knowledge-based systems and **expert systems** are specialized types of computer software. Knowledge-based systems use facts and rules you might find in a manual or textbook. An example of a knowledge-based system is a chess-playing program. Expert systems provide you with the knowledge of human experts in a particular field. The programs are based on a combination of textbook knowledge and the decision-making processes supplied by experts. One example is the medical consultation system called CADUCEUS. This system is like many expert systems in that it is interactive. The user works with the system to solve a problem through a series of questions and answers, which eventually leads to a solution.

Activity ❶ Knowledge-Based Systems

1. Work with your classmates to list the knowledge-based systems that you know. You are probably familiar with more than you think.

2. Describe what each knowledge-based system is used for.

Activity 2 Expert Systems

1. Describe what you would expect to find in an expert system called *Bird Species Identification*.

2. Use your research skills to find 5 areas in which expert systems have already been developed. Write a brief description of what each system does.

3. Predict 3 areas in which you think expert systems will be developed in the future. Describe how you think the systems will be used.

4. In what area would you like to see an expert system developed? What would you want the program to tell you?

5. There are many areas in which it is very difficult to create expert systems. A system that tells you how to predict the weather is one example. Why is an expert system difficult to create in this area? In what other areas do you think it would be difficult to have an expert system?

2.10 Expressions with Integers

Activity: Use a Formula

To raise money for charity, the Bilko Company held a paper airplane contest. The planes were flown from a 50th-floor window of a skyscraper. The formula for the height of the winning plane at any given time was

$$h = \frac{700 + 15t - t^2}{4}$$

where h was the height of the plane in metres and t was the flight time in seconds.

Inquire

1. Substitute 1 for t to find the height after 1 s.
2. What was the height after 2 s? 3 s?
3. What was the height after 10 s? 15 s? 20 s?
4. Describe the plane's motion so far.
5. What was the height after 35 s? 40 s? Explain.
6. After how many seconds did the plane hit the ground? Explain.

In many formulas and expressions, variables are replaced by integers and then evaluated.

Example 1

Evaluate the expressions for $x = -2$.

a) $-7x$ **b)** $8 - 5x$ **c)** $-x^2 + 3x - 5$

Solution

a)
$$\begin{aligned} &-7x \\ &= -7(-2) \\ &= 14 \end{aligned}$$

b)
$$\begin{aligned} &8 - 5x \\ &= 8 - 5(-2) \\ &= 8 + 10 \\ &= 18 \end{aligned}$$

c)
$$\begin{aligned} &-x^2 + 3x - 5 \\ &= -(-2)^2 + 3(-2) - 5 \\ &= -4 - 6 - 5 \\ &= -15 \end{aligned}$$

Example 2

Evaluate for $s = -3$ and $t = 4$.

a) $-6st$ **b)** $5s - 2t$ **c)** $4s^2 - 2st - t$

Solution

a)
$$\begin{aligned} &-6st \\ &= -6(-3)(4) \\ &= 72 \end{aligned}$$

b)
$$\begin{aligned} &5s - 2t \\ &= 5(-3) - 2(4) \\ &= -15 - 8 \\ &= -23 \end{aligned}$$

c)
$$\begin{aligned} &4s^2 - 2st - t \\ &= 4(-3)^2 - 2(-3)(4) - (4) \\ &= 4(9) - 2(-3)(4) - (4) \\ &= 36 + 24 - 4 \\ &= 56 \end{aligned}$$

Practice

1. Evaluate for $x = 2$.

a) $4x$ **b)** $x + 2$ **c)** $x - 2$
d) $-2x$ **e)** $7 - x$ **f)** $2x + 4$

2. Evaluate for $x = -3$.

a) $2x$ **b)** $x + 5$ **c)** $-4x$
d) $x - 1$ **e)** $5 - x$ **f)** $2x + 3$

3. Evaluate for $s = -2$ and $t = 3$.

a) $5s$ **b)** $4st$ **c)** $s + t$
d) $t - s$ **e)** $-3st$ **f)** $-s - t$

4. Evaluate for $r = -6$.

a) $-4r$ **b)** $3(r + 1)$ **c)** $7 - 3r$
d) $-2 + 3r$ **e)** $-(r - 4)$ **f)** $-r - 4$
g) $2r + 1$ **h)** $-7r - 2$ **i)** $5r - 8$

5. Evaluate for $m = -2$.

a) $m^2 + 1$ **b)** $6 + m^3$
c) $-4m^3$ **d)** $m^2 + 3m - 1$
e) $-m^3 - 4m^2$ **f)** $m^2 - 2m - 3$
g) $(7 - m)^2$ **h)** $2m^2 + 5m + 6$

6. Evaluate for $x = -4$ and $y = -1$.

a) $3x + 2y$ **b)** $7 - 3xy$
c) $x^2 + y^2$ **d)** $4x + 3y - 6$
e) $2x^2 - y^2$ **f)** $5(x + y)^2$
g) $-3(x - y)^3$ **h)** $x^2 - 3xy + y^2$
i) $2x^2 - xy - y^2$ **j)** $(x + 2)(y - 4)$

Problems and Applications

7. Copy and complete the tables.

a)

x	$2x - 1$
2	
1	
0	
−1	
−2	

b)

t	$t^2 + 3$
2	
1	
0	
−1	
−2	

c)

y	$y + 1$
2	
1	
0	
−1	
−2	

d)

m	$m - 2$
2	
1	
0	
−1	
−2	

8. The formula for the height of a model rocket is

$$h = -t^2 + 20t$$

where h is the height in metres and t is the time in seconds after lift-off.

a) What is the rocket's height after 7 s? 10 s?

b) Find the rocket's height after 20 s. Explain your answer.

9. Suppose a rock is fired upward at a speed of 20 m/s from the edge of a 60-m cliff beside the ocean. The height of the rock above the edge of the cliff is given by the formula

$$h = 20t - 5t^2$$

where h is the height in metres and t is the time in seconds since the rock was fired upward.

a) What is the height of the rock after 1 s? 2 s? 3 s?

b) What is the height of the rock after 4 s? Explain.

c) What is the height of the rock after 5 s? Explain.

d) After how many seconds does the rock hit the water?

10. A weather station in Montreal records the temperature at noon every day. The formula

$$\frac{a + b + c + d}{4}$$

calculates the average of 4 readings. Find the average of the following.

a) $a = -2$°C, $b = -6$°C, $c = -5$°C, $d = +1$°C

b) $a = -1$°C, $b = +2$°C, $c = -3$°C, $d = -6$°C

2.11 Use a Data Bank

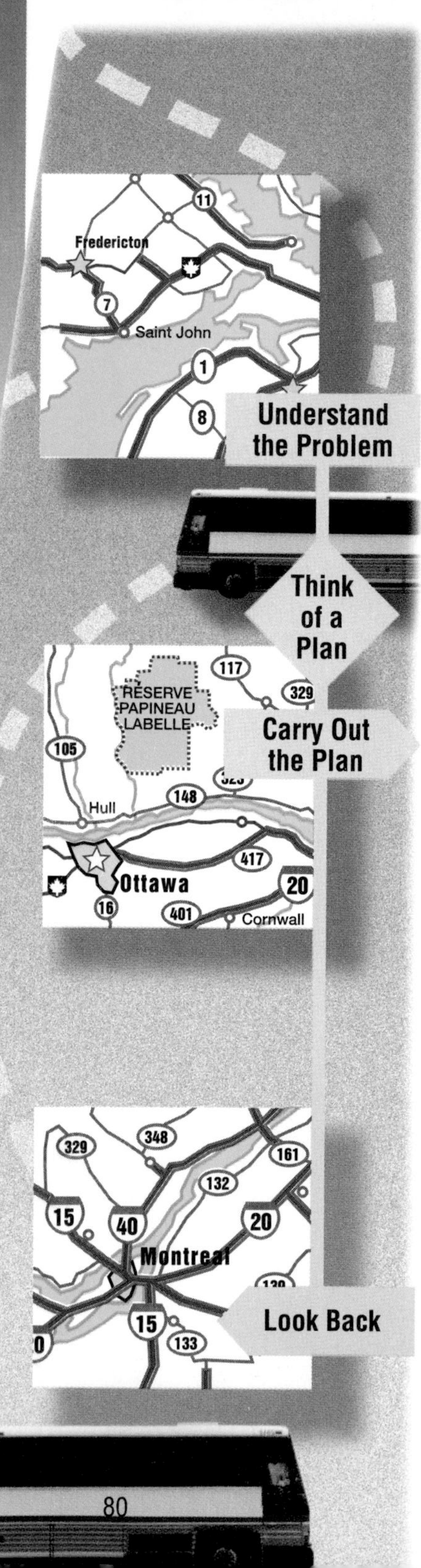

You must locate information to solve some problems. There are many sources of information, including computer files, libraries, newspapers, magazines, atlases, experts, and data banks.

A grade 9 history class is going from Saint John to Ottawa, then to Montreal, and back to Saint John. The students will spend 3 days in Ottawa and 2 days in Montreal. Their bus can average 70 km/h. The students plan to leave Saint John on a Monday morning. Draw up a schedule for the trip.

1. What information are you given?

2. What are you asked to find?

3. What information do you need to locate?

4. Do you need an exact or approximate answer?

You need the driving distances between the cities. You could use an atlas, almanac, or road map, or ask at a tourist information centre. To calculate the time required for each part of the trip, divide the distance by the average speed.

Use the Data Bank on pages 564 to 569 in this book. Use the chart of "Driving Distances Between Cities."

Saint John to Ottawa	1130 km
Ottawa to Montreal	190 km
Montreal to Saint John	940 km

Saint John to Ottawa takes $1130 \div 70 \doteq 16$ h. Allow 2 days.
Ottawa to Montreal takes $190 \div 70 \doteq 3$ h. Allow $\frac{1}{2}$ a day.
Montreal to Saint John takes $940 \div 70 \doteq 13$ h. Allow $1\frac{1}{2}$ days.

Schedule: Leave Saint John Monday morning. Arrive in Ottawa Tuesday evening. Spend Wednesday, Thursday, and Friday in Ottawa. Leave Saturday morning for Montreal. Arrive in Montreal at noon on Saturday. Spend the rest of Saturday and Sunday in Montreal. Leave Montreal for Saint John on Monday morning. Arrive in Saint John on Tuesday around noon.

Does the schedule seem reasonable?

Use a Data Bank

1. Look up the information you need.
2. Solve the problem.
3. Check that the answer is reasonable.

Problems and Applications

Use the Data Bank on pages 564 to 569 of this book to solve the problems.

1. a) What is the flying distance between Vancouver and Regina?
b) What is the driving distance between Vancouver and Regina?
c) How much longer is the driving distance than the flying distance?

2. The St. Joseph's High School Band is planning a tour. The band will leave from Halifax and stop in Fredericton, Quebec City, Toronto, and Montreal, in that order. The band will play one concert at a high school in each city and will then return to Halifax. The band will travel on a bus that can average 80 km/h. The band will leave Halifax on a Monday morning. Draw up a schedule for the trip.

3. Sandra flew from Vancouver to Halifax to compete in a gymnastics competition. She left Vancouver at 11:00. The flight to Toronto took 4 h. She took 1 h and 15 min to change planes in Toronto. The flight to Halifax took 1 h and 30 min. What time was it in Halifax when she landed?

4. What is the wind chill temperature if the thermometer reading is −23°C and the wind speed is 32 km/h?

5. If it is 09:00 on July 1 in Paris, France, what is the time and the date in Sydney, Australia?

6. Justine drove from Halifax to Vancouver. She drove on the Trans Canada Highway for the entire trip.
a) How far did she drive?
b) If she averaged 90 km/h, how many hours did she spend driving?
c) What major cities did she drive through?

7. The world's longest covered bridge is at Hartland, New Brunswick. The world's longest cantilevered bridge is the Pont de Quebec across the St. Lawrence River. How many times longer is the Pont de Quebec?

8. The population density of a country, city, or region is the average number of people living in each square kilometre of land. Use the area and the population of each Canadian province to work out the population density of each. Rank the population densities from highest to lowest.

9. a) You are the manager of a music group that wants to tour cities in Canada for two weeks to promote its new release. Make up a travel schedule for the group.
b) Compare your schedule with your classmates'. Decide the best features of each schedule.

NUMBER POWER

This puzzle appeared in the work of a Chinese mathematician, Sun Tzu, who lived in the 4th or 5th century A.D.

Pick a whole number less than 60 and greater than 5. Divide the whole number by 3. Call the remainder x. Divide the whole number by 4. Call the remainder y. Divide the whole number by 5. Call the remainder z.

Evaluate the expression
$40 \times x + 45 \times y + 36 \times z$
and divide the value by 60.

a) What do you notice about the remainder?
b) Compare your finding with your classmates'.

Radiocarbon Dating

One theory about how people first entered North and South America is that they crossed the Bering Strait over 40 000 years ago. One method that archaeologists use to establish such dates is called *radiocarbon dating*.

All living plants and animals contain the same amount of radioactive carbon–14, C–14, per kilogram of mass. When they die, the C–14 slowly changes to nitrogen–14, N–14, which is not radioactive. The level of radioactivity of the dead plant or animal slowly decreases with time. By measuring the level of radioactivity, archaeologists can find how much C–14 is left.

It takes about 5700 years for half the C–14 to change to N–14. Another way of saying this is that C–14 has a "half-life" of about 5700 years.

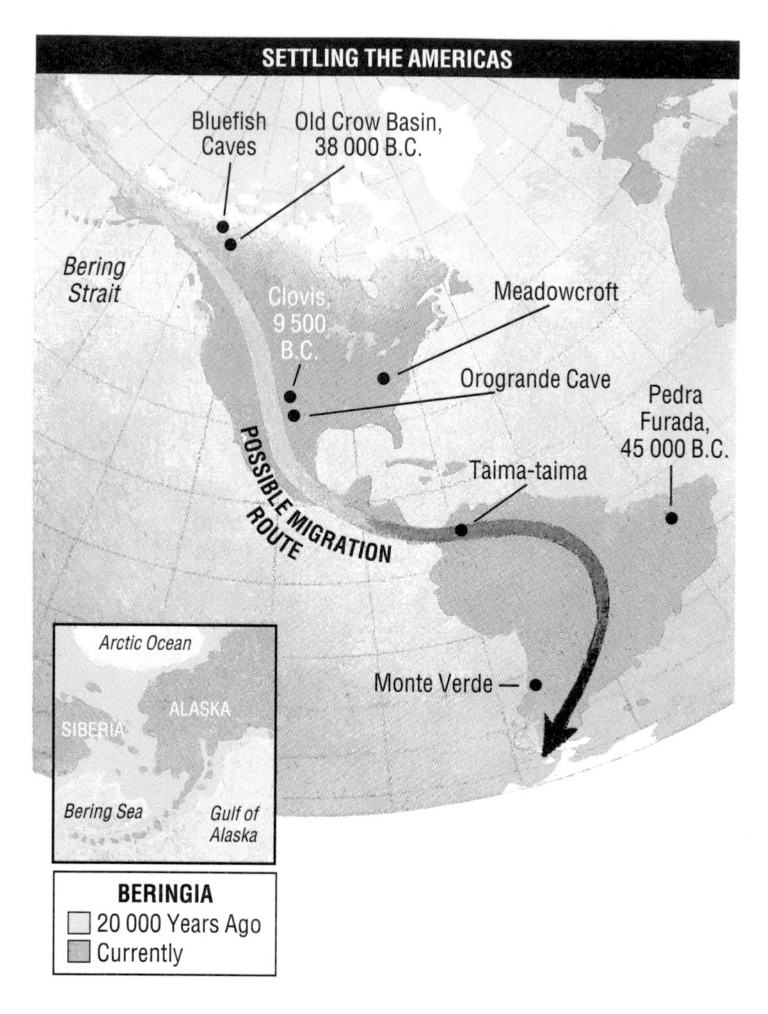

Activity ❶

Copy and complete the table. Graph the percent of C–14 remaining versus time after death.

Time After Death (Years)	Percent of C–14 Remaining
0	100
5700	50
11 400	25
17 100	
22 800	
28 500	
34 200	
39 900	

Percent of C–14 Remaining: 0, 10, 20, 30, 40, 50, 60, 70, 80, 90, 100
Time After Death (Thousands of Years): 0, 10, 20, 30, 40

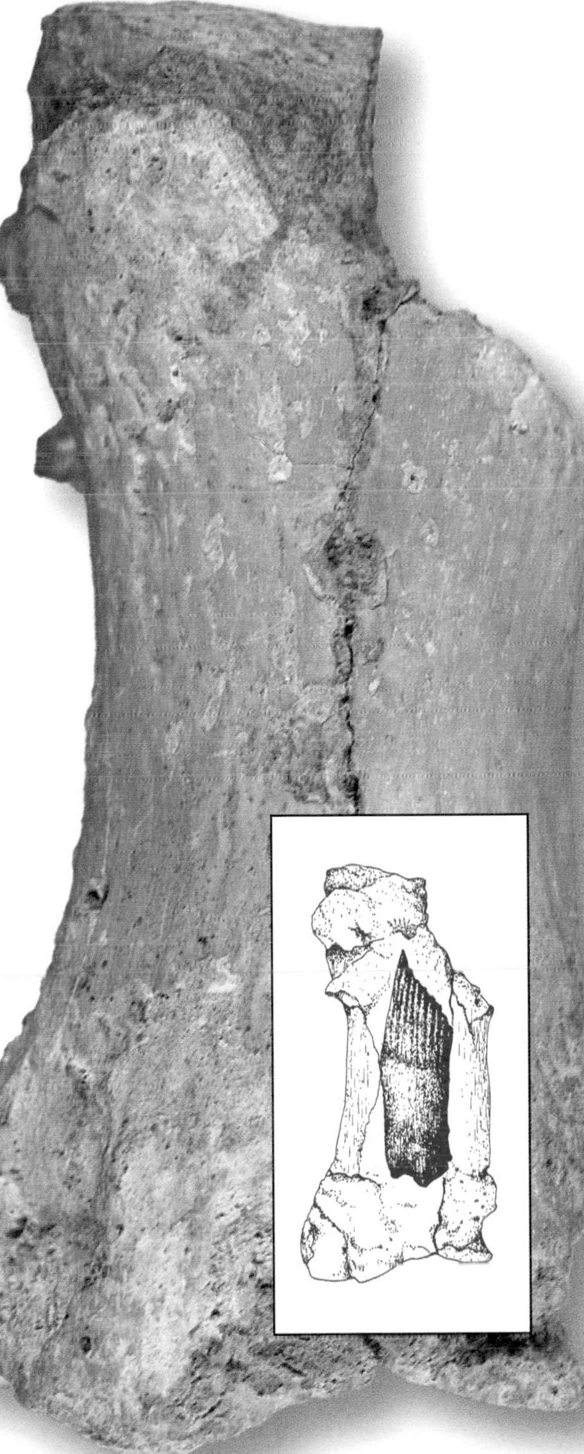

Courtesy of the Cooley-Dickinson Hospital

Spear point embedded in toe bone of a horse discovered in New Mexico.

Activity ❷

The map shows where archaeological digs have discovered evidence of human activity.

1. At the Bluefish Caves in the Yukon, Jacques Cinq-Mars of the Archaeological Survey of Canada found a caribou bone that was cut and shaped to form a tool. Six percent of the C–14 remained in the bone. How old is the bone? What was the date of the settlement at the Bluefish Caves?

2. At Monte Verde in southern Chile, dead plants from 15 species still used today for medicinal purposes were found under a peat bog. The plants came from the coast, 100 km away. The settlers probably trekked to the coast to get them. The plants had 20% of their C–14 remaining. How old are they?

3. At Taima-taima in Venezuela, the remains of a slain mastodon were found. The remains had 21% of their C–14 remaining. How old are the remains?

4. At Orogrande Cave in New Mexico, a horse's toe bone with a spear point in it had 5% of its C–14 remaining. How old is the bone?

5. In southwestern Pennsylvania, a mat woven from bark was found in a settlement that has been named Meadowcroft. The bark had 10% of its C–14 remaining. How old is the mat? What was the date of the settlement?

Review

1. State the opposite of each integer.

a) 6 **b)** −4 **c)** −7 **d)** 9

2. Write in order from largest to smallest.

a) −1, 5, −4, −6 **b)** 7, 9, −1, −5, 0

c) −9, −7, −8, −2 **d)** 0, −1, 1, 2, −2

3. Identify each integer.

a) 6 more than −1 **b)** 7 less than 3

c) 5 more than −8 **d)** 6 less than −3

4. Graph on a number line.

a) integers less than or equal to 3

b) integers greater than −4

5. List and graph these integers.

a) less than 4 and greater than −2

b) greater than −7 and less than −1

Add.

6. $(+5) + (+7)$ **7.** $(-4) + (+2)$

8. $(-2) + (-9)$ **9.** $(+6) + (-6)$

10. $(+11) + (-12)$ **11.** $(-13) + (-9)$

Simplify.

12. $(-5) + (-3) + (+2)$

13. $(+5) + (-2) + (-3) + (-9)$

14. $(-5) + (-6) + (-7) + (-8)$

Find the value of ■.

15. $(+5) + (■) = (+3)$

16. $(-3) + (■) = (-7)$

17. $(■) + (-2) = (-1)$

18. $(■) + (-3) = (-5)$

Add.

19. $-6 + 8$ **20.** $-8 + (-3)$

21. $-9 + 6$ **22.** $4 + (-11)$

Subtract.

23. $(+5) - (+2)$ **24.** $(+7) - (-4)$

25. $(-3) - (-6)$ **26.** $(-2) - (+5)$

27. $(+8) - (+9)$ **28.** $(+5) - (-5)$

Find the value of ■.

29. $(+7) - (■) = (-3)$

30. $(-6) - (■) = (+3)$

31. $(■) - (+3) = (+2)$

32. $(■) - (+5) = (-4)$

Simplify.

33. $(+6) + (+4) - (+2)$

34. $(+9) - (-3) + (-2)$

35. $(-3) - (-4) - (-1)$

36. $(+7) + (-2) - (-5)$

Subtract.

37. $5 - 4$ **38.** $8 - (-4)$ **39.** $-7 - 6$

40. $7 - 9$ **41.** $-2 - (-3)$ **42.** $-4 - 8$

Simplify.

43. $-3 + 4 - 6 + 2$

44. $-2 - 3 - 5 - 7$

45. $7 - 6 + 4 - (-2)$

46. $-3 + 7 - (-9) - 1$

Multiply.

47. $(+7) \times (+2)$ **48.** $(-3) \times (-5)$

49. $(-10) \times (-11)$ **50.** $(+9)(-7)$

51. $(+6)(+12)$ **52.** $(-8)(+4)$

Find the value of ■.

53. $(■)(-4) = 8$ **54.** $(+5)(■) = -10$

55. $(■)(-7) = 21$ **56.** $(-2)(■) = 0$

Simplify.

57. -11×9 **58.** $9 \times (-4)$

59. $-6 \times (-7)$ **60.** $-2(-6)$

61. $-3 \times 4 \times (-2)$ **62.** $(-4)^2$

63. -4^2 **64.** $(-2)^2 \times 3$

Divide.

65. $(+12) \div (+4)$

66. $(-20) \div (-5)$

67. $(+18) \div (-2)$

68. $(-15) \div (+3)$

Find the value of ■.

69. $(-9) \div ($■$) = (+3)$

70. $(-27) \div ($■$) = (-9)$

71. $($■$) \div (-2) = (-7)$

72. $($■$) \div (+5) = (+2)$

Divide.

73. $20 \div (-2)$

74. $-14 \div 2$

75. $-16 \div (-4)$

76. $21 \div 7$

77. $0 \div (-2)$

78. $-36 \div (-9)$

Simplify.

79. $5 + (-3) + (-2)$

80. $-6 \times 2 + 8$

81. $10 + -3(-4)$

82. $15 \div (-5) + 6 - (-1)$

83. $3 \times (-2) + 4 \times (-3)$

84. Evaluate for $t = -2$.

a) $3t + 7$

b) $t^2 + 2t - 1$

c) $t(5t - 4)$

d) $2t^2 - t + 8$

e) $t^3 - t^2 - t$

f) $-t^3 + t^2 + t$

85. Evaluate for $x = -2$ and $y = -3$.

a) $3x + 2y$

b) $6xy$

c) $x^2 + y^2$

d) $7(x + y)^2 - 1$

e) $-3(x - y)$

f) $x - 2xy + y$

g) $2x^2 - y^2$

h) $4xy - x^2 - y^2$

86. The formula for the height of a model rocket is

$$h = -t^2 + 40t$$

where h is the height in metres and t is the time in seconds after lift-off. Find the height of the rocket at these times after lift-off.

a) 10 s **b)** 20 s **c)** 40 s

87. A plane is flying at 30 000 m. The temperature outside the plane is −27°C. As the plane descends, the temperature increases by 1°C for every 1000-m drop in altitude. What is the temperature on the ground?

Group Decision Making
Advertising in the Media

In this activity, your group will do one of the following.

- write and act out a 30-s television commercial
- write and act out a 30-s radio commercial
- design a full-page newspaper advertisement

1. Meet as a class and choose 6 or 7 products you would like to advertise. Decide as a class which group will advertise which product.

2. In your home group, decide which type of advertising best suits your product. List the reasons for your decision.

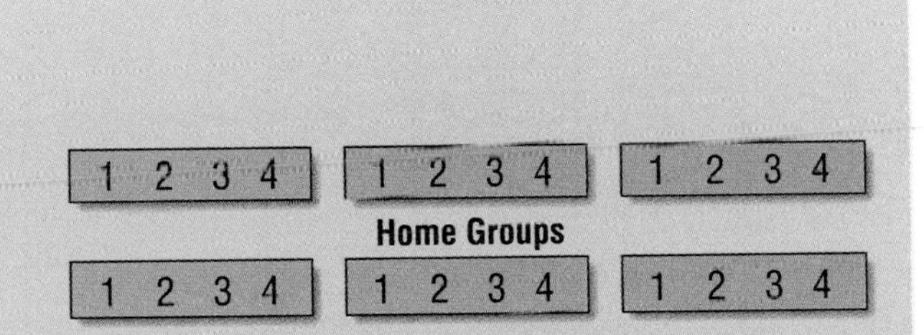

3. Prepare your advertisement in your home group.

4. Make a presentation to the class. Include a description of how you used math in preparing the advertisement.

5. In your home group, discuss your group work. Decide what worked well and what you would do differently next time.

6. Meet as a class to compare your opinions on your group work.

Chapter Check

1. Write the integers in order from largest to smallest.

a) $-1, 2, 0, -2, -3, 4$

b) $-3, -6, -9, -7, -5$

2. Identify each integer.

a) 2 less than 1

b) 5 more than -2

c) 4 less than -8

d) 6 more than -9

3. List and graph.

a) integers less than 3 and greater than -4

b) integers greater than -9 and less than -1

Simplify.

4. $5 + (-8)$

5. $-4 - 7$

6. $-3 - (-6)$

7. $6 \times (-2)$

8. $(-5)(-3)$

9. $-3 \times (-4)$

10. $-21 \div (-7)$

11. $-9 \div 3$

12. $16 \div (-2)$

Simplify.

13. $+12 - 13 - 7$

14. $-3 - 4 - 5$

15. $(-2)(-6) \div (-4)$

16. $30 \div (-5) \times (-2)$

17. $-4(-5) - 2$

18. $-7 + (-1)(-6)$

19. $(-2)^2 + 5$

20. $-3^2 - 7$

Find the value of ■ .

21. $(-6) + (■) = -2$

22. $(■) - (-4) = +5$

23. $(-7) \times (■) = +21$

24. $(■) \div (-3) = +6$

Simplify.

25. $(-5)^2 - 4 + 5$

26. $(-3)(-8) \div (-6)$

27. $-7 + 10 \div (-5)$

28. $5 \times (-2) - 8 \times 2$

29. Evaluate for $s = -4$ and $t = -5$.

a) $4s + 2t$

b) $3t - s$

c) $-2st + s^2 + t^2$

d) $2s^2 + 3st - t^2$

30. At 09:00, the temperature in Brockville, Ontario, was 1°C. The temperature dropped 2°C/h for the next 3 h. What was the temperature at noon?

31. Canadian nuclear reactors, known as CANDU reactors, use a type of water called "heavy water." The freezing point of "normal" water is 4°C lower than the freezing point of heavy water. What is the freezing point of heavy water?

3-18
LET'S SEE... THAT'S ABOUT 2,800 IN MOTH YEARS
don addis

BENT OFFERINGS by Don Addis
By permission of Don Addis and Creators Syndicate.

Using the Strategies

1. Determine the pattern and write the next 3 rows.

1	1	1						
1	2	3	2	1				
1	3	6	7	6	3	1		
1	4	10	16	19	16	10	4	1

2. The perimeter of an isosceles triangle is 8 cm. The length of each side is a whole number. How long is the shortest side?

3. The number 72 can be factored as follows.

$$\begin{aligned} 72 &= 36 \times 2 \\ &= 18 \times 2 \times 2 \\ &= 9 \times 2 \times 2 \times 2 \\ &= 3 \times 3 \times 2 \times 2 \times 2 \end{aligned}$$

Use this technique to find the age of each person in a group of teenagers if the product of their ages is 661 500.

4. The first 4 triangular numbers are shown.

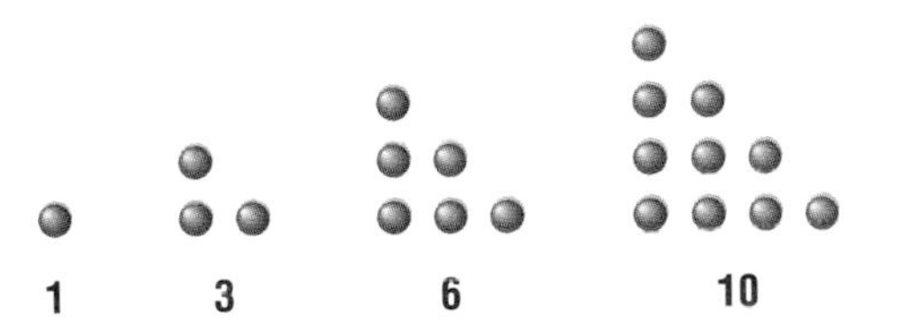

What are the next 3 triangular numbers?

5. There are 20 students in the gym. The teacher wants them in 3 groups, with an even number of students in each group. In how many ways can the teacher put them into groups?

6. Your train leaves at 08:15. The bus trip to the train station takes 25 min. The bus stop is a 5-min walk from your place. You should get to the train station to buy your ticket 15 min before the train leaves. It will take you 55 min to get dressed, eat breakfast, and pack. For what time should you set your alarm clock?

7. Determine the pattern and predict the next 2 lines.

$101 \times 101 = ■$
$202 \times 202 = ■$
$303 \times 303 = ■$

8. The number 8 is a perfect cube.
$2^3 = 2 \times 2 \times 2 = 8$ or $2^3 = 8$
The number 27 is also a perfect cube.
$3^3 = 3 \times 3 \times 3 = 27$ or $3^3 = 27$
We can write 8 as the sum of 2 consecutive odd numbers.

$$3 + 5 = 8$$

We can write 27 as the sum of 3 consecutive odd numbers.

$$7 + 9 + 11 = 27$$

a) The next perfect cube is 64 because $4^3 = 64$. Write 64 as the sum of 4 consecutive odd numbers.
b) The next perfect cube is 125. Write 125 as the sum of 5 consecutive odd numbers.
c) The next perfect cube is 216. Write 216 as the sum of 6 consecutive odd numbers.
d) How does the square of the number that is cubed fit into the sum of the consecutive odd numbers?
e) Use this pattern to write 7^3 as the sum of 7 consecutive odd numbers.

9. Sketch a graph to show the length of time you spend watching TV versus the day of the week.

Use the Data Bank on pages 564 to 569 to find the information you need.

1. How much higher above sea level is Mexico City than Calgary?

2. Express Venezuela's annual gold production as a percent of Canada's annual gold production.

Square Roots and Powers

A square engraving has sides of length 4 cm.

The sides of a photographic enlargement of the engraving are 3 times as long as the sides of the engraving. How do the areas of the photograph and the engraving compare?

A reduced image of the engraving is used on a postage stamp. The area of the image on the stamp is 4 times smaller than the area of the engraving. What is the length of a side of the image on the stamp?

Activity ❶ Prime Numbers

The number 10 has 4 **factors**, 1, 2, 5, and 10. Each factor divides 10 evenly. A **prime number** is a whole number with exactly 2 factors, itself and 1. The number 1 is not considered to be prime. The first 4 prime numbers are 2, 3, 5, and 7.

1. The numbers 5 and 7 are called **twin primes**. They are consecutive odd numbers that are also prime. Find all the twin primes less than 100.

2. Prime triplets are 3 consecutive odd numbers that are also prime. What is the only set of prime triplets?

3. In 1742, the mathematician Goldbach stated that every even number, except 2, can be written as the sum of 2 prime numbers.

$$8 = 3 + 5$$

$$18 = 7 + 11 \text{ or } 18 = 5 + 13$$

Write each of the following as the sum of 2 prime numbers.

a) 24 **b)** 30 **c)** 42 **d)** 100

4. Until 1903, it was thought that $2^{67} - 1$ was a prime number. That year, a mathematician named Cole received a standing ovation at a meeting of the American Mathematical Association. He multiplied out 2^{67} and subtracted 1, then moved to another chalkboard and multiplied the following numbers.

$$\begin{array}{r} 761\ 838\ 257\ 287 \\ \times\ 193\ 707\ 721 \\ \hline \end{array}$$

He got the same result from both calculations. Cole had factored a number previously thought to be prime.

a) Estimate $2^{67} - 1$. Compare your method with your classmates'.

b) Investigate why mathematicians find prime numbers so interesting.

Activity ❷ Squares, Cubes, and Factors

Study the following steps.

- List the factors of 6.

  1, 2, 3, 6

- Find the number of factors each factor of 6 has.

Factor of 6	1	2	3	6
Number of Factors	1	2	2	4

- Find the sum of the cubes of the numbers of factors.

Number of Factors	1	2	2	4
Number of Factors Cubed	1	8	8	64

$$1 + 8 + 8 + 64 = 81$$

- Find the square of the sum of the numbers of factors.

$$1 + 2 + 2 + 4 = 9 \text{ and } 9^2 = 81$$

1. Write a statement about the relationship between "the sum of the cubes of the numbers of factors" and "the square of the sum of the numbers of factors" for the factors of a number.

2. Test your statement, starting with these numbers.

a) 12 **b)** 20

Activity ❸ Amazing Square

Copy the square onto grid paper.

Cut out the 4 pieces and rearrange them to form a rectangle.

1. What was the area of the original square?
2. What is the area of the rectangle?
3. Can you explain the difference in the areas, or is there some magic in squares?

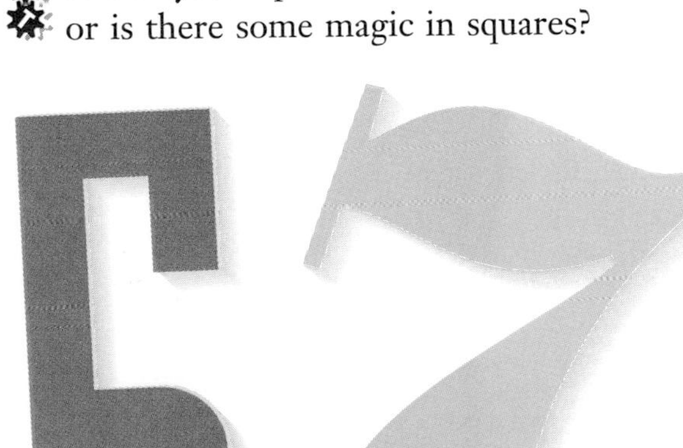

Mental Math

Calculate.

1. 5×23 **2.** 16×9
3. 7^2 **4.** 11^2
5. 66×5 **6.** $5 \times 2 \times 8$
7. 19×7 **8.** 103×4
9. 100×16 **10.** 2323×3

Calculate.

11. $85 \div 5$ **12.** $123 \div 10$
13. $217 \div 7$ **14.** $366 \div 6$
15. $6684 \div 2$ **16.** $5000 \div 100$
17. $950 \div 1000$ **18.** $1200 \div 20$
19. $1500 \div 50$ **20.** $800 \div 25$

Calculate.

21. $356 + 222$ **22.** $140 + 555$
23. $199 + 299$ **24.** $234 + 587$
25. $435 - 225$ **26.** $810 - 105$
27. $367 - 202$ **28.** $287 - 369$
29. $450 + (-301)$ **30.** $-215 - 680$

Calculate.

31. $17 + 23 - 8$ **32.** $9 - 8 - 3$
33. $27 + 14 + 38$ **34.** $15 - 16 + 20$
35. $3 + 22 - 16$ **36.** $-4 + 15 - 18$
37. $22 + (-15) + 13$ **38.** $-10 + 35 + 12$
39. $42 + 16 + (-10)$ **40.** $-17 + 26 - 35$

Calculate.

41. 2×5^2 **42.** $100 \div (6 + 4)$
43. $5 \times 2 \times 8$ **44.** $73 - 7 \times 5$
45. 4×7^2 **46.** $5^2 + 3^2$
47. $(17 - 11)^2$ **48.** $(11 - 17)^2$
49. $15 \times (65 - 55)$ **50.** $10 \times 5 - 70$

Perfect Squares

The Pythagoreans were a secret society of Greek mathematicians in the sixth century B.C. They were interested in special types of numbers, including prime numbers, triangular numbers, square numbers, and pentagonal numbers. The Pythagoreans used lines, triangles, and squares of pebbles to represent different types of numbers. Our word "calculate" comes from the Latin word *calculus*, which means "pebble." Perfect square numbers could be represented by squares of pebbles. Perfect squares are found by squaring whole numbers.

1 4 9 16

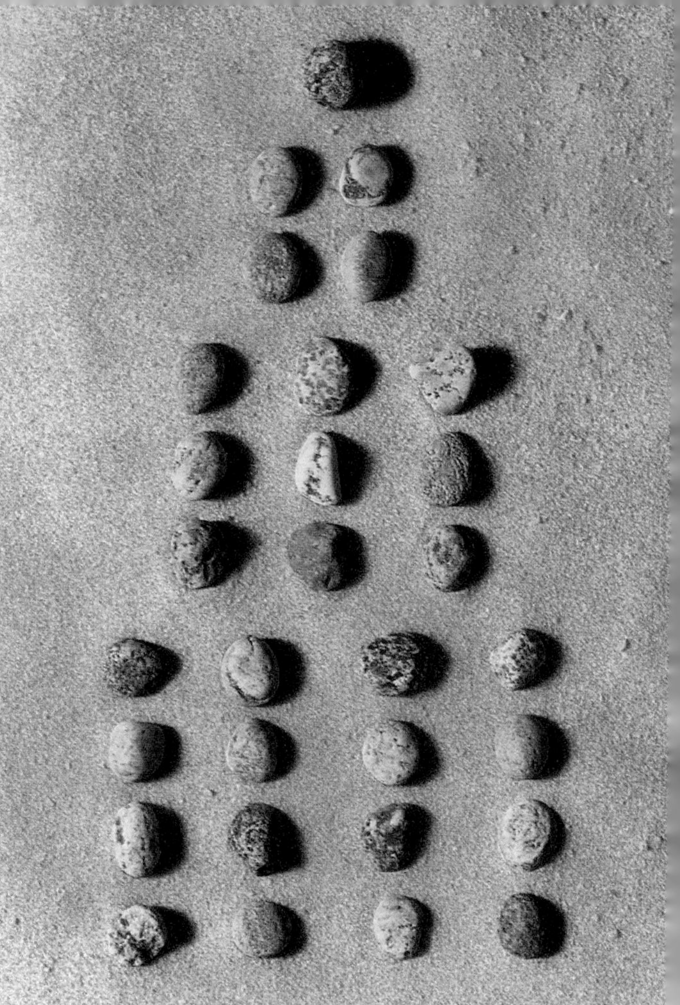

Activity ❶ Odd Numbers and Perfect Squares

Copy and complete the table.

Odd Numbers	Sum	Diagram
First one	1 = 1	■
First two	1 + 3 = 4	□□ □■
First three	1 + 3 + 5 = ■	
First four		
First five		

What is the sum of

a) the first 8 odd numbers?

b) the first 10 odd numbers?

c) the first 1000 odd numbers?

Activity ❷ Triangular Numbers and Perfect Squares

The first 4 triangular numbers are shown.

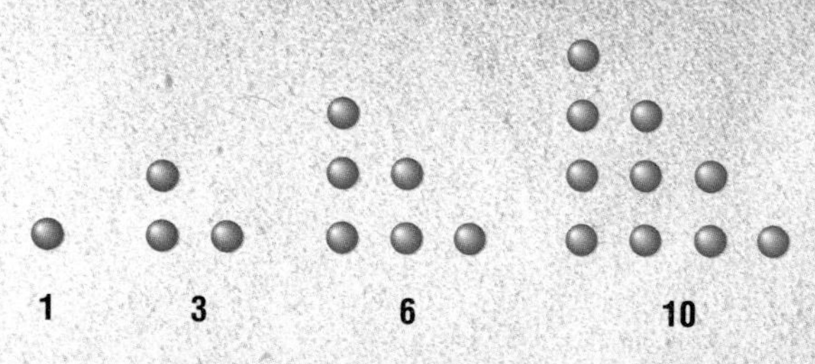

Use diagrams to illustrate your solutions to the following.

1. Find the next 3 triangular numbers.

2. Why are they called triangular numbers?

3. What is the sum of any 2 consecutive triangular numbers?

Activity ❸ The Locker Problem

Southridge School has a very long hallway, with 1000 lockers along one side. They are numbered from 1 to 1000.

On April Fool's Day, each of Southridge's 1000 students walks down the hallway to leave the school. Every student walks in the same direction, past locker 1 first and locker 1000 last.

The first student closes every locker door. The second student opens every second locker door. The third student changes the state of every third locker door. This means that, if the locker door is open, the student closes it. If it is closed, the student opens it. The fourth student changes the state of every fourth locker door.

This pattern continues until the thousandth student leaves and changes the state of the thousandth locker door.

What are the numbers of the locker doors that are closed after the last student leaves?

Hint: Solve a simpler problem. You may want to make a table to show the states of some locker doors after the first few students walk down the hall. Use the table to predict which of the 1000 doors are closed after 1000 students leave. Another way is to use cards or counters as lockers and act out the problem.

	Locker Number																			
Student	1	2	3	4	5	6	7	8	9	10	11	12	13	14	15	16	17	18	19	20
1	C	C	C	C	C	C	C	C	C	C	C	C	C	C	C	C	C	C	C	C
2		O		O		O		O		O		O		O		O		O		O
3			O			C			O			C			O			C		
4																				
5																				

Activity ❹ Fermat and Perfect Squares

French mathematician Pierre de Fermat (1601–1675) stated that any whole number can be written as the sum of, at most, 4 perfect squares.

$21 = 4^2 + 2^2 + 1^2$ or $21 = 3^2 + 2^2 + 2^2 + 2^2$

$39 = 5^2 + 3^2 + 2^2 + 1^2$ or $39 = 6^2 + 1^2 + 1^2 + 1^2$

1. Write each of the following as the sum of, at most, 4 perfect squares.

a) 33 **b)** 42 **c)** 77 **d)** 88

e) 153 **f)** 212 **g)** 208 **h)** 903

2. Compare your answers with a classmate's.

3.1 Estimating and Calculating Square Roots

About 1 in every 5 Canadian adults plays chess. The world's top chess players are called Grandmasters. Thirteen-year-old Alexandre Lesiège of Quebec beat a Russian Grandmaster at the World Open in 1989.

The playing surface of a chess board is a square. It is covered by 64 smaller squares that are equal in area. There are 8 squares along each side of the board. We can say that 64 is the square of 8, but how can we describe 8 in relation to 64?

Activity: Use the Diagrams

Draw figures A, B, C, and D on 1-cm grid paper. Determine the area of each square.

Inquire

1. The length of each side of a square is the **square root** of its area. What is the square root of area A? area B? area C?

2. Count squares to determine the area of square D. Measure the side of square D. What is the approximate square root of area D?

3. Write your definition of the square root of a number.

4. a) What is the square root of 64?

b) Is any other integer also the square root of 64? Explain.

5. A Snakes and Ladders game board is a square with 100 smaller squares marked on it.

a) How many of the smaller squares lie along each side of the board?

b) What are the square roots of 100?

Since $5 \times 5 = 25$ and $(-5) \times (-5) = 25$, then 5 and −5 are the square roots of 25.

Since $1.2 \times 1.2 = 1.44$ and $(-1.2) \times (-1.2) = 1.44$, then 1.2 and −1.2 are the square roots of 1.44.

The **radical sign**, $\sqrt{\ }$, is used to represent the positive square root of a number.

$$\sqrt{25} = 5, \text{ and } \sqrt{1.44} = 1.2$$

The positive square root is also called the **principal square root**. To avoid confusion, mathematicians do not use the radical sign for negative square roots.

Example 1

Evaluate.

a) $\sqrt{81}$

b) $\sqrt{0.36}$

c) $-\sqrt{6400}$

Solution

a) $9 \times 9 = 81$
so $\sqrt{81} = 9$

b) $0.6 \times 0.6 = 0.36$
so $\sqrt{0.36} = 0.6$

c) $80 \times 80 = 6400$
so $-\sqrt{6400} = -80$

Many numbers, such as $\sqrt{2}$ and $\sqrt{3}$, cannot be written as a fraction or a terminating decimal. These square roots are non-terminating, non-repeating decimals, or **irrational numbers**. To determine the approximate values of such square roots, we estimate or use a calculator.

Example 2

Estimate $\sqrt{356}$.

Solution

Start with numbers whose squares you know.
$30 \times 30 = 900$ (too high)
$20 \times 20 = 400$ (too high)
$10 \times 10 = 100$ (too low)
Since 356 is between 100 and 400, and is closer to 400, a good estimate for $\sqrt{356}$ is $\sqrt{400}$ or 20.

Another way to estimate the square root of a number is to divide the number into groups of 2 digits, starting at the decimal point. Then, estimate the square root of the group furthest from the decimal point and add 1 zero for each other group. Thus, for $\sqrt{1535}$, we consider the two groups 15 and 35.
$\sqrt{15} \doteq 4 \leftarrow$ square root of the group furthest from the decimal point.
There is one other group of digits, so we add one zero.
So, $\sqrt{1535} \doteq 40$.

Example 3

Estimate the square roots.

a) $\sqrt{567}$

b) $\sqrt{12\ 300}$

c) $\sqrt{0.45}$

d) $\sqrt{0.0067}$

Solution

a) $\sqrt{567}$
$\sqrt{5}\quad 67$
$\downarrow\quad\downarrow$
$2\quad 0$
$\sqrt{567} \doteq 20$

b) $\sqrt{12\ 300}$
$\sqrt{1}\quad 23\quad 00$
$\downarrow\quad\downarrow\quad\downarrow$
$1\quad 0\quad 0$
$\sqrt{12\ 300} \doteq 100$

c) $\sqrt{0.45}$
$0\ .\ \sqrt{45}$
$\downarrow\quad\downarrow$
$0\ .\ 7$
$\sqrt{0.45} \doteq 0.7$

d) $\sqrt{0.0067}$
$0\ .\ 00\ \sqrt{67}$
$\downarrow\quad\downarrow\quad\downarrow$
$0\ .\ 0\quad 8$
$\sqrt{0.0067} \doteq 0.08$

The square root key on your calculator may appear as [√] or [√x].
To find the square root of a number, enter the number, then press the square root key.

CONTINUED

Example 4

Evaluate to the nearest tenth.

a) $\sqrt{42}$

b) $-\sqrt{164}$

Solution

a) Press 42 [√]
Display 42 6.4807407

EST: $6 \times 6 = 36$, so $\sqrt{42} \doteq 6$

So, $\sqrt{42}$ is 6.5 to the nearest tenth.

b) Press 164 [√]
Display 164 12.806248

EST: $13 \times 13 = 169$, so $-\sqrt{164} \doteq -13$

So, $-\sqrt{164}$ is -12.8 to the nearest tenth.

Example 5

Evaluate to the nearest tenth.

a) $5\sqrt{3} + 7\sqrt{2}$

b) $\frac{6\sqrt{11}}{7}$

Solution

a) $5\sqrt{3}$ means $5 \times \sqrt{3}$ and $7\sqrt{2}$ means $7 \times \sqrt{2}$

[C] 5 [×] 3 [√] [+] 7 [×] 2 [√] [=] 18.559749

So, $5\sqrt{3} + 7\sqrt{2} = 18.6$ to the nearest tenth.

EST: $5\sqrt{3} + 7\sqrt{2}$
$\doteq 5 \times 2 + 7 \times 1$
$= 17$

b) [C] 6 [×] 11 [√] [÷] 7 [=] 2.8428212

So, $\frac{6\sqrt{11}}{7} = 2.8$ to the nearest tenth.

EST: $\frac{6\sqrt{11}}{7} \doteq \frac{6 \times 3}{6}$
$= 3$

Practice

Find the square roots of each number.

1. 49 **2.** 81 **3.** 121 **4.** 625
5. 0.64 **6.** 0.01 **7.** 1.96 **8.** 0.25

Evaluate.

9. $\sqrt{25}$ **10.** $\sqrt{100}$ **11.** $\sqrt{225}$
12. $\sqrt{256}$ **13.** $\sqrt{169}$ **14.** $\sqrt{0.36}$
15. $\sqrt{0.04}$ **16.** $\sqrt{1.21}$ **17.** $\sqrt{0.81}$

Estimate.

18. $\sqrt{30}$ **19.** $\sqrt{66}$ **20.** $\sqrt{92}$
21. $\sqrt{765}$ **22.** $\sqrt{989}$ **23.** $\sqrt{3245}$
24. $\sqrt{7800}$ **25.** $\sqrt{56\ 000}$ **26.** $\sqrt{880\ 000}$

Estimate.

27. $\sqrt{0.8}$ **28.** $\sqrt{0.77}$
29. $\sqrt{0.05}$ **30.** $\sqrt{0.067}$

31. $\sqrt{0.0382}$ **32.** $\sqrt{0.0023}$
33. $\sqrt{0.009}$ **34.** $\sqrt{0.0006}$
35. $\sqrt{0.000\ 22}$ **36.** $\sqrt{0.000\ 34}$

Estimate. Then, calculate to the nearest tenth.

37. $\sqrt{31}$ **38.** $\sqrt{44}$
39. $\sqrt{62}$ **40.** $\sqrt{79}$
41. $\sqrt{101}$ **42.** $\sqrt{206}$
43. $\sqrt{1123}$ **44.** $\sqrt{20\ 183}$
45. $\sqrt{86\ 003}$ **46.** $\sqrt{202\ 183}$

Evaluate to the nearest tenth.

47. $\sqrt{3} + \sqrt{7}$ **48.** $\sqrt{10} - \sqrt{5}$
49. $3\sqrt{11}$ **50.** $6\sqrt{23}$
51. $(\sqrt{3})(\sqrt{2})$ **52.** $7\sqrt{12} - 6$
53. $\sqrt{14} \div \sqrt{5}$ **54.** $\frac{\sqrt{20} - \sqrt{10}}{\sqrt{31}}$

Problems and Applications

55. Evaluate for $a = 5$ and $b = -2$.

a) $\sqrt{4a + 2b}$ **b)** $\sqrt{\dfrac{125}{a}}$

c) $\sqrt{-18b}$ **d)** $\sqrt{a^2 - 12b}$

e) $-\sqrt{(ab)^2}$ **f)** $\sqrt{10a - 5b + 4}$

g) $\sqrt{-10ab^3}$ **h)** $-3\sqrt{a^2 + 2ab + b^2}$

i) $7.3\sqrt{a^2 - 2ab + b^2}$

56. Calculate the perimeter of each figure to the nearest tenth of a unit.

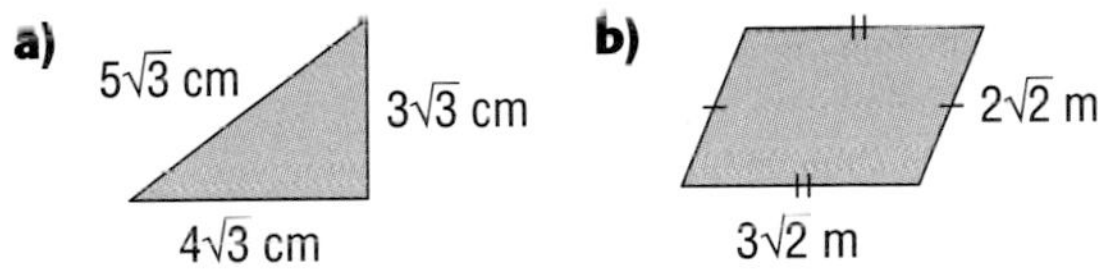

57. The Greek mathematician Heron found the following formula for the area of a triangle.

$$A = \sqrt{s(s-a)(s-b)(s-c)}$$

where a, b, and c are the side lengths, and s is half the perimeter.

$$s = \frac{a + b + c}{2}$$

Calculate the area of each triangle.

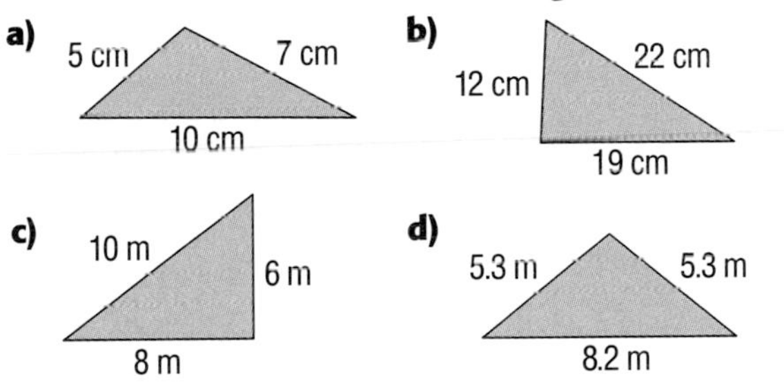

58. The area, a, of an equilateral triangle is given by the formula

$$a \doteq 0.43s^2$$

where s is the side length.
For each area, find the side length to the nearest tenth of a unit.

a) $15.6\ \text{cm}^2$ **b)** $44\ \text{m}^2$

c) $346.9\ \text{mm}^2$ **d)** $876.4\ \text{cm}^2$

59. The world's largest city square is Tiananmen Square in Beijing, China. It has an area of $0.396\ \text{km}^2$. What is the length of a side of the square to the nearest metre?

60. a) Evaluate $2\sqrt{2}$. **b)** Evaluate $\sqrt{8}$.

c) Compare your answers and explain your findings.

61. How are the square roots of a perfect square related?

62. A square has an area of $169\ \text{cm}^2$. What is the radius of the largest circle you can fit inside it? Explain.

63. Try to evaluate $\sqrt{-9}$ on your calculator. What is the result? Why?

64. To find the square root of a perfect square by subtraction, subtract the odd numbers in increasing order until the result is 0.

25 − 1 = 24
24 − 3 = 21
21 − 5 = 16
16 − 7 = 9
9 − 9 = 0

We need 5 subtractions to reach 0 from 25, so $\sqrt{25} = 5$.

a) Use this method to find the following.

$\sqrt{289}$ $\sqrt{576}$ $\sqrt{784}$ $\sqrt{1089}$

b) Explain why this method works. Compare your explanation with your classmates'.

LOGIC POWER

There are 14 cubes in the structure, which sits on a table. The face of each cube has an area of $1\ \text{m}^2$. If you paint the exposed surface, how many square metres do you paint?

3.2 Applying Square Roots

Activity: Use a Formula

The distance you can see out to sea when you stand on a beach is given by the formula

$$d = 0.35\sqrt{h}$$

In this formula, h is the height, in centimetres, of your eyes above the water, and d is the distance you can see in kilometres.

Inquire

1. The height of Sandra's eyes is 144 cm. How far can she see from a beach?

2. If Sandra sits on a lifeguard tower, and her eyes are 350 cm above the water, how far can she see?

3. Explain why the formula may not apply if you are not standing on a beach.

4. If you stand on a beach and watch an approaching sailboat, why do you see the mast of the boat before the hull?

Example

For many years, people believed Aristotle's theory of gravity, which stated that the heavier an object, the faster it would fall. Galileo disproved this theory by dropping 2 objects of different masses from the Leaning Tower of Pisa. They hit the ground at the same time.
The formula for calculating the time it takes an object to fall to the ground is

$$t = 0.45 \times \sqrt{h}$$

where h is the height in metres and t is the time in seconds. The Leaning Tower of Pisa is 54.5 m tall. How long would it take an object dropped from the top to hit the ground? Give your answer to the nearest tenth of a second.

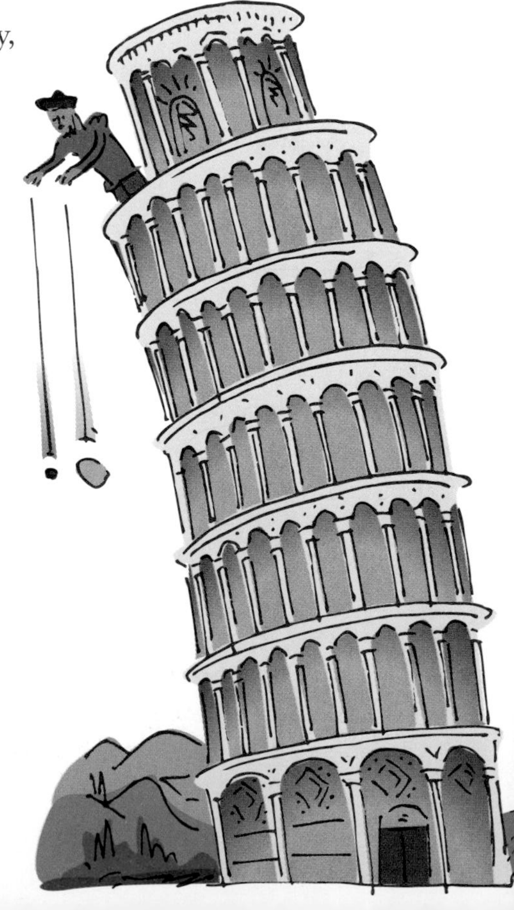

Solution

$$\begin{aligned} t &= 0.45\sqrt{h} \\ &= 0.45 \times \sqrt{54.5} \\ &\doteq 0.45 \times 7.38 \\ &= 3.321 \end{aligned}$$

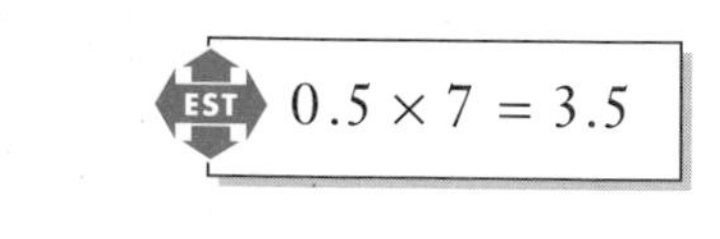

It would take 3.3 s, to the nearest tenth of a second, for the object to hit the ground.

Practice

Estimate. Then, evaluate to the nearest tenth.

1. $\sqrt{46}$
2. $\sqrt{303}$
3. $-\sqrt{90}$
4. $5\sqrt{66}$
5. $-7\sqrt{88}$
6. $\sqrt{13} + \sqrt{79}$
7. $3\sqrt{7} + 6\sqrt{91}$
8. $16\sqrt{56} - 4\sqrt{11}$
9. $\dfrac{2\sqrt{45}}{3}$
10. $\dfrac{5\sqrt{10} + 6\sqrt{21}}{4}$
11. $\dfrac{\sqrt{123}}{\sqrt{60}}$
12. $\dfrac{\sqrt{112}}{\sqrt{13}}$

Problems and Applications

Given the area of each square, determine to the nearest tenth of a unit

a) *the length of its side*
b) *its perimeter*

13. 25 cm^2
14. 225 m^2
15. 55 m^2
16. 90 cm^2
17. 200 m^2
18. 800 cm^2

19. Land areas are measured in hectares (ha).

$$1 \text{ ha} = 10\ 000 \text{ m}^2$$

a) If a square field has an area of 1 ha, what are its dimensions?
b) A house lot in a big city may be only about 0.25 ha in area. If this amount of land were square, what would its dimensions be?

20. The water at the surface of a river moves faster than the water near the bottom. The formula that relates these two speeds is

$$\sqrt{b} = \sqrt{s} - 1.3$$

where b is the speed near the bottom and s is the speed at the surface. Both speeds are in kilometres per hour.
a) What is the speed of the water near the bottom of a river, to the nearest tenth of a kilometre per hour, if the surface speed is 16 km/h? 20 km/h?
b) What is the surface speed of the water, to the nearest tenth of a kilometre per hour, if the speed near the bottom is 9 km/h? 7 km/h?

21. The velocity of sound changes with temperature and with the material the sound passes through. The formula to calculate the velocity in air is

$$V = 20\sqrt{273 + T}$$

where V is the velocity in metres per second and T is the temperature in degrees Celsius. Calculate the velocity of sound in air, to the nearest metre per second, at
a) 16°C **b)** 0°C **c)** −10°C

22. For a satellite to stay in orbit, its speed must satisfy the formula

$$s = \sqrt{\frac{5.15 \times 10^{12}}{d}}$$

where s is the speed in kilometres per hour and d is the distance in kilometres from the satellite to the centre of the Earth. If a satellite is in orbit 35 840 km above the Earth, it hovers over one spot on the Earth. Research the radius of the Earth and determine the speed of a satellite that hovers over one spot on the Earth. Compare your answer with a classmate's.

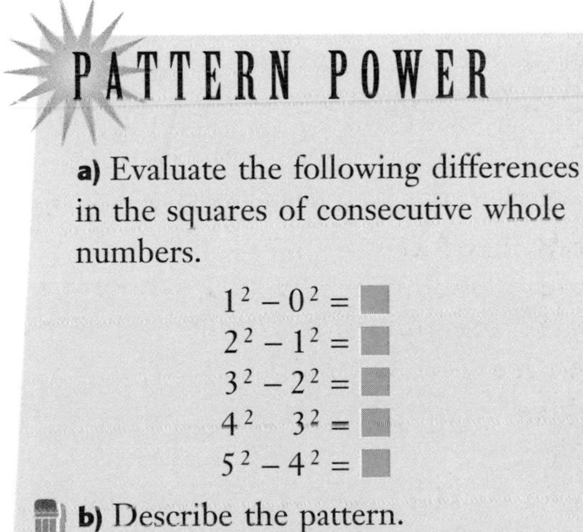

PATTERN POWER

a) Evaluate the following differences in the squares of consecutive whole numbers.

$1^2 - 0^2 = \blacksquare$
$2^2 - 1^2 = \blacksquare$
$3^2 - 2^2 = \blacksquare$
$4^2 - 3^2 = \blacksquare$
$5^2 - 4^2 = \blacksquare$

b) Describe the pattern.
c) Use the pattern to evaluate mentally $245^2 - 244^2$.

3.3 Use Logic

Problems that can be solved using logic do not require any special skills in mathematics. These problems sharpen your deductive thinking skills.

A cat, dog, monkey, and elephant are named Trixie, Wags, Snow, and Boots. Use the clues to name each animal.

- Trixie is a friend of the cat and the elephant.
- The dog and Boots enjoy popcorn.
- Snow and Boots play golf with the monkey.
- The elephant does not play outdoor sports.

Understand the Problem

1. What information are you given?
2. What are you asked to find?

Think of a Plan

Make a table and fill in the facts from the clues.

Carry Out the Plan

Trixie is a friend of the cat and the elephant. So, Trixie is not the cat or the elephant.

	C	D	M	E
Trixie	n			n
Wags				
Snow				
Boots				

The dog and Boots enjoy popcorn. So, the dog is not Boots.

	C	D	M	E
Trixie	n			n
Wags				
Snow				
Boots		n		

Snow and Boots play golf with the monkey. So, Snow and Boots are not the monkey.

	C	D	M	E
Trixie	n			n
Wags				
Snow			n	
Boots		n	n	

The elephant does not play outdoor sports. So, the elephant does not play golf and is not Snow or Boots. The table now shows that the Elephant must be Wags, and the cat must be Boots. We can now complete the table.

	C	D	M	E
Trixie	n			n
Wags				
Snow			n	n
Boots		n	n	n

	C	D	M	E
Trixie	n	n	y	n
Wags	n	n	n	y
Snow	n	y	n	n
Boots	y	n	n	n

Trixie is the monkey, Wags is the elephant, Snow is the dog, and Boots is the cat.

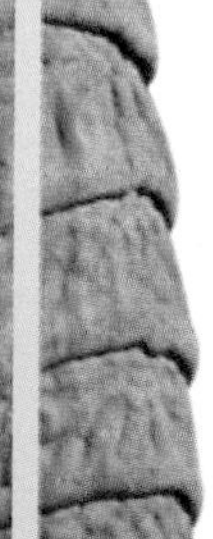

Look Back

Check that the answer agrees with the given facts.

Use Logic

1. Organize the information.
2. Draw conclusions from the information.
3. Check that your answer is reasonable.

Problems and Applications

1. Maria, Paula, and Shelly attend the same school. One is in grade 9, one is in grade 10, and one is in grade 11. Use the clues to find which grade each is in.

- Maria and the grade 9 student eat lunch with Shelly.
- Maria is not in grade 10.

2. Four students named Al, Bjorn, Carl, and Don each have a favourite sport. The sports are swimming, running, bowling, and golf. Use the clues to match each student with his sport.

- The runner met Bjorn and the golfer for lunch.
- Neither Bjorn's sport nor Carl's sport requires a ball.
- Don is in the same math class as the golfer's sister.

3. Susan, Irina, Traci, and Debbie are a pilot, a dentist, a doctor, and a writer. Use the clues to match each person with her profession.

- Debbie is a friend of the doctor.
- Susan and the writer sail with Traci.
- Neither Susan nor Irina has patients.

4. Four people are running a marathon. Manuel is 30 m behind Margaret. Coreen is 20 m behind Tom. Margaret is 75 m ahead of Coreen. How far ahead of Tom is Manuel?

5. Sonya had $1.19 in change. None of the coins was a dollar. Greg asked her for change for a dollar, but Sonya could not make change. What coins did she have?

6. The odometer on Anitha's car showed 25952 km. The number 25952 is called a **palindrome**. It reads the same forward and backward. Anitha drove for 2 h, and the odometer showed the next palindrome. How far did she drive in the 2 h?

7. A building has 5 doors. In how many ways can you enter the building by one door and leave by another?

8. A photographer wants to take a picture of the curling team. In how many ways can the 4 curlers be arranged side by side?

9. In a bacterial culture, the number of bacteria doubled every minute. At midnight, the glass jar was full and contained 1 000 000 bacteria. At what time was the jar half full?

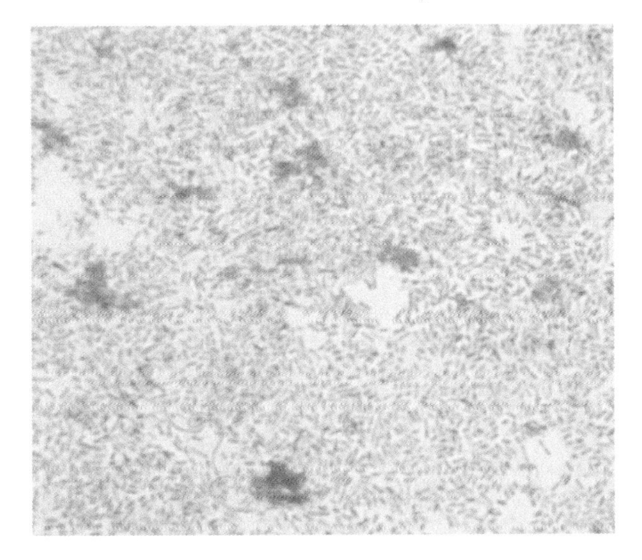

10. Evans, Thompson, Smith, and DiMaggio are a teacher, plumber, artist, and banker. Use the clues to match each person with a profession.

- Thompson met the artist when he hired her to paint a picture of his horse.
- The plumber and DiMaggio are friends who have never had any business dealings.
- Neither Evans nor the plumber has ever met Smith.
- DiMaggio asked the banker for a loan.

11. Make your own logic puzzle like those in questions 2 and 3. First, set up a chart and mark the answers you want. Then, write the clues. Test your puzzle before asking a classmate to solve it.

3.4 Working with Exponents

The leg span of the giant spider from South America is 25 cm. Like all spiders, it has 8 legs.

We can write the number 8 as the power 2^3. In this power, the exponent is 3 and the base is 2.

base → 2^3 ← exponent
↑
power

The base of a power can also be a negative number. We can extend the exponent rules to simplify powers with integers as bases.

Activity: Look for a Pattern

Copy and complete the table.

Exponential Form	Result as a Repeated Multiplication	Result in Exponential Form	Result in Standard Form
$(-2)^2 \times (-2)^3$	$(-2) \times (-2) \times (-2) \times (-2) \times (-2)$	$(-2)^{2+3}$ or $(-2)^5$	-32
$(-2)^3 \times (-2)^4$			
$(+3)^2 \times (+3)^3$			
$(-2)^8 \div (-2)^3$			
$(-3)^4 \div (-3)^2$			
$(+5)^3 \div (+5)$			
$(-4)^3 \times (-4)^0$			
$(-3)^4 \div (-3)^0$			

Inquire

1. Write the exponent rule for multiplying powers with the same integral base.

2. Write the exponent rule for dividing powers with the same integral base.

3. Describe the difference between

a) $(1 + 3)^2$ and $1^2 + 3^2$?

b) $(1 - 3)^2$ and $1^2 - 3^2$?

c) $(-3)^2$ and -3^2?

Example 1

Simplify.

a) $(-4)^3 \times (-4)^5$

b) $(-y)^2 \times (-y)^3$

Solution

To multiply powers with the same integral base, add the exponents.

$x^m \times x^n = x^{m+n}$

a) $(-4)^3 \times (-4)^5$
$= (-4)^{3+5}$
$= (-4)^8$

b) $(-y)^2 \times (-y)^3$
$= (-y)^{2+3}$
$= (-y)^5$

Example 2

Simplify.

a) $(-6)^5 \div (-6)^2$

b) $(-t)^3 \div (-t)$

Solution

To divide powers with the same integral base, subtract the exponents.

$x^m \div x^n = x^{m-n}$

a) $(-6)^5 \div (-6)^2$
$= (-6)^{5-2}$
$= (-6)^3$

b) $(-t)^3 \div (-t)$
$= (-t)^{3-1}$
$= (-t)^2$

Example 3

Simplify.

a) $((-2)^2)^3$

b) $((-m)^4)^2$

Solution

a) $((-2)^2)^3$
$= (-2)^2 \times (-2)^2 \times (-2)^2$
$= (-2)^{2+2+2}$
$= (-2)^6$

b) $((-m)^4)^2$
$= (-m)^4 \times (-m)^4$
$= (-m)^{4+4}$
$= (-m)^8$

Example 3 suggests the following rule. To raise a power with an integral base to a power, multiply the exponents.

$$(x^m)^n = x^{m \times n}$$

Example 4

Evaluate $(-5)^4(-5)^3$.

Solution

$(-5)^4(-5)^3 = (-5)^{4+3}$
$= (-5)^7$
$= -78\ 125$

C 5 +/– Yˣ 7 =   – 78 125.

Example 5

Evaluate $(-4)^2 \times (-3)^3$.

Solution

Because the bases are not the same, evaluate each power before multiplying.

$(-4)^2 \times (-3)^3$
$= 16 \times (-27)$
$= -432$

EST
20×-25
$= -500$

C 4 +/– Yˣ 2 = × 3 +/– Yˣ 3 =   – 432.

Example 6

Evaluate for $x = -2$ and $y = -1$.

a) $x^2 + y^3$

b) $3x^3 - 2y^2$

Solution

a) $x^2 + y^3$
$= (-2)^2 + (-1)^3$
$= 4 - 1$
$= 3$

b) $3x^3 - 2y^2$
$= 3(-2)^3 - 2(-1)^2$
$= 3(-8) - 2(+1)$
$= -24 - 2$
$= -26$

CONTINUED

Practice

State the base.

1. 2^6 **2.** $(-5)^2$ **3.** -1^4 **4.** $(-9)^3$

State the exponent.

5. -2^5 **6.** 4^2 **7.** $(-4)^0$ **8.** -5

Write in exponential form.

9. $(-4)(-4)(-4)$

10. $(-3)(-3)(-3)(-3)(-3)$

11. $p \times p \times p \times p \times p$

12. $(-n)(-n)(-n)(-n)$

13. $3 \times 3 \times 3 \times (-2) \times (-2) \times 3 \times (-2)$

Write as a repeated multiplication.

14. $(-2)^5$ **15.** -2^5 **16.** $(-x)^3$

Write in standard form.

17. 3^2 **18.** $(-3)^2$ **19.** $(-1)^4$

20. -1^5 **21.** $(-5)^3$ **22.** -5^3

23. $(-0.5)^3$ **24.** 1.1^4 **25.** $(-2.5)^2$

Simplify.

26. $5^3 \times 5^6$ **27.** $(-8)^2 \times (-8)^3$

28. $(-2)^3(-2)^4$ **29.** $2^2 \times 2^3 \times 2$

30. $(-2.1)^5(-2.1)^3$ **31.** $(-0.2)^3(-0.2)^2$

Simplify.

32. $5^4 \div 5^3$ **33.** $6^8 \div 6^2$

34. $\frac{(-4)^5}{(-4)^3}$ **35.** $\frac{(-9)^7}{(-9)^2}$

Simplify.

36. $(2^3)^2$ **37.** $((-3)^7)^4$

38. $((-5)^2)^3$ **39.** $((-6)^5)^3$

40. $((-4)^6)^7$ **41.** $((-2.3)^3)^4$

Simplify.

42. $x^4 \times x^2$ **43.** $y^{12} \div y^5$

44. $z^8 \div z$ **45.** $(-m)^6(-m)^4$

46. $(s^2)^4$ **47.** $((-r)^3)^2$

Simplify, then calculate.

48. $(-5)^2 \times (-5)^3$ **49.** $6^2 \times 6^5$

50. $(-2)^3(-2)^5(-2)^2$ **51.** $(-1)^5(-1)^7$

52. $(-3.1)^5(-3.1)^3$ **53.** $(-3)^6 \div (-3)^4$

54. $(-10)^5 \div (-10)$ **55.** $(-4)^6 \div (-4)^5$

Calculate.

56. $2^8 \div 2^4$ **57.** $(-3)^7 \div (-3)$

58. $(-5)^2 \times (-5)^3$ **59.** $(3^2)^3$

60. $\frac{(-4)^3 \times (-4)^5}{(-4)^5}$ **61.** $\frac{4^9}{4^3 \times 4^2}$

62. $(-2)^3(-2)^5$ **63.** $(-3)^0(-3)^5$

64. $((-2)^3)^2$ **65.** $(6^2)^3 \div (6^2)^2$

Evaluate.

66. $(-8)^2$ **67.** $6(-4)^3$

68. $(-3)^2(6)^2$ **69.** $(-1)^5 + 3^3$

70. $4^5 - 3^5$ **71.** $(-2)^5 \times (-3)^4$

72. $9^2 \div (-2)^3$ **73.** $(-5)^2(-4)^4$

74. $(1.3)^2(-2)^4$ **75.** $(1.5)^2 \div (-5)^3$

76. Evaluate for $n = 3$.

a) $5n^2$ **b)** $-6n^3$

c) $1 + 7n^5$ **d)** $n^3 - 6n$

77. Evaluate for $x = -2$ and $y = 3$.

a) x^3 **b)** $5y^4$

c) $x^2 + y^2$ **d)** $3x^3y^3$

e) $(x - y)^3$ **f)** $(y - x)^2$

g) $-6xy$ **h)** $-4x^2y^3$

i) $4x^3 - 5y$ **j)** $(3x^2)(-2y^2)$

Problems and Applications

78. Find the value of x.

a) $3^x = 81$ **b)** $(-2)^x = -512$

c) $x^5 = 1024$ **d)** $(-x)^3 = -1000$

e) $-5^x = -625$ **f)** $-x^2 = -1.69$

g) $(0.2)^x = 0.0016$ **h)** $x^3 = -0.216$

79. The formula for the volume, V, of a cube is

$$V = s^3$$

where s is the side length. Copy the chart and use the formula to complete it.

Side Length (cm)	Volume (cm³)
5	
7	
4.2	
	1000
	512

80. There are 10 bacteria in a culture at the beginning of an experiment. The number of bacteria doubles every 24 h. The table shows the number of bacteria present over the first few days.

Elapsed Time (days)	Number of Bacteria
1	10×2 $= 10 \times 2^1$ $= 20$
2	$10 \times 2 \times 2$ $= 10 \times 2^2$ $= 40$
3	$10 \times 2 \times 2 \times 2$ $= 10 \times 2^3$ $= 80$

a) Complete the table up to the end of day 8. How many bacteria are there after 8 days?
b) Use the pattern to calculate the number of bacteria after 2 weeks.
c) How many bacteria are there after 40 days?

81. If the base of a power is negative, and the exponent is even, is the standard form of the number positive or negative? Explain.

82. If the base of a power is negative, and the exponent is odd, is the standard form of the number positive or negative? Explain.

83. Any positive number has 2 square roots, but you can calculate only 1 side length if you are given the area of a square. Explain why.

84. The exponent rules can be applied to powers with fractional bases.

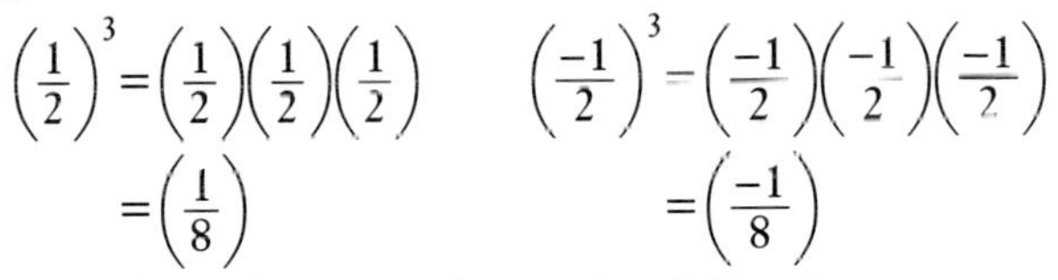

$$\left(\frac{1}{2}\right)^3 = \left(\frac{1}{2}\right)\left(\frac{1}{2}\right)\left(\frac{1}{2}\right) = \left(\frac{1}{8}\right) \qquad \left(\frac{-1}{2}\right)^3 = \left(\frac{-1}{2}\right)\left(\frac{-1}{2}\right)\left(\frac{-1}{2}\right) = \left(\frac{-1}{8}\right)$$

Use the rules to evaluate the following.

a) $\left(\frac{-2}{3}\right)^5$ **b)** $\left(\frac{11}{2}\right)^4$ **c)** $\left(\frac{-3}{2}\right)^3$

d) $4\left(\frac{-1}{2}\right)^2$ **e)** $\left(\frac{1}{2}\right)^2\left(\frac{-1}{2}\right)^3$ **f)** $\left(\frac{-1}{4}\right)^2 - 1$

85. a) Are the products $(-4)^3 \times (-4)^2$ and $(-4)^2 \times (-4)^3$ the same? Explain.
b) Are the quotients $(-4)^3 \div (-4)^2$ and $(-4)^2 \div (-4)^3$ the same? Explain.
c) Are the powers $((-4)^3)^2$ and $((-4)^2)^3$ the same? Explain.

86. A new student in your class does not know how to evaluate such powers as $(-2)^4$ and -2^4. Write an explanation and compare it with your classmates'.

PATTERN POWER

A triangular pyramid has 3 layers of golf balls. There are 10 golf balls altogether. How many golf balls will there be in a triangular pyramid with 5 layers? 10 layers?

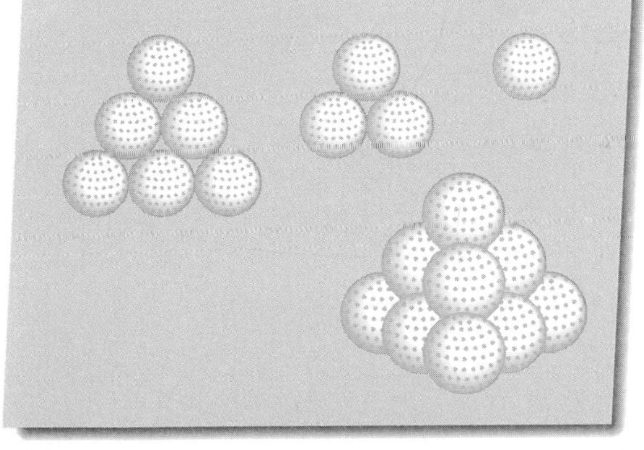

Patterns and Powers

What is the units digit of 2^{95}? To solve this problem, you might try using a calculator. What answer does your calculator give for 2^{95}? Another way to find the answer is to look for a pattern.

Activity 1

To find the units digit of 2^{95}, complete the table and draw the graph.

$2^1 = 2$
$2^2 = 4$
$2^3 = ■$
$2^4 = ■$
$2^5 = ■$
$2^6 = ■$
$2^7 = ■$
$2^8 = ■$
$2^9 = ■$
$2^{10} = ■$
$2^{11} = ■$
$2^{12} = ■$

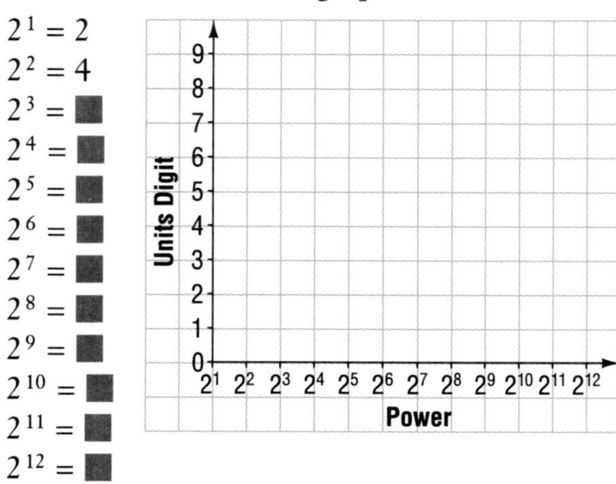

1. Describe the pattern in the units digits.

2. What is the units digit for the powers with exponents that are divisible by 4?

3. What is the units digit of

a) 2^{92}? **b)** 2^{93}? **c)** 2^{94}?

4. What is the units digit of 2^{95}?

Activity 2

To find the units digit of 3^{100}, complete the table and draw the graph.

$3^1 = ■$
$3^2 = ■$
$3^3 = ■$
$3^4 = ■$
$3^5 = ■$
$3^6 = ■$
$3^7 = ■$
$3^8 = ■$
$3^9 = ■$

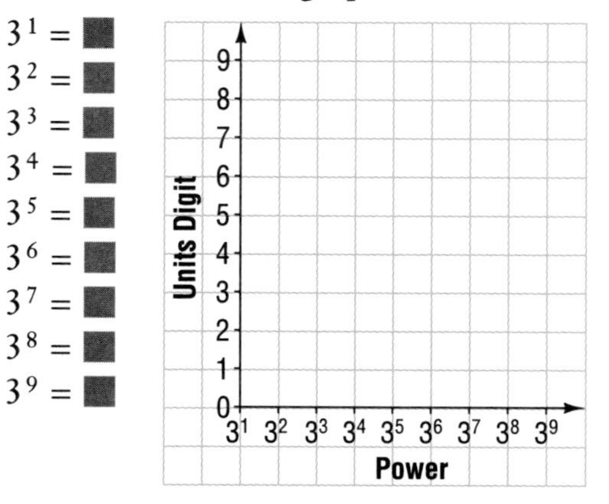

1. Describe the pattern in the units digits.

2. What is the units digit of 3^{100}?

Activity 3

To find the units digit of 7^{111}, complete the table and draw the graph.

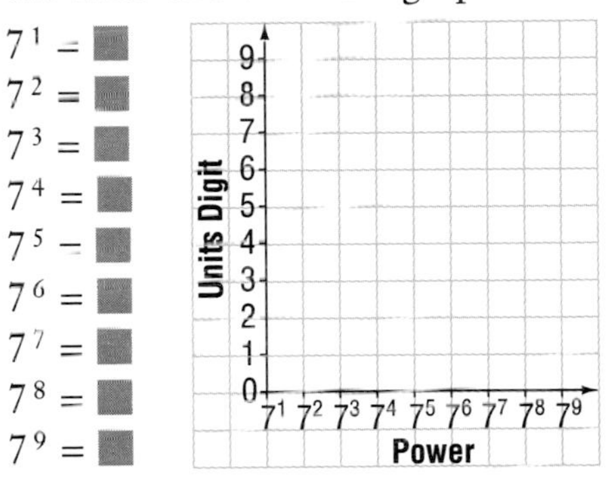

Activity 4

1. Without drawing a graph, determine the units digit of 6^{80}.

2. What would the graph of the units digits versus the powers of 6 look like?

3. What other bases give similar graphs?

Activity 5

1. Without drawing a graph, determine the units digit of

a) 4^{92} **b)** 4^{93}

2. What would the graph of the units digits versus the powers of 4 look like?

3. What other bases give similar graphs?

Activity 6

1. Write a problem using a power of 8.

2. Have a classmate solve your problem.

3.5 Work Backward

Working backward can be a very useful problem solving technique. To solve some crimes, a detective will start at the scene of the crime and retrace the criminal's steps, or work backward, until the crime is solved.

The departure time for the Orient Express train from Calais to Paris is 21:20. Sam Orwell, a private detective, plans to board the train 35 min before departure. Orwell's hotel, the Rampant Lion, is a 25-min cab ride from the train station.

Before leaving the hotel, Orwell needs 30 min to eat dinner. After dinner, he needs 10 min to pack and 15 min to phone Interpol from his room. On the way to the train station, Orwell needs to stop for 15 min at the ambassador's residence to get a copy of a set of fingerprints. To reach the train on time, when should Orwell

a) leave his hotel for the train? **b)** go to dinner?

Understand the Problem

1. What information are you given?
2. What are you asked to find?
3. Do you need an exact or approximate answer?

Think of a Plan

Start at the departure time of the train and work backward.

Carry Out the Plan

Train departure time	21:20
Time to arrive before departure	35 min
Orwell's arrival at the train	20:45
Time for cab ride and ambassador	25 + 15 = 40 min
When to leave hotel for the train	20:05
Time for dinner, packing, and phone call	30 + 10 + 15 = 55 min
When to go to dinner	19:10

a) Orwell should leave the hotel at 20:05.
b) Orwell should go to dinner at 19:10.

Look Back

Is the answer reasonable?
Can you think of another way to solve the problem?

Work Backward	1. Start with what you know. 2. Work backward to get an answer. 3. Check that the answer is reasonable.

Problems and Applications

1. Clara's plane leaves at 08:15. She has to be at the airport 40 min early. The cab ride to the airport is about 55 min. Clara wants to spend 30 min at her office on her way to the airport. When should she leave home?

2. Michiko went to the county fair. She spent half her money on rides. She spent half of what was left on food. Of the remainder, she spent half at the arcade. She had $6 left. How much money did she start with?

3. Sandra bought a gold chain for $150. The jeweller added one-half of his cost price to get the price of $150. What did the jeweller pay for the chain?

4. The exam starts at school at 09:00. You want to get there 15 min early. It takes 20 min to 30 min to get to school, depending on the buses. You plan to study for 1 h at home. It will take you 1 h to shower, dress, and eat breakfast. For what time should you set your alarm?

5. Jennifer wants a jacket that costs $230, including tax. She has saved $70. If she saves the rest in equal amounts over the next 4 months, how much will she save each month?

6. Ahmed's plane leaves at 12:30. Ahmed can drive 120 km to the airport at an average speed of 80 km/h. He needs to be at the airport 45 min before departure. When should he leave home?

7. Tanya saved for a band trip costing $725. She worked part-time at a convenience store and earned $7.50/h. Then, her aunt gave her $125 toward the trip. For how many hours did Tanya have to work to afford the trip?

8. Dino multiplied his age by 3. Then he subtracted 3, divided by 7, and added 12. The result was 18. How old was Dino?

9. The train left Sutton for Bent Creek. At Brownsville, 35 people got on and 27 got off. Next came Liberty, where 24 people got on and 11 got off. At Long Lake, 55 people got on and 42 people got off. At Bent Creek, 105 people got off and 10 got on. There were 203 people on the train when it left Bent Creek. How many people were on the train when it left Sutton?

10. Mike works for an art gallery. An art dealer offers him a painting for $72 000. The dealer says the painting has increased in value by one-half of the previous year's value in each of the past 2 years.

a) What was its value 2 years ago?

b) If the painting's value continues to increase at this rate, what will it be 3 years from now?

11. The Canadian population was about 27 300 000 in 1991. It had increased by about 3 000 000 since 1981. The 1981 figure was about 7 times the figure at Confederation in 1867. What was the Canadian population at Confederation to the nearest 100 000?

12. Write a problem that can be solved by working backward. Have a classmate solve your problem.

WORD POWER

In an anagram, the letters can be in any order except the right one. The word TEN has 5 anagrams.

NET NTE ETN ENT TNE

1. How many anagrams are there for

a) FOUR? **b)** NINE?

2. Why are your answers different, even though FOUR and NINE both contain 4 letters?

3.6 Zero and Negative Exponents

There are 8 A notes on a piano. They can be named as shown in the diagram.

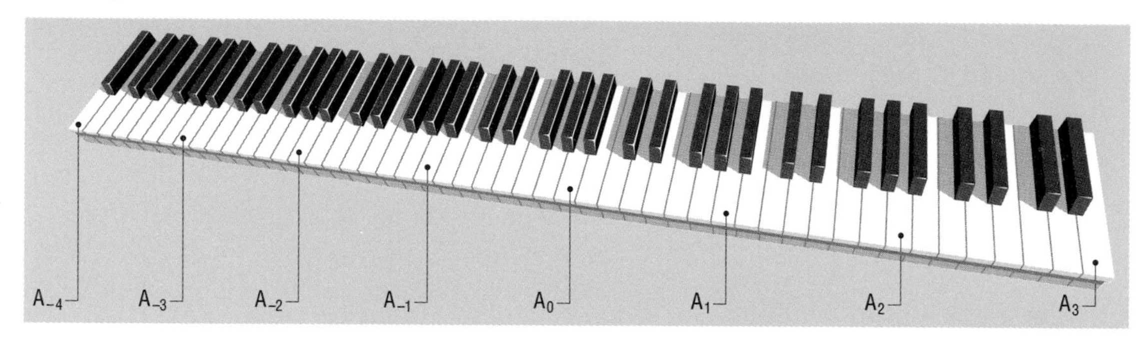

When tuning a piano, a tuner first adjusts the note A_0. This note, called "concert A," is the A above middle C. The formulas to determine the frequencies of the A notes are as follows. The units of frequency are hertz, symbol Hz.

$A_{-4} = 440 \times 2^{-4}$ $A_{-3} = 440 \times 2^{-3}$ $A_{-2} = 440 \times 2^{-2}$ $A_{-1} = 440 \times 2^{-1}$

$A_0 = 440 \times 2^0$ $A_1 = 440 \times 2^1$ $A_2 = 440 \times 2^2$ $A_3 = 440 \times 2^3$

Calculate the frequencies of notes A_1, A_2, and A_3.
To calculate the other frequencies, we need to know about zero and negative exponents.

Activity: Discover the Relationship

Copy and complete the table by doing each division twice. First expand and divide, and then use the exponent rule for division.

Division	Expand and Divide	Exponent Rule
$2^4 \div 2^4$	$\frac{2 \times 2 \times 2 \times 2}{2 \times 2 \times 2 \times 2} = ?$	$\frac{2^4}{2^4} = 2^{4-4} = 2^?$
$3^2 \div 3^2$		
$4^3 \div 4^3$		
$10^4 \div 10^4$		
$x^5 \div x^5$		

Inquire

1. How do the 2 answers in each row compare?

2. What is the value of 5^0? 7^0? y^0?

3. What is the value of any number raised to the exponent 0?

4. Calculate the frequency of A_0 on the piano.

Activity: Discover the Relationship

Copy and complete the table by doing each division twice. First expand and divide, and then use the exponent rule for division.

Division	Expand and Divide	Exponent Rule
$2^3 \div 2^5$	$\frac{2 \times 2 \times 2}{2 \times 2 \times 2 \times 2 \times 2} = \frac{1}{2^?}$	$\frac{2^3}{2^5} = 2^{3-5} = 2^?$
$3^2 \div 3^6$		
$5 \div 5^3$		
$10^4 \div 10^5$		
$x^5 \div x^9$		

Inquire

1. How do the 2 answers in each row compare?
2. State another way to write
   a) 3^{-2} **b)** 4^{-3} **c)** 1^{-4}
3. Write a rule for writing a number with a negative exponent as a number with a positive exponent.
4. Evaluate.
   a) 2^{-4} **b)** 5^{-2} **c)** 3^{-3} **d)** 4^{-2} **e)** 1^{-7}
5. Calculate the frequencies of the following notes on the piano.
   a) A_{-4} **b)** A_{-3} **c)** A_{-2} **d)** A_{-1}

Example 1

Evaluate.

a) 4^0
b) $(-5)^0$
c) -2^0

Solution

$x^0 = 1$, where x can be any number except 0

a) $4^0 = 1$ **b)** $(-5)^0 = 1$ **c)** $-2^0 = -1$

Example 2

Evaluate.

a) 4^{-3}
b) $(-3)^{-3}$
c) $2^0 - 2^{-2}$

Solution

$x^{-m} = \frac{1}{x^m}$, where x can be any number except 0

a) $4^{-3} = \frac{1}{4^3} = \frac{1}{64}$

b) $(-3)^{-3} = \frac{1}{(-3)^3} = -\frac{1}{27}$

c) $2^0 - 2^{-2} = 1 - \frac{1}{2^2} = 1 - \frac{1}{4} = \frac{3}{4}$

CONTINUED

Example 3

Evaluate.

a) $3^{-3} \times 3^{5} \times 3^{-4}$

b) $(-2)^{-5} \div (-2)^{-2}$

c) $(4^{-1})^{2}$

d) $\frac{1}{5^{-2}}$

Solution

a) $3^{-3} \times 3^{5} \times 3^{-4}$
$= 3^{-3+5-4}$
$= 3^{-2}$
$= \frac{1}{3^{2}}$
$= \frac{1}{9}$

b) $(-2)^{-5} \div (-2)^{-2}$
$= (-2)^{-5-(-2)}$
$= (-2)^{-3}$
$= \frac{1}{(-2)^{3}}$
$= -\frac{1}{8}$

c) $(4^{-1})^{2}$
$= 4^{(-1)\times 2}$
$= 4^{-2}$
$= \frac{1}{4^{2}}$
$= \frac{1}{16}$

d) $\frac{1}{5^{-2}}$
$= \frac{1}{\frac{1}{5^{2}}}$
$= \frac{1}{\frac{1}{25}}$
$= 1 \times \frac{25}{1}$
$= 25$

Practice

1. Write each power with a positive exponent.

a) 2^{-3} **b)** 4^{-1} **c)** 10^{-7}

d) 9^{-8} **e)** 1^{-4} **f)** $(0.5)^{-6}$

g) $(-7)^{-6}$ **h)** $(-2)^{-3}$ **i)** $\frac{1}{2^{-3}}$

j) $\frac{1}{4^{-2}}$ **k)** $\frac{1}{5^{-4}}$ **l)** $\frac{1}{(-3)^{-5}}$

2. Write in exponential form.

a) $\frac{1}{8 \times 8}$ **b)** $\frac{1}{7 \times 7 \times 7}$ **c)** $\frac{1}{9 \times 9 \times 9 \times 9}$

3. Write in exponential form.

a) $\frac{1}{4}$ **b)** $\frac{1}{27}$ **c)** $\frac{1}{64}$ **d)** $\frac{1}{343}$

Evaluate.

4. 5^{0} **5.** 2^{6} **6.** 3^{-1}

7. 4^{-2} **8.** $(-1)^{7}$ **9.** 10^{-3}

10. $(-6)^{0}$ **11.** 8^{-1} **12.** $(-3)^{-4}$

13. $(-10)^{-3}$ **14.** $(0.1)^{-3}$ **15.** $(-1)^{-6}$

16. $\frac{1}{3^{-1}}$ **17.** $\frac{1}{4^{-2}}$ **18.** $\frac{1}{(-3)^{-1}}$

Write each expression as a power.

19. $7^{4} \times 7^{5}$ **20.** $9^{6} \times 9^{-4}$

21. $8^{-3} \times 8^{-5}$ **22.** $6^{7} \div 6^{3}$

23. $5^{-7} \div 5^{-2}$ **24.** $4^{-2} \div 4^{6}$

25. $(3^{3})^{4}$ **26.** $(9^{-2})^{4}$

27. $(8^{-1})^{-5}$ **28.** $-(2^{-3})^{-2}$

Write each expression as a power.

29. $2^{4} \times 2^{-3} \times 2^{2}$ **30.** $3^{-5} \times 3^{-3} \times 3^{2}$

31. $5^{6} \times 5^{-9} \times 5$ **32.** $8^{4} \times 8^{-5} \div 8^{-2}$

33. $(-2)^{-4} \times (-2)^{-3} \div (-2)^{-1}$

34. $(-3)^{-6} \div (-3)^{-2} \times (-3)^{4}$

Evaluate.

35. $3^{2} \times 3^{2}$ **36.** $4^{7} \div 4^{5}$

37. $2^{4} \times 2^{-3}$ **38.** $3^{2} \div 3^{-1}$

39. $5^{2} \times 5^{-4}$ **40.** $6^{-2} \div 6^{0}$

41. $7^{-4} \div 7^{-5}$ **42.** $6^{-3} \div 6^{-3}$

43. $\frac{(2^{2})^{2}}{(5^{3})^{-1}}$ **44.** $\frac{(3^{-1})^{3}}{(2^{-3})^{-2}}$

45. $\frac{(6^{0})^{-4}}{(10^{2})^{-2}}$ **46.** $\frac{(7^{-3})^{0}}{(0.1^{-1})^{-2}}$

Evaluate.

47. $2^{3} \times 2^{-2} \times 2^{-2}$ **48.** $3^{4} \div 3^{5} \times 3$

49. $3^{0} + 3^{3}$ **50.** $4^{2} - 2^{-1}$

51. $5^{3} + 3^{2}$ **52.** $2^{-2} + 5^{0}$

53. $2^{3} \times 2^{-1} + 5$ **54.** $(6 - 3)^{-2}$

55. $(9^{0} + 2^{0})^{-1}$ **56.** $\frac{1}{2^{-2}} + \frac{1}{3^{-1}}$

57. $\frac{2^{-1}}{3^{-1}}$ **58.** $\frac{-3^{-2}}{4^{-1}}$

Evaluate.

59. $\left(\frac{1}{3}\right)^{-1}$ **60.** $\left(\frac{-1}{5}\right)^{2}$ **61.** $\left(\frac{-7}{8}\right)^{0}$

62. $\left(\frac{1}{10}\right)^{-2}$ **63.** $\left(\frac{-2}{3}\right)^{-3}$ **64.** $\left(\frac{-3}{4}\right)^{-2}$

Simplify.

65. $x^4 \times x^3$

66. $x^{-2} \times x^3$

67. $y^{-1} \times y^{-3}$

68. $t^8 \div t^4$

69. $m^6 \div m^{-2}$

70. $b^{-2} \div b^{-4}$

71. $(m^1)^2$

72. $(t^{-2})^4$

73. $(y^{-5})^{-2}$

74. $m^3 \times m^{-2} \times m^4$

75. $\frac{a^{-4} \times a^{-2}}{a^2}$

76. $\frac{t^4 \times t^3}{t^9}$

77. $y^{-3} \times y^7 \times y^{-5}$

78. $t^{-4} \div t^{-6} \times t^6$

79. $n^5 \times n^{-6} \div n$

Problems and Applications

80. State whether or not all the numbers in each row are equal. If they are not, state which number does not equal the other 3. Explain.

a)	$\frac{1}{5 \times 5}$	0.25	$\frac{1}{25}$	5^{-2}
b)	$\frac{1}{3 \times 3 \times 3}$	0.027	$\frac{1}{27}$	3^{-3}
c)	$\frac{1}{2 \times 2 \times 2 \times 2}$	0.0625	$\frac{1}{16}$	2^{-3}
d)	$\frac{1}{10 \times 10 \times 10}$	0.001	$\frac{1}{1000}$	10^{-2}

81. Evaluate for $a = 2$ and $b = 3$.

a) a^3 **b)** b^4 **c)** a^0

d) a^{-3} **e)** b^{-2} **f)** a^{-4}

g) $(a \times b)^{-2}$ **h)** $(a + b)^{-1}$

i) $(a - b)^{-1}$ **j)** $(b^2 - a^2)^{-2}$

82. Find the value of x.

a) $2^x = 8$ **b)** $3^x = 81$

c) $4^x = 1$ **d)** $3^x = \frac{1}{9}$

e) $5^x = \frac{1}{125}$ **f)** $x^3 = 27$

g) $x^{-2} = \frac{1}{4}$ **h)** $x^{-3} = \frac{1}{1000}$

83. Evaluate.

a) $\left(\frac{1}{2}\right)^0$ **b)** $\left(\frac{2}{3}\right)^{-3}$

c) $\left(\frac{1}{3}\right)^2 \left(\frac{-1}{2}\right)^3$ **d)** $\left(\frac{4}{5}\right)^0 - \left(\frac{3}{4}\right)^2$

e) $\left(\frac{4}{5}\right)^6 \left(\frac{4}{5}\right)^{-8}$ **f)** $\left(\frac{5}{3}\right)^3 \div \left(\frac{5}{3}\right)^5$

84. The radioactivity of a sample of carbon-14 drops to $\frac{1}{2}$ or 2^{-1} of its original value in about 5700 years. After 11 400 years, the radioactivity is $\frac{1}{4}$ or 2^{-2} of its original value.

a) What fraction of the radioactivity remains after 28 500 years?

b) Write the fraction as a power with a negative exponent.

c) Write the fraction as a power with a positive exponent.

d) After how long is the radioactivity $\frac{1}{128}$ or 2^{-7} times its original value?

85. Is each statement always true, sometimes true, or never true? Explain.

a) The value of a power with a negative exponent is less than 0.

b) The value of a power with a fractional base is less than 1.

c) Two powers in which the exponents are both 0 have equal values.

86. On a test, 3 students evaluated $2^{-2} \times 2^0$ as follows.

Terry $2^{-2} \times 2^0 = 4^{-2}$

$= \frac{1}{4^2}$

$= \frac{1}{16}$

Sean $2^{-2} \times 2^0 = 2^0$

$= 1$

Michel $2^{-2} \times 2^0 = 4^0$

$= 1$

a) What errors did each student make? What did each do correctly?

b) What is the correct answer?

87. Describe how to evaluate such powers as $(-3)^{-2}$ and -3^{-2}. Compare your description with your classmates'.

Virtual Reality

Virtual reality is the use of computers and graphics software to create sights and sounds from real-world data. The result is an impression of the real world. Soon, even touch and smell will be included in the "world" created by virtual reality.

Virtual reality is seen as a powerful training tool for people who take hazardous jobs or who must learn to use expensive equipment.

A Canadian company uses virtual reality in the systems it designs to train air traffic controllers in several countries. The controllers practise on a virtual reality system before tackling the real job at an airport.

Activity ❶

Virtual reality may be used to train doctors.

1. List 5 skills that trainee doctors might practise on a virtual reality system.

2. Describe the advantages of having doctors train in this way.

3. Describe any disadvantages of having doctors train in this way.

Activity ❷

Firefighters may be trained on a virtual reality system. What could the system simulate for a firefighter?

Activity ❸

Choose a situation in which you think a virtua system could be used for training. Describe what your system would do.

Activity ❹

Describe a computer game that could be played with a virtual reality system.

3.7 Use a Flow Chart

A flow chart provides a way of describing or checking many processes. This flow chart shows how you might try to solve a problem.

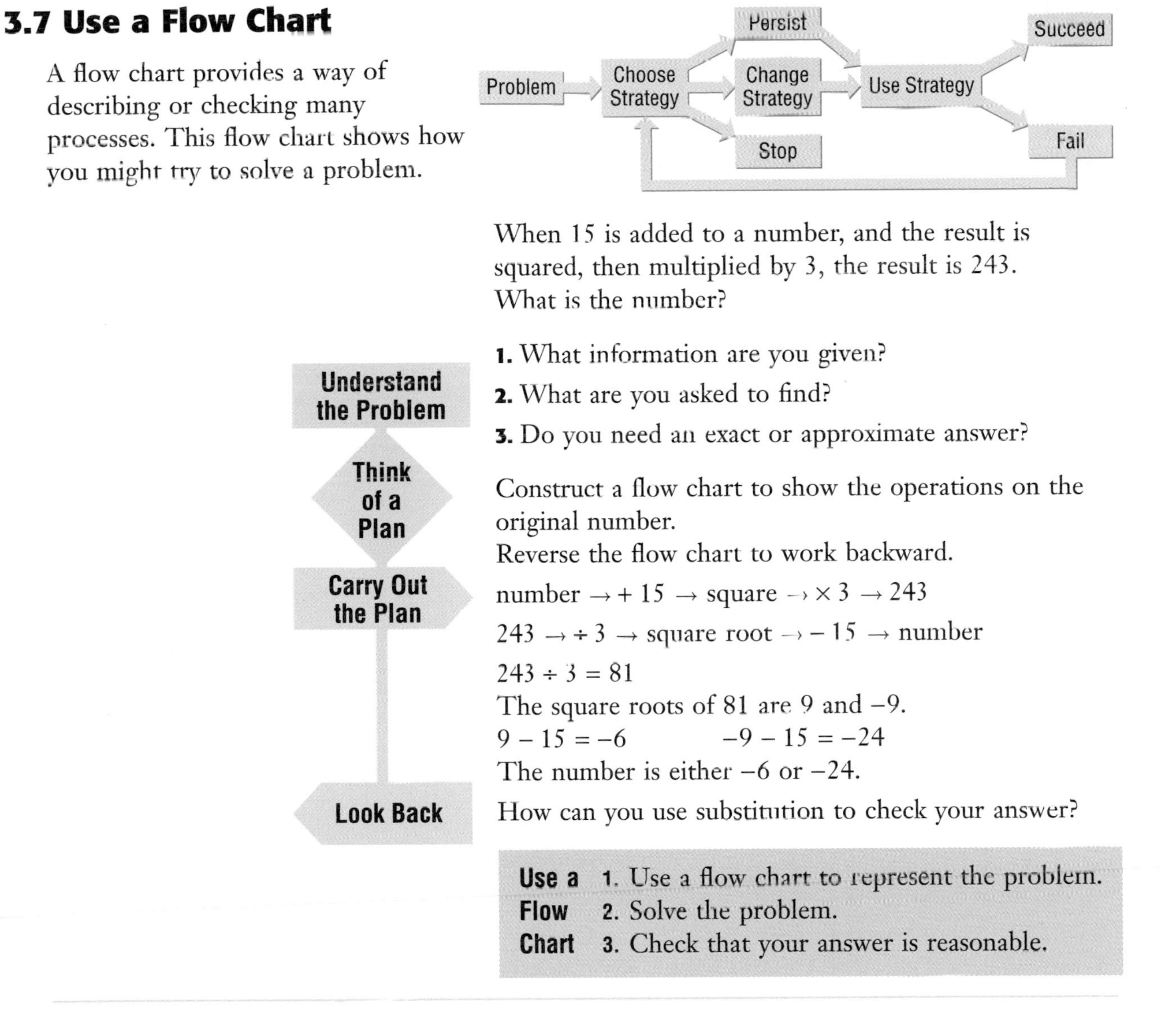

When 15 is added to a number, and the result is squared, then multiplied by 3, the result is 243. What is the number?

Understand the Problem

1. What information are you given?
2. What are you asked to find?
3. Do you need an exact or approximate answer?

Think of a Plan

Construct a flow chart to show the operations on the original number.
Reverse the flow chart to work backward.

Carry Out the Plan

number → + 15 → square → × 3 → 243

243 → ÷ 3 → square root → − 15 → number

$243 \div 3 = 81$

The square roots of 81 are 9 and −9.

$9 - 15 = -6$ $\qquad$ $-9 - 15 = -24$

The number is either −6 or −24.

Look Back

How can you use substitution to check your answer?

Use a Flow Chart

1. Use a flow chart to represent the problem.
2. Solve the problem.
3. Check that your answer is reasonable.

Problems and Applications

1. The cube of a number is divided by 20, then 36 is subtracted. The result is 14. What is the number?

2. Saskatchewan's population is about 1 000 000. This is about 6 000 000 less than Quebec's population, which is about 1.4 times smaller than Ontario's population. Find Ontario's population to the nearest million.

3. Troy has read 104 pages of a 496-page book. If he reads 32 pages a day, how many more days will he take to finish the book?

4. Draw a flow chart to represent each process. Compare your flow charts with a classmate's.

a) shopping for a new jacket

b) phoning a friend from your home

c) writing an essay

3.8 Scientific Notation: Small Numbers

The wavelength is the distance from one peak of a wave to the next. The wavelength of red light is 0.000 07 cm. The wavelength of violet light is 0.000 04 cm. This is the range of wavelengths we can see. To work with these small numbers, we need to extend our knowledge of scientific notation.

Activity: Look for a Pattern

Copy and complete the table.

Decimal Form	Product Form	Scientific Notation
0.52	$\frac{5.2}{10}$ or $5.2 \times \frac{1}{10^1}$	5.2×10^{-1}
0.052	$\frac{5.2}{?}$ or $5.2 \times \frac{1}{10^?}$	$5.2 \times 10^?$
0.0052	$\frac{5.2}{?}$ or $5.2 \times \frac{1}{10^?}$	$5.2 \times 10^?$
0.000 52	$\frac{5.2}{?}$ or $5.2 \times \frac{1}{10^?}$	$5.2 \times 10^?$

Inquire

1. Write a rule for writing numbers less than 1 in scientific notation.

2. Write in scientific notation the wavelength of
a) red light **b)** violet light

3. Why would you not write 0.2347 in scientific notation?

Example 1

Write in scientific notation.

a) 5100
b) 0.0051

Solution

Write the number in the form $x \times 10^n$, where x is greater than or equal to 1 but less than 10.

a) $5100 = 5.1 \times 1000$
$= 5.1 \times 10^3$

b) $0.0051 = 5.1 \times \frac{1}{1000}$
$= 5.1 \times \frac{1}{10^3}$
$= 5.1 \times 10^{-3}$

For 5100, the decimal point is moved 3 places to the left, and the exponent is 3. $5100 = 5.1 \times 10^3$

For 0.0051, the decimal point is moved 3 places to the right and the exponent is −3. $0.0051 = 5.1 \times 10^{-3}$

Example 2

Evaluate $(3.2 \times 10^{-2}) \times (5 \times 10^{-4})$.

Solution

$$(3.2 \times 10^{-2}) \times (5 \times 10^{-4}) = 3.2 \times 5 \times 10^{-2} \times 10^{-4}$$
$$= 16 \times 10^{(-2)+(-4)}$$
$$= 16 \times 10^{-6}$$
$$= 1.6 \times 10^{-5}$$

Practice

Write in scientific notation.

1. 4 500 000
2. 0.089
3. 0.2
4. 0.000 055

Write in scientific notation.

5. 45×10^7
6. 0.34×10^6
7. 33×10^{-8}
8. 10^{-8}
9. 0.06×10^{-7}
10. 100×10^{-14}

Write in decimal form.

11. 2.3×10^8

12. 4.7×10^{-6}

13. 7×10^{-9}

14. 10^{-6}

Write the number that is 10 times as large as each of the following.

15. 2.3×10^6

16. 4.5×10^{-9}

17. 5×10^{-7}

18. 10^{-11}

Write the number that is one-tenth as large as each of the following.

19. 7.8×10^9

20. 6.8×10^{-6}

21. 8×10^{-10}

22. 10^{-7}

Estimate, then calculate. Write each answer in scientific notation.

23. $(2.5 \times 10^{-3}) \times (5 \times 10^{-4})$

24. $(8 \times 10^{-5}) \times (1.2 \times 10^{-1})$

25. $(3.2 \times 10^{-3}) \times (4.1 \times 10^{-2})$

26. $(5.2 \times 10^{3}) \div (2 \times 10^{-2})$

27. $(3.5 \times 10^{-4}) \div (7 \times 10^{-3})$

Problems and Applications

28. Order from highest to lowest.

a) 4.3×10^{-3}, 4.35×10^{-3}, 8.4×10^{-4}, 10^{-3}

b) 5.6×10^{-9}, 10^{-9}, $\frac{1}{10^8}$, 5.6×10^{-8}

c) $\frac{1}{1000}$, 2.1×10^{-3}, 10^{-2}, 2.12×10^{-3}

29. One water molecule has a mass of about 3×10^{-26} kg. Calculate the number of water molecules in Lake Superior, which holds about 1.2×10^{16} kg of water.

30. Energy is measured in joules, symbol J. If you tap your finger 10 times on your desk, you use about 1 J of energy. Express the following quantities of energy in scientific notation.

a) If you run for 1 h, you use about 2 000 000 J of energy.

b) When a cricket chirps, it uses about 0.0008 J of energy.

c) The food energy in a slice of apple pie is about 1 500 000 J.

d) The energy released by splitting 1 uranium atom is about 0.000 000 000 08 J.

31. Why is 0.23×10^{-9} not in scientific notation?

32. Is it possible to write the number 5 in scientific notation? Explain.

33. a) The waves that bombard us every day have wavelengths that range from 0.000 000 000 000 01 cm to 1 000 000 000 cm. Write these numbers in scientific notation.

b) Draw a number line showing the wavelengths in powers of 10.

10^{-3} 10^{-2} 10^{-1} 1 10 10^2 10^3

Wavelength (cm)

Mark the range of wavelengths visible to humans.

c) Use your research skills to find out if there are creatures that can see outside the human range.

CALCULATOR POWER

1. On your calculator, multiply 0.000 000 4 × 0.000 055. How does your calculator display the answer?

2. Scientific calculators display [2.2 – 11]. What does this answer mean?

3. Many calculators have an [EE] key. To input 5.2×10^{-8}, press 5 [.] 2 [EE] 8 [+/–]

Use the [EE] key to evaluate

a) $(3.5 \times 10^{-4}) \times (4.8 \times 10^{-7})$

b) $(1.5 \times 10^{-8}) \div (5 \times 10^{-12})$

c) $\frac{(3.6 \times 10^{6}) \times (2.4 \times 10^{-8})}{2.7 \times 10^{-3}}$

3.9 Guess And Check

You can solve many problems by guessing the answer and then testing it to see if it is correct. If it is wrong, you can keep guessing until you get the right answer.

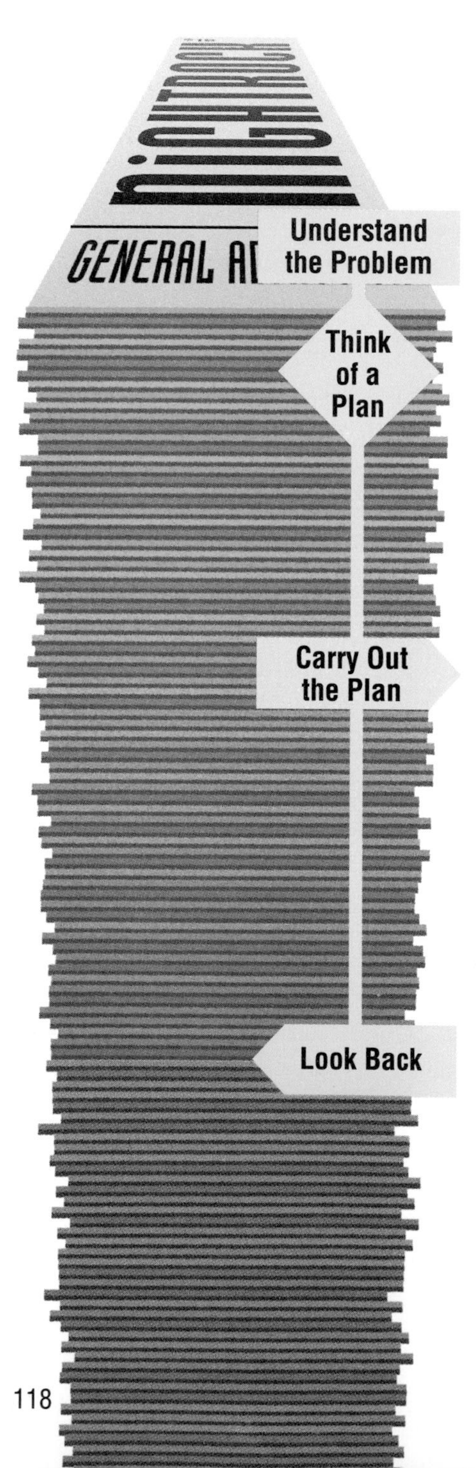

Jeannine wanted to raise \$10 000 for charity by holding a massed bands concert in a 400-seat auditorium. She decided to sell reserved seats for \$40 each and general admission seats for \$16 each. How many of each type of ticket did she have printed?

1. What information are you given?

2. What are you asked to find?

Guess at the number of reserved tickets to be printed. This number will give the number of general admission tickets. Then, calculate the total receipts to see if they equal \$10 000. If not, try another guess.

GUESS **CHECK**

Number of Reserved Tickets	Number of General Admission Tickets	Total Revenue (\$)	Is the total \$10 000?
100	300	$100 \times 40 + 300 \times 16$ $= 8800$	Too low
200	200	$200 \times 40 + 200 \times 16$ $= 11\,200$	Too high
150	250	$150 \times 40 + 250 \times 16$ $= 10\,000$	10 000 = 10 000

CHECKS!

Jeannine had 150 reserved seating tickets and 250 general admission tickets printed.

Check the answer against the given information.

Does the answer seem reasonable?

Guess and Check	1. Guess an answer that fits one of the facts. 2. Check the answer against the other facts. 3. If necessary, adjust your guess and check again.

Problems and Applications

1. Multiplying a number by 7 and adding 13 gives 55. What is the number?

2. Multiplying a number by 14 and subtracting 12 gives 114. What is the number?

3. What 3 consecutive numbers have a sum of 237?

4. What 4 consecutive even numbers have a sum of 196?

5. The square of the number of students in the band is close to 3000. About how many students are in the band?

6. The 7 Trudeaus went to the fair. Adult tickets cost \$8. Children's tickets cost \$5. The total cost was \$47. How many adults were there?

7. In how many ways could 4 darts hit the board to give a total score of 19?

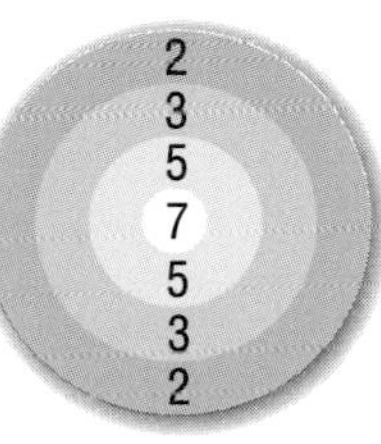

8. The perimeter of a rectangle is 60 m. The length is 4 m more than the width. What are the dimensions of the rectangle?

9. Karen wants to cut 56 m of television cable into 2 pieces, so that one is 6 m longer than the other. How long should she make each piece?

10. The cube of the number of people on the swim team is close to 4000. About how many are on the swim team?

11. Justine has some rare quarters and nickels. She has 8 more nickels than quarters. The face value of the coins is \$2.50. How many are quarters?

12. The dimensions of a rectangular solid are whole numbers. The areas of the faces are 42 cm^2, 48 cm^2, and 56 cm^2. What are the dimensions of the solid?

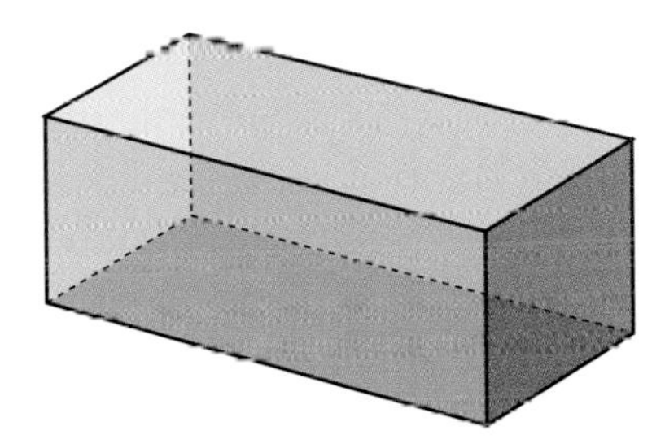

13. Piero works at the Snack Shack. He has been asked to make a mixture of 24 kg of cashews and peanuts that will cost \$7.25/kg. Cashews cost \$14/kg and peanuts cost \$5/kg. Copy and complete the table to find how many kilograms of cashews and how many kilograms of peanuts should be mixed together so that the cost of the mixture is \$7.25/kg.

GUESS | CHECK

Mass of Cashews (kg)	Mass of Peanuts (kg)	Total Cost of Mixture (\$)	Cost per Kilogram (\$)	Does cost equal \$7.25/kg?

14. Canada's top honour is the Order of Canada. There are 3 categories — Companion, Officer, and Member. A total of up to 153 awards can be given in any year. If all the awards are given, the number of new Members is twice the number of new Officers. There are 31 more new Officers than new Companions. How many awards are made in each category?

15. Write a problem that can be solved using the guess and check strategy. Ask a classmate to solve the problem.

Rolling Balls and Inductive Reasoning

Inductive reasoning is the process of drawing conclusions from a number of specific cases. In the following activities, you will consider a number of cases in which a ball rebounds from the sides of a table.

When a ball moves on a flat table with raised sides, the angle at which the ball strikes a side equals the angle at which the ball rebounds.

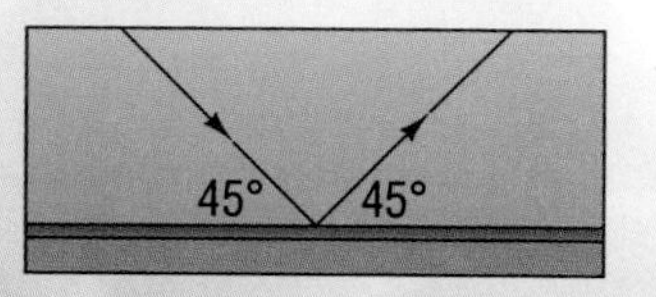

In the following activities, assume that:

a) The corners of the table are named top right (TR), top left (TL), lower right (LR), and lower left (LL).
b) The ball always starts from the lower left (LL) corner at an angle of 45° to the sides.
c) The ball keeps rolling until it hits a corner. Then, the ball stops.
d) The table dimensions are stated as height first and then width. A 3-by-2 table is shown here.

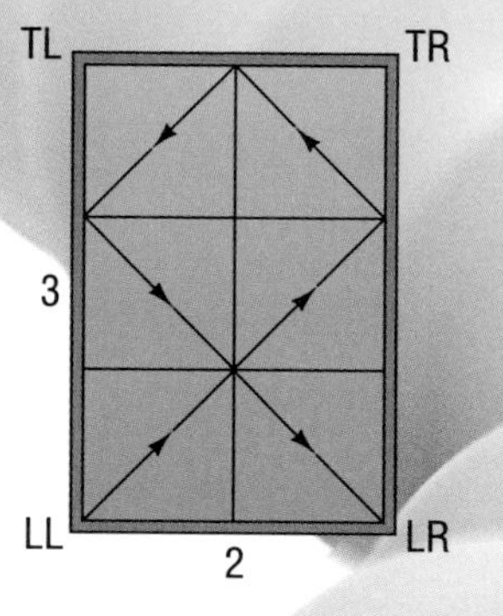

Activity 1

The diagram shows the path of a ball on a 3-by-2 table. The ball stops at the lower right (LR) corner.

1. Draw tables with the following dimensions on grid paper. For each grid, draw the path of a ball that starts at the LL corner.

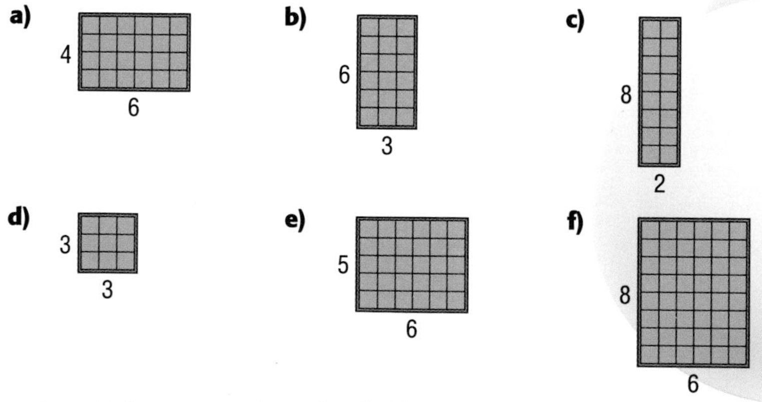

2. In which corner does the ball stop on each table?

3. What are the dimensions of the table that gives
a) the most complicated path? **b)** the simplest path?

4. On which table does the ball cross every square?

5. Will a ball ever stop at the LL corner? Explain.

Activity ❷

1. Draw tables with the following dimensions and draw the path of the ball.

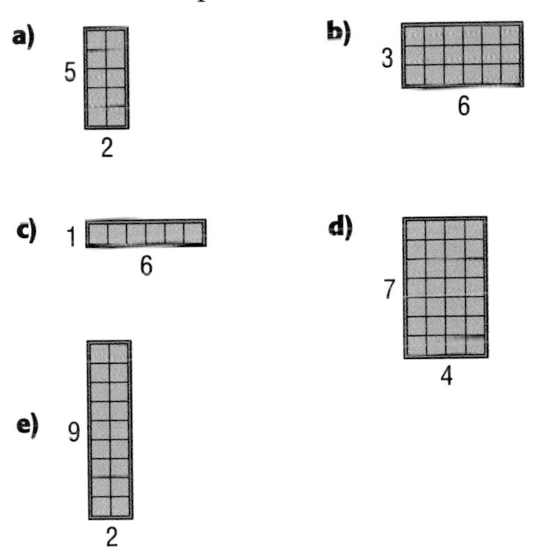

2. What kind of number is the height of each table?

3. What kind of number is the width of each table?

4. Where does the ball always stop?

Activity ❸

1. Draw each table and the path of the ball.

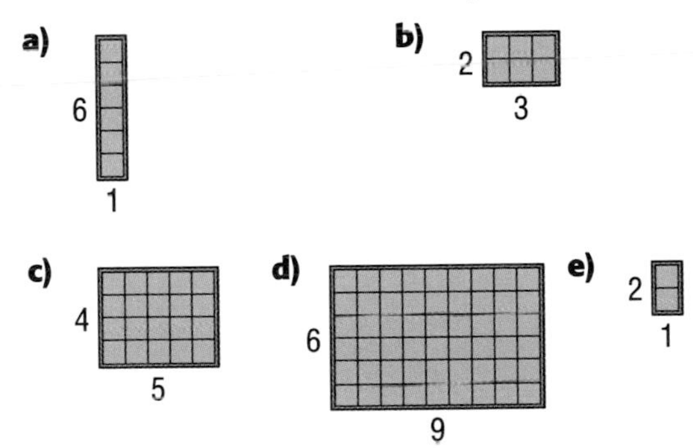

2. What kind of number is the height of each table?

3. What kind of number is the width of each table?

4. Where does the ball always stop?

Activity ❹

1. Draw tables with the following dimensions and draw the path of the ball.

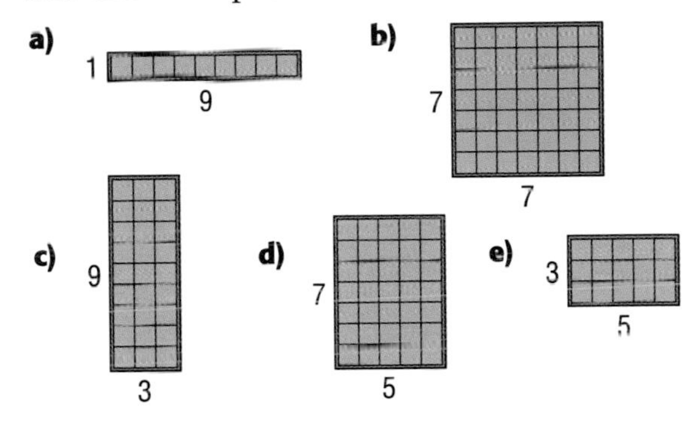

2. What kind of number is the height of each table?

3. What kind of number is the width of each table?

4. Where does the ball always stop?

Activity ❺

1. Use inductive reasoning to write a rule that states where the ball stops if the dimensions are

a) odd-by-even **b)** even-by-odd

c) odd-by-odd

2. State where the ball stops on tables with the following dimensions.

a) 32-by-13 **b)** 37-by-37

c) 29-by-28 **d)** 101-by-101

e) 244-by-133 **f)** 79-by-88

Review

Write as a power.

1. $5 \times 5 \times 5 \times 5 \times 5 \times 5$
2. $(-3)(-3)(-3)$
3. $\left(\frac{1}{2}\right)\left(\frac{1}{2}\right)\left(\frac{1}{2}\right)\left(\frac{1}{2}\right)$
4. $2 \times 2 \times 2 \times 2 \times 2$

Evaluate.

5. 3^5 **6.** 12^2 **7.** $(-5)^3$

8. $5(2)^5$ **9.** 6^0 **10.** $(-5)^0$

11. 2^{-1} **12.** 3^{-2} **13.** $\left(\frac{2}{3}\right)^3$

14. $\left(\frac{3}{5}\right)^{-2}$ **15.** $\frac{1}{3^{-2}}$ **16.** $\frac{5}{2^{-3}}$

Evaluate.

17. $3^{-1} - 3^0$ **18.** $2^{-3} + 2^{-2}$

Simplify.

19. $2^3 \times 2^{-5}$ **20.** $n^{-4} \div n^{-6}$

21. $((-3)^{-2})^4$ **22.** $((-x)^5)^{-1}$

23. $(-0.4)^{-5}(-0.4)^{-3}$

24. $4^2 \times 4^{-3} \times 4^{-1}$

Evaluate.

25. $5^3 \times 5^2 \times 5$ **26.** $2^0 \times 2^3$

27. $6^5 \div 6^3$ **28.** $(2^3)^4$

29. $2^{-3} \times 2^2$ **30.** $(4^2)^3 \div 4^4$

31. $(2^3)^{-2}$ **32.** $((-3)^2)^{-1}$

33. Evaluate for $x = 3$ and $y = -2$.

a) x^2y^2 **b)** y^5

c) $-6xy$ **d)** $(x - y)^3$

e) $6x^3 - 5y$ **f)** $(-x)^2(-y)^2$

Calculate. Write your answer in decimal notation.

34. 4.716×100 **35.** 16.93×10^3

36. $0.63 \div 10$ **37.** $123.94 \div 10^3$

38. $\frac{2.95}{100}$ **39.** $\frac{26.95}{10^4}$

40. 463.9×10^{-2} **41.** 1000×10^{-2}

42. $\frac{1.93}{10^{-1}}$ **43.** $\frac{10^{-3}}{10^{-1}}$

Write in scientific notation.

44. 27 300 000 **45.** 0.000 000 019 3

Write in decimal notation.

46. 2.53×10^7 **47.** 9.71×10^{-4}

Calculate. Write your answer in scientific notation.

48. $(2.5 \times 10^3)(3.1 \times 10^5)$

49. $(6.2 \times 10^{-4})(3.7 \times 10^2)$

50. $(8.6 \times 10^{21}) \div (2.5 \times 10^7)$

Calculate.

51. 5^2 **52.** 7^2 **53.** 12^2

54. 31^2 **55.** $\sqrt{0.16}$ **56.** $\sqrt{81}$

57. $\sqrt{121}$ **58.** $\sqrt{400}$

59. $6\sqrt{36}$ **60.** $3^2 + 4^2 + 7^2$

61. $-8\sqrt{49}$ **62.** $\frac{6\sqrt{25} - 8}{2}$

Write both square roots of each number.

63. 25 **64.** 36 **65.** 64 **66.** 144

Estimate. Then, calculate to the nearest tenth.

67. $\sqrt{59}$ **68.** $\sqrt{372}$ **69.** $\sqrt{1273}$

70. $\sqrt{41\ 093}$ **71.** $\sqrt{4.93}$ **72.** $\sqrt{0.0187}$

Calculate to the nearest tenth.

73. $\sqrt{5} - \sqrt{7}$ **74.** $3\sqrt{17}$

75. $10\sqrt{7} - 6\sqrt{19}$ **76.** $(\sqrt{23})(\sqrt{8})$

77. $\sqrt{18} \div \sqrt{5}$ **78.** $\frac{\sqrt{30} - \sqrt{20}}{\sqrt{3}}$

Determine the side of a square with each area. Round to the nearest tenth of a unit if necessary.

79. 36 m² **80.** 80 cm²

81. 156 cm² **82.** 28.4 m²

83. A box of mirror tiles contains 25 square tiles. Each tile covers 400 cm^2.

a) What are the dimensions of each tile?

b) What is the total area covered by the 25 tiles?

c) What length of edging is needed around the mirror?

edging

84. The velocity of sound in air may be found from the formula

$$V = 20\sqrt{273 + T}$$

where V is the velocity in metres per second and T is the temperature in degrees Celsius. Calculate the velocity of sound in air when the temperature is

a) 18°C **b)** 30°C

c) the temperature on Canada's coldest day ever

85. A sound whose frequency is twice the frequency of another sound is said to be an octave higher. A sound of frequency 400 Hz is an octave higher than a sound of frequency 200 Hz and 2 octaves higher than a sound of frequency 100 Hz.

a) How many octaves higher than a sound of frequency 500 Hz is a sound of frequency 4000 Hz?

b) How many octaves lower than a sound of frequency 6400 Hz is a sound of frequency 100 Hz?

86. Quebec City has a population of about 6.4×10^5 and an area of about 3.2×10^3 km^2.

a) How many square kilometres are there per person in Quebec City? Write your answer in scientific notation.

b) If each person's area was a square, what would its side be to the nearest metre?

Group Decision Making

Researching Education Careers

1. Brainstorm as a class some careers in education you want to know about. They might include high school teacher, elementary school teacher, university professor, librarian, or business administrator.

2. Choose a career to research on your own.

3. In your home group, decide on some questions that each individual might try to answer.

4. Before you start your research, list the questions you want to answer. Make sure that you include a question about how math is used in the career.

5. Research the career and prepare a report.

6. Evaluate your own research process and your report. Identify what went well and what you would do differently next time.

7. Have other students in the class evaluate your report.

Chapter Check

Evaluate.

1. $(-3)^4$

2. $(-1)^{14}$

3. 6^0

4. 5^{-1}

5. $(3^{-2})^2$

6. $(-3)^{-2}$

7. -2^3

8. $(-0.7)^2$

9. $2^5 \div (-2)^3$

10. $5^{-2} \times 5^3 \times 5^{-2}$

11. $7^0 - 4^{-1}$

12. $(1-3)^4$

Write in scientific notation.

13. 0.0004

14. 0.000 000 000 002 31

Write in decimal form.

15. 4.6×10^{-3}

16. 3.21×10^{-6}

Estimate, then calculate. Express your answer in scientific notation.

17. $(9.5 \times 10^{-2}) \times (5.1 \times 10^{-6})$

18. $(6.3 \times 10^{-4}) \div (1.9 \times 10^{-7})$

Calculate.

19. $\sqrt{36}$

20. $5\sqrt{16}$

21. $-\sqrt{121}$

Estimate. Then, calculate to the nearest tenth.

22. $\sqrt{93}$

23. $(\sqrt{8})(\sqrt{39})$

Simplify.

24. $(-3)^4 \div (-3)^5$

25. $s^2 \times s^{-5}$

26. $((2)^{-3})^{-1}$

27. $3^{-1} \div 3^{-2} \times 3^{-3}$

28. Evaluate for $m = -2$.

a) $3m^4$

b) $4m^3 + 5$

29. Forty percent of Canada's cropland is in Saskatchewan. The total area of cropland in Saskatchewan is about 134 000 km^2. What are the dimensions of a square that is big enough to hold all this land? Give your answer to the nearest kilometre.

THESE ARE CALLED "RADICAL" SIGNS...
I'M AS RADICAL AS THEY COME..
10-5
© 1991 United Feature Syndicate, Inc.

Using the Strategies

1. Copy the diagram. Place the digits from 1 to 5 in the squares so that the product is 3542.

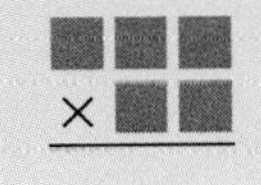

2. What is the side of a cube that can be made with 294 cm^2 of cardboard? What assumptions have you made?

3. Jason chose a whole number less than 10. He multiplied the number by 6 and added 1. The result was a perfect square. What numbers could he have chosen?

4. Find 3 consecutive whole numbers whose sum is 144.

5. How many Friday the thirteenths will there be in the year you turn 21?

6. Assume that the following pattern continues.

A, BBB, CCCCC, DDDDDDD, ...

a) How many letter Ms will there be?
b) How many letter Zs will there be?

7. In the figure, the side of each small square represents 1 cm.

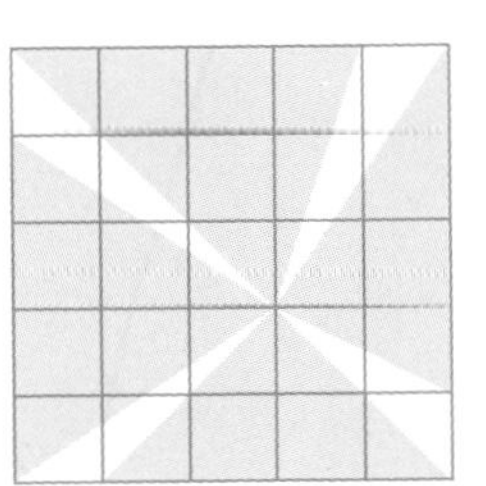

Find the area of
a) the shaded region.
b) the unshaded region.

8. How many different combinations of coins have a value of $0.28? Copy and complete the table to find out.

	Coins			
Combinations	**$0.25**	**$0.10**	**$0.05**	**$0.01**
1	1	0	0	3
2	0	1	2	8
3				

9. a) On a digital clock, how many times a day are the digits all the same?

b) Write another question about the numbers on a digital clock. Have a classmate solve your problem.

10. Draw a flow chart to represent each process. Compare your flow charts with a classmate's.
a) getting a driver's permit
b) writing a computer program

11. Sketch a graph that shows the distance you walk in a school day. Label the vertical axis "Distance Walked" and the horizontal axis "Time of Day." Represent a complete 24-h period, starting at midnight. Compare your graph with a classmate's and explain any differences.

1. If the Canadian province with the greatest area had the shape of a square, what would be the length of each side to the nearest kilometre?

2. Would you feel colder at a temperature of −7 °C with the wind at 32 km/h or at a temperature of −18 °C with the wind at 8 km/h? Explain.

CHAPTER 4

Rational Numbers

About 100 lightning bolts blast the surface of the Earth every second. How many is that a day? a year?

St. Hubert, Quebec, has thunderstorms on 27 days of the year, on average. Fort Smith, Northwest Territories, has thunderstorms on 12 days of the year, on average. How many times more frequent are thunderstorms in St. Hubert than in Fort Smith?

Lightning starts about a third of all forest fires. About how many forest fires does lightning start each year in Canada?

Activity ❶ Fraction Diagrams

Write the fraction that represents the shaded part of each diagram.

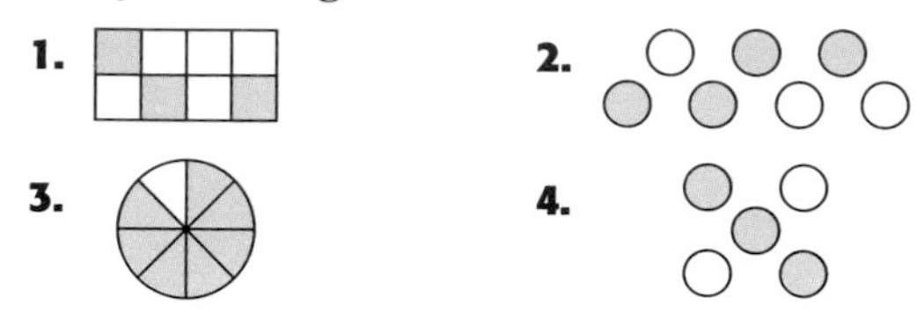

The above diagrams show fractions as part of a whole or part of a group. Draw diagrams to represent each of the following fractions as part of a whole and as part of a group.

5. $\frac{3}{4}$ **6.** $\frac{1}{6}$ **7.** $\frac{4}{5}$

8. $\frac{2}{3}$ **9.** $\frac{5}{8}$ **10.** $\frac{3}{10}$

Activity ❷ Equivalent Fractions

Equivalent fractions have the same value. Write 2 fractions equivalent to each fraction.

1. $\frac{1}{2}$ **2.** $\frac{2}{5}$ **3.** $\frac{3}{4}$

4. $\frac{5}{6}$ **5.** $\frac{1}{3}$ **6.** $\frac{7}{10}$

Write each fraction in simplest form.

7. $\frac{3}{6}$ **8.** $\frac{10}{12}$ **9.** $\frac{8}{10}$

10. $\frac{6}{24}$ **11.** $\frac{40}{100}$ **12.** $\frac{10}{25}$

Activity ❸ Dominoes and Fractions

There are 28 pieces in a set of dominoes, going from [blank domino] to [6/6 domino].

1. Sketch all the dominoes that make a fraction equivalent to $\frac{1}{2}$, like [1/2 domino].

2. Sketch all the dominoes that make a fraction equivalent to $\frac{1}{3}$.

3. Find as many other pairs of equivalent fractions as you can in a set of dominoes and sketch the solutions.

Activity ❹ Fractions and Words

The first $\frac{1}{2}$ of BASEBALL is BASE.

1. What is the last $\frac{1}{2}$ of BASEBALL?

2. What is the middle $\frac{1}{2}$ of NINE?

3. What is the last $\frac{3}{8}$ of ELEPHANT?

4. What is the first $\frac{2}{3}$ of NUMBER?

5. What is the last $\frac{2}{3}$ of RECTANGLE?

6. What is the first $\frac{3}{8}$ of PENTAGON?

7. What is the middle $\frac{3}{5}$ of STARS?

Activity ❺ Championship Fractions

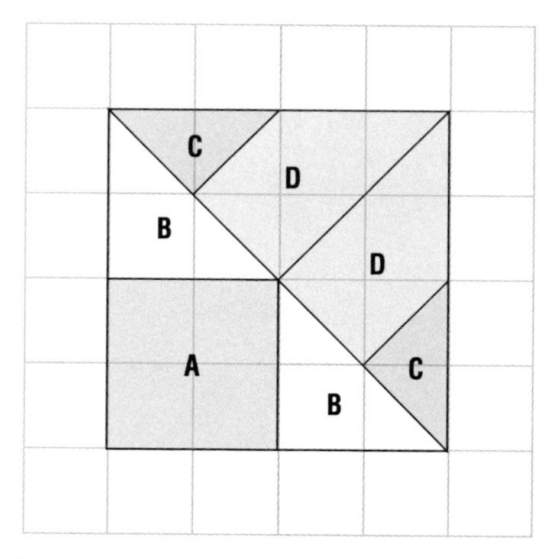

These 7 pieces, similar to a tangram, were used in the first World Puzzle Championships. The pieces are drawn on grid paper.

1. Find the area of each piece.

2. What fraction of A is B?

3. What fraction of A is C?

4. What fraction of A is D?

5. What fraction of D is C?

6. What fraction of D is B?

7. What fraction of the whole square is A? B? C? D?

Warm Up

Add. Write each answer in simplest form.

1. $\frac{1}{4} + \frac{2}{4}$
2. $\frac{2}{5} + \frac{2}{5}$
3. $\frac{3}{8} + \frac{1}{8}$
4. $\frac{2}{3} + \frac{1}{3}$
5. $\frac{1}{2} + \frac{1}{6}$
6. $\frac{3}{8} + \frac{1}{4}$
7. $\frac{2}{5} + \frac{3}{10}$
8. $\frac{2}{3} + \frac{1}{4}$
9. $\frac{1}{4} + \frac{5}{6}$
10. $\frac{2}{5} + \frac{1}{2}$

Subtract. Write each answer in simplest form.

11. $\frac{3}{4} - \frac{1}{4}$
12. $\frac{4}{5} - \frac{2}{5}$
13. $\frac{5}{8} - \frac{3}{8}$
14. $\frac{5}{7} - \frac{3}{7}$
15. $\frac{1}{2} - \frac{1}{4}$
16. $\frac{7}{8} - \frac{1}{4}$
17. $\frac{9}{10} - \frac{3}{5}$
18. $\frac{3}{4} - \frac{3}{8}$
19. $\frac{1}{2} - \frac{1}{3}$
20. $\frac{3}{4} - \frac{2}{3}$
21. $\frac{1}{2} - \frac{1}{5}$
22. $\frac{5}{6} - \frac{3}{4}$

Multiply. Write each answer in simplest form.

23. $\frac{1}{2} \times \frac{1}{2}$
24. $\frac{2}{3} \times \frac{1}{3}$
25. $\frac{1}{2} \times \frac{3}{4}$
26. $\frac{5}{6} \times \frac{2}{3}$
27. $\frac{2}{5} \times \frac{3}{4}$
28. $\frac{3}{8} \times \frac{1}{2}$
29. $\frac{3}{5} \times \frac{2}{3}$
30. $\frac{5}{6} \times \frac{4}{5}$

Divide. Write each answer in simplest form.

31. $\frac{1}{2} \div \frac{1}{2}$
32. $\frac{1}{4} \div \frac{3}{4}$
33. $\frac{1}{3} \div \frac{2}{3}$
34. $\frac{3}{5} \div \frac{2}{5}$
35. $\frac{3}{4} \div \frac{1}{2}$
36. $\frac{3}{8} \div \frac{1}{2}$
37. $\frac{1}{3} \div \frac{1}{2}$
38. $\frac{1}{2} \div \frac{1}{3}$
39. $\frac{3}{8} \div \frac{5}{8}$
40. $\frac{2}{3} \div \frac{5}{6}$

Mental Math

Add. State each answer in simplest form.

1. $\frac{1}{2} + \frac{1}{2}$
2. $3 + \frac{1}{2}$
3. $1\frac{1}{2} + 1$
4. $1\frac{1}{4} + \frac{3}{4}$
5. $\frac{3}{8} + \frac{5}{8}$
6. $\frac{2}{7} + \frac{3}{7}$
7. $\frac{3}{10} + \frac{1}{10}$
8. $2\frac{1}{2} + 3\frac{1}{2}$

Subtract. State each answer in simplest form.

9. $\frac{1}{4} - \frac{1}{4}$
10. $1 - \frac{1}{4}$
11. $2\frac{1}{2} - \frac{1}{2}$
12. $\frac{7}{8} - \frac{5}{8}$
13. $2\frac{3}{4} - 2$
14. $\frac{9}{10} - \frac{7}{10}$
15. $1\frac{5}{6} - 1\frac{4}{6}$
16. $1 - \frac{2}{3}$

Multiply. State each answer in simplest form.

17. $\frac{1}{3} \times \frac{1}{3}$
18. $\frac{1}{2} \times \frac{1}{4}$
19. $\frac{1}{5} \times \frac{1}{2}$
20. $\frac{1}{6} \times \frac{1}{2}$
21. $2 \times \frac{1}{2}$
22. $\frac{1}{3} \times 3$
23. $\frac{5}{6} \times 0$
24. $4 \times \frac{1}{2}$

Copy and complete the statements.

25. $1\frac{3}{4} + ■ = 3$
26. $2\frac{2}{3} + ■ = 5$
27. $\frac{3}{8} + ■ = 2$
28. $\frac{7}{10} + ■ = 3$
29. $\frac{1}{2}$ of $20 = ■$
30. $\frac{1}{3}$ of $60 = ■$
31. $\frac{1}{10}$ of $70 = ■$
32. $\frac{1}{2}$ of $22 = ■$

4.1 Rational Numbers

A corporation may sell shares to individuals. These individuals become part-owners of the corporation. Shares in the ownership of a business are called **stock**. The Toronto Stock Exchange publishes a daily report on the changing values of stocks.

If the report shows that the value of a stock has changed by $+\frac{3}{4}$, it has gone up by $\frac{3}{4}$ of a dollar or \$0.75.

If the report shows a change in value of $-\frac{1}{2}$, it has gone down by $\frac{1}{2}$ of a dollar or \$0.50.

Recall that the fractions $\frac{3}{4}$ and $-\frac{1}{2}$ are written in **lowest terms** or **simplest form**. This means that the numerator and denominator can only be divided evenly by 1.

Activity: Interpret the Data

The value of one McDonald's share was up $1\frac{1}{4}$ dollars or \$1.25 from the previous day.

The value of one Sony share was down $1\frac{1}{2}$ dollars or \$1.50 from the previous day.

The value of one IBM share was down $4\frac{3}{4}$ dollars, and the value of one A&W Foods share was up 2 dollars.

McDonald's	$+1\frac{1}{4}$
Sony	$-1\frac{1}{2}$
IBM	$-4\frac{3}{4}$
A&W Foods	+2

Inquire

1. Write $+1\frac{1}{4}$ as the quotient of two integers in lowest terms. In how many different ways can you do this?

2. Write $-1\frac{1}{2}$ as the quotient of two integers in lowest terms. In how many different ways can you do this?

3. Write $-4\frac{3}{4}$ as the quotient of two integers in lowest terms. In how many different ways can you do this?

4. Write +2 as the quotient of two integers in lowest terms. In how many different ways can you do this?

Rational numbers are numbers that can be written as the quotient of two integers, that is, in the form $\frac{a}{b}$, where a is any integer and b is any integer except 0.

This means that all integers, fractions, terminating decimals, and repeating decimals are rational numbers. Numbers that cannot be written as the quotient of two integers are called **irrational numbers.** These numbers are non-repeating, non-terminating decimals. Examples of irrational numbers are $\sqrt{2}$, $\sqrt{3}$, and π. **Real numbers** include all of the rational and irrational numbers.

Rational numbers can be written in many equivalent forms. For example, the rational number 2 can be written as $\frac{2}{1}, \frac{4}{2}, \frac{-2}{-1}, \frac{-4}{-2}, -\frac{-4}{2}, -\frac{4}{-2}$, and so on.

Example 1

Express each rational number in an equivalent form.

a) $\frac{-4}{7}$ **b)** -6 **c)** $-\frac{-6}{-8}$ **d)** $\frac{5}{-7}$

Solution

a) $\frac{-4}{7} = \frac{-4}{7} \times 1$
$= \frac{-4 \times -1}{7 \times -1}$
$= \frac{4}{-7}$

b) $-6 = \frac{-6}{1} \times 1$
$= \frac{-6 \times 2}{1 \times 2}$
$= \frac{-12}{2}$

c) $-\frac{-6}{-8} = -\frac{-6}{-8} \div 1$
$= -\frac{-6 \div 2}{-8 \div 2}$
$= -\frac{-3}{-4}$
$= -\frac{3}{4}$

d) $\frac{5}{-7} = \frac{5}{-7} \times 1$
$= \frac{5 \times -3}{-7 \times -3}$
$= \frac{-15}{21}$

The results of Example 1 suggest the following general statement.

$\frac{-a}{b} = \frac{a}{-b} = -\frac{a}{b} = -\frac{-a}{-b}$, where $b \neq 0$.

Rational numbers may be located on a number line.

Recall that numbers increase in value as you go from left to right on the number line.

For example, $-1\frac{3}{4}$ is less than -1.

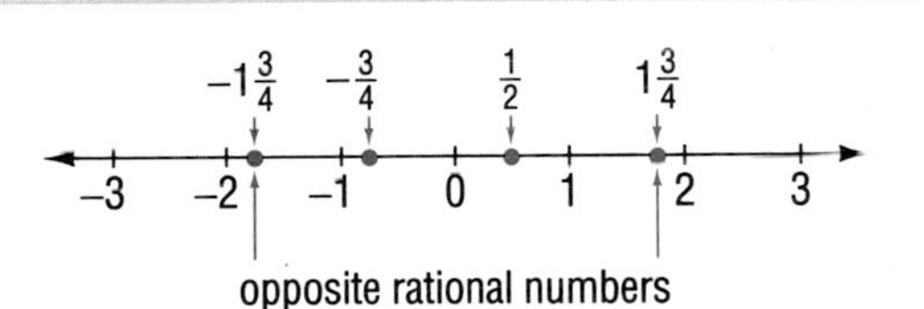

Example 2

Which is larger, $\frac{-2}{3}$ or $\frac{-4}{5}$?

Solution

The LCD of 3 and 5 is 15.

$\frac{-2 \times 5}{3 \times 5} = \frac{-10}{15}$ $\frac{-4 \times 3}{5 \times 3} = \frac{-12}{15}$

Compare numerators.

$-10 > -12$

Therefore, $\frac{-2}{3} > \frac{-4}{5}$.

CONTINUED

Example 3

Write each of the following as the quotient of two integers in lowest terms.

a) $1\frac{1}{2}$ **b)** -0.8 **c)** 3 **d)** -2.51

Solution

a) $1\frac{1}{2} = \frac{3}{2}$

b) 0.8 means 8 tenths

so $-0.8 = -\frac{8}{10}$

$= -\frac{4}{5}$

c) $3 = \frac{3}{1}$

d) 2.51 is $2\frac{51}{100}$

or $\frac{251}{100}$

so $-2.51 = -\frac{251}{100}$

Practice

State which are rational numbers.

1. $\frac{1}{2}$ **2.** $\frac{2}{5}$ **3.** $-\frac{1}{3}$ **4.** $\frac{-3}{0}$

5. $\frac{-0}{4}$ **6.** 0 **7.** -7 **8.** $2\sqrt{2}$

9. π **10.** $\frac{-8}{-1}$ **11.** $\frac{18}{-19}$ **12.** $\frac{7}{0}$

Express each rational number in an equivalent form.

13. $\frac{-2}{5}$ **14.** 7 **15.** $\frac{-6}{12}$ **16.** $\frac{8}{-3}$

17. $\frac{-3}{4}$ **18.** -5 **19.** $\frac{2}{-3}$ **20.** $\frac{-2}{-5}$

Write each rational number in lowest terms.

21. $\frac{4}{8}$ **22.** $\frac{-3}{6}$ **23.** $\frac{-2}{8}$ **24.** $\frac{-14}{2}$

25. $\frac{-18}{21}$ **26.** $\frac{-8}{10}$ **27.** $\frac{-3}{9}$ **28.** $\frac{12}{15}$

Write each of the following as the quotient of two integers in lowest terms.

29. $2\frac{3}{4}$ **30.** 0.4 **31.** -7 **32.** -1.2

33. $-3\frac{1}{2}$ **34.** 2.25 **35.** -0.3 **36.** $\frac{-12}{18}$

Which is larger?

37. $\frac{-1}{2}$ or $\frac{3}{4}$ **38.** $\frac{3}{5}$ or $\frac{8}{3}$

39. $\frac{-3}{4}$ or $\frac{5}{-4}$ **40.** 0 or $\frac{-4}{-5}$

Replace each ● by <, >, or = to make each statement true.

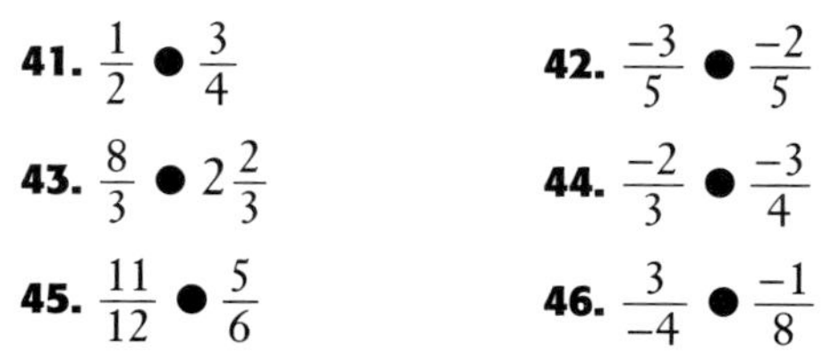

41. $\frac{1}{2}$ ● $\frac{3}{4}$ **42.** $\frac{-3}{5}$ ● $\frac{-2}{5}$

43. $\frac{8}{3}$ ● $2\frac{2}{3}$ **44.** $\frac{-2}{3}$ ● $\frac{-3}{4}$

45. $\frac{11}{12}$ ● $\frac{5}{6}$ **46.** $\frac{3}{-4}$ ● $\frac{-1}{8}$

Arrange the rational numbers in order from largest to smallest.

47. $\frac{1}{2}, \frac{4}{5}, 1\frac{2}{5}, 0, \frac{-1}{2}, \frac{-7}{-10}, 2$

48. $1\frac{7}{8}, 1\frac{1}{2}, 1\frac{2}{3}, \frac{5}{4}, \frac{11}{6}, \frac{9}{4}$

49. $\frac{-5}{3}, \frac{11}{9}, \frac{-11}{-9}, \frac{4}{-3}$

50. Mark the location of each of the following rational numbers on a number line.
$\frac{7}{4}, -2, \frac{-6}{8}, \frac{9}{2}, \frac{8}{-3}, \frac{-6}{-2}, -3\frac{1}{2}$

Problems and Applications

51. Write an example of a rational number
a) with a numerator of 1 and a denominator greater than 5
b) with a numerator less than −3 and a denominator greater than 0
c) with a denominator of −16 and a numerator that makes a rational number in lowest terms

52. How many times more provinces are there to the east of Manitoba than there are to the west of Manitoba? Write your answer as a quotient of two integers in lowest terms.

53. Divide the number of consonants by the number of vowels in each name. Express each answer as a quotient of two integers in lowest terms.

a) Montreal **b)** Red Deer
c) Nanaimo **d)** Saskatoon
e) Barrie **f)** Truro
g) Fredericton **h)** Gander

54. a) List your answers from question 53 in order from largest to smallest.
b) Which name has the greatest quotient?

55. The following game report was published by the operators of an animal safari.

Safari Game Report

	Population 5 Years Ago	Population Now
Elephants	18	14
Monkeys	6	12
Giraffes	14	19
Rhinoceroses	3	1
Hyenas	8	8

The growth factor for each type of animal can be calculated using the following formula.

$$GF = \frac{\text{Pop. now} - \text{Pop. 5 years ago}}{\text{Pop. 5 years ago}}$$

Calculate the growth factor for each type of animal. Express each answer as the quotient of 2 integers in lowest terms.

56. Between 1951 and 1966, the Canadian population increased from about 14 000 000 to about 20 000 000. How many times greater was the population in 1966 than it was in 1951? Express your answer as a quotient of two integers in lowest terms.

57. Is $\frac{-3}{4}$ the opposite of $\frac{4}{-3}$? Explain.

58. Since $4 \times 2 = 8$, $\frac{8}{2} = 4$. Use this idea to show why $\frac{8}{0}$ is not possible.

59. State whether each statement is always true, sometimes true, or never true. Explain.
a) The quotient of two integers has a value greater than 1 if the numerator is greater than the denominator.
b) The value of the quotient of two integers decreases as the denominator increases.
c) An integer is a rational number.
d) A rational number is a real number.
e) A real number is a rational number.

LOGIC POWER

Assume that there are no cubes missing from the back of this stack.

1. How many cubes are in the stack?

2. If you painted the outside of the stack green, including the bottom, what fraction of the cubes would have the following numbers of green faces?

a) 1 **b)** 2 **c)** 3 **d)** 4

Writing Decimals as Fractions

Some decimals, like 0.5 and 0.64, terminate or end. Others repeat and never end, like $0.\overline{3}$ or 0.3333....

Activity ❶ Terminating Decimals to Fractions

To write a terminating decimal as a fraction, use your knowledge of our number system. 0.7 means "seven tenths," so $0.7 = \frac{7}{10}$. 0.45 means "forty-five hundredths," so $0.45 = \frac{45}{100} = \frac{9}{20}$.

1. Write each decimal as a fraction or mixed number in lowest terms. Check your answer with a calculator.

a) 0.3 **b)** −0.6 **c)** 1.5
d) 3.75 **e)** −2.06 **f)** −1.875

Activity ❷ Non-Terminating Repeating Decimals to Fractions

For these types of decimals, eliminate the repeating part of the decimal by subtraction. Study the steps used to change $0.\overline{4}$ and $0.\overline{63}$ to fractions.

Let $F = 0.\overline{4}$ or 0.4444...

Multiply by 10	$10F = 4.\overline{4}$
and	$F = 0.\overline{4}$
Subtract	$9F = 4$
Divide by 9	$\frac{9F}{9} = \frac{4}{9}$
	$F = \frac{4}{9}$

Let $G = 0.\overline{63}$ or 0.636363...

Multiply by 100	$100G = 63.\overline{63}$
and	$G = 0.\overline{63}$
Subtract	$99G = 63$
Divide by 99	$\frac{99G}{99} = \frac{63}{99}$
	$G = \frac{63}{99}$ or $\frac{7}{11}$

1. Write each decimal as a fraction or mixed number in lowest terms. Check your answers with a calculator.

a) $0.\overline{3}$ **b)** $0.\overline{4}$ **c)** $0.\overline{23}$
d) $0.\overline{46}$ **e)** $3.\overline{7}$ **f)** $5.\overline{36}$

2. Write each decimal as a fraction or mixed number in lowest terms. Check your answers with a calculator.

a) $-0.\overline{2}$ **b)** $-0.\overline{7}$ **c)** $-0.\overline{33}$
d) $-0.\overline{52}$ **e)** $-3.\overline{1}$ **f)** $-4.\overline{18}$

3. Write each decimal as a fraction or mixed number in lowest terms. Check your answers with a calculator.

a) $0.2\overline{6}$ **b)** $0.1\overline{5}$ **c)** $0.3\overline{9}$ **d)** $3.8\overline{9}$

Activity ❸

1. Make up a repeating decimal and convert it to a fraction or mixed number in lowest terms.

2. Have a classmate convert your decimal to a fraction or mixed number in lowest terms.

3. Check your classmate's answer.

4.2 Decimal Forms of Rational Numbers

The prices for shares on the stock exchange are stated as dollars and fractions of dollars instead of dollars and cents.

If a stock increased by $1\frac{7}{8}$ dollars, you might want to know how many dollars and cents that is.

When we write an amount of money such as $1.25, the cents are expressed as a decimal.

To write a fraction as a decimal, divide the numerator by the denominator.

Activity: Complete the Table

Copy the table. Use a calculator to complete the table. The first two rows have been done for you.

Fraction	Division	Calculator Display	Decimal Form
$\frac{3}{8}$	$3 \div 8$	0.375	0.375
$\frac{1}{3}$	$1 \div 3$	0.3333333	$0.\overline{3}$
$\frac{7}{8}$			
$\frac{13}{20}$			
$\frac{3}{16}$			
$\frac{5}{11}$			
$\frac{5}{6}$			
$\frac{4}{9}$			

Inquire

1. There are 2 kinds of decimals in the table. Describe the difference between the 2 kinds.
2. What kind of decimals do we often write using a bar, $\overline{}$? Explain why this notation is often used.

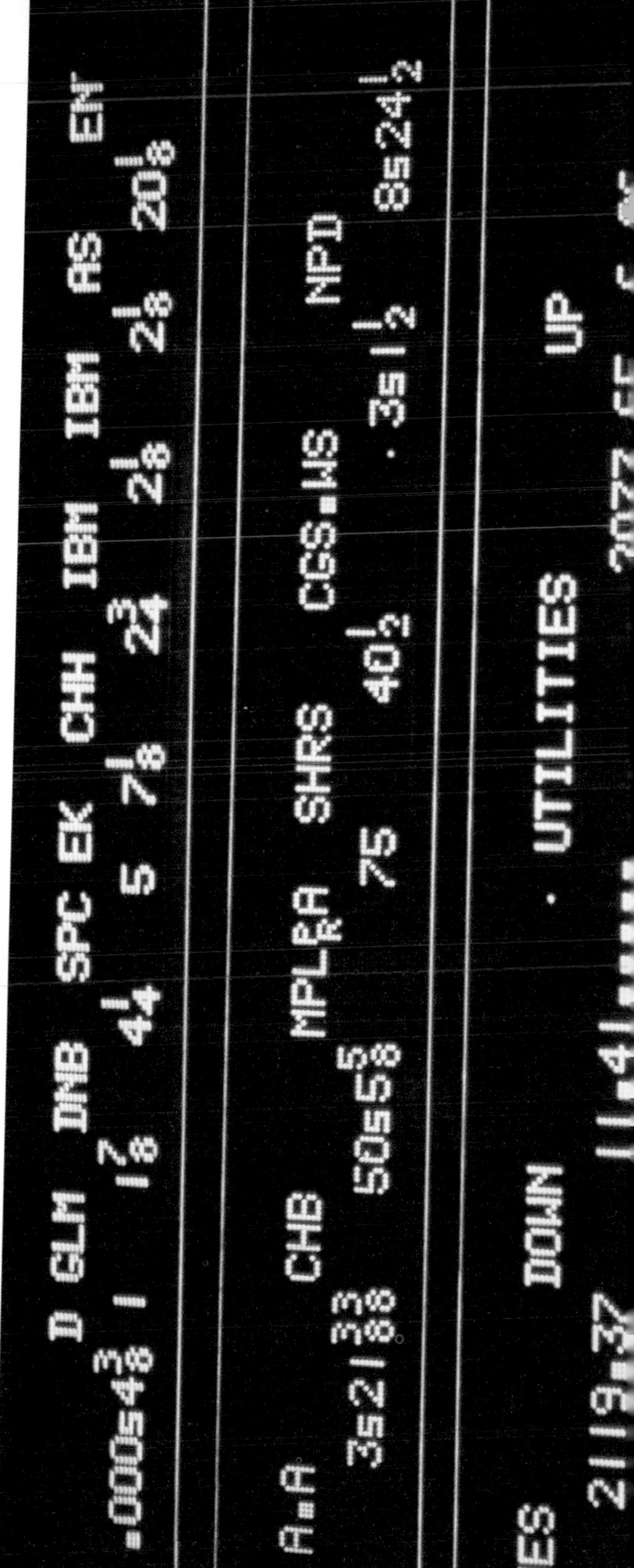

The following are examples of rational numbers.

$\frac{1}{2} = 0.5$ — **terminating** decimal

$\frac{3}{11} = 0.27272727\ldots$ — **non-terminating** but **repeating** decimal

The repeating decimal 0.27272727... is also written as $0.\overline{27}$.
The digits that repeat are called the **period**.
The number of digits that repeat is called the **length of the period.**
Here, the period is 27, and the length of the period is 2.

Example 1

Express each decimal as a fraction in simplest form.

a) 0.4
b) 0.85
c) −0.26
d) −1.825

Solution

a) $0.4 = \frac{4}{10} = \frac{4 \div 2}{10 \div 2} = \frac{2}{5}$

b) $0.85 = \frac{85}{100} = \frac{85 \div 5}{100 \div 5} = \frac{17}{20}$

c) $-0.26 = -\frac{26}{100} = -\frac{26 \div 2}{100 \div 2} = -\frac{13}{50}$

d) $-1.825 = -\frac{1825}{1000} = -\frac{1825 \div 25}{1000 \div 25} = -\frac{73}{40}$

Practice

Write true or false for each statement.

1. $\frac{31}{9} = 3.4$ **2.** $\frac{182}{5} = 36.4$

3. $\frac{9}{11}$ has 2 repeating digits

Match each number on the left with an equivalent number on the right.

4. 5.67 **a)** $\frac{4}{5}$

5. 0.8 **b)** $128\frac{3}{10}$

6. 1.283 **c)** $\frac{1283}{1000}$

7. 128.3 **d)** $5\frac{67}{100}$

e) $\frac{8}{100}$

Write each repeating decimal using bar notation.

8. 0.31313... **9.** 0.666...

10. 0.9555... **11.** 5.6161...

12. 32.42874287... **13.** −18.5252...

Write as a decimal.

14. $\frac{2}{3}$ **15.** $-\frac{7}{8}$ **16.** $\frac{9}{2}$

17. $-\frac{5}{12}$ **18.** $-\frac{11}{5}$ **19.** $\frac{4}{11}$

20. $\frac{3}{7}$ **21.** $-\frac{7}{9}$ **22.** $\frac{-7}{20}$

Identify the period and the length of the period in each rational number.

23. $\frac{2}{9}$ **24.** $\frac{7}{11}$ **25.** $-\frac{1}{6}$

26. $-\frac{1}{3}$ **27.** $\frac{5}{6}$ **28.** $-\frac{7}{12}$

Write as a fraction.

29. -1.1 **30.** 6.51 **31.** 0.07

32. 1.003 **33.** -3.37 **34.** -2.79

Express in the form $\frac{a}{b}$ and reduce to lowest terms.

35. 0.5 **36.** -0.4 **37.** 1.4

38. 0.8 **39.** -2.75 **40.** 1.25

41. -0.21 **42.** 3.25 **43.** 1.62

44. -7.4 **45.** -2.32 **46.** 3.1

Replace each ● with >, < or = to make each statement true.

47. $0.3 \bullet 0.\overline{3}$ **48.** $0.007 \bullet 0.07$

49. $1.21 \bullet 1\frac{21}{100}$ **50.** $0.\overline{61} \bullet \frac{3}{5}$

51. $\frac{7}{8} \bullet -0.875$ **52.** $5\frac{1}{3} \bullet 5.\overline{3}$

53. $-\frac{7}{11} \bullet -0.\overline{71}$ **54.** $6.\overline{001} \bullet 6.00\overline{1}$

55. $-1\frac{1}{6} \bullet 1.6$ **56.** $6\frac{3}{4} \bullet 6.75$

Write in order from smallest to largest.

57. $\frac{1}{2}, \frac{1}{3}, \frac{3}{5}, 0.59$

58. $0.71, 0.7\overline{1}, \frac{711}{1000}, 0.\overline{7}$

59. $-9.01, -8.93, -9.\overline{1}, -8.\overline{9}$

60. $0.113, 0.\overline{113}, 0.1\overline{13}, 0.\overline{1}$

61. $\frac{3}{8}, -\frac{4}{7}, 0.3\overline{7}, 0.3\overline{75}$

Problems and Applications

62. Write a rational number in fraction form to represent each of the following.

a) a loss of $12.25

b) a profit of $6.50

c) 35.2 m below sea level

d) 3.1°C above zero

e) 0.6 s slower

63. How many times taller is a 60-storey building than a 50-storey building? Write your answer as a decimal. What assumptions have you made?

64. At the 1992 Summer Olympics in Barcelona, Spain, Canadians won 6 gold medals, 5 silver medals, and 7 bronze medals.

a) State in simplest terms the fraction of the Canadian medals of each kind.

b) Write each fraction as a decimal.

65. Canadian mines produce about 0.24 of the world's nickel ore. Express this decimal in fraction form in lowest terms.

66. a) Express each of the following as a repeating decimal.

$\frac{1}{11}$ $\frac{2}{11}$ $\frac{3}{11}$

b) Describe the pattern.

c) Use the pattern to predict the following.

$\frac{7}{11}$ $\frac{10}{11}$

67. The Celsius temperature scale uses positive and negative rational numbers to represent temperatures. The Kelvin temperature scale does not use negative rational numbers. Research the Kelvin temperature scale and explain why it does not use negative rational numbers.

LOGIC POWER

The 2 squares intersect at 2 points.

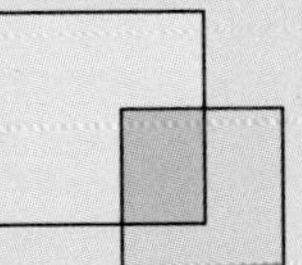

Draw diagrams to show 2 squares intersecting at 3, 4, 5, and 6 points.

Multiplying and Dividing Fractions

Activity ❶ Multiplying Fractions

What is $\frac{1}{4}$ of $\frac{1}{2}$? In mathematics, we write $\frac{1}{4}$ of $\frac{1}{2}$ as $\frac{1}{4} \times \frac{1}{2}$.
We can shade grid paper to multiply a fraction by a fraction.

1. Draw a rectangle of 8 squares on grid paper.

2. Shade $\frac{1}{2}$ of the rectangle.

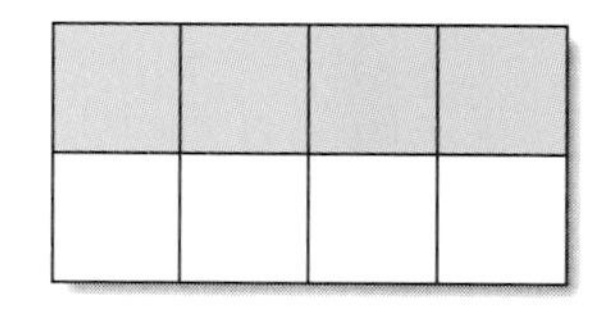

3. Show $\frac{1}{4}$ of the shaded area by shading over the first shading.

4. To describe the fraction with the double shading, we write $\frac{1}{4} \times \frac{1}{2} = \frac{1}{8}$.
Use grid paper to show the following and write a multiplication sentence for each.

a) $\frac{1}{3}$ of $\frac{1}{2}$ **b)** $\frac{2}{3}$ of $\frac{3}{4}$ **c)** $\frac{1}{2}$ of $\frac{4}{5}$ **d)** $\frac{1}{2}$ of $\frac{2}{3}$

5. Write a rule for multiplying fractions.
Compare your rule with your classmates'.

Activity ❷ Dividing Fractions

1. When we write $\frac{3}{4} \div \frac{1}{4}$, we mean: "How many $\frac{1}{4}$s are in $\frac{3}{4}$?"
Complete the division sentence $\frac{3}{4} \div \frac{1}{4} = ■$
Complete the multiplication sentence $\frac{3}{4} \times \frac{4}{1} = ■$

2. How many $\frac{1}{10}$s are in $\frac{3}{5}$?
Write the division sentence.
Complete the multiplication sentence $\frac{3}{5} \times \frac{10}{1} = ■$

3. Draw a diagram to show each of the following.
Write each division sentence.
Write a multiplication sentence for each.

a) How many $\frac{1}{8}$s are in $\frac{5}{8}$?

b) How many $\frac{1}{12}$s are in $\frac{5}{6}$?

c) How many $\frac{1}{10}$s are in $\frac{1}{2}$?

4. Write a rule for dividing fractions.
Compare your rule with your classmates'.

4.3 Multiplying Rational Numbers

Activity: Study the Table

The value of one share of a company's stock changes depending on what other people are willing to pay. The following table shows the change in value of some stocks in one day.

Stock	Change
Sony	-1
Coca-Cola	$+\frac{1}{2}$
Nike	$-\frac{1}{4}$
Reebok	$-\frac{3}{4}$
Disney	$-\frac{1}{2}$

Inquire

1. By how much did the value of 500 shares of Coca-Cola increase?

2. By how much did the value of 1000 shares of Disney decrease?

3. By how much did the value of 400 shares of Reebok decrease?

4. Did the value of 500 shares of Nike stock increase or decrease during the day? By how much?

When multiplying rational numbers in decimal form, multiply as you did with decimals and use the *sign rules* you used with integers.

Example 1

Multiply $0.67 \times (-4.3)$.

Solution

$0.67 \times (-4.3)$
$= -2.881$

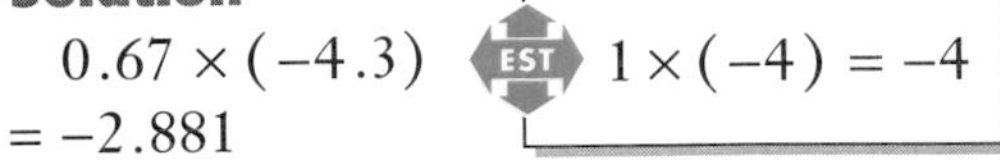

When multiplying rational numbers in fraction form, use the rule you used to multiply fractions and the *sign rules* you used with integers.

Example 2

Multiply.

a) $\frac{1}{3} \times \frac{3}{5}$ **b)** $\frac{-4}{7} \times (-3)$ **c)** $\frac{-3}{-4} \times \frac{2}{-5}$ **d)** $-2\frac{1}{4} \times 1\frac{2}{3}$

Solution

a) $\frac{1}{3} \times \frac{3}{5}$
$= \frac{3}{15}$
$= \frac{1}{5}$

b) $\frac{-4}{7} \times (-3)$
$= \frac{12}{7}$

c) $\frac{-3}{-4} \times \frac{2}{-5}$
$= \frac{-6}{20}$
$= \frac{-3}{10}$

d) $-2\frac{1}{4} \times 1\frac{2}{3}$
$= \frac{-9}{4} \times \frac{5}{3}$
$= \frac{-45}{12}$
$= \frac{-15}{4}$

You can use a calculator to multiply rational numbers.

Example 3

Multiply $2\frac{1}{4} \times \left(-1\frac{1}{2}\right)$.

Solution

$2\frac{1}{4} \times \left(-1\frac{1}{2}\right)$
$= \frac{9}{4} \times \frac{-3}{2}$
$= \frac{-27}{8}$
$= -3\frac{3}{8}$

C 2 . 25 × 1 . 5 +/− = −3.375

EST $2 \times (-2) = -4$

Some calculators have fraction keys.

To multiply $2\frac{1}{4} \times \left(-1\frac{1}{2}\right)$ on this type of calculator, press,

2 a b/c 1 a b/c 4 × 1 a b/c 1 a b/c 2 +/− = −3⌟3⌟8

Practice

Write a multiplication sentence to represent each diagram.

1. **2.** **3.**

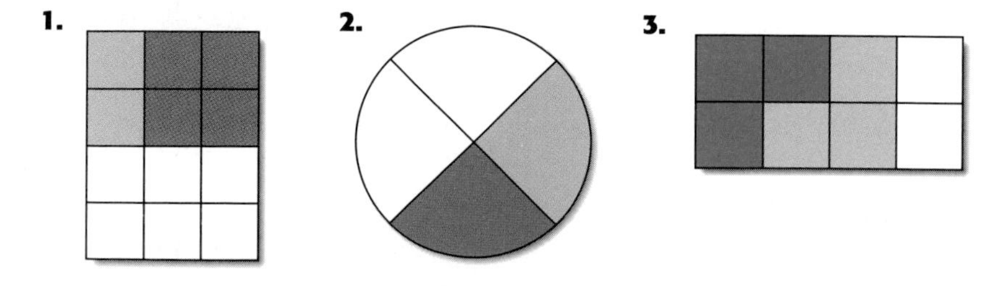

Estimate each product.

4. $-8\frac{1}{2} \times \left(-1\frac{1}{4}\right)$

5. $-6\frac{2}{3} \times 3\frac{1}{2}$

6. $5\frac{1}{2} \times 2\frac{1}{4}$

7. $3\frac{1}{5} \times \left(-2\frac{5}{6}\right)$

Simplify.

8. $-\frac{1}{4} \times \frac{3}{5}$

9. $-\frac{2}{7} \times \left(-\frac{7}{2}\right)$

10. $\frac{3}{4} \times \left(-\frac{1}{5}\right)$

11. $\frac{1}{8} \times \frac{2}{3}$

12. $\frac{1}{3} \times \left(-\frac{1}{4}\right)$

13. $\frac{1}{2} \times \left(-\frac{3}{4}\right)$

14. $-\frac{-18}{5} \times \left(-\frac{1}{2}\right)$

15. $-\frac{2}{3} \times \left(-\frac{3}{4}\right)$

16. $\frac{4}{4} \times \frac{1}{4}$

Simplify.

17. $1\frac{1}{2} \times 3\frac{1}{4}$

18. $0 \times \left(-\frac{3}{4}\right)$

19. $-\frac{3}{4} \times 2\frac{1}{2}$

20. $-1\frac{1}{2} \times \left(-1\frac{2}{3}\right)$

21. $\frac{7}{2} \times 1\frac{1}{4}$

22. $\frac{1}{10} \times 100$

23. $1 \times \left(-\frac{7}{8}\right)$

24. $18 \times \left(-\frac{2}{9}\right)$

25. $2\frac{2}{3} \times \frac{1}{2}$

26. Evaluate when $p = \frac{1}{2}$, $q = \frac{3}{5}$, and $r = -\frac{3}{4}$.

a) $p \times q$ **b)** $q \times r$ **c)** $p \times q \times r$

Problems and Applications

27. The price of General Motors stock fell by $\frac{3}{8}$ one day. What was the change in value of 400 shares of General Motors stock?

28. Martina can ride 11.5 km/h on her mountain bike. What distance can she ride in 3.25 h?

29. The temperature in Halifax changed by −1.3°C/h overnight. What was the temperature change from 01:00 to 06:00?

30. How many hours are there in $3\frac{1}{4}$ weeks?

31. In 1993, about 6600 satellites were orbiting the Earth. There were also about $2\frac{3}{11}$ times that number of pieces of junk in orbit. The junk included used rockets and debris from explosions in space. About how many pieces of junk were orbiting the Earth?

32. It is estimated that the loon population today is about $\frac{1}{2}$ what it was 10 years ago.

a) If the population 10 years ago was approximately one million, what is the loon population today?

b) What factors might be responsible for this decline in the loon population?

33. Decide whether each statement is always true, sometimes true, or never true. Explain.

a) The square of a non-zero rational number is negative.

b) The product of 2 negative rational numbers is greater than either of them.

c) The product of 2 rational numbers is not zero.

34. Work with a classmate. Research the cost of advertising space in your local newspaper or your favourite magazine. Determine the cost of a $\frac{1}{4}$-page, $\frac{1}{2}$-page, and $\frac{1}{8}$-page advertisement. List any factors other than size that might affect the cost of an advertisement.

DESIGN POWER

How many triangles can you find in this figure?

Make up a similar type of design using a different geometric shape. Ask a classmate to solve it.

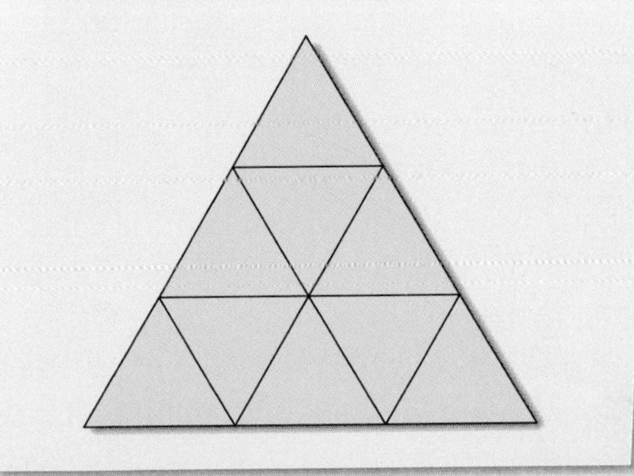

4.4 Dividing Rational Numbers

Every day, the newspaper compares the Canadian dollar with the U.S. dollar. If the Canadian dollar is quoted as 80.03¢ U.S., one Canadian dollar (100 cents) will buy approximately 80 cents of U.S. money.

Activity: Complete the Table

Suppose you want to buy some U.S. money to take on vacation. You would like to know how much Canadian money buys 1 U.S. dollar. To do this, you divide 100 Canadian cents by the number of U.S. cents they will buy.

Copy the table in your notebook and use your calculator to complete it. Round your answers to the nearest cent.

Value of 100 Canadian Cents in U.S. Cents	Division	Cost of 1 U.S. Dollar in Canadian Dollars
80.03	100 ÷ 80.03	
83.45		
91.34		
101.56		
77.34		
106.42		

Inquire

1. How much does it cost to buy $200 U.S. if the value of 1 Canadian dollar is 80.03¢ U.S.?

2. How much does it cost to buy $500 U.S. if the value of 1 Canadian dollar is 91.34¢ U.S.?

3. How much does it cost to buy $800 U.S. if the value of 1 Canadian dollar is 106.42¢ U.S.?

4. If you have $80 Canadian and the value of 1 Canadian dollar is 83.45¢ U.S., do you have enough money to buy $70 U.S.?

When dividing rational numbers in fraction form, use the same rule you used to divide fractions. Replace the divisor by the reciprocal and multiply. Use the same *sign rules* you used to divide integers.

Example 1

Divide.

a) $\frac{1}{4} \div \frac{1}{2}$ **b)** $3 \div \frac{1}{3}$ **c)** $-\frac{2}{5} \div \frac{3}{10}$ **d)** $1\frac{1}{4} \div (-3)$

Solution

a)
$$\frac{1}{4} \div \frac{1}{2} = \frac{1}{4} \times \frac{2}{1} = \frac{2}{4} = \frac{1}{2}$$

b)
$$3 \div \frac{1}{3} = 3 \times \frac{3}{1} = 9$$

c)
$$-\frac{2}{5} \div \frac{3}{10} = -\frac{2}{5} \times \frac{10}{3} = -\frac{20}{15} = -\frac{4}{3}$$

d)
$$1\frac{1}{4} \div (-3) = \frac{5}{4} \div \frac{-3}{1} = \frac{5}{4} \times \frac{1}{-3} = -\frac{5}{12}$$

You can use a calculator to divide rational numbers.

Example 2

Divide $-1\frac{1}{2} \div \frac{4}{7}$.

Solution

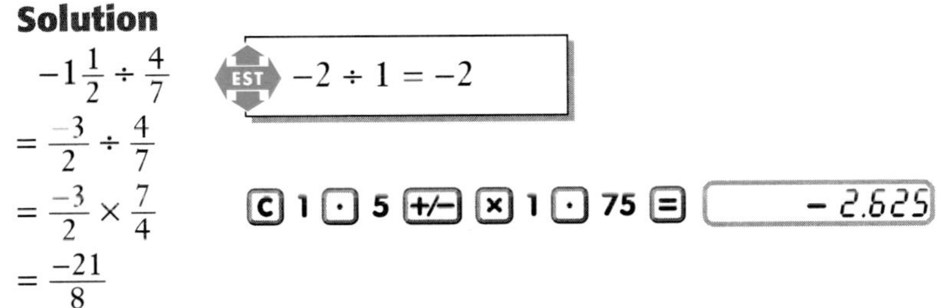

$$-1\frac{1}{2} \div \frac{4}{7} = \frac{-3}{2} \div \frac{4}{7} = \frac{-3}{2} \times \frac{7}{4} = \frac{-21}{8}$$

EST $-2 \div 1 = -2$

C 1 . 5 +/− × 1 . 75 = −2.625

To divide $-1\frac{1}{2} \div \frac{4}{7}$ using a calculator with a fraction key, press,

1 a b/c 1 a b/c 2 +/− ÷ 4 a b/c 7 = − 2⌟5⌟8

Practice

Write a division statement to represent each diagram.

1.

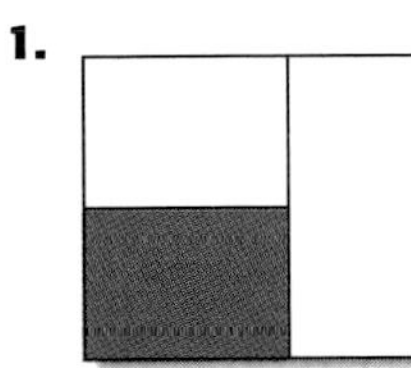

2.

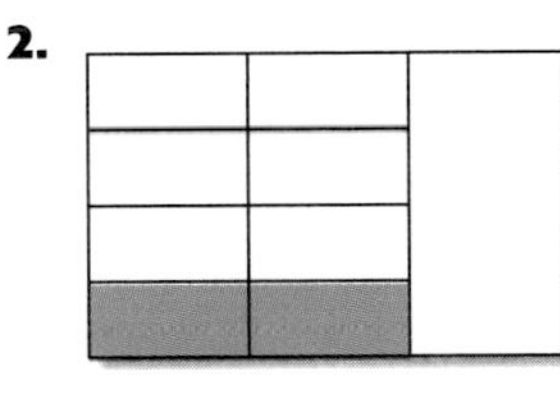

Estimate each quotient.

3. $3\frac{1}{2} \div 2\frac{1}{4}$ **4.** $-4\frac{1}{4} \div 1\frac{7}{8}$

5. $1\frac{2}{5} \div \left(-5\frac{1}{2}\right)$ **6.** $-2\frac{2}{3} \div \left(-1\frac{1}{2}\right)$

State the reciprocal.

7. $\frac{1}{4}$ **8.** $-\frac{1}{2}$ **9.** $\frac{-5}{3}$ **10.** $\frac{7}{2}$

11. 3 **12.** $1\frac{3}{4}$ **13.** 0 **14.** $-2\frac{1}{5}$

State whether each answer will be positive or negative.

15. $-\frac{1}{3} \div \frac{1}{2}$ **16.** $6 \div \frac{1}{3}$

17. $\frac{4}{5} \div 5$ **18.** $\frac{7}{9} \div \left(-\frac{1}{3}\right)$

19. $-4\frac{1}{2} \div \left(-1\frac{1}{2}\right)$ **20.** $\frac{2}{5} \div \left(-\frac{1}{6}\right)$

CONTINUED

Write a division statement and solve.

21. How many $\frac{1}{2}$s are in $\frac{3}{4}$?

22. How many $\frac{1}{4}$s are in $1\frac{3}{4}$?

23. How many $\frac{1}{3}$s are in $\frac{2}{3}$?

Simplify.

24. $2 \div \frac{2}{7}$ **25.** $-\frac{7}{8} \div \left(-\frac{1}{4}\right)$ **26.** $-10.5 \div 0.25$

27. $-50 \div 0.4$ **28.** $\frac{1}{2} \div 0$ **29.** $-\frac{4}{5} \div \left(-\frac{8}{10}\right)$

30. $2\frac{1}{3} \div \left(-\frac{1}{2}\right)$ **31.** $\frac{-7}{4} \div 4$ **32.** $-6.75 \div (-3)$

Problems and Applications

33. Cecil had $1\frac{2}{3}$ pizzas. He wanted each person to have $\frac{1}{3}$ of a pizza. How many friends could he share with?

34. Kerry drove the 299 km from Calgary to Edmonton in $3\frac{1}{4}$ hours. What was her average speed?

35. There are 100 pennies in a British pound. If a Canadian dollar is worth 49.32 British pennies, how many Canadian dollars does it cost to buy

a) a British pound?

b) 250 British pounds?

36. If the Swiss franc is worth 80.27¢ Canadian, how many Swiss francs can you buy for $100.00 Canadian?

37. A chemist is measuring the acid needed for an experiment. If she has $2\frac{1}{5}$ cylinders of acid and needs $\frac{1}{5}$ of a cylinder for each experiment, how many experiments can she do?

38. If $\frac{1}{2}$ a loaf of bread is cut into 8 equal slices, what fraction is each slice of the original loaf?

39. If a car travels 12.5 km on 1 L of gasoline, how many litres of gasoline does it use when it travels 100 km?

40. The value of an Inco share changed by $-1\frac{1}{4}$ one day. If Lisa's Inco shares decreased in value by $325.00 that day, how many Inco shares did she have?

41. The national debt is the amount of money the federal government owes. The debt stood at about $388 000 000 000 in 1991. The Canadian population was about 27 300 000. If all the people in Canada had shared equally in paying off the debt, how much would each of them have paid? Round your answer to the nearest $100.

42. What is the value of the quotient of 2 opposite rational numbers? Explain.

43. Write the following numbers. Compare your answers with a classmate's.

a) 2 rational numbers with a quotient of $\frac{5}{4}$

b) 2 rational numbers with a quotient of $-\frac{1}{2}$

c) 2 rational numbers whose quotient is an integer

44. Write a problem that requires division of rational numbers to solve it. Have a classmate solve your problem.

WORD POWER

Change the word READ to the word BOOK by changing one letter at a time. You must form a real word each time you change a letter. The best solution contains the fewest steps.

4.5 Adding and Subtracting Rational Numbers

The **opening price** of a stock tells you how much one share of a stock cost when the stock market opened for the day. The **closing price** of a stock tells you what one share was worth when the stock market closed for the day.

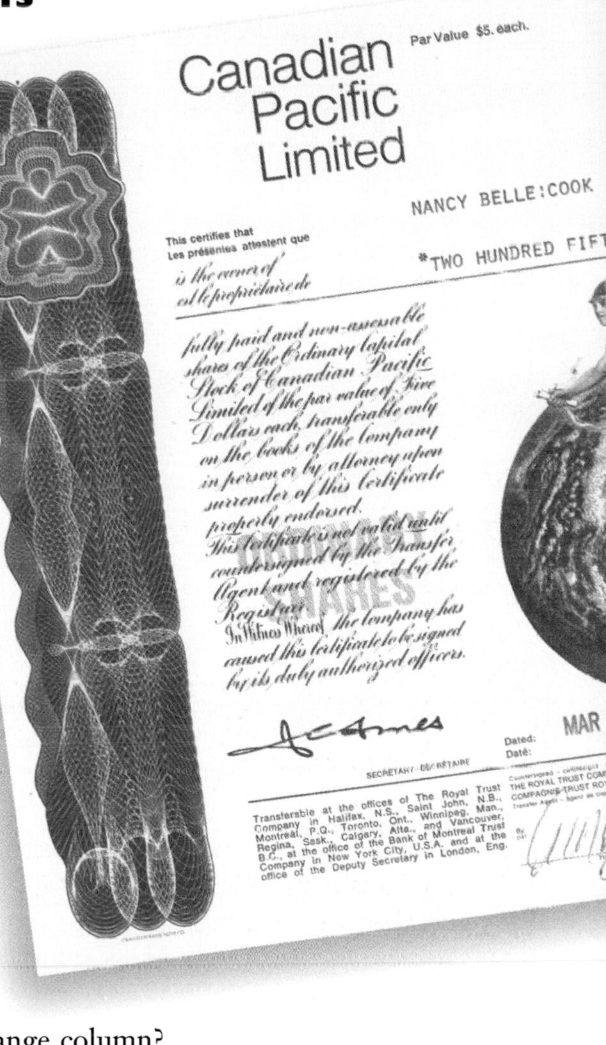

Activity: Complete the Table

The table shows the opening price for TCBY on a Monday and the daily change in price for 5 days. Copy and complete the table. The closing price on Monday becomes the opening price on Tuesday, and so on.

	Opening Price	Change	Closing Price
Monday	9	$-1\frac{1}{2}$	
Tuesday		$-\frac{3}{4}$	
Wednesday		$+\frac{1}{4}$	
Thursday		$-\frac{1}{4}$	
Friday		$-\frac{1}{2}$	

Inquire

1. What was the closing price on Friday?

2. What is the sum of the Change column for the week?

3. What number do you get when you subtract the closing price on Friday from the opening price on Monday?

4. How does this number compare with the sum of the Change column?

5. What number do you get when you subtract the opening price on Monday from the closing price on Friday?

6. How does this number compare with the sum of the Change column? Explain.

When adding or subtracting rational numbers in decimal form, add or subtract as with decimals and use the *sign rules* you used for integers.

Example 1

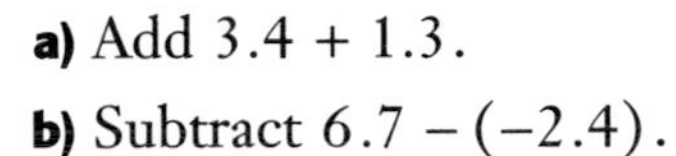

a) Add $3.4 + 1.3$.

b) Subtract $6.7 - (-2.4)$.

Solution

a) $3.4 + 1.3$
$= 4.7$

EST $3 + 1 = 4$

b) $6.7 - (-2.4)$
$= 6.7 + 2.4$
$= 9.1$

EST $7 + 2 = 9$

CONTINUED

When adding or subtracting rational numbers in fraction form, use the same rules you used for adding or subtracting fractions and use the *sign rules* you used for integers.

Example 2

Simplify.

a) $-\frac{7}{8} + \frac{3}{4}$

b) $-3 - 1\frac{2}{3}$

Solution

a) $-\frac{7}{8} + \frac{3}{4}$

$= -\frac{7}{8} + \frac{6}{8}$

$= -\frac{1}{8}$

EST $-1 + 1 = 0$

b) $-3 - 1\frac{2}{3}$

$= -\frac{9}{3} - \frac{5}{3}$

$= -\frac{14}{3}$

EST $-3 - 2 = -5$

You can use a calculator to add and subtract rational numbers.

Example 3

Simplify.

$1\frac{1}{4} - 2\frac{3}{8}$

Solution

$1\frac{1}{4} - 2\frac{3}{8}$

$= \frac{5}{4} - \frac{19}{8}$

$= \frac{10}{8} - \frac{19}{8}$

$= -\frac{9}{8}$

C 1 . 25 − 2 . 375 = −1.125

EST $1 - 2 = -1$

To subtract $1\frac{1}{4} - 2\frac{3}{8}$ on a calculator with a fraction key, press,

1 ab/c 1 ab/c 4 − 2 ab/c 3 ab/c 8 = −1⌟1⌟8

Practice

State the lowest common denominator for each pair of fractions.

1. $\frac{1}{2}, \frac{1}{5}$

2. $-\frac{3}{8}, -\frac{1}{4}$

3. $1\frac{2}{3}, -2\frac{3}{4}$

Estimate.

4. $1\frac{1}{3} + 2\frac{3}{4}$

5. $-1\frac{1}{2} + 3\frac{1}{8}$

6. $4\frac{2}{3} - 6\frac{1}{4}$

7. $-\frac{5}{6} - 2\frac{1}{4}$

8. $2\frac{1}{8} + (-2\frac{3}{4})$

9. $4\frac{1}{3} - (-2\frac{1}{2})$

Write true or false for each number sentence.

10. $\frac{1}{4} - \frac{7}{8} = -\frac{5}{8}$

11. $-2\frac{1}{3} + \frac{5}{6} = -\frac{2}{6}$

12. $\frac{7}{10} + \frac{11}{20} = \frac{18}{20}$

13. $-\frac{11}{30} - \frac{7}{6} = -\frac{23}{15}$

Simplify.

14. $\frac{1}{4} - 2\frac{1}{3}$

15. $-2.55 + 6.3$

16. $\frac{7}{9} + \frac{1}{3}$

17. $\frac{9}{10} - \frac{3}{10}$

18. $8.83 - 9.75$

19. $\frac{3}{5} - \frac{1}{4}$

20. $-1\frac{4}{5} - \frac{1}{10}$

21. $-\frac{5}{8} + 2\frac{3}{16}$

22. $-4\frac{8}{9} - 2\frac{2}{3}$

23. $-\frac{1}{4} + \frac{21}{100}$

24. $6.23 - (-3.65)$

25. $-\frac{3}{10} + \frac{53}{100}$

26. $1 - \frac{1}{100}$

27. $-8.93 + (-5.63)$

28. $-4.61 - 1.84$

29. $-3.65 + 2.99$

30. $2\frac{1}{4} - (-1\frac{1}{2})$

31. $3\frac{1}{2} - 5\frac{1}{3}$

Problems and Applications

32. The nightly low temperatures one week in Winnipeg were $-1.5°C$, $+2.1°C$, $-3.8°C$, $-2.2°C$, $+4.3°C$, $-5.3°C$, and $-6.9°C$. What was the average low temperature that week?

33. Shares of ABC Foods opened at $9\frac{1}{8}$ and closed at $7\frac{5}{8}$. What was the change that day?

34. Six friends bought 3 pizzas. Each person ate $\frac{3}{8}$ of a pizza. How much pizza was left over?

35. On their vacation, the Reikos spent $\frac{1}{4}$ of their money on gas, $\frac{1}{4}$ on food and lodging, and $\frac{1}{8}$ on tourist attractions. What fraction of their money did they spend? If they started with \$840.00, how much money did they have left?

36. In February 1993, Canada's Kate Pace won the gold medal at the World Alpine Ski Championships. Her winning time was 1 min 27.38 s. She finished $\frac{28}{100}$ of a second ahead of Astrid Loedemal of Norway. What was Astrid's time?

37. About $\frac{2}{5}$ of Canada's gold production comes from Ontario. About $\frac{3}{10}$ comes from Quebec and $\frac{1}{10}$ from British Columbia. What fraction of Canada's gold production comes from the rest of the country? Write your answer in lowest terms.

38. a) If the sum of 2 different rational numbers is zero, what can you say about the numbers?

b) If the sum of 2 equal rational numbers is zero, what can you say about the numbers?

39. Decide whether each statement is always true, sometimes true, or never true. Explain.

a) The difference between 2 negative rational numbers is negative.

b) The sum of 2 non-zero rational numbers is greater than either of the numbers.

c) The difference between 2 rational numbers is an integer.

40. Write the following numbers. Compare your answers with a classmate's.

a) 2 rational numbers with a sum of -0.6

b) 3 rational numbers with a sum of $-2\frac{1}{4}$

c) 2 rational numbers with a difference of $\frac{5}{8}$

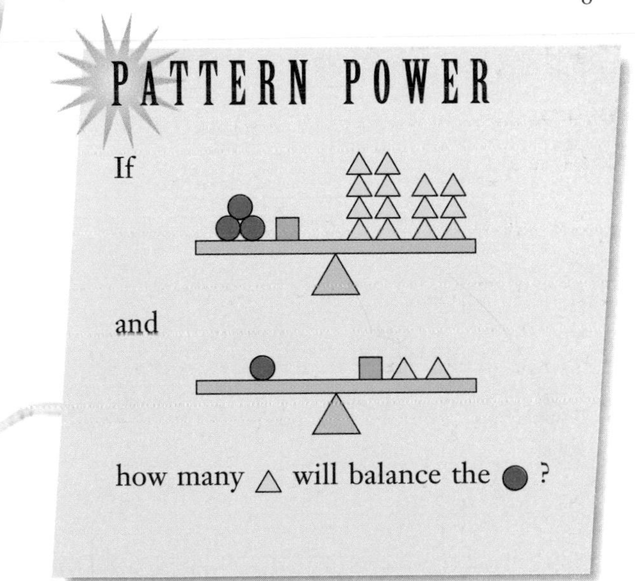
PATTERN POWER
If
and
how many △ will balance the ● ?

Solving Problems Using Spreadsheets

A spreadsheet consists of rows and columns. It allows you to quickly analyze and change data. A spreadsheet can be an important problem solving tool.

The following problem solving activities show both the power and simplicity of a spreadsheet.

Activity 1

A collection of 48 coins has a total value of $7.32. The collection consists of pennies, nickels, dimes, quarters, and half dollars. How many coins of each type are there?

1. Set up the following spreadsheet. Each formula in Column C is typed into that particular cell as indicated.

	A	B	C
1	**Coin**	**Number**	**Value**
2	Penny		0.01*B2
3	Nickel		0.05*B3
4	Dime		0.1*B4
5	Quarter		0.25*B5
6	Half Dollar		0.5*B6
7	Totals	@SUM(B2. .B6)	@SUM(C2. .C6)

2. Enter the number of each type of coin into Column B of the spreadsheet until the total number of coins is 48, and the total value of all the coins is $7.32.

3. What strategy did you use to solve the problem?

4. Is there more than one solution?

Activity ❷

You can use a spreadsheet to help select a store for shopping.

1. Obtain advertising flyers from several stores.

2. Find several items that are common to all stores.

3. Set up a spreadsheet similar to the one shown to determine which store gives you the best value for your money.

	A	B	C	D
1	ITEM	STORE 1	STORE 2	STORE 3
2		Cost	Cost	Cost
3	Toothpaste			
4	Shampoo			
.				
.				
	Totals			

Activity ❸

1. Make up a problem of your own that can be solved using a spreadsheet.

2. Give your problem to a classmate to solve.

4.6 Order of Operations

The order in which people perform some operations in their daily activities is important. For example, a person cannot drive a car before starting the engine or make a call on a pay phone without first depositing a quarter.

The order in which operations are performed in mathematics is also important.

Activity: Solve the Problem

Suni worked 10 h at Champions Fitness Centre on the holiday weekend. She worked at the reception desk for $4\frac{1}{2}$ h and earned \$8.00/h. She worked for $5\frac{1}{2}$ h as manager for the centre and earned \$10.50/h. She received a bonus of \$20.00 for working on the holiday weekend. How much did she earn altogether?

Inquire

1. How much did she earn at the reception desk?

2. How much did she earn as manager?

3. What were her total earnings?

4. The computer calculated her earnings by evaluating the expression $4.5 \times 8.00 + 5.5 \times 10.50 + 20.00$. Which operation does the computer do first?

5. Does adding the hourly rates, multiplying by 10 h, and then adding \$20.00 give the same result? Explain.

The order of operations followed for whole numbers, decimals, and integers also applies to rational numbers. Remember the order with the acronym **BEDMAS**.

Example

Evaluate.

a) $\frac{7}{10} - \frac{1}{2} \times \frac{2}{5}$

b) $-6.2 \div (0.5)^2 + 1.6$

c) $\frac{1}{2} - 3(\frac{1}{3} + 2)$

Solution

a)

$$\begin{aligned} &\frac{7}{10} - \frac{1}{2} \times \frac{2}{5} && \textbf{multiply} \\ &= \frac{7}{10} - \frac{2}{10} && \textbf{subtract} \\ &= \frac{5}{10} \\ &= \frac{1}{2} \end{aligned}$$

b)

$$\begin{aligned} &-6.2 \div (0.5)^2 + 1.6 && \textbf{exponent} \\ &= -6.2 \div 0.25 + 1.6 && \textbf{divide} \\ &= -24.8 + 1.6 && \textbf{add} \\ &= -23.2 \end{aligned}$$

c)

$$\begin{aligned} &\frac{1}{2} - 3\left(\frac{1}{3} + 2\right) && \textbf{brackets} \\ &= \frac{1}{2} - 3\left(\frac{7}{3}\right) && \textbf{multiply} \\ &= \frac{1}{2} - 7 && \textbf{subtract} \\ &= -6\frac{1}{2} \end{aligned}$$

Practice

Evaluate.

1. $8 + 6 \times 2$

2. $-7 \times 3 - 8$

3. $18 + 6 \div 2$

4. $\frac{18 + 6}{5 + 1} - 4$

5. $9.8 - 1.4 \div 0.7$

6. $7(3 - 2 + 6)$

7. $-4[8 + 7(6 - 3)]$

8. $\frac{7}{10} - \frac{1}{10} \div \frac{1}{10}$

9. $-3 + \left(\frac{3}{4} - \frac{1}{4}\right)^2$

10. $-3 + (6 - 2)^2$

Add brackets to make each statement true.

11. $\frac{2}{3} + 4 \times \frac{1}{2} + \frac{1}{4} \div 3 = \frac{5}{3}$

12. $0.5^2 - 0.1 \times 8 \div 2 = 0.6$

13. $-2 \times 18.5 - 6.3 \div 4 = -6.1$

14. Evaluate each expression for $a = 3$, $b = 6$, and $c = -\frac{1}{2}$.

a) $a(b + c)$ **b)** $ab + ac$

c) $b^2 + a$ **d)** $a \times b \times c$

e) $b \div c \times a$ **f)** $b \times c + 3$

g) $a \div b \div c$ **h)** $a \div (b \div c)$

15. Evaluate each expression for $x = 0.5$, $y = 7.2$, and $z = -1.8$.

a) $x(y + z)$ **b)** $xy + xz$

c) $y \div z \times x$ **d)** $x \times y \times z$

e) $x^2 - z$ **f)** $x^2 + y^2$

g) $x^2 - y^2$ **h)** $y \div x + z$

Problems and Applications

16. Copy the cross-number puzzle into your notebook. Evaluate the following expressions to complete the puzzle.

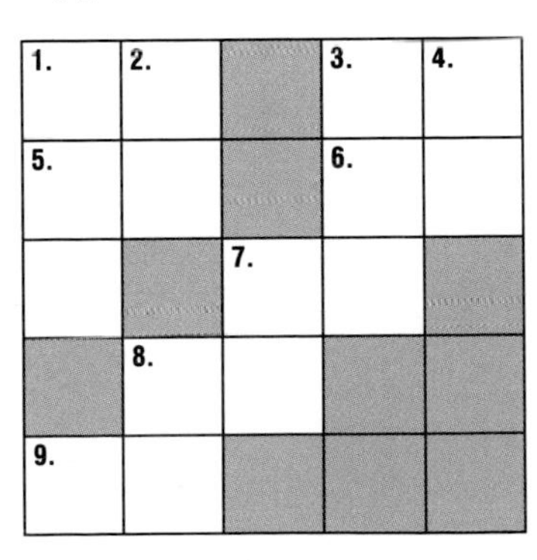

Across

1. $17 + 5 - 3$

3. $[(6 - 3)^2 - 4]^2$

5. $5^2 - 16 + 3$

6. $(75 - 10) - (23 - 8)$

7. $96 - 3(4.2 - 0.2)$

8. $9 + 4 \div \frac{1}{6} \times 3$

9. $(19.5 - 6.5)^2 - (8.4 - 4.4) - 7 \times 10$

Down

1. $323 - 2(121 - 16)$

2. $(12 - 2)^2 + 6 - 14$

3. $2^8 - 2$

4. $10 \div \frac{1}{2} \times 2.5$

7. $(6 \div 2 \times 3)^2$

8. $5(18 - 3) + 10$

17. Two of the following skill-testing questions were answered incorrectly. Identify the incorrect answers and correct them.

a) $8 \div 2 \times 3 + (6 - 1) \div 5 = 13$

b) $7 - 8 \times 0.5 - 0.5 = 7$

c) $2^2 \times 3 - 1 = 8$

18. Make up your own skill-testing question. Have a classmate solve your question.

19. For each part, compare your expressions with a classmate's.

a) Here is a way of using four $\frac{1}{2}$s and the order of operations to make the number 1.

$$\frac{\frac{1}{2} + \frac{1}{2}}{\frac{1}{2} + \frac{1}{2}} = 1$$

Write 3 other ways of using four $\frac{1}{2}$s to make the number 1.

b) Write 3 ways of using four $-\frac{1}{2}$s to make the number 1.

c) Use four $\frac{1}{2}$s to make the number 3.

d) Use four $-\frac{1}{2}$s to make the number 3.

NUMBER POWER

On what day in which month and year will you have lived for at least 1 billion seconds?

4.7 Rational Numbers and Formulas

Hailstones are balls of ice that grow as they are held up in the clouds by thunderstorm updrafts. While they are held up, supercooled water drops hit them and freeze, causing the hailstones to grow. Large hailstones can fall at speeds of up to 150 km/h and can have masses of up to 2 kg. Because of the damage that can occur during a hailstorm, weather forecasters try to warn people when hail is probable.

Activity: Use a Formula

Weather forecasters can use a formula to estimate the size of hailstones if the speed of the thunderstorm updraft is known. The formula is

$$d = 0.05s$$

where s is the speed of the updraft in km/h, and d is the diameter of a hailstone in centimetres.

Inquire

1. What is the predicted diameter of a hailstone when the updraft speed is 30 km/h? 80 km/h? 100 km/h?

2. The smallest hailstone possible is about 0.5 cm in diameter. What is the approximate updraft speed to produce this hailstone?

3. How might weather forecasters determine the speed of an updraft?

Example

Tidal waves are caused by underwater earthquakes or large storms at sea. Tidal waves can be very destructive if they crash into populated areas. The speed of a tidal wave in metres per second can be found using the formula

$$s = 3.1 \times \sqrt{d}$$

where d is the depth of the ocean in metres. To the nearest tenth of a metre per second, what is the speed of a tidal wave in 300 m of water? 50 m of water?

Solution

$s = 3.1 \times \sqrt{d}$

In 300 m of water

$s = 3.1 \times \sqrt{300}$

$\doteq 53.7$

In 300 m of water, the speed is about 53.7 m/s.

In 50 m of water

$s = 3.1 \times \sqrt{50}$

$\doteq 21.9$

In 50 m of water, the speed is about 21.9 m/s.

Problems and Applications

1. The stopping distance a car requires in good road conditions is given by the formula

$$d = 0.4s + 0.002s^2$$

where d represents the approximate stopping distance in metres and s represents the speed in kilometres per hour. Calculate the distance required to stop at the following speeds.

a) 50 km/h **b)** 80 km/h

c) 100 km/h **d)** 120 km/h

2. The drama club decided to raise money by performing in a drama marathon. Pledges averaged $5.50 per actor for every hour of acting. Write a formula to calculate the amount each member could earn for the club. Let h represent the number of hours each member performs and v represent the amount of money earned.

3. The Purple Cab Company needs a formula to program into its computers to calculate fares. The manager has decided that a minimum fare of $3.50 must be charged. There is also a charge of $0.70 per half kilometre.

a) Write the formula for the total fare.

b) Determine the cost of a 7.5 km ride.

4. A rental company charges for lawn ornament rentals according to the formula

$$C = 2.40n + 2.00$$

where n is the number of ornaments rented and C is the cost in dollars. Calculate the cost of renting 12 pink flamingos for a birthday party.

5. The face value, f dollars, of a collection of dimes and nickels is given by the equation

$$f = 0.1d + 0.05n$$

where d is the number of dimes and n is the number of nickels. What is the face value of 35 dimes and 48 nickels?

6. The following formula calculates the time, t seconds, it takes for an object to fall from a height, h metres.

$$t = \sqrt{h \div 4.9}$$

The High Level Bridge in Lethbridge, Alberta, is the highest railway bridge of its type in the world. It stands 96 m above the river valley below. How long does it take for an object to fall from the top of the High Level Bridge to the ground below it?

7. The period of revolution of a planet around the sun is the time it takes for the planet to complete one orbit of the sun. The period, P years, is given by Kepler's third law

$$P^2 = D^3$$

where D is the average distance of the planet from the sun in astronomical units (AU). One astronomical unit is the average distance of the Earth from the sun. Use the average distance from the sun to find the period of revolution for each of the following planets to the nearest tenth of a year.

a) Jupiter; 5.20 AU

b) Mercury; 0.387 AU

c) Uranus; 19.2 AU

8. Write a problem that requires a formula. Have a classmate solve your problem.

CALCULATOR POWER

Here is a way to write the digits from 1 to 9 in order, so that when + or − signs are included the answer is 100.

$$123 + 45 - 67 + 8 - 9 = 100$$

Find 4 other ways to do this.

Sound

Sound waves or vibrations are measured by their frequency, period, amplitude, and wavelength.

The **amplitude** is the height of a wave. The **wavelength** is the length of a complete wave. The **period** is the time taken for 1 complete wave or vibration to pass a fixed point. The period is measured in fractions of a second. The **frequency** is the number of waves or vibrations that pass a fixed point in 1 s. Frequency is measured in vibrations per second.

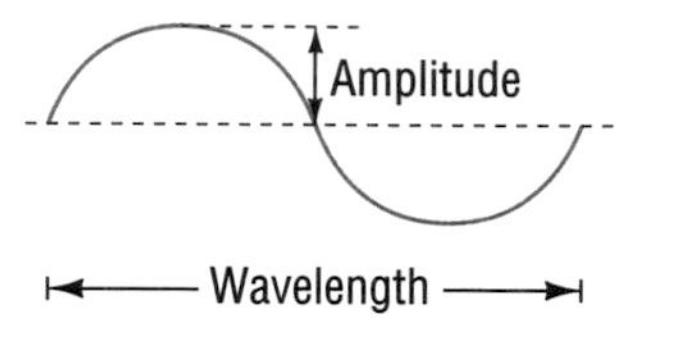

Activity ❶ Period and Frequency

The diagram shows the number of waves that pass a fixed point in 1 s. The period is $\frac{1}{2}$s because it takes $\frac{1}{2}$s for 1 complete wave to pass a fixed point. The frequency is 2 vibrations/s because 2 complete waves pass a fixed point every second.

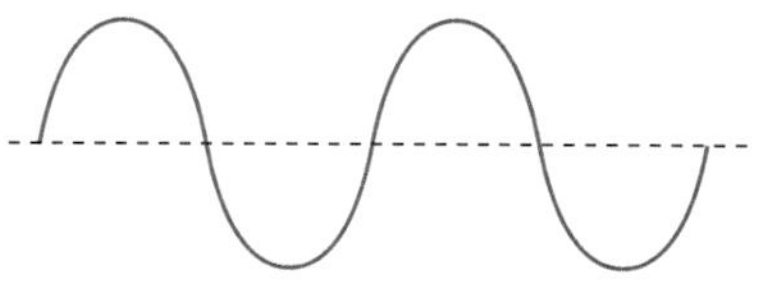

1. The diagram below shows the number of waves that pass a fixed point in 1 s. What is the period and frequency of this wave?

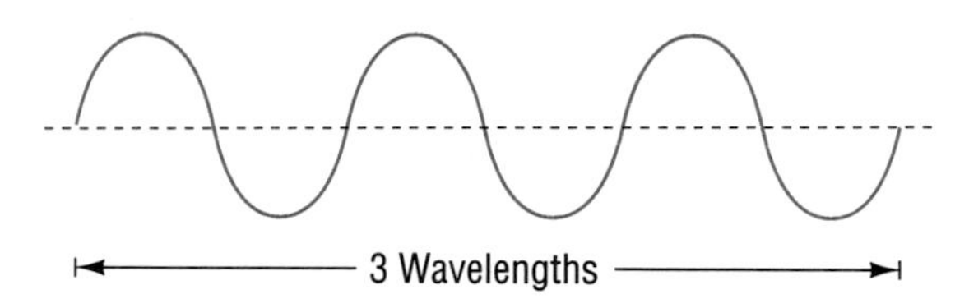

2. A wave has a period of $\frac{1}{4}$s. Draw a diagram to show the number of waves that pass a fixed point in 1 s.

3. A wave has a frequency of 5 vibrations/s. Draw a diagram to show the number of waves that pass a fixed point in 1 s.

4. A wave has a period of $\frac{1}{10}$s. What is the frequency?

5. A wave has a frequency of 8 vibrations/s. What is the period?

6. "Concert A," the A above middle C on a piano, has a period of $\frac{1}{440}$s. What is the frequency?

7. If you know the period of a wave, how do you find the frequency?

8. If you know the frequency of a wave, how do you find the period?

9. Let C be the period and f be the frequency. Write a formula to calculate each of the following.

a) the period from the frequency **b)** the frequency from the period

Activity ❷ Radio Broadcasting

Commercial radio stations broadcast on 2 bands, AM and FM.

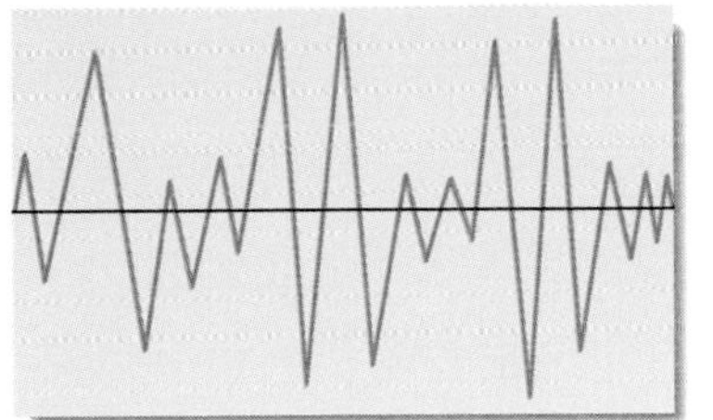

Amplitude Modulation (AM)

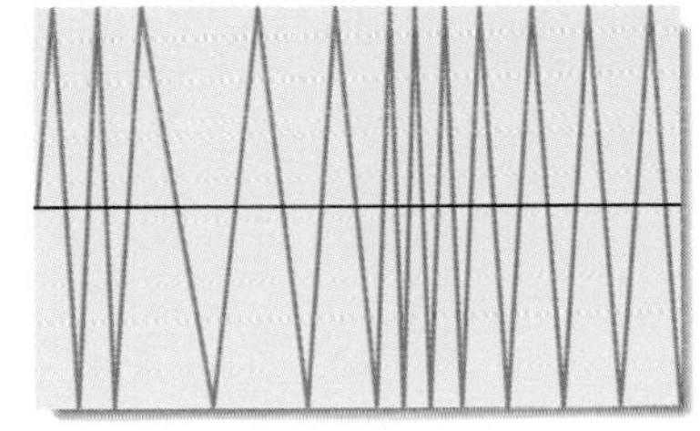

Frequency Modulation (FM)

AM stations broadcast on frequencies between 535 000 vibrations/s and 1 705 000 vibrations/s. The frequency of an AM station is found by multiplying the station number by 1000.

For example, 1050 AM broadcasts at 1050×1000 or 1 050 000 vibrations/s.

FM stations broadcast on frequencies between 88 000 000 and 108 000 000 vibrations/s. The frequency of an FM station is found by multiplying the station number by 1 000 000. For example, 99.9 FM broadcasts at $99.9 \times 1\ 000\ 000$ or 99 900 000 vibrations/s.

1. What are the frequency and period of the radio waves transmitted by radio station 900 AM? 101.2 FM?

2. What are the frequency and period of the radio waves transmitted by a radio station in your area?

3. A radio station is transmitting waves that have a period of $\frac{1}{1\ 320\ 000}$ s. What is the frequency? Is it an AM or FM station? What is the station number?

4. A radio station is transmitting waves that have a period of $\frac{1}{101\ 300\ 000}$ s. What is the frequency? Is it an AM or FM station? What is the station number?

Activity ❸ The Sounds You Hear

1. Which notes on a piano have a higher frequency, high sounding notes or low sounding notes?

2. There are some whistles that dogs can hear, but you cannot. Why? Are there sounds you can hear that a dog cannot?

3. Research the range of sounds that humans, other mammals, fish, and birds can hear. Display your findings on a graph.

Review

Which numbers are rational?

1. $\frac{8}{7}$ **2.** $\frac{0}{3}$ **3.** -1.2

4. -3.3 **5.** $1\frac{5}{7}$ **6.** π

7. $7.2121...$ **8.** $\frac{-7}{0}$ **9.** $\sqrt{2}$

10. 8.076 **11.** 0.6 **12.** $\frac{51}{100}$

Write each of the following as the quotient of 2 integers in lowest terms.

13. -0.7 **14.** 0.45 **15.** -3

16. $-1\frac{1}{3}$ **17.** $\frac{2}{3}$ **18.** 0.54

Write in order from smallest to largest.

19. $\frac{1}{2}, \frac{7}{8}, \frac{3}{4}$ **20.** $\frac{-3}{10}, \frac{-1}{2}, \frac{7}{-5}$

21. $0.71, \frac{7}{10}, 0.\overline{7}$ **22.** $0.2, 0.02, -0.2$

Replace each ● by <, >, or = to make each statement true.

23. $\frac{3}{8}$ ● $\frac{9}{24}$ **24.** -2 ● $\frac{-4}{3}$

25. $\frac{3}{4}$ ● $\frac{2}{3}$ **26.** 0.3 ● $\frac{1}{3}$

27. $\frac{11}{6}$ ● 1.9 **28.** $\frac{-3}{8}$ ● $\frac{-2}{3}$

29. $1\frac{1}{2}$ ● $\frac{3}{2}$ **30.** $\sqrt{5}$ ● 2.413

Write as a decimal.

31. $\frac{-3}{5}$ **32.** $\frac{-11}{3}$ **33.** $\frac{4}{5}$

34. $\frac{5}{8}$ **35.** $\frac{1}{7}$ **36.** $\frac{-3}{4}$

Write in the form $\frac{a}{b}$ and reduce to lowest terms.

37. 0.5 **38.** -0.24 **39.** 1.2

40. -0.06 **41.** 0.625 **42.** 0.6

Solve only the questions in which the answers are from 1 to 5, inclusive.

43. $\frac{1}{2} \times 7$ **44.** $-\frac{3}{4} \div (-4)$

45. $18 - 12\frac{1}{2}$ **46.** $\frac{3}{4} + \frac{7}{8}$

47. $21 \div \frac{7}{3}$ **48.** $\frac{11}{3} + \frac{1}{6}$

49. $\frac{3}{100} \div 1$ **50.** $\frac{5}{6} - \frac{1}{2}$

51. $-3.6 + 5.4$ **52.** -7×0.5

53. $9.1 - 3.7$ **54.** $-6 - \frac{1}{2}$

Simplify.

55. $\left(\frac{7}{2} + \frac{1}{4}\right) \times \frac{1}{2}$ **56.** $\frac{8}{3} - 3 \times \frac{1}{6}$

57. $7.16 - 1.16 \div 2$ **58.** $9^2 - 6 + 2.5$

59. $11.1 \times 100 \div 10$ **60.** $(18 - 10)^2 \div \left(-\frac{1}{8}\right)$

61. $\frac{1}{3} \times \frac{2}{5} \div 2$ **62.** $-7.2 - 7\frac{1}{5}$

63. $(1.14 - 2.14)^2$ **64.** $3 \div \frac{1}{3} \times \frac{1}{27}$

65. Evaluate when $m = -\frac{1}{2}$, $n = \frac{2}{3}$, and $p = -1$.

a) $m \times n \times p$ **b)** $n + p$

c) $m + n + p$ **d)** $m \times n \div p$

e) $m \div n$ **f)** $m \times p + n$

66. In Alberta's Kananaskis Country, there is a very small lake called Wedge Pond. This lake was created by hauling 200 000 m^3 of mud from the pond to the golf course. It took 33 000 truck loads to haul the mud. Each trip to the golf course was 3.5 km, one way. How many kilometres in total were driven to haul the mud?

67. Janelle is paid according to the formula

$$p = t \times r + 50$$

where t is the number of hours worked and r is the hourly rate of pay. How much does she earn if she works 3.5 h at $5.75/h?

68. Calculate the area of each figure.

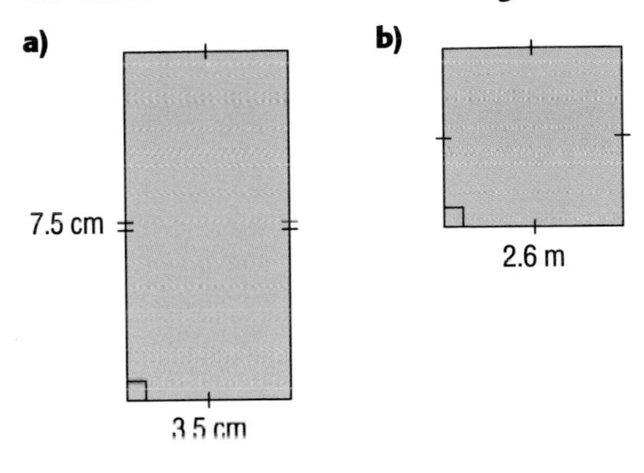

69. The formula that the Safety Taxi Co. uses to determine fares is

$$F = 2.50 + 0.125d$$

where F is the fare in dollars and d is the distance travelled in kilometres. Find the fare from Hyde Street to 34 Busy Road if the distance is 5.6 km.

70. Find today's closing price for ABC Magazine Publishing stock if it closed yesterday at $64\frac{1}{4}$ and the net change today was $-7\frac{1}{8}$.

Group Decision Making
The SCAMPER Technique

The writer B. Eberle used the mnemonic SCAMPER to help people expand their thinking during brainstorming sessions. The letters in SCAMPER each represent a different idea to think about.

S : Substitute – What if a thing or a person takes another's place?

C : Combine – What if you put things together or combine purposes?

A : Adapt – What if you adjust something? What else is this?

M : Modify, Magnify, Minify – What if you change the purpose? the size? the colour? the sound? the speed?

P : Put to Other Uses – What other uses are there?

E : Eliminate – What if you get rid of a part? a whole?

R : Rearrange – What if you change the order? turn something around? backwards? upside down?

Suppose you were designing a home entertainment centre. You might consider the following questions.

a) What if you *substitute* 8 television screens for 1?

b) What if you *combine* the room and a gym so you can be actively involved in the music or movies?

c) What if you *adapt* the ceiling to become television screens?

d) What if you *modify* the roof of your house so that it holds 4 satellite dishes?

e) What if you *eliminate* the walls of one room in the house?

f) What if you *rearrange* the floors in the house so that the top floor becomes the bottom floor?

1. In home groups, use this technique to design a classroom or a movie theatre.

2. Present your design to the class and explain how you used the SCAMPER technique.

3. Evaluate the technique.

Chapter Check

Which numbers are rational numbers?

1. -3.2 **2.** $\frac{22}{7}$ **3.** $\sqrt{13}$

4. $1\frac{1}{2}$ **5.** 0.75 **6.** π

Write in order from largest to smallest.

7. $\frac{1}{2}, \frac{4}{5}, \frac{7}{10}$ **8.** $-1\frac{1}{2}, -1\frac{1}{4}, -2$

9. $1.1, 1.12, \frac{8}{7}$ **10.** $\frac{4}{3}, \frac{3}{2}, 1.7$

Write as a fraction in lowest terms.

11. -0.35 **12.** 1.4

13. -1.3 **14.** 0.6

Write as a decimal.

15. $\frac{2}{7}$ **16.** $-\frac{13}{5}$

17. $3\frac{1}{3}$ **18.** $-2\frac{6}{11}$

Simplify.

19. $\frac{1}{2} + \frac{5}{6}$ **20.** -1.8×0.2

21. $-\frac{8}{3} \div \left(-\frac{1}{3}\right)$ **22.** $1\frac{1}{4} \div 2\frac{1}{3}$

Simplify.

23. $9 - \left(7 \times 2 \times \frac{1}{2}\right)$ **24.** $(8 - 2)^2 \div \frac{1}{6}$

25. $7.6 + 2 \times \left(\frac{1}{2} + \frac{1}{4}\right)$

26. Evaluate $x + y \div z$ when $x = \frac{1}{2}$, $y = \frac{1}{4}$, and $z = -\frac{1}{3}$.

27. Evaluate $s \times r \div t$ when $s = 0.75$, $r = -1.3$, and $t = 2.6$.

28. The cost of renting a car from a car rental agency is given by the formula

$$C = 50n + 0.25d$$

where C is the cost in dollars, n is the number of days the car is rented for, and d is the total distance driven in kilometres.

a) Calculate the cost of renting a car for 3 days and driving 350 km per day.

b) How much money do you have left for food and accommodations if you have $750 budgeted for the trip?

Reprinted by permission: Tribune Media Services

Using the Strategies

1. In 1 year, December had exactly 4 Tuesdays and 4 Saturdays. On what day did December 1 fall that year?

2. An apartment building has 6 floors. The top floor has 1 apartment. Each of the other floors has twice the number of apartments as the floor above it. How many apartments are there?

3. Fill in the squares in the diagram with the numbers 1 to 8 so that consecutive numbers are not adjacent in any direction—horizontally, vertically, or diagonally.

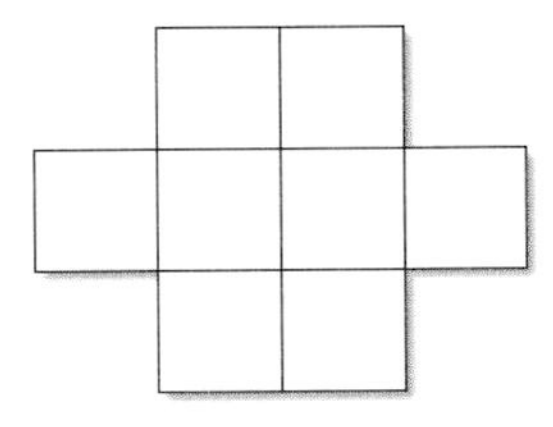

4. There are 6 h of lifeguard work available for the weekend. Alicia, Martin, and Rica have each agreed to work a whole number of hours. In how many different ways can the 6 h be divided so that each person works at least 1 h?

5. If you open a book and the product of the page numbers on the 2 facing pages is 45 156, what are the page numbers?

6. A rectangular floor is tiled with 36 square tiles. The tiles around the outside edge of the rectangle are red, and the tiles on the inside are white. How many red tiles are there? Is there more than 1 possible answer?

7. Use each of the digits 1, 3, 4, 6, 7, 8 only once to make the smallest possible sum. What is the sum?

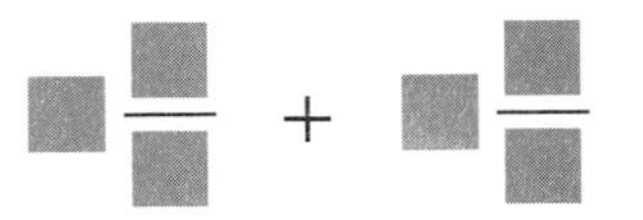

8. A coffee shop on the corner of a busy intersection is open 24 h a day. Sketch a graph of the number of customers in the shop versus the time of day.

9. Tania has a collection of compact discs. When she puts them in piles of 2, she has 1 left over. She also has 1 left over when she puts them in piles of 3 and piles of 4. She has none left over when she puts them in piles of 7. What is the smallest number of compact discs she can have?

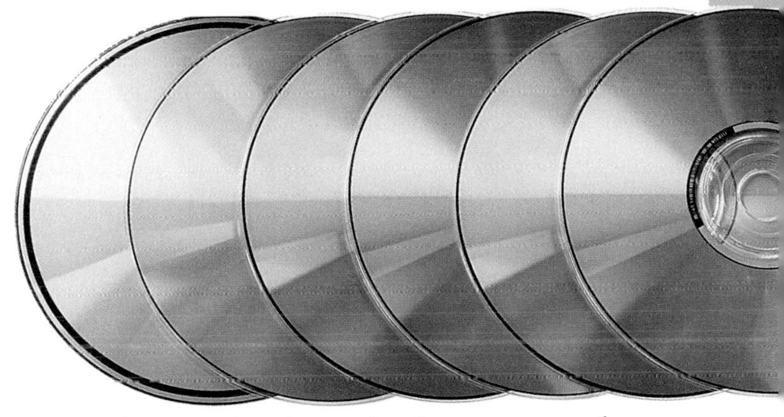

10. A shopper went into a hardware store and asked the clerk how much 1 cost. The clerk said "$1.00." The shopper then asked how much 10 would cost, and the clerk replied "$2.00." "Fine," the shopper replied, "I'll take 1035." The shopper then paid the clerk $4.00. What did the shopper buy?

DATA BANK

1. How many hours ahead of the time in St. John's, Newfoundland, is the time in Tokyo, Japan? Express your answer in decimal form.

2. How many times more moons does Saturn have than Neptune? Express your answer as a fraction in lowest terms and as a decimal.

Chapter 1

1. Terri is a police officer. Write 2 questions her supervisor might ask that would need
a) an estimate **b)** an exact answer

2. Write a problem where you would use
a) paper and pencil **b)** a calculator
c) a computer **d)** mental math

Estimate, then calculate.

3. $234 + 676 + 99$ **4.** $34.6 + 0.89 - 12.3$
5. 87×61 **6.** 3.4×120
7. $2400 \div 48$ **8.** $13.2 \div 5.5$
9. $0.7 - 0.003$ **10.** $340 - 239$

Evaluate.

11. $4.65 \times 10\ 000$ **12.** $320 \div 1000$
13. $23 \div 10\ 000$ **14.** 63×0.01
15. 123×1000 **16.** $0.4 \div 0.001$

Write in scientific notation.

17. 23 000 000 **18.** 8 800 000

Write in standard form.

19. 3.4×10^8 **20.** 4.0×10^6
21. 8.88×10^3 **22.** 3.01×10^{10}

Evaluate.

23. 3^3 **24.** 2^4 **25.** 5^2 **26.** 10^4

27. Evaluate for $x = 3$ and $y = 4$.
a) $3x + 5y$ **b)** $4xy - 2x - y$
c) $x^2 + 2xy + y^2$ **d)** $y - x - 1$

28. Evaluate for $x = 1.2$ and $y = 0.4$.
a) $4x - 2y$ **b)** $10xy - y$
c) $6y - x$ **d)** $4x + 6y - xy$

29. Sam bought 6 baseball cards for \$0.75 each, 3 hockey cards for \$1.50 each, and a card magazine for \$4.25. How much change should he get from a twenty dollar bill?

Chapter 2

1. List the integers from smallest to largest.
$-2, 4, -6, 0, -1, -7, 5$

2. Identify each integer.
a) 5 less than 3 **b)** 6 more than -1
c) 4 less than 2 **d)** 7 less than 0

3. List and graph.
a) integers less than 2 and greater than -1
b) integers less than -3 and greater than -8

Simplify.

4. $3 + (-7)$ **5.** $-4 - 3$
6. $5 \times (-2)$ **7.** $-3 \times (-6)$
8. $15 \div (-5)$ **9.** $7 - 9$
10. $-18 \div 6$ **11.** -4×7

Simplify.

12. $-3 + 4 - 7$ **13.** $-2 - 5 - 6$
14. $-2 \times (-2) + 4$ **15.** $-6 + (-3)(-4)$
16. $-10 \div 5 - 7$ **17.** $(-3)^2 - 5$
18. $-3^2 - 5$ **19.** $-6 - 12 \div (-6)$

Find the value of ■.

20. $-7 +$ ■ $= 2$ **21.** ■ $- (2) = 1$
22. $-24 \div$ ■ $= 3$ **23.** $-2 \times (-2) + 4 =$ ■
24. ■ $- (-2) = 4$ **25.** ■ $- 5 = 1$
26. ■ $\times (-2) = -8$ **27.** ■ $\div (-3) = 2$

Simplify.

28. $-4(-6) \div (-2)$ **29.** $(-3)(-2) + (-1)^2$
30. $3 \times (-2) - 4 \div 2$ **31.** $-5 \times (-1) - 4 \times (-3)$

32. Evaluate for $x = -2$ and $y = -3$.
a) $4x - 3y$ **b)** $5x + 2y$
c) $x - y$ **d)** $y - x$
e) $x^2 - y^2$ **f)** $3xy - 2x + 3y$

33. The temperature in Caledonia at noon was $-4\,°C$ and it fell to $-13\,°C$ by midnight. What was the change in temperature?

Chapter 3

Calculate.

1. $3^4 + 3^0$
2. 2^{-1}
3. $3^{-1} - 3^{-2}$
4. $(2^3)^2$
5. $(-1)^{12}$
6. $(-1)^{21}$
7. -5^3
8. $(3-2)^{-1}$

Calculate

9. 42.57×1000
10. $\frac{2.357}{100}$

Write in scientific notation.

11. 37 300 000
12. 0.000 000 000 015 4

Estimate the square root.

13. 80
14. 291

Calculate.

15. $\sqrt{81}$
16. $3\sqrt{9} - \sqrt{4}$
17. $\sqrt{3^2 + 4^2 + 12^2}$
18. $\frac{5\sqrt{16} - 8}{3}$

19. The approximate diameter of a circle can be found using the formula

$$d = \sqrt{\frac{4A}{3.14}}$$

where A is the area. Determine the diameter of circles with the following areas. Round each answer to the nearest tenth of a unit.

a) 78 cm^2 b) 1000 m^2

20. The world's smallest known spider is found in Western Samoa. About 550 spiders, placed end-to-end, would take up the length of this page. Calculate the length of 1 spider to the nearest thousandth of a centimetre. Express your answer in scientific notation.

Chapter 4

Which numbers are rational numbers?

1. $0.\overline{7}$ 2. 3.14 3. $1\frac{3}{4}$ 4. 0.6 5. $\sqrt{5}$

Write in order from largest to smallest.

6. $\frac{-1}{3}, \frac{2}{6}, \frac{1}{6}$
7. $1.7, \frac{-3}{2}, \sqrt{3}$
8. $3.3, 3\frac{1}{3}, \frac{10}{3}$
9. $-1\frac{5}{6}, -1\frac{1}{4}, -1\frac{1}{3}$

Write as a fraction in lowest terms.

10. 0.62
11. 1.4
12. -7.25

Write as a decimal.

13. $-\frac{22}{3}$
14. $\frac{13}{7}$
15. $-4\frac{7}{8}$

Calculate.

16. $\frac{7}{5} \times \frac{1}{7}$
17. $2\frac{1}{2} \div \frac{1}{6}$
18. $\frac{1}{2} + \frac{1}{6}$
19. $-1.5 - 0.8$
20. $3\frac{1}{4} \times \frac{2}{3}$
21. $-\frac{4}{5} \div \frac{6}{10}$
22. $-3.75 + 11.23$
23. 3.2×4.6

Calculate.

24. $5.5 + 4(3 + 2) \div 1.5$
25. $7.36 + 6 \times 8.1 - 3$
26. $(6 - 5)^2 + (8 - 7) \div \frac{1}{2}$

27. A swinging pendulum is often used to keep time. The period of a pendulum is defined as the time it takes for a complete swing to and fro. At sea level on the surface of the Earth, the period, T seconds, is given by the formula

$$T = 2.01\sqrt{l}$$

where l is the length of the pendulum in metres. What is the period of a pendulum of length 1 m? 2 m? 3 m? Write each answer to the nearest hundredth of a second.

CHAPTER 5

Algebra

Crime writers have used codes many times. "The Shadow" was a crime fighter in the 1930s. He was created by Walter B. Gibson. Here is a cipher used by "The Shadow."

A B C D E F G H I J K L M N O P Q R S T U V W X Y Z

Extra Symbols: 1 2 3 4

The extra symbols tell the decoder how far the letter symbols have been rotated to the right in the message in order to disguise them. The symbol means they have been rotated 90° to the right.

Decode this message.

Sir Arthur Conan Doyle used a code in the Sherlock Holmes story "The Adventure of the Dancing Men." Research how he used the code.

Activity ❶ Patterns in Tables

The patterns in the tables are found in the columns. Copy and complete the tables. Write the rules for completing each table.

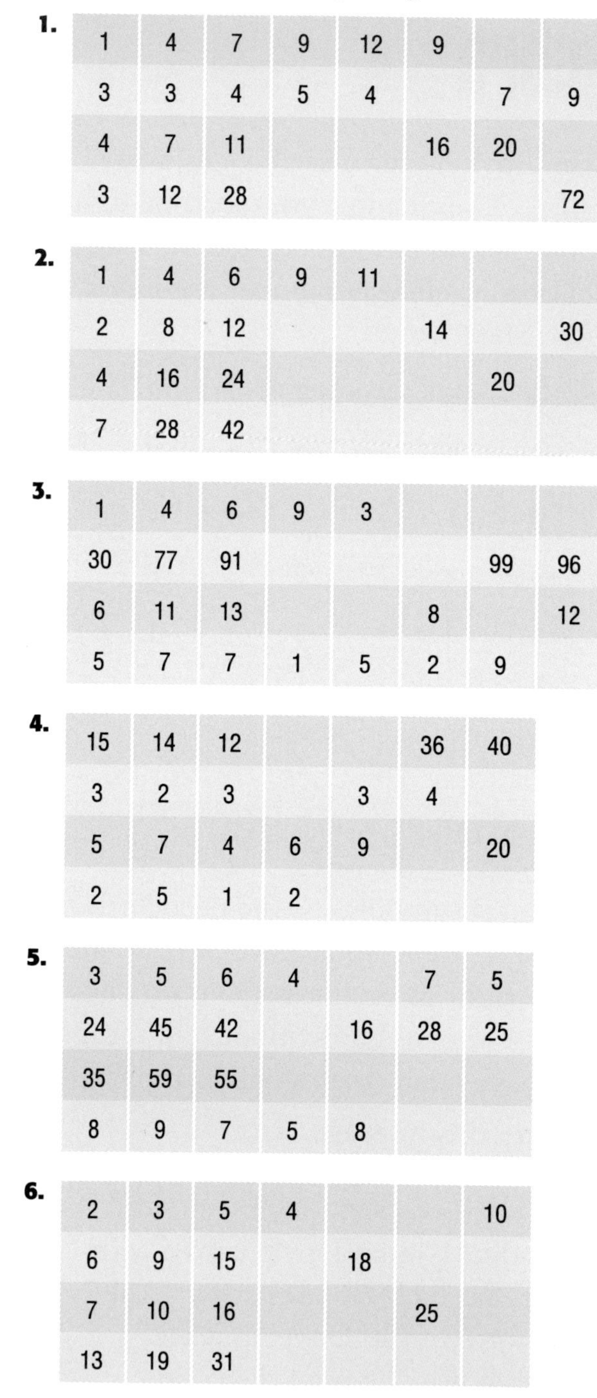

1.

1	4	7	9	12	9		
3	3	4	5	4		7	9
4	7	11			16	20	
3	12	28					72

2.

1	4	6	9	11			
2	8	12			14		30
4	16	24				20	
7	28	42					

3.

1	4	6	9	3			
30	77	91				99	96
6	11	13			8		12
5	7	7	1	5	2	9	

4.

15	14	12			36	40
3	2	3		3	4	
5	7	4	6	9		20
2	5	1	2			

5.

3	5	6	4		7	5
24	45	42		16	28	25
35	59	55				
8	9	7	5	8		

6.

2	3	5	4			10
6	9	15		18		
7	10	16			25	
13	19	31				

Activity ❷ Stamp Patterns

Two stamps can be attached to form two rectangular shapes.

1. In how many ways can 12 stamps be attached to form rectangular shapes? Sketch your solutions.

2. How many rectangular shapes can be made with 10 attached stamps?

3. How many rectangular shapes can be made with 24 attached stamps?

4. Without drawing the stamps, how can you determine how many rectangular shapes can be made with 100 attached stamps?

Activity ❸ Tile Patterns

A floor is being tiled using the following pattern. Each yellow tile is surrounded by white tiles.

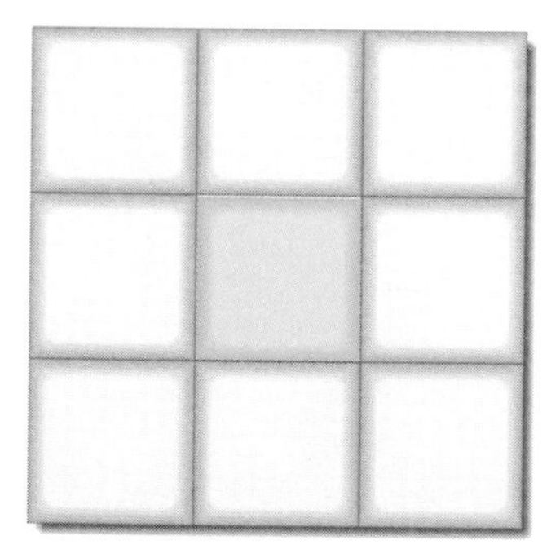

1. How many tiles of each colour do you need to tile each of the following areas?

a) 3 tiles by 3 tiles
b) 5 tiles by 3 tiles
c) 5 tiles by 5 tiles
d) 7 tiles by 3 tiles
e) 7 tiles by 5 tiles
f) 7 tiles by 7 tiles

Activity 4 Patterns in Drawings

Determine the pattern and then sketch the next figure in the sequence.

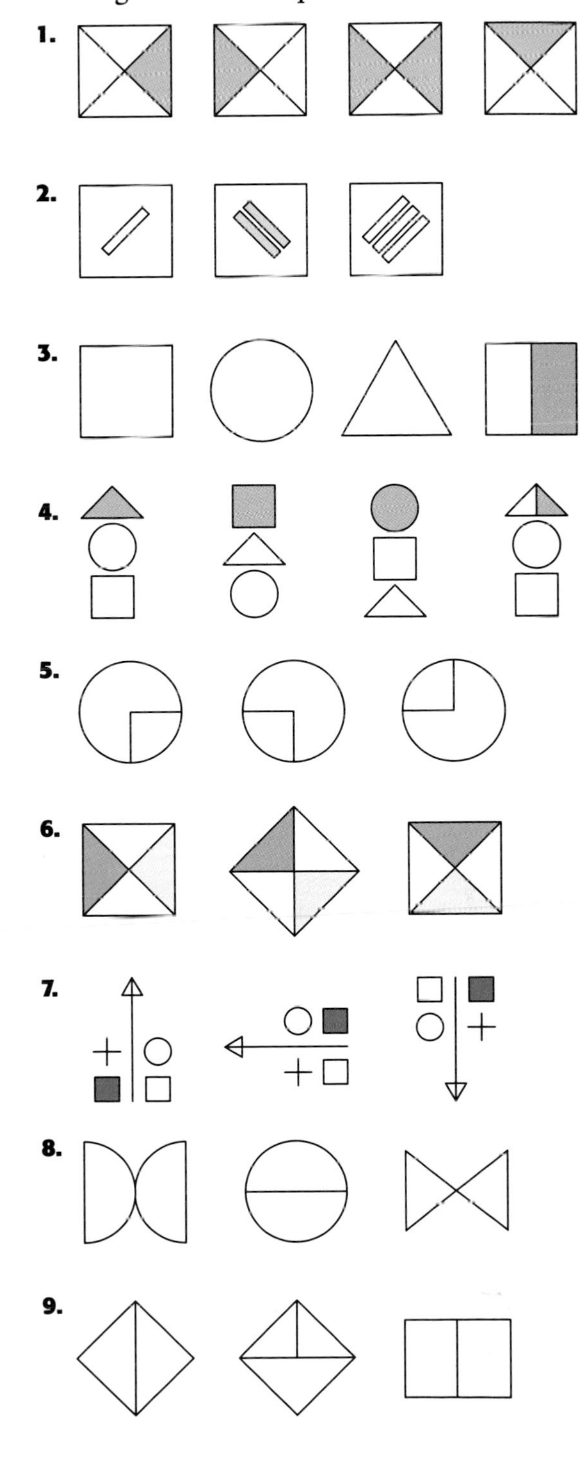

Mental Math

Simplify.

1. $10 + 20 - 30$ **2.** $10 - 10 + 10$
3. $-10 + 2 + 2$ **4.** $15 - 5 - 4$
5. $-15 + 2 + 1$ **6.** $-10 - 10 - 10$
7. $-10 - 10 - 20$ **8.** $-15 + 5 - 15$
9. $-50 + 25 + 5$ **10.** $100 - 80 - 10$

Simplify.

11. $(2)^3$ **12.** $(-3)^2$ **13.** $(-1)^5$
14. $(-1)^8$ **15.** -5^2 **16.** -5^0
17. $(4)^2$ **18.** $(-2)^2$ **19.** -10^2
20. $(3)^3$ **21.** -1^3 **22.** -1^6
23. $(-8)^0$ **24.** $(-3)^3$ **25.** -3^2

Multiply.

26. $-2 \times (-3) \times 5$ **27.** $-2 \times (-2) \times (-5)$
28. $5 \times (-5) \times 2$ **29.** $5 \times 4 \times (-3)$
30. $-4 \times 3 \times 2$ **31.** $6 \times (-5) \times (-2)$
32. $10 \times 10 \times (-10)$ **33.** $-10 \times (-10) \times 2$

Divide.

34. $\frac{-18}{9}$ **35.** $\frac{21}{-7}$ **36.** $\frac{-42}{6}$
37. $\frac{-14}{-7}$ **38.** $\frac{-20}{-5}$ **39.** $\frac{-24}{-6}$
40. $\frac{21}{3}$ **41.** $\frac{-49}{7}$ **42.** $\frac{-54}{-6}$
43. $\frac{-33}{11}$ **44.** $\frac{44}{-4}$ **45.** $\frac{-50}{-5}$

Simplify. State each answer in exponential form.

46. 3×3^0 **47.** 2×2^2 **48.** $3^5 \times 3^{-2}$
49. $6^4 \times 6^{-2}$ **50.** $4^{-2} \times 4^3$ **51.** $5^{-3} \times 5^5$
52. $3^7 \div 3^5$ **53.** $4^{10} \div 4^8$ **54.** $2^8 \div 2^7$
55. $2 \div 2^{-2}$ **56.** $3 \div 3^{-1}$ **57.** $4 \div 4^{-1}$
58. $\frac{5^5}{5^3}$ **59.** $\frac{6^3}{6^2}$ **60.** $\frac{2}{2^{-2}}$

Algebra Tiles

Each red tile represents $+1$.
Each white tile represents -1.
Each long green tile represents $+x$ or x.
Each long white tile represents $-x$.
Each square green tile represents $+x^2$ or x^2.
Each square white tile represents $-x^2$.

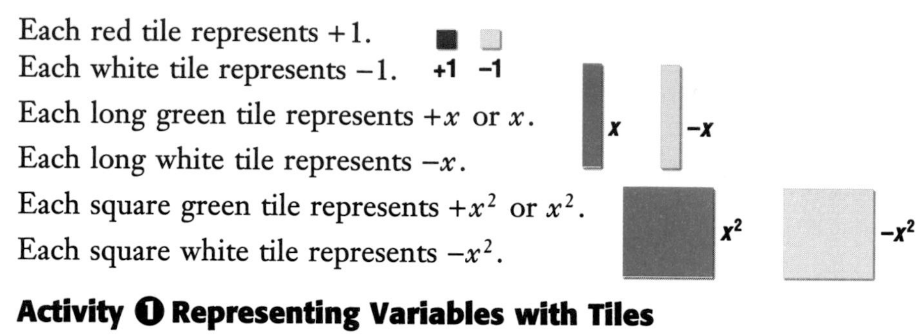

Activity 1 Representing Variables with Tiles

1. Write the expression represented by each group of tiles and then evaluate each expression for $x = 2$ and $x = -3$.

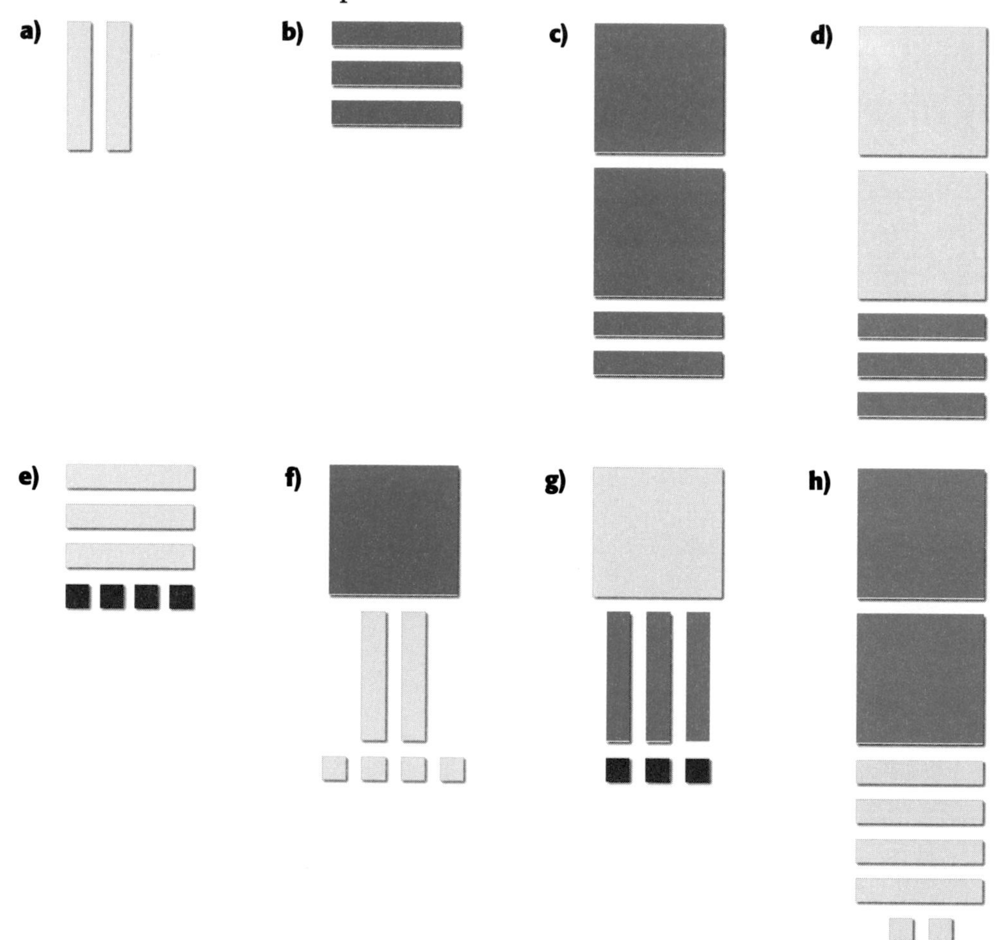

2. Use tiles to model the following expressions. Sketch your solutions.

a) $4x$ **b)** $2x^2$ **c)** $-4x$ **d)** $3x^2 - 2x$

e) $2x - x^2$ **f)** $-3x^2 + 4x + 3$ **g)** $-2x^2 - 2x - 1$ **h)** $3x^2 - 2x + 4$

Activity ❷ Representing Zero with Variables

Each pair represents zero.

1. Copy and complete the table. The first row has been done for you.

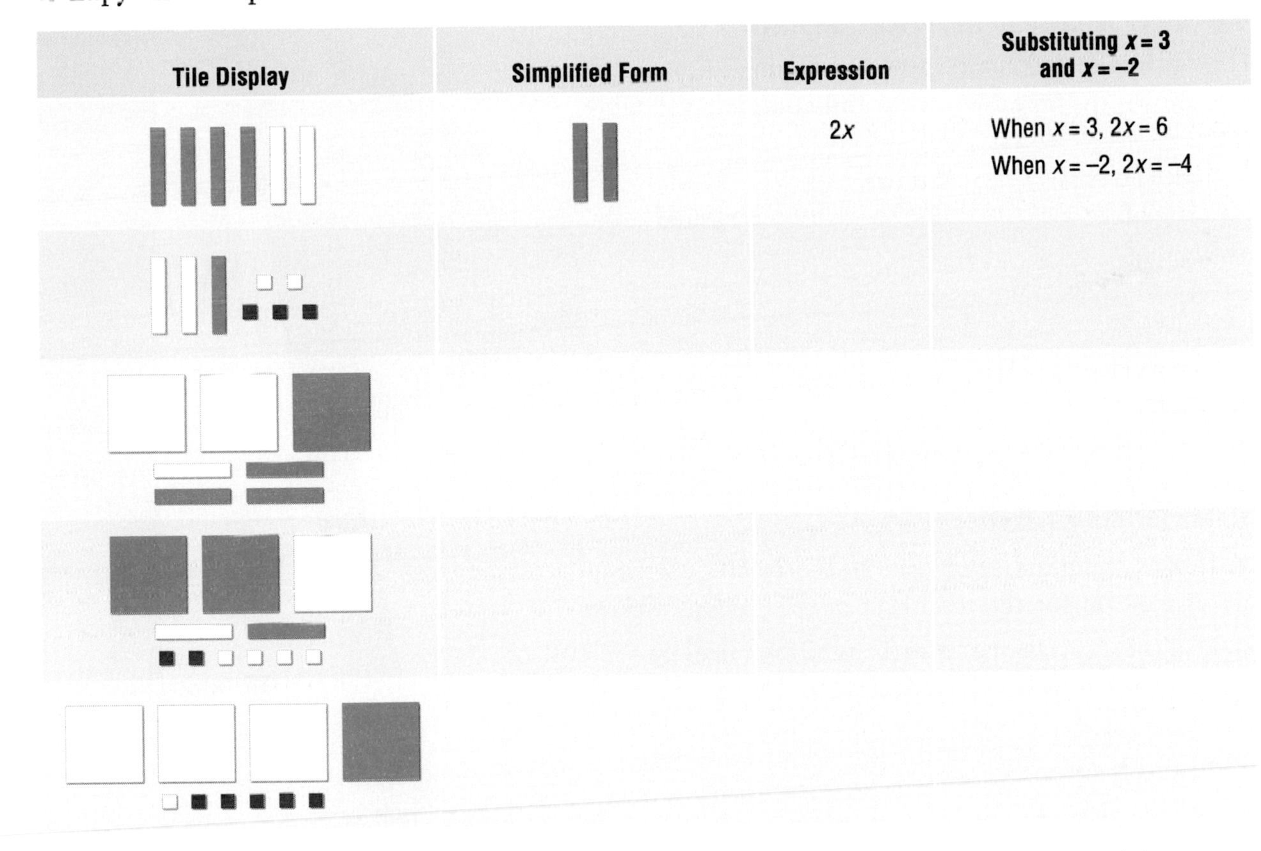

Tile Display	Simplified Form	Expression	Substituting $x = 3$ and $x = -2$
		$2x$	When $x = 3$, $2x = 6$ When $x = -2$, $2x = -4$

2. What is the smallest number of tiles you can add to each group to make zero? Explain.

a) b) c) d)

e) f) g) h)

3. Represent each integer using exactly 10 tiles. Use only tiles that represent $+1$ and -1.

a) $+8$ **b)** -6 **c)** -4 **d)** -8 **e)** $+2$ **f)** -2

5.1 Evaluating Expressions

At the 1993 Juno Awards, the theme song from the movie *Beauty and the Beast* was voted best single of the year. It was sung by Quebec's Celine Dion and an American, Peabo Bryson.

Activity: Discover the Relationship

The prices of tickets for a recent Celine Dion concert are shown in the table. Copy and complete the table.

Number of Tickets Purchased	Main Floor	Balcony
1	$32.50	$28.50
2		
3		
4		
5		

Inquire

1. If n is the number of seats on the main floor, write an algebraic expression for the total cost of the main-floor tickets.

2. If m is the number of seats in the balcony, write an expression for the total cost of the balcony tickets.

3. Use the expressions from 1 and 2 to calculate the cost of tickets for everyone in your class to
a) sit on the main floor.
b) sit in the balcony.

A letter that represents one or more numbers is called a *variable*. Expressions such as $x + 2y + 10$, which include numbers and variables, are called *algebraic expressions*.

The expression $x + 2y + 10$ has 3 **terms**. The number 10 is called a **constant**.

Algebraic expressions can have many different values, depending on the value assigned to each variable. When we assign a specific value to a variable in an algebraic expression, the process is known as **substitution**.

Example 1

Evaluate for $x = 4$.
a) $2x + 1$
b) $-7x$
c) $x - 11.5$

Solution

a) $2x + 1 = 2(4) + 1$
$= 8 + 1$
$= 9$

b) $-7x = -7(4)$
$= -28$

c) $x - 11.5 = 4 - 11.5$
$= -7.5$

Example 2

Evaluate for $x = 3$, $y = 4$, and $z = 5$.

a) $x + y - z$

b) $5x - 2y + 2z$

Solution

a) $x + y - z = 3 + 4 - 5$
$= 2$

b) $5x - 2y + 2z = 5(3) - 2(4) + 2(5)$
$= 15 - 8 + 10$
$= 17$

Practice

Name the variable in each expression.

1. $a + 8$ **2.** $c - 3$ **3.** $\frac{1}{4} + q$

4. $b - 5$ **5.** $-k - 0.1$ **6.** $t + 1$

Name each variable and constant.

7. $c + 10$ **8.** $a + b - 6$

9. $x - y - 3$ **10.** $xy + z + 0.5$

State the number of terms in each expression.

11. $a + b$ **12.** $s + t - 5$

13. $1 - 5c + 4d$ **14.** $jkl + mn$

15. The value of the green x-tile is 2. What is the value of the expression?

16. The value of the green x-tile is 6. Evaluate the expression.

17. Evaluate $5x$ for each value of x.

a) 2 **b)** -5 **c)** 0 **d)** -4

18. Evaluate $3y + 1$ for each value of y.

a) 0 **b)** 1 **c)** 5 **d)** -3

19. Evaluate $2z - 1$ for each value of z.

a) -1 **b)** 2 **c)** -5 **d)** 10

20. Evaluate for $x = -2$.

a) $x + 3$ **b)** $3x + 2$

c) $2x - 6$ **d)** $8 - 2x$

21. Evaluate for $x = 2$, $y = 3$, and $z = -4$.

a) $x + y + z$ **b)** xyz

c) $x + 2z$ **d)** $xy - z$

22. Evaluate for $a = 1.5$, $b = 2.5$, and $c = 3.5$.

a) $a + b + c$ **b)** $ab - 0.75$

c) $5a - 3b$ **d)** $1.2a + 3.2b$

e) $5.65 - ab$ **f)** $2.4b - a$

g) $ac - 2b$ **h)** $ab - bc - ca$

Problems and Applications

23. a) In a basketball game, team A scored $7x + 46y + 15z$ points and team B scored $2x + 27y + 31z$ points. In these expressions, x is the value of a 3-point field goal, y is the value of a 2-point field goal, and z is the value of a 1-point free throw. Which team won?

b) At the end of the 1991–1992 season, the Chicago Bulls' Michael Jordan had the highest all-time scoring average in the NBA at 32.3 points per game. His total points were described by the expression $206x + 6881y + 4620z$. The variables have the same values as in part a). How many points had Michael Jordan scored?

24. An apple contains about 290 kJ of energy, a banana 360 kJ, and a grapefruit 210 kJ.

a) Write an expression for the energy in a mixture of apples, bananas, and grapefruit.

b) Calculate the energy in a fruit salad made from 2 apples, 3 bananas, and $1\frac{1}{2}$ grapefruit.

5.2 Like Terms

Activity: Use a Tangram

The tangram square is made up of 7 pieces. Copy the square into your notebook. Label the pieces with letters starting at A. Label pieces with equal areas with the same letter.

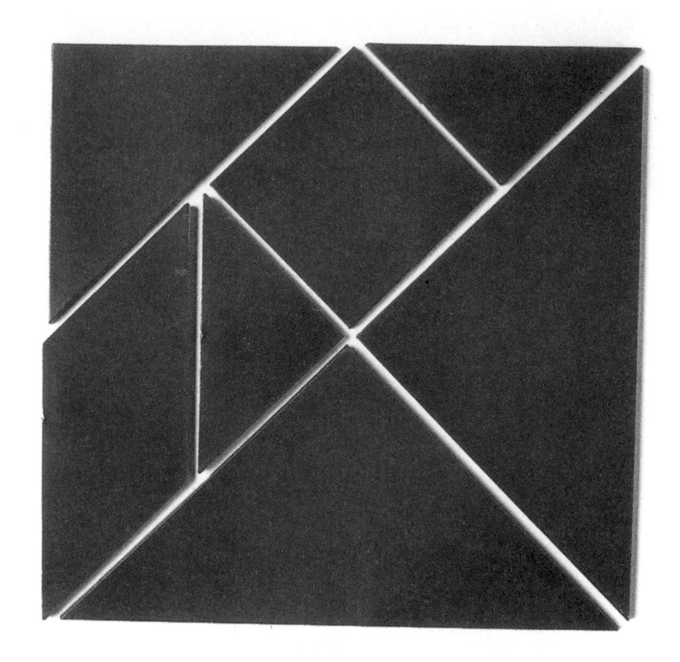

Inquire

1. How many different areas make the square?

2. Which areas are identical or *like*?

3. Which areas are not identical or *unlike*?

4. Write 2 different expressions that give the area of the tangram square.

5. Which of the 2 expressions in question 4 is simpler?

Like terms in algebra have the same variables raised to the same exponents. For example, the terms a, $3a$, and $7a$ are like terms. So are $2x^2$, $3x^2$, and $7x^2$.

Unlike terms in algebra have different variables, or the same variable but different exponents. For example, the terms $7b$, $3b^2$, $10q$, $5t$, $3x$, $4x^3$, and $17y$ are unlike terms.

Example 1

Simplify.

a) $x + 2x$ **b)** $3b - b$

c) $2y + y - 4$

Solution

Combine like terms.

a) $x + 2x = 3x$ **b)** $3b - b = 2b$

c) $2y + y - 4 = 3y - 4$

Example 2

Simplify.

a) $5x - 6y - 2x + 2y + 1$

b) $-11 + a - 2c + 16 + 3a - b - 5c$

Solution

a) $5x - 6y - 2x + 2y + 1$
$= 5x - 2x - 6y + 2y + 1$
$= 3x - 4y + 1$

b) $-11 + a - 2c + 16 + 3a - b - 5c$
$= a + 3a - b - 2c - 5c - 11 + 16$
$= 4a - b - 7c + 5$

Practice

Simplify.

1. $3x + 5x$
2. $11t - t$
3. $-10b + 3b$
4. $-12y - y$
5. $11m + 10m$
6. $15p - 9p$
7. $-21s + 12s$
8. $-13a - 10a$

Simplify.

9. $6r + 4r - r$
10. $2t + 3t - t$
11. $9p - p - 6p$
12. $-w + 6w - w$
13. $11a - 10a - 5a$
14. $-5y - 2y + 9y$
15. $-8q - 9q + 10q$
16. $16n + n - 17n$

Simplify.

17. $3t + 5 - 2t$
18. $2x - 5x - 7$
19. $2a - b + 3a$
20. $-4x + y - 6x$
21. $8y - 2z + 7y$
22. $-9p - 10q + 8p$
23. $-2r - 3s + 6r$
24. $-j + 7k - 3j$

Simplify.

25. $5c + d - 2c - d$
26. $7p - q + p - 2q$
27. $3j + k - 5j - 2k$
28. $8a - 2b - 6a - 3b$
29. $-r + s + 2r - 2s$
30. $-5x + 5y + 5x - y$

Simplify.

31. $5x + 10 + 5y - 3x + 1$
32. $3a - 4b - 5 + a - b$
33. $-7t + 2 + 8r + 9r - 8t$
34. $-8 - 3z + 9x - 11 - 6z$
35. $8r - 11 - 18q + 5p + 7q$
36. $-4w + 7c - 8x - 9w - 3c$
37. $13j - 18d - 5d + 2j + 2c$
38. $-q + 7q + 11n + 11p - 8q$

Simplify.

39. $-p + 5q - 8r - q + 3p + 9r + 1$
40. $-2z + 3y - 10x - 4y + 4z - 3x$
41. $5q - 9s - 8r + 8q - 7r + 13s$
42. $-10 - a + 6c - 11 - 9c + 8a - d$

Problems and Applications

43. Simplify, then evaluate.

a) $2a + 8a$, when $a = 2$
b) $7t - 3t$, when $t = 3$
c) $-6k - 8k$, when $k = -2$
d) $-6y - 9y - y$, when $y = -3$
e) $3 + 4x - 2x$, when $x = 5$
f) $-5p - 3p + 1 - 6$, when $p = -1$

CONTINUED

44. Write an expression for each perimeter in 2 different ways.

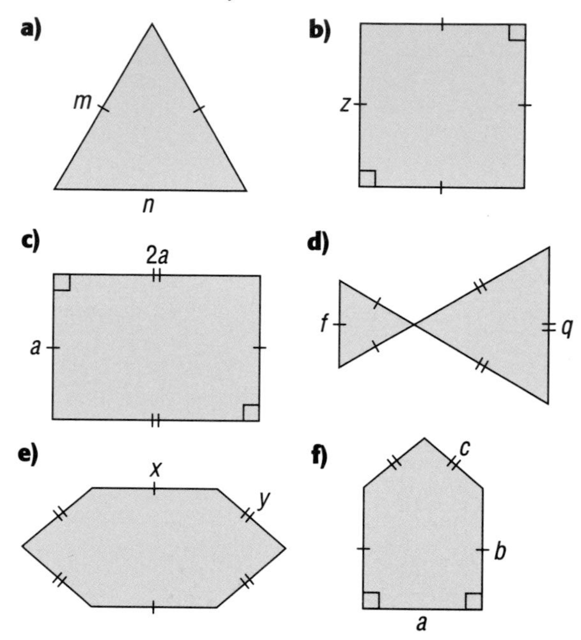

45. Write and simplify an expression for each perimeter.

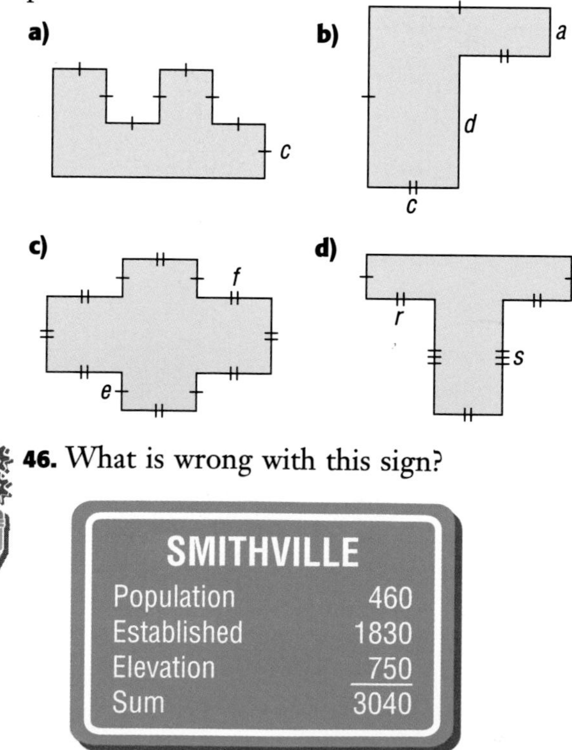

46. What is wrong with this sign?

47. Work with a partner to calculate the perimeter of this figure. Describe the simplest method you can find.

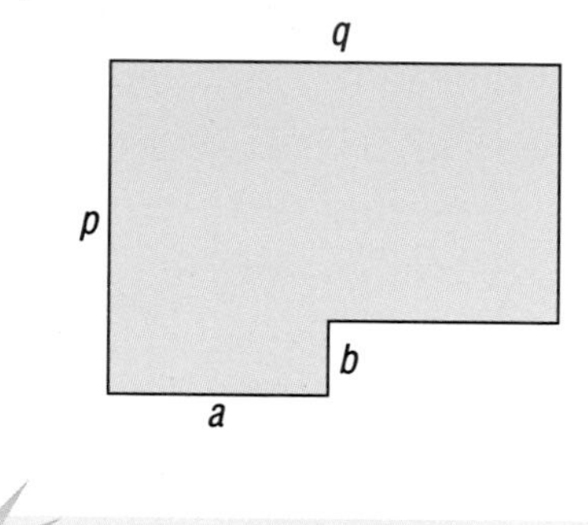

LOGIC POWER

The locomotive, L, is on the main track. Two boxcars, A and B, are on sidings that meet at point P. Where the sidings meet, there is room on the track for 1 boxcar but not the locomotive.

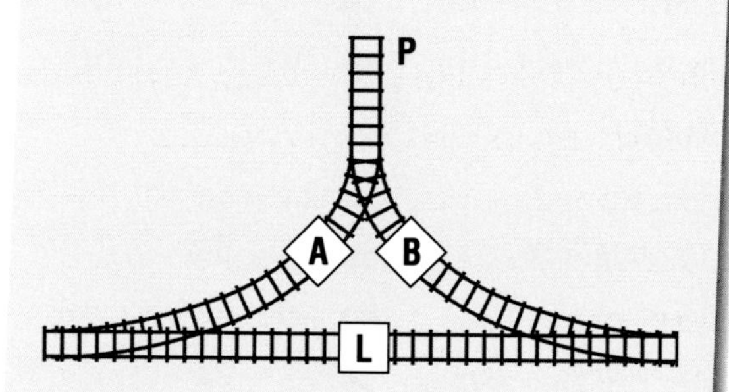

The problem is to switch the positions of the 2 boxcars and leave the locomotive in its original position.

The locomotive can push or pull the boxcars. It can push or pull 2 at a time. It can go between them and push one and pull the other.

Hint: Draw the diagram and use different coins for the locomotive and the boxcars.

5.3 Polynomials

The Olympic biathlon consists of 2 events. The events are cross-country skiing and shooting. At one winter Olympics, Canada's Miriam Bedard won the bronze medal in the biathlon.

Just like combination Olympic events, algebraic expressions have special names. A monomial is a number or variable or the product of numbers and variables. A polynomial is a monomial or the sum of monomials. A polynomial with 2 terms is called a **binomial**. A polynomial with 3 terms is called a **trinomial**.

Activity: Use the Diagrams

Copy the diagrams into your notebook.

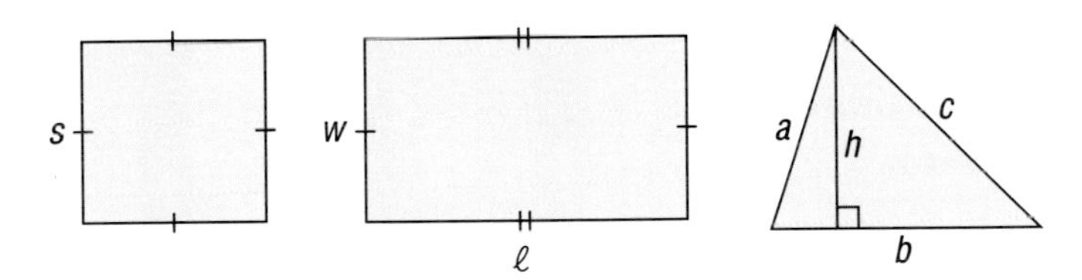

Inquire

1. Write the formulas for the perimeter and area of each figure. Combine like terms where possible.

2. Which expressions are monomials?

3. Which expression is a binomial?

4. Which expression is a trinomial?

5. Explain why it is appropriate to call expressions with 1, 2, and 3 terms monomials, binomials, and trinomials.

Example 1

Classify each of the following as a monomial, binomial, or trinomial.

a) $a - 2b$ **b)** $x^2 + y^2 + z^2$ **c)** 10

Solution

a) The expression $a - 2b$ is a binomial because it contains 2 terms.

b) The expression $x^2 + y^2 + z^2$ is a trinomial because it contains 3 terms.

c) The number 10 is a monomial because it contains 1 term.

CONTINUED

The **degree of a monomial** is the sum of the exponents of its variables.

Monomial	Degree
$4x^3$	3
$5a^2b^3c$	$2 + 3 + 1 = 6$

$3 + 2 + 4 = 9$

$2x^3y^2z^4$

The **degree of a polynomial in one variable** is the highest power of the variable in any one term.

Polynomials	Degree
$6x^2 + 3x$	2
$x^5 + 7x^2 - 3$	5

The **degree of a polynomial in two variables or more** is the largest sum of the exponents in any one term.

Polynomials	Degree
$x^2y^3 + xy^4 + xy^5$	6
$3x^3y^4 + 7xy^3 - 2xy$	7

Example 2

Classify each polynomial and state its degree.

a) $3abc$ **b)** $2x^2 + x$ **c)** $3xy + 5x^2y^2 - 3$

Solution

a) $3abc$ is a monomial. The sum of the exponents is $1 + 1 + 1 = 3$. It is a third-degree polynomial.
b) $2x^2 + x$ is a binomial. The highest power, 2, is contained in the term $2x^2$. It is a second-degree polynomial.
c) $3xy + 5x^2y^2 - 3$ is a trinomial. The largest exponent sum is contained in the term $5x^2y^2$. It is a fourth-degree polynomial.

The terms of a polynomial are usually arranged so that the powers of one variable are in either ascending order or descending order.

Descending Order	Ascending Order
$x^3 + 2x^2 - 5x + 7$	$7 - 5x + 2x^2 + x^3$
(in x) $5x^2 + 7xy + 3y^2$	(in x) $3y^2 + 7xy + 5x^2$

Practice

Identify as a monomial, binomial, or trinomial.

1. $5xyz$ **2.** $x + 2y$ **3.** $a - 2b + 3c$

4. $x^2 + y^2$ **5.** 23 **6.** $x - y + 2$

State the degree of each monomial.

7. $25x$ **8.** $25x^2y^2$ **9.** 17

10. $2x^2y^3$ **11.** $-5x^3y^4$ **12.** $-6xy^4z$

State the degree of each polynomial.

13. $5x^2y^2 + 3xy^3$

14. $3x + 2y - 5z$

15. $x^4 + 2x^3 + 3x^2 + 4$

16. $4x^4y^2 + 2x^3y^5 - 23$

17. $3x - 2y + z^2$

18. $25m^3n + 36m^3n^3$

19. $-5x^4y^2z + 2x^2y^2z^2$

Arrange the terms in each polynomial in descending powers of x.

20. $1 + x^3 + x^2 + x^5$

21. $5 - 3x^3 + 2x$

22. $5y^2 + 2xy - x^2$

23. $25xy^2 - 5x^2y + 3xy^3 - 4x^4$

24. $5ax + 7b^2x^4 - 3x^3 + 4abx^2$

Arrange the terms in each polynomial in ascending powers of x.

25. $3x^2 - 2x^3 + 5x^5 + x - 2$

26. $4x^4 + x^2 - 3x^3 + 5 - x$

27. $4xy^2 - 2x^2y^2 - 3x^4 + 2x^3y$

28. $5x^2yz^2 + 2xy^4z + 3x^3y^4z^2 - 3$

29. $z - xy + x^2$

30. $x^2 - 2xy - 3x^3 + 16$

31. $2x^3y + 3xy - x^5$

32. $3x^3y^2 + x^4y + xy - 1$

Problems and Applications

33. Identify each type of polynomial.

a) $\frac{4\pi r^3}{3}$ **b)** $\pi r^2 + 2\pi rh$ **c)** $4\pi r$

34. What type of polynomial is represented by the perimeter of each of these figures?

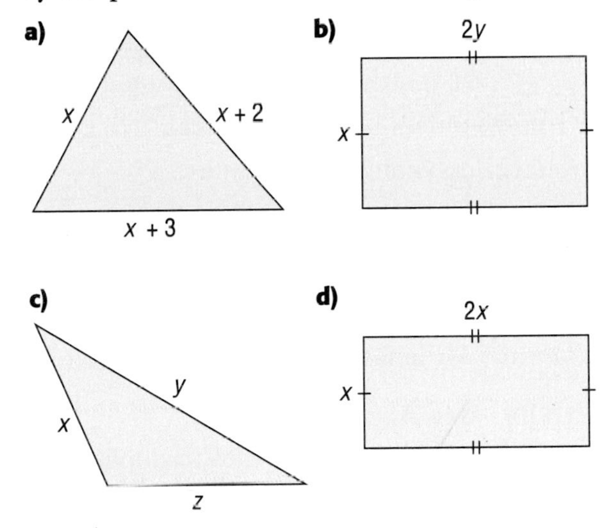

35. a) Calculate the area of each face of this box.

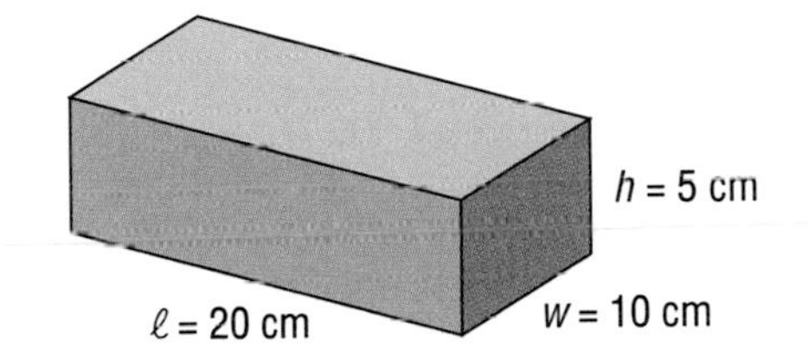

b) What is the total area of the box?

c) Write a polynomial that can be used to calculate the answer you gave in part b).

36. The formula for the volume of a rectangular jewellery box is lwh. Its dimensions are 25 cm × 18 cm × 17 cm. It has 2 cm thick walls. What is the volume of the box's interior to the nearest cubic centimetre?

37. Write a problem that can be solved with a polynomial. Have a classmate solve your problem.

5.4 Adding Polynomials

A sports car has 4 wheels, 2 seats, and 1 engine. To build 2 sports cars, we need

$4 + 4$ or 8 wheels,
$2 + 2$ or 4 seats,
and $1 + 1$ or 2 engines.

We have added the like parts of the 2 cars. Algebra tiles model the addition of like parts of 2 polynomials.

To model polynomials, we can use x^2-tiles, x-tiles, and 1-tiles. When the x^2 or x-tile is green side up, it means x^2 or x. When the x^2 or x-tile is white side up, it means $-x^2$ or $-x$. When the 1-tile is red side up, it means $+1$. When the 1-tile is white side up it means -1.

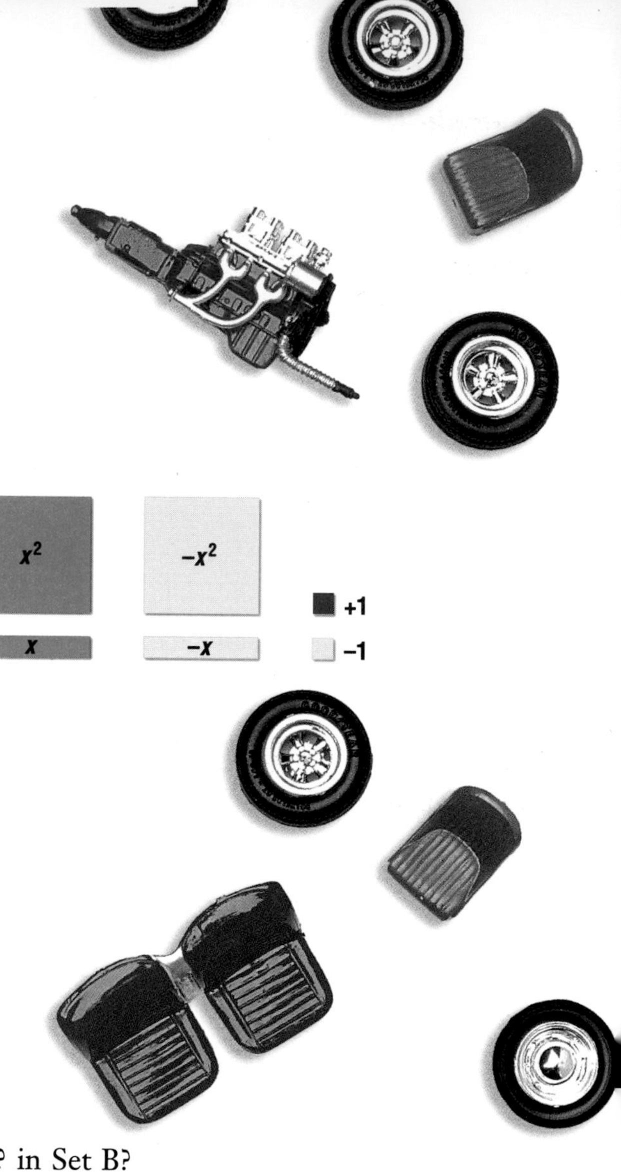

Activity: Use Algebra Tiles

Use x^2-tiles, x-tiles, and 1-tiles to duplicate the algebraic expressions in Set A and Set B.

Inquire

1. How many x^2-tiles, x-tiles, and 1-tiles are in Set A? in Set B?

2. Write an algebraic expression to represent
a) Set A **b)** Set B

3. Combine the tiles in Set A and Set B into 1 set, so that like tiles are placed together.

4. Write the algebraic expression that represents the combined set in 3.

5. Use tiles to represent the following pairs of expressions.

a) $\begin{array}{r} x^2 + 2x + 3 \\ \underline{2x^2 + 3x + 2} \end{array}$ **b)** $\begin{array}{r} x^2 - 2x + 6 \\ \underline{3x^2 + 5x - 2} \end{array}$ **c)** $\begin{array}{r} 3x^2 + 4x - 2 \\ \underline{5x^2 - 2x + 3} \end{array}$

6. Combine each pair in 5 and write its algebraic sum.

7. Write a rule for adding polynomials.

Example 1

Add $(x^2 + 4x + 3) + (2x^2 + 5x + 1)$.

Solution

Remove the brackets, collect like terms, and add.

$$\begin{aligned}(x^2 + 4x + 3) + (2x^2 + 5x + 1) &= x^2 + 4x + 3 + 2x^2 + 5x + 1\\ &= x^2 + 2x^2 + 4x + 5x + 3 + 1\\ &= 3x^2 + 9x + 4\end{aligned}$$

Note that the answer to Example 1, $3x^2 + 9x + 4$, is in **simplest form**. A polynomial is in simplest form when it contains no like terms.

Example 2

Use column form to add these polynomials.

a) $(5x^2 - 3x + 5) + (2x^2 - 7x - 10)$

b) $(3y^3 - 3y + 2y^2 + 8) + (-5 - 6y^2 + 2y)$

Solution

a) Write the like terms of the polynomials in columns, then add.

$$\begin{array}{r} 5x^2 - 3x + 5 \\ +2x^2 - 7x - 10 \\ \hline 7x^2 - 10x - 5 \end{array}$$

b) Rewrite the terms in descending powers of y before arranging them in columns and adding.

$$\begin{array}{rcr} 3y^3 - 3y + 2y^2 + 8 & \longrightarrow & 3y^3 + 2y^2 - 3y + 8 \\ -5 - 6y^2 + 2y & \longrightarrow & - 6y^2 + 2y - 5 \\ \hline & & 3y^3 - 4y^2 - y + 3 \end{array}$$

Practice

Find the sums of the expressions, A and B, represented by the algebra tiles.

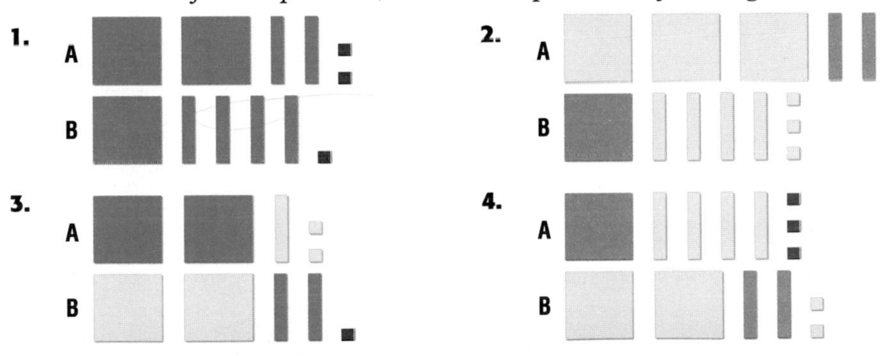

Identify the like terms in each expression.

5. $2x + 3y - 4xy + 5x - 2y + 6xy$

6. $2a + 5a - 6b + 8b - 2c + 3c$

7. $3s^2 + 5s - 2 + 7s^2 + s - 3$

Add.

8. $(3x + 1) + (4x - 2)$

9. $(3x^2 + 5x - 4) + (x^2 - 7x + 2)$

10. $(-y^2 + 7y - 5) + (2y^2 + 7y - 4)$

11. $(2y^3 - 3y^2 - 1) + (-5y^2 - 4y^3 + 3)$

Add.

12. $\begin{array}{r} x + 7 \\ +5x + 2 \\ \hline \end{array}$

13. $\begin{array}{r} 3y^2 + 2y + 8 \\ +4y^2 + 7y + 11 \\ \hline \end{array}$

14. $\begin{array}{r} 5x - 2y + 6 \\ +3x - 6y + 9 \\ \hline \end{array}$

15. $\begin{array}{r} 5x^2 - 3x + 7 \\ +2x^2 - 5x - 12 \\ \hline \end{array}$

16. $\begin{array}{r} 5x^2 + 7x - 9 \\ +4x^2 - 8x + 11 \\ \hline \end{array}$

17. $\begin{array}{r} 3y^2 - 8y + 3 \\ +2y^2 + 8y - 9 \\ \hline \end{array}$

Simplify.

18. $(5z + 6 - 3z^2) + (4 - 7z + 2z^2)$

19. $(3x^2 + 2y^2 - 5) + (4x^2 + 3y^2 - 11)$

20. $(2x^4 + 7x - 5x^2 + 3) + (2x^3 - 7)$

Add.

21. $(5x^2 + 7x - 7) + (4x^2 - 8x + 12)$

22. $(3y^2 - 8y + 3) + (2y^2 + 8y - 9)$

23. $(m^3 + 5m^2 + 3) + (4m^2 + 7)$

24. $(x^2 + x + 3) + (x^2 - 6) + (x^2 - 2x - 3)$

Simplify.

25. $(4x^2 + 3xy - 2y^2) + (-x^2 - 5xy + 7y^2)$

26. $(5y^2 + 3y - 7) + (-2y^2 - 5y + 8)$

27. $(3x^2y - 2xy + 4y^2) + (x^2y + y^2)$

Problems and Applications

28. a) Write an expression in simplest form for the perimeter of the figure.

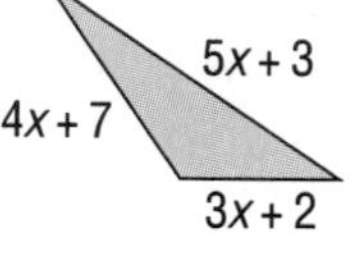

b) If $x = 4$ cm, what is the perimeter?

29. a) Write an expression in simplest form for the perimeter of the figure.

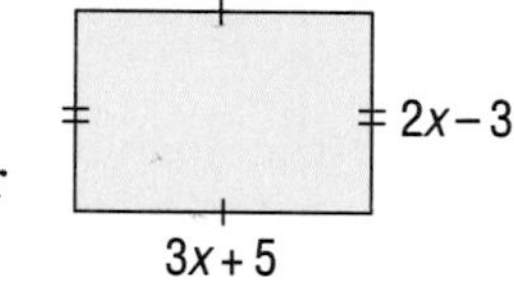

b) If $x = 7$ cm, what is the perimeter?

30. a) The perimeter of historic Fort York is expressed by the following polynomial.

$$x + (x + 171) + (x + 156)$$

What geometric shape is the perimeter of Fort York?

b) If $x = 153$ m, what is the perimeter?

31. Add $(3x^2 + x - 1) + (x^2 - 3x - 2)$. How can you use $x = 1$ to check your solution?

32. Work with a partner to write a monomial that describes the perimeter of this figure.

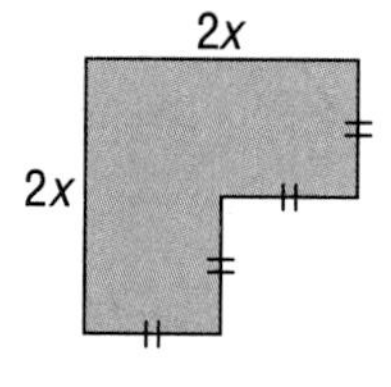

LOGIC POWER

If the hole passes right through the stack, how many cubes are there?

Budgets from a Spreadsheet

A spreadsheet is a useful tool for keeping track of expenses and the amount of money that remains in a budget.

Activity 1

1. Work with a partner. Use magazines, catalogues, and newspapers to choose 5 prizes of different values. They will be 1st, 2nd, 3rd, 4th, and 5th prizes for the students who raise the most in a fund-raising project. Keep the total cost below $500.00.

2. Copy the following spreadsheet into your notebook. Calculate and record the entries.

You will need to enter the rate of provincial tax for your province. You may need to correct the rate of the GST.

		B	C	D	E	F
1		Prize	Price	Provincial Tax	GST	Item Cost
2	1st			+C2* ■	+C2*.07	+C2+D2+E2
3	2nd					
4	3rd					
5	4th					
6	5th					
7					Total Cost	@SUM(F2..F6)

3. Is the calculation method shown in the spreadsheet correct for all provinces? Explain.

Activity 2

1. Set up the spreadsheet from Activity 1 on a computer and have the computer calculate the total cost.

2. Adjust your prizes until you have spent no more than $350.00 and no less than $300.00.

3. Challenge another group of students to come closer to $350.00 than your total cost does.

Activity 3

1. Redo the spreadsheet to allow for 8 prizes, but leave the total cost at no more than $350.00.

2. Challenge another group of students to come closer to $350.00 than your total cost does.

5.5 Subtracting Polynomials

Activity: Use Algebra Tiles

Use tiles to duplicate Set A and Set B.

Inquire

1. Write the algebraic expression represented by each group of tiles.

2. How many x^2-tiles, x-tiles, and 1-tiles are there in each set?

3. If the x^2-tiles, x-tiles, and 1-tiles in Set B are subtracted from those in Set A, how many of each type of tile are left?

4. Turn Set B over to make its opposite. Write the sum of Set A and the opposite of Set B. Compare with your answer in 3.

5. Use algebra tiles to represent each pair of expressions.

a) $\begin{array}{r} x^2 + 5x - 7 \\ \underline{-2x + 3} \end{array}$ **b)** $\begin{array}{r} 3x^2 - 4x + 3 \\ \underline{x^2 + 3x - 5} \end{array}$ **c)** $\begin{array}{r} -2x^2 + 4x - 5 \\ \underline{-x^2 + x - 3} \end{array}$

6. In each pair, add the opposite of the second polynomial to the first polynomial.

7. Write a rule for subtracting polynomials.

Example 1

Use column form to subtract $2x^2 + 2x + 5$ from $5x^2 - 7x + 4$.

Solution

When we subtract 1 polynomial from another, we add the opposite of the polynomial that is being subtracted.

The opposite of $2x^2 + 2x + 5$ is $-2x^2 - 2x - 5$.

$$\begin{array}{r} 5x^2 - 7x + 4 \\ \underline{-2x^2 - 2x - 5} \\ 3x^2 - 9x - 1 \end{array}$$

How can you check your answer by addition?

Example 2

Simplify $(4x^2 - 5x + 7) - (3x^2 + 2x - 5)$.

Solution

Write the opposite of $3x^2 + 2x - 5$ and add.

$$\begin{aligned} &(4x^2 - 5x + 7) - (3x^2 + 2x - 5) \\ &= (4x^2 - 5x + 7) + (-3x^2 - 2x + 5) \\ &= 4x^2 - 5x + 7 - 3x^2 - 2x + 5 \\ &= 4x^2 - 3x^2 - 5x - 2x + 7 + 5 \\ &= x^2 - 7x + 12 \end{aligned}$$

Practice

Write the opposite.

1. $x^2 + 4x + 1$
2. $x^2 - 2x - 3$
3. $2x^2 + x - 5$
4. $-3x^2 - 7x + 2$

Subtract.

5. $(3x - 5) - (x + 2)$
6. $(x + 5) - (3x - 1)$
7. $(x + 4) - (-x - 3)$
8. $(3x - 5) - (x + 4)$

Subtract.

9. $\begin{array}{r} 5x^2 + 3x - 5 \\ \underline{2x^2 - 5x - 4} \end{array}$

10. $\begin{array}{r} -3x^2 + 5x - 7 \\ \underline{2x^2 + 3x - 3} \end{array}$

11. $\begin{array}{r} -4x^2 - 4x + 3 \\ \underline{-3x^2 + 4x - 8} \end{array}$

12. $\begin{array}{r} x^2 - 5x + 1 \\ \underline{x^2 - 5x + 6} \end{array}$

13. $\begin{array}{r} x^2 + 7x - 1 \\ \underline{x^2 + 4x + 1} \end{array}$

14. $\begin{array}{r} 12x^3 + 3x^2 - 5x \\ \underline{9x^3 + 4x^2 - 4x} \end{array}$

Subtract.

15. $(2y^2 + 3y - 5) - (2y^2 + 4y + 6)$
16. $(4s^2 + s - 2) - (-3s^2 + s - 5)$
17. $(y^2 - 5y + 3) - (-2y^2 + 7y + 5)$

Subtract.

18. $3x^2 + 7x - 3$ from $2x^2 - 2x + 3$
19. $5y^2 + 7y - 5$ from $-2y^2 + 3y - 2$
20. $-t^2 + 5t - 1$ from $2t^2 + 3t + 6$

Simplify.

21. $(-5n^2 - n - 8) - (-2n^2 + 7n - 3)$
22. $(4 + 2x - x^2) - (3 - 7x^2 + 5x)$
23. $(-t^2 + 4t - 7) - (3t^2 + 4t - 2)$
24. $(x^2 + 5x + 3) - (-x^2 - 7x + 11)$
25. $(3m^2 + 7m - 8) - (-m^2 + m - 1)$
26. $(-5y^2 + 7y - 12) - (-3y^2 + 4y - 2)$

Problems and Applications

27. Find the length of PQ.

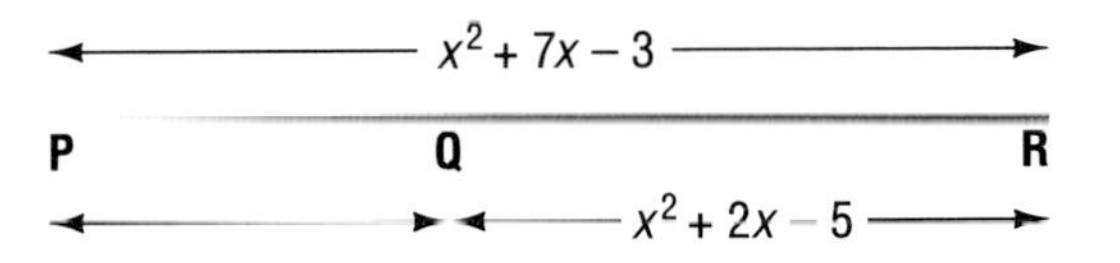

28. Given the perimeter, P, of a figure and 2 or 3 sides, find the missing length.

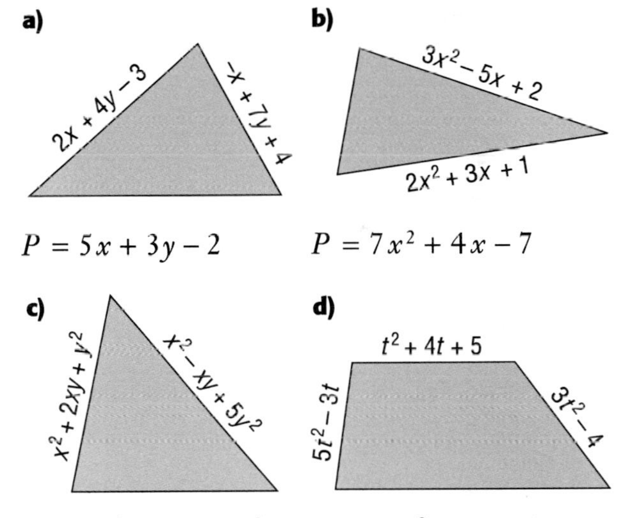

a) $P = 5x + 3y - 2$

b) $P = 7x^2 + 4x - 7$

c) $P = 5x^2 + xy + 5y^2$

d) $P = 11t^2 + 6t + 9$

29. The area of the large rectangle is $3x^2 + 2x + 1$. What is the area of the shaded region?

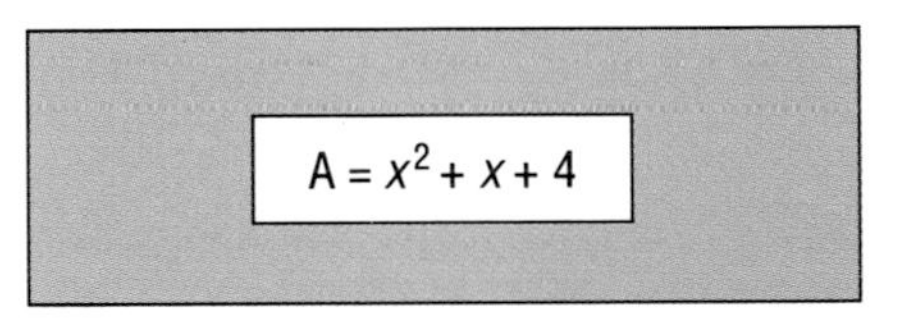

30. What is the result of adding a polynomial and its opposite? Explain why.

31. a) If you add 2 polynomials, A and B, does it matter if you add A + B or B + A? Explain and give an example.

b) If you subtract 2 polynomials, A and B, does it matter if you subtract A – B or B – A? Explain and give an example.

5.6 The Distributive Property

Activity: Use Algebra Tiles

An x-tile has an area of x square units.
A 1-tile has an area of 1 square unit.

How many units long are the sides of a 1-tile?
How many units long is each side of an x-tile?

Use x-tiles and 1-tiles to model these rectangles.

a)
b)

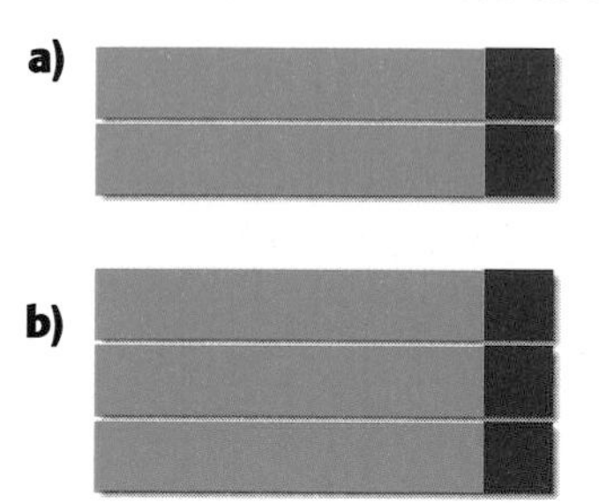

c)

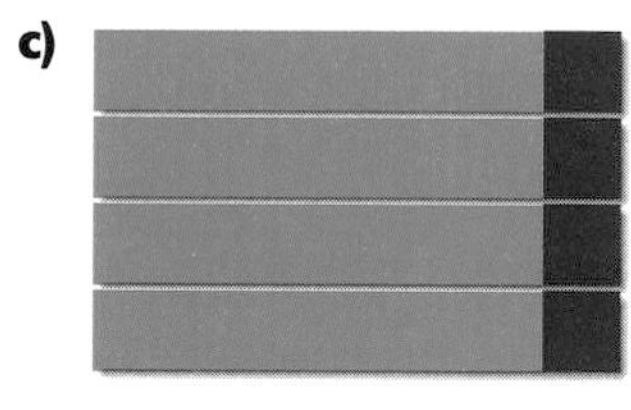

Inquire

1. What is the area of each rectangle?
2. What are the length and width of each rectangle?
3. Describe a method for multiplying the length by the width to give the area of each rectangle.

To **expand** an expression means to remove the brackets and simplify. To do this, we use the **distributive property**.
$a(b + c) = ab + ac$

Example 1

Expand using the distributive property.

a) $3(x + 5)$
b) $-(2 - x)$

Solution

Multiply each term inside the brackets by the term outside the brackets.

a)
$$\begin{aligned} 3(x + 5) &= 3(x + 5) \\ &= 3(x) + 3(5) \\ &= 3x + 15 \end{aligned}$$

b)
$$\begin{aligned} -(2 - x) &= -1(2 - x) \\ &= -1(2) - 1(-x) \\ &= -2 + x \end{aligned}$$

Example 2

Expand.

a) $4(3x - 2y + 7)$
b) $-2(x + 3) - (x - 5)$

Solution

a)
$$\begin{aligned} &4(3x - 2y + 7) \\ &= 4(3x - 2y + 7) \\ &= 12x - 8y + 28 \end{aligned}$$

b)
$$\begin{aligned} &-2(x + 3) - (x - 5) \\ &= -2(x + 3) - 1(x - 5) \\ &= -2x - 6 - x + 5 \\ &= -2x - x - 6 + 5 \\ &= -3x - 1 \end{aligned}$$

Practice

Expand.

1. $5(x+1)$ **2.** $3(x-2)$ **3.** $4(x+2)$

4. $2(x-3)$ **5.** $7(x-1)$ **6.** $5(x+3)$

7. $2(x+6)$ **8.** $4(x-5)$ **9.** $7(x+3)$

10. $3(x-4)$ **11.** $10(x+2)$ **12.** $9(x-3)$

Expand.

13. $2(3x+2)$ **14.** $3(3x+1)$

15. $5(2x+1)$ **16.** $4(2x+3)$

17. $6(2x-1)$ **18.** $5(3x-2)$

19. $7(2x+1)$ **20.** $6(3x+2)$

Expand.

21. $-3(x+2)$ **22.** $-4(2x+1)$

23. $-2(5x-2)$ **24.** $-3(3x-2)$

25. $-5(2x-1)$ **26.** $-2(5x-3)$

27. $-4(x+5)$ **28.** $-(2x-1)$

Expand.

29. $-2(3x-2y)$ **30.** $-3(5x+3y)$

31. $-4(5x+2y)$ **32.** $-(2x+y)$

33. $-5(-x+y)$ **34.** $-7(x-3y)$

35. $-2(3x+7)$ **36.** $-4(2x-y)$

Expand.

37. $-4(2x+3y+z)$ **38.** $3(5x-2y+2)$

39. $6(-x+3y+4)$ **40.** $-(2x-3y+5)$

Expand and simplify.

41. $4(y-3)+2(y+4)$

42. $2(5x-1)-(3x+2)$

43. $4(x+7)-(3x+2)$

44. $3(2x-3)-3(2x-5)$

45. $3(2x-5)+3(2x+1)$

46. $4(-x+3)-2(3x+1)$

47. $-2(x-3y+2)-3(x-2y)$

Problems and Applications

Expand and simplify.

48. $3(x^2+4x-3)+2(x^2+5)$

49. $2(3x^2+7x-3)+4(x-3)$

50. $5(y^2+3y-4)+2(y^2-4y+1)$

51. $-(2x+3y-5)+5(2x-3y+6)$

52. $-(2x+3)+3(x-1)-4(x-2)$

53. $(2x+1)-4(2x-3)+2(3x-1)$

Expand and simplify.

54. $4(4x-3y+2)-(2x+5y-4)$

55. $-(3x-2y+7)+4(x+y-2)$

56. $2(x^2+y^2)-3x^2+4y^2+7$

57. $4(x^2+2x-3)-2(x^2+4y-1)$

58. $5(2y^2+3y-2)-2(y^2-4y+1)$

For each large rectangle, state

a) *the length and width.*

b) *the area in expanded form.*

59. **60.**

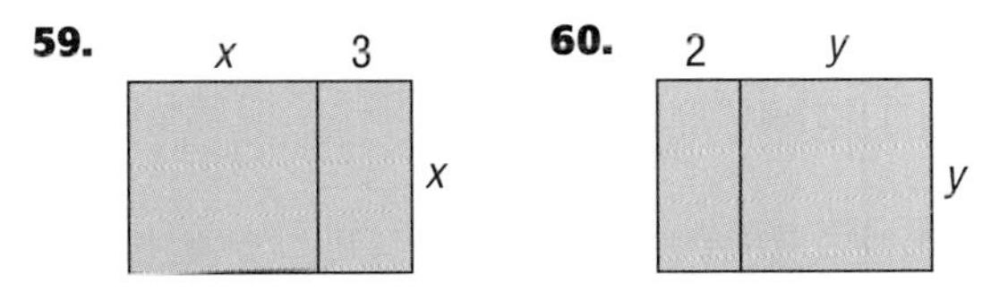

Write the area of the shaded rectangle in expanded form.

61. **62.**

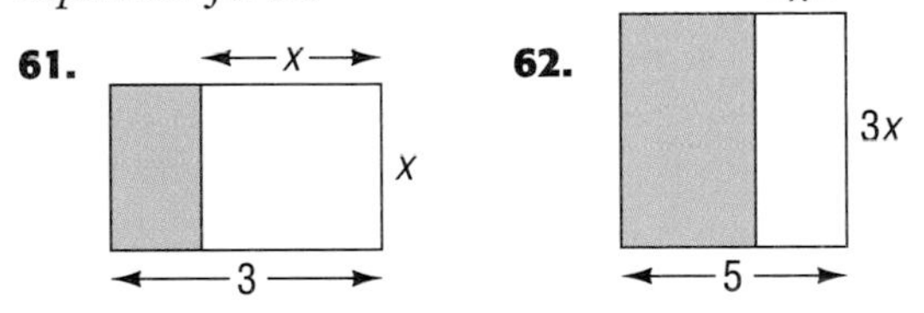

63. The money you deposit in a bank is called the principal, P. After your money earns interest, the new principal is given by the expression $P(1+rt)$, where r is the interest rate and t is the length of time your money was deposited.

a) Expand the expression $P(1+rt)$.

b) Evaluate the expression for $P = \$300$, $r = 0.05$, and $t = 2$ years.

5.7 Multiplying Monomials by Monomials

The first male killer whale born in captivity in North America was born at Marineland in Niagara Falls. The volume of the tank in which *Splash* lives can be expressed as the product of 3 monomials, $l \times w \times h$, where l is the length, w is the width, and h is the height of the tank.

Activity: Use Algebra Tiles

Arrange six x^2-tiles to form a rectangle as shown.

Inquire

1. What is the area of the rectangle?

2. What is the length of the rectangle?

3. What is the width of the rectangle?

4. Write the area of the rectangle as a product of its dimensions.

5. Use x^2-tiles to model these products. State the product in each case.

a) $2x \times 2x$ **b)** $x \times 2x$ **c)** $3x \times x$

6. Write a rule for multiplying a monomial by a monomial.

Example

Find the product.

a) $(30x)(4y)$ **b)** $-5x^2(-3yz)$

Solution

a)
$$\begin{aligned}&(30x)(4y)\\&= 30 \times x \times 4 \times y\\&= 30 \times 4 \times x \times y\\&= 120xy\end{aligned}$$

b)
$$\begin{aligned}&-5x^2(-3yz)\\&= -5 \times x^2 \times (-3) \times y \times z\\&= -5 \times (-3) \times x^2 \times y \times z\\&= 15x^2yz\end{aligned}$$

Practice

Multiply.

1. $5x \times 3y$ **2.** $2m \times 3n$ **3.** $5s \times 7t$

4. $4a \times 6b$ **5.** $3x^2 \times 2y$ **6.** $4a \times 5b^2$

7. $4b \times 3c$ **8.** $3a \times 2b^2$ **9.** $6s \times 3t$

Multiply.

10. $(3x)(2y)$ **11.** $(3a)(4b)$ **12.** $(5x^2)(2y^2)$

13. $(5ab)(3c)$ **14.** $(4x)(3y)$ **15.** $(6xy)(5z)$

16. $(2a^2)(3b^2)$ **17.** $(3a)(3b)$ **18.** $(7a)(5b)$

Multiply.

19. $(3x^2)(-5y^2)$ **20.** $-2t^3(-4a)$

21. $(6ab)(-2c^2)$ **22.** $-8a^2(3y^2)$

23. $(5x)(5yz)$ **24.** $-12x^2(4y^2)$

Multiply.

25. $(-2xy)(-7xy)$ **26.** $-5m^2(-2mn)$

27. $4s^2t^3(-3st)$ **28.** $-3abx(-2a^2b^4y)$

29. $-2s^2t^3(-5s^4t^2)$ **30.** $-5x^2y^2(4c^2x^3y^8)$

Multiply.

31. $-3x^2yz(5yz)$ **32.** $-2x^2(-3cy^2z^3)$

33. $-2x^2(-2y^2)(z)$ **34.** $-5x^2(-7y)(-2z)$

35. $-5x(-7y)(-2z)$ **36.** $-6xy(-5z^2)(3t^2)$

Multiply.

37. $(4a^2x^3z)(-2x^3y^2z^2)$

38. $(2b^2xy^4z^2)(-3xyz)$

39. $-5a^2b^3(-2a^2b^2)$

Multiply.

40. $(3abc)(-4abc)(2abc)$

41. $(-x^2yz)(2xy^2z)(-2xyz^2)$

42. $(-2jkl)(-3jkl)(-4jkl)$

Problems and Applications

43. Find the area of each figure.

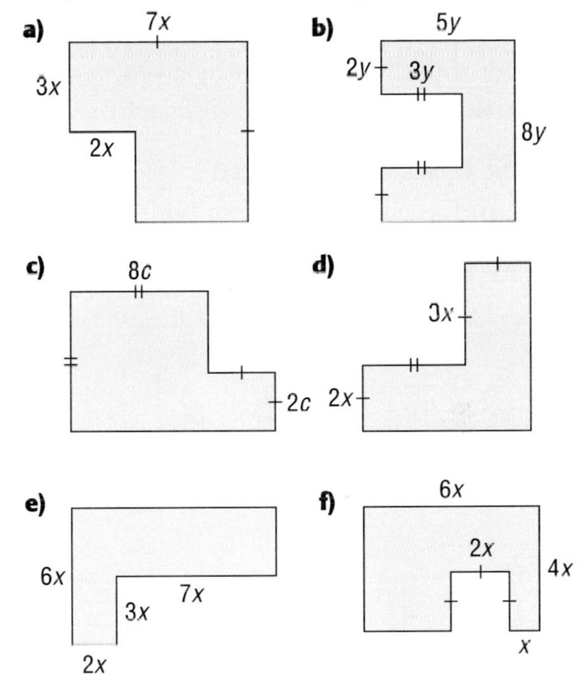

44. a) If x is the time needed in minutes to make a hat, and y is the cost of production in dollars per minute, what does the expression $24xy$ mean?

b) Use $x = 3$ and $y = 0.5$ to evaluate your expression in part a).

45. Write an expression for the volume of each prism.

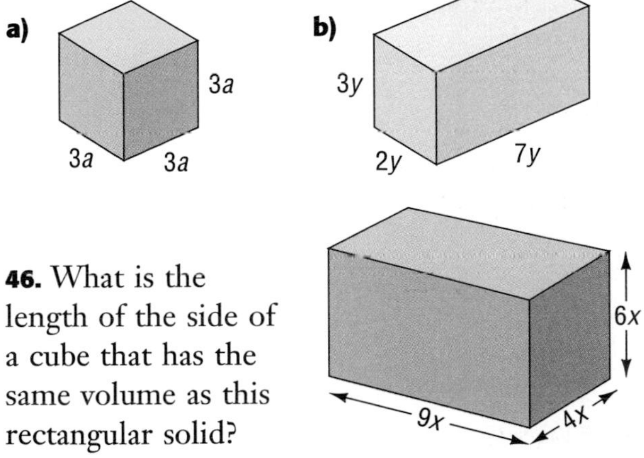

46. What is the length of the side of a cube that has the same volume as this rectangular solid?

5.8 Powers of Monomials

There are 35 major pyramids in Egypt. The typical Egyptian pyramid has a square base. The area of the base is given by a power of a monomial, s^2, where s is the side length.

Activity: Look for a Pattern

The following cubes have been built from interlocking cubes.

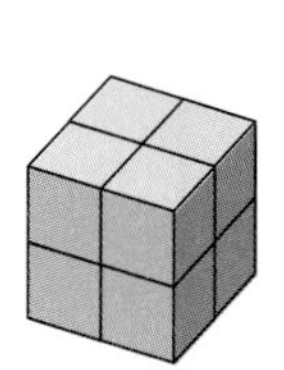

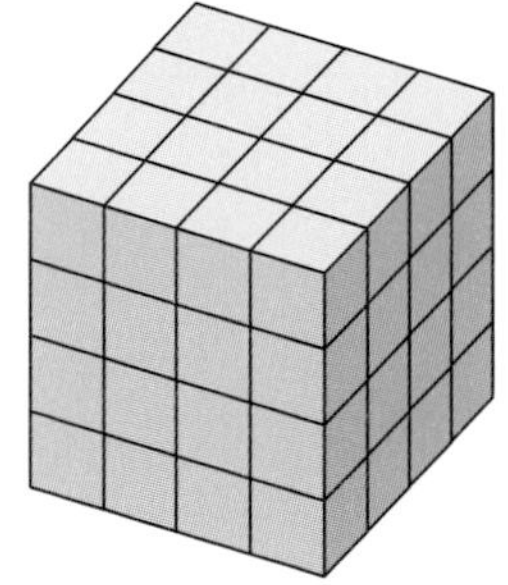

Inquire

1. Copy the table. Work with a partner to complete it. Consider each interlocking cube to have a side length of 1 unit.

Side Length of Large Cube	Total Number of Interlocking Cubes
2	
3	
4	
5	
10	
100	

2. To what exponent is the side length of each cube raised to express the volume?

3. What algebraic expression represents the volume of a cube with side length x?

4. What algebraic expression represents the volume of a cube with side length $2x$?

5. The volume of a cube with side length $2x$ can also be found by multiplying the lengths of the sides.
$2x \times 2x \times 2x = 8x^3$
How does the resulting expression compare with the expression you wrote in question 4?

6. Write a rule to find powers of a monomial.

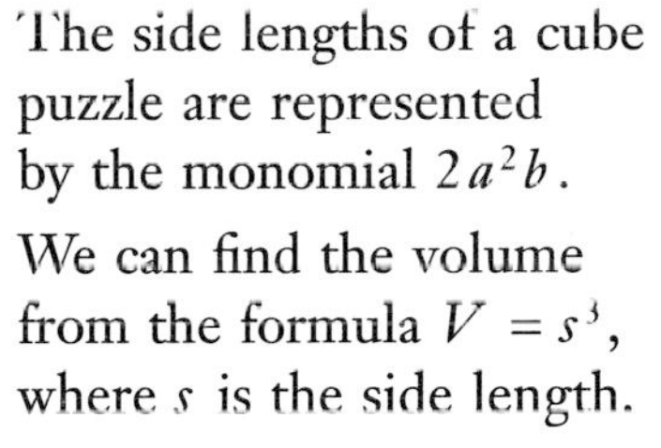

The side lengths of a cube puzzle are represented by the monomial $2a^2b$.

We can find the volume from the formula $V = s^3$, where s is the side length.

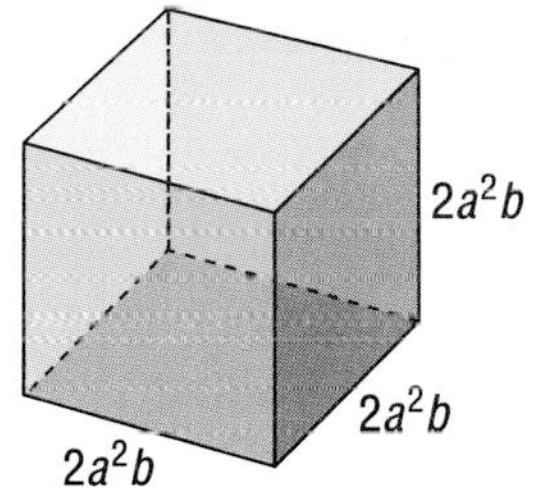

For the given cube, the volume is

$$\begin{aligned} V &= (2a^2b)^3 \\ &= (2a^2b)(2a^2b)(2a^2b) \\ &= (2 \times 2 \times 2)(a^2 \times a^2 \times a^2)(b \times b \times b) \\ &= 8a^6b^3 \end{aligned}$$

We obtain the same result by finding the third power of each factor in the monomial.

$$\begin{aligned} (2a^2b)^3 &= (2)^3(a^2)^3(b)^3 \\ &= 8a^6b^3 \end{aligned}$$

In general, $(x^my^n)^a = x^{am}y^{an}$

Example 1

Simplify.

a) $(x^3)^2$

b) $(y^3)^4$

Solution

a) $(x^3)^2 = x^{3\times2}$
$= x^6$

b) $(y^3)^4 = y^{3\times4}$
$= y^{12}$

Example 2

Simplify.

a) $(5x^2y^4)^3$

b) $(-2x^3y)^4$

c) $(3x^4y^2)(-x^2y^3)^2$

Solution

a)
$$\begin{aligned} &(5x^2y^4)^3 \\ &= (5)^3(x^2)^3(y^4)^3 \\ &= 125x^6y^{12} \end{aligned}$$

b)
$$\begin{aligned} &(-2x^3y)^4 \\ &= (-2)^4(x^3)^4(y)^4 \\ &= 16x^{12}y^4 \end{aligned}$$

c)
$$\begin{aligned} &(3x^4y^2)(-x^2y^3)^2 \\ &= (3x^4y^2)(-1)^2(x^2)^2(y^3)^2 \\ &= (3x^4y^2)(1)(x^4)(y^6) \\ &= (3 \times 1)(x^{4+4})(y^{2+6}) \\ &= 3x^8y^8 \end{aligned}$$

CONTINUED

Practice

Simplify.

1. $(x)^2$
2. $(a)^3$
3. $(p)^5$
4. $(n^2)^2$
5. $(-t^3)^2$
6. $(-y^2)^3$

Simplify.

7. $(x^2)^3$
8. $(y^3)^2$
9. $(m^2)^2$
10. $(n^3)^4$
11. $(x^3)^3$
12. $(y^2)^3$
13. $(z^4)^3$
14. $(m^4)^5$
15. $(p^{18})^2$
16. $(-s^{10})^2$
17. $(-x)^{31}$
18. $-(-b^0)^3$

Simplify.

19. $(xy)^2$
20. $(ab)^3$
21. $(-xy)^2$
22. $(mn)^4$
23. $(pq)^3$
24. $(-2xt)^2$
25. $(4xy)^2$
26. $(-2ax)^3$
27. $-(3rs)^3$

Simplify.

28. $(x^2y^2)^3$
29. $(x^2y^3)^2$
30. $(a^2b)^3$
31. $(ab^3)^2$
32. $(mn)^3$
33. $(-ab^2)^2$
34. $(-j^3k^4)^2$
35. $(x^2y)^2$
36. $-(s^3t^2)^0$

Simplify.

37. $(2x^2)^3$
38. $(3y^3)^2$
39. $(4x^4)^2$
40. $(5y^2)^2$
41. $(-m^2)^2$
42. $(-n^2)^3$
43. $(-2n^2)^3$
44. $(-3y^2)^2$
45. $(3pqr)^2$
46. $(-3yz)^3$
47. $(-4x^2y^3)^3$
48. $-(3xy^0)^2$

Simplify.

49. $(2x^2y^3)^2(x^2y)$
50. $(-3xy)(-2xy)^2$
51. $(2xy^2)^3(3x^2y^2)$
52. $(2a^4b^3)(-3ab)^3(10a^2b^2)$
53. $(10abc)^2(-2a^2bc)$

Problems and Applications

54. Write and simplify an expression for the area of each square.

a) **b)**

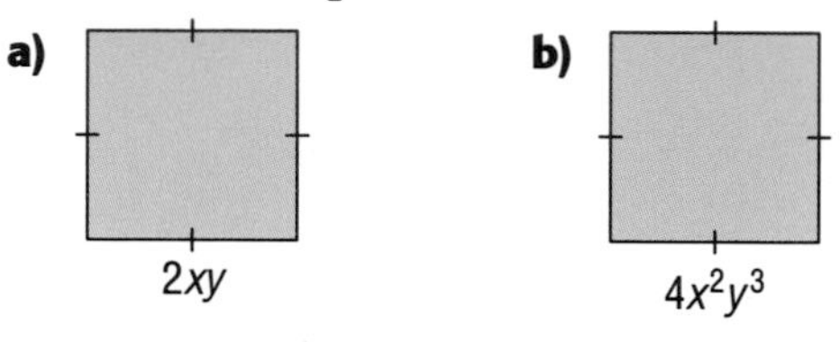

55. Write and simplify an expression for the volume of each cube.

a)

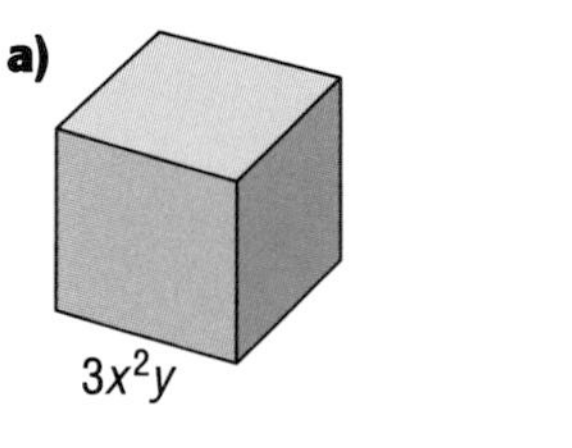

b)

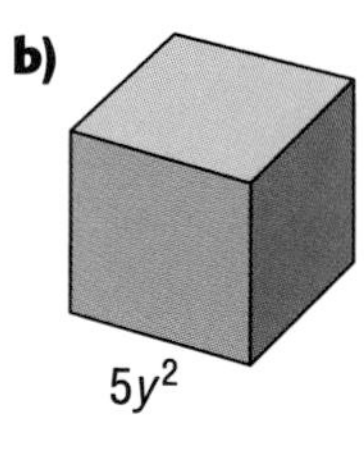

56. Find the volume of this cube if $a = 1$, $b = 2$, and $c = 3$.

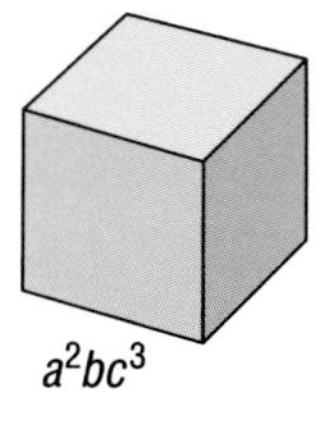

57. Does replacing the variables by their opposites, x by $-x$ and y by $-y$, result in the opposite of each monomial? Explain.

a) $(3x^2y^3)^2$ **b)** $(3x^2y^3)^3$

NUMBER POWER

Copy the diagram and place the numbers from 1 to 9 in the circles so that the consecutive sums of each side differ by 4.

Electronic Spreadsheets in Sports

Sports statisticians use spreadsheets to keep track of individual and team statistics during a season so that they can be reported to fans. A spreadsheet allows simple daily updates of the data.

Activity ❶ Scoring-Leader Statistics

The spreadsheet gives the partial statistics for several hockey players for a recent season.

GP = Games Played
G = Goals
A = Assists
PTS = Points (1 for each goal and assist)
OWN = Number of even-strength and shorthanded goals scored by a player's team while the player was on the ice.
OPP = Number of goals scored by the opponents while the player was on the ice.
+/− = **(OWN − OPP)**
S = Shots on Goal
Pct = Percent of shots that scored goals.

	A	B	C	D	E	F	G	H	I	J
1	Player	GP	G	A	PTS	OWN	OPP	+/−	S	Pct
2										
3	Hull	73	70	39		112	114		408	
4	Messier	79	35	72		132	101		212	
5	Stevens	80	54	69		90	82		325	
6	Gretzky	74	31	90		87	99		215	
7	Lemieux	64	11	87		97	70		289	
8	Robitaille	80	44	63		124	128		240	
9	Roenick	80	53	50		119	96		234	

1. Enter the data and formulas on a spreadsheet that will calculate the missing statistics.

2. Rank the players from best to worst. Give your reasons and compare them with a classmate's.

Activity ❷ Other Sports

1. With a partner, choose a sport of your preference.

2. With your partner, set up a spreadsheet that will calculate team statistics or individual player statistics.

5.9 Multiplying a Polynomial by a Monomial

Northern Dancer was one of the fastest thoroughbred horses that ever raced. He was born, raised, and trained at Winfields Farm in Oshawa. The 500 ha of Winfields Farm contains many different sizes and shapes of rectangular fields attached to each other, like the rectangles shown below.

Activity: Explore the Pattern

Copy or trace this diagram onto a blank sheet of paper.

Inquire

1. In simplest form, what is the area of

a) the square? **b)** the rectangle attached to the square?

2. Write the sum of the areas for the square and the rectangle.

3. Consider the larger rectangle that includes the square.

a) What is the length of the larger rectangle?

b) What is the width of the larger rectangle?

c) Use your answers to parts a) and b) to write an expression for the area.

d) How can you multiply the monomial in your expression by the binomial in your expression to arrive at the answer you gave in question 2?

4. Write a rule for multiplying a polynomial by a monomial.

Example 1

Expand and simplify.

a) $2x(x^2 - 2x + 5)$

b) $4x(x - 3) - 2(x + 3)$

Solution

Use the distributive property to expand each expression. Then, collect like terms and simplify.

a) $2x(x^2 - 2x + 5)$

$$= 2x(x^2 - 2x + 5)$$
$$= 2x^3 - 4x^2 + 10x$$

b) $4x(x - 3) - 2(x + 3)$

$$= 4x(x - 3) - 2(x + 3)$$
$$= 4x^2 - 12x - 2x - 6$$
$$= 4x^2 - 14x - 6$$

Example 2

Expand and simplify $2(3x^2-4x+5)-2x(x-3)$.

Solution

$2(3x^2 - 4x + 5) - 2x(x - 3)$

$$= 2(3x^2 - 4x + 5) - 2x(x - 3)$$
$$= 6x^2 - 8x + 10 - 2x^2 + 6x$$
$$= 4x^2 - 2x + 10$$

Practice

Expand.

1. $x(x+2)$ **2.** $x(x-3)$ **3.** $a(a+1)$
4. $t(t-1)$ **5.** $y(y+4)$ **6.** $m(m+5)$
7. $x(x-5)$ **8.** $y(y-7)$ **9.** $a(a-10)$

Expand.

10. $3x(x+2)$ **11.** $4b(b-11)$
12. $5t(t+3)$ **13.** $2x(3+x)$
14. $7y(y-5)$ **15.** $-2x(x+4)$
16. $-x(x+2)$ **17.** $-y(y-3)$

Expand and simplify.

18. $x(x+3)-x(x-2)$
19. $y(2+y)+y(y-1)$
20. $m(m-1)+m(m-1)$
21. $x(x+2) \quad (2x-2)$
22. $y(y-4)-y(3-2y)$
23. $a(2a-1)+a(a+1)$
24. $x(x-2)-x(x+1)$

Expand and simplify.

25. $3x(x+2)+2x(x+5)$
26. $2x(x-3)-x(x-5)$
27. $3x(2x+1)+x(3x+2)$
28. $-2y(y-3)-y(y+1)$
29. $2a(a+3)+3a(a-2)$
30. $-x(3x-4)-2x(1-x)$
31. $4x(x+2)+2x(7-2x)$

Expand.

32. $x(x^2+2x+3)$ **33.** $3(x^2+2x-5)$
34. $5x(x^2+2x-7)$ **35.** $-(x^2-3x-1)$
36. $4m(m^2-5m+6)$ **37.** $3y(2y^2-4y+3)$
38. $-3b(3b^2-5b+1)$ **39.** $-5z(z^2-2z-5)$

Problems and Applications

Expand and simplify.

40. $3(x^2+2x-5)-x(x+1)$
41. $5(x^2+2x-7)+3x(x+1)$
42. $-(x^2-3x-1)+x(3x+2)$
43. $4(2x+3)+3x(x^2-x+3)$
44. $3m(m-2)+4(m^2-5m+6)$
45. $5y(1-y)+3(2y^2-4y+3)$
46. $-3x(x+2)+2x(2x-1)$

47. Write, expand, and simplify an expression for the area of each figure.

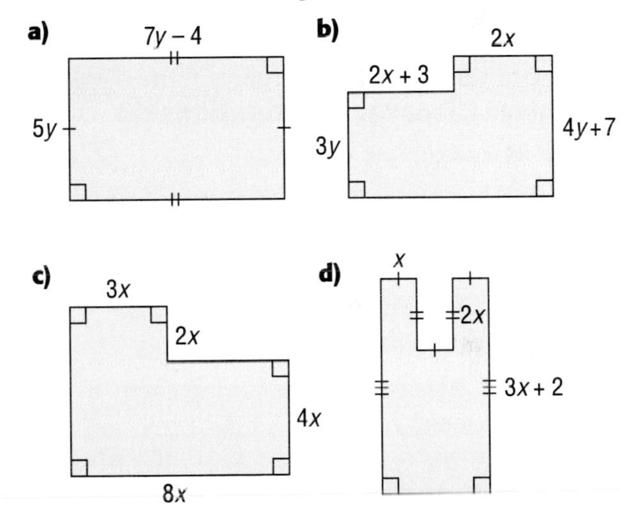

48. Write a problem similar to those in question 47. Have a classmate solve your problem.

WORD POWER

Find a three-letter word that can go in front of each of the following words to form a new word.

TON PET ROT

5.10 Dividing Monomials by Monomials

When Patricia Hy became Canadian Women's Tennis Champion, she played on a regulation court. If we know the area of the court and its length or width, we can calculate the other dimension. Dividing the area, A, by the length, l, or width, w, involves the division of a monomial by a monomial.

Recall that for the multiplication statement $3 \times 2 = 6$, there are 2 division statements.

$6 \div 2 = 3$ $\qquad$ $6 \div 3 = 2$

Activity: Discover the Relationship

Copy and complete.

a) $2a \times 3b$ **b)** $3x \times 5x$ **c)** $4x^2 \times 2x$

Inquire

1. Write 2 division statements for each multiplication statement above.

2. Write 2 division statements for each of the following.

a) $3g \times 4h = 12gh$ **b)** $5y \times 2y^2 = 10y^3$

c) $2x \times 3y = 6xy$ **d)** $x^2 \times 2y^3 = 2x^2y^3$

3. Write a rule for dividing monomials by monomials.

Example

Divide.

a) $10x^3y^4 \div 2y^3$ **b)** $\dfrac{-8x^6y^7}{4x^5y^5}$ **c)** $\dfrac{60a^4b^3c^2}{30a^2b^2c^2}$

Solution

a)
$$
\begin{aligned}
&10x^3y^4 \div 2y^3 \\
&= \frac{10x^3y^4}{2y^3} \\
&= \left(\frac{10}{2}\right)\left(\frac{x^3}{1}\right)\left(\frac{y^4}{y^3}\right) \\
&= 5x^3y
\end{aligned}
$$

b)
$$
\begin{aligned}
&\frac{-8x^6y^7}{4x^5y^5} \\
&= \left(\frac{-8}{4}\right)\left(\frac{x^6}{x^5}\right)\left(\frac{y^7}{y^5}\right) \\
&= -2xy^2
\end{aligned}
$$

c)
$$
\begin{aligned}
&\frac{60a^4b^3c^2}{30a^2b^2c^2} \\
&= \left(\frac{60}{30}\right)\left(\frac{a^4}{a^2}\right)\left(\frac{b^3}{b^2}\right)\left(\frac{c^2}{c^2}\right) \\
&= 2a^2b
\end{aligned}
$$

Practice

Divide.

1. $\frac{6x}{3}$
2. $\frac{-15a}{5}$
3. $\frac{24y}{8}$
4. $\frac{36m}{9}$
5. $\frac{-30x}{6}$
6. $\frac{-25y}{5}$
7. $\frac{12x}{4x}$
8. $\frac{-18y}{3y}$
9. $\frac{24a}{a}$
10. $\frac{-32b}{-b}$
11. $\frac{-40x}{40}$
12. $\frac{-32m}{-8}$

Divide.

13. $\frac{15xyz}{5xy}$
14. $\frac{-18ab}{6a}$
15. $\frac{12pqr}{4pqr}$
16. $\frac{36abc}{-4a}$
17. $\frac{28xy}{7y}$
18. $\frac{-15rst}{3rt}$

Divide.

19. $25xy \div 5xy$
20. $22ab \div 11ab$
21. $21xyz \div 3xyz$
22. $9amn \div amn$
23. $36rst \div 3rs$
24. $39jkl \div 13kl$
25. $52pqrs \div 13ps$
26. $51defg \div 3eg$

Simplify.

27. $5x^4y^2 \div x^3y$
28. $-3a^3b^4 \div ab$
29. $18j^7k^7 \div (-9j^4)$
30. $-20x^5y^{15} \div 4x^2y$
31. $7a^3b^2c \div (-7ab)$
32. $-8x^2y^9 \div (-2xy^8)$

Simplify.

33. $\frac{10a^6b^3}{5a^4b^2}$
34. $\frac{-15x^6y^8}{-5x^5y^7}$
35. $\frac{-12m^6n^2}{4m^3n^2}$
36. $\frac{22x^2y^4z^3}{11xy^3z}$
37. $\frac{20a^4b^6c}{15a^4b^6c}$
38. $\frac{-18x^5y^8}{12x^3y^6}$

Simplify.

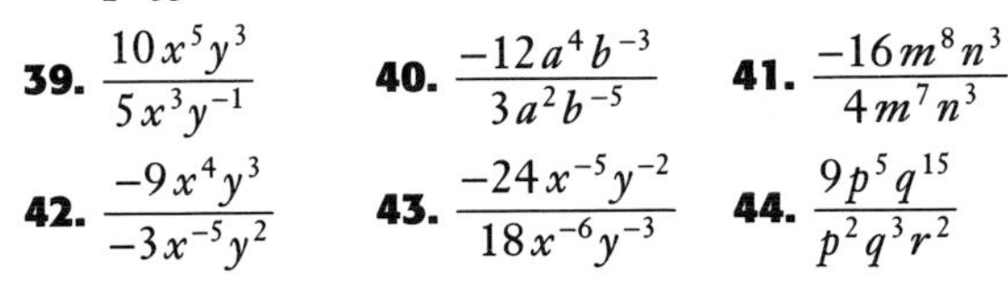

39. $\frac{10x^5y^3}{5x^3y^{-1}}$
40. $\frac{-12a^4b^{-3}}{3a^2b^{-5}}$
41. $\frac{-16m^8n^3}{4m^7n^3}$
42. $\frac{-9x^4y^3}{-3x^{-5}y^2}$
43. $\frac{-24x^{-5}y^{-2}}{18x^{-6}y^{-3}}$
44. $\frac{9p^5q^{15}}{p^2q^3r^2}$

Problems and Applications

45. Find the missing dimension in each rectangle.

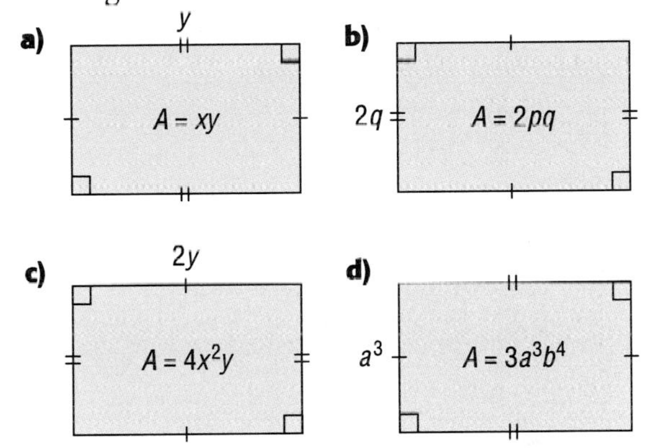

46. What are the dimensions of each rectangle if each area is 160 cm²?

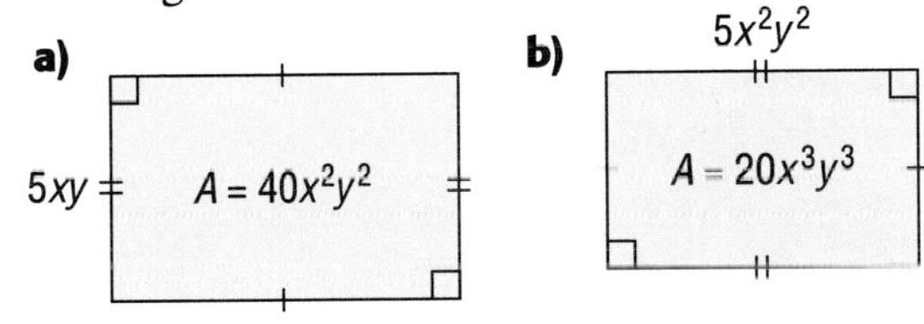

47. Find the quotient of the volume of this box divided by its area.

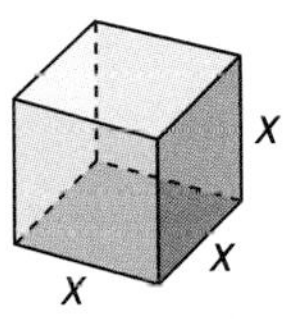

48. A and B represent monomials. When you multiply A × B, the result is A. When you divide A ÷ B, the result is A. What is the monomial B?

49. Write a problem similar to problem 46. Have a classmate solve your problem.

PATTERN POWER

There are x cars in a bumper-to-bumper lineup. The cars' bumpers are touching. How many bumpers are touching?

5.11 Dividing Polynomials by Monomials

An x^2-tile has an area of x^2 square units.
An x-tile has an area of x square units.

Activity: Use Algebra Tiles

Use an x^2-tile and an x-tile to model this rectangle.

Inquire

1. What is the area of the rectangle?

2. What is the length of the rectangle?

3. What is the width of the rectangle?

4. Describe how you can divide the area by the width to give the length.

5. Write a rule for dividing a polynomial by a monomial.

The distributive property applies to division as well as to multiplication.

$$\frac{a+b}{c} = \frac{1}{c}(a+b) = \frac{a}{c} + \frac{b}{c}$$

Example 1

Simplify.

a) $\dfrac{5xyz + 10xy}{5xy}$

b) $\dfrac{8x^5 + 12x^3 - 4x^2}{4x}$

Solution

Divide each term of the polynomial by the monomial.

a)
$$\frac{5xyz + 10xy}{5xy} = \frac{5xyz}{5xy} + \frac{10xy}{5xy}$$
$$= z + 2$$

b)
$$\frac{8x^5 + 12x^3 - 4x^2}{4x} = \frac{8x^5}{4x} + \frac{12x^3}{4x} - \frac{4x^2}{4x}$$
$$= 2x^4 + 3x^2 - x$$

Example 2

Simplify.

a) $\dfrac{15y^4 - 12y^5 + 9y^3}{-3y^2}$

b) $\dfrac{16a^5b^3 - 20a^4b^4 + 24a^5b^3}{4a^3b^3}$

c) $\dfrac{9a^4b^7z^5 - 15ab^3z^2}{-3ab^3z}$

Solution

a)
$$\frac{15y^4 - 12y^5 + 9y^3}{-3y^2}$$
$$= \left(\frac{15y^4}{-3y^2}\right) - \left(\frac{12y^5}{-3y^2}\right) + \left(\frac{9y^3}{-3y^2}\right)$$
$$= -5y^2 + 4y^3 - 3y$$
$$= 4y^3 - 5y^2 - 3y$$

b)
$$\frac{16a^5b^3 - 20a^4b^4 + 24a^5b^3}{4a^3b^3}$$
$$= \frac{16a^5b^3}{4a^3b^3} - \frac{20a^4b^4}{4a^3b^3} + \frac{24a^5b^3}{4a^3b^3}$$
$$= 4a^2 - 5ab + 6a^2$$

c)
$$\frac{9a^4b^7z^5 - 15ab^3z^2}{-3ab^3z}$$
$$= \frac{9a^4b^7z^5}{-3ab^3z} - \left(\frac{15ab^3z^2}{-3ab^3z}\right)$$
$$= -3a^3b^4z^4 - (-5z)$$
$$= -3a^3b^4z^4 + 5z$$

Practice

Divide.

1. $\frac{12xy}{4xy}$

2. $\frac{24ab}{-4ab}$

3. $\frac{-12mn}{3m}$

4. $\frac{-30xy}{-5xy}$

5. $\frac{11ab}{ab}$

6. $\frac{5xy}{5x}$

7. $\frac{24x^2y}{6xy}$

8. $\frac{15ab^3}{-5ab}$

9. $\frac{36x^3y^7}{-6x^2y^2}$

Divide.

10. $\frac{12xy - 15y^2 + 24y}{3y}$

11. $\frac{5x^3 + 10x^2 - 15x}{5x}$

12. $\frac{7y^4 + 7y^3 - 21y}{-7y}$

13. $\frac{4m^3 + 8m^2 - 12m}{4m}$

14. $\frac{9x^3 - 24x^2 - 15x}{-3x}$

15. $\frac{6j^5 + 12j^4 + 18j^3}{-6j}$

Divide.

16. $\frac{10x^4 + 5x^3 - 15x^2}{-5x^2}$

17. $\frac{-21m^2 + 14m^3 - 21m^4}{-7m^2}$

18. $\frac{10p^2q^2 - 15pq^3 + 25p^3q^4}{-5pq^2}$

19. $\frac{-12a^3b^2 + 9a^2b^3 + 24a^4b^4}{3a^2b^2}$

Divide.

20. $\frac{-20x^3yz + 30x^2y^2z - 40xy^3z}{-10xyz}$

21. $\frac{8a^3b^2c^3 - 12a^2b^2c^2 + 16a^2b^3c}{4a^2b^2}$

22. $\frac{-12x^4y^6 - 16x^5y^5 - 24x^6y^4}{-4x^4y^4}$

23. $\frac{30m^3n^5 - 36m^4n^4 - 30m^5n^3}{6m^3n^3}$

24. $\frac{25a^3b^3c^5 - 40a^4b^3c^4 + 35a^6b^4c^3}{-5a^2b^2c^2}$

Problems and Applications

25. Determine the length, l, of the unknown side, given the area and the length of one side.

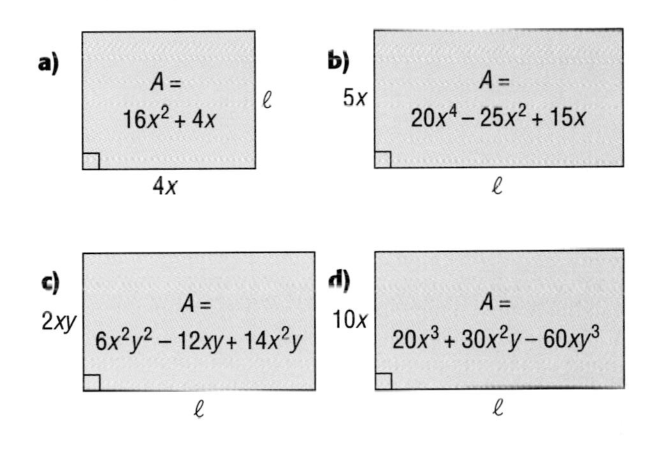

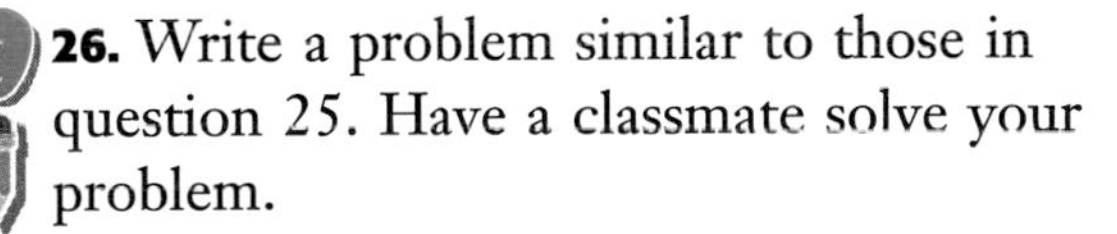

26. Write a problem similar to those in question 25. Have a classmate solve your problem.

LOGIC POWER

Four students entered a problem solving contest. Each student represented a different zone of the town. Use the clues to determine which zone each student represented and in which order the students finished.

1. David came second, just behind the student from the west zone.

2. Petra represented neither the east nor the west zones.

3. The student from the north zone finished second last, just ahead of Frank.

4. David and Jarvi represented opposite zones in the town.

The Solar System

The table shows the diameters of the star and planets in the solar system relative to the Earth's diameter.

Planet or Star	Diameter (Earth = 1)
Sun	109
Mercury	0.38
Venus	0.95
Earth	1.00
Mars	0.53
Jupiter	11.2
Saturn	9.4
Uranus	4.0
Neptune	3.8
Pluto	0.2

Activity ❶ Diameters in the Solar System

1. Work with a partner to explain what is meant by the heading "Earth = 1" in the diameter column of the table.

2. How were the diameters of the other planets calculated? Compare your answer to a classmate's.

3. a) Estimate how many times the smallest planet could fit across the diameter of the sun.
b) Use your calculator to check your estimate. How close was your estimate?

4. Prepare some arguments that would support the statement, "Some planets are approximately the same size."

5. If the diameter of the Earth is approximately 12 756 km, what are the diameters of the other planets?

6. Use circles, or parts of circles, to make a scale drawing that shows how the sizes of the planets and the sun compare.

7. Write three English words that are derived from *sol*, the Latin word for sun.

Activity 2 Distances in the Solar System

This table shows how far from the sun each planet is relative to the Earth.

Planet	Average Distance from Sun (Earth = 1)
Mercury	0.39
Venus	0.72
Earth	1.00
Mars	1.52
Jupiter	5.2
Saturn	9.5
Uranus	19.2
Neptune	30.1
Pluto	39.6

1. Work with a partner to explain why the numbers are increasing from the beginning to the end of this table.

2. Compare the meaning of the heading "Earth = 1" in this table to its meaning in the previous table.

3. If the average distance of the Earth from the sun is 149.6 million kilometres, calculate the average distance from the sun for the rest of the planets.

4. What sort of orbit does Pluto make as it revolves around the sun?

5. Make a scale drawing to show how the distances of the planets from the sun compare.

Activity 3 Researching a Space Probe

1. Research the latest space probe. Then, write a press release to describe what was supposed to happen.

2. List 3 points supporting the view that the money spent on all the space probes to date is well worth it.

3. List 3 points supporting the view that the money spent on all the space probes to date is not worth it.

4. a) Work with a classmate to compare the lists that each of you prepared in questions 2 and 3 of this activity.
b) On the basis of your lists, discuss whether space probes should continue.

5. Divide the class into small groups to prepare a debate on this topic.

6. Choose a side, elect a speaker for each side of the argument, and hold the debate.

Review

Evaluate $3x - 2$ for each value of x.

1. 3 **2.** 5 **3.** 7

4. 0 **5.** -3 **6.** 1

7. -5 **8.** 2 **9.** -10

Evaluate for $y = -3$.

10. $y + 5$ **11.** $y - 2$ **12.** $y + 7$

13. $y - 3$ **14.** $2y$ **15.** $-3y$

16. y^2 **17.** y^0 **18.** $2(y^0)$

Evaluate $3x + 2y$ for the following values of the variables.

19. $x = 1,\ y = 3$ **20.** $x = 4,\ y = 2$

21. $x = 7,\ y = 1$ **22.** $x = 3,\ y = 3$

23. $x = -4,\ y = 0$ **24.** $x = -2,\ y = -1$

25. $x = 3,\ y = -5$ **26.** $x = -1,\ y = -3$

Evaluate each expression for $a = 2.5$, $b = -3.5$, and $c = 5.5$.

27. ab **28.** $a + b + c$

29. $ab + c$ **30.** $a - bc$

31. $2a + 3b$ **32.** $c - 5b$

33. $a + 2b - 3c$ **34.** $2a - b + 4c$

Simplify.

35. $7x + 10 + 7y - 4x + 1$

36. $2a - 3b - 5 + a - b$

37. $-q + 4q - 9r - q + 2r + 1$

Identify each polynomial type and state its degree.

38. xyz **39.** $3x^2y^3$

40. $5xy^2 - 5xy$ **41.** $x^2 + y$

42. $ab^3 + b^2 + 1$ **43.** $3m^2n^3 + 5mn$

44. $x^2 + 2x^4 + 5x^2$ **45.** $3x^3 - 5x + 3$

46. $x^2yz + 2yz$ **47.** $b^2x^3 - 2x^3 - 4xy^2$

Arrange each polynomial with its powers in descending order.

48. $x^2 + 3x - 5x^3$

49. $5 + 2y^2 - 3y + y^4$

50. $2m + 6 - 3m^2 - m^4 + 6m^3$

51. $3 + x + x^2 + x^3 + x^4$

52. $y^5 - 2y^7 + 3y - 4y^2 + 5y^6$

Add.

53. $(m^2 + 3m - 5) + (2m^2 - 5m + 7)$

54. $(5a + 3a^2 - 2) + (a^2 + 3a + 4)$

55. $(b + 5 - 2b^2) + (1 - 2b + b^2)$

Add.

56. $\begin{array}{r} 2x^3 - 3x^2 + 5x \\ \underline{-\ x^2 + 3x} \end{array}$ **57.** $\begin{array}{r} 3x^2 - 7x + 5 \\ \underline{-x^2 - x - 3} \end{array}$

58. $\begin{array}{r} -5a^2 - 2a - 7 \\ \underline{6a^2 + 4a + 3} \end{array}$ **59.** $\begin{array}{r} 7t^2 + 8t - 9 \\ \underline{2t^2 - 10t - 5} \end{array}$

Subtract.

60. $(5a^2 - 3a + 6) - (2a^2 + 3a + 7)$

61. $(3m^2 + 5m + 1) - (5m^2 + 2m - 1)$

62. $(3x^2 + 2x - 5) - (x^2 + x + 1)$

Subtract.

63. $\begin{array}{r} 5x^2 - 3x + 2 \\ \underline{-x^2 - 2x + 4} \end{array}$ **64.** $\begin{array}{r} 4x^2 - 7x + 5 \\ \underline{7x^2 - 9x + 3} \end{array}$

65. $\begin{array}{r} 2x^2 + 8x - 4 \\ \underline{-3x^2 + 13x - 11} \end{array}$ **66.** $\begin{array}{r} 8x^2 - 2x - 1 \\ \underline{3x^2 + x + 7} \end{array}$

Multiply.

67. $(10x)(4y)$ **68.** $(-30y)(5xy)$

69. $(3ax)(4bx)$ **70.** $(-2ap)(-5ab)$

71. $(3xy)(-5xy)$ **72.** $-4a^2(-ab)$

73. $4x^2(-2xy^2z)$ **74.** $-s^2t^2(3s^3t^3)$

Simplify.

75. $(x^3)^2$ **76.** $(y^4)^3$ **77.** $(xy)^3$

78. $(3y^3)^2$ **79.** $(x^2y^3z)^2$ **80.** $(-r^3s^4t^2)^3$

81. $(-2x^2y^3)^3$ **82.** $(-5x^2y^2)^2$ **83.** $(-2abc^4)^3$

Simplify.

84. $(2x^2yz^2)(x^2y^2z^2)^3$

85. $(3a^3b)^3(-a^3bx^4)$

86. $(-2j^3k^2l^2)^3(-j^2k^3l^3)$

87. $(a^4b^3)(-2ab)^3(5a^2b^2)$

88. $(-abc)^2(-a^2b^3c^3)(-a^2b^2c)$

Expand and simplify.

89. $2(3x+1)+3(x+4)$

90. $4(a+5)+3(a-2)$

91. $3y(2y-7)+2(y-5)$

92. $2(m+1)-m(m+1)$

93. $-z(2z+3)-(z-5)$

94. $5x(2x-7)-3x(3x-5)$

Expand.

95. $2y(-y^2+3y-7)$

96. $-3t(1-2t-t^2)$

97. $4m(m^2+2m-3)$

98. $-3x(2x^2-4x+2)$

Divide.

99. $\frac{20x^5y^8}{5x^4y^6}$ **100.** $\frac{36x^3y^4}{4x^2y^3}$

101. $\frac{45a^3b^6}{9ab^4}$ **102.** $\frac{-20a^4b^4}{10a^3b^{-3}c^2}$

103. $\frac{15abc+5ab}{5ab}$ **104.** $\frac{4x^5+8x^3-2x^2}{2x^2}$

105. $\frac{8a^3b^5-16a^4b^4+4a^5b^3}{4a^3b^3}$

106. $\frac{6a^4b^5z^5-12a^4b^5z^4}{-3a^3b^4z^2}$

Group Decision Making
Researching Computer Careers

1. Brainstorm the computer careers that you would like to investigate. They could include careers such as programming, systems design, sales, marketing, and hardware technician. As a class, select six careers to research. Number the careers from 1 to 6.

2. Go to home groups. As a group, decide which career each group member will research.

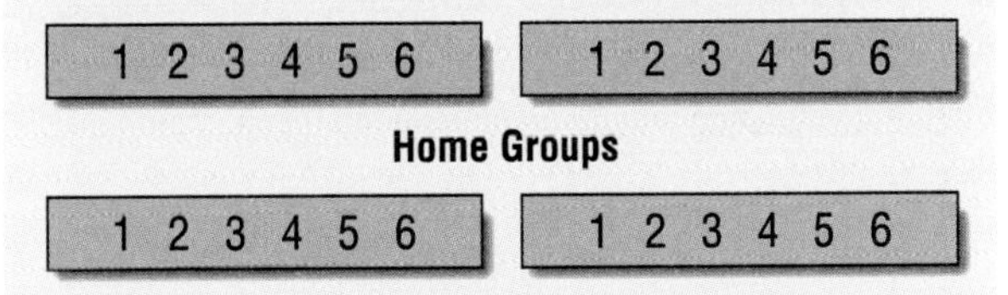

3. Form an expert group with students who have the same career to research. In your expert group, decide what questions you want to answer about the career. Then, research the answers.

4. Return to your home group and tell the others in the group what you learned in your expert group.

5. In your home group, prepare a report on the 6 careers. As a group, decide what form the report will take. It could be written, acted out, presented as a video, displayed on a poster, or presented in any other appropriate form.

6. In your home group, evaluate the process and your presentation.

Chapter Check

1. Evaluate $ab + bc - ca$ for $a = 2$, $b = 3$, and $c = 4$.

Collect like terms and simplify.

2. $3x - 2y - 3x + 3y + 2$

3. $-15 + 2a - 3c + 10 + 4a - b - 7c$

Classify each polynomial and state its degree.

4. $x^2 + 3x + 7$ **5.** $3x^3 + x^2$

6. $3x^2y$

Arrange each of the following in descending powers of x.

7. $1 + x^3 + x^5 + x^4$

8. $5x^3 + 2xy - yx^4$

9. $x^3y - 3x^4 - 2x^2y + 10xy^2$

10. $3ax - 5b^2x^3 - 1 + 5abx^2$

Simplify.

11. $(5a^2 - x + 9) + (a^2 + x - 13)$

12. $(b^3 + b^2 - 2y + 7) + (3y - 8 - 2b^2)$

Simplify.

13. $(6y^2 - 6x + 3) - (y^2 + 6x - 3)$

14. $(-8t^2 + 4t - 1) - (4 - 7t^2 + 5t)$

Simplify.

15. $(10x^2 - 3x + 8) + (-11x^2 - 4x - 2)$

16. $(8y^2 + 6y - 1) - (3y^2 - 8y - 2)$

Add.

17.
$$\begin{array}{r} 3x^2 + 2x - 7 \\ \underline{+5x^2 - 11x - 5} \end{array}$$

18.
$$\begin{array}{r} 10x^2 + 4x - 5 \\ \underline{-12x^2 + 6x + 6} \end{array}$$

Multiply.

19. $-10x(5y)$ **20.** $(-15x^2y)(-3xy^2)$

21. $-2x^5(3b^2yz)$ **22.** $3x^3yz(-5xyz^2)$

Simplify.

23. $(-3x^4y^3)^3$ **24.** $(-ab^2c^2)^5$

25. $(x^2y^2z)(xy)^3$ **26.** $(x^2y^4)(-3x^3y^2)^2$

Expand and simplify.

27. $2x(x - 2) + 3(x + 3)$

28. $4x(x + 1) - 5x(x - 1)$

29. $-3x(x^2 - 2x + 1)$

30. $2m(2m^2 - 6m - 10)$

Divide.

31. $\dfrac{-6a^8b^5}{3a^5b^3}$ **32.** $\dfrac{-39j^7k^5l^5}{-13j^5k^4l^3}$

Divide.

33. $\dfrac{5y^4 - 10y^5 + 15y^6}{5y^4}$

34. $\dfrac{12a^5b^3 - 8a^4b^3 + 4a^3b^3}{4a^3b^3}$

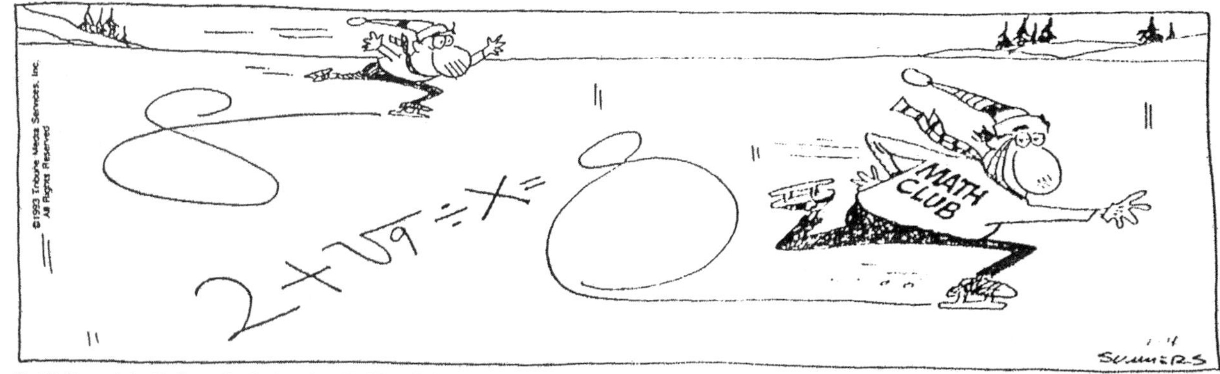

Using the Strategies

1. How many hours are in 1000 s?

2. a) What is the largest 2-digit number that is a perfect square?
b) What is the largest 3-digit number that is a perfect square?

3. Find 3 consecutive numbers that have a sum of 111.

4. There are 4 boys, Barry, Bob, Hans, and Joe and 5 girls, Marie, Sue, Paola, Natasha, and Oba, who wish to enter a badminton mixed doubles tournament. How many different 2 player teams can these nine players make?

5. A picture 20 cm by 20 cm is to be bordered by 1-cm squares. How many squares are needed for the border?

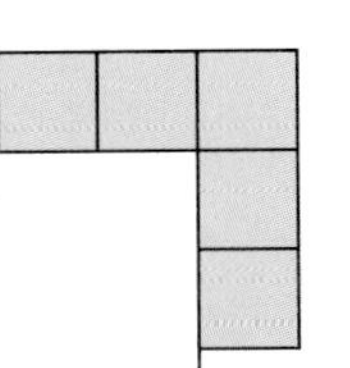

6. The sides of a triangular field are 40 m, 44 m, and 52 m. A fence is to be built around the field with a post in each corner and the posts 4 m apart. How many fence posts are needed?

7. Sue is 4 years older than Alex. In four years, the sum of their ages will be 36. How old are they today?

8. A restaurant offers a choice of how to pay for a class lunch.

Plan A: $8 per person
Plan B: $20 service charge plus $7 per person

a) For what number of students is Plan A better?
b) For what number of students is Plan B better?
c) For what number of students are the 2 plans the same?

9. In a bean toss game, each person throws 2 bags.

Scoring A and B gives 18 points.
Scoring A and C gives 15 points.
Scoring B and C gives 13 points.

What will you score if you toss both bags in B?

10. Two grade 6 students and two grade 9 students want to cross a river in a canoe. The canoe is big enough to hold the two grade 6 students or one grade 6 student and one grade 9 student. How many times must the canoe cross the river to get all the students to the other side?

11. Find the 4-digit perfect square that is a pair of 2-digit perfect squares written side by side.

12. Chicken pieces come in boxes of 6, 9, and 20. You can buy 21 pieces by buying 2 boxes of 6 pieces and 1 box of 9 pieces.

$$6 + 6 + 9 = 21$$

a) How can you buy 44 pieces?
b) How can you buy 41 pieces?
c) How can you buy 42 pieces?
d) Can you buy 43 pieces?

1. You leave Halifax at 09:00 on a Friday and drive to Vancouver. You average 80 km/h and drive for 9 h/day, starting at 09:00. On what day and at about what time will you arrive in Vancouver?

2. The Canadian city with the greatest elevation is 561 m higher than Saskatoon. What is the elevation of Saskatoon?

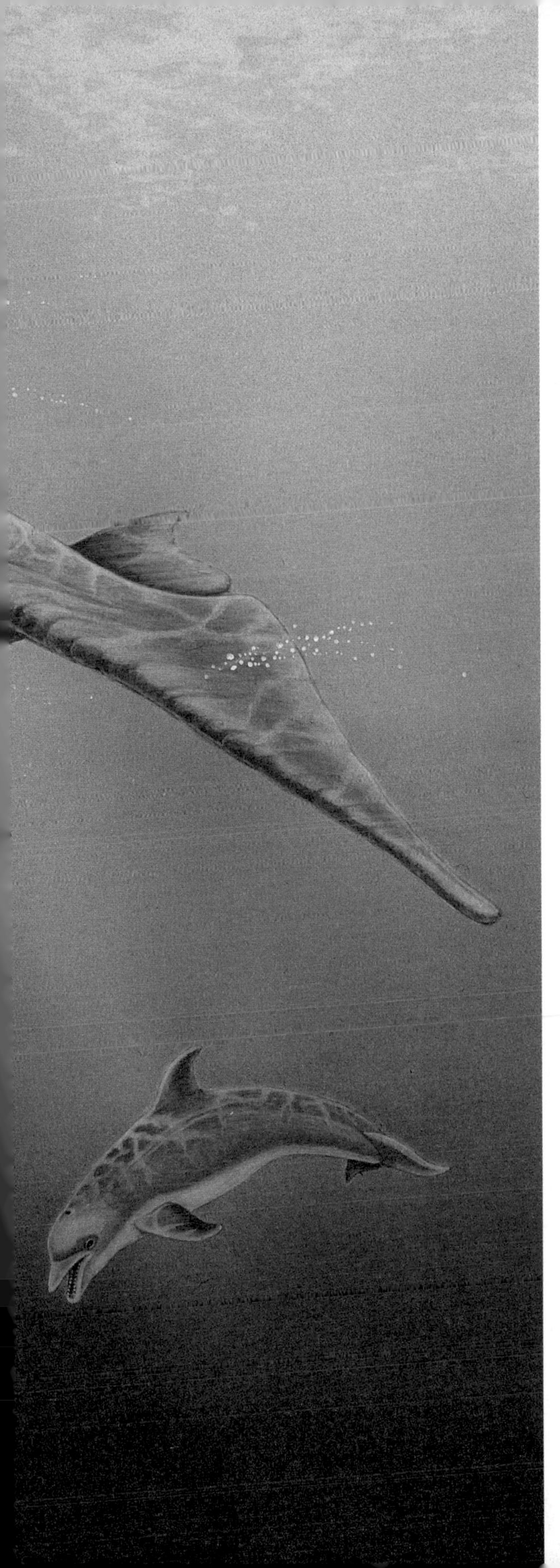

CHAPTER 6

Equations and Inequalities

Whales, sharks, and dolphins have shapes that minimize water resistance. The ideal shape for minimum water resistance is a torpedo shape, with a width that is one-quarter the length.

The equations below show how the widths of a blue whale, a shark, and a dolphin are related to their lengths.

Blue Whale	$w = 0.21l$
Shark	$w = 0.26l$
Dolphin	$w = 0.25l$

About how wide is a shark that is 18 m long?

The blue whale is the world's largest mammal. Is a 30-m long blue whale wider than your classroom?

What objects made by humans have torpedo shapes? Explain why they are made in this shape.

Activity ❶ Balancing Act One

Suppose you have masses of 1 kg, 3 kg, and 9 kg, and you want to add 13 kg of sand to a container. Place the 1 kg, 3 kg, and 9 kg masses on one side of the balance scale. Pour sand into the pan on the other side until the scale balances.

To measure 6 kg of sand, place the 9 kg mass on one side and the 3 kg mass on the other. Now add sand until the scale balances.

Draw balance scales to show how you can measure any unit mass of sand from 1 kg to 12 kg using the 1 kg, 3 kg, and 9 kg masses.

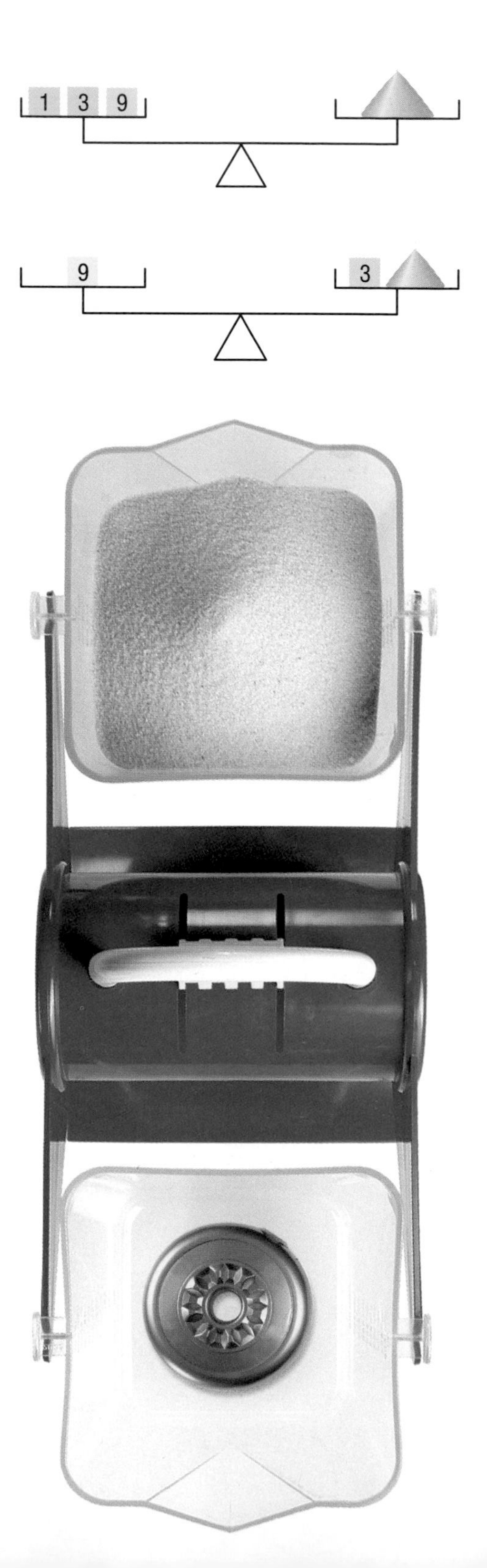

Activity ❷ Balancing Act Two

In this activity you have masses of 1 kg, 3 kg, 9 kg, and 27 kg available to you.

To measure 30 kg of sand, you put the 27 kg mass and the 3 kg mass on one side of the balance scale. Now pour sand until the scale balances.

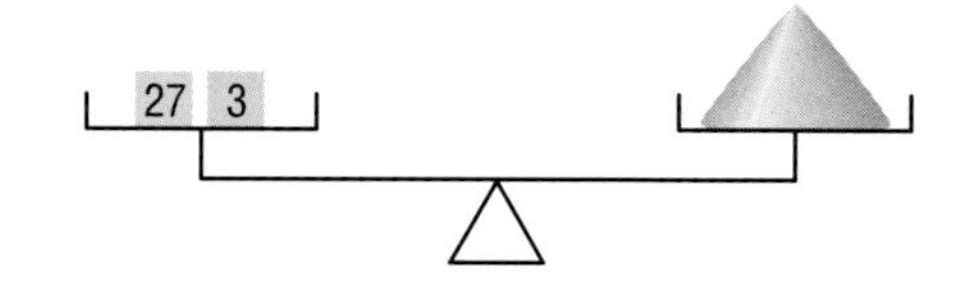

To measure 24 kg of sand place the 27 kg mass on one side and the 3 kg mass on the other. Pour sand until the scale balances. Why is there 24 kg of sand on the scale?

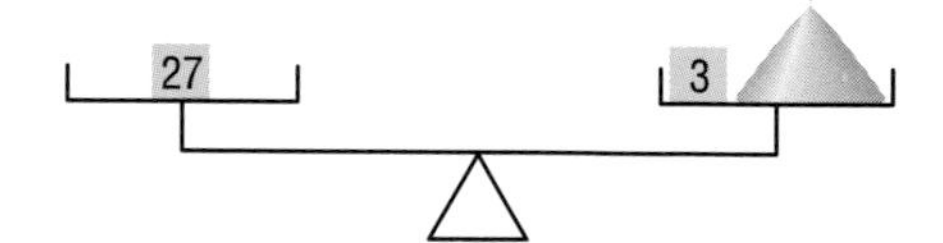

You have seen in Activity 1 how 1 kg, 3 kg, and 9 kg masses let you measure unit masses of sand from 1 kg to 13 kg. You now have the 27 kg mass as well as the 1 kg, 3 kg, and 9 kg masses. Draw balance scales to show how you can now measure any unit mass of sand from 14 kg to 40 kg.

Warm Up

Simplify.

1. $3x + 4x$
2. $5x - 2x$
3. $6m - 2m + 5m$
4. $6t - 7t + 5t + 2t$
5. $2a - 3a - 4a$
6. $-5n + 2n - 3n$
7. $8s - 4s - 7s + 5s$
8. $3y + 6y - 10y + 2y$

Simplify.

9. $3x + 2y - 2x + 5y$
10. $3a - 4a - 2b + 5b$
11. $4x - 7 + 3x - 6$
12. $-7t + 4s - 5s - 8t$
13. $3x - 4y + 5 - 6x + 2$
14. $-3a - 2b - 7b + 5a$
15. $10 - 7x - 8x + 5y - 11y$

Expand.

16. $2(x + 7)$
17. $5(2x - 5)$
18. $3(6x - 1)$
19. $7(5x - 4)$
20. $4(2a - 3b)$
21. $6(2p - q)$
22. $-2(3y - 7)$
23. $-(4x - 9)$
24. $-5(2x - 3y - 4)$
25. $-(4x + 5y - 1)$

Evaluate for $x = -2$ and $y = 3$.

26. $x + y$
27. $4x + 5y$
28. $x - y$
29. $y - x$
30. $4y - 3x - 7$
31. $-4 - 2x - 4y$
32. $3xy - 4y - 3x$
33. $-5xy - 5x - 11$
34. $y - 2xy - x$
35. $10 - y - x - xy$

Evaluate for $x = 3$ and $y = -2$.

36. $x + 2y$
37. $7x - 4y$
38. $2x + y$
39. $y - 3x$
40. $2y - 3x - 1$
41. $5 - y - x - 6$
42. $5 - y - x + xy$
43. $-2xy + 12 - x$
44. $x - 5xy - 2y$
45. $y - x - xy - 1$

Mental Math

Add.

1. $4 + (-3) + 2$
2. $-1 + (-2) + (-3)$
3. $5 + 6 + (-10)$
4. $-3 + 4 + 5 + (-3)$
5. $-6 + (-4) + (-3)$
6. $-3 + (-6) + (-2) + 8$

Simplify.

7. $6 - (-7)$
8. $7 - (-6) + 2$
9. $-8 - (-8) - (-8)$
10. $11 - 13 + 14$
11. $0 - (-4) + 6 + 5$

Simplify.

12. $5 - (-7) + 1$
13. $8 - 4 - (-3) + 8$
14. $-3 + (-5) - 7 + 5$
15. $-2 - (-6) + 1 + 10$
16. $5 - (-5) - 3 - 1$

Simplify.

17. $6 \times (-3) \times (-2) \times (-1)$
18. $-4 \times (-2) \times (3) \times (-2)$
19. $12 \div (-6) \div (-2)$
20. $-10 \div 2 \div 5 \times (-1)$
21. $14 \div 7 \div 2 - 1$

Solve by inspection.

22. $x + 4 = 11$
23. $x - 5 = 9$
24. $5x = 20$
25. $2x = -10$
26. $x - 3 = 0$
27. $x + 1 = 0$
28. $4x = 24$
29. $x - 3 = 8$
30. $5 + x = 12$
31. $3x = 21$

6.1 Solving Equations

Many horses take part in sports and shows in Canada. In the RCMP Musical Ride, 32 horses and their riders perform complex movements to music.

Activity: Solve the Problem

Suppose that you own a horse. The monthly cost to board your horse at a public stable is $100. You have $225 a month to spend, and you want to take riding lessons that cost $25 each. Use the equation $25n + 100 = 225$ to find the number of riding lessons you can take per month.

Inquire

1. How did you solve the equation $25n + 100 = 225$?

2. What is the value of x in each of the following equations?

a) $x + 4 = 9$ **b)** $x - 7 = 11$ **c)** $3x = 27$ **d)** $4x + 11 = 31$

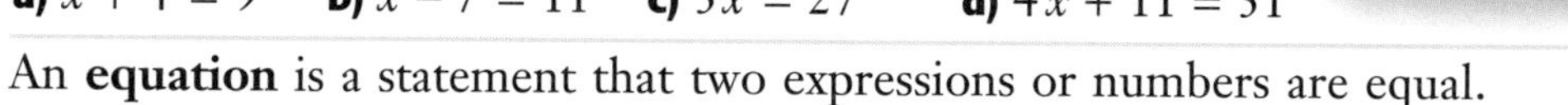

An **equation** is a statement that two expressions or numbers are equal.

This equation is true. $5 + 3 = 8$

This equation is false. $7 - 4 = 2$

This equation is neither true nor false. $x + 5 = 9$

The equation $x + 5 = 9$ is called an **open sentence**, because it is not possible to say whether it is true or false. It is true if x is replaced by 4. Other replacements make the statement false. A value of the variable that makes the statement true is called the **solution** or **root** of the equation. This value is said to *satisfy* the equation.

Example

Solve.

a) $x + 5 = 2$

b) $7x + 13 = 55$

Solution

a) Simple equations like $x + 5 = 2$ can be solved mentally. This process is called **solving by inspection**. The solution is $x = -3$.

b) Use **systematic trial**. Substitute different values for the variable until you find the solution.

Substitute 3 for x. $7x + 13 = 7(3) + 13$
$= 21 + 13$
$= 34$ When $x = 3$, $7x + 13$ is less than 55.

Substitute 7 for x. $7x + 13 = 7(7) + 13$
$= 49 + 13$
$= 62$ When $x = 7$, $7x + 13$ is greater than 55.

Substitute 6 for x. $7x + 13 = 7(6) + 13$
$= 42 + 13$
$= 55$ The solution is $x = 6$.

Practice

Solve by inspection.

1. $x + 3 = 7$
2. $x + 5 = 4$
3. $x + 5 = 5$
4. $x + 5 = 3$
5. $2 + m = 9$
6. $n + 1 = 6$
7. $p + 10 = 3$
8. $4 = 7 + s$
9. $y + 8 = 4$
10. $z + 8 = 12$

Solve by inspection.

11. $x - 5 = -7$
12. $x - 3 = 7$
13. $4 = -10 + y$
14. $y + 5 = -3$
15. $z - 1 = 6$
16. $t - 12 = 10$
17. $n + 5 = -2$
18. $4 - x = 0$
19. $m + 3 = 0$
20. $2 = x + 7$

Is the number in brackets the correct solution to the given equation?

21. $x + 7 = 10$ (3)
22. $x - 5 = 0$ (−5)
23. $3x = -18$ (−6)
24. $5x = 20$ (−4)
25. $\frac{x}{5} = \frac{4}{5}$ (4)
26. $\frac{2x}{7} = -\frac{6}{7}$ (−3)
27. $2x + 1 = 7$ (−3)
28. $10 = -2 + 3x$ (4)

Solve by systematic trial.

29. $6x + 17 = 65$
30. $73 - 7x = 45$
31. $14 + 9x = 86$
32. $35 = 83 - 8t$
33. $8k - 37 = 59$
34. $5b + 41 = 76$
35. $4 - 3s = -11$
36. $2m - 15 = -11$
37. $-15 = 7 - x$
38. $21 = 5b + 31$

Problems and Applications

Solve by inspection.

39. $x + 2.5 = 3.5$
40. $t - 4.3 = 1.7$
41. $6.6 - m = 2.6$
42. $2x = 4.8$
43. $0.5t = 16$
44. $7.1 + x = 10.2$
45. $5 = m - 3.7$
46. $12 = x + 4.6$
47. $9 - x = 8.2$
48. $1.2 + x = -5.3$

49. Solve the equation $9w = 36$ to find the width of the rectangle.

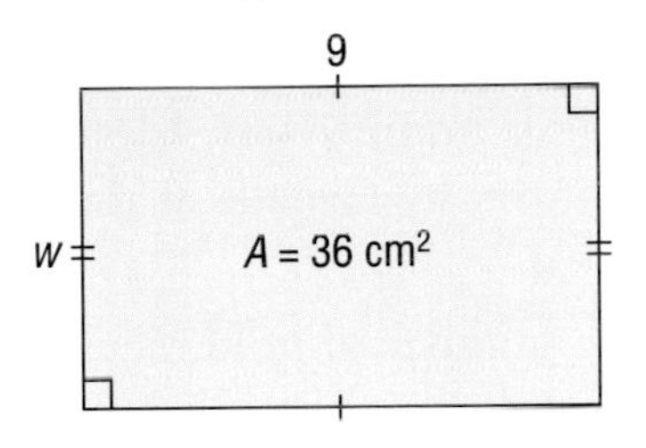

50. Point Pelee National Park has an area of 15 km², which is 7 km² less than the area of Prince Edward Island National Park. Solve the equation $x - 7 = 15$ to find the area of Prince Edward Island National Park.

51. London, Ontario, has thunderstorms on 36 days a year, the highest number in Canada. Prince Rupert, B.C., has the fewest days of thunderstorms. If you multiply the number for Prince Rupert by 10 and add 6, you get the number for London. Solve the equation $10d + 6 = 36$ to find the number of days of thunderstorms in Prince Rupert per year.

52. Use x, 2, and 6 to write equations with the following solutions. Check your answers by substitution.

a) 3 **b)** −4 **c)** 4
d) 8 **e)** 12 **f)** $\frac{1}{3}$

53. Find 2 solutions to each equation.

a) $x^2 = 16$ **b)** $t^2 - 6 = 19$

54. Write an equation that can be solved by inspection and in which the solution is $x = -4$. Have a classmate solve it.

55. Write an equation that can be solved by systematic trial and in which the solution is $x = 8$. Have a classmate solve it.

56. Write an equation that can be solved by systematic trial and in which the solution is $x = -15$. Have a classmate solve it.

6.2 Using Addition and Subtraction to Solve Equations

One year, Canada won 2 gold medals at the Winter Olympics. This number was 4 less than the number of gold medals won by Canada at the Summer Olympics that same year.

If we let the number of gold medals Canada won at the Summer Olympics be x, we can represent the above information with the equation $x - 4 = 2$.

Activity: Use Algebra Tiles

Write the equation represented by the balanced scale.

Inquire

1. How many red 1-tiles must be added to the left side of the balance so that only the value of the x-tile remains?

2. What must you do to the right side of the scale to keep it balanced?

3. What is the value of x?

4. How many gold medals did Canada win at the Summer Olympics?

5. Model the following equations using algebra tiles. Then, solve for x. Compare your answers with a classmate's.

a) $x - 2 = 5$ **b)** $x - 1 = 2$

c) $x - 4 = 2$ **d)** $x - 5 = -4$

6. Write a rule for solving equations by addition.

To solve an equation, isolate the variable on one side of the equation.

Example 1

Solve the equation $x - 2 = 6$.

Solution

Add 2 to both sides of the equation.

$$x - 2 = 6$$
$$x - 2 + 2 = 6 + 2$$
$$x = 8$$

The solution is $x = 8$.

Activity: Use Algebra Tiles

Write the equation represented by the balanced scale.

Inquire

1. How many red 1-tiles must be subtracted from the left side of the balance so that only the x-tile remains?

2. What must you do to the right side of the scale to keep it balanced?

3. What is the value of x?

4. Model the following equations using algebra tiles. Then, solve for x. Compare your answers with a classmate's.

a) $x + 1 = 5$ **b)** $x + 3 = 2$ **c)** $x + 4 = -2$ **d)** $x + 2 = -3$

5. Write a rule for solving equations by subtraction.

Example 2

Solve and check $x + 7 = -2$.

x +7 = -2

Solution

To isolate the variable, subtract 7 from both sides of the equation.

x +7 −7 = −2 −7

$$x + 7 = -2$$
$$x + 7 - 7 = -2 - 7$$
$$x = -9$$

x = −9

Substitute $x = -9$ into the left side of the equation. The solution is correct if the value of the left side equals the value of the right side.

Check: **L.S.** $= x + 7$ **R.S.** $= -2$

$= -9 + 7$

$= -2$

The solution is $x = -9$.

CONTINUED

Practice

Write and solve the equations shown by the tiles.

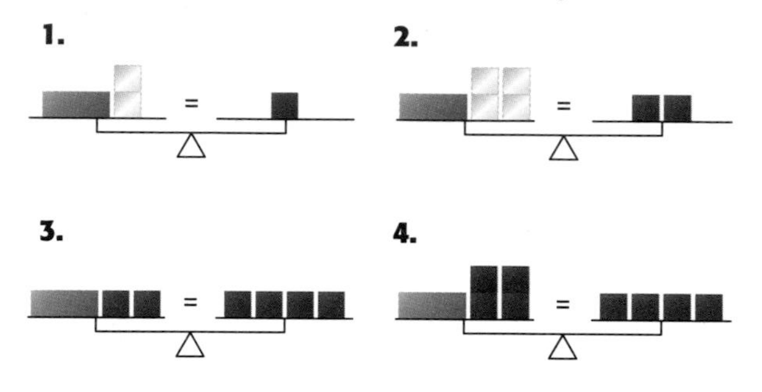

What number would you add to both sides to solve each equation?

5. $x - 3 = 11$
6. $x - 1 = 5$
7. $n - 7 = -8$
8. $-8 = -4 + m$
9. $3 = y - 10$
10. $z - 5 = -11$

What number would you subtract from both sides of each equation to solve it?

11. $x + 6 = 13$
12. $8 = x + 1$
13. $y + 2 = -7$
14. $m + 5 = -5$
15. $x + 3 = 9$
16. $-10 = 7 + z$

Solve.

17. $m - 5 = -4$
18. $2 = -3 + n$
19. $p - 7 = -3$
20. $r + 7 = -9$
21. $-3 = s + 5$
22. $t - 8 = 10$
23. $2 + x = -4$
24. $7 = x - 5$
25. $x + 4 = -8$
26. $-11 = t - 1$

Problems and Applications

Solve and check.

27. $x + 1.5 = 3.5$
28. $m - 3.2 = 4.8$
29. $4.6 = t - 1.4$
30. $5.7 = r + 3.6$
31. $4.3 + m = 1.3$
32. $y - 2.4 = -1.4$
33. $9 = 8.2 + x$
34. $3.7 + t = 6.4$
35. $-1.4 + q = 4.4$
36. $10.3 = x - 5.4$
37. $-8.8 = w - 1.1$
38. $x + 15.7 = -18.1$

In questions 39–41, find the equation that represents the problem and solve it.

39. Three more than a number, x, is eight. What is the number?

a) $x - 3 = 8$ **b)** $x + 8 = 3$
c) $x + 3 = 8$ **d)** $x - 8 = 3$

40. How many boxes are left to unload if there are 195 boxes in the shipment and 72 boxes have been unloaded?

a) $x + 72 = 195$ **b)** $x + 195 = 72$
c) $x - 72 = 195$ **d)** $x - 195 = 72$

41. In a triangle, the sum of the sides is 9.7 cm. The lengths of two of the sides are 3.2 cm and 4.5 cm. What is the length of the unknown side?

a) $x - 4.5 = 9.7 - 3.2$
b) $9.7 - x = 4.5 - 3.2$
c) $4.5 + x - 3.2 = 9.7$
d) $x + 3.2 + 4.5 = 9.7$

42. The number of moons around Uranus is 5 more than the number of rings. Uranus has 15 moons. Solve the equation $r + 5 = 15$ to find the number of rings around Uranus.

43. A nurse walks about 6.3 km a day which is 0.8 km less than a letter carrier walks. Solve the equation $x - 0.8 = 6.3$ to find out how far a letter carrier walks in a day.

44. What is the result if you add 0 to both sides of an equation?

45. Find 2 solutions for each equation.

a) $x^2 - 1.8 = 7.2$ **b)** $82.2 = t^2 + 18.2$

46. Write 1 equation that can be solved by addition and 1 equation that can be solved by subtraction. Each equation should have $x = -3$ as its solution. Have a classmate solve your equations.

6.3 Using Division and Multiplication to Solve Equations

In 1992, Canada won the Women's World Hockey Championships. In the semi-final game against Finland, Canada scored 6 goals, 3 times as many as Finland. If we let x be the number of goals Finland scored, we can represent the above information with the equation $3x = 6$.

Activity: Use Algebra Tiles

Write the equation represented by the balanced scale.

Inquire

1. How many pairs of 1-tiles are shown?

2. How many x-tiles are shown?

3. How many 1-tiles does an x-tile represent?

4. Divide both sides of the equation $3x = 6$ by 3. Compare your result with your answer to question 3.

5. How many goals did Finland score against Canada?

6. Represent the following equations using algebra tiles. Then, solve for x.

a) $2x = 10$ **b)** $3x = -6$ **c)** $-2x = 4$ **d)** $-4x = -4$

7. Write a rule for solving equations by division.

Example 1

Solve the equation $2x = 6$.

Solution

Divide both sides of the equation by 2.

$$2x = 6$$
$$\frac{2x}{2} = \frac{6}{2}$$
$$x = 3$$

The solution is $x = 3$.

CONTINUED

Activity ❷ Interpret the Diagram

The balanced scale represents the equation $\frac{x}{2} = 1$. What part of an x-tile is shown on the scale?

$$\frac{x}{2} = 1$$

Inquire

1. How many 1-tiles are shown on the scale?
2. How many 1-tiles does a whole x-tile represent?
3. Multiply both sides of the equation by 2. What is the value of an x-tile? Compare this result with your answer to question 2.
4. Write a rule for solving equations by multiplication.

Example 2

Solve the equation $\frac{x}{2} = -5$.

$\frac{x}{2}$ = -5

Solution

Multiply both sides of the equation by 2.

$$\frac{x}{2} = -5$$
$$2 \times \frac{x}{2} = 2 \times (-5)$$
$$x = -10$$

The solution is $x = -10$.

$\frac{x}{2}$ $\frac{x}{2}$ = -5 -5

x = -10

Example 3

Solve and check $-2x = 18.4$.

Solution

Divide both sides of the equation by -2.

$$-2x = 18.4$$
$$\frac{-2x}{-2} = \frac{18.4}{-2}$$
$$x = -9.2$$

Check: **L.S.** $= -2x$ **R.S.** $= 18.4$

$= -2(-9.2)$

$= 18.4$

The solution is $x = -9.2$.

Practice

Write and solve the equation represented by each balanced scale.

1.

2.

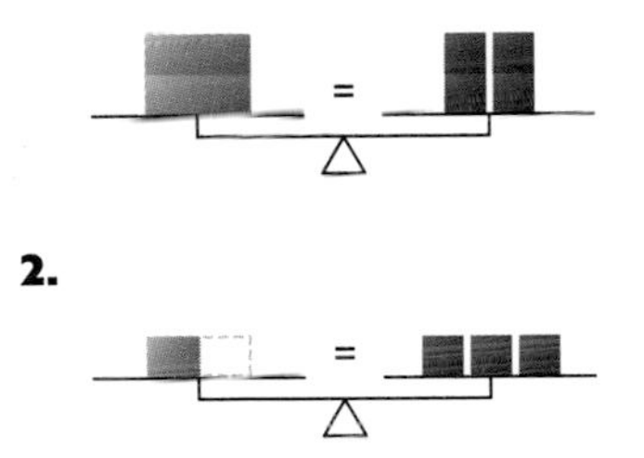

By what number would you divide both sides to solve each equation?

3. $6x = 12$ **4.** $2x = -8$ **5.** $7z = -14$

6. $11r = 22$ **7.** $6n = -18$ **8.** $8y = 64$

By what number would you multiply both sides to solve each equation?

9. $\frac{x}{3} = 9$ **10.** $\frac{x}{2} = -4$ **11.** $\frac{y}{7} = 4$

12. $\frac{m}{5} = -6$ **13.** $\frac{t}{4} = -1$ **14.** $\frac{x}{6} = 0$

Solve.

15. $3x = 15$ **16.** $2y = -12$ **17.** $-8x = 16$

18. $-15 = 5p$ **19.** $-6t = -24$ **20.** $12 = 3s$

21. $2a = 24$ **22.** $-25 = 5j$ **23.** $-7x = -21$

Solve.

24. $\frac{x}{2} = 2$ **25.** $\frac{y}{2} = -4$ **26.** $\frac{x}{3} = 7$

27. $-4 = \frac{p}{3}$ **28.** $\frac{m}{4} = 8$ **29.** $-3 = \frac{n}{5}$

30. $\frac{s}{10} = 5$ **31.** $-2 = \frac{t}{12}$ **32.** $\frac{u}{11} = 4$

Solve and check.

33. $2x = 8.4$ **34.** $3m = -6.3$

35. $\frac{x}{5} = 2.4$ **36.** $\frac{x}{2} = -1.5$

37. $-4x = -2.8$ **38.** $1.6x = 4.8$

39. $4 = \frac{m}{1.5}$ **40.** $-8.4 = 0.2x$

Problems and Applications

In questions 41 and 42, find the equation that represents the problem and solve it.

41. Six dollars is to be divided equally among three people. How much money does each person receive?

a) $3x = 6$ **b)** $6x = 3$

c) $\frac{x}{3} = 6$ **d)** $\frac{x}{6} = 3$

42. One-third of the sum of the side lengths of an equilateral triangle is 4.2 cm. What is the sum?

a) $3x = 4.2$ **b)** $\frac{1}{3}x = 4.2$

c) $\frac{1}{3}x + 4.2 = 0$ **d)** $x = \frac{1}{3}(4.2)$

43. The average annual snowfall in Vancouver is about one-sixth of the average value for St. John's. Vancouver averages 60 cm of snow a year. Solve the equation $\frac{s}{6} = 60$ to find the average annual snowfall in St. John's.

44. The mass of a white-tailed deer is 1.4 times the mass of a cougar. White-tailed deer average 98 kg in mass. Solve the equation $1.4x = 98$ to find the mass of a cougar.

45. What is the result if you multiply both sides of an equation by 0?

46. Describe the solutions to these equations.

a) $3x = 0$ **b)** $0y = 2$

47. Find 2 solutions to each equation.

a) $2x^2 = 72$ **b)** $\frac{m^2}{4} = 1$

48. Write one equation that can be solved by division and one that can be solved by multiplication. Each equation should have $x = 7$ as its solution. Have a classmate solve your equations.

Equations and Number Tiles

Prepare a set of 10 number tiles with the numbers from 0 to 9 written on them.

Activity ❶

1. Use each of the tiles only once to solve the equations. The x-value for each equation is shown in brackets. Three of the tiles have been placed for you.

$x - \blacksquare = 2$	$(x = 5)$
$x - \blacksquare = \blacksquare - 3$	$(x = 5)$
$0 + x - \blacksquare = \blacksquare + 1$	$(x = 10)$
$x + \blacksquare - 8 = 9 + \blacksquare$	$(x = 11)$

Activity ❷

Use each of the tiles only once to solve the equations. The x-value for each equation is shown in brackets. Three of the tiles have been placed for you.

$x + \blacksquare = 7$	$(x = 4)$
$x + \blacksquare = \blacksquare + 2$	$(x = 3)$
$\blacksquare + x + 6 = \blacksquare + 3$	$(x = 5)$
$x - 2 + \blacksquare = \blacksquare - 0$	$(x = 3)$

Activity ❸

Use each of the tiles only once to solve the equations. The x-value for each equation is shown in brackets. Three of the tiles have been placed for you.

$$\blacksquare x = 8 \qquad (x = 2)$$

$$\blacksquare x + \blacksquare = 8 \qquad (x = 1)$$

$$\blacksquare + 2x = \blacksquare \qquad (x = 3)$$

$$0 + \blacksquare x = \blacksquare + 9 + 1 \qquad (x = 3)$$

Activity ❹

Use each of the tiles only once to solve the equations. The x-value for each equation is shown in brackets. Two of the tiles have been placed for you.

$$\frac{x}{\blacksquare} = 2 \qquad (x = 12)$$

$$\frac{x}{\blacksquare} + \blacksquare = 5 \qquad (x = 5)$$

$$\blacksquare + \frac{x}{2} = \blacksquare + 2 \qquad (x = 6)$$

$$\frac{x}{\blacksquare} + \blacksquare = 7 - \blacksquare \qquad (x = 18)$$

Activity ❺

Create a set of equations similar to those in Activities 1, 2, 3, and 4. Have a classmate solve your equations.

6.4 Solving Equations Using More Than One Step

A cow sleeps 7 h a day. This is 1 h less than twice the amount an elephant sleeps a day. If we let the hours that an elephant sleeps be x, we can represent this information with the equation $2x - 1 = 7$.

Activity: Use Algebra Tiles

Write the equation represented by the balanced scale.

Inquire

1. How many red 1-tiles must be added to both sides of the balance so that only the value of the 2 x-tiles remains on the left?

2. How many red 1-tiles does an x-tile represent?

3. For how many hours does an elephant sleep each day?

4. Represent the following equations using algebra tiles. Solve the equations.

a) $3x + 2 = 8$ **b)** $4x - 1 = 11$ **c)** $2x + 3 = -9$ **d)** $5x - 1 = -11$

5. Write the steps for solving each equation in question 4.

6. The flow charts show the order of the steps to solve the equation $5x + 2 = 17$. First, isolate x on the left side.

Left Side START → $5x - 2$ → Subtract 2 → Divide by 5 → x → STOP

Then, apply the same steps to the right side.

Right Side START → 17 → Subtract 2 → Divide by 5 → 3 → STOP

Use flow charts to solve these equations.

a) $2x + 3 = 11$ **b)** $3x - 1 = 17$

Example 1

Solve the equation $5x + 4 = -16$.

Solution

$$5x + 4 = -16$$

Subtract 4 from both sides: $5x + 4 - 4 = -16 - 4$

$$5x = -20$$

Divide both sides by 5: $\frac{5x}{5} = -\frac{20}{5}$

$$x = -4$$

The solution is $x = -4$.

Example 2

Solve and check $7.2 = 3x - 4.2$.

Solution

$$7.2 = 3x - 4.2$$

Add 4.2 to both sides: $7.2 + 4.2 = 3x - 4.2 + 4.2$

$$11.4 = 3x$$

Divide both sides by 3: $\frac{11.4}{3} = \frac{3x}{3}$

$$3.8 = x$$

Substitute $x = 3.8$.

Check: **L.S.** $= 7.2$ **R.S.** $= 3(3.8) - 4.2$

$= 11.4 - 4.2$

$= 7.2$

The solution is $x = 3.8$.

Example 3

Solve and check $\frac{9x}{2} - 6 = 21$.

Solution

$$\frac{9x}{2} - 6 = 21$$

Add 6 to both sides: $\frac{9x}{2} - 6 + 6 = 21 + 6$

$$\frac{9x}{2} = 27$$

Multiply both sides by 2: $2 \times \frac{9x}{2} = 2 \times 27$

$$9x = 54$$

Divide both sides by 9: $\frac{9x}{9} = \frac{54}{9}$

$$x = 6$$

Substitute $x = 6$.

Check: **L.S.** $= \frac{9x}{2} - 6$ **R.S.** $= 21$

$= \frac{9(6)}{2} - 6$

$= \frac{54}{2} - 6$

$= 27 - 6$

$= 21$

The solution is $x = 6$.

CONTINUED

Example 4

Solve the equation $7x - 9x - 6 = 21 - 5$.

Solution

Simplify both sides of the equation. Then, apply the rules for solving equations.

$$7x - 9x - 6 = 21 - 5$$

$$-2x - 6 = 16$$

Add 6 to both sides: $-2x - 6 + 6 = 16 + 6$

$$-2x = 22$$

Divide both sides by -2: $\frac{-2x}{-2} = \frac{22}{-2}$

$$x = -11$$

The solution is $x = -11$.

Example 5

Solve $3x - 2 = 5$. Round your answer to the nearest tenth.

Solution

$$3x - 2 = 5$$

Add 2 to both sides: $3x - 2 + 2 = 5 + 2$

$$3x = 7$$

Divide both sides by 3: $\frac{3x}{3} = \frac{7}{3}$

$$x \doteq 2.33$$

The solution is $x = 2.3$ to the nearest tenth.

Practice

Draw flow charts to show the solution steps for each equation.

1. $5x + 2 = 22$

2. $2x + 5 = 25$

Solve.

3. $3x = 11 + 1$

4. $2y - 5 = 9$

5. $5n + 2n = -14$

6. $3m + 7 = 19$

7. $4x + 2x = -18$

8. $4y - 9y = 35$

9. $5w + w = 7 + 23$

10. $6n + 3n = -18$

Solve.

11. $3y - 5y = 4$

12. $3t + 7t = -30$

13. $3x + 2x = -20$

14. $5 + 11 = 4y$

15. $15 = 2n + 3n$

16. $5 + n = -15$

17. $3 + m = -7$

18. $4 + x = 20$

Solve and check.

19. $\frac{x}{3} = 4 + 2$

20. $\frac{y}{2} = 6 - 3$

21. $\frac{n}{3} + 3 = 5$

22. $\frac{m}{5} - 5 = -9$

23. $4 + \frac{x}{3} = -7$

24. $\frac{y}{2} + 5 = 3$

25. $\frac{x}{4} - \frac{1}{4} = -\frac{3}{4}$

26. $\frac{y}{2} + \frac{1}{4} = \frac{3}{4}$

Solve and check.

27. $4x + 3x + 7 = 21$

28. $2y - 5y - 5 = 13$

29. $4t + 7t = 15 - 4$

30. $6s + 2s + s = 18$

31. $4t - 7t = 8 + 4$

32. $3y = 10 - 6 - 7$

33. $4 + 3t = -6 - 2$

34. $x + 2x = -15 - 6$

Solve and check.

35. $5n + 3n - 2n = 17 - 2 + 9$

36. $2m + 5m + m = -10 - 5 - 1$

37. $3x + 5x + 1 = 11 - 2$

38. $4 + 3t - 5t = 12$

39. $2s + 3s - 5 = 18 + 12$

40. $4y - 7 + 2y = -24 - 1$

Solve.

41. $2x + 1.4 = 7.8$

42. $5t - 2.1 = 8.9$

43. $1.4m - 3.6 = 3.4$

44. $9.2 = 1.5t + 1.7$

45. $6 - 1.2x = 8.4$

46. $9.3 + 2.5k = 1.3$

47. $2 - 1.8r = 11$

48. $4.2 + 0.5y = 8.1$

Solve. Round your answer to the nearest tenth.

49. $7x + 5 = -3$

50. $2.4n - 3.8 = 1.1$

51. $3 = 3y + 8$

52. $4 = -6z + 14$

Problems and Applications

53. The equation $(x + 12) + 3x + 8 = 40$ represents the perimeter of the triangle. Find the lengths of the sides.

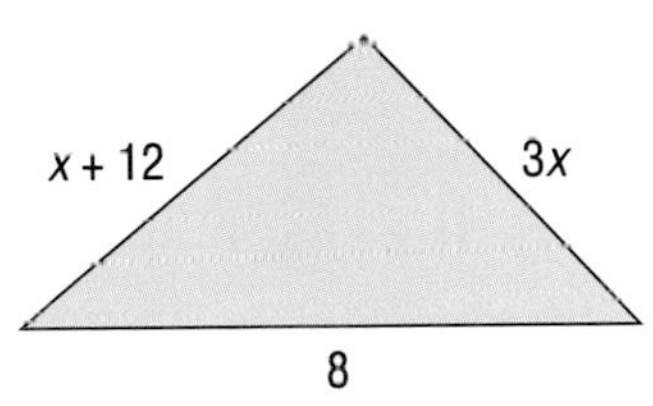

54. Pierre Trudeau was Prime Minister of Canada for 15 years. This was 3 years longer than twice the number of years that John Diefenbaker was Prime Minister. Solve the equation $2y + 3 = 15$ to find out how long John Diefenbaker was Prime Minister.

55. a) Solve the equation $2x + 5 = 11$, using the division rule first.

b) Is the result the same when you use the subtraction rule first?

c) Is one method better than the other? Explain.

56. Write an equation in which the solution requires at least 2 steps and is $x = -15$. Have a classmate solve your equation.

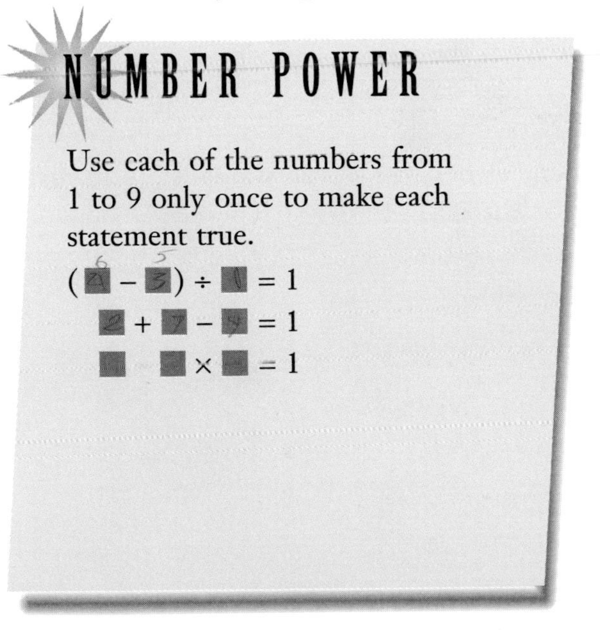

NUMBER POWER
Use each of the numbers from 1 to 9 only once to make each statement true.
(■ − ■) ÷ ■ = 1
■ + ■ − ■ = 1
■ ■ × ■ = 1

6.5 Solving Equations with Variables on Both Sides

At the 1992 Summer Olympics, Canadians won medals in 19 events, including the demonstration sport of taekwondo. Marcia King won a silver medal for Canada in this event.

Canadians were more successful in rowing than in any other sport. If we let x be the number of rowing events in which Canadians won medals, we can write the equation $x + 9 = 19 - x$. This is an example of an equation with a variable on both sides.

Activity: Use Algebra Tiles

Write the equation represented by the balanced scale.

Inquire

1. Solve the equation for x. Describe your method.
2. Solve the equation $x + 7 = 2x + 3$ by isolating the variable on the right side of the equation.
3. Solve the equation $x + 7 = 2x + 3$ by isolating the variable on the left side of the equation.
4. Compare your results in questions 2 and 3.
5. Write the steps for solving equations with variables on both sides.
6. Solve the equation $x + 9 = 19 - x$ to find the number of rowing events in which Canadians won medals at the 1992 Summer Olympics.

Example

Solve the equation $5x - 8x = x + 8$.

Solution

Simplify before solving.

$$5x - 8x = x + 8$$
$$-3x = x + 8$$

Subtract x from both sides:

$$-3x - x = x + 8 - x$$
$$-4x = 8$$

Divide both sides by -4:

$$\frac{-4x}{-4} = \frac{8}{-4}$$
$$x = -2$$

The solution is $x = -2$.

Practice

Write and solve the equations shown by the tiles.

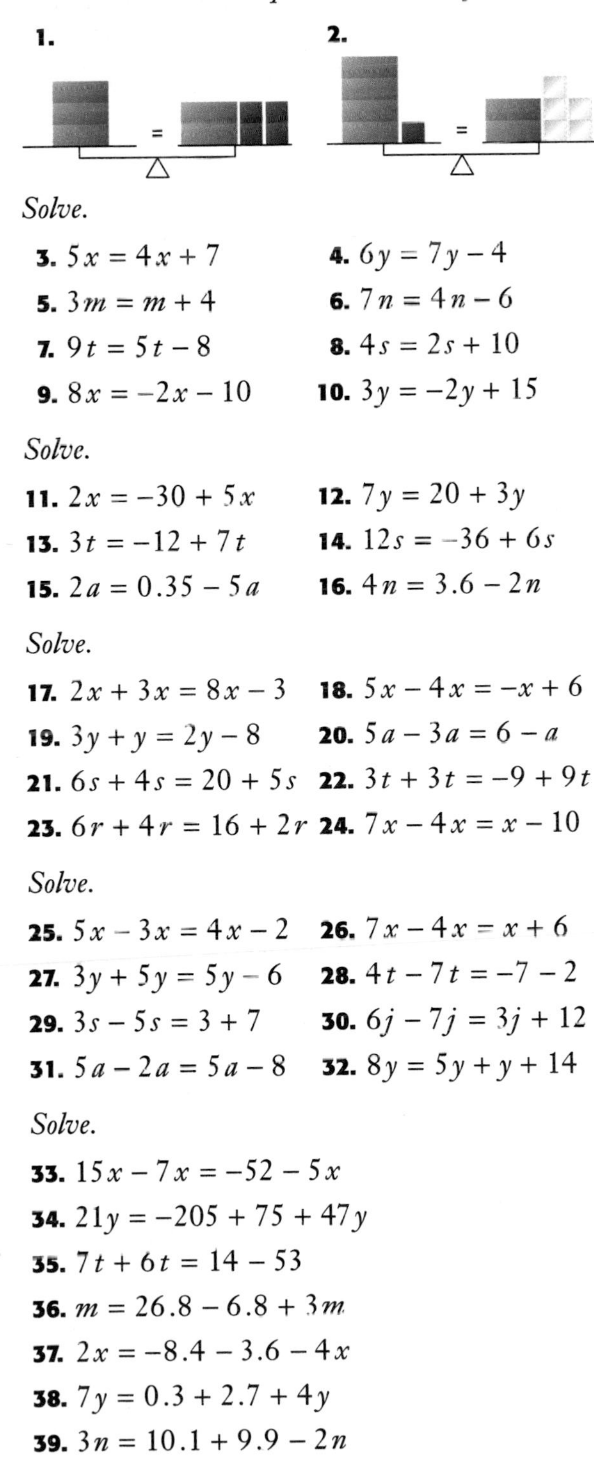

1. **2.**

Solve.

3. $5x = 4x + 7$ **4.** $6y = 7y - 4$

5. $3m = m + 4$ **6.** $7n = 4n - 6$

7. $9t = 5t - 8$ **8.** $4s = 2s + 10$

9. $8x = -2x - 10$ **10.** $3y = -2y + 15$

Solve.

11. $2x = -30 + 5x$ **12.** $7y = 20 + 3y$

13. $3t = -12 + 7t$ **14.** $12s = -36 + 6s$

15. $2a = 0.35 - 5a$ **16.** $4n = 3.6 - 2n$

Solve.

17. $2x + 3x = 8x - 3$ **18.** $5x - 4x = -x + 6$

19. $3y + y = 2y - 8$ **20.** $5a - 3a = 6 - a$

21. $6s + 4s = 20 + 5s$ **22.** $3t + 3t = -9 + 9t$

23. $6r + 4r = 16 + 2r$ **24.** $7x - 4x = x - 10$

Solve.

25. $5x - 3x = 4x - 2$ **26.** $7x - 4x = x + 6$

27. $3y + 5y = 5y - 6$ **28.** $4t - 7t = -7 - 2$

29. $3s - 5s = 3 + 7$ **30.** $6j - 7j = 3j + 12$

31. $5a - 2a = 5a - 8$ **32.** $8y = 5y + y + 14$

Solve.

33. $15x - 7x = -52 - 5x$

34. $21y = -205 + 75 + 47y$

35. $7t + 6t = 14 - 53$

36. $m = 26.8 - 6.8 + 3m$

37. $2x = -8.4 - 3.6 - 4x$

38. $7y = 0.3 + 2.7 + 4y$

39. $3n = 10.1 + 9.9 - 2n$

Solve.

40. $2x + 10 = 9 + 11 - 3x$

41. $11s + 25 = 54 + 25 + 2s$

42. $65 + n = 85 - 16 + 3n$

Problems and Applications

43. Bill has x baseball cards. Abbas has $4x$ baseball cards. The number of cards Abbas has is equal to the number of cards Bill has plus 60. Solve the equation $x + 60 = 4x$ to find the number of cards Bill has.

44. Marlene scored y ringette goals. Lynn scored $3y$ goals. The number of goals that Marlene scored equals the number of goals Lynn scored minus 12. Solve the equation $y = 3y - 12$ to find the number of goals Marlene scored.

45. Solve for x. Assume the letters a, b, c, and y represent non-zero integers.

a) $x + a = b + c + y$ **b)** $a + (x - b) = c - y$

c) $ax + b = c - y$ **d)** $a(x - b) + y = c$

e) $b + y = cx - a$ **f)** $\frac{x}{a} + b - y = -c$

46. The average wind speed in Bonavista, Newfoundland, is twice the average wind speed in Whitehorse, Yukon Territory. The difference between the wind speeds is 14 km/h. Solve the equation $2x - 14 = x$ to find the average wind speed in Whitehorse.

47. The number of states in India is 1 more than 4 times the number of states in Australia. There are 19 more states in India than in Australia. Solve the equation $x + 19 = 4x + 1$ to find the number of states in Australia.

48. Write an equation with a variable on both sides and a solution of $x = -2$. Have a classmate solve your equation.

6.6 Solving Equations with Brackets

Activity: Solve the Equation

Canada became a country in 1867. If we take the number of provinces in Canada at that time, add 1, and double the sum, the result is Canada's present number of provinces. Thus, we can find the number of provinces in 1867 by solving the equation $2(x + 1) = 10$.

Inquire

1. Expand to remove the brackets in the equation $2(x + 1) = 10$.

2. Solve for x to find the number of provinces in 1867.

3. Solve these equations.

a) $3(x - 1) = 12$

b) $-2(x + 1) = 6$

c) $2(x + 3) = -4$

d) $-(x + 1) = -7$

4. Write a rule for solving equations with brackets.

5. Research and list Canada's provinces in 1867.

Example 1

Solve the equation $2(x + 2) = -6$ and check.

Solution

Expand then solve.

$$2(x + 2) = -6$$
$$2x + 4 = -6$$

Subtract 4 from both sides: $2x + 4 - 4 = -6 - 4$

$$2x = -10$$

Divide both sides by 2: $\frac{2x}{2} = \frac{-10}{2}$

$$x = -5$$

Substitute $x = -5$.

Check: **L.S.** $= 2(x + 2)$ $= 2(-5 + 2)$ $= 2(-3)$ $= -6$ **R.S.** $= -6$

The solution is $x = -5$.

Example 2

Solve the equation $2(x-3)-5=13-4x$ and check.

Solution

To follow the order of operations, first expand to remove the brackets.

$$
\begin{aligned}
2(x-3)-5 &= 13-4x \\
2x-6-5 &= 13-4x \\
2x-11 &= 13-4x \\
2x-11+11 &= 13-4x+11 \\
2x &= 24-4x \\
2x+4x &= 24-4x+4x \\
6x &= 24 \\
\frac{6x}{6} &= \frac{24}{6} \\
x &= 4
\end{aligned}
$$

Substitute $x = 4$.

Check: **L.S.** $= 2(x-3)-5$
$= 2(4-3)-5$
$= 2(1)-5$
$= 2-5$
$= -3$

R.S. $= 13-4x$
$= 13-4(4)$
$= 13-16$
$= -3$

The solution is $x = 4$.

Example 3

Solve and check.

$3(2x-5)-(x+3)=2(x+1)+4$

Solution

$$
\begin{aligned}
3(2x-5)-(x+3) &= 2(x+1)+4 \\
6x-15-x-3 &= 2x+2+4 \\
6x-x-15-3 &= 2x+6 \\
5x-18 &= 2x+6 \\
5x-2x &= 6+18 \\
3x &= 24 \\
x &= 8
\end{aligned}
$$

Check: **L.S.** $= 3(2x-5)-(x+3)$
$= 3[2(8)-5]-(8+3)$
$= 3(16-5)-(11)$
$= 3(11)-11$
$= 33-11$
$= 22$

R.S. $= 2(x+1)+4$
$= 2(8+1)+4$
$= 2(9)+4$
$= 18+4$
$= 22$

The solution is $x = 8$.

CONTINUED

Practice

Solve.

1. $2(x + 1) = 4$
2. $2(x - 3) = 2$
3. $3(x + 1) = 6$
4. $2(x + 3) = -6$
5. $3(x + 2) = -9$
6. $2(x + 5) = -4$

Solve and check.

7. $2(3x + 4) = 14$
8. $14 = 2(3x - 2)$
9. $3(x + 5) = 18$
10. $3(2x + 3) = -3$
11. $-24 = 4(x + 3)$
12. $5(2x + 3) = -15$

Solve.

13. $2(x + 3) - 3 = 8 - 3x$
14. $3(x + 1) + 10 = 8 - 2x$
15. $8 - 3x = 4(x - 3) + 6$
16. $5(2x - 3) + 6 = -35 - 3x$

Solve.

17. $5(2x - 3) = 2(x - 2) + 5$
18. $2(x + 1) = (3x - 2) + 1$
19. $3(y + 4) = 5(y - 3) + 23$
20. $4(n + 7) = -44 + 2(n + 6)$
21. $5(2x - 3) = 2(x + 7) + 11$

Solve and check.

22. $2(x - 3) + (x + 3) = 6x$
23. $5(x + 4) - (x + 2) = 8x + 2$
24. $4(m - 2) - (m + 3) = m - 1$
25. $4(n - 7) - 2(n + 3) = -15n$
26. $4(y + 2) - 5(y + 1) = y - 1$

Solve.

27. $3(2x + 1) - (x - 2) = 2(x + 4)$
28. $12(s - 1) - 4(2s - 1) = 2(s + 1)$
29. $2(x - 8) - (x - 4) = 3(x + 5) + 3$
30. $7(x - 1) - 2(x - 6) = 2(x - 5) + 6$
31. $3(4n - 1) = 4(2n + 9) - 7$

Problems and Applications

32. The equation $5(x + 1) = 20$ represents the area of the following rectangle.

a) Solve the equation.

b) Find the rectangle's width.

5

$x + 1$

$A = 20$

33. The equation $2(3x + 2) + 2(x + 4) = 36$ represents the perimeter of the rectangle shown below.

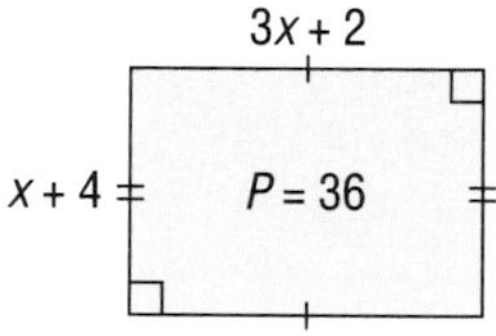

a) Solve the equation.

b) Calculate the rectangle's dimensions.

34. The Pacific Ocean accounts for 46% of the area of the water on the Earth's surface. If we take the percent that the Atlantic Ocean accounts for, subtract 1, and double the difference, the result is the percent that the Pacific Ocean accounts for. Solve the equation $2(x - 1) = 46$ to find the percent that the Atlantic Ocean accounts for.

35. Write an equation that has 1 bracketed expression on its left side and 1 bracketed expression plus a number on its right side. The equation should have a solution of $x = -2$. Have a classmate solve your equation.

WORD POWER

Change the word BARN to the word DOOR by changing one letter at a time. You must form a real word each time you change a letter. The best solution has the fewest steps.

6.7 Solving Equations with Fractions and Decimals

In 1990, a survey of Canadian households found that the percent with camcorders was 2.1% less than one-half the percent with compact disc players. The percent of households with camcorders was 5.6%.

We can write the equation $0.5x - 2.1 = 5.6$, where x is the percent of households with compact disc players.

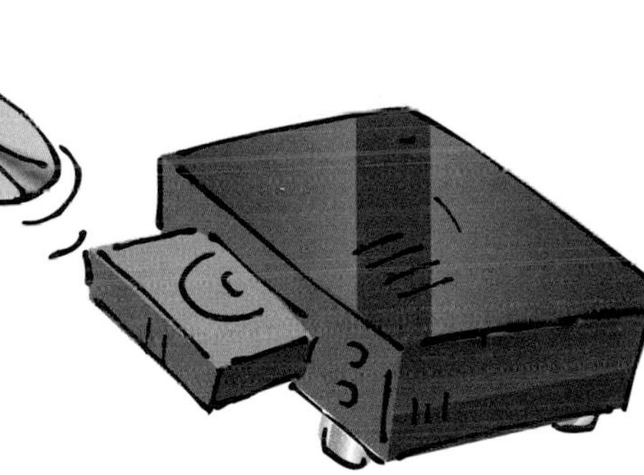

Activity: Solve the Equations

Solve these equations.

a) $1.2x - 0.4 = 2$ **b)** $0.25x - 5 = 2.75$

c) $3.14n + 2 = 12.99$ **d)** $2.3t - 1.47 = 6.85 + 0.7t$

Inquire

1. Multiply the terms in the equations a) to d) by a power of 10 to produce an equation with whole-number coefficients.

2. Solve each equation you wrote in question 1.

3. Compare the 2 solution methods you used. Which method do you prefer? Explain.

4. Solve the equation $0.5x - 2.1 = 5.6$ to find the percent of Canadian households with compact disc players in 1990.

5. Research the percent of Canadian households with compact disc players today. Compare your findings with your answer to question 4.

Example 1

Solve the equation $5.84 - 2.2y = 18.6$.

Solution 1

The number 5.84 has the greatest number of decimal places. To obtain whole-number coefficients, multiply each term by 100.

$$\begin{aligned} 5.84 - 2.2y &= 18.6 \\ 100 \times (5.84 - 2.2y) &= 100 \times 18.6 \\ 100 \times 5.84 - 100 \times 2.2y &= 100 \times 18.6 \\ 584 - 220y &= 1860 \\ -220y &= 1276 \\ y &= -5.8 \end{aligned}$$

EST $1000 \div (-200) = -5$

The solution is $y = -5.8$.

Solution 2

Keep the decimals and solve with a calculator.

$$\begin{aligned} 5.84 - 2.2y &= 18.6 \\ -2.2y &= 18.6 - 5.84 \\ -2.2y &= 12.76 \\ y &= -5.8 \end{aligned}$$

EST $10 \div (-2) = -5$

CONTINUED

Example 2

Solve $\frac{x}{3} - \frac{3x}{2} = \frac{1}{6} - x$.

Solution

To eliminate the fractions, multiply each term in the equation by the lowest common denominator, 6.

$$\frac{x}{3} - \frac{3x}{2} = \frac{1}{6} - x$$

$$6 \times \left(\frac{x}{3}\right) - 6 \times \left(\frac{3x}{2}\right) = 6 \times \left(\frac{1}{6}\right) - 6 \times x$$

$$2x - 9x = 1 - 6x$$

$$-7x = 1 - 6x$$

$$-7x + 6x = 1$$

$$-x = 1$$

$$x = -1$$

The solution is $x = -1$.

Example 3

Solve and check $\frac{x-2}{3} + \frac{x+1}{2} = 4$.

Solution

Multiply all terms by the lowest common denominator, 6.

$$\frac{x-2}{3} + \frac{x+1}{2} = 4$$

$$6 \times \frac{(x-2)}{3} + 6 \times \frac{(x+1)}{2} = 6 \times 4$$

The division bar is a grouping symbol. It acts like a bracket.

$$2(x-2) + 3(x+1) = 24$$

$$2x - 4 + 3x + 3 = 24$$

$$5x - 1 = 24$$

$$5x = 25$$

$$x = 5$$

Check: **L.S.** $= \frac{x-2}{3} + \frac{x+1}{2}$ **R.S.** $= 4$

$$= \frac{5-2}{3} + \frac{5+1}{2}$$

$$= \frac{3}{3} + \frac{6}{2}$$

$$= 1 + 3$$

$$= 4$$

The solution is $x = 5$.

Practice

Solve.

1. $x + 0.2 = 0.8$
2. $y - 0.5 = 1.2$
3. $s + 1.2 = 1.5$
4. $t + 4.5 = -3.0$
5. $m + 5.2 = 2.2$
6. $-2.3 = 4.6 + n$

Solve.

7. $0.3x = 0.3$
8. $-5 = 0.5y$
9. $1.2m = -3.6$
10. $0.3n = 1.2$
11. $4.8 = 2.0n$
12. $1.1s = -4.4$

Solve and check.

13. $3x - 0.4 = 0.8$ **14.** $3.64 = 2m + 4.7$

15. $4y + 0.88 = 5.24$ **16.** $2.5 = 3.7 + 0.2x$

17. $0.3r + 0.54 = -3$ **18.** $1.2s + 3.6 = 4.8$

Solve.

19. $\frac{x}{4} = \frac{1}{2}$ **20.** $\frac{y}{12} = \frac{1}{3}$ **21.** $\frac{8}{10} = \frac{n}{5}$

22. $\frac{m}{6} = -\frac{1}{3}$ **23.** $-\frac{1}{9} = \frac{y}{27}$ **24.** $\frac{x}{8} = -\frac{1}{4}$

Solve and check.

25. $\frac{y}{2} = \frac{y}{3} - 1$ **26.** $\frac{y}{4} = \frac{y}{5} + 1$

27. $\frac{5n}{2} = \frac{4n}{3} - \frac{7}{6}$ **28.** $\frac{p}{2} - \frac{3p}{4} = \frac{3}{4} - p$

29. $\frac{n}{3} + 2 = \frac{n}{5} + 4$ **30.** $\frac{(x+1)}{3} = \frac{(x-1)}{5}$

31. $\frac{(3-y)}{5} = \frac{(-2-3y)}{4}$ **32.** $\frac{(2x-3)}{2} = \frac{(-x-1)}{4}$

Solve and check.

33. $\frac{1-x}{4} - \frac{x}{2} = 7$

34. $\frac{x-1}{3} + \frac{x+2}{6} = 7$

35. $\frac{x+1}{2} - \frac{x-7}{6} = 3$

36. $\frac{2n+10}{4} - \frac{n}{3} = 1$

37. $\frac{4x+5}{3} - \frac{3x}{2} = -x$

38. $4 - \frac{x+1}{3} + \frac{x+5}{5}$

39. $\frac{x+1}{3} + \frac{2-3x}{2} = -1$

40. $\frac{x+1}{3} + \frac{x-2}{7} = 1$

Problems and Applications

41. On the average, the mass of a raccoon is 1 kg more than $\frac{1}{4}$ of the mass of a coyote. The mass of a raccoon is 9.5 kg. To find the mass of a coyote, solve the equation $\frac{m}{4} + 1 = 9.5$.

42. On the average, a Canadian adult spends 35% or 0.35 of the time sleeping. Solve the equation $9t - 0.01 = 0.35$ to find the fraction of the time spent shopping.

PATTERN POWER

1. Take a 20 cm by 20 cm square piece of paper and cut out a 1-by-1 square from each corner. Fold up the sides to make a box. Calculate the volume of the box.

2. Cut out a 2-by-2 square from each corner and calculate the volume of the new box.

3. Repeat the procedure for a 3-by-3 square, a 4-by-4 square, and so on. Copy and complete the table.

Dimensions of Corners (cm)	Volume of Box (cm³)
1 × 1	
2 × 2	
3 × 3	
...	

4. What are the dimensions of the box with the greatest volume?

5. What was the size of the square you removed from each corner to give the box with the greatest volume?

6. To get the box with the greatest volume, what size of square must you remove from each corner of each of the following squares?

a) 14 cm by 14 cm

b) 25 cm by 25 cm

The Binary Number System

The decimal system that we use has ten digits.

0, 1, 2, 3, 4, 5, 6, 7, 8, 9

The **binary system** consists of 2 digits, 0 and 1.

A computer stores information in a **bit**, which is a short form of "binary digit." The electrical current in a bit can be either off or on. Off is represented by 0 and on by 1. Bits are combined into groups of 8 called **bytes**.

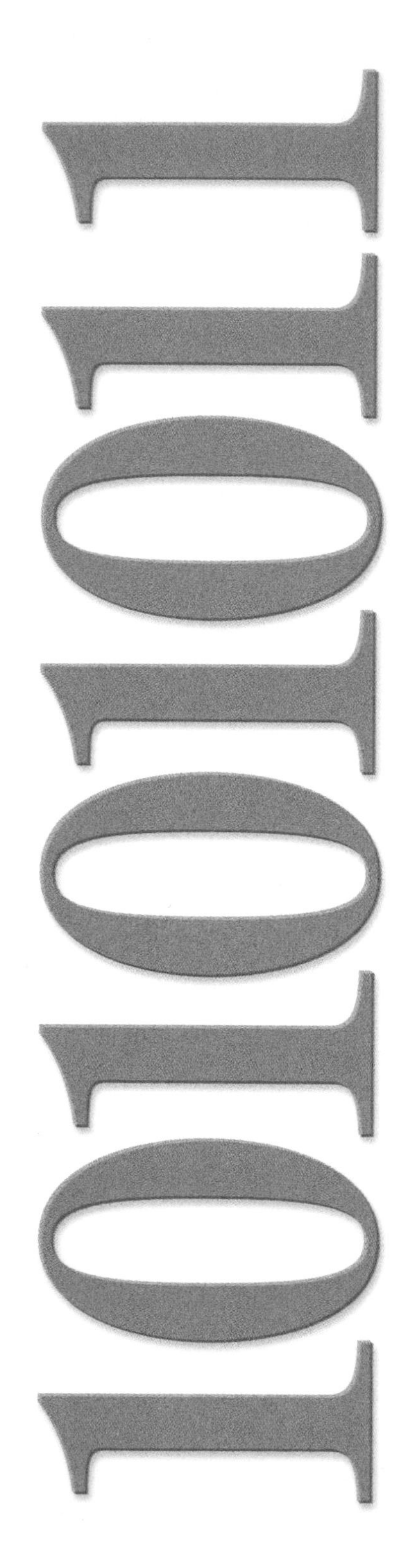

Activity ❶ Binary Numbers

Copy and complete the table to write the first 16 decimal numbers in binary. The first 5 have been done for you.

Decimal Form	Binary Form
0	0
1	1
2	10
3	11
4	100

Activity ❷ Powers of Two

1. Copy the following table. To complete it, use your results from Activity 1 and look for a pattern.

Decimal Number	Power of 2	Binary Number
1	2^0	1
2	2^1	10
4	2^2	100
8		
16		
32		
64		
128		

2. How is the exponent in the power of 2 related to the number of zeros in the binary number?

Activity ❸ Whole Numbers in Binary

You can use the results of Activity 2 to help you write whole numbers in binary form, because whole numbers can be written as sums of powers of 2. To write 43 as the sum of powers of 2, successively add the largest possible powers of 2.

$$\begin{aligned}43 &= 32 + 11\\ &= 32 + 8 + 3\\ &= 32 + 8 + 2 + 1\\ &= 2^5 + 2^3 + 2^1 + 2^0\end{aligned}$$

In the binary system $2^5 + 2^3 + 2^1 + 2^0$

$$= 100000 + 1000 + 10 + 1$$
$$= 101011$$

Write each of these decimal numbers in binary form.

1. 20 **2.** 37 **3.** 56 **4.** 78 **5.** 147

Activity ❹ Binary Numbers to Whole Numbers

To write binary numbers as whole numbers, reverse the steps in Activity 3.

$$\begin{aligned}110110 &= 100000 + 10000 + 100 + 10\\ &= 2^5 + 2^4 + 2^2 + 2^1\\ &= 32 + 16 + 4 + 2\\ &= 54\end{aligned}$$

Write these binary numbers as whole numbers.

1. 10101 **2.** 11001

3. 101010 **4.** 110011

Activity ❺ Binary Codes

Some television stations scramble their signals so that viewers cannot use unauthorized satellite dishes. Authorized users receive a code to activate their descramblers. The code is a binary number.

If you had 1 bit to use, you could use one of 2 codes. 0 or 1

If you had 2 bits, you could use one of 4 codes.

00 01 10 11

1. How many different codes could you use if you had 3 bits? 4 bits? 5 bits?

2. Television stations have 2^{56} bits available to them. Approximately how many codes can they use? If you set about finding the code by trying 1 000 000 codes a day, about how many years would it take you to find the code if you succeeded on the last try?

6.8 Solving Inequalities

Dr. Roberta Bondar was the second Canadian in space. She wore a space suit that protected her against pressure and temperature changes. The mass of a space suit is 23 kg. The combined mass of an astronaut plus space suit must be less than 110 kg.

Activity: Complete the Table

Copy and complete the table.

Astronaut's Mass (kg)	Mass of Astronaut Plus Space Suit (kg)	Astronaut's Mass (kg)	Mass of Astronaut Plus Space Suit (kg)
71		91	
76		96	
81		101	
86		106	

Inquire

1. From the table, what is the maximum mass that an astronaut can have without exceeding the mass restriction?

2. Let the mass of an astronaut be x. Write an equation to show that the mass of an astronaut plus space suit can equal 110 kg.

3. What is the difference between the equation you wrote in question 2 and the statement about combined mass given in the introduction?

4. Which symbol, greater than, "$>$," or less than, "$<$," should you use to replace the "=" sign in the equation you wrote in question 2?

5. Rewrite your equation using the symbol you chose in question 4.

An **inequality** is a mathematical sentence that contains one of these symbols: $>$, $<$, $\geq$, or $\leq$. The following table shows each symbol and its meaning.

Symbol	Meaning
$<$	is less than
$>$	is greater than
$\leq$	is less than or equal to
$\geq$	is greater than or equal to

Activity: Discover the Relationships

Copy the table. Complete it by writing the correct inequality after each operation.

Inequality	Operation on Both Sides	New Inequality
$6 > -4$	Add 2	
$6 > -4$	Subtract 2	
$6 > -4$	Multiply by 2	
$6 > -4$	Divide by 2	

Inequality	Operation on Both Sides	New Inequality
$6 > -4$	Add (–2)	
$6 > -4$	Subtract (–2)	
$6 > -4$	Multiply by (–2)	
$6 > -4$	Divide by (–2)	

Inquire

1. Repeat the table for the inequality $-8 < -2$.

2. Does the direction of the inequality sign stay the same when you do each of the following?

a) add the same positive or negative number to both sides
b) subtract the same positive or negative number from both sides
c) multiply both sides by the same positive number
d) divide both sides by the same positive number
e) multiply both sides by the same negative number
f) divide both sides by the same negative number

Example 1

Solve the inequality $5x - 2 \geq 2x + 4$ and graph the solution.

Solution

Add 2 to both sides:
$$5x - 2 + 2 \geq 2x + 4 + 2$$
$$5x \geq 2x + 6$$

Subtract $2x$ from both sides:
$$5x - 2x \geq 2x + 6 - 2x$$
$$3x \geq 6$$

Divide both sides by 3:
$$\frac{3x}{3} \geq \frac{6}{3}$$
$$x \geq 2$$

The solution is $x \geq 2$.

–3 –2 –1 0 1 2 3 4 5 6 7 8

The closed dot at $x = 2$ means that 2 is included in the solution.

Example 2

Solve the inequality $2(x - 3) > 4x + 2$ and graph the solution.

Solution

$$2(x - 3) > 4x + 2$$
$$2x - 6 > 4x + 2$$
$$2x - 4x > 2 + 6$$
$$-2x > 8$$

Divide both sides by -2:
$$\frac{-2x}{-2} < \frac{8}{-2}$$
$$x < -4$$

Remember to reverse the symbol when multiplying or dividing by a negative number.

–10 –9 –8 –7 –6 –5 –4 –3 –2 –1 0 1 2

The open dot at $x = -4$ means that -4 is not included in the solution.

CONTINUED

Practice

1. Which inequalities are true?

a) $5 > 4$ **b)** $10 > 6$ **c)** $15 \ge 17$
d) $6 < 8$ **e)** $7 \le -7$ **f)** $11 < 13$
g) $-3 > -2$ **h)** $-1 > -5$ **i)** $-1 \ge -2$
j) $-4 \le -4$ **k)** $-5 < -8$ **l)** $-6 < -5$

2. Which of the x values shown make each inequality true?

a) $x + 2 < 8$ (3, 6) **b)** $-5 > x - 2$ (−5, 0)
c) $x + 5 < 7$ (4, 1) **d)** $x - 3 > 3$ (0, 8)
e) $x + 3 \le 8$ (7, −1) **f)** $x - 5 \ge -9$ (−5, 1)
g) $6 \le -1 + x$ (−4, 7) **h)** $x + 2 \le -2$ (−2, −4)

Solve each inequality and graph its solution.

3. $x - 2 > 0$ **4.** $10 > x + 5$
5. $x - 3 < 1$ **6.** $x - 5 > 2$
7. $y - 4 > -3$ **8.** $y + 3 < 4$
9. $1 < -2 + z$ **10.** $z - 5 < -2$

Solve.

11. $4x < 8$ **12.** $2y > -8$ **13.** $4m < 20$
14. $5n < -20$ **15.** $3s > 0$ **16.** $21 \ge 7y$
17. $6p \le -12$ **18.** $-15 \ge 5t$ **19.** $8b \le 24$

Solve.

20. $-3m < 9$ **21.** $-4n > 12$
22. $-15 > -5x$ **23.** $-6 > -2y$
24. $-3t \ge -18$ **25.** $-10y \le -50$
26. $21 \ge -7x$ **27.** $4x \ge -12$

Solve. Graph the solution.

28. $2x < 10$ **29.** $2y > -8$ **30.** $-8 > 4m$
31. $2n > 6$ **32.** $5t \le -15$ **33.** $4 \le 4s$
34. $3y \le -6$ **35.** $8x \le 40$ **36.** $2x \ge -2$

Solve.

37. $4x + 2 < 3x + 5$ **38.** $2x - 4 > x + 2$
39. $7y - 4 \ge 6y + 3$ **40.** $5y + 7 \le 4y - 2$
41. $2t + 7 < t - 1$ **42.** $10a + 4 > 9a + 2$
43. $3n + 4 < 2n + 2$ **44.** $5n + 2 > 4n - 1$

Solve.

45. $4(x - 3) > 3x + 1$
46. $3y + 9 < 2y + 12$
47. $4(2m - 1) \ge 7m - 3$
48. $3x - 10 > 2x - 9$
49. $3y + 2 \le 2y + 1$
50. $3m + 14 < 2(m + 6)$
51. $17 - 6x > 12 - 7x$
52. $20m - 7 < 19m - 2$
53. $2(5 - 3b) \ge -7b + 2$
54. $5w - 7 \le 4w - 3$
55. $7p + 11 \ge 6p + 17$
56. $10x + 10 \le 9x + 15$
57. $19 - 13x \ge -14x + 16$

Solve. Graph the solution.

58. $7x + 4 < 5x + 8$
59. $5y - 2 \le 2y + 7$
60. $2m - 3 < 9 - 2m$
61. $3x + 10 < -2 - 3x$
62. $4y + 5 \ge 6y - 1$
63. $6t - 5 \le 8t + 3$
64. $7 + y < 4y + 13$
65. $6x - 13 \ge 8x - 15$

Solve.

66. $6(y - 2) + 7 > 8y - 25$
67. $4(t - 1) < 8t + 20$
68. $8(x - 2) > 8 - 4x$
69. $24x + 18 > 12 + 7(3x - 3)$

70. $3(5 - 5x) + 3 \geq 34 - 7x$

71. $2(7x - 11) \leq 9x - 32$

72. $15 + 4x \geq 5(2x - 1) - 10$

73. $2(8x - 13) + 4 < 18x + 26$

74. $3t + 45 \geq 6(t - 4)$

75. $6(m - 4) - 2(m + 2) < 7(m - 4) - 6$

76. $5(y + 1) - 2(y + 3) \leq 5(y - 1)$

77. $3(2x - 5) \geq 2(1 + 2x) + 5$

Problems and Applications

78. In her last math test, Giselle got 10 marks more than on any other test this year. Her mark was less than 100. Solve the inequality $x + 10 < 100$ to find her highest possible mark on any other test.

79. On Saturdays, Campus Clothes sells at least 25 more jackets than on any other day of the week. The store has never sold more than 84 jackets in one day. Solve the inequality $84 \geq n + 25$ to find the number of jackets that could have been sold on any other day of the week.

80. Matt wants to keep his annual travel expenses under \$4680. Solve the inequality $52x < 4680$ to find how much Matt can spend each week.

81. Canada' s top honour is the Order of Canada. There are 3 types of appointments to the Order: Member, Officer, and Companion. In any year, up to 46 new Officers can be appointed. This number is 1 more than 3 times the maximum number of Companions who can be appointed. Solve the inequality $3x + 1 \leq 46$ to find the number of Companions appointed in any year.

82. Bianca must keep her phone bill below \$55/month. The basic charge is \$15, and it costs her \$2/min to phone her friend in Taiwan. Solve the inequality $2t + 15 < 55$ to determine how long Bianca can spend talking to her friend each month.

83. A peregrine falcon can dive at up to 350 km/h. This speed is 50 km/h faster than 3 times the top speed of the fastest land animal, the cheetah.

a) Solve the inequality $350 \geq 3x + 50$ to determine the speed of a cheetah.

b) Does a cheetah have a minimum speed? Explain.

84. a) Solve the inequality $x + 15 + 18 \leq 50$ to find the values of x that give this triangle a perimeter of no more than 50.

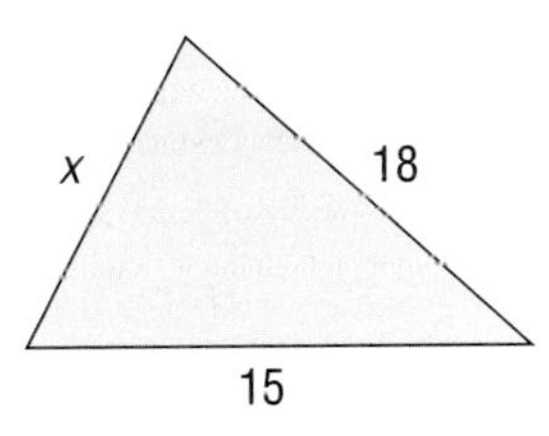

b) Does x have a minimum value? Explain.

85. a) Solve the inequality $10(x + 1) < 50$ to find values of x that give this rectangle an area of less than 50.

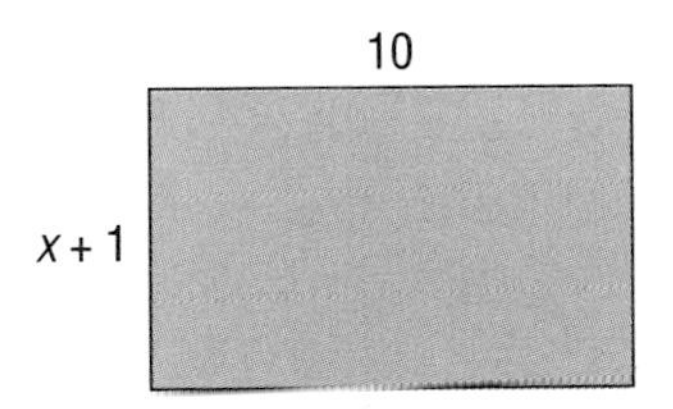

b) Does x have a minimum value? Explain.

86. Write an inequality that has variables and numbers on both sides and has a solution of $x \leq -1$. Have a classmate solve your inequality.

The Möbius Strip

Activity ❶ Making a Möbius Strip

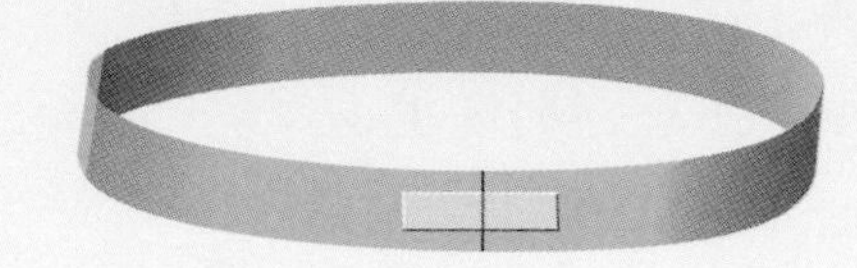

1. Take a strip of paper about 30 cm long and 5 cm wide and tape it together to make a ring. You could paint this ring with 2 colours, red on the inside and blue on the outside, because the ring has 2 sides. To get from one side to the other, you have to cross an edge.

2. Take another strip of paper about 30 cm long and 5 cm wide. Turn 1 end of the paper to make a half twist. Then, tape the 2 ends together. The result is called a Möbius strip after the German mathematician, Augustus Möbius, who discovered it.

3. Use a coloured pencil and start "painting" the strip. You will notice you can "paint" the whole strip with just one colour. You never have to cross an edge. The strip has only one side.

Activity ❷ Cutting a Möbius Strip in Half

If you cut a paper ring around the middle, you get 2 paper rings, each one-half the width of the original.

1. Predict what you will get when you cut a Möbius strip around the middle.

2. Cut the Möbius strip around the middle and describe the results. How many sides does the resulting figure have? Is it a Möbius strip? Explain.

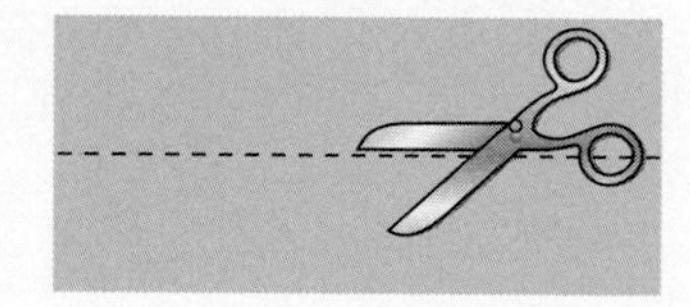

Activity ❸ Cutting a Möbius Strip into Thirds

If you cut a paper ring into thirds, you get 3 paper rings, each one-third the width of the original.

1. Predict what you will get if you cut a Möbius strip into thirds.

2. Cut a Möbius strip starting one-third the way in from its edge. The scissors will make 2 trips around the strip. Describe the result. How many sides does each of the resulting figures have? Are they both Möbius strips? Explain.

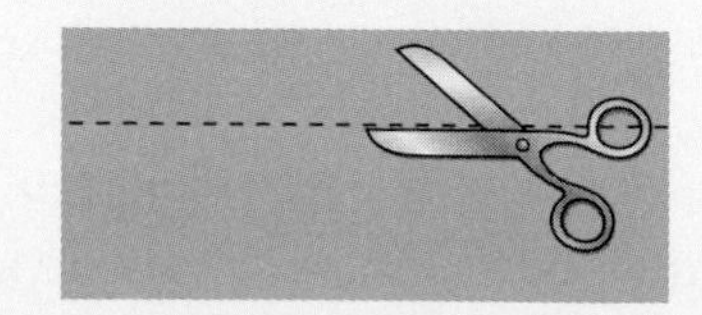

Sprouts

Sprouts is a game for 2 players. Start with any number of points on a piece of paper. Three is a good number to begin with.

The first player makes a move by drawing a line that joins one point to another or a point to itself, as shown in the diagram below. The first player then marks a new point anywhere on the line.

Examples of some first moves are shown.

Lines are drawn according to the following rules.

1. The line can be of any shape but must not cross itself, cross another line, or go through a previously drawn point.

2. No point can have more than 3 lines drawn from it.

Players take turns drawing lines, adding a new point on each turn. The last player to make a move is the winner.

The following is a sample game of Sprouts, starting with 3 points.

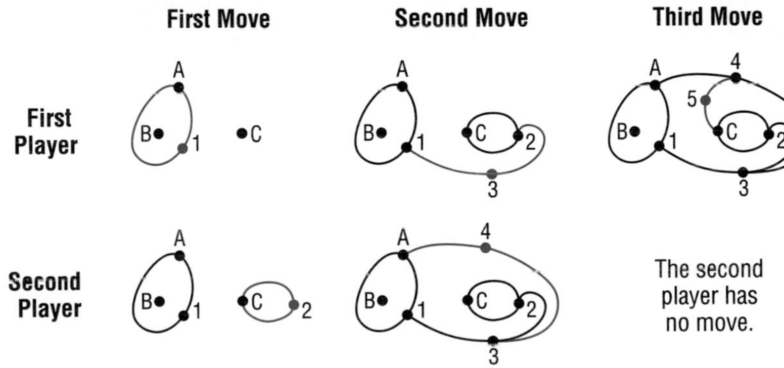

Play several games of Sprouts with a classmate. Describe the most effective strategies for winning the game.

Review

1. Which bracketed values are solutions of the equations shown?

a) $x + 7 = 11$ (4) **b)** $3x = 12$ (9)
c) $x - 5 = 3$ (7) **d)** $5y = -10$ (−2)
e) $x + 3 = 11$ (14) **f)** $2x - 10 = 4$ (7)
g) $x + 5 = -7$ (−4) **h)** $3x + 2 = -7$ (3)

Solve.

2. $3x = 12$
3. $y + 7 = 12$
4. $2 = -4 + s$
5. $3 + z = 7$
6. $x + 1 = -2$
7. $5m = 40$
8. $3n = -21$
9. $-24 = 12x$
10. $5p = 3.0$
11. $7z = -2.8$

Solve.

12. $\frac{x}{2} = 5$
13. $\frac{y}{2} = -5$
14. $\frac{b}{3} = -3$
15. $\frac{z}{4} = 6$

Solve.

16. $\frac{x}{3} + 2 = 4$
17. $\frac{y}{2} - 3 = 2$
18. $4 + \frac{b}{3} = -1$
19. $2 - \frac{z}{4} = -1$

Solve and check.

20. $2x - 3 = 9$
21. $3y - 2 = 13$
22. $5y - 7 = 3$
23. $-22 = 7y - 8$
24. $3x + 2x = -25$
25. $4x + 5x = -81$
26. $3 + 2q = 21$
27. $2.9 + 6x = 1.1$

Solve.

28. $3x - 17 = 13$
29. $56 = 5y + 11$
30. $2x + 3.1 = 1.1$
31. $5m + 25 = -65$
32. $3y + 10 = 82$
33. $0.3x + 2.6 = 4.4$
34. $-39 = 25 + 4a$
35. $-54 = 6z - 18$
36. $-33 + 7t = -61$
37. $8q - 42 = -26$

Solve.

38. $6a - 3a + 2a = 10 - 2 + 7$
39. $2p + 3p - p = 9 - 15 - 6$
40. $10j - 5j - 8j = -6 + 15 + 12$
41. $12k - 8k - 6k = -18 + 20 - 6$

Solve and check.

42. $2x + 7 = 5x + 13$
43. $5y - 12 = y - 4$
44. $3x - 4 = 8 - x$
45. $6x - 6 = 4x - 10$
46. $7 - x = -7 + 6x$
47. $4 + x = 36 - 7x$
48. $2 + 3j = 5j - 24$
49. $-3 + 2r = -30 - 7r$

Solve.

50. $5x + 6x - 2x = 3x - 18 + 12 - 6$
51. $2x + 15 = -12 - 13 - 3x$
52. $b - 17 = 15 + 3b + 18$
53. $21 + 4q = q + 13 + 17$

Solve and check.

54. $7 + 5(x - 3) = 3(x + 2)$
55. $4(y - 4) + 3(y + 7) = -30$
56. $-38 = 5(2y - 1) + y$
57. $6(3x - 4) - (1 - 3x) = -67$

Solve.

58. $9m - 0.8 = 7m + 4.8$
59. $5y + 0.4 - 3y = -6$
60. $1.6m - 2.4 + 0.1m = 0.1m + 0.8$
61. $8.4x - 0.8 = 4.4x - 4$
62. $12.2x + 0.3 = 20.7 + 5.4x$

Solve.

63. $\frac{y}{2} - \frac{1}{2} = \frac{1}{2}$
64. $\frac{y}{3} + \frac{1}{3} = -\frac{2}{3}$
65. $\frac{y}{2} + \frac{y}{3} = \frac{5}{6}$
66. $\frac{5y}{4} - \frac{y}{2} = \frac{-3}{4}$
67. $\frac{2y + 1}{3} = -5$
68. $\frac{x - 2}{6} = \frac{x + 2}{2}$
69. $\frac{x + 2}{4} = \frac{2x - 5}{2}$
70. $\frac{x - 2}{2} - \frac{x}{6} = -2$

Solve and check.

71. $\frac{n}{2} - \frac{1 + n}{5} = -\frac{1}{2}$

72. $\frac{2x - 1}{3} - \frac{x + 2}{4} = -5$

73. $\frac{3y + 5}{3} - \frac{y - 3}{6} = -2$

74. $\frac{2y + 1}{3} - \frac{y + 4}{5} = 7$

Solve. Graph each solution.

75. $2x < 10$ **76.** $3y > -15$ **77.** $4 + x > 8$

78. $8 > 4x$ **79.** $3x \geq 12$ **80.** $-3 \leq 3x$

81. $-2x < 6$ **82.** $-5x > -5$ **83.** $-8x \leq 16$

Solve these inequalities.

84. $3(x - 2) \leq 3$ **85.** $18 \geq 10 + 4x$

86. $4t - 3 \geq -15$ **87.** $4 - 2x \geq 8$

88. $7x - 8 < 4x + 1$ **89.** $7t + 7 \leq 10t - 14$

90. $6t - 5 > 2t - 1$ **91.** $3(s + 6) \leq -9$

Solve.

92. $4y - 2 \leq 2y + 8$

93. $5(x + 6) - 2(x - 1) > 2$

94. $2(x + 4) \leq 5 - 11$

95. $4(x - 5) - (x + 1) < 3$

96. $7 - 3(2x - 5) - x > 1$

97. The number of Members of Parliament (MPs) elected from Ontario is 99. This number is 5 less than 4 times the number elected from Alberta. Solve the equation $4n - 5 = 99$ to find the number of MPs elected from Alberta.

98. The world's largest rodent is the capybara of South America. The mass of a capybara can be up to 113 kg. This mass is 8 kg more than 5 times the greatest known mass of a domestic cat. Solve the inequality $5m + 8 \leq 113$ to find the masses of domestic cats.

Group Decision Making

Designing a Rube Goldberg Invention

The cartoonist Rube Goldberg is famous for his impractical inventions, which are shown as one-scene cartoons. This cartoon is like a Rube Goldberg cartoon. It shows a different way of pouring water into a sink.

Sources of power for Goldberg inventions might include:

- the sun melting a block of ice that releases a rope that was frozen into it
- the rays of the sun being focused through a magnifying glass
- a cat chasing a dog

1. Work in home groups. Decide on a Rube Goldberg invention designed to do something in an original way. Some tasks might include sharpening a pencil, travelling to school, or cleaning the chalkboard.

2. As a group, design a flow chart to show the sequence of events in your invention before you start to draw the cartoon.

3. In your group, draw and discuss rough sketches of the cartoon. Draw the final version and present it to the class.

4. As a class, evaluate the cartoons for originality and humour.

Chapter Check

Solve and check.

1. $x - 12 = -4$

2. $x + 15 = 12$

3. $x + 2 = 8$

4. $x - 10 = -19$

5. $5 = x - 3$

6. $15 - x = -5$

Solve.

7. $6x = -54$

8. $4x = 44$

9. $\frac{x}{5} = 7$

10. $\frac{x}{3} = 4$

Solve.

11. $5x + 8 = 3x + 2$

12. $5x + 12 = 7x + 16$

13. $6z + 8 = 9z - 7$

14. $2y - 6 = 10 + 11y + 2$

Solve.

15. $3(2x - 4) = 9x + 3$

16. $5(2x - 1) + 9 = 2(x - 2)$

Solve.

17. $10x - 0.4 = 2x - 3.6$

18. $7.6x - 0.7 = 5.6x - 6.5$

Solve.

19. $\frac{2x}{3} = -6$

20. $\frac{y}{3} + \frac{1}{3} = -\frac{2}{3}$

21. $\frac{y}{3} - \frac{5y}{6} = -\frac{1}{2}$

Solve.

22. $\frac{3y + 1}{2} = 5$

23. $\frac{x + 2}{2} = \frac{x - 1}{5}$

24. $\frac{x + 2}{2} + \frac{x - 2}{5} = 2$

Solve and graph.

25. $-3x > 21$

26. $-2x + 5 < 9$

27. $4r + 4 > r + 1$

Solve.

28. $2(3x + 1) - 2(x - 1) \geq -16$

29. $4(x - 1) - 6(x + 1) < 10$

30. The average Canadian spends 23% or 0.23 of the time on leisure activities. This fraction is 0.01 less than 4 times the fraction of the time spent eating. Solve the equation $4t - 0.01 = 0.23$ to find the fraction of the time spent eating.

31. A Boeing 747 is about 60 m long. This length is about 4 m longer than 8 times the length of the longest known earthworm. Solve the inequality $8x + 4 \leq 60$ to find the lengths of earthworms.

Using the Strategies

1. What day of the week will it be 1000 days from today?

2. What are the numbers of the wheels that turn clockwise?

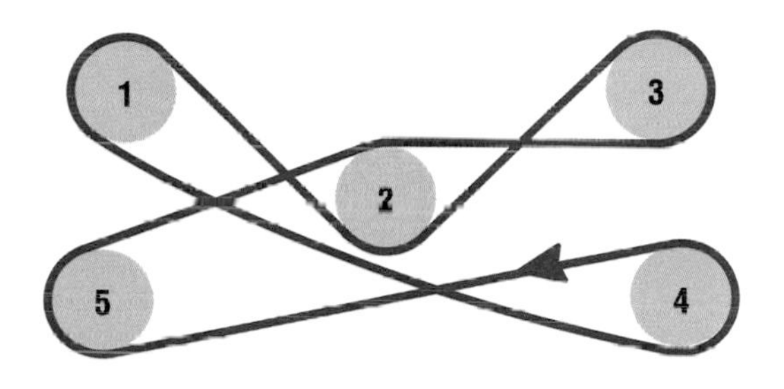

3. Find 5 consecutive whole numbers whose sum is 405.

4. The map shows the locations of 4 houses, the train station, and the airport.

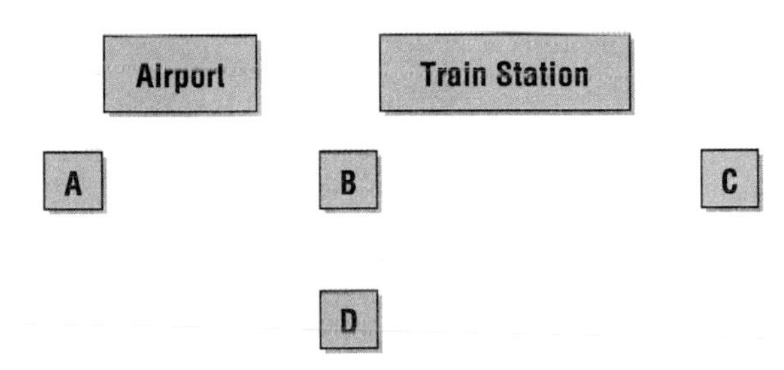

Match each house with the correct number on the distance graph. Give reasons for your answer.

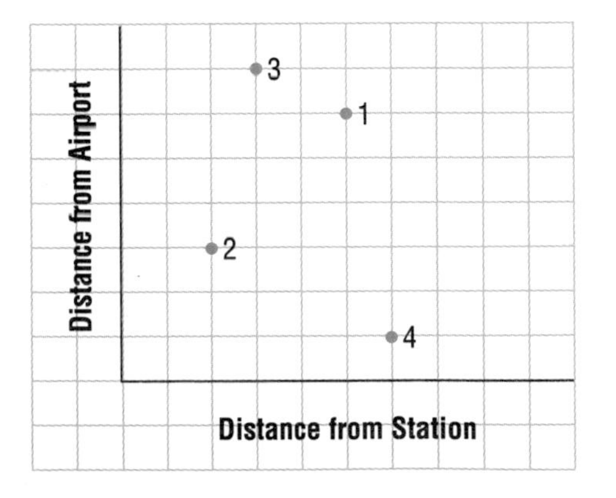

5. Place the digits 1, 2, 3, 4, 5, 6, and 7 in the circles so that the sum of each line of connected circles is the same.

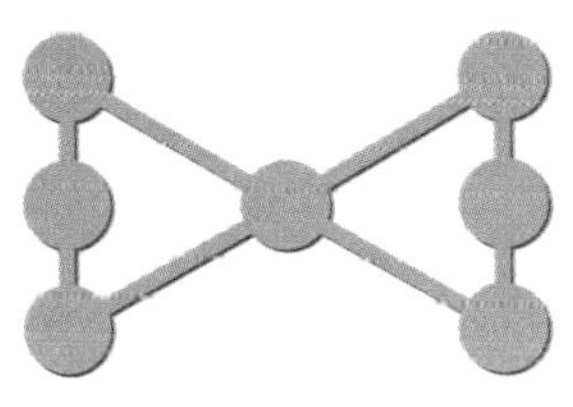

6. In how many ways can you use only dimes and quarters to make change for a dollar?

7. How many days are there between July 1 and December 25?

8. In a collection of coins, the numbers of dimes and nickels add to 5. The numbers of dimes and quarters add to 7. The numbers of nickels and quarters add to 8. How many of each type of coin are there?

9. The diagram shows a Latin Square. The numbers 1, 2, and 3 have been placed so that each number appears only once in each row and column.

1	2	3
2	3	1
3	1	2

There are 11 other different Latin Squares that use the numbers 1, 2, and 3. Draw them.

1. The flying distance, in kilometres, from Toronto to Chicago is given by the expression $\frac{d}{2} + 35$, where d is the flying distance, in kilometres, from Vancouver to Regina. What is the flying distance from Toronto to Chicago?

2. Use information from the Data Bank on pages 564 to 569 to write a problem. Have a classmate solve your problem.

CHAPTER 7

Using Equations to Solve Problems

The following are the times for a typical space shuttle launch.

At $t - 3.8$ s, the shuttle's 3 main engines are ignited.

At $t + 2.88$ s, the twin boosters are ignited, and the hold-down bolts are released.

At $t + 2$ min 12 s, the boosters burn out and are jettisoned.

At $t + 8$ min 32 s, just before the shuttle reaches orbital velocity, the engines are shut down.

The huge external tank is jettisoned at $t + 8$ min 50 s.

The shuttle's own orbit-adjust engines put the shuttle into orbit.

If t is 08:00, at what time does each of the above events take place? For how long is the shuttle bolted to the lift-off platform while the engines are on? Why?

Activity ❶ A Scavenger Hunt

When you solve problems in mathematics, you often replace words with symbols.

1. The following are some symbols that have been used to replace words. State what each symbol means and, if possible, give an example of where you have seen it.

2. Design an international symbol of your own. Ask your classmates to tell you what they think it means.

Activity ❷ Equations

Solve.

1. $x - 4 = -6$

2. $y - 5 = -8$

3. $8 = m + 3$

4. $n + 5 = 8$

5. $s - 7 = 7$

6. $t + 7 = 7$

7. $x + 5 = 11$

8. $-9 = -3 + y$

9. $q + 9 = -1$

Solve.

10. $3y = 12$

11. $5x = -20$

12. $-m = 6$

13. $-n = -3$

14. $18 = 6s$

15. $-45 = 9t$

16. $\frac{x}{3} = 5$

17. $\frac{y}{4} = 1$

18. $\frac{y}{2} = -3$

Solve and check.

19. $2x + 7 = -13$

20. $1 = -5 + 3y$

21. $m + 6 = 4m + 18$

22. $4n - 5 = 2n + 1$

23. $6 = 2(x - 1)$

24. $3(y + 2) + 1 = y + 9$

25. $2(m + 3) + 4 = 3(m - 7) + 19$

Activity ❸ Perimeter and Area

Calculate each perimeter.

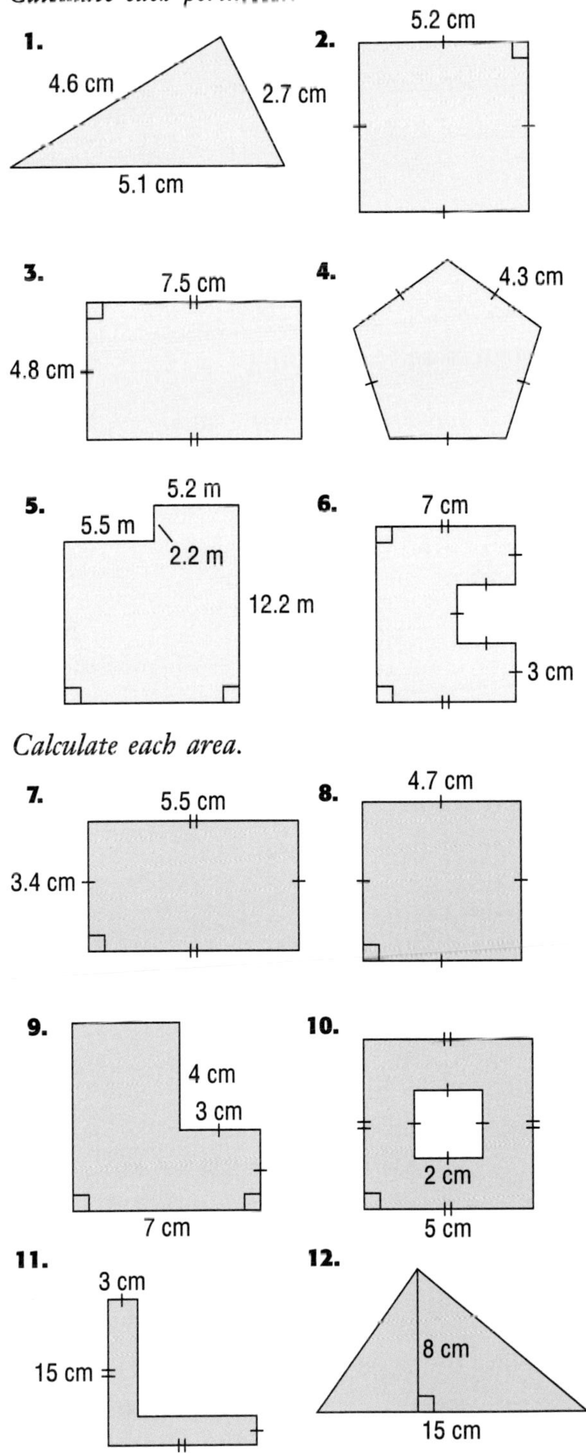

Mental Math

Add.

1. $18 + 2$ **2.** $18 + 12$ **3.** $25 + 5$

4. $25 + 25$ **5.** $46 + 5$ **6.** $46 + 15$

7. $54 + 6$ **8.** $54 + 16$ **9.** $54 + 26$

Subtract.

10. $36 - 6$ **11.** $36 - 16$ **12.** $54 - 4$

13. $54 - 24$ **14.** $65 - 6$ **15.** $65 - 16$

16. $71 - 2$ **17.** $71 - 12$ **18.** $71 - 22$

Multiply.

19. $(25)(4)$ **20.** $(25)(40)$ **21.** $(4)(5)$

22. $(400)(5)$ **23.** $(5)(8)$ **24.** $(5)(800)$

25. $(4)(9)$ **26.** $(40)(9)$ **27.** $(400)(9)$

Divide.

28. $28 \div 4$ **29.** $280 \div 4$ **30.** $35 \div 7$

31. $3500 \div 7$ **32.** $2000 \div 5$ **33.** $2000 \div 50$

Simplify.

34. $(3)(2) + 5$ **35.** $4 + (2)(5)$

36. $5(2 + 6)$ **37.** $(8 - 3) \times 4$

38. $5 + 9 - 2$ **39.** $24 - (8 - 2)$

40. $27 - 9 - 3$ **41.** $30 - (10 + 5)$

Simplify.

42. $4 + 6 + 5 + 7$ **43.** $7 - 3 + 4 - 5$

44. $(2)(6) - 3 + 4$ **45.** $12 - 3 + 6 - 5$

46. $24 - (4)(2)(1)$ **47.** $20 - 5 - (4)(3)$

48. $(6 + 4)(3 - 1)$ **49.** $5 + 7 - (7 - 4)$

Simplify.

50. $\frac{4}{8} + \frac{1}{8}$ **51.** $\frac{3}{9} + \frac{1}{9}$ **52.** $\frac{5}{10} + \frac{2}{10}$

53. $\frac{5}{10} - \frac{1}{5}$ **54.** $\frac{5}{6} - \frac{1}{6}$ **55.** $\frac{3}{10} - \frac{1}{5}$

Words to Symbols

Activity 1

The longest parade in the world runs from March to July. The "floats" have masses between 500 000 t and 10 000 000 t. They are up to 10 000 years old. The parade consists of hundreds of icebergs that pass Newfoundland's east coast.

The part of an iceberg above the water is called the tip. The height, t, of the tip is $\frac{1}{7}$ of the total height of an iceberg. So, an expression for the total height of an iceberg is $7t$. Write an expression in terms of t for the height of the part of an iceberg that is under water.

Activity 2

Match the words with the symbols.

1. a number increased by five	$5x$
2. a number decreased by one	$w + 3$
3. a number divided by nine	$n - 1$
4. two less than a number	$b + 5$
5. three more than a number	$6 - t$
6. five times a number	$m - 2$
7. six decreased by a number	$2 + y$
8. two increased by a number	$q \div 9$

If n represents a number, write an algebraic expression using numbers and symbols for each of the following statements.

9. three times a number

10. a number increased by one

11. a number decreased by five

12. the product of six and a number

13. a number divided by five

14. three more than twice a number

15. ten decreased by a number

16. seven divided by a number

In each case, let the number be x. Write an expression for each statement.

17. a number increased by four

18. seven times a number

19. a number multiplied by five

20. three divided by a number

21. seven more than a number

22. half of a number

23. a number diminished by eight

24. the product of a number and two

25. Write 3 new statements. Have a classmate write an expression for each.

Activity ❸

Copy and complete the tables.

1. The first number is 23 more than the second.

First Number	45	56		x		$3m$	$4t + 21$	
Second Number			18		y			$6r - 5$

2. Francine is 4 years younger than Terry.

Francine's Age		19		m		$4l$	
Terry's Age	16		x		$5y$		$3w + 8$

3. There are 6 times as many cars as motorcycles.

Number of Cars		72		y		
Number of Motorcycles	13		x		$m + 7$	$3t + 8$

Activity ❹

1. The length of a rectangle is 5 cm more than its width.

a) Which dimension is smaller?

b) Let the smaller dimension be x. Then, write the larger dimension in terms of x.

c) Draw and label a diagram of the rectangle.

d) Write an expression for the perimeter of the rectangle in terms of your variable.

2. The length of a rectangle is 4 m more than its width. Write an expression for its area.

3. The side length of a square is x metres. What is its perimeter?

4. a) Copy and complete the table for the dimensions of a rectangle. Its length is 3 m more than its width.

Width (m)	10	15	x	$10x$	$4x + 1$
Length (m)					

b) Write expressions for the perimeter and area of each rectangle. Compare your expressions with a classmate's.

5. Repeat question 4 for the following table. The length of the rectangle is now 7 m more than the width.

Width (m)	10		x		$3t$		$5z + 4$
Length (m)		24		y		$4n$	

7.1 Writing Equations

The Canadian Coast Guard patrols Canada's rivers and coastline. The Coast Guard is mainly involved in search-and-rescue operations.

Activity: Study the Information

A Coast Guard boat that travels at a speed of 15 km/h in still water was patrolling up a river with a current of 6 km/h. The boat's actual speed, s kilometres per hour relative to the river bank, was given by the equation $s = 15 - 6$.

When it patrolled down the river, the boat's actual speed, s kilometres per hour relative to the river bank, was given by the equation $s = 15 + 6$.

Inquire

1. Let the speed of the boat in still water be x kilometres per hour. Write an equation to find x if the boat patrols up a river where

a) the current is 6 km/h and the boat's actual speed is 11 km/h.

b) the current is 4 km/h and the boat's actual speed is 10 km/h.

2. Write an equation to find x if the boat patrols down a river where

a) the current is 5 km/h and the actual speed is 17 km/h.

b) the current is 7 km/h and the actual speed is 20 km/h.

Example 1

It costs 4 times as much to go to the movie theatre as it does to rent a movie. Together, both ways of seeing a movie cost $10. Write an equation to find each cost.

Solution

Let the smaller cost, a movie rental, be represented by x.
Then, the cost of going to a movie theatre is $4x$.
Together, both ways cost $10.
The total cost of both ways equals the sum of x and $4x$.

$$x + 4x = 10$$
$$5x = 10$$

Example 2

Lake Ontario is 77 km shorter than Lake Erie. The lengths of the 2 lakes total 699 km. Write an equation to find the length of Lake Ontario.

Solution

Let the length of Lake Ontario be represented by x.
Then, the length of Lake Erie is $x + 77$.
The sum of x and $x + 77$ is equal to 699.

$$x + (x + 77) = 699$$
$$2x + 77 = 699$$

Practice

Match each sentence with the correct equation.

1. Three times a number is equal to eighteen.	$x - 4 = 10$
2. A number decreased by six is four.	$3x = 18$
3. Kim's age in four years will be eighteen.	$\frac{m}{4} = 18$
4. Shuji's age four years ago was ten.	$y - 6 = 4$
5. A number divided by four is eighteen.	$x + 4 = 18$

Write an equation for each sentence.

6. Four times a number is twenty.

7. A number divided by two equals five.

8. Six more than a number is fifteen.

9. A number increased by five is twelve.

10. A number decreased by six is ten.

11. Four less than a number is seven.

12. The square of a number is twenty-five.

13. Ten decreased by a number is two.

14. Three more than a number is eighteen.

15. Five less than a number is nine.

16. A number increased by four is twenty-one.

17. A number multiplied by three is nine.

18. A number divided by five is ten.

19. A number decreased by six is negative eight.

Problems and Applications

Write an equation for each statement in questions 20–25.

20. There are 16 more white keys than black keys on a full-sized piano keyboard. There are 88 keys on a piano.

21. Brad has $12 more than Pietro. Together they have $84.

22. Mike is seven years older than Carol. The sum of their ages is 29.

23. The area of the Pacific Ocean is twice the area of the Atlantic. The sum of their areas is 250 000 000 km^2.

24. The population of Japan is about 4.5 times the population of Canada. The sum of the populations is about 150 000 000.

25. Niagara Falls has 2 parts. The American Falls are 2 m higher than the Horseshoe Falls. Their average height is 58 m.

26. The perimeter of each rectangle is 36 cm. Write an equation to find the dimensions of each rectangle.

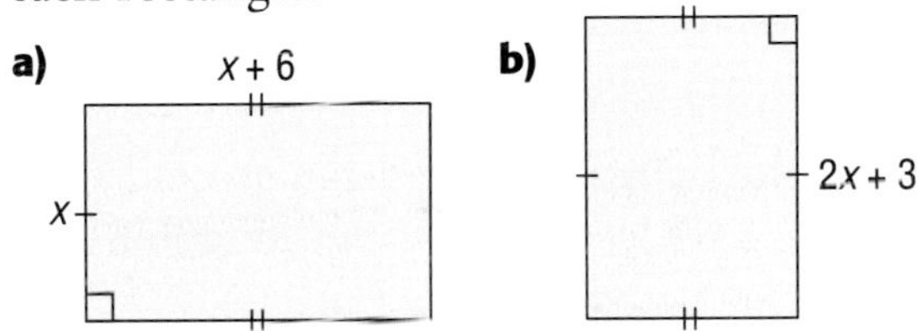

27. The perimeter of each triangle is 39 m. Write an equation to find the dimensions of each triangle.

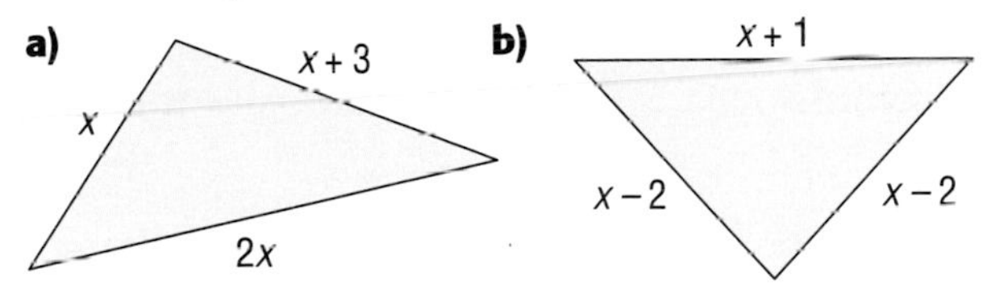

WORD POWER

Change the word WARM to the word COLD by changing 1 letter at a time. You must form a real word each time you change a letter. The best solution has the fewest steps.

7.2 Solving Problems Using Equations

The ground speed of a jet depends on its air speed and the speed of the headwind or tailwind.

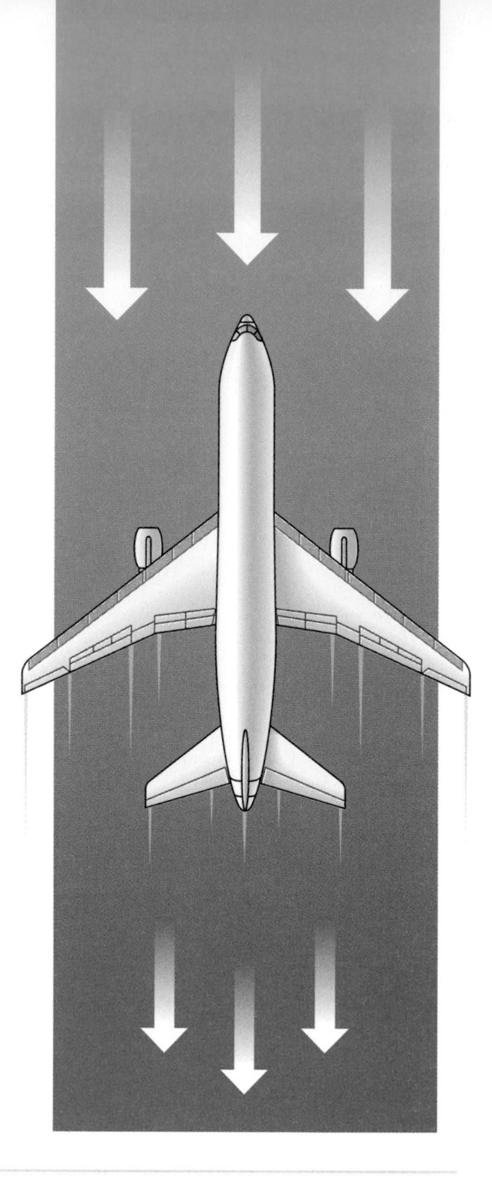

Activity: Study the Information

The equation for the ground speed of a jet, y, flying at 450 km/h with a headwind, x, is:

$$y = 450 - x$$

If there is a tailwind, the equation is:

$$y = 450 + x$$

Inquire

1. What is the ground speed when there is
a) a headwind of 80 km/h?
b) a tailwind of 120 km/h?

2. What is the headwind if the ground speed is 390 km/h?

3. What is the tailwind if the ground speed is 515 km/h?

4. Write the equation to find the ground speed when the headwind is 60 km/h. Solve the equation.

5. Write the equation to find the ground speed when the tailwind is 100 km/h. Solve the equation.

6. Write the equation to find the headwind when the ground speed is 400 km/h. Solve the equation.

7. Write the equation to find the tailwind when the ground speed is 475 km/h. Solve the equation.

Example 1

A compact disc player costs $75 more than a tape deck. Together, they cost $725. How much does each cost?

Solution

Let the cost of the tape deck be x dollars.
Then, the cost of the compact disc player is $(x + 75)$ dollars.
The sum of the costs is $725.

$$\begin{aligned} x + (x + 75) &= 725 \\ 2x + 75 &= 725 \\ 2x &= 725 - 75 \\ 2x &= 650 \\ x &= 325 \end{aligned}$$

The cost of the tape deck is $325.
The cost of the compact disc player is $325 + $75 or $400.
Check: $325 + $400 = $725

Example 2

The sides of a triangle are 3 consecutive whole numbers of centimetres. The perimeter of the triangle is 48 cm. How long is each side?

Solution

Let x represent the length of the shortest side in centimetres. Then, the lengths of the other 2 sides are $x + 1$ and $x + 2$. The sum of the 3 sides is 48 cm.

$$
\begin{aligned}
x + (x + 1) + (x + 2) &= 48 \\
x + x + x + 1 + 2 &= 48 \\
3x + 3 &= 48 \\
3x &= 45 \\
x &= 15
\end{aligned}
$$

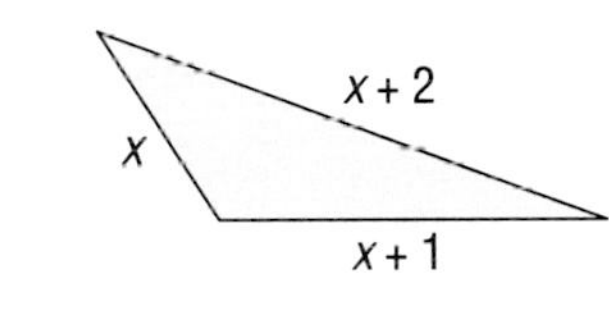

If $x = 15$, $x + 1 = 16$, and $x + 2 = 17$.
The lengths of the sides are 15 cm, 16 cm, and 17 cm.
Check: The numbers 15, 16, and 17 are consecutive whole numbers, and $P = 15 + 16 + 17$ or 48 cm.

Example 3

A parking meter contains $27.05 in quarters and dimes. There are 146 coins. How many quarters are there?

Solution

Let x represent the number of quarters.
Then, $(146 - x)$ is the number of dimes.
If there are x quarters, then the value of the quarters is $25x$ cents.
If there are $(146 - x)$ dimes, the value of the dimes is $10(146 - x)$ cents.
The total value of the coins is $27.05 or 2705 cents.

$$
\begin{aligned}
10(146 - x) + 25x &= 2705 \\
1460 - 10x + 25x &= 2705 \\
1460 + 15x &= 2705 \\
15x &= 1245 \\
x &= 83
\end{aligned}
$$

The number of quarters is 83.
The number of dimes is $146 - 83$ or 63.
Check: The value of the coins in dollars is $(0.25)(83) + (0.10)(63)$ or $27.05.

CONTINUED

Example 4

One number is 2 times another number. If you subtract 10 from each number, the sum is 40. What are the numbers?

Solution

Let the smaller number be x. Then, the larger number is $2x$. Subtract 10 from each number to give $x - 10$ and $2x - 10$. The sum equals 40.

$$\begin{aligned} (x - 10) + (2x - 10) &= 40 \\ x + 2x - 10 - 10 &= 40 \\ 3x - 20 &= 40 \\ 3x &= 60 \\ x &= 20 \end{aligned}$$

If $x = 20$, $2x = 40$. The numbers are 20 and 40.

Check: $20 - 10 = 10$ $\quad 40 - 10 = 30$ $\quad 10 + 30 = 40$

Practice

Each statement has two unknowns. Represent both in terms of x.

1. The sum of two numbers is 35.

2. There are 50 nickels and dimes.

3. There are 125 quarters and dimes.

4. The length and width of a rectangle total 36 cm.

5. There is a total of 32 males and females in the class.

6. The cafeteria sold 758 hamburgers and hot dogs.

7. Jim and Janice sold 468 kg of cheese.

8. The parking meter had 246 coins in quarters and dimes.

Find the length of each side.

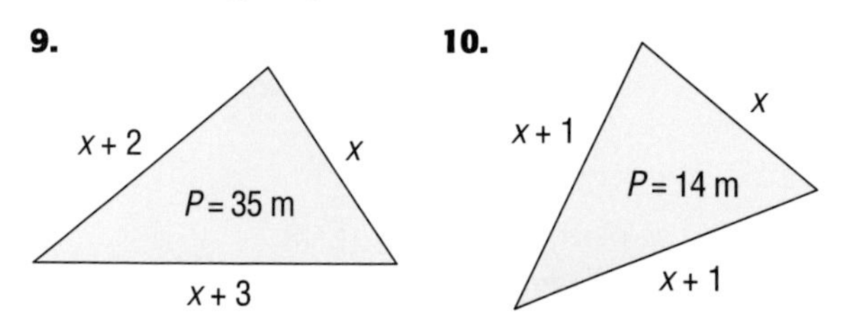

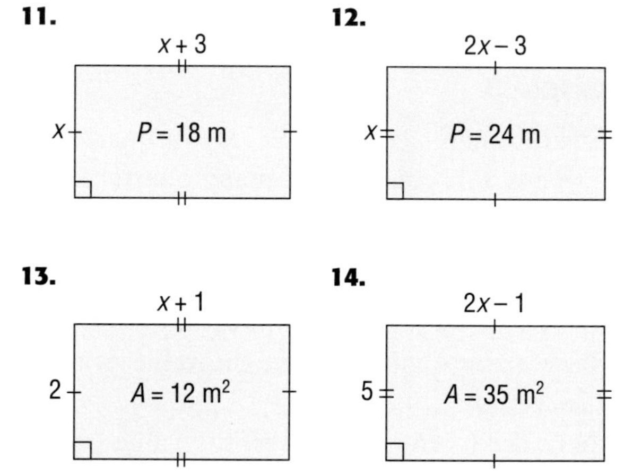

Problems and Applications

15. The sum of 2 numbers is 46. One number is 12 more than the other. What are the numbers?

16. The Mackenzie river is 1183 km longer than the St. Lawrence. The sum of their lengths is 7299 km. How long is each river?

17. The length of a rectangle is 5 m more than its width. Its perimeter is 90 m. What are its dimensions?

18. The sum of 3 consecutive numbers is 105. Find the numbers.

19. The number of frost-free days per year in Quebec City is 18 less than the number in Halifax. The total number in both cities is 292. How many frost-free days are there per year in Halifax?

20. The maximum life span of a brown bear is 10 times the maximum life span of a mouse. The sum of their maximum life spans is 33 years. What is the maximum life span of a mouse?

21. When Canada won the Women's World Hockey Championships, the Canadian team scored 35 more goals than were scored against it. The total number of goals scored in the games that Canada played was 41.

a) How many goals were scored against Canada?

b) How many goals did Canada score?

22. There are 2.25 times as many days of fog per year in Vancouver as in Winnipeg. The total number of days of fog in both cities is 65. How many days of fog does Winnipeg have per year?

23. One of the equal sides of an isosceles triangle is 3 m less than twice its base. The perimeter is 44 m. Find the lengths of the sides.

24. The cost of a pen is 3 times the cost of a pencil. The cost of 4 pencils and 3 pens is $9.75. What is the cost of a pencil?

25. Sally has twice as many dimes as nickels. The total value is $3.50. How many nickels does she have?

26. Aretha has $0.85 in nickels and dimes. She has 2 more nickels than dimes. How many nickels and dimes does she have?

27. A box contains 140 dimes and nickels. The total value is $11.15. How many dimes and how many nickels are there?

28. A picture is 5 cm longer than it is wide. The perimeter of the picture is 90 cm.

a) What is the width of the picture?

b) The picture is surrounded by a border, which is 6 cm wide. What are the outside dimensions of the border?

29. The sum of 2 numbers is 39. Twice the first number plus 3 times the second number is 101. Find the numbers.

30. One number is 5 more than another number. Three times the first plus twice the second is 30. Find the numbers.

31. A jar contains $18.50 in dimes and quarters. There are 110 coins in the jar. How many quarters are in the jar?

32. Pietra sells tickets at the theatre. At the end of the evening, she had $524.00 in $2.00 and $5.00 bills. The total number of bills was 145. How many of each kind of bill were there in total?

33. Large pizzas cost $12.50 and small pizzas cost $9.00. The pizza parlor sold 38 pizzas with a total value of $415.50. How many of each type of pizza did the parlor sell?

34. Tickets to the concert cost $9.00 for adults and $6.50 for students. A total of 950 people paid $7675.00 to attend. How many students attended the concert?

35. A garden is 20 m by 25 m. It is surrounded by a walk-way. The outside perimeter of the walk-way is 114 m. What is the width of the walk-way?

36. A picture measures 40 cm by 30 cm. The outside perimeter of the frame around the picture is 156 cm.

a) What is the width of the frame?

b) What is the area of the frame?

37. Write a problem that can be solved using an equation. Have a classmate solve your problem.

7.3 Working with Formulas

A squid has 8 arms and 2 tentacles. The arms and the ends of the tentacles are covered with tooth-rimmed suction pods. The squid uses its tentacles to pull prey into its arms. Giant squids prey on sharks and, in turn, are preyed upon by sperm whales.

Activity: Use the Formula

Scientists can determine the length of a giant squid by measuring the diameter of the suction-pod scars left on its prey. The formula $l = 180d$ gives the length of a squid in centimetres from the diameter of its suction pods in centimetres.

Inquire

1. How long is a squid whose suction pods have a diameter of 0.5 cm?

2. Squids found off the coast of Newfoundland have suction pods with diameters of 5 cm. How long are these squids in metres?

3. The longest squid, to date, was found in New Zealand. The suction pods were 5.4 cm in diameter. How long was this squid in metres?

4. A sperm whale was found with suction-pod scars that measured 35 cm in diameter. How many metres long was the squid that made these scars?

Example

a) Given $P = 2(l + w)$, solve for w.

b) Evaluate w for $P = 350$ cm and $l = 120$ cm.

Solution

a) Since we want the value of w, isolate this variable.

$$P = 2(l + w)$$
$$P = 2l + 2w$$

Subtract $2l$ from both sides: $P - 2l = 2l + 2w - 2l$

$$P - 2l = 2w$$

Divide both sides by 2: $\frac{P - 2l}{2} = \frac{2w}{2}$

$$\frac{P - 2l}{2} = w$$

b) Substitute $P = 350$ and $l = 120$ into the formula.

$$\begin{aligned} w &= \frac{P - 2l}{2} \\ &= \frac{350 - 2(120)}{2} \\ &= \frac{350 - 240}{2} \\ &= \frac{110}{2} \\ &= 55 \end{aligned}$$

So, $w = 55$ cm.

Check: Substitute into $P = 2(l + w)$.

$$350 = 2(120 + 55)$$
$$350 = 2(175)$$
$$350 = 350$$

Practice

1. For the formula $A = lw$,
a) find A if $l = 8$ cm and $w = 5$ cm.
b) find w if $A = 40$ m^2 and $l = 10$ m.
c) find l if $A = 238$ m^2 and $w = 14$ m.

2. Assume $\pi = 3.14$. For the formula $C = 2\pi r$,
a) find C if $r = 10$ cm.
b) find r if $C = 628$ cm.

3. For the formula $A = \frac{1}{2}bh$,
a) find A if $b = 6$ cm and $h = 8$ cm.
b) find h if $A = 40$ cm^2 and $b = 4$ cm.
c) find b if $A = 60$ m^2 and $h = 20$ m.

4. For the formula $P = 2(l + w)$,
a) find P if $l = 9$ m and $w = 6$ m.
b) find w if $P = 60$ m and $l = 16$ m.
c) find l if $P = 84$ m and $w = 5$ m.

Solve each formula for the indicated variable.

5. $A = lw$ for w
6. $A = \frac{1}{2}bh$ for b
7. $I = Prt$ for P
8. $C = 2\pi r$ for r
9. $E = mc^2$ for m
10. $A = \frac{1}{2}h(a + b)$ for b

Problems and Applications

11. The maximum desirable pulse rate for a person exercising can be found using the formula

$$m = 220 - a$$

where m is the pulse rate in beats per minute and a is the person's age in years.
a) Copy and complete the table.

Age (years)	Maximum Desirable Pulse Rate (beats/min)
20	
27	
39	
44	
61	

b) To what ages do these maximum desirable pulse rates correspond?
198 171 183

12. The distance travelled by a spaceship is given by the formula

$$d = 40\ 000t$$

where d is the distance in kilometres and t is the time in hours.
a) How far does a spaceship travel in 12 h? in 17.5 h?
b) How long does it take a spaceship to travel 130 000 km?

13. The amount of food energy required per day by military personnel on active duty is given by the formula

$$E = -125T + 15\ 250$$

where E is the amount of food energy in kilojoules (kJ) and T is the outside temperature in degrees Celsius. Copy and complete the table.

Temperature (°C)	Energy (kJ)
40	
23	
0	
−10	
−45	

14. Shawna is at an outdoor concert in Vancouver. She is sitting 100 m from the band. The formula that gives the length of time for the band's sound to reach her is

$$t = \frac{d}{330}$$

where t is the time in seconds and d is Shawna's distance from the band in metres. Paul is listening to the same concert on a radio in Halifax. The formula that gives the length of time for the band's sound to reach him is

$$t = \frac{d}{300\ 000}$$

where t is the time in seconds and d is the distance from Halifax to Vancouver in kilometres. Who hears each sound first, Shawna or Paul?

Developing Pick's Formula for Area

Activity ❶ No Points in the Interior

Each polygon has been constructed so that there are no points in its interior.

1. Construct the polygons on grid paper or on a geoboard.

2. Copy and complete this table. The first line has been completed for you.

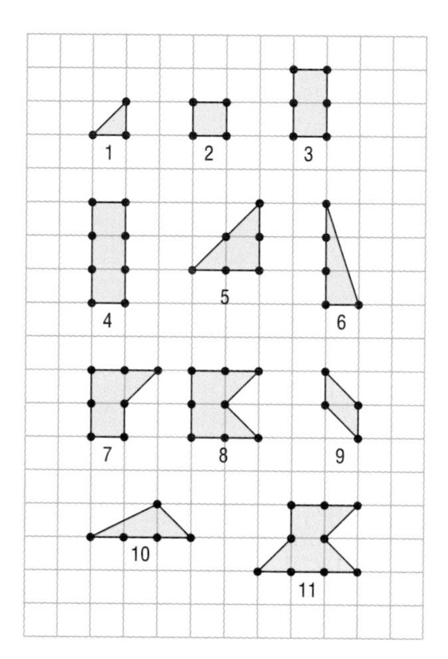

Polygon	Points on the Perimeter (P)	Area (square units)
1	3	$\frac{1}{2}$
2		
3		
4		
5		
6		
7		
8		
9		
10		
11		

3. If 2 figures have the same number of points on the perimeter, what can you say about their areas?

4. List the areas from smallest to largest. Do not repeat areas. The first two are:
3 points, 0.5 square units
4 points, 1 square unit

5. Predict the area of a figure with 10 points on the perimeter and no points in its interior. Check your prediction by making 2 such shapes and finding their areas.

6. Use the pattern in the table to write a formula for the area of a polygon with no points in the interior if you know the number of perimeter points.

7. Draw 3 different figures with no interior points. Verify your formula.

8. Use your formula to determine the area of a figure with 99 points on the perimeter and no interior points.

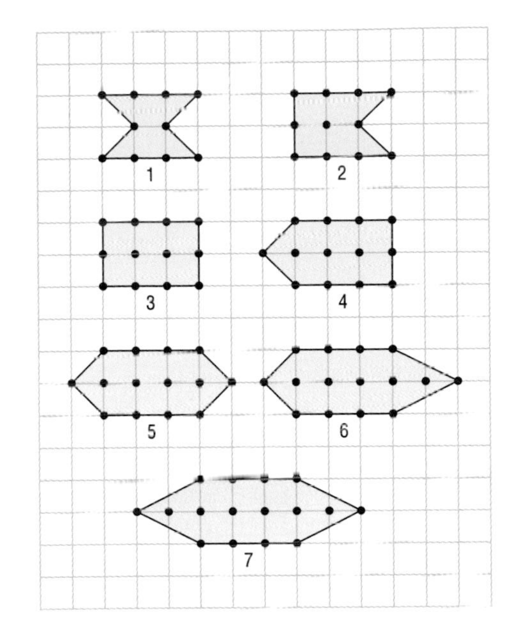

Activity ❷ Points in the Interior

Each polygon has 10 points on the perimeter. The number of points in the interior, I, increases from 0 to 6.

1. Calculate the area of each polygon.

2. Copy and complete the table, calculating each area in square units.

Polygon	Points on the Perimeter (P)	Interior Points (I)	Area (square units)
1			
2			
3			
4			
5			
6			
7			

3. What happens to the area of the polygon as the number of interior points increases by 1?

4. Predict the area of figures with 10 points on the perimeter and 7 and 8 points in the interior. Check your prediction by making 2 such shapes and finding their areas.

5. Use the pattern in the table and the formula for area you found in Activity 1. Write a formula for the area of a polygon if you know the number of perimeter points and interior points. This formula is called **Pick's formula** after the mathematician who discovered it.

6. Draw 3 different figures with points in the interior. Verify your formula.

7. Use your formula to determine the areas of the following figures.

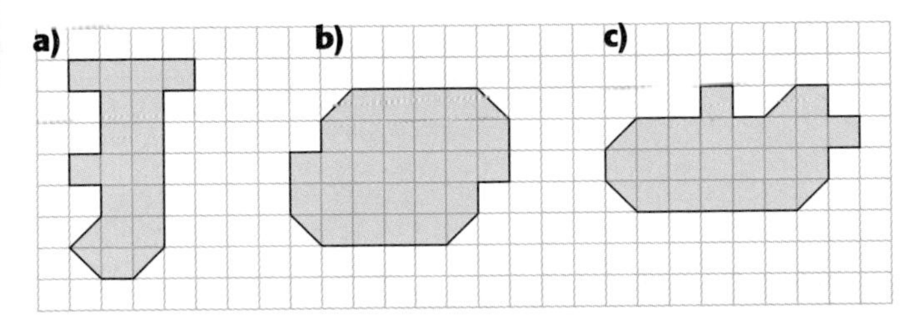

7.4 Developing Formulas

The *Bluenose* was Canada's most famous sailing ship. It was launched in Lunenburg, Nova Scotia, in 1921. It was the fastest racing schooner of its time, winning many championships until its last race in 1938.

Activity: Study the Information

In the early days of sailing ships, navigators estimated the speed of their ship by tying a rope to a log and tossing the log into the water behind the ship. After 30 s, the length of the rope was measured, showing how far the ship had travelled in this amount of time.

Inquire

1. If the length of the rope after 30 s was 90 m, what was the speed of the ship in metres per second?

2. Let the length of the rope after 30 s be l. Write a formula for the speed of the ship in metres per second.

3. If the length of the rope after 30 s was 240 m, what was the speed of the ship in metres per second?

4. Use the formula to calculate the speed of the ship if the length of the rope after 30 s is 210 m.

5. Rewrite the formula to calculate the speed of the ship in kilometres per hour.

Example

Many early navigators also estimated the speed of a ship by measuring the number of seconds it took a piece of wood to float from one end of the ship to the other.

a) Write a formula to estimate the speed of a galleon in metres per second.

b) Use the formula to find the speed of a 40 m long galleon if $t = 8$ s.

Solution

a) Let the time in seconds be t. Let the speed in metres per second be s.

The distance travelled in metres is the length of the galleon, l.

$$\text{Speed} = \frac{\text{Distance}}{\text{Time}}$$

$$s = \frac{l}{t}$$

b) $s = \frac{40}{8}$

$= 5$

The speed of the galleon was 5 m/s.

Practice

Complete the table and state a rule for each pattern.

1.

a	1	2	3	4	5	6	7
b	3	6	9	12			

2.

m	2	4	6	8			
n	5	7	9	11	13	15	17

3.

t	25	24	23	22			
a	100	96	92	88	84	80	76

Complete the table. Then, use the variables to write a formula for each pattern.

4.

Number of Books (n)	1	2	3	4
Cost (c)	4	8	12	

5.

Hours (h)	5	10	15	20
Wages (w)	37.50	75.00	112.50	

6.

Selling Price (s)	300	400	500	600
Profit (p)	60	80	100	

Problems and Applications

7. The cost to rent a bus is $100 plus a certain amount per kilometre. The table gives the cost of 3 bus trips. Write a formula to calculate the cost of a bus trip in terms of distance.

Trip	Distance, d (km)	Cost, C ($)
1	50	250
2	100	400
3	150	550

8. This table shows the cost of a taxi ride. Write a formula that relates kilometres driven, d, to the cost of the ride, C.

d (km)	0	5	10	15	20
C ($)	3.00	10.50	18.00	25.50	33.00

9. The map shows an area where there are 6 towns, 9 roads, and 4 regions.

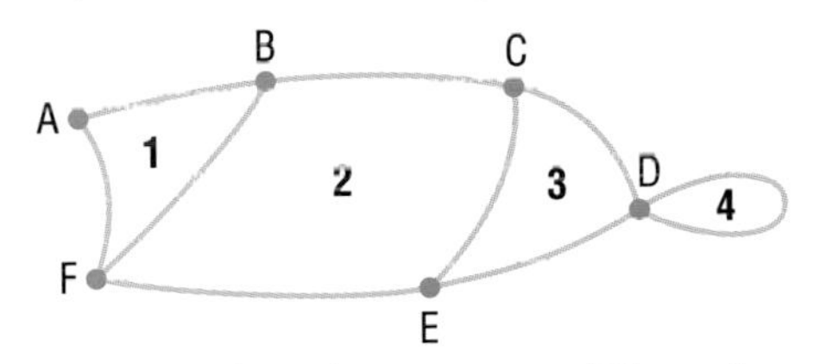

Roads start and end at a town. If roads cross, there is a town where they cross. A road can connect a town to itself. Regions are completely surrounded by roads. Each town is connected to at least 1 other town.
In such maps, there is a relationship between the numbers of towns, regions, and roads.

a) Make at least 5 maps, each with 4 regions. Complete a table like the one shown.

Map	Regions	Towns	Roads
1	4		
2	4		
3	4		
4	4		
5	4		

Look for a pattern in the Towns and Roads columns.
Write a formula that lets you determine the number of regions if you know the numbers of towns and roads.
Regions = ▬
Test your formula on another map.

b) Make at least 5 maps, each with 6 towns. Complete a table like the one in part a). Write a formula that lets you determine the number of towns if you know the numbers of roads and regions. Test your formula on another map.

c) Make at least 5 maps, each with 6 roads. Complete a table like the one in part a). Write a formula that lets you determine the number of roads if you know the numbers of towns and regions. Test your formula on another map.

7.5 Uniform Motion Problems

Problems involving **uniform motion** are those in which the speeds of the objects do not change. The formula $D = r \times t$ is used to solve these problems, where D is the distance, r is the rate of speed, and t is the time.

Activity: Study the Information

Frank left the ranch for town on horseback. He rode at 15 km/h. One hour later, his brother Jesse left the ranch and tried to catch up with Frank. Jesse rode at 20 km/h. How long did it take Jesse to catch up with Frank?

Inquire

1. Let x be the time in hours that Frank rode before Jesse caught up with him. Write an expression for how far Frank rode in x hours.

2. If x is the time Frank rode, write an expression in terms of x for the time Jesse rode.

3. Use your expression from question 2 to write an expression for the distance Jesse rode.

4. When Jesse caught up with Frank, the distances they had ridden were the same. Write an equation stating the 2 distances were equal.

5. Solve the equation for x.

6. a) How long had Frank been riding when Jesse caught up with him?
b) How long had Jesse been riding?

Example

A plane left Montreal for Calgary, a distance of 3000 km, travelling at 550 km/h. At the same time, a plane left Calgary for Montreal travelling at 450 km/h. How long after take-off did the planes pass each other?

Solution

Set up a table and let x be the time from take-off until the planes passed. Write expressions for D using r and t.

	t (h)	r (km/h)	D (km)
Montreal to Calgary	x	550	$550x$
Calgary to Montreal	x	450	$450x$

The sum of the distances is 3000 km.

$$550x + 450x = 3000$$
$$1000x = 3000$$
$$x = 3$$

The planes passed each other 3 h after take-off.

Check:
In 3 h, the plane from Montreal flew 3×550 or 1650 km.
In 3 h, the plane from Calgary flew 3×450 or 1350 km.
1650 km + 1350 km = 3000 km

Practice

Calculate the distance travelled.

1. 3 h at 60 km/h **2.** 2 h at 85 km/h

3. $\frac{1}{2}$ h at 90 km/h **4.** $\frac{3}{4}$ h at 60 km/h

How long does each trip take?

5. 40 km at 80 km/h

6. 400 km at 50 km/h

7. 20 km at 100 km/h

8. 360 km at 80 km/h

Calculate each speed.

9. 300 km in 3 h **10.** 400 km in 5 h

11. 360 km in 4 h **12.** 40 km in $\frac{1}{2}$ h

Complete the table.

	Distance (km)	Rate (km/h)	Time (h)
13.	450	100	
14.		65	3
15.	600		8
16.		80	x
17.		90	$x + 1$
18.		85	$x - 1$
19.	200	x	
20.	400		x
21.		r	t
22.	D	r	
23.	D		t

Problems and Applications

24. A cruise ship left Halifax for Bermuda at 20 km/h. A private boat left for Bermuda 1 h later and travelled at 25 km/h. After how long did the private boat overtake the cruise ship?

25. Two cars left a service centre at 16:30. One car travelled in one direction at 75 km/h. The other car travelled in the opposite direction at 85 km/h.

a) After how long were they 600 km apart?

b) At what time were they 600 km apart?

26. Two friends, one living in Winnipeg and one living in Edmonton, decided to meet on the TransCanada Highway. The distance from Edmonton to Winnipeg is about 1360 km. They both left home at 08:00 Winnipeg time. The friend from Winnipeg drove at 80 km/h, and the friend from Edmonton drove at 90 km/h.

a) After how long did they meet?

b) What was the time in Winnipeg when they met?

c) What assumptions have you made?

27. A plane left Vancouver for Los Angeles at 08:30 and flew at 600 km/h. Fifteen minutes later, another plane left Vancouver for Los Angeles and flew at 700 km/h.

a) How long did it take the second plane to overtake the first one?

b) At what time did it happen?

28. A car left a garage on the highway at 100 km/h. Fifteen minutes later, a police cruiser left the same garage at 120 km/h in pursuit of the car. How long did it take the cruiser to catch up with the car?

29. Write a problem involving uniform motion and ask a classmate to solve it.

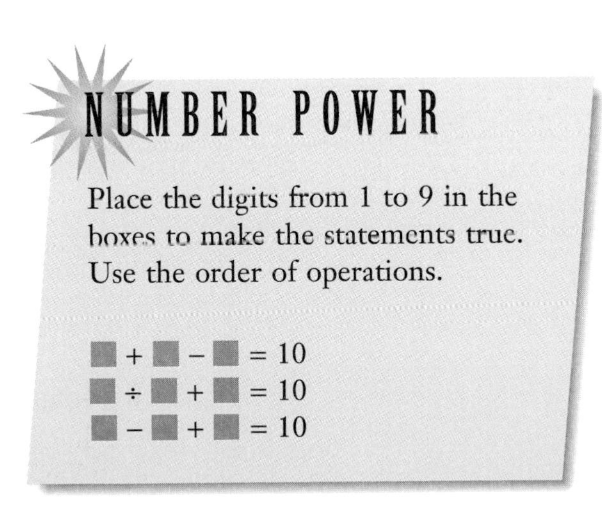

NUMBER POWER
Place the digits from 1 to 9 in the boxes to make the statements true. Use the order of operations.
■ + ■ − ■ = 10
■ ÷ ■ + ■ = 10
■ − ■ + ■ = 10

7.6 Rate of Work Problems

Activity: Study the Process

Tania and Justine own a cruise boat that takes people out to the Atlantic Ocean to show them how lobsters are caught. At the end of the day, Tania and Justine take turns cleaning the boat for the next day's cruise. Tania is older and cleans the boat in 2 h. It takes Justine 3 h. We want to know how long it would take them to clean the boat together. Problems of this type are known as **rate of work problems**.

Inquire

1. What fraction of the boat does Tania clean in 1 h?

2. What fraction of the boat does Justine clean in 1 h?

3. What is the sum of these fractions?

4. If they clean $\frac{5}{6}$ of the boat in 1 h, how long will it take to clean the last $\frac{1}{6}$?

5. What is the total time to clean the boat?

Example

After the restaurant closes, Roberto takes 2 h to clean it. Bill is new to the job and takes 4 h. How long would it take them if they worked together?

Solution

One way to solve this type of problem is to use an equation. Let x hours represent the time the job takes if they work together. Then, $\frac{x}{2}$ represents the part cleaned by Roberto and $\frac{x}{4}$ the part cleaned by Bill. Together, the 2 parts must add to the whole job.

$$\frac{x}{2} + \frac{x}{4} = 1$$

Multiply by 4: $4 \times \frac{x}{2} + 4 \times \frac{x}{4} = 4 \times 1$

$$2x + x = 4$$

$$3x = 4$$

$$x = \frac{4}{3} \text{ or } 1\frac{1}{3}$$

Roberto and Bill can clean the restaurant in $1\frac{1}{3}$ h by working together.

Check:

In $\frac{4}{3}$ h, the fraction of the restaurant cleaned by Roberto is $\frac{\frac{4}{3}}{2}$ or $\frac{4}{3} \times \frac{1}{2} = \frac{2}{3}$.

In $\frac{4}{3}$ h, the fraction of the restaurant cleaned by Bill is $\frac{\frac{4}{3}}{4}$ or $\frac{4}{3} \times \frac{1}{4} = \frac{1}{3}$.

$\frac{2}{3} + \frac{1}{3} = 1$ or the entire restaurant

Practice

Write 2 fractions for each question.

1. Ahmed takes 2 h to mow his lawn. His brother, Sami, takes 4 h. What fraction of the lawn does each mow in 1 h?

2. Athena takes 6 h to paint an apartment. Helena takes 8 h to do the same job. What fraction of the apartment do they each paint in 1 h?

Problems and Applications

Solve. Round answers to the nearest tenth, where necessary.

3. Julio can fill a water tank in 4 min using a large hose. He takes 6 min using a smaller hose. How long will he take if he uses both hoses?

4. Andrea can deliver 500 handbills in 2 h. Althea can deliver the same number in 3 h. How long will they take to deliver 500 handbills if they work together?

5. Murray can tile a floor in one hour. His partner can do the same job in half the time. How long will it take them to tile the floor if they work together?

6. Mario can take inventory at the store in 30 min. His partner, Carmen, can take inventory in 20 min. If they work together, how long will the inventory take?

7. Ken and Milan are office cleaners. Ken earns $10/h and takes 8 h to clean an office. Milan earns $8/h and takes 10 h to clean it.
a) How long will it take Ken and Milan to clean the office together?
b) What is the cost of cleaning the office using only Ken? only Milan? Ken and Milan together?

8. Uri, Max, and Boris work for a company that installs carpet tiles in offices. Uri can install 2000 tiles in 10 h. It takes Max 12 h to do the same job. Boris installs 2000 tiles in 15 h. How long would it take them to install 2000 tiles if they work together?

9. Mary takes 3 h to complete a task. Mary and Jim together take 2 h to complete the same task. How long will it take Jim to complete the task working alone?

10. Dan and Brad are brothers who attend the same school. Brad can walk to school in 15 min. Dan takes 20 min.
a) How long will it take them to walk to school together?
b) What makes this problem different from the others in this section?

11. Write a rate of work problem. Have a classmate solve your problem.

LOGIC POWER

Draw the grid and place 3 pennies, a nickel, and a dime on it as shown. By sliding one coin at a time into a neighbouring empty square, make the nickel and the dime change places. You can move horizontally or diagonally. Make the switch in as few moves as possible.

Database Managers

A database is a collection of data. A database manager organizes the data so that they can be retrieved easily.

Activity ❶ Non-Computerized Database Managers

There are databases and database managers that are not computerized. A dictionary is an example. The words and their meanings make up the database. The book that they are in is the manager.

List some non-computerized databases and database managers that you have used. Describe how you used them.

Activity ❷ Computerized Database Managers

If a database has the names, addresses, ages, and phone numbers of the students in your class, this collection of data is called a **file**. Each line about one student is called a **record**. Each piece of information in a record is called a **field**. One phone number is an example of a field.

Lastname	Firstname	Address	Age	Phone
Bevan	Sarah	83 1st St.	14	527–0101
Brooks	Thomas	129 Elm Ave.	13	575–9919

The principal features of electronic database managers are as follows.

Retrieve and Display

Records can be located easily and displayed on the screen for updating or viewing.

Sort

The user can sort through records and rearrange them in different ways.

Calculate and Format

Math formulas can be used to manipulate data. Data can be printed out in different formats.

1. Describe how a computerized database manager might be used by

a) a police officer **b)** a pharmacist

2. List 3 other occupations in which computerized database managers would be useful. Describe how they would be used.

Energy and Exercise

Recent studies suggest that people who exercise frequently are healthier and live longer than people who do not exercise.

One way to determine how much exercise is required to stay fit is to measure the amount of energy a specific activity uses. The recommended amount of exercise per week should use a total of 6300 kJ to 8400 kJ of energy.

The table lists some common activities, times, and approximate amounts of energy.

Activity	Time (min)	Energy Used (kJ)
Stair Climbing (Normal Speed)	5	147
Fast Stair Climbing (Gym Class)	10	357
Rowing Machine	10	273
Treadmill Running	10	399
Stationary Cycling (16 km/h)	15	441
Swimming	20	756
Cycling	20	567
Gardening	20	462
Housecleaning	20	336
Mowing the Lawn	20	630
Raking Grass and Yard Work	20	567
Washing the Car	20	273
Brisk Walking	30	672

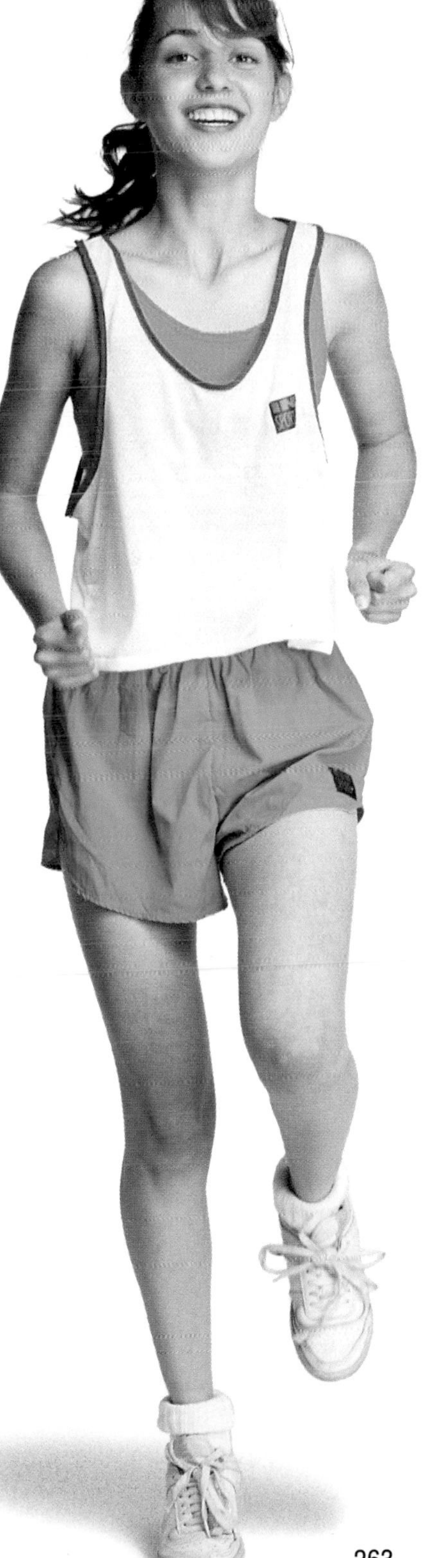

Activity ❶

1. Plan a fitness program for someone who needs to burn 7000 kJ/week. Decide how many days per week the person should exercise to keep fit. Design a week's activity plan.

2. The amount of energy that a person needs to burn per day to keep fit is called the Daily Kilojoule Target.

a) Determine the Daily Kilojoule Target for the activity plan that you designed in step 1.

b) Choose any 4 activities in the table. For each activity you choose, find the amount of time needed to meet the Daily Kilojoule Target.

Activity ❷

1. a) Form a group of 3 or 4 people. Describe a new way of reporting the information in the above table.

b) Compare your new way of reporting the information with that of another group.

Review

In questions 1–5, write an equation for each statement.

1. Eight more than a number is twenty.

2. A number multiplied by six is seventy-two.

3. Hadib is eight years older than Cam. The sum of their ages is twenty-four.

4. Sonya has ten dollars less than Paula. Together they have ninety dollars.

5. Carl has three times as many stamps as Tim. Together they have 2400 stamps.

6. The perimeter of each figure is 64 m. Write an equation to find the dimensions of each figure.

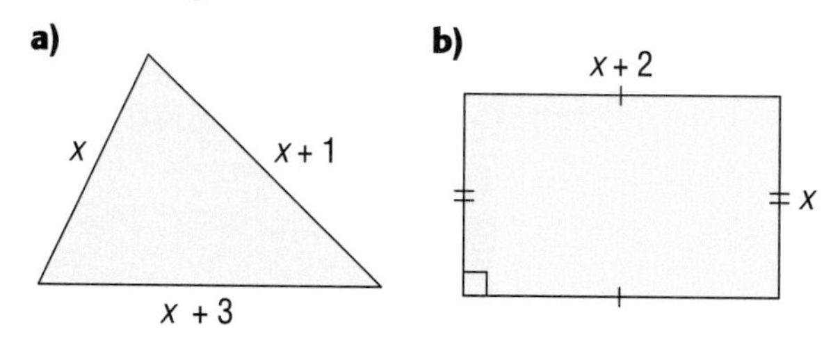

Solve.

7. One number is five times another and their sum is 36. What are the numbers?

8. The length of a rectangle is 3 m greater than the width. The perimeter is 26 m. What are the dimensions of the rectangle?

9. The sum of three consecutive numbers is 183. Find the numbers.

10. The length of a rectangular field is three times the width. The perimeter is 1688 m. What are the dimensions of the field?

11. The lengths of the sides of a triangle are 3 consecutive numbers. The perimeter of the triangle is 126 m. How long is each side?

12. One number is 4 times another. If you subtract 5 from each number, the sum is 50. What are the numbers?

13. The mass of synthetic material in a hockey puck is 9 times the mass of natural rubber. A regulation hockey puck has a mass of 170 g. What mass of natural rubber is there in a hockey puck?

14. Find the length of each side.

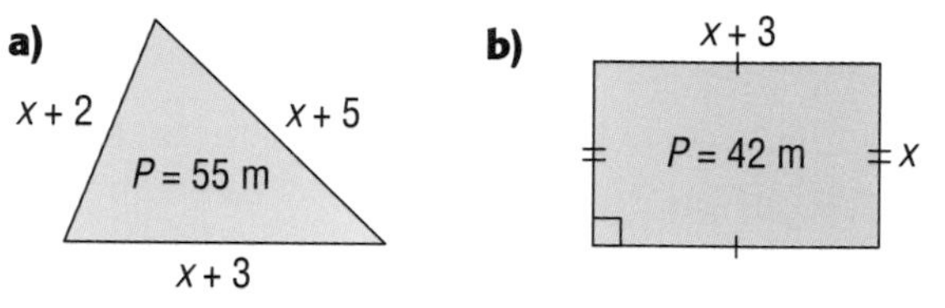

15. The sum of two numbers is 51. Twice the first plus 4 times the second is 128. What are the numbers?

16. One number is 5 more than another number. Four times the larger number plus 3 times the smaller is 97. Find the numbers.

17. A jar contains $36.25 in dimes and quarters. There are 250 coins in the jar. How many quarters are in the jar?

18. In a pile of coins, there are 15 more quarters than loonies. The total value of the coins is $21.25. How many quarters are there?

19. The formula $Q = 12t + 35$ gives the amount of water in a tank after t minutes. The tank originally contained 35 L of water, and water runs into it at a rate of 12 L/min.

a) How much water will be in the tank after 5 min? 8 min?

b) How long will it take to fill the tank if it holds 203 L?

20. A hot water tank holds 200 L of water. When the tap is opened, the water drains at a rate of 15 L/min. The amount of water left in the tank after t minutes is given by $A = 200 - 15t$.

a) How much water will be in the tank after 5 min? 10 min?

b) How long, to the nearest minute, will it take to drain the tank?

21. The formula for the perimeter of a rectangle is $P = 2(l + w)$. Solve for l.

22. Shandra is a math tutor. She charges $20 a visit plus $25/h.
a) Write a formula that determines her earnings from the hours she tutors on a visit.
b) How much will she earn if she tutors for 3 h? 2.5 h?

23. Marie started jogging at 08:30 at 9 km/h. Heather started jogging 15 min later in the same direction. Heather jogged at 12 km/h.
a) How long did it take Heather to catch up to Marie?
b) At what time did she catch up to her?

24. Two cars left the same highway restaurant at the same time but in opposite directions. One travelled at 65 km/h, the other at 55 km/h. After how long were they 600 km apart?

25. Heidi and Kelly drove in an antique car rally. Kelly left a checkpoint at 09:00 travelling at 45 km/h. One hour later, Heidi left the same checkpoint travelling at 50 km/h.
a) After how long did Heidi overtake Kelly?
b) At what time did Heidi overtake Kelly?

26. A plane left Darwin for Beijing, a distance of 6000 km, travelling at 450 km/h. At the same time, a plane left Beijing for Darwin travelling at 550 km/h. How long after take-off did the planes pass each other?

27. It takes Renate 3 h to cut the lawn. It takes her older sister Adrianna 2 h to cut the same lawn. How long will it take them to cut the lawn if they cut it together?

28. A small gas barbecue will cook for 4 h on a tank of propane. A larger barbecue will only cook for 3 h on a tank of propane. For how long can you cook if both barbecues are attached to the same tank?

Group Decision Making
Researching Medical Careers

1. Brainstorm as a class possible careers to investigate. They could include careers like dentist, doctor (general practitioner), doctor (specialist), nurse, X-ray technician, psychologist, dietitian, dental hygienist, or physiotherapist. Decide as a class on 6 careers.

2. Go to home groups. Decide as a group what career each member will research.

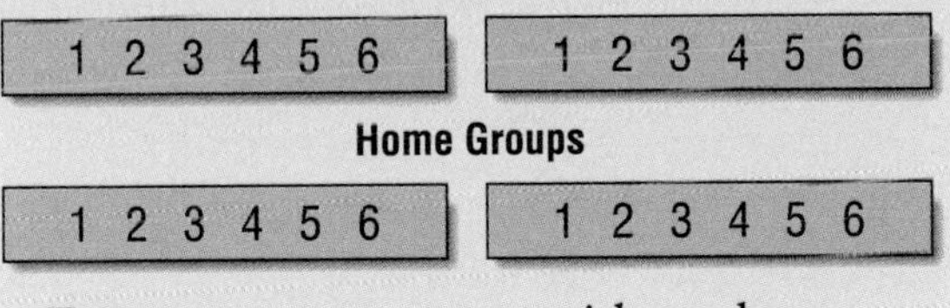

3. Form an expert group with students who have the same career as you to research. Decide on the questions you want answered, including how math is used in the career. Research the career in your expert group.

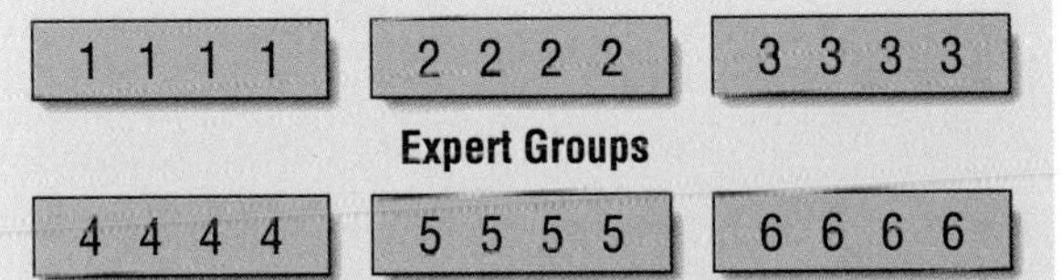

4. Return to your home group and tell the others what you found out in your expert group. Ask for questions and comments from your home group.

5. Return to your expert group and report what you have learned from your home group. In your expert group, prepare a report on the career. The form of the report is to be decided by the group.

6. In your expert group, evaluate the process and your report.

Chapter Check

1. Burt is four years older than Anna and the sum of their ages is 36. Write the equation to find their ages.

2. When 47 is subtracted from a certain number, the result is 34. Find the number.

3. The length of a rectangle is 15 m longer than the width. The perimeter of the rectangle is 74 m. Find the length and width.

4. The lengths of the sides of a triangle are 3 consecutive whole numbers. The perimeter of the triangle is 102 m. Find the lengths of the sides.

5. The formula for the perimeter of a rectangle is $P = 2(l + w)$. Find w if $P = 200$ m and $l = 56$ m.

6. The formula for the area of a triangle is $A = \frac{1}{2}bh$. Solve for h.

7. A cargo ship left Montreal for Halifax at 15 km/h. One hour later, a patrol boat left Montreal at 20 km/h, trying to overtake the cargo ship. After how long did the patrol boat overtake the cargo ship?

8. You see lightning flash before you hear thunder because light travels much faster than sound. The light reaches you almost instantaneously. The sound travels at a speed of 330 m/s. Write a formula to find your distance, d metres, from a thunderstorm in terms of the number of seconds between the lightning flash and when you hear the thunder.

9. Mark lived 350 km from his agent. Mark had to sign a contract his agent had. Mark and his agent left their apartments at the same time in their cars and drove along the same road toward each other. Mark drove at 65 km/h and his agent drove at 75 km/h. After how long did they meet?

10. Sunil and Kim both do volunteer work at Ronald McDonald House. It takes Sunil 6 h to cut the lawns and trim the hedges, and it takes Kim 4 h to do the same. How long would it take them to cut the lawns and trim the hedges if they worked together?

THE FAR SIDE By GARY LARSON

"Yes, yes, I *know* that, Sidney ... *everybody* knows *that*! ... But look: Four wrongs *squared*, minus two wrongs to the fourth power, divided by this formula, *do* make a right."

Using the Strategies

1. If the pattern continues, find the following product.

$\left(1-\frac{1}{2}\right)\left(1-\frac{1}{3}\right)\left(1-\frac{1}{4}\right)\ldots\left(1-\frac{1}{24}\right)$

2. The figure is made up of 11 identical squares. The area of the figure is 539 cm^2.

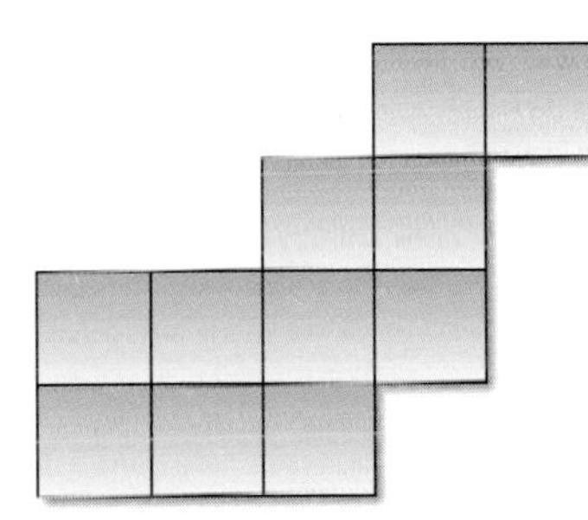

What is the perimeter of the figure?

3. Your plane is scheduled to leave for Casablanca at 18:15. You must arrive at the airport 25 min before the flight to pick up your tickets. It takes a taxi 30 min to drive to the airport from your hotel. Before you leave for the airport, the naval attaché is delivering some documents to you in the lobby of your hotel. It will take the attaché 10 min to explain the documents to you. Before meeting the attaché, you need 1 h 15 min to shower, pack, change clothes, and eat. It is now noon. You want to spend some time sightseeing. What is the latest time you can return to the hotel?

4. A train travelling at 90 km/h passed a car sitting at a railway crossing. It took 45 s for the train to pass. How long was the train in metres?

5. An engine plant built 485 automobile engines. There were 2 models of engines, 4-cylinder and 6-cylinder. The total number of cylinders in the engines was 2224. How many of each model of engine did the plant build?

6. A 2-digit number is changed by decreasing the tens digit by 1 and increasing the units digit by 1. When the new number is added to the original number, the result is 59. What was the original number?

7. If you disregard the year on a calendar, only a limited number of calendars is necessary. For example, you need only one calendar for any year in which January 1 is on a Monday, providing the year is not a leap year.

a) How many different calendars are necessary?

b) Start with the year 2001 and determine how many years must pass before each of the calendars you found in part a) is used at least once.

8. The graphs show the motion of Xenia's car. The vertical axis shows the distance from her house and the horizontal axis shows time. Write a story to explain each graph.

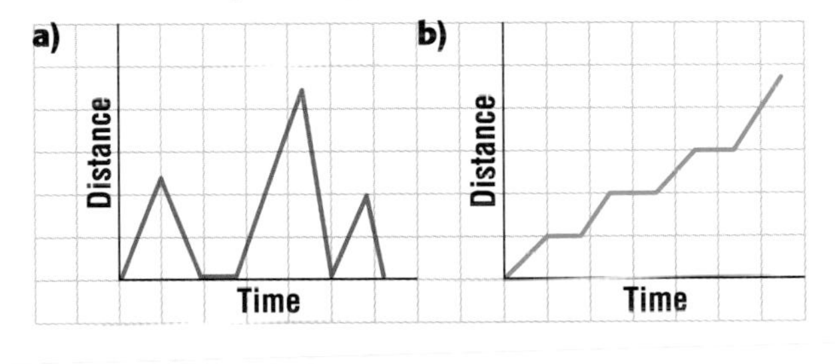

1. Before he left to sail down the Nile River, Livingstone, who was in Cairo, called his friend Stanley, who was in Edmonton. Livingstone called at 10:00 Cairo time. What time was it in Edmonton?

2. Use the Data Bank on pages 564 to 569 to write a uniform motion problem. Have a classmate solve it.

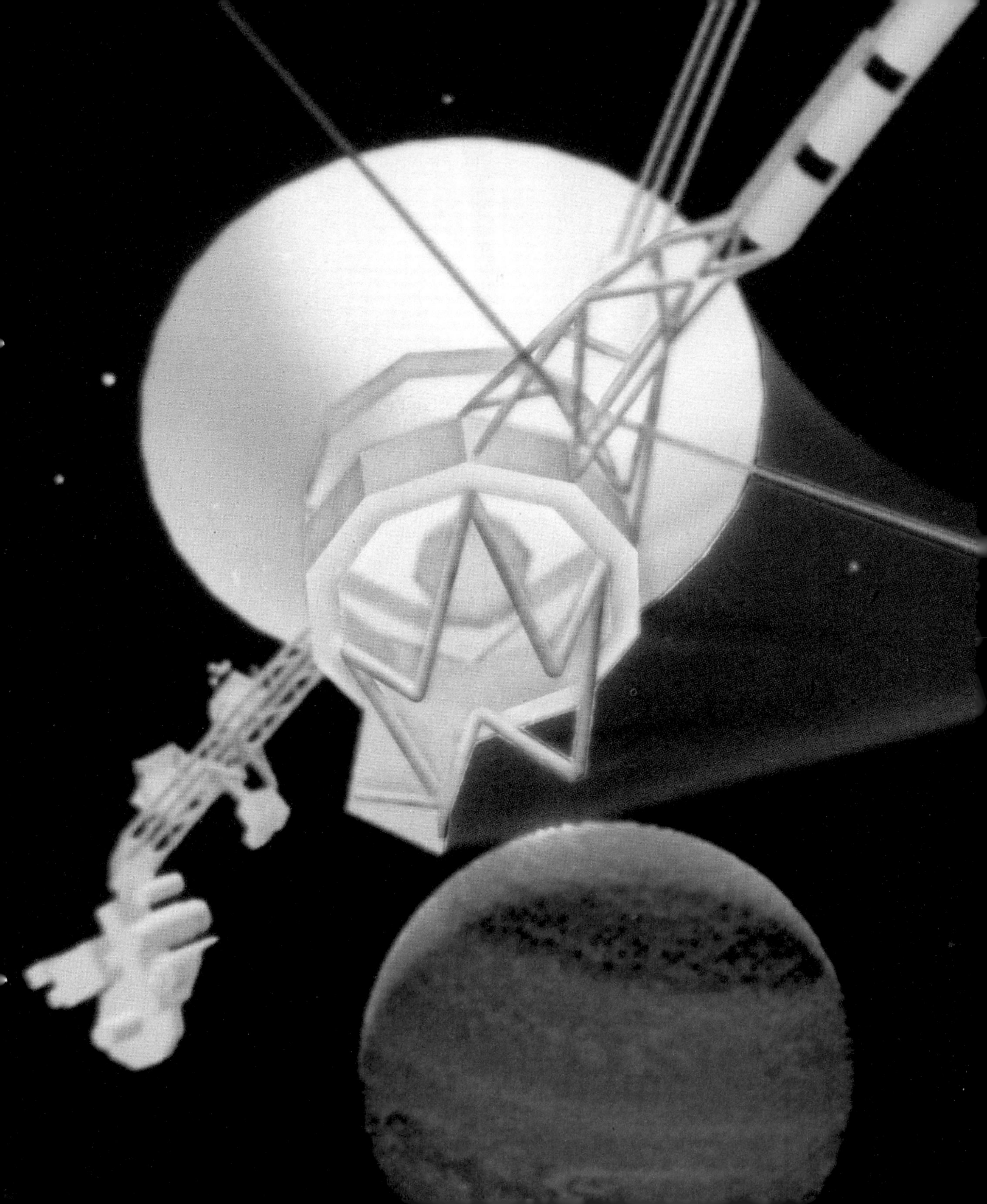

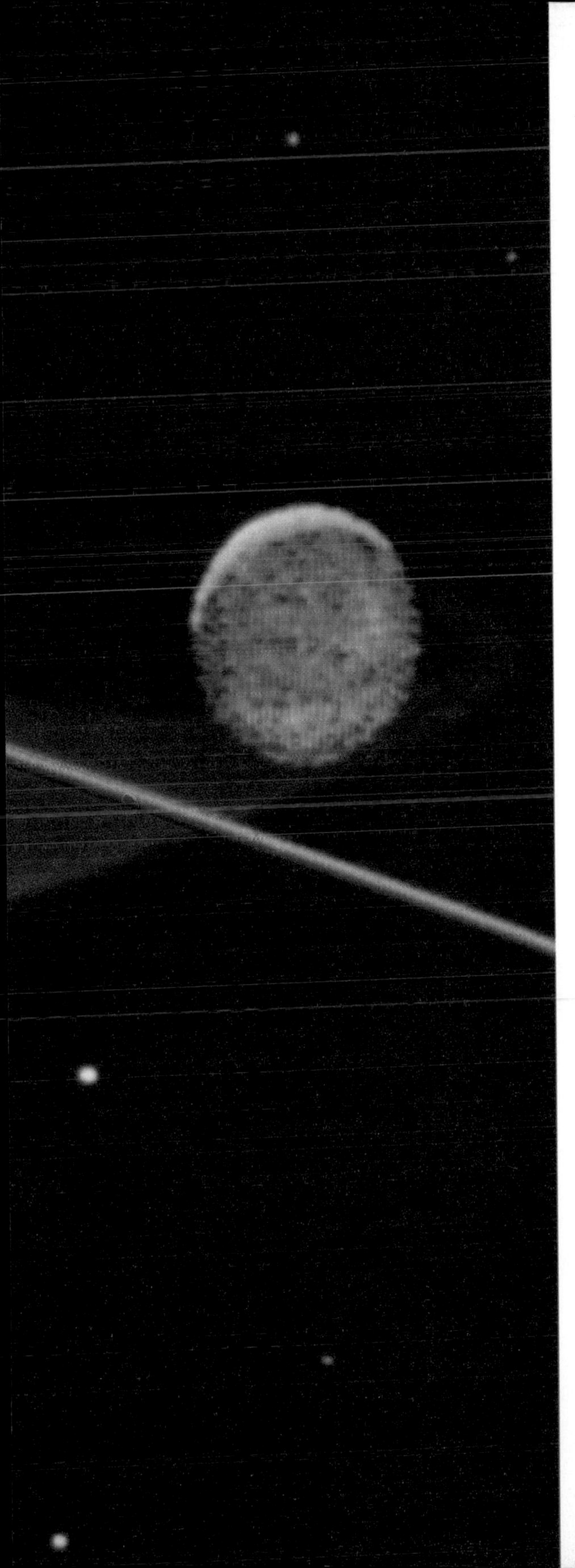

Special Products and Factoring

The formula $v = 1.15 \times 10^{-5} \times \sqrt{\frac{m}{r}}$ gives the minimum velocity that space probes, such as *Voyager 1* and *Voyager 2*, must have to escape the Earth's gravitational pull. In this formula, v is the minimum velocity in metres per second, m is the mass of the Earth in kilograms, and r is the radius of the Earth in metres.

The mass of the Earth is 5.98×10^{24} kg, and the radius of the Earth is 6.38×10^{6} m.

Use these data to calculate the escape velocity in metres per second. What is the escape velocity in kilometres per hour?

Activity 1 Factors

List the whole-number factors for these numbers.

1. 12 **2.** 15 **3.** 21

4. 8 **5.** 45 **6.** 36

7. 52 **8.** 17 **9.** 100

10. 24 **11.** 80 **12.** 39

Write each number as a product of prime factors. For example, $12 = 2 \times 2 \times 3$.

13. 8 **14.** 15 **15.** 24

16. 20 **17.** 36 **18.** 42

19. 28 **20.** 19 **21.** 26

22. Copy and complete this table.

First Number	Second Number	Sum	Product
3	4	7	12
–2	–3		
7	–5		
–9	2		
–5		–9	
	–6		–18
	10		–10
–3		1	
		9	20
		–7	12
		12	20
		–11	30

Determine the missing factor in each of the following expressions.

23. $6a = (3a)(\blacksquare)$ **24.** $-7ab = (-7a)(\blacksquare)$

25. $4x = (\blacksquare)(2x)$ **26.** $4x^2 = (-2x)(\blacksquare)$

27. $15a^2 = (5a)(\blacksquare)$ **28.** $-18xy = (9y)(\blacksquare)$

29. $12xy = (3x)(\blacksquare)$ **30.** $-4x^2 = (x)(\blacksquare)$

31. $12xy = (\blacksquare)(6xy)$ **32.** $36xy^3 = (\blacksquare)(-6y)$

33. $7x^2 = (\blacksquare)(7x^2)$ **34.** $20abc = (\blacksquare)(-5bc)$

Activity 2 Division by Monomials

Divide.

1. $x^5 \div x^2$ **2.** $p^3 \div p^2$

3. $a^7 \div a$ **4.** $x^5 \div x$

5. $\frac{-5t^5}{t^2}$ **6.** $\frac{a^4b^2}{a^2b^2}$

7. $\frac{-12x^4}{4x}$ **8.** $\frac{8a^4}{-2a}$

9. $\frac{-30x^3y^2}{-6xy^2}$ **10.** $\frac{72m^2n^2}{12mn}$

11. The area of a rectangle is $45a^2b^2$ and its length is $9ab$. Find an expression for its width.

Use this example to simplify each expression.

$$\frac{4a^2b - 6ab^2}{2ab} = \frac{4a^2b}{2ab} - \frac{6ab^2}{2ab} = 2a - 3b$$

12. $\frac{6a+12}{6}$ **13.** $\frac{9x-6y}{3}$

14. $\frac{6t-18}{6}$ **15.** $\frac{33xy-22y}{11y}$

16. $\frac{27m-18n+9}{9}$ **17.** $\frac{4x^3-10x^2+6x}{2x}$

18. $\frac{6a^2b+4ab^2}{2ab}$ **19.** $\frac{12xy^4-9x^3y}{3xy}$

Activity 3 Square Roots and Squares

State each square root. Assume that variables are positive.

1. $\sqrt{25}$ **2.** $\sqrt{121}$

3. $\sqrt{36}$ **4.** $\sqrt{x^2}$

5. $\sqrt{4a^2}$ **6.** $\sqrt{49x^2}$

Square.

7. 2^2 **8.** $(-5)^2$

9. 9^2 **10.** $(2x)^2$

11. $(-3a)^2$ **12.** $(5x^3)^2$

Activity 4 Area

Write an expression for each area.

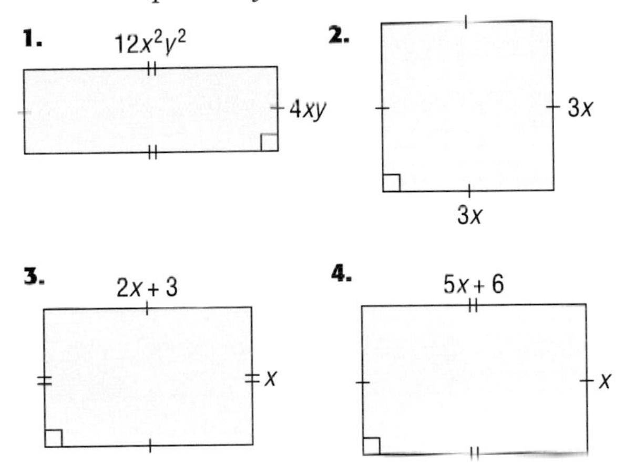

Add the areas of the smaller rectangles to write an expression for the area of each complete figure.

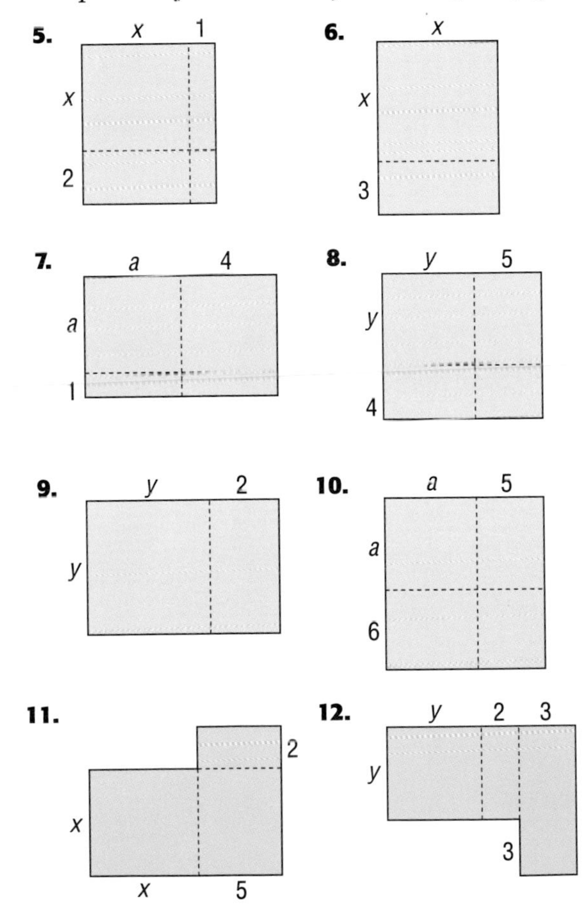

Mental Math

Calculate.

1. $12 + 6 - 3$
2. $20 - 12 + 6$
3. $24 \div 6 \times 4$
4. $18 \times 2 \div 4$
5. $12 \times 3 - 4 \div 2$
6. $24 - 18 \div 3 \times 4$
7. $3 + 8 \times 3 \div 4$
8. $(10)(3^2)$
9. $(2 + 3)^2$
10. $(4)(5^2)$
11. $3^2 + 4^2$
12. $2^3 + (4 + 1)^2$

Calculate.

13. $(8 + 4) \div 2$
14. $(20 - 8) \div 4$
15. $6 \times (5 + 4)$
16. $3 \times 4 + 6 \times 4$
17. $8 \times (12 - 9) + 2$
18. $15 \times (10 - 8) + 1$
19. $(20 + 10) \div 2 - 3 + 1$
20. $40 + 24 \div 8 - 3 + 1$
21. $(40 + 24) \div 8 - 3 + 1$

Simplify.

22. $\frac{1}{2} \times \frac{1}{3}$
23. $\frac{1}{4} \times \frac{1}{5}$
24. $\frac{3}{4} \times 4$
25. $\frac{2}{3} \times 3$
26. $\frac{1}{3} + \frac{1}{3}$
27. $\frac{5}{8} + \frac{1}{8}$
28. $\frac{1}{2} \div \frac{1}{2}$
29. $\frac{1}{2} \div \frac{1}{4}$
30. $\frac{2}{3} - \frac{1}{3}$
31. $\frac{5}{8} - \frac{1}{8}$
32. $\frac{3}{4} + \frac{3}{4}$
33. $\frac{3}{5} + \frac{4}{5}$
34. $\frac{1}{4} \times \frac{1}{3}$
35. $\frac{1}{2} \div 2$
36. $4 \div \frac{1}{2}$
37. $\frac{1}{4} + \frac{1}{2}$
38. $\frac{1}{2} - \frac{1}{4}$
39. $\frac{7}{8} - \frac{3}{4}$
40. $\frac{1}{3} - \frac{1}{5}$
41. $\frac{1}{4} - \frac{1}{8}$
42. $\frac{2}{5} - \frac{1}{10}$
43. $3 \div \frac{1}{3}$
44. $\frac{1}{2} \times \frac{2}{3}$
45. $\frac{1}{3} + \frac{1}{6}$

Activity ❶ Patterns on a Calendar

Examine the following monthly calendar. What month might it be?

S	M	T	W	T	F	S
		1	2	3	4	5
6	7	8	9	10	11	12
13	14	15	16	17	18	19
20	21	22	23	24	25	26
27	28	29	30	31		

1. Copy any 3-by-3 block, such as the one shaded in the table.

2. Choose 3 numbers by circling only 1 number from each column and each row of the block.

3. Record the sum of the 3 numbers.

4. Choose 3 other numbers in the same block. Again, choose only 1 number from each column and each row. Record the sum of the 3 numbers.

5. Compare your results from steps 3 and 4.

6. Describe how you can use 1 number in the block to find the sum of the 3 numbers you chose.

7. Compare your findings with your classmates'.

8. How many 3-by-3 blocks can you make from the above calendar month?

9. How many 3-by-3 blocks could you make if the first of the month was a Sunday?

10. Complete these 3-by-3 calendar blocks to show the patterns.

a)

x	$x+1$	$x+2$
$x+7$		

b)

	x	

c)

x		

11. Investigate the historical development of the calendar.

Activity 2 Patterns in Tables

1. Copy and complete the table.

Addition				
+	3	5	7	9
2				
4				
6				
8				

2. Choose 4 sums by circling only 1 sum in each row and each column of the table.

3. Add the 4 numbers you circled.

4. Compare the sum of your 4 numbers with that of a classmate. Is the result always the same?

5. Copy and complete the following table. Then carry out steps 2 and 3 above. Explain why the sum is always the same.

+	b	d	f	h
a				
c				
e				
g				

6. Work with a partner to rewrite steps 2, 3, and 4 to deal with multiplication, instead of addition.

7. Copy and complete the table.

Multiplication				
×	1	2	3	4
1				
2				
3				
5				

8. Use your instructions to investigate the products in the table. Compare your results with your classmates'.

8.1 Common Factors and the GCF

On the average, Yellowknife has blowing snow on 10 days of the year, and Halifax has blowing snow on 14 days of the year.

The number 10 has 4 factors: 1, 2, 5, and 10. The number 14 also has 4 factors: 1, 2, 7, and 14.

The **greatest common factor** (GCF) of the numbers 10 and 14 is 2.

Activity: Discover the Relationship

Write all the factors of each number.

a) 6 **b)** 12 **c)** 18

Inquire

1. Which prime factors are common to all 3 numbers?

2. What is the GCF of the 3 numbers?

3. Write a rule for finding the GCF of a set of numbers from their prime factors.

4. Write 3 different numbers that have a GCF of

a) 3 **b)** 8 **c)** 10

5. Write 4 different numbers that have a GCF of

a) 4 **b)** 5 **c)** 12

Example 1

Find the common factors of 15 and 18.

Solution

Write each number as a product of its prime factors.

$15 = 3 \times 5$
$18 = 2 \times 3 \times 3$

The numbers 15 and 18 have only 1 common factor, 3.

Note that the factor 1 is common to all numbers. Therefore, we do not include it as a common factor.

Example 2

Determine the GCF of each pair of numbers.

a) 24 and 48 **b)** 36 and 42

Solution

a) Write each number as a product of its prime factors.

$24 = 2 \times 2 \times 2 \times 3$

$48 = 2 \times 2 \times 2 \times 2 \times 3$

Multiply the common factors to calculate the GCF.
The GCF of 24 and 48 is $2 \times 2 \times 2 \times 3$ or 24.

b) $36 = 2 \times 2 \times 3 \times 3$

$42 = 2 \times 3 \times 7$

The GCF of 36 and 42 is 2×3 or 6.

Example 3

a) Determine the common factors of the monomials $2x^2$ and $4x$.

b) Write their GCF.

Solution

For algebraic expressions, the variable must also be factored. Write each expression as a product. Write the factors common to both.

a) $2x^2 = 2 \times x \times x$

$4x = 2 \times 2 \times x$

The common factors of $2x^2$ and $4x$ are 2 and x.

b) The GCF of $2x^2$ and $4x$ is $2x$.

Example 4

Find the GCF of $2x^3y$, $4x^2y^2$, and $2x^2y$.

Solution

$2x^3y = 2 \times x \times x \times x \times y$

$4x^2y^2 = 2 \times 2 \times x \times x \times y \times y$

$2x^2y = 2 \times x \times x \times y$

The GCF of $2x^3y$, $4x^2y^2$, and $2x^2y$ is $2 \times x \times x \times y$ or $2x^2y$.

Practice

Write the prime factors of each number.

1. 12 **2.** 16 **3.** 28

4. 63 **5.** 144 **6.** 225

Factor fully.

7. $4xy^2$ **8.** $18a^2b^3$ **9.** $36x^2yz^2$

10. $10x^2y$ **11.** $54x^5$ **12.** $125a^4b^2$

Determine the GCF of each pair.

13. 15, 20 **14.** 16, 24 **15.** 27, 36

16. 28, 42 **17.** 48, 72 **18.** 64, 96

Determine the GCF of each pair.

19. $4a$, $6a$ **20.** $2x^2$, $3x$

21. $12m^3$, $10m^2$ **22.** $12abc$, $3abc$

23. $2x$, $4y$ **24.** $14a$, $7b$

25. $5x^2$, $10x$ **26.** $4xy$, $5xy$

27. $9mn^2$, $8mn$ **28.** $2a^3$, $8a^2$

29. $15bc$, $25b^2c$ **30.** $6x^2y^2$, $9xy$

Determine the GCF of each set.

31. $5xyz$, $10abc$, $25pqr$

32. $20x$, $10x^3$, $8x^2$

33. $12abc$, $18ab$, $6ac$

34. $10x^2y$, $15xy^2$, $25xyz$

35. $21a^2b$, $35a^2b^2c$, $49ab^2c$

36. $12xy$, $16x^2y$, $20xyz$

37. $56abc$, $64a^2b$, $36ab^2c$

Find the GCF.

38. x^2y^2, x^2y^3, x^3y^4

39. $2x^3y$, $4x^2y^4$, $2x^2y^4$

40. $3x^2y^3$, $3x^3y^2$, $6xy^2$

41. $4a^3b^3$, $8a^2b^3$, $16ab^3$

42. $10s^4t^5$, $5s^5t^4$, $15s^3t^4$

Problems and Applications

43. Montreal has 162 wet days a year, whereas Beijing has 66. What is the GCF of these numbers?

44. Two rectangles are attached as shown.

$A = xy$	$A = 2xy$

What is the length of the common side? Explain.

45. Suppose students can get to school by walking at 5 km/h, riding a bike at 10 km/h, or being driven in a car at 30 km/h. Bob rides a bike, as does Karin. Bob lives the same distance from the school as Collette, who walks. Gustav and Shirley come by car. Karin and Gustav live the same distance from school. The graph gives the times the five students take to get to school one day.

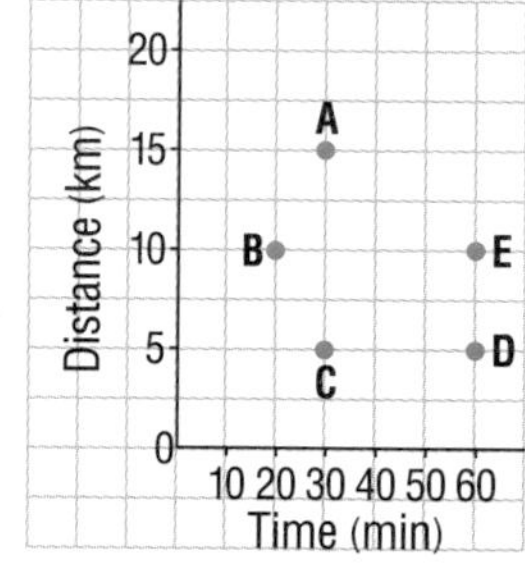

a) Which person does each point represent? Give reasons for your answers.

b) Draw a map of the area, showing where each student could live. Compare your map with a classmate's.

LOGIC POWER

If the outside of the large cube is painted red, how many of the smaller cubes have the following numbers of red faces?

a) 3 **b)** 2

c) 1 **d)** 0

8.2 Factoring Expressions with Common Factors

The world's narrowest commercial building is in Vancouver, British Columbia. The Sam Kee building is 30 m long by 1.8 m wide. In how many ways can you determine the perimeter?

Activity: Discover the Relationship

Examine the diagram of an x-tile.

1

x

Inquire

1. What is the sum of the length and width of the x-tile?

2. Write an expression for twice the sum of the length and width. Do not expand.

3. Add the sides of the x-tile to find the perimeter. Collect like terms.

4. How do the quantities represented in questions 2 and 3 compare?

5. Write a rule for removing a common factor from the terms in a polynomial.

6. Use your rule to remove a common factor from each of the following.

a) $3x + 6$ **b)** $4x + 4$ **c)** $2x - 8$

Example

a) Factor the expression $6x^2 - 14x$.
b) Check your answer by expanding.

Solution

a) Determine the GCF of both terms.
Then, divide both terms by the GCF.

$6x^2 = 2 \times 3 \times x \times x$
$14x = 2 \times 7 \times x$

The GCF is $2x$.

The second factor is $\frac{6x^2}{2x} - \frac{14x}{2x}$ or $3x - 7$.

The factors of $6x^2 - 14x$ are $2x$ and $3x - 7$.

Therefore, $6x^2 - 14x = 2x(3x - 7)$.

b) $2x(3x - 7) = 2x(3x - 7)$
$= 6x^2 - 14x$

Checks!

CONTINUED

Practice

State the missing factor.

1. $12x + 18y = (\blacksquare)(2x + 3y)$

2. $3x^2 - 5x = (\blacksquare)(3x - 5)$

3. $4ab + 3ac = (\blacksquare)(4b + 3c)$

4. $5x^2 + 10x = (\blacksquare)(x + 2)$

5. $8abc - 12ab = (\blacksquare)(2c - 3)$

Copy and complete.

6. $3y^2 + 18y = 3y(y + \blacksquare)$

7. $14a - 12b = 2(\blacksquare - 6b)$

8. $4a^3 - 8a^2 = 4a^2(\blacksquare - 2)$

9. $10x^3 - 5x^2 + 15x = 5x(2x^2 - \blacksquare + \blacksquare)$

Copy and complete.

10. $33ab - 22b = 11b(\blacksquare - \blacksquare)$

11. $4a^3 - 10a^2 + 6a = 2a(\blacksquare - \blacksquare + \blacksquare)$

12. $27a^2b^2 - 18ab + 9b = 9b(\blacksquare - \blacksquare + \blacksquare)$

13. $6x^2y - 4xy^2 = 2xy(\blacksquare - \blacksquare)$

14. $9a^3b - 12ab^4 = 3ab(\blacksquare - \blacksquare)$

Factor each binomial.

15. $10x + 15$

16. $28y - 14$

17. $2mn - n$

18. $5x^2 + 10x$

19. $8x^2 + 4x^3$

20. $9a^3b^2 - 6a^2b$

21. $4x^2y^2 - 6xy^2z^2$

22. $14a^2b^4 - 21b^2c^2$

23. $6x^2y^3z + 12xy^2z$

24. $15a^2b^5 - 9b^4c^5$

Problems and Applications

Factor each trinomial.

25. $9a - 6b + 3$

26. $4a - 8b + 16$

27. $12x^3 - 6x^2 + 24x$

28. $10x^3 - 5x^2 + 15x$

29. $24x^4y - 18x^3y + 12x^2y^2$

30. $8a^2b + 16ab - 24a$

31. $25m^3n - 15m^2n^2 + 5mn^3$

32. a) Write the rectangle's perimeter as the sum of 2 different products and as the product of a number and a sum.

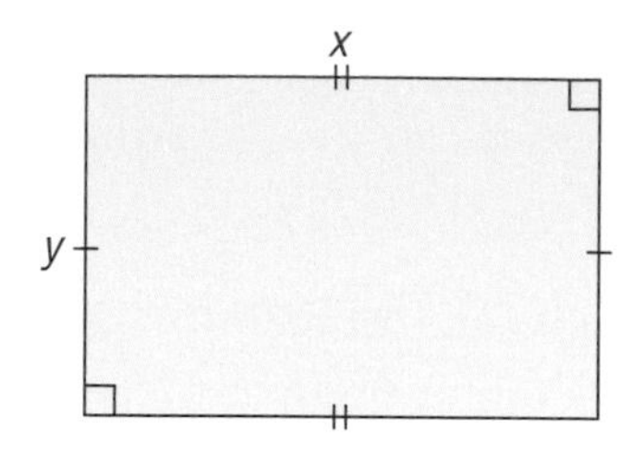

b) Which of the 2 forms in part a) is the factored form?

33. The perimeter of a rectangle is 46 cm. The length is 1 cm longer than the width. What are the rectangle's dimensions?

34. Find the GCF of each expression and factor fully.

a) $(a + b)x + (a + b)y$

b) $x(x - 2) + 3(x - 2)$

c) $x(2x - 3) - 5(2x - 3)$

d) $2a(a - b) + b(a - b)$

35. If you stand on the Earth and jump up at 5 m/s, your approximate height in metres above the ground after t seconds is given by the expression $5t - 5t^2$.

a) Factor the binomial.

b) Evaluate for $t = 0.4$ s.

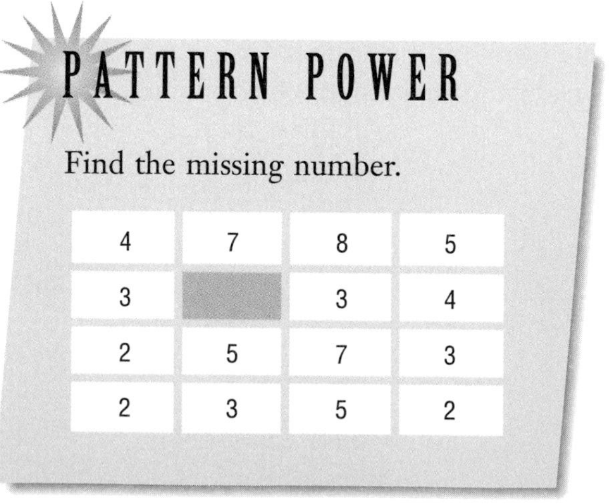

PATTERN POWER

Find the missing number.

4	7	8	5
3		3	4
2	5	7	3
2	3	5	2

8.3 Binomial Products

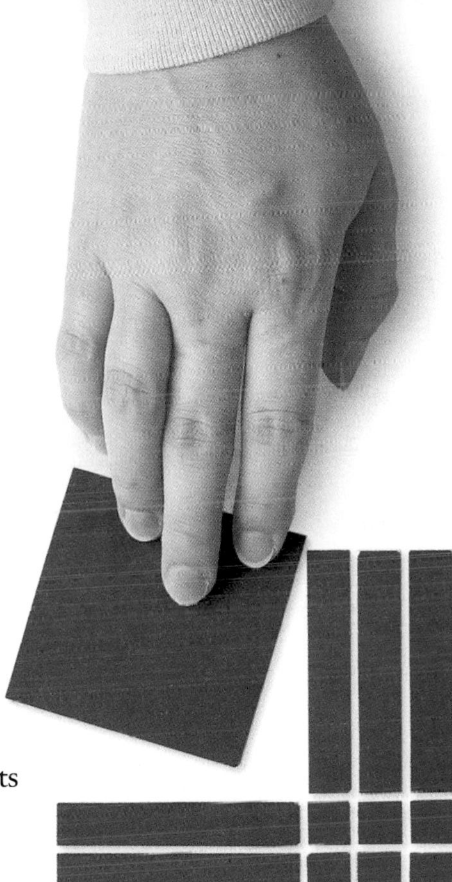

Activity: Use Algebra Tiles

Use algebra tiles to model this rectangle.

Inquire

1. Write an expression for the area of the rectangle by counting the numbers of x^2-tiles, x-tiles, and 1-tiles. Collect like terms and write the expression in descending powers of x.
2. Write an expression for the length of the rectangle.
3. Write an expression for the width of the rectangle.
4. Use the expressions for the length and width to write an expression for the area of the rectangle. Do not expand.
5. How do the quantities you represented in questions 1 and 4 compare?
6. Write a rule for multiplying 2 binomials.
7. Use your rule to multiply the following. Check your results by modelling with algebra tiles.
   a) $(x + 1)(x + 3)$ **b)** $(x + 2)(x + 4)$
8. Check your results in question 7 by substituting 1 for x.

Example

Find the product of these binomials.

a) $(x + 1)(x + 2)$ **b)** $(3x - 4)(5x + 1)$

Solution

To expand a binomial, use the distributive property.

a)
$$\begin{aligned}(x + 1)(x + 2) &= x(x + 2) + 1(x + 2)\\ &= x^2 + 2x + x + 2\\ &= x^2 + 3x + 2\end{aligned}$$

b)
$$\begin{aligned}(3x - 4)(5x + 1) &= 3x(5x + 1) - 4(5x + 1)\\ &= 15x^2 + 3x - 20x - 4\\ &= 15x^2 - 17x - 4\end{aligned}$$

The same result can be obtained by multiplying each term in the first binomial by each term in the second binomial.
You can remember this method with the acronym FOIL, which stands for First terms, Outside terms, Inside terms, and Last terms.

$$\begin{aligned}(x + 1)(x + 2) &= (x + 1)(x + 2)\\ &= \underset{F}{x^2} + \underset{O}{2x} + \underset{I}{x} + \underset{L}{2}\\ &= x^2 + 3x + 2\end{aligned}$$

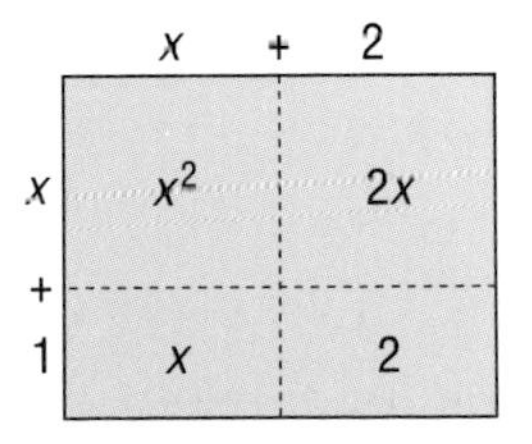

CONTINUED

Practice

Express each area as a product and in expanded form.

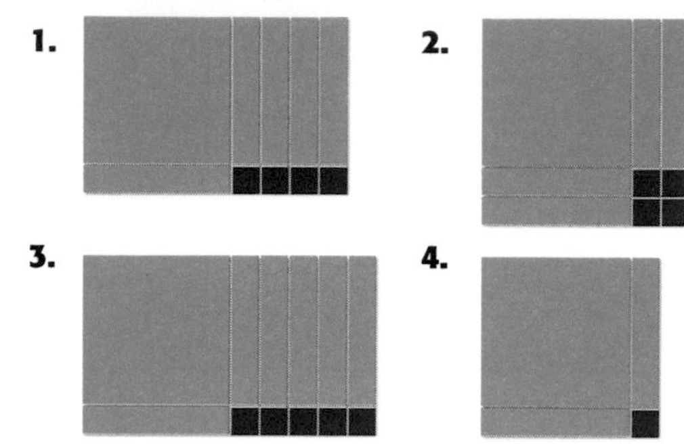

Expand.

5. $3(x-5)$ **6.** $x(2x+3)$

7. $-7(2x-6)$ **8.** $4x(3x-1)$

9. $5a^2(3a-4b)$ **10.** $-2x(3x+5y)$

Find the product.

11. $(x+1)(x+2)$ **12.** $(x+4)(x+3)$

13. $(a+4)(a+4)$ **14.** $(y+5)(y+6)$

15. $(x-4)(x-3)$ **16.** $(a-4)(a-2)$

17. $(b-1)(b-5)$ **18.** $(y-9)(y-9)$

19. $(x-6)(x+3)$ **20.** $(c+2)(c-8)$

21. $(t+10)(t-10)$ **22.** $(q-2)(q+5)$

Expand.

23. $(c+3)(c-4)$ **24.** $(x+2)(x-5)$

25. $(y+6)(y-2)$ **26.** $(a+9)(a-5)$

27. $(x-3)(x+3)$ **28.** $(b-7)(b+10)$

29. $(y-12)(y+3)$ **30.** $(x-7)(x+1)$

Multiply.

31. $(x+5)(2x+1)$ **32.** $(3y+1)(y+2)$

33. $(x-1)(2x-1)$ **34.** $(a-3)(2a-5)$

35. $(5y-7)(y+3)$ **36.** $(x-5)(4x+3)$

37. $(2x+3)(2x+1)$ **38.** $(5y-2)(3y-4)$

39. $(4x+1)(3x-5)$ **40.** $(2y-9)(5y+2)$

41. $(7y-3)(2y-7)$ **42.** $(3x-2)(8x+5)$

Problems and Applications

Multiply.

43. $(x+0.5)(x+2)$ **44.** $(x-1.2)(x+3)$

45. $(x-2.5)(x-10)$ **46.** $(x-3)(x+2.1)$

Expand the following.

47. $2(x+3)(x+5)$ **48.** $4(x-9)(x+5)$

49. $-1(a+3)(a-2)$ **50.** $10(x+7)(x-5)$

51. $3(2x-1)(3x-2)$ **52.** $2x(x+7)(x-10)$

53. $0.5(x-1)(x+3)$ **54.** $1.8(x+1)(x+1)$

55. Evaluate each area in 2 different ways for $x = 3$ cm.

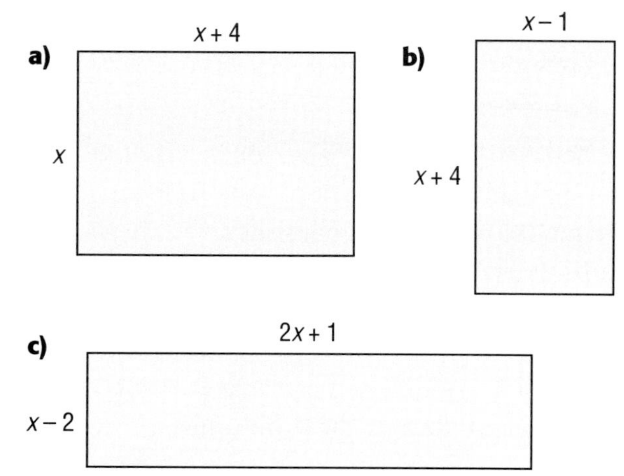

56. a) Verify that $(x+6)(x+2) \neq x^2 + 12$ by substituting 1 for x.
b) Expand $(x+6)(x+2)$ correctly.

57. A square building of side x metres is extended by 10 m on one side and 5 m on the other side to form a rectangle.
a) Express the new area as the product of 2 binomials.
b) Evaluate the new area for $x = 20$.

58. After 2 years, an investment of $1000 compounded annually at an interest rate r grows to the amount $1000(1+r)^2$ in dollars.
a) Expand the expression.
b) Evaluate for $r = 0.08$.

8.4 Factoring Trinomials: $x^2 + bx + c$

Many trinomials can be written as a product of 2 binomials.

Activity: Use Algebra Tiles

Use algebra tiles to model this rectangle.

Inquire

1. Write an expression for the area of the rectangle by counting tiles. Collect like terms and write the expression in descending powers of x.

2. Use expressions for the length and width to write an expression for the area of the rectangle. Do not expand.

3. How are the quantities represented by your 2 expressions related?

4. How is the numerical term in the expression from question 1 related to the numerical terms in the expression from question 2?

5. How is the coefficient of the x-term in the expression from question 1 related to the numerical terms in the expression from question 2?

6. How is the x^2-term in the expression from question 1 related to the x-terms in the expression from question 2?

7. Use your findings to factor each of the following.

a) $x^2 + 5x + 6$ **b)** $x^2 - 3x + 2$ **c)** $x^2 - x - 2$

Expanding 2 general binomials gives the following results.

$$\begin{aligned}(x + m)(x + n) &= x^2 + nx + mx + mn \\ &= x^2 + mx + nx + mn \\ &= x^2 + (m + n)x + mn\end{aligned}$$

Compare the above result to the general trinomial $x^2 + bx + c$.

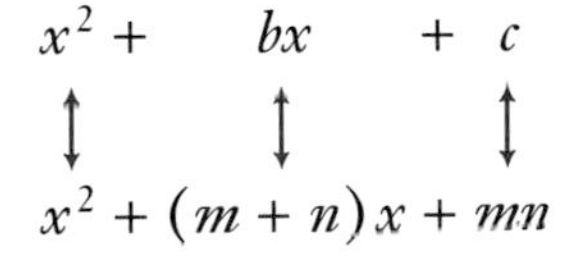

$$\begin{array}{ccc} x^2 + & bx & + \ c \\ \updownarrow & \updownarrow & \updownarrow \\ x^2 + & (m + n)x & + \ mn \end{array}$$

When we make this comparison, we see that $b = m + n$ and $c = mn$.

Therefore, the general trinomial $x^2 + bx + c = x^2 + (m + n)x + mn$
$= (x + m)(x + n)$

CONTINUED

Example 1

Factor $x^2 - 10x + 16$.

Solution

To factor $x^2 - 10x + 16$, find m and n so that $m + n = -10$ and $mn = 16$.
This means that both factors are negative. List pairs of factors of 16 and choose the correct pair.

Pairs of Factors of 16	Sum of Pairs of Factors
1, 16	17
−1, −16	−17
2, 8	10
−2, −8	−10
4, 4	8
−4, −4	−8

Therefore, $m = -2$ and $n = -8$.

$$\begin{aligned} x^2 - 10x + 16 &= x^2 + (-2 - 8)x + (-2)(-8) \\ &= (x - 2)(x - 8) \end{aligned}$$

Example 2

Factor $x^2 + 4x - 21$.

Solution

Find m and n so that $m + n = 4$ and $mn = -21$.

Pairs of Factors of −21	Sum of Pairs of Factors
1, −21	−20
−1, 21	20
3, −7	−4
−3, 7	4

This means that one factor is positive and one factor is negative.

Therefore, $m = -3$ and $n = 7$.

$x^2 + 4x - 21 = (x - 3)(x + 7)$

Example 3

Factor $2x^2 - 4x - 70$.

Solution

Factor the trinomial first to remove the GCF, which is 2.

$2x^2 - 4x - 70 = 2(x^2 - 2x - 35)$

Find m and n so that $m + n = -2$ and $mn = -35$.
One value is positive and one value is negative.

Pairs of Factors of −35	Sum of Pairs of Factors
1, −35	−34
−1, 35	34
5, −7	−2
−5, 7	2

Therefore, $m = 5$ and $n = -7$.

$2x^2 - 4x - 70 = 2(x + 5)(x - 7)$

Practice

Express the area of each rectangle in expanded and factored form.

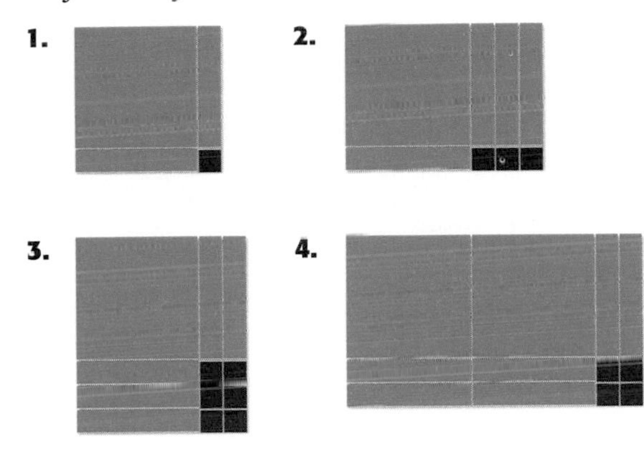

State the values of m and n that satisfy the given conditions.

5.

	Sum $m + n$	Product mn
a)	7	12
b)	8	15
c)	13	12
d)	18	77
e)	−8	15
f)	−10	25
g)	−7	12
h)	−7	6

6.

	Sum $m + n$	Product mn
a)	−1	−12
b)	1	−12
c)	−3	−40
d)	25	150
e)	4	−5
f)	−1	−42
g)	−7	−60
h)	2	−48

Factor.

7. $x^2 + 7x + 10$ **8.** $y^2 - 8y + 15$
9. $w^2 - w - 56$ **10.** $z^2 + 3z - 40$
11. $x^2 - x - 30$ **12.** $a^2 - 17a + 16$
13. $x^2 - 9x - 10$ **14.** $x^2 + 12x + 20$
15. $x^2 + 10x + 25$ **16.** $m^2 - 9m + 18$
17. $a^2 - 6a + 9$ **18.** $y^2 + 11y + 30$
19. $x^2 + 10x + 9$ **20.** $x^2 - 15x - 16$
21. $a^2 + 6a - 16$ **22.** $x^2 + 9x + 20$
23. $a^2 - 25a + 24$ **24.** $y^2 - 9y + 14$
25. $y^2 - 7y - 18$ **26.** $x^2 - x - 72$
27. $s^2 - 2s - 80$ **28.** $a^2 - 18a + 81$

Remove the GCF and factor fully.

29. $3x^2 - 21x + 36$ **30.** $5x^2 - 5x - 10$
31. $7x^2 + 35x + 42$ **32.** $4x^2 - 24x + 32$
33. $6b^2 + 12b - 90$ **34.** $bx^2 - 28bx + 75b$
35. $5jx^2 - 40jx + 75j$ **36.** $3tx^2 + 12tx + 12t$
37. $t^3 + t^2 - 12t$ **38.** $3k^3 + 15k^2 - 18k$

Problems and Applications

Factor, if possible.

39. $x^2 + x + 1$ **40.** $a^2 - 7a \quad 8$
41. $b^2 + 14b + 48$ **42.** $y^2 + 7y - 12$
43. $z^2 - 20z + 100$ **44.** $m^2 - 2m + 5$
45. $x^2 - 4x - 4$ **46.** $y^2 + 8y - 20$

47. The area of a rectangle is represented by the expression $x^2 + 9x + 20$.
a) Factor the expression.
b) A smaller rectangle is 1 unit shorter on each side than the first rectangle. Write a factored expression for the area of the smaller rectangle.
c) Expand the expression for the area of the smaller rectangle.

48. Write 4 trinomials in the form $x^2 + 12x + ■$ that can be factored.

49. a) Complete the factoring by supplying the missing terms.
$x^2 + 6x + ■ = (x + ■)(x + ■)$
$x^2 - 5x + ■ = (x - ■)(x - ■)$
$x^2 + ■x + 12 = (x + ■)(x + ■)$
$x^2 - ■x + 5 = (x - ■)(x - ■)$
$x^2 - ■x - 12 = (x - ■)(x + ■)$
b) Compare your answers with a classmate's. Which cases have more than one solution? Explain.

50. Make up 5 trinomials in the form $x^2 + bx + c$. Make 3 factorable and 2 impossible to factor. Exchange trinomials with a classmate. Try to factor each other's trinomials.

8.5 Special Product: $(a - b)(a + b)$

How are the binomials $x + 4$ and $x - 4$ the same and how are they different?

Activity: Look for a Pattern

Copy and complete this table.

Binomials	Product	Simplified Product
$(x + 4)(x - 4)$	$x^2 + 4x - 4x + 16$	
$(a - 7)(a + 7)$		
$(k + 5)(k - 5)$		
$(x - y)(x + y)$		
$(2a - 1)(2a + 1)$		

Inquire

1. a) What do you notice about the 2 middle terms in each product?
b) What is the algebraic sum of the middle terms in each product?

2. How many terms are there in each simplified product?

3. Write a rule for expanding 2 binomials that are identical except for the signs in the middle.

Example

Expand.

a) $(x + 3)(x - 3)$ **b)** $(7x - 1)(7x + 1)$

Solution

a) $(x + 3)(x - 3) = (x + 3)(x - 3)$
$= x^2 - 3x + 3x - 9$
$= x^2 - 9$

b) $(7x - 1)(7x + 1) = (7x - 1)(7x + 1)$
$= 49x^2 + 7x - 7x - 1$
$= 49x^2 - 1$

These expanded forms represent a special case of the product of 2 binomials, called the **difference of squares**.

$(a + b)(a - b) = a^2 - b^2$

Practice

Write the missing factor.

1. $(a-7)(\blacksquare) = a^2 - 49$
2. $(x+2)(\blacksquare) = x^2 - 4$
3. $(\blacksquare)(3m-7) = 9m^2 - 49$
4. $(9x+8)(\blacksquare) = 81x^2 - 64$
5. $(x-y)(\blacksquare) = x^2 - y^2$
6. $(\blacksquare)(2a+3b) = 4a^2 - 9b^2$

Expand.

7. $(x+1)(x-1)$
8. $(a+5)(a-5)$
9. $(p+6)(p-6)$
10. $(x-9)(x+9)$
11. $(y-8)(y+8)$
12. $(t-10)(t+10)$

Multiply.

13. $(2x-1)(2x+1)$
14. $(3y+1)(3y-1)$
15. $(4a+3)(4a-3)$
16. $(6t-5)(6t+5)$
17. $(x-y)(x+y)$
18. $(10t-3)(10t+3)$
19. $(p+q)(p-q)$
20. $(3a-b)(3a+b)$
21. $(x-6y)(x+6y)$
22. $(j-10r)(j+10r)$
23. $(x^2+3)(x^2-3)$
24. $(k^2+9)(k^2-9)$

Problems and Applications

25. Factor each of the numbers as shown and complete the table.

Numbers	$(a+b)(a-b)$	Product
33×27	$(30+3)(30-3)$	891
24×16		
47×53		
62×58		

Multiply each of the following, using the method in question 25.

26. $(10+2)(10-2)$
27. $(15+3)(15-3)$
28. $(20-2)(20+2)$
29. 14×6
30. 17×23
31. 32×28

Expand.

32. $(x-1)(x+1)(x^2+1)$
33. $(a+2)(a-2)(a^2+4)$
34. $(x-9)(x+9)(x^2+81)$
35. $(3x-2)(3x+2)(9x^2+4)$
36. $(x-1)(x+1)(x^2+1)(x^4+1)$
37. If a square field is made into a rectangle by shortening 2 opposite sides by 50 m each and lengthening the other 2 sides by 50 m each, how do the areas of the original field and the new field compare? Explain.
38. Which of the following numbers of terms are not possible as the product of 2 binomials?
    1 term
    2 terms
    3 terms
    4 terms
    5 terms
    Give reasons for your answers.

LOGIC POWER

Use your knowledge of geography to find out what province this area is in. Then find the area on a complete map.

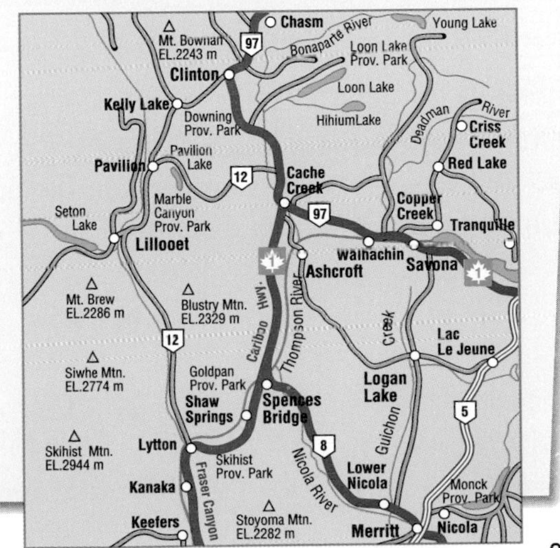

8.6 Factoring the Difference of Squares

Activity: Complete the Table

Copy and complete the table.

Factored Form	Product
$(x-5)(x+5)$	x^2-25
$(b+8)(b-8)$	
$(z+9)(z-9)$	
	a^2-36
	y^2-100
	r^2-81

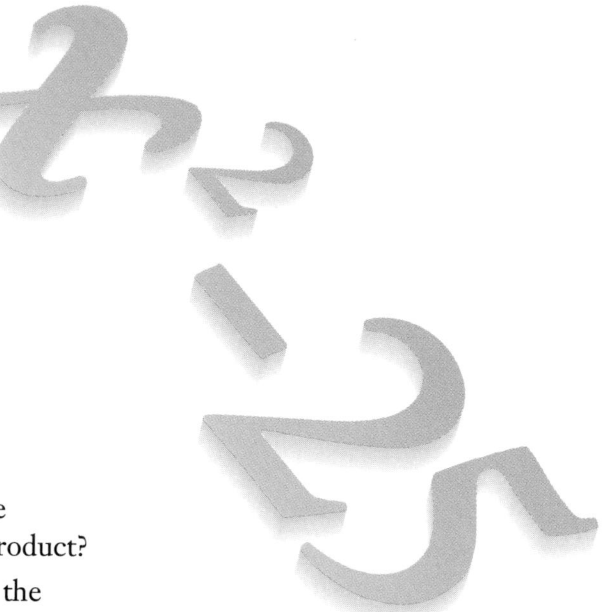

Inquire

1. How is the first term in each binomial of the factored form related to the first term in the product?

2. How is the second term in each binomial of the factored form related to the second term in the product?

3. Write a rule for working backward from the product to the factored form.

4. Use your rule to factor each of the following.

a) x^2-4 **b)** y^2-49 **c)** $4z^2-25$

Example

Factor.

a) x^2-49 **b)** $25a^2-81b^2$ **c)** $9y^2-81$ **d)** $12b^2-27c^2$

Solution

Remove any common factors.
Take the square root of each term in the difference of squares.
Write 2 binomial factors that are identical except for the signs in the middle.

a) x^2-49
$=(x+7)(x-7)$

b) $25a^2-81b^2$
$=(5a-9b)(5a+9b)$

c) $9y^2-81z^2$
$=9(y^2-9z^2)$
$=9(y+3z)(y-3z)$

d) $12b^2-27c^2$
$=3(4b^2-9c^2)$
$=3(2b-3c)(2b+3c)$

Practice

Determine the missing factor.

1. $x^2 - 25 = (x - 5)(■)$

2. $w^2 - 100 = (w + 10)(■)$

3. $k^2 - 81 = (k - 9)(■)$

4. $4a^2 - 121 = (2a - 11)(■)$

5. $9 - 16x^2 = (3 + 4x)(■)$

6. $x^4 - 36 = (x^2 + 6)(■)$

Factor, if possible, using a difference of squares.

7. $t^2 - 4$ **8.** $x^2 - 16$

9. $b^2 - 9$ **10.** $m^2 - 49$

11. $p^2 - 25$ **12.** $w^2 - 36$

13. $a^2 - 81$ **14.** $q^2 - 100$

15. $y^2 + 144$ **16.** $1 - c^2$

17. $64 - x^2$ **18.** $121 + w^2$

19. $s^2 - t^2$ **20.** $z^2 - x^2$

21. $h^2 - g^2$ **22.** $p^2 - q^2$

Factor.

23. $4a^2 - 9b^2$ **24.** $16p^2 - 81$

25. $25a^2 - 49$ **26.** $9b^2 - 25$

27. $16 - x^2$ **28.** $25 - 36b^2$

29. $16 - 49x^2$ **30.** $81 - 4a^2$

31. $16x^2 - 1$ **32.** $144 - 121x^2$

33. $169 - 100t^2$ **34.** $225 - 49w^2$

Factor, if possible.

35. $6x^2 - 25$ **36.** $16y^2 - 49$

37. $9 - 4z^2$ **38.** $25a^2 - 36$

39. $x^2y^2 - 4$ **40.** $m^2 + 64$

41. $(a + b)^2 - (a - b)^2$

42. $25 - 81p^2q^2$

Remove the common factor, then factor using a difference of squares.

43. $2m^2 - 50$ **44.** $9x^2 - 36$

45. $20r^2 - 45$ **46.** $8a^2 - 50$

47. $10y^2 - 1000$ **48.** $x^3 - x$

49. $50y^2 - 72$ **50.** $x^3 - 9x$

51. $8y^2 - 8$ **52.** $4p^2 - 16$

53. $27k^2 - 12$ **54.** $16t^2 - 16$

55. $64 - x^2$ **56.** $50 - 18x^2$

57. $12x^2 - 75y^2$ **58.** $50x^2 - 98y^2$

Problems and Applications

59. Cut a 10-by-10 square out of grid paper. Remove a 3-by-3 square from one corner.

a) The area of the new figure is $10^2 - 3^2$. What is the area?

b) Cut off one of the 7-by-3 rectangles and add it to the larger rectangle to make one rectangle. What are the dimensions of this rectangle?

c) How do the dimensions of the rectangle compare with the result of factoring $10^2 - 3^2$?

d) Repeat parts a) to c) for a square that is 6-by-6, with a 2-by-2 square removed. The area of the resulting figure is $6^2 - 2^2$.

e) Repeat parts a) to c) for a starting square of side length x, with a square of side length y removed.

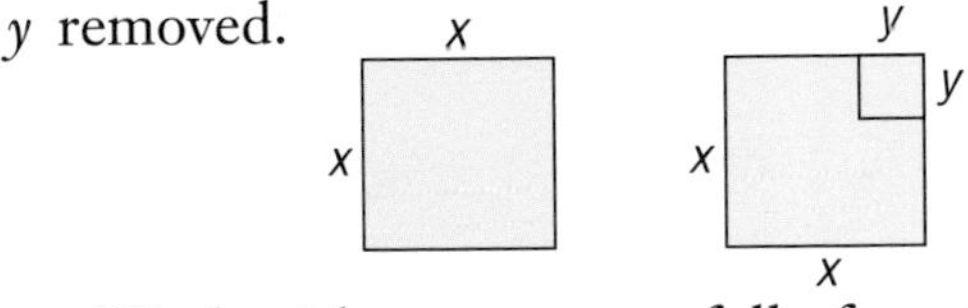

60. Work with a partner to fully factor the following expressions.

a) $x^4 - 1$ **b)** $x^8 - 1$ **c)** $x^4 - 625$

8.7 Special Products: Perfect Squares

An Austrian monk, Gregor Mendel (1822–1884), began the modern study of genetics. He found that the seeds from some tall pea plants produced dwarf plants, as well as tall ones.

We can represent Mendel's findings with a diagram called a Punnett square. Each parent plant, represented outside the square, has a dominant gene (T) for tallness and a recessive gene (t) for dwarfism. Because the tallness gene is dominant, each parent is tall.

The genes inherited by the offspring appear inside the square. Offspring with the combinations TT, Tt, and tT are tall plants. Offspring with 2 recessive genes (tt) are dwarf plants. If both tall parents have a recessive gene for dwarfism, what fraction of their offspring are dwarf plants?

	T	t
T	TT	Tt
t	tT	tt

A Punnett square is an example of a perfect square. Many algebraic expressions are also perfect squares.

Activity: Use Algebra Tiles

Use algebra tiles to model these squares.

a) b) c)

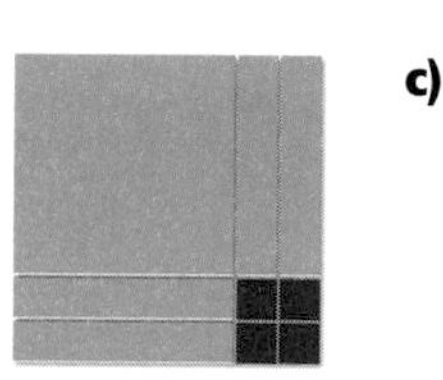

Inquire

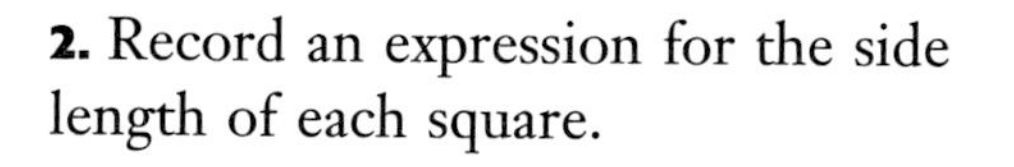

1. Count tiles to write a polynomial that represents the area of each square. Collect like terms and write each polynomial in descending powers of x.

2. Record an expression for the side length of each square.

3. Use your expression for the side length to write an expression for the area of each square. Do not expand.

4. How are the quantities represented by your expressions from questions 1 and 3 related?

5. Write a rule for finding the square of a binomial.

6. Use your rule to expand each of the following.

a) $(x+4)^2$ **b)** $(x-1)^2$ **c)** $(x-2)^2$

Example

Expand the following binomials.

a) $(x+6)^2$ **b)** $(x-5)^2$ **c)** $(2x-7)^2$

Solution

A **perfect square** consists of 2 identical binomial factors.
Expand, collect like terms, and simplify.

a) $(x + 6)^2$
$= (x + 6)(x + 6)$
$= x^2 + 6x + 6x + 36$
$= x^2 + 12x + 36$

b) $(x - 5)^2$
$= (x - 5)(x - 5)$
$= x^2 - 5x - 5x + 25$
$= x^2 - 10x + 25$

c) $(2x - 7)^2$
$= (2x - 7)(2x - 7)$
$= 4x^2 - 14x - 14x + 49$
$= 4x^2 - 28x + 49$

In general, the following are true for the product of 2 identical binomials.

$$\begin{aligned}(a + b)^2 &= (a + b)(a + b)\\ &= a^2 + ab + ab + b^2\\ &= a^2 + 2ab + b^2\end{aligned}$$

$$\begin{aligned}(a - b)^2 &= (a - b)(a - b)\\ &= a^2 - ab - ab + b^2\\ &= a^2 - 2ab + b^2\end{aligned}$$

Practice

Square.

1. $(-7)^2$ **2.** $(-9)^2$ **3.** $(-6)^2$
4. $(-12)^2$ **5.** $(2x)^2$ **6.** $(-3a)^2$
7. $(11y)^2$ **8.** $(-x)^2$ **9.** $(-4y)^2$

What is the first term in each product?

10. $(x + 7)^2$ **11.** $(a - 9)^2$ **12.** $(2x - 1)^2$
13. $(9t + 5)^2$ **14.** $(10b - 3)^2$ **15.** $(3y + 6)^2$
16. $(7p - 2)^2$ **17.** $(4j + 1)^2$ **18.** $(6q - 8)^2$

What is the middle term, including its sign, in each product?

19. $(x - 3)^2$ **20.** $(y + 8)^2$
21. $(x + y)^2$ **22.** $(a - b)^2$
23. $(2x + 3)^2$ **24.** $(4a - 5)^2$
25. $(3x + 2y)^2$ **26.** $(6p - 7)^2$

Square.

27. $(y - 10)^2$ **28.** $(3a - 1)^2$
29. $(5x + 2)^2$ **30.** $(3 - x)^2$
31. $(5 - y)^2$ **32.** $(5a + b)^2$
33. $(3x + y)^2$ **34.** $(4x - 3y)^2$
35. $(7a - 2b)^2$ **36.** $(4m + 5n)^2$

Problems and Applications

Rewrite in the form $(a + b)^2$ or $(a - b)^2$.

37. $x^2 + 14x + 49$ **38.** $x^2 - 16x + 64$
39. $4a^2 + 12a + 9$ **40.** $9b^2 - 24b + 16$
41. $64m^2 - 32m + 4$ **42.** $81n^2 + 90n + 25$

None of the following trinomials is a perfect square. Change 1 term to make the trinomial a perfect square.

43. $x^2 + 12x + 18$ **44.** $a^2 + 7a + 16$
45. $y^2 - 9y + 9$ **46.** $m^2 - 4m + 16$
47. $4x^2 - 4x + 2$ **48.** $9y^2 + 10y + 4$

49. Verify that $(x + 3)^2 \neq x^2 + 9$ by substituting 1 for x.

50. Use the diagram to expand $(a + b + c)^2$. Collect like terms.

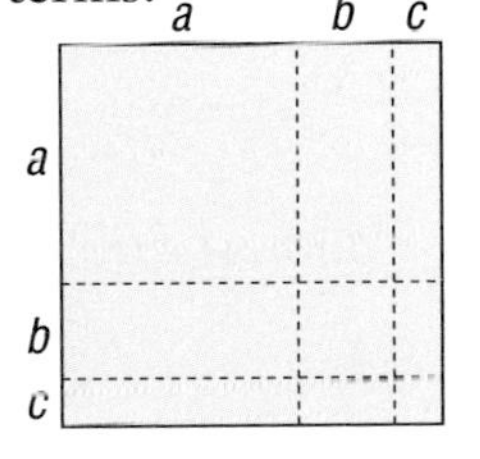

51. Explain how you can recognize a perfect square trinomial. You might use a sketch. Discuss terms and coefficients.

8.8 Products of Polynomials

Activity: Complete the Table

Copy and complete this table.

Product Form	Expanded Using the Distributive Property	Simplified Form
$(x+2)(x+3)$	$(x+2)x+(x+2)3$	
$(x+3)(x^2+x+2)$	$(x+3)x^2+(x+3)x+(x+3)2$	
$(x-1)(x^2-2x+4)$	$(x-1)x^2+(x-1)(-2x)+(x-1)4$	

Inquire

1. Find the products in the first column by multiplying each term of the first factor by each term of the second factor and simplifying.

2. How do the answers you found in 1 compare with the answers in the simplified form in the table?

3. Write a rule for multiplying polynomials.

Example

Expand.

a) $(x+1)(x^2+2x+1)$ **b)** $(x-2)(x^2-3x-5)$ **c)** $(x-5)(2x^2-x+2)$

Solution

Multiply each term in the first expression by each term in the second expression and simplify.

a)
$$\begin{aligned}(x+1)(x^2+2x+1) &= (x+1)(x^2+2x+1)\\ &= x^3+2x^2+x+x^2+2x+1\\ &= x^3+2x^2+x^2+x+2x+1\\ &= x^3+3x^2+3x+1\end{aligned}$$

	x^2	$2x$	1
x	x^3	$2x^2$	x
1	x^2	$2x$	1

b)
$$\begin{aligned}(x-2)(x^2-3x-5) &= (x-2)(x^2-3x-5)\\ &= x^3-3x^2-5x-2x^2+6x+10\\ &= x^3-3x^2-2x^2-5x+6x+10\\ &= x^3-5x^2+x+10\end{aligned}$$

c)
$$\begin{aligned}(x-5)(2x^2-x+2) &= (x-5)(2x^2-x+2)\\ &= 2x^3-x^2+2x-10x^2+5x-10\\ &= 2x^3-x^2-10x^2+2x+5x-10\\ &= 2x^3-11x^2+7x-10\end{aligned}$$

Practice

Expand.

1. $3(x^2 + 3x - 5)$
2. $2(3a^2 - 5a + 7)$
3. $2(x^2 + 7x + 12)$
4. $3(a^2 + 7a + 10)$
5. $5(y^2 + 4y + 10)$
6. $x(x^2 - 10x + 25)$
7. $x(x^2 + 11x + 28)$
8. $a(2a^2 + 8a + 15)$
9. $x(4x^2 - 3x - 4)$
10. $x(6x^2 - 23x + 20)$

Expand and simplify.

11. $(x + 1)(x^2 + 2x + 3)$
12. $(x + 2)(x^2 - 3x + 1)$
13. $(x + 3)(x^2 + 2x - 3)$
14. $(x - 3)(x^2 + 6x + 5)$
15. $(x - 4)(x^2 - 8x - 7)$
16. $(x - 2)(x^2 - 5x + 6)$
17. $(x - 5)(x^2 + 10x - 11)$

Expand and simplify.

18. $(a - 1)(3a^2 + a + 5)$
19. $(b - 7)(5b^2 - b - 2)$
20. $(x - 8)(2x^2 - 3x + 3)$
21. $(x - 5)(4x^2 - 3x - 7)$
22. $(x - 2)(3x^2 - 5x - 4)$
23. $(a - 1)(7a^2 - a + 1)$
24. $(y - 5)(2y^2 + 5y + 2)$

Expand and simplify.

25. $(2x^2 + 3x - 2)(x + 5)$
26. $(3a^2 - 5a - 6)(a + 3)$
27. $(5b^2 - 7b + 10)(b + 1)$
28. $(7w^2 - w - 1)(w - 1)$
29. $(2y^2 + 3y + 2)(y - 2)$
30. $(4t^2 - 2t - 1)(t - 5)$
31. $(x^2 + 2xy + y^2)(x + 1)$
32. $(x^2 - 3xy + y^2)(x - 2)$
33. $(b^2 - 3by - y^2)(x + 3)$

Problems and Applications

Expand and simplify.

34. $(3x + 2)(5x^2 - 2xy + y^2)$
35. $(2y - 5)(2y^3 - 3y^2 + 7)$
36. $(4x + 9)(7x^2 + x - 3)$
37. $(3a - 7)(2a^2 - 3a + 5)$
38. $(2y + 1)(3y^2 + 4y + 2)$
39. $(3a^2 - 6a - 7)(3a - 5)$
40. Determine the area of each figure.

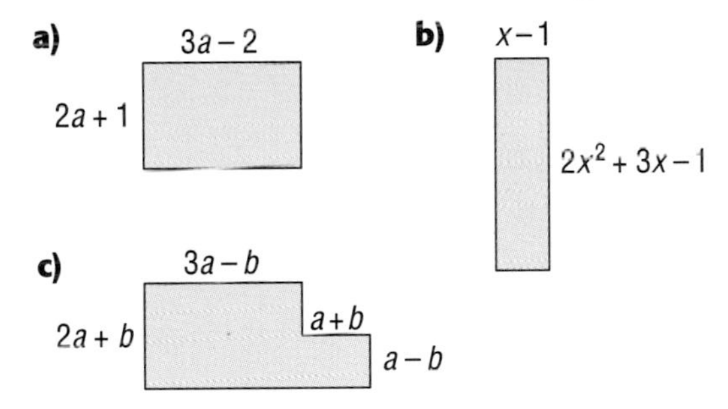

41. Determine the volume of each box.

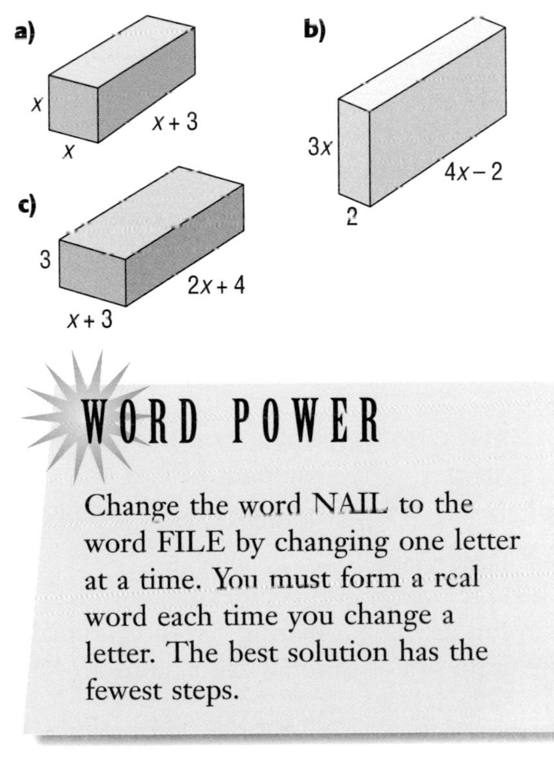

WORD POWER

Change the word NAIL to the word FILE by changing one letter at a time. You must form a real word each time you change a letter. The best solution has the fewest steps.

8.9 Rational Expressions

Duane Ward had the lowest earned run average of all the Blue Jays' pitchers the first year they won the World Series. His ERA was 1.95, meaning that, on average, about 2 runs were scored against him for every 9 innings he pitched. The ERA is a fraction, calculated by dividing 9 times the number of runs, r, by the number of innings pitched, i. The expression $\frac{9r}{i}$ is an example of a rational expression.

Activity: Discover the Relationship

Copy and complete the table.

Fraction	$\frac{2}{5}+\frac{1}{5}$	$\frac{2}{5}-\frac{1}{5}$	$\frac{2}{5}\times\frac{1}{5}$	$\frac{2}{5}\div\frac{1}{5}$
Simplified Fraction				

Inquire

1. Substitute x for the 2 in each expression and perform the operations.

2. Substitute x for 2 and y for 1 in each expression and perform the operations.

3. Substitute x for 5 in each expression and perform the operations.

4. Substitute x for 5 and y for 2 in each expression and perform the operations.

5. Use your findings to simplify the following.

a) $\frac{3}{x}+\frac{2}{x}$ **b)** $\frac{3}{x}-\frac{2}{x}$ **c)** $\frac{3}{x}\times\frac{2}{x}$ **d)** $\frac{3}{x}\div\frac{2}{x}$

Recall that rational numbers are numbers that can be written as the quotient of two integers, $\frac{a}{b}$, where a is any integer, and b can be any integer except zero.

A **rational expression** is an expression that can be written as the quotient of two polynomials. Some examples of rational expressions are shown.

$$\frac{a}{2} \quad \frac{4}{y} \quad \frac{x}{x-2}$$

Since division by zero is not defined, restrictions must sometimes be placed on the variables.

For the expression, $\frac{4}{y}$, y cannot be zero. We write $\frac{4}{y}$, $y \neq 0$.

For the expression, $\frac{x}{x-2}$, x cannot be 2. We write $\frac{x}{x-2}$, $x \neq 2$.

Example 1

Simplify. State the restrictions on the variables.

a) $\frac{3x^2y}{4x} \times \frac{2xy^3}{5y}$ **b)** $\frac{-8x^5y^8}{4x^3y^5}$ **c)** $\frac{3x^2y}{4} \div \frac{2x}{5}$

Solution

a)
$$\begin{aligned} &\frac{3x^2y}{4x} \times \frac{2xy^3}{5y} \\ &= \frac{6x^3y^4}{20xy} \\ &= \left(\frac{6}{20}\right)\left(\frac{x^3}{x}\right)\left(\frac{y^4}{y}\right) \\ &= \frac{3}{10}x^2y^3 \\ &= \frac{3x^2y^3}{10},\ x \neq 0,\ y \neq 0 \end{aligned}$$

b)
$$\begin{aligned} &\frac{-8x^5y^8}{4x^3y^5} \\ &= \left(\frac{-8}{4}\right)\left(\frac{x^5}{x^3}\right)\left(\frac{y^8}{y^5}\right) \\ &= -2x^2y^3,\ x \neq 0,\ y \neq 0 \end{aligned}$$

c)
$$\begin{aligned} &\frac{3x^2y}{4} \div \frac{2x}{5} \\ &= \frac{3x^2y}{4} \times \frac{5}{2x} \\ &= \frac{15x^2y}{8x} \\ &= \left(\frac{15}{8}\right)\left(\frac{x^2}{x}\right)\left(\frac{y}{1}\right) \\ &= \frac{15xy}{8},\ x \neq 0 \end{aligned}$$

Example 2

Simplify.

a) $\frac{x-1}{2} + \frac{x-2}{3}$ **b)** $\frac{(x+1)}{2} - \frac{2(x-1)}{5}$

Solution

To add or subtract rational expressions with different denominators, find equivalent fractions with a common denominator.

a)
$$\begin{aligned} &\frac{x-1}{2} + \frac{x-2}{3} \\ &= \frac{3(x-1)}{3 \times 2} + \frac{2(x-2)}{2 \times 3} \\ &= \frac{3x - 3 + 2x - 4}{6} \\ &= \frac{3x + 2x - 3 - 4}{6} \\ &= \frac{5x - 7}{6} \end{aligned}$$

b)
$$\begin{aligned} &\frac{(x+1)}{2} - \frac{2(x-1)}{5} \\ &= \frac{5(x+1)}{5 \times 2} - \frac{2[2(x-1)]}{2 \times 5} \\ &= \frac{5(x+1) - 2(2x-2)}{10} \\ &= \frac{5x + 5 - 4x + 4}{10} \\ &= \frac{5x - 4x + 5 + 4}{10} \\ &= \frac{x + 9}{10} \end{aligned}$$

CONTINUED

Practice

For each pair of fractions, write the lowest common denominator.

1. $\frac{1}{4}, \frac{3}{8}$
2. $\frac{5}{12}, \frac{7}{24}$
3. $\frac{1}{3}, \frac{1}{6}$
4. $\frac{5}{3x}, \frac{1}{7x}$
5. $\frac{6}{5a}, \frac{2}{3a}$
6. $\frac{7}{8x}, \frac{5}{6x}$

Simplify. State the restrictions on the variables.

7. $\frac{12x}{4xy}$
8. $\frac{20}{10y}$
9. $\frac{9x^2}{3x}$
10. $\frac{6abc}{3ab}$
11. $\frac{-10x^3y^4}{5x^2y^2}$
12. $\frac{20a^2b^2}{-5ab}$

Simplify.

13. $\frac{x}{2} \times \frac{3}{x}$
14. $\frac{2}{m} \times \frac{m}{4}$
15. $\frac{x}{y} \times \frac{y}{x}$
16. $\frac{m}{3} \div \frac{m}{4}$
17. $\frac{2}{y} \div \frac{4}{y}$
18. $\frac{x}{y} \div \frac{x}{y}$

Multiply.

19. $\frac{x^2y}{x} \times \frac{xy}{y}$
20. $\frac{2xy^2}{2x} \times \frac{5x^2y}{3x}$
21. $\frac{3p^2q^2}{2p^2} \times \frac{4pq}{3qr}$
22. $\frac{7a^2b}{2a} \times \frac{2ab}{2}$

Divide these monomials.

23. $\frac{x^2y^3}{3} \div \frac{xy}{3}$
24. $\frac{2x^2y}{5} \div \frac{2xy}{5}$
25. $\frac{xy^2}{10} \div \frac{2xy^5}{5x^3y^3}$
26. $\frac{3x^3y^2}{2x^2y^2} \div \frac{3}{xy}$

Simplify these fractions.

27. $\frac{7}{y} + \frac{12}{y}$
28. $\frac{x}{a} - \frac{11}{a}$
29. $\frac{7}{12x} - \frac{4}{12x}$
30. $\frac{45xy}{4z} - \frac{9xy}{4z}$

Simplify.

31. $\frac{2}{x} + \frac{3-x}{x}$
32. $\frac{x+3}{3x^2} + \frac{4-x}{3x^2}$
33. $\frac{3}{x^2} - \frac{(7-x)}{x^2}$
34. $\frac{2-x}{5x^2} - \frac{1}{5x^2}$

Simplify.

35. $\frac{x-2}{2} + \frac{x+5}{4}$
36. $\frac{x-2}{3} - \frac{x-6}{4}$
37. $\frac{x+2}{2} - \frac{4-x}{3}$
38. $\frac{2-x}{5} - \frac{3-x}{3}$
39. $\frac{x-1}{4} - \frac{2(x-3)}{6}$
40. $\frac{2(x-2)}{5} + \frac{x+1}{2}$

Problems and Applications

41.

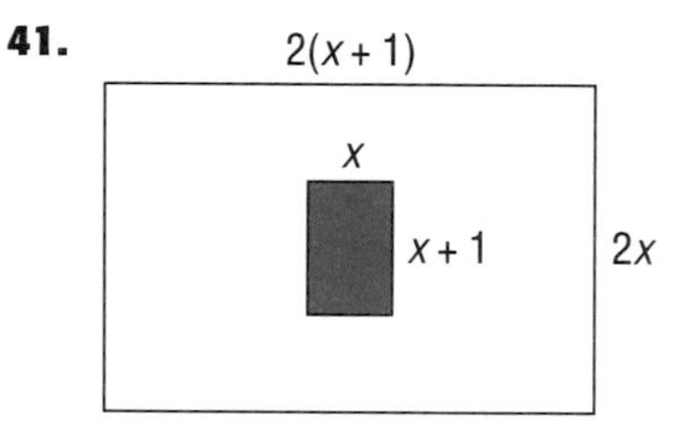

a) What is the total area of the figure? Leave your answer in factored form.
b) What is the area of the red rectangle in factored form?
c) What is the quotient of the area of the red rectangle divided by the total area of the figure? Write your answer in simplest form.
d) Check your answer by substituting 1 for x.

42.

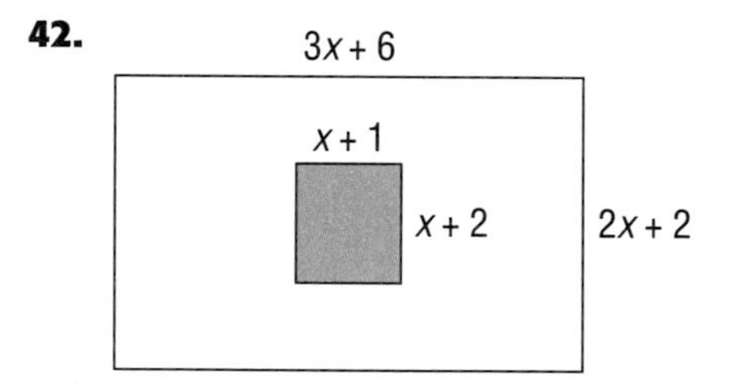

a) What is the total area of the figure in factored form?
b) What is the area of the green rectangle in factored form?
c) What is the quotient of the area of the green rectangle divided by the total area of the figure? Write your answer in simplest form.
d) Verify your answer by substituting 2 for x.

Fitness Centres

Activity ❶

Many Canadians are members of local fitness centres. There are many examples of modern technology in fitness centres. Describe some examples of how technology is used. Consider the office, weight rooms, pool area, aerobics room, and so on.

Activity ❷

Design a fitness centre that has all of the technology that you would like to have available. Create your own machines and explain what physical activities they would measure or involve. Describe how the machines would make use of mathematics.

Activity ❶ Designing a Theatre

You have been asked to submit a design for the seating in a theatre that will be located in an old movie house. The floor slopes to the stage and orchestra pit, so all you need to do is to place the seats and the aisles. The stage and orchestra pit are centred and measure 30 m across.

Each seat measures 50 cm by 50 cm, including arm rests. There is to be a 50-cm space in front of each seat for leg room and access. The seating area measures 40 m wide by 50 m deep. Aisles must be at least 2 m wide.

There must be an aisle across the front and the back of the seating area, plus aisles to get customers to their seats. Aisles may be curved or straight.

There should be no more than 20 seats in a row. There are 2 doors evenly spaced at the back of the theatre and a door in the middle of each side. There is no balcony.

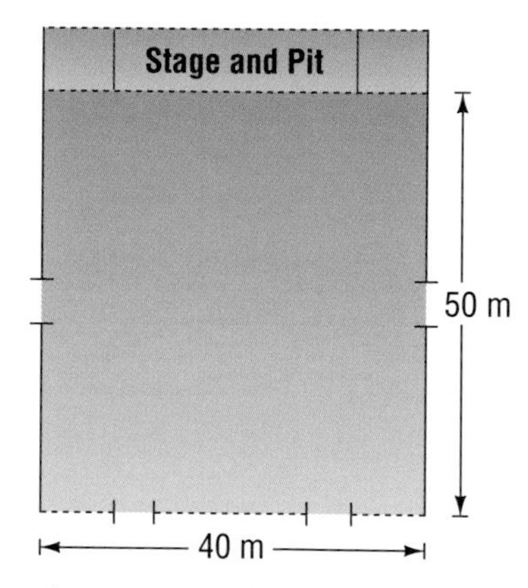

Design the placement of the seats and aisles so that people have easy access to their seats. The owners of the theatre want you to decide how many seats to put in. The owners want a pleasant, comfortable arrangement, but keep in mind that they are in the business of selling tickets. If you include too few seats, your design may not be accepted. You must label each aisle and each seat for the purpose of ticket sales.

Activity ❷ Designing a Circular Track

Suppose there is to be a race between 5 people on a circular track. The track has an inside radius of 50 m. Each lane of the track is 1 m wide. The lanes are numbered 1 to 5, as shown. The start and finish line is also shown.

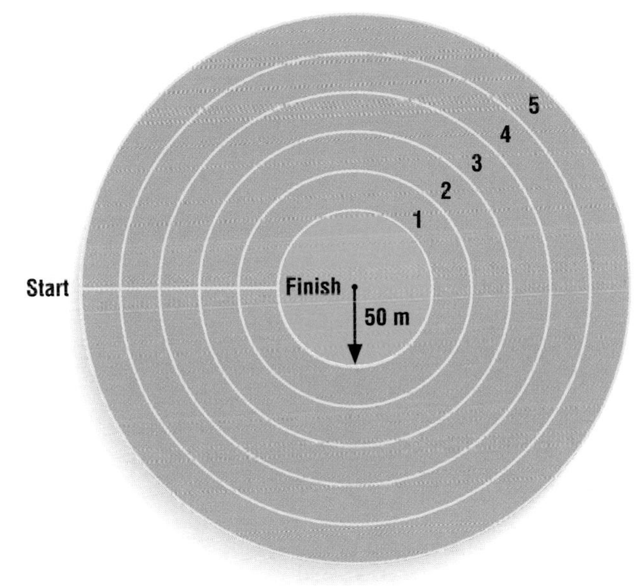

1. Suppose all 5 runners begin at the start line and run once around the track. Calculate the distance that each runner runs. Assume that $\pi = 3.14$.

2. To have all runners run the same distance, the start line for each lane must be staggered. Make the necessary calculations and mark a different start line for each lane. Each of the other runners must cover the same distance as the runner in lane 1 covers in 1 lap of the track from the start to the finish line.

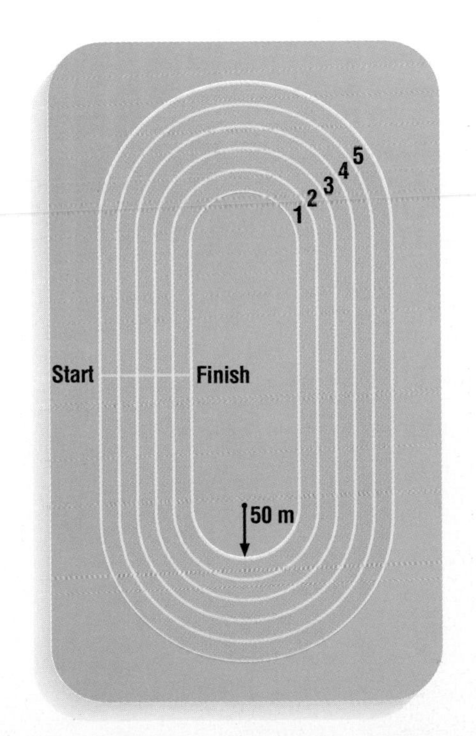

Activity ❸ Designing an Oval Track

1. Design an oval track like the one used in the Olympic Games. The ends of the track have an inside radius of 50 m. The sides of the track are straight. There are 5 lanes and the width of each lane is 1 m. The runner in lane 1 must run 400 m in 1 lap from the start to the finish.

2. Calculate the distance each runner would run if all the runners started at the same start line and finished after 1 lap of the track.

3. Make the necessary calculations and mark the start line for each lane. Each runner must run 400 m in a lane, and the finish line is the same for all runners.

Review

1. Write the GCF of each pair.

a) $35, 40$ **b)** $21, 28$

c) $34, 51$ **d)** $120, 96$

e) $10a^2, 5a$ **f)** $16xy, 12xz$

g) $10ab^2, 18ab$ **h)** $15xy, 25x^2y^2$

2. Write the GCF of each set.

a) $21x^2y, 15xy^2, 9x^2y^2$

b) $24x^2, -16xy, 32y^2$

c) $18xy^2, -27x^2y^2, 36x^2y$

Factor.

3. $5x - 15$ **4.** $6x^2 - 18x$

5. $5ab + 10ac$ **6.** $7a^2 + 35a^3$

7. $8abc - 12bc$ **8.** $3x^2 + 9y^2$

9. $3a^2 - 6ab + a$ **10.** $2x + 6y - 10z$

Expand.

11. $(x + 2)(x - 3)$ **12.** $(x - 4)(x + 7)$

13. $(x + 5)(x + 2)$ **14.** $(x - 2)(x - 3)$

15. $(x + 10)(x - 2)$ **16.** $(2x - 5)(3x - 2)$

17. $(2a - 3)(5a + 2)$ **18.** $(4x - 7)(3x + 2)$

19. $(3x - 5)(7x + 6)$ **20.** $(5a - b)(2a + 3b)$

Factor.

21. $x^2 + 8x + 7$ **22.** $x^2 - 6x + 5$

23. $y^2 + 8y + 15$ **24.** $a^2 + 8a + 12$

25. $b^2 + 10b + 24$ **26.** $x^2 - 7x + 6$

27. $x^2 - 11x + 28$ **28.** $a^2 - 7a + 12$

Factor.

29. $a^2 - a - 20$ **30.** $x^2 - x - 30$

31. $x^2 - 5x - 14$ **32.** $m^2 - 6m - 40$

33. $x^2 + 4x - 21$ **34.** $x^2 + 10x - 24$

35. $x^2 + 2x - 35$ **36.** $x^2 - 2x - 15$

Factor fully.

37. $2x^2 + 24x + 40$ **38.** $5a^2 - 40a + 80$

39. $4w^2 - 4w - 120$ **40.** $3r^2 - 21r + 30$

41. $2j^2 - 6j + 8$ **42.** $3t^2 + 18t - 21$

43. $7y^2 + 7y - 140$ **44.** $3z^2 - 39z + 126$

Expand.

45. $(m + 3)(m - 3)$ **46.** $(x - 5)(x + 5)$

47. $(2c + 1)(2c - 1)$ **48.** $(2x - 3)(2x + 3)$

49. $(4s - 7t)(4s + 7t)$ **50.** $(3j - 5q)(3j + 5q)$

Factor.

51. $x^2 - 1$ **52.** $y^2 - 4$ **53.** $4a^2 - 9$

54. $a^2 - 4b^2$ **55.** $4x^2 - y^2$ **56.** $4a^2 - 9b^2$

57. $9 - x^2$ **58.** $25 - 49x^2$ **59.** $2a^2 - 50$

60. $5x^2 - 20$ **61.** $4x^2 - 36$ **62.** $16a^2 - 36$

Identify the expressions that are perfect squares.

63. $x^2 - 2x + 1$ **64.** $x^2 + 9x + 3$

65. $x^2 - 8x - 16$ **66.** $x^2 + 10x + 25$

Expand.

67. $(x + 2)^2$ **68.** $(x - 3)^2$

69. $(y + 6)^2$ **70.** $(m - 5)^2$

Expand.

71. $(x - 2)(x^2 - 3x + 2)$

72. $(x - 3)(x^2 - 5x - 3)$

73. $(x + 1)(2x^2 - 2x + 1)$

74. $(x + 5)(3x^2 - x - 1)$

75. $(x^2 - x + 1)(x + 1)$

76. $(x^2 + 2x + 3)(x - 1)$

77. $(5x^2 - 5x - 3)(x - 4)$

78. $(4x^2 - 3x - 7)(x + 2)$

Divide.

79. $\frac{6x^3y^2}{3xy^2}$

80. $\frac{-30a^2b^2}{15a^2b}$

81. $\frac{-10x^8y^9}{-5x^6y^6}$

82. $\frac{4x^5y^7}{-8x^3y^3}$

Simplify.

83. $\frac{2pq^2}{3} \div \frac{2p}{3q}$

84. $\frac{2x^3y}{3} \div \frac{4}{3y}$

85. $\frac{x^3y^2}{3} \div \frac{3}{xy}$

86. $\frac{3x^3y^4}{7} \div \frac{2xy}{7}$

Multiply.

87. $\frac{x^2y}{2x} \times \frac{xy^3}{2y}$

88. $\frac{2a^3y^2}{4a} \times \frac{2ay}{3y}$

89. $\frac{20a^3b^2}{2a^2} \times \frac{ab}{5b^3}$

90. $\frac{5a^3b^8}{2ab^6} \times \frac{2a}{b}$

Simplify.

91. $\frac{x-1}{2} + \frac{x-4}{3}$

92. $\frac{x-2}{3} + \frac{1-x}{5}$

93. $\frac{3-x}{3} + \frac{2(x-2)}{4}$

94. $\frac{x-5}{6} + \frac{2(x+3)}{7}$

95. $\frac{1}{x^2} + \frac{1-x}{x^2}$

96. $\frac{3x+1}{7x^2} + \frac{2-x}{7x^2}$

Simplify.

97. $\frac{x-5}{2} - \frac{x+1}{3}$

98. $\frac{x+2}{3} - \frac{x-3}{4}$

99. $\frac{2(x+3)}{2} - \frac{x-1}{7}$

100. $\frac{x-1}{6} - \frac{2-x}{7}$

101. $\frac{1-x}{7} - \frac{2-x}{5}$

102. $\frac{x-2}{4} - \frac{x-3}{7}$

103. $\frac{2x+1}{2x} - \frac{x-1}{2x}$

104. $\frac{7-x}{x^2} - \frac{2-x}{x^2}$

105. The area of a rectangle is represented by the expression $x^2 + 10x + 16$.

a) Factor the expression.

b) A larger rectangle is 2 units longer on each side. Write a factored expression for the area of the larger rectangle.

c) Expand and simplify the expression for the area of the larger rectangle.

Group Decision Making

Mirror, Mirror on the Wall

Solve this problem in home groups.

1 2 3 4	1 2 3 4	1 2 3 4

Home Groups

1 2 3 4	1 2 3 4	1 2 3 4

The problem is this: When you are standing in front of a mirror and you want to see more of yourself, should you move forward or backward?

1. As a group, discuss the solution to the problem. Try to find a solution that everyone agrees with. Use a mirror to test your group's solution.

2. Next, make a drawing of the problem. Use a stick figure for the person in front of the mirror and a point on the figure for the person's eyes. Remember that the reflection appears at the same distance behind the mirror as the person is in front of the mirror.

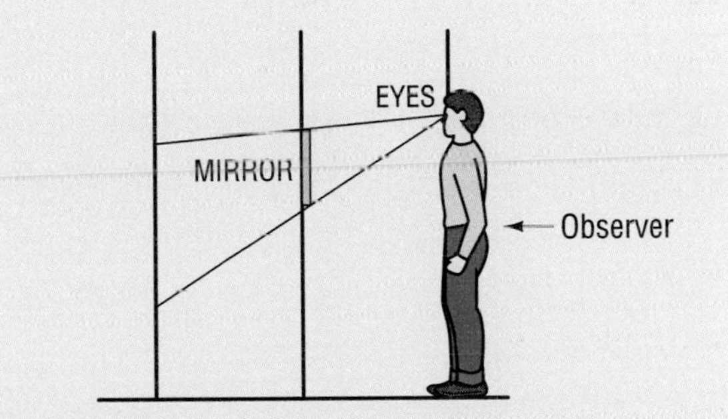

3. Make other drawings on top of the first one, with the stick figure moved forward and backward. Discuss the results.

4. Does a change in the size of the mirror change the problem?

5. Meet as a class. Present your group's solution to the problem.

Chapter Check

Factor fully.

1. $3xy - 6xz$

2. $24xy^2 + 12x^2y$

Expand.

3. $(x - 2)(x + 4)$

4. $(x - 3)(2x + 7)$

5. $(4a - 3b)(2a - 3b)$

Factor.

6. $x^2 + 7x + 10$

7. $x^2 - 9x + 18$

8. $x^2 - 3x - 10$

9. $x^2 + 2x - 35$

10. $x^2 + 8x + 16$

11. $x^2 - 18x + 17$

Factor fully.

12. $2x^2 - 8x + 8$

13. $5x^2 - 5x - 100$

Expand.

14. $(x - 2)^2$

15. $(w + 7)^2$

Factor.

16. $a^2 - 4$

17. $4x^2 - 25$

18. $81 - x^2$

19. $1 - 4b^2$

Factor.

20. $2t^2 - 200$

21. $3x^2 - 12$

22. $100 - 16t^2$

23. $2 - 18y^2$

State whether each trinomial is a perfect square.

24. $x^2 - 10x + 25$

25. $a^2 - 14a - 49$

26. $x^2 - 12x + 36$

27. $y^2 - 16y - 64$

Simplify.

28. $\frac{25xy}{5y} \times \frac{2x^2y^3}{2y}$

29. $\frac{10a^2b^3}{5ab^4} \div \frac{2ab}{5a^2b^2}$

Simplify.

30. $\frac{x + 3}{2} + \frac{5 + x}{3}$

31. $\frac{2(x - 3)}{5} - \frac{x - 1}{7}$

32. A rectangle has an area of $x^2 + 13x + 30$.

a) Express the area as a product of 2 binomials.

b) The length of a second rectangle is 2 units more than the length of the first rectangle. The width of the second rectangle is 1 unit less than the width of the first rectangle. Write an expression in factored form for the area of the second rectangle.

c) Expand and simplify your expression from part b).

BENT OFFERINGS by Don Addis.
By Permission of Don Addis and Creators Syndicate.

Using the Strategies

1. Using different fractions with 10s as denominators, fill in the 3-by-3 grid below so that the sum vertically, horizontally, and diagonally is $1\frac{1}{2}$.

2. a) Write any 2-digit number.
b) Keep summing the digits in your number until you obtain a single digit. For example, for 67 the sum of the digits is 6 + 7 or 13. The sum of the digits for 13 is 1 + 3 or 4.
c) Divide the 2-digit number you wrote by 9 and note the remainder. What do you notice?
d) Repeat parts a) to c) for other 2-digit numbers. Describe your results.
e) Do the results for numbers with more than 2 digits follow the same pattern?

3. In the following division problem, a and b represent missing single-digit numbers. Find values for a and b.

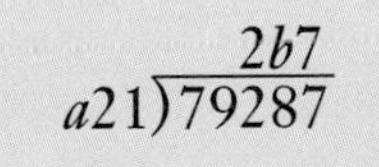

4. The greatest common factor of 2 numbers, m and n, is 14. If $m = 2 \times 5 \times 7^2$, name 3 numbers that could be n.

5. Kim has 2 chores at home. Every 4 days, she must clean the gerbil cage. Every 6 days, she must clean the canary cage. Last Monday, she did both jobs. On what day of the week will she next do both jobs?

6. To hang a picture on a bulletin board, Masao uses 4 thumbtacks, 1 in each corner. For 2 pictures of the same size, Masao can overlap the corners and hang 2 pictures with only 6 tacks.

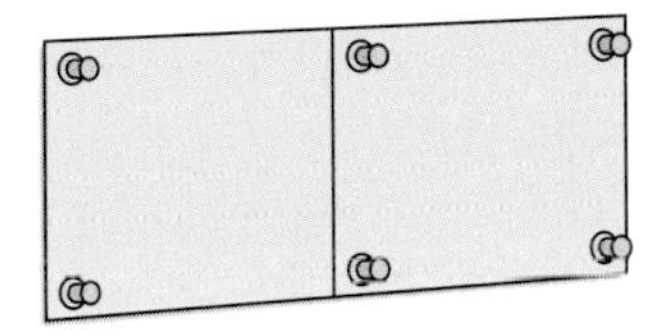

a) What is the minimum number of tacks Masao needs to hang 6 pictures of the same size in a row?
b) Write an expression for finding the number of tacks needed to hang any number of pictures in a row.

7. The graph shows the speed of a car for 10 min. Write a story to explain the graph.

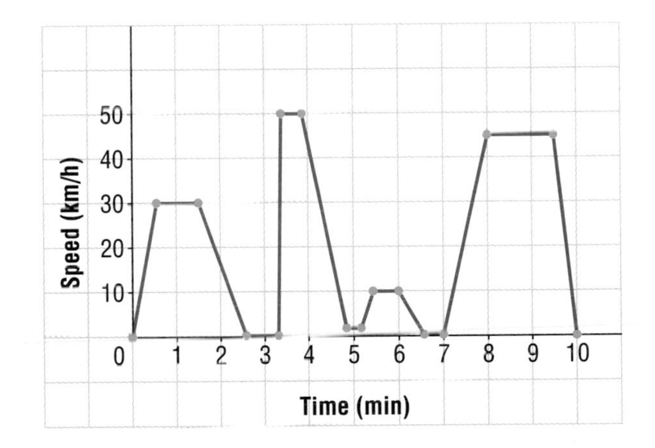

1. Ontario and Quebec are Canada's 2 largest provinces. What percent of Canada's total area is covered by these 2 provinces combined?

2. Use information from the Data Bank on pages 564 to 569 to write a problem. Have a classmate solve your problem.

Chapter 5

Evaluate for $x = 3$.

1. $5x - 4$ **2.** $3(x + 2)$
3. $3(x + 1) - (x - 3)$ **4.** $2(x + 1) + 3(x - 2)$

Expand and simplify.

5. $3(x + 1)$ **6.** $2(x - 5)$
7. $4(2x + 1)$ **8.** $5(3 - x)$
9. $-3(2x - 2)$ **10.** $-4(3 - 2x)$

Expand and simplify.

11. $4(2x - 3y + 5)$
12. $2(x - 3) - (x - 2)$
13. $-3(x - 2) + 2(x + 4)$
14. $3(x + 4) - 2(x + 5)$

Simplify.

15. $(3x^2 + 2x - 3) + (x^2 + x + 1)$
16. $(2y^2 - 3y + 1) + (3y^2 - y - 1)$
17. $(x^2 + x - 7) - (2x^2 + 3x + 2)$
18. $(2x^2 - x - 1) - (3x^2 - x + 2)$

Simplify.

19. $3x(2x - 3)$ **20.** $-2x(3 - x)$
21. $(3x)(5y)$ **22.** $(-6s)(2t)$

Simplify.

23. $\begin{array}{r} 3x^2 - 5x + 2 \\ \underline{+\ \ x^2 + 2x + 1} \end{array}$ **24.** $\begin{array}{r} 4x^2 + \ \ x - 2 \\ \underline{-\ -3x^2 - 4x + 4} \end{array}$

Divide.

25. $45x \div 9$ **26.** $36y \div 4$
27. $\dfrac{20xy}{4x}$ **28.** $\dfrac{-32xy}{8y}$
29. $\dfrac{4x - 8x^2 + 12x^3}{4x}$ **30.** $\dfrac{5x^3 - 15x^2 - 10x}{5x}$

Chapter 6

Solve.

1. $x - 2 = 10$ **2.** $y - 9 = 3$
3. $a - 3 = 3$ **4.** $b - 5 = 6$
5. $y + 3 = 4$ **6.** $x + 4 = 8$
7. $n + 7 = 10$ **8.** $t + 3 = 15$

Solve.

9. $7y = 21$ **10.** $3x = 12$
11. $4a = 24$ **12.** $6p = 6$
13. $\frac{x}{2} = 8$ **14.** $\frac{y}{3} = 5$
15. $\frac{a}{10} = 5$ **16.** $\frac{b}{9} = 5$

Solve.

17. $2x = 10 + 2$ **18.** $5x = 15 - 10$
19. $\frac{x}{2} - 3 = 5 + 2$ **20.** $\frac{x}{5} - 1 = 4 + 6$

Solve.

21. $6x + 12 = 3x$ **22.** $4x + 3x = 5x + 2$
23. $\frac{4x}{3} - 3 = \frac{x}{3} + 2$ **24.** $\frac{x}{6} - 2 = -\frac{x}{6} + 1$

Solve.

25. $2(x - 1) - 2 = 11 - 3x$
26. $2(x + 3) - (2x - 5) = 2(x + 1) + 1$

Solve.

27. $\dfrac{x - 1}{2} = \dfrac{x + 3}{3}$ **28.** $\dfrac{3n + 1}{2} - \dfrac{n}{3} = 4$

Solve each inequality and graph the solution on a number line.

29. $7x - 3 \geq 5x - 1$ **30.** $3x - 2 < 2x - 4$
31. $3x - 1 > 5x - 3$ **32.** $6x - 4 \leq 9x + 5$

Chapter 7

Replace the words with symbols.

1. A number is increased by four.

2. A number is divided in half.

3. A number is multiplied by ten.

4. A number is decreased by one-half.

Write an equation for each statement.

5. Three times a number is fifteen.

6. Five more than a number is six.

7. Two times a number is equal to the number increased by 10.

8. A number increased by 2 and then multiplied by 5 is 20.

Solve.

9. The sides of a triangle are 3 consecutive numbers. Its perimeter is 54 cm. How long is each side?

10. a) Copy and complete the table for the areas of 4 triangles, each with a fixed height of 10 cm.

Base (cm)	4	6	8	10
Area (cm)^2				

b) Write a formula for the areas of the triangles with a fixed height of 10 cm.
c) Use the formula to find the area of a triangle with a height of 10 cm and a base of 15 cm.

11. One painter paints a room in 1 h. A second paints it in 2 h. How long will they take to paint a room together?

12. An aircraft flew at 400 km/h to its destination and returned at 500 km/h. If the total trip took 9 h, how far did the aircraft fly?

13. The sum of 2 numbers is 26. Four times the first number plus twice the second is 70. What are the numbers?

Chapter 8

Find the GCF.

1. $42, 56$ **2.** $22, 55$

3. $10t^2, 2t$ **4.** $14xy, 35x^2y^2$

Factor.

5. $3x - 21$ **6.** $5x^2y - 15xy^2$

7. $18x^2 - 27x$ **8.** $2a - 8b - 10$

Expand.

9. $(x-3)(x+1)$ **10.** $(x-4)(x-5)$

11. $(2x-1)(3x-2)$ **12.** $(5x-2)(7x+9)$

Factor.

13. $x^2 - 7x - 8$ **14.** $a^2 + 10a - 11$

15. $b^2 - 2b + 1$ **16.** $y^2 - 10y - 56$

Factor fully.

17. $2x^2 - 20x + 50$ **18.** $3x^2 + 3x - 36$

19. $4a^2 - 4a - 80$ **20.** $5a^2 - 25a + 30$

21. $2y^2 - 20y + 42$ **22.** $3y^2 - 21y - 54$

Expand.

23. $(x+7)(x-7)$ **24.** $(3-t)(3+t)$

25. $(5p-1)(5p+1)$ **26.** $(6-9y)(6+9y)$

Factor.

27. $x^2 - 4$ **28.** $x^2 - 36$ **29.** $x^2 - 25$

30. $4p^2 - 49$ **31.** $9q^2 - 64$ **32.** $81 - 25x^2$

33. $8x^2 - 72$ **34.** $50x^2 - 8$ **35.** $16 - 36t^2$

Simplify.

36. $\frac{12x^5y^6}{4x^3y^2}$ **37.** $\frac{25x^2y}{5} \times \frac{2x^4y^2}{5x^3y^3}$

38. $\frac{2+x}{2} + \frac{5-x}{5}$ **39.** $\frac{x-3}{7} - \frac{7-x}{3}$

CHAPTER 9

Ratio, Rate, and Percent

The map shows Canada's national highway network. There are 24 500 km of highway.

How many hours would it take you to drive along all of Canada's national highway network at a speed of 80 km/h?

Could the population of Canada make a hand-to-hand chain 24 500 km long?

One highway accounts for almost one-third of the length of Canada's highway network. Name this highway. In which cities does it begin and end?

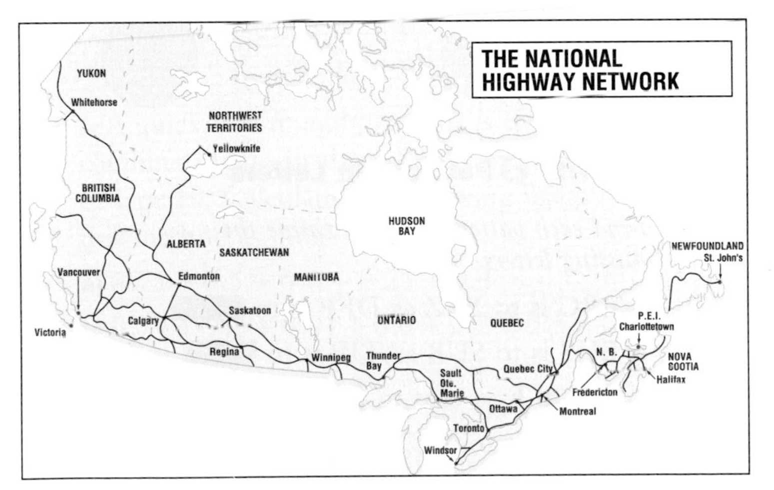

Ratios and Rates

A **ratio** is a fraction that compares 2 numbers with the same units. A **rate** compares 2 numbers with different units.

Activity ❶ Writing Ratios

We can write the ratio of pennies to dimes in words as "two to five", in ratio form as 2:5, or in fraction form as $\frac{2}{5}$.

1. Write the ratio of dimes to nickels in words.
2. Write the ratio of nickels to pennies in fraction form.
3. Write the ratio of nickels to dimes in ratio form.
4. Write the ratio of dimes to all the coins in fraction form.

Activity ❷ Ranking Baseball Pitchers

There are many ways to compare pitchers in baseball. The simplest way is to compare the numbers of games they win. Another way is to calculate the numbers of runners per inning they allow in a season. The smaller the number of runners a pitcher allows, the fewer runs the opposition can score. To calculate the runners per inning rate, add the number of hits and walks a pitcher gives up, then divide by the number of innings pitched. Round the quotient to the nearest hundredth.

The table gives the numbers of innings pitched and the numbers of hits and walks for some Toronto Blue Jays pitchers in the year their team first won the World Series.

Pitcher	Innings Pitched	Hits	Walks	$\frac{\text{Hits + Walks}}{\text{Innings Pitched}}$
Cone	53	39	29	
Henke	56	40	22	
Key	217	205	59	
Guzman	181	135	72	
Morris	241	222	80	
Stottlemyre	174	175	63	
Ward	101	76	39	
Wells	120	138	36	

1. Copy the table. Complete it by calculating each pitcher's runners per inning rate.
2. Rank the pitchers from best to worst rate.

Activity ❸ Ranking Baseball Hitters

One way to rank baseball hitters is to use the ratio of their number of hits to their number of at bats. This ratio is called the batting average.

The table shows the numbers of hits and at bats for some team members of the Toronto Blue Jays in the year they first won the World Series.

Player	At Bats	Hits	Ratio of Hits to At Bats (fraction form)	Ratio of Hits to At Bats (decimal form)
Alomar	571	177		
Bell	25	6		
Borders	480	116		
Carter	622	164		
Gruber	446	102		
Lee	396	104		
Olerud	458	130		
White	641	159		
Winfield	583	169		

1. Copy the table. Complete it by first writing the ratio of the hits to at bats in fraction form.

2. Write each ratio as a decimal to the nearest thousandth.

3. Rank the players from highest average to lowest.

Activity ❹ Other Uses of Ratios and Rates

1. List sports or other types of activities in which ratios and rates are used to compare teams or individuals.

2. Display the results of your research in a chart.

3. Why are sports fans so interested in statistics?

9.1 Ratios

Halifax is the capital of Nova Scotia. The word "Halifax" is made up of 4 consonants and 3 vowels.

A ratio is a comparison of 2 numbers with the same units. The ratio of consonants to vowels in the word "Halifax" is "four to three."

We can write a ratio in different ways. The ratio of consonants to vowels is written in ratio form as 4:3, in fraction form as $\frac{4}{3}$, or in words as "four to three."

The ratio 4:3 has 2 **terms**. The **first term** is 4, and the **second term** is 3.

Activity: Complete the Table

Copy the table. Complete it by finding the ratio of consonants to vowels for the capital city of each of Canada's other provinces and territories.

Ratio of Consonants to Vowels

City	In Words	In Ratio Form	In Fraction Form
Halifax	four to three	4:3	$\frac{4}{3}$
St. John's			
Quebec			

Inquire

1. List the capitals that have the same ratios of consonants to vowels. Explain your reasoning.

2. Which capital has the highest ratio of consonants to vowels?

A ratio is in **lowest terms** or **simplest form** when the greatest common factor of the terms is 1.

Example 1

Of the Members of Parliament (MPs) in Ottawa, 99 are elected from Ontario, and 11 are elected from Nova Scotia. Write the ratio of the number of MPs elected from Ontario to the number elected from Nova Scotia. Express the ratio in lowest terms.

Solution

The ratio of MPs elected from Ontario to MPs elected from Nova Scotia is 99:11.

$$
\begin{aligned}
99:11 &= \frac{99}{11} \\
&= \frac{99 \div 11}{11 \div 11} \\
&= \frac{9}{1}
\end{aligned}
$$

Divide the numerator and denominator by the greatest common factor.

To write a ratio of 2 quantities with different units, first make the units the same. For example, the ratio of 1 week to 1 day is not 1:1. The correct ratio is 7 days to 1 day or 7:1.

Example 2

Write each ratio in simplest form in 3 ways.

a) 1 min to 30 s **b)** 2 years to 15 months

Solution

The units of the quantities being compared must be the same.

a) 1 min = 60 s

$$
\begin{aligned}
60:30 &= \frac{60}{30} \\
&= \frac{60 \div 30}{30 \div 30} \\
&= \frac{2}{1}
\end{aligned}
$$

The ratio is $\frac{2}{1}$, or 2:1, or 2 to 1.

b) 1 year = 12 months

so 2 years = 24 months

$$
\begin{aligned}
24:15 &= \frac{24}{15} \\
&= \frac{24 \div 3}{15 \div 3} \\
&= \frac{8}{5}
\end{aligned}
$$

The greatest common factor is 3.

The ratio is $\frac{8}{5}$, or 8:5, or 8 to 5.

Practice

1. Use the diagram to write each ratio in 3 ways.

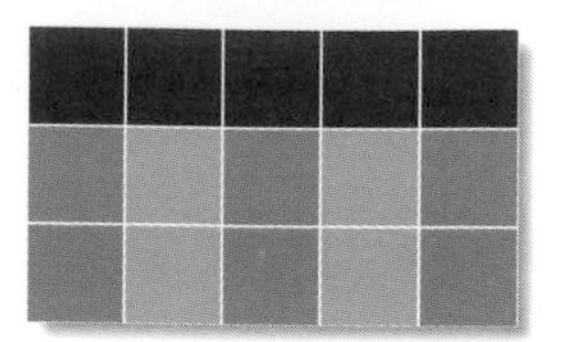

a) red squares to blue squares
b) red squares to green squares
c) green squares to blue squares
d) red squares to all squares
e) all squares to blue squares
f) blue squares to all squares
g) green squares to red squares

2. Draw four 5-by-5 squares on grid paper. Use a different diagram to illustrate each ratio.

a) 17:8 **b)** 3:2 **c)** 4:21 **d)** 16:9

3. Write each ratio in 3 ways.
a) 3 cups of sugar to 5 cups of fruit
b) 3 cans of water to 1 can of concentrate
c) 50 L of gasoline to 1 L of oil
d) 25 drops of paint to 4 drops of solvent
e) $2 saved to $3 spent

4. Write each ratio in simplest form.

a) $\frac{6}{15}$ **b)** 4:12 **c)** $\frac{8}{16}$
d) 3 to 18 **e)** 6:10 **f)** 14:7
g) 12 to 2 **h)** 8:18 **i)** $\frac{24}{10}$
j) 18:4 **k)** 10:25 **l)** 56:28

5. Write as a ratio in simplest form.

a) 1 min to 45 s **b)** 2 kg to 250 g
c) 73 days to 1 year **d)** 1.5 L to 750 mL
e) 35¢ to $1.05 **f)** 1 m to 175 cm

Problems and Applications

6. One year, 78% of Canadian households owned cars and 24% owned air conditioners. Write the ratio of households with cars to households with air conditioners as a ratio in lowest terms.

7. For the name of each country, express the ratio of the number of consonants to the number of vowels in lowest terms.

a) Canada **b)** Australia
c) France **d)** Germany
e) Iceland **f)** Bolivia
g) Afghanistan **h)** Mozambique
i) Cuba **j)** Madagascar

8. A food guide reported that half a cup of orange juice contained 49 mg of Vitamin C and half a cup of grapefruit juice contained 42 mg of Vitamin C. Write the ratio of the mass of Vitamin C in orange juice to the mass of Vitamin C in grapefruit juice. Express your answer in simplest form.

9. Naji took 15 pictures on a 36-exposure roll of film. Express each ratio in lowest terms.
a) the number of pictures he took to the number of exposures remaining
b) the total number of exposures on the roll to the number of pictures he took
c) the number of exposures remaining to the total number of exposures on the roll

10. The power used by a light bulb is indicated by the power rating in watts (W). Express the following power ratings as ratios in lowest terms.
a) 100 W to 60 W **b)** 40 W to 60 W
c) 40 W to 150 W **d)** 150 W to 100 W

11. During a volleyball game, one team scored its points as follows.

4 points from serves
3 points from spikes
6 points from rallies

Write a ratio in simplest form to compare
a) points from serves to points from spikes
b) points from spikes to points from rallies
c) points from rallies to points from serves

12. The same ballpoint pen cost 90¢ in one store and $1.20 in another store. Write the ratio of the lower cost to the higher cost in lowest terms.

13. One term of a two-term ratio is a prime number. What can you state about the other term if it is possible to write the ratio in simpler form?

14. a) Without looking at a newspaper, estimate as a group the ratio of advertising space to non-advertising space in a local newspaper.
b) Get a copy of the local paper and determine the ratio of advertising space to non-advertising space. You need to estimate and add fractions of pages to get this ratio.
c) Compare your solution and your strategy with those of other groups.
d) Is the ratio different for a national newspaper?
e) Is the ratio different for a magazine?

15. Estimate, then determine the ratio of "DJ talk minutes" to "music minutes" on your favourite radio station.

16. Estimate, then determine the ratio of "advertising minutes" to "non-advertising minutes" for a 1-h TV show of your choice.

17. a) Estimate the ratio of the height to the width of a TV screen. Is the ratio constant for all TV screens? Compare your findings with a classmate's.
b) Repeat part a) for a door instead of a TV screen.

18. Research these ratios for your school. Write them in lowest terms.
a) grade 9 girls to grade 9 boys
b) grade 9 students to all students
c) all students to all teachers

19. The Buffon needle experiment was performed by Count Buffon in the eighteenth century. To repeat the experiment, draw parallel lines on a piece of paper so that the distance between adjacent lines equals the length of a toothpick.
a) Drop 10 toothpicks onto the paper and count the toothpicks that touch the lines. Repeat this step 9 more times. If your results agree with Count Buffon's, you will find that

$$\frac{2 \times \text{number dropped}}{\text{number touching lines}} \doteq \pi$$

How close is your result to the value of π?
b) Combine your data with your classmates'. Compare the result with the value of π. Is the result from the combined data closer to π than the result you obtained in part a)?

LOGIC POWER

In the 1880s, some companies put their advertising on puzzle cards to get people's attention. The following farmer's puzzle was very popular. Copy the diagram into your notebook. Draw 3 paths, one from the house to gate A, another from the barn to gate C, and the third from the well to gate B. Make sure that none of the paths cross and that all the paths stay inside the fence.

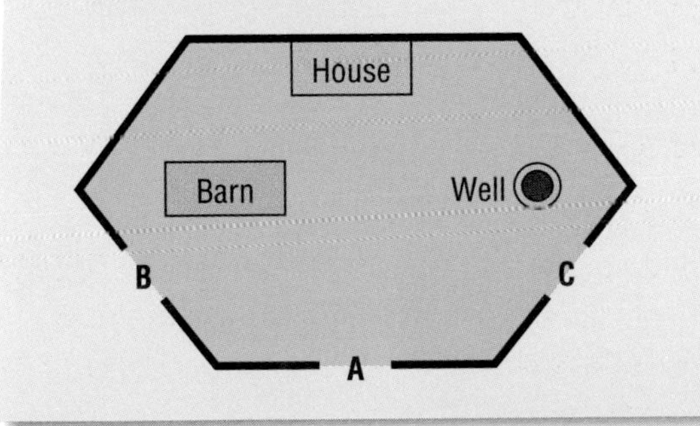

The Golden Ratio

Activity 1 The Golden Rectangle

The golden rectangle was discovered by the Greeks in the fifth century B.C. Many people find the shape of the rectangle very appealing.

To construct a golden rectangle, follow these steps.

1. Construct square ABCD of side length 4 cm.

2. Find the midpoint of BC and name it E.

3. With centre E and radius ED, draw an arc to intersect the extension of BC at F.

4. Construct a perpendicular to BF at F to meet the extension of AD at G. The rectangle ABFG is a golden rectangle.

5. Measure AG and AB.

6. Use your calculator to divide AG by AB. Write the ratio of AG to AB with the second term equal to 1. The result is an approximation of the golden ratio.

7. The golden rectangle often appears in nature. Verify that the rectangles shown are golden rectangles.

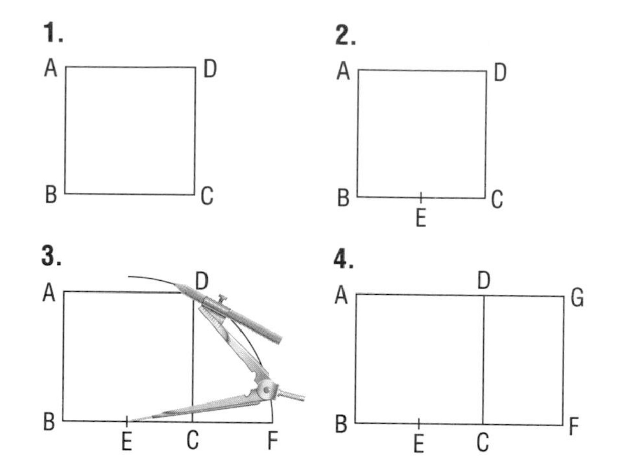

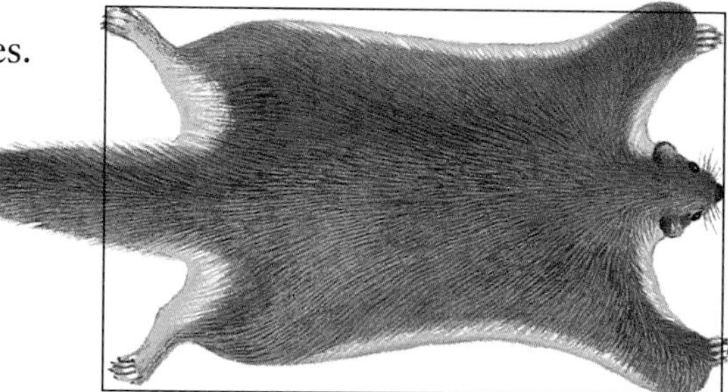

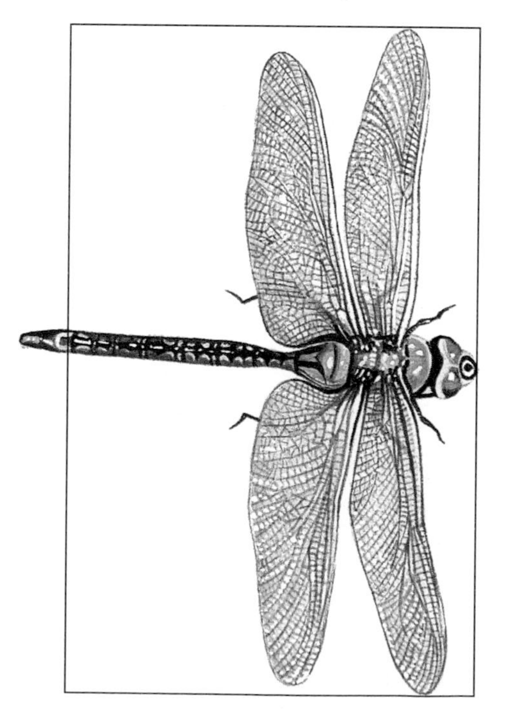

Activity ❷ The Golden Ratio and You

Some people claim that the golden ratio is found in many measurements of the human body. Find the following ratios for each member of your group. Express each ratio with a second term of 1. Are any of the ratios close to the golden ratio?

a) the distance from your shoulder to your fingertips to the distance from your elbow to your fingertips

b) the length of your head to the distance from the top of your ears to your chin

c) the length of your index finger to the length from the second joint to the fingertip

Activity ❸ The Golden Ratio and the Fibonacci Sequence

The golden ratio is about 1.618.
The Fibonacci sequence is
1, 1, 2, 3, 5, 8, 13, 21, 34, 55, ...

1. Use your calculator to find the values of the ratios of successive terms in the sequence.

$\frac{1}{1} = \blacksquare$ $\frac{2}{1} = \blacksquare$ $\frac{3}{2} = \blacksquare$ $\frac{5}{3} = \blacksquare$

$\frac{8}{5} = \blacksquare$ $\frac{13}{8} = \blacksquare$ $\frac{21}{13} = \blacksquare$ $\frac{34}{21} = \blacksquare$

2. What do the ratios seem to do with respect to the golden ratio?

3. Take any 2 numbers, like 4 and 7, and construct a sequence in the same way as the Fibonacci sequence is made.

4, 7, 11, 18, ...

4. Use your calculator to find the ratios of successsive terms in the sequence. Compare your results with those in question 2.

Activity ❹ Researching the Golden Ratio

Use your research skills to find examples of the golden ratio in architecture and nature.

9.2 Equivalent Ratios and Proportions

Painter is to *Brush* as *Bowler* is to *Ball.* The above statement is an analogy. The word "analogy" comes from the Greek word *analogous*, which means "in due ratio." Analogy questions are very common on aptitude tests. These questions ask you to find a relationship between 2 words, then find a similar relationship in a different pair of words. The analogy above is usually written as Painter:Brush as Bowler:Ball. The colon symbol, :, means "is to."

Activity: Complete the Analogies

Copy and complete the analogies. Compare your solutions with a classmate's.

a) Teacher:Classroom as Mechanic: ▇
b) Lukewarm:Boiling as Flurry: ▇
c) Knife:Bread as Pen: ▇
d) Letter:Alphabet as Finger: ▇
e) Speak:Mumble as Walk: ▇
f) Hug:Affection as Wince: ▇
g) Acrobat:Cartwheel as Magician: ▇

Inquire

1. Write 3 more word analogies and ask a classmate to solve them.

2. Copy and complete these number analogies.

a) 1:2 as 2:▇ **b)** 4:12 as 5:▇ **c)** 10:5 as 16:▇

3. How do word analogies and number analogies compare?

Such ratios as 1:2, 2:4, 3:6, and 4:8 make the same comparison. They are known as **equivalent ratios** or **equal ratios**.
The table shows the number of cans of water and concentrated lime juice needed to make limeade.

Cans of Water	4	8	12	16	20
Cans of Concentrate	1	2	3	4	5

The ratios of cans of water to cans of concentrate are 4:1, 8:2, 12:3, 16:4, and 20:5, or
$\frac{4}{1}, \frac{8}{2}, \frac{12}{3}, \frac{16}{4}$, and $\frac{20}{5}$.
These are all examples of equivalent ratios.

We can find equivalent ratios by dividing or multiplying by the same whole number, except 0.

Example 1

Find an equivalent ratio.

a) 3:4
b) 18:12

Solution

a) $3{:}4 = \frac{3}{4}$

$= \frac{3 \times 2}{4 \times 2}$

$= \frac{6}{8}$

b) $18{:}12 = \frac{18}{12}$

$= \frac{18 \div 3}{12 \div 3}$

$= \frac{6}{4}$

Example 2

Are the ratios in each pair equivalent?

a) 12:8, 3:2
b) 3:5, 50:75

Solution

a) $\frac{12}{8} = \frac{12 \div 4}{8 \div 4}$

$= \frac{3}{2}$

$\frac{3}{2} = \frac{3}{2}$

So, 12:8 and 3:2 are equivalent ratios.

b) $\frac{3}{5} = \frac{3 \times 15}{5 \times 15}$

$= \frac{45}{75}$

$\frac{50}{75} \neq \frac{45}{75}$

So, 3:5 and 50:75 are not equivalent ratios.

A statement that ratios are equal is called a **proportion**.

$\frac{1}{3} = \frac{2}{6}$ or $1{:}3 = 2{:}6$ are 2 ways of writing the same proportion.

Both forms are read as "one is to three as two is to six."

Notice that, if $\frac{1}{3} = \frac{2}{6}$, then $1 \times 6 = 3 \times 2$.

In general, if $\frac{a}{b} = \frac{c}{d}$ Multiply both sides by *bd*.

$$(bd) \times \frac{a}{b} = (bd) \times \frac{c}{d}$$

and $ad = bc$

So, if $\frac{a}{b} = \frac{c}{d}$, then $ad = bc$, $b \neq 0$, $d \neq 0$.

This is known as the **cross-product rule**.

Example 3

There was a fantastic archeological find in Shensi Province in China in 1974. A life-sized army made of pottery was found buried near the tomb of the Emperor Ch'in Shih Huang Ti, who died about 210 B.C. The tomb and the army were built by about 1 000 000 workers. To estimate the number of soldiers in the army, archeologists excavated a 10-m strip of the 200 m long rectangle. They found 270 of the pottery soldiers. How many soldiers did they estimate to be in the 200-m long rectangle?

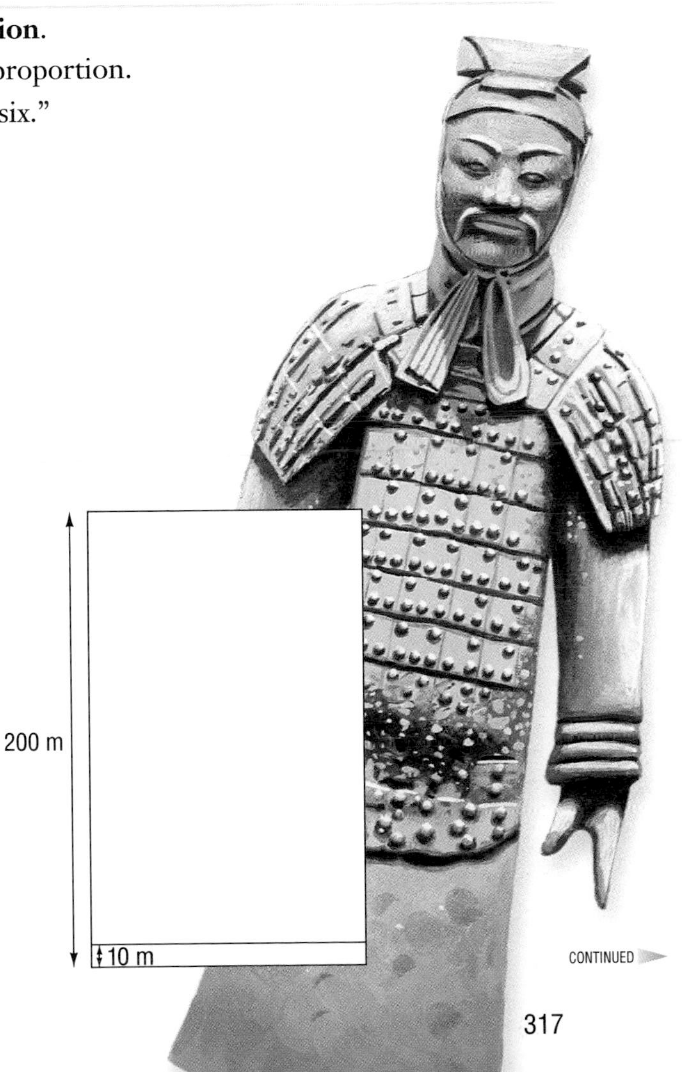

CONTINUED

Solution

Estimate the number of soldiers by solving a proportion.
Let x represent the total number of soldiers.

$$\frac{\text{Total number of soldiers}}{\text{Length of rectangle}} = \frac{\text{Number of soldiers found}}{\text{Number of metres excavated}}$$

So $$\frac{x}{200} = \frac{270}{10}$$

Method 1

Use equivalent ratios.

$$\frac{x}{200} = \frac{270 \times 20}{10 \times 20}$$
$$\frac{x}{200} = \frac{5400}{200}$$
$$x = 5400$$

Method 2

Multiply both sides by 200.

$$200 \times \frac{x}{200} = 200 \times \frac{270}{10}$$
$$x = \frac{200 \times 270}{10}$$
$$= \frac{54\ 000}{10}$$
$$= 5400$$

Method 3

Use the cross-product rule.

$$\frac{x}{200} = \frac{270}{10}$$
$$10x = 200 \times 270$$
$$10x = 54\ 000$$
$$x = 5400$$

EST $300 \times 200 = 60\ 000$
$60\ 000 \div 10 = 6000$

About 5400 soldiers were in the rectangle.

Example 4

Find the unknown terms.

a) $30{:}24 = x{:}8$

b) $\frac{4}{1.5} = \frac{10}{n}$

Solution

a) Multiply both sides by 8.

$$\frac{30}{24} = \frac{x}{8}$$
$$8 \times \frac{30}{24} = 8 \times \frac{x}{8}$$
$$\frac{240}{24} = x$$
$$10 = x$$

b) Use the cross-product rule.

$$\frac{4}{1.5} = \frac{10}{n}$$
$$4n = 10 \times 1.5$$
$$4n = 15 \quad \text{Divide both sides by 4.}$$
$$\frac{4n}{4} = \frac{15}{4}$$
$$n = 3.75$$

EST $16 \div 4 = 4$

To solve some problems with proportions, consider the number of equal parts a ratio represents.

Example 5

To break in a new motor bike, the recommended ratio of gas to oil is 25:1, by volume. How much gas and how much oil are needed to fill a 13-L tank on a new motor bike?

Solution

The total number of parts in the mixture is 25 + 1 or 26.
Gas makes up $\frac{25}{26}$ of the mixture. Oil makes up $\frac{1}{26}$ of the mixture.
The mixture has a volume of 13 L.
Let the volume of gas in the mixture be x.

$$\frac{x}{13} = \frac{25}{26}$$
$$26x = 25 \times 13$$
$$26x = 325 \quad \text{Divide both sides by 26.}$$
$$x = 325 \div 26$$
$$= 12.5$$
$$13 - 12.5 = 0.5$$

EST $30 \times 10 = 300$
$300 \div 30 = 10$

So, 12.5 L of gas and 0.5 L of oil are needed.

Practice

Write an equivalent ratio.

1. 6:5 **2.** 12:1 **3.** 1:5

4. 25:40 **5.** $\frac{2}{8}$ **6.** $\frac{18}{10}$

7. 9:6 **8.** $\frac{16}{2}$ **9.** $\frac{4}{7}$

State whether the ratios in each pair are equivalent.

10. $\frac{2}{3}, \frac{4}{6}$ **11.** $\frac{4}{5}, \frac{8}{10}$

12. $\frac{5}{15}, \frac{1}{4}$ **13.** $\frac{6}{15}, \frac{8}{10}$

14. 9:5, 18:15 **15.** 5:10, 6:12

16. 8:6, 4:3 **17.** 16:12, 12:9

Solve for x.

18. $\frac{x}{15} = \frac{3}{5}$ **19.** $\frac{5}{1} = \frac{x}{3}$

20. $\frac{3}{x} = \frac{18}{24}$ **21.** $\frac{4}{3} = \frac{8}{x}$

22. x:2 = 10:5 **23.** 6:x = 10:25

24. 8:x = 6:3 **25.** 5:3 = 45:x

Solve for x.

26. 2.5:x = 2:7 **27.** 10:6.5 = x:2.6

28. $\frac{4.9}{1.75} = \frac{3.36}{x}$ **29.** $\frac{x}{12.5} = \frac{0.8}{4}$

Problems and Applications

30. Neptune orbits the sun 3 times in the time it takes Pluto to orbit the sun twice.
a) How many orbits does Neptune complete while Pluto orbits 8 times?
b) How many orbits does Pluto complete while Neptune orbits 15 times?

31. The Sam Kee Building in Vancouver is the narrowest commercial building in the world. The ratio of the building's length to its width is 50:3. The building is 30 m long. How wide is it?

32. A camera costs $175. You and a friend share the cost in the ratio 2:3. What is your share?

33. To make liquid skim milk from powder, you use powder and water in the ratio 1:3, by volume. How much powder and how much water do you need to make 1 L of liquid skim milk? Give your answers in millilitres.

34. To make an environmentally-safe furniture polish, you can combine lemon oil and mineral oil in the ratio 3:200, by volume. How much lemon oil do you need to make 10.15 L of polish?

35. The dimensions of a rectangle are in the ratio 5:3. If the longer side is 18 cm, what is the shorter side?

36. Two squares have side lengths in the ratio 3:2.
a) What is the ratio of the areas of the squares?
b) If the smaller square has an area of 24 cm^2, what is the area of the larger square?

37. Two squares have areas in the ratio 49:25.
a) What is the ratio of the side lengths of the squares?
b) If the larger square has a side length of 17.5 m, what is the side length of the smaller square?

38. Bronze is an alloy that contains the metals copper and tin in the ratio 23:2, by mass.
a) What mass of each metal is present in 100 g of the alloy?
b) How many grams of copper are there for 100 g of tin?
c) How many grams of tin are there for 100 g of copper? Write your answer to the nearest tenth of a gram.

39. Work with a partner to try to make 2 equivalent ratios from 4 different prime numbers. Describe and explain your findings.

9.3 Rates

In May of 1992, Canadian rower Silken Laumann broke her right leg in a rowing accident. Doctors told her that she would not row for at least 6 months. Ten weeks after her accident, Silken won a bronze medal in the women's single sculls at the Olympic Games in Barcelona. The race was 2 km long, and Silken rowed at an average rate, or average speed, of 16 km/h.

A **rate** is a comparison of 2 numbers expressed in different units. A rate is usually written as a **unit rate**, in which the second term is 1. Silken's average speed is an example of a unit rate. It could be written as 16 km:1 h.

Activity: Solve the Problem

A Greek cycling champion named Kanellos Kanellopoulos set a world distance record for human-powered flight when he repeated the mythical flight of Daedalus. Kanellos pedalled an ultralight aircraft 5 m above the sea for 120 km, from the island of Crete to the island of Santorini. The trip took about 3 h. We want to know the rate at which he flew.

Inquire

1. Write a rate that compares the distance Kanellos flew in kilometres to the time his flight took in hours.

2. Express his speed as a unit rate in kilometres per hour.

3. a) Who travelled faster, Silken or Kanellos?
b) How many times faster?

Example 1

Tania earned $43.50 for working for 6 h at a supermarket checkout. What was her rate of pay?

Solution

$$\text{Rate of pay} = \frac{\text{amount earned}}{\text{time worked}} = \frac{\$43.50}{6\text{ h}} = \$7.25/\text{h}$$

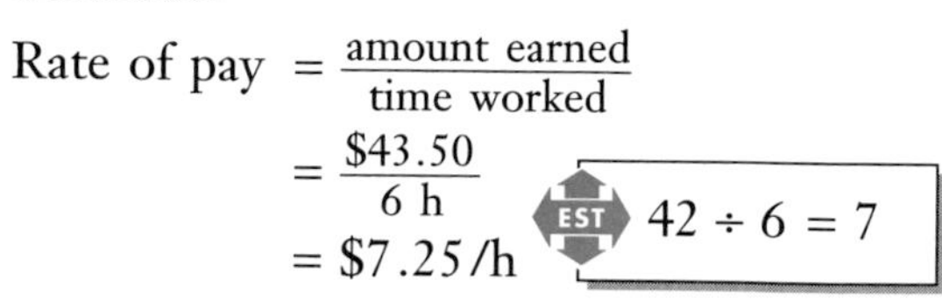

Example 2

A printer can print 213 characters/s.
How many characters can it print in 5 min?

Solution

Let the number of characters printed in 5 min be x.
Write a proportion, making sure that the units agree.
The rate is $\frac{213 \text{ characters}}{1 \text{ s}}$.

$$\frac{213}{1} = \frac{x}{300}$$

5 min = 300 s

$$\frac{213 \times 300}{1 \times 300} = \frac{x}{300}$$

EST $200 \times 300 = 60\ 000$

$$\frac{63\ 900}{300} = \frac{x}{300}$$

$$63\ 900 = x$$

The printer can print 63 900 characters in 5 min.

How could you work backward to check the answer?

Recall the steps

Understand the Problem
Think of a Plan
Carry Out the Plan
Look Back

Practice

Write as a unit rate.

1. 48 hot dogs for 16 people

2. 120 words typed in 3 min

3. driving 240 km in 3 h

4. \$22.80 for 3 h of work

5. \$9.48 for 3 cans of juice

6. \$126 for a 6-day car rental

7. 432 heartbeats in 6 min

8. jogging 1.8 km in 12 min

Express the unit rate in the units shown in brackets.

9. 66 h of sunshine in 12 days (h/day)

10. \$98 tax on 14 barrels of oil (\$/barrel)

11. \$66 for 16.5 m of fabric (\$/m)

12. 120 km:1 h (km/min)

13. 9 km in 6 h (m/min)

Problems and Applications

14. The wing beat of a tiny insect is the fastest action of any animal. The common midge can beat its wings at up to about 132 000 times per minute. Express this rate in beats per second.

15. Copy and complete the table to show the earnings of 4 newspaper carriers.

Carrier	Wages for 4 Weeks	Wages Per Week
Bob	\$39.00	■
Terry	■	\$19.50
Susan	■	\$18.75
Yurko	\$59.40	■

16. Paolo earned \$300/week by working 5 days/week and 8 h/day. What was his hourly rate of pay?

17. The space shuttle *Discovery* carried Dr. Roberta Bondar to an altitude of 300 km in about 8.5 min after liftoff. At what speed did the shuttle climb

a) to the nearest kilometre per minute?

b) to the nearest hundred kilometres per hour?

CONTINUED

18. a) Write a pay rate of \$10/h as a ratio with units hours per dollar.
b) How are the 2 ways of writing the ratio related?
c) What is the product of the 2 ways of writing the ratio?

19. The table shows the buffalo populations of 4 national parks and the areas of the parks in hectares.

National Park	Place	Area (ha)	Number of Buffalo
Elk Island	Alberta	19 400	750
Wind Cave	S. Dakota	11 450	325
Wood Buffalo	Alberta – NWT	4 480 400	3375
Yellowstone	Idaho – Montana – Wyoming	898 300	2850

a) Find the area per buffalo in each park to the nearest tenth of a hectare.
b) Rank the parks from greatest to smallest area per buffalo.

20. An advertisement stated that 4 L of paint would cover 40 m^2. How much paint is needed to paint 70 m^2?

21. The Earth spins 360° about its axis in 24 h. How many degrees does it spin in 10 h?

22. Eating 2 fresh peaches gives you about 294 kJ of energy. How much energy would 3 fresh peaches give you?

23. The world's most accurate mechanical clock is in Copenhagen Town Hall in Denmark. The clock will lose or gain no more than 0.5 s in 300 years. How long would the clock take to lose or gain 1 min? What assumptions have you made?

24. Fuel consumption for cars with 5-speed transmissions is reported in litres per 100 km (L/100 km) by Transport Canada. The smaller the value, the more fuel efficient the car is.
a) Copy the table. Complete it by calculating the fuel consumption of each car to the nearest tenth of a litre per hundred kilometres.

Car	Distance Travelled (km)	Fuel Used (L)	Fuel Consumption (L/100 km)
A	500	26.5	
B	700	40.5	
C	600	33	
D	1000	74	
E	1500	99	

b) Rank the cars from highest to lowest fuel consumption.
c) Another way to determine fuel efficiency is to find the number of kilometres travelled per litre of fuel (km/L). Convert the fuel consumption in L/100 km to the units km/L. Round each answer to the nearest tenth of a kilometre per litre.
d) Research the fuel consumptions of cars. List 5 cars that could be called "gas guzzlers" and 5 cars you regard as fuel efficient. Compare your lists with your classmates'.

25. a) Measure your heart beat for 10 s. At this rate, how many times does your heart beat in 1 min? 1 day? 1 year? Compare your results with a classmate's.
b) List some factors you must take into account when comparing heart rates.
c) Research how doctors and nurses measure heart rates.

WORD POWER

A certain vowel appears in the names of 8 of the 10 Canadian provinces. Which vowel?

Computers and Connectivity

Connectivity means connecting a computer by telephone lines or other methods to other computers or other sources of information. The connected system can be as simple as the computer at your desk and a regular telephone. A simple system for a car is a laptop computer and a cellular telephone. Both systems allow you to send and receive information. They give you access to many resources.

Activity ❶ Electronic Bulletin Boards

Like the public bulletin boards you sometimes see in grocery stores, electronic bulletin boards are usually open to everyone. There are bulletin boards for almost every subject.

1. How would sports cards collectors use an electronic bulletin board?

2. How would mystery novelists use one?

3. What kind of service could a newspaper provide on an electronic bulletin board?

4. One of the biggest electronic bulletin boards is called CompuServe. What kind of service does it supply? How much does it cost to use the service?

Activity ❷ Commercial Services

Some businesses offer services for computer users. One type of service is travel reservations. You can get information on airline schedules and fares, then order your tickets and charge the cost to a credit card.

1. Describe how home banking would work on a computer.

2. Describe how teleshopping would work.

3. What other commercial services do you think would be useful?

Activity ❸ Electronic Mail

Electronic mail is also called E-mail. It is used within companies, so that employees can send messages to each other. To send a message, you type it into your computer and send the message to the receiver's electronic mailbox. You can also put the message in several mailboxes at once, or broadcast it to everyone on the E-mail system. To open your own mailbox, you type your password into your computer and "open" your mail.

1. What advantages does E-mail have over a telephone?

2. Why do E-mail users have passwords?

3. Would E-mail be useful in homes?

9.4 Making Comparisons: Rates and Unit Pricing

We often compare rates. For example, we might want to know which of 2 speeds is faster.

Activity: Solve the Problem

Canada's Mark McKoy won the 110-m hurdles at the 1992 Olympic Games in a time of 13.12 s. American Kevin Young won the 400-m hurdles in 46.78 s. We want to know which athlete ran faster.

Inquire

1. For each gold medal performance, write a rate with distance in metres over time in seconds.

2. Calculate the unit rate for each performance to the nearest hundredth of a metre per second.

3. Which athlete ran faster?

4. Why do we write rates as unit rates before we compare them?

To compare prices of various goods that come in different sizes, we use unit rates, known as **unit prices**.

Example 1

A 200-g bag of popcorn costs \$1.00. A 750-g bag of popcorn costs \$2.70. Which size is the better value?

Solution

Let the unit price be the cost per gram or $\frac{\text{cost}}{1\text{ g}}$.

Let the $\frac{\text{cost}}{1\text{ g}}$ be $\frac{x}{1}$. Convert dollars to cents.

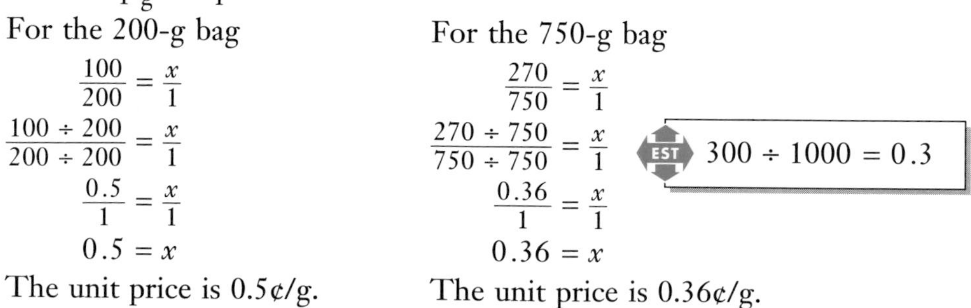

For the 200-g bag

$$\frac{100}{200} = \frac{x}{1}$$

$$\frac{100 \div 200}{200 \div 200} = \frac{x}{1}$$

$$\frac{0.5}{1} = \frac{x}{1}$$

$$0.5 = x$$

The unit price is 0.5¢/g.

For the 750-g bag

$$\frac{270}{750} = \frac{x}{1}$$

$$\frac{270 \div 750}{750 \div 750} = \frac{x}{1}$$

$$\frac{0.36}{1} = \frac{x}{1}$$

$$0.36 = x$$

The unit price is 0.36¢/g.

EST $300 \div 1000 = 0.3$

Since $0.36 < 0.5$, the 750-g size is the better value.

Since the price per unit of mass or volume is often very small, we may express the unit price as the price per 100 units.

Example 2

The price of a 1-L bottle of shampoo was \$4.20, while a 750-mL bottle cost \$3.00.

a) Express each unit price as the cost per 100 mL.

b) Which size is the better value?

Solution

Let the $\frac{\text{cost}}{100 \text{ mL}}$ be $\frac{x}{100}$.

Recall that 1 L = 1000 mL.

For the 1-L bottle

$$\frac{4.20}{1000} = \frac{x}{100}$$

$$4.20 \times 100 = 1000x$$

$$\frac{420}{1000} = \frac{1000x}{1000}$$

$$0.42 = x$$

The unit price is \$0.42/100 mL.

For the 750-mL bottle

$$\frac{3.00}{750} = \frac{x}{100}$$

$$3.00 \times 100 = 750x$$

$$\frac{300}{750} = \frac{750x}{750}$$

$$0.4 = x$$

EST $300 \div 1000 = 0.3$

The unit price is \$0.4/100 mL.

Since \$0.4/100 mL < \$0.42/100 mL, the 750-mL bottle is the better value.

SHAMPOO

750 mL

Practice

Express as a unit rate.

1. 68 students for 2 buses

2. 120 hamburgers for 60 people

3. driving 165 km in 3 h

4. 750 mL of juice for 6 people

5. skiing 14 km in 4 h

6. \$35 for 4 h of work

7. 900 g of batter for 4 cakes

Find the unit price. Round to the nearest tenth of a cent if necessary.

8. \$15.30 for 3 small pizzas

9. \$4.20 for 30 coloured markers

10. \$1.29 for 10 pens

11. \$32.00 for 25 party hats

12. \$5.00 for 6 sandwiches

13. \$0.70 for 40 g of potato chips

14. \$42.00 for 2.5 h of sailing

Problems and Applications

15. Nicole earned \$46.00 for 8 h of babysitting. Leon earned \$67.20 for 12 h of babysitting. Who had the higher rate of pay and by how much?

16. In Halifax, Nova Scotia, there are about 320 000 people in an area of 80 km^2. In Kingston, Ontario, there are about 60 000 people in an area of 30 km^2. In which city are there more people per square kilometre?

17. One year, British Columbia had 6149 police officers and a total population of 3 282 000. Nova Scotia had 1542 police officers and a total population of 900 000. Which province had the greater number of people for each police officer?

18. Each of your fingernails grows at about 0.05 cm/week. Each of your toenails grows at about 0.65 cm/year. Do your toenails or your fingernails grow faster?

CONTINUED

19. Great Britain's Linford Christie won the men's 100-m run at the 1992 Summer Olympics in 9.96 s. The Canadian women's team won the 3000-m speed skating relay at the 1992 Winter Olympics in 4 min 36.62 s. Who travelled faster, the sprinter or the speed skaters?

20. The density of a material is defined as its mass per unit volume. A 15-cm^3 lump of uranium has a mass of 285.75 g. A 25-cm^3 lump of gold has a mass of 472 g.

a) Express the density of each of these metals in grams per cubic centimetre.

b) State which metal has the higher density.

21. Decide which is the better value.

a) $25.00 for concert tickets for 2 people or $36.00 for concert tickets for 3 people

b) $340.00 for a bus for 35 people or $432.00 for a bus for 40 people

c) $4.30 for 4 L of milk or $2.40 for 2 L of milk

d) 28 g of mixed nuts for $0.98 or 35 g of mixed nuts for $1.40

e) ski rentals at $72.00 for 5 h or $52.50 for 3 h

f) 1.5 L of juice for $1.65 or 2.5 L of juice for $2.25

g) 500 sheets of computer paper for $9.50 or 2000 sheets of computer paper for $36.00

h) 18 L of gas for $9.36 or 8 L of gas for $4.08

22. Calculate the rate per 100 units to find the better value.

a) 450 g of peanuts for $1.80 or 250 g of peanuts for $0.75

b) 700 g of cheese for $4.20 or 0.5 kg of cheese for $3.50

c) 90 mL of hand lotion for $4.05 or 0.6 L of hand lotion for $21.00

d) 50 drinking straws for $0.69 or 110 drinking straws for $1.49

23. Loose-leaf paper costs $1.49 for 200 sheets or $3.49 for 500 sheets. Find

a) the least you can pay for 1000 sheets.

b) the least you can pay for 1600 sheets.

24. List reasons why the unit price is often lower for a larger size of a product than for a smaller size of the same product.

25. With a classmate, list the advantages and disadvantages of buying 1 L of milk for $2.00 or 3 L of milk for $5.75.

26. Collect store flyers that advertise food products. With a classmate, compare prices of 10 products in different stores. When product sizes differ, calculate the best value.

27. Write a problem that involves 2 unit rates. Have a classmate solve your problem.

PATTERN POWER

a) Find these differences in the squares of consecutive whole numbers.

$1^2 - 0^2 = ■$

$2^2 - 1^2 = ■$

$3^2 - 2^2 = ■$

$4^2 - 3^2 = ■$

$5^2 - 4^2 = ■$

b) Describe the pattern.

c) Use the pattern to evaluate mentally $245^2 - 244^2$.

9.5 Scale Drawings

Models are often used to represent real objects. Many models are not the same size as the objects they represent. These models are called **scale models**. A **scale** is a ratio of the size of a model to the size of the object it represents. Scales are usually written in simplest terms.

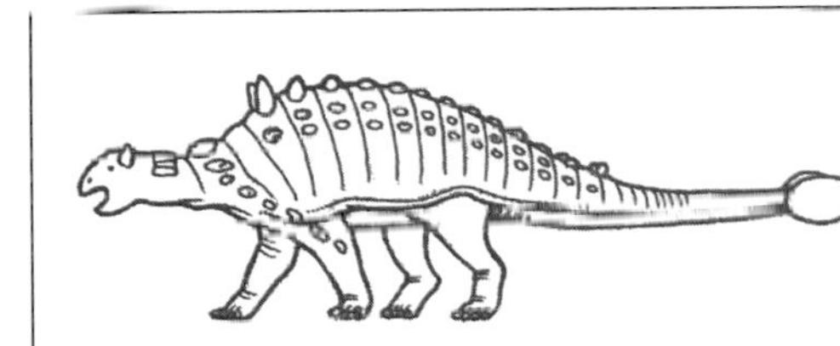

Scale 1 cm represents 1 m

Activity: Interpret the Model

The dinosaur Euoplocephalus once roamed Alberta. Above is a scale drawing of the dinosaur.

Inquire

1. Measure the height and length of the scale drawing.

2. Determine the actual height and length of the dinosaur.

3. List situations in which you use scale models to represent real objects. Compare your list with a classmate's.

4. Describe the advantages and disadvantages of using scale models.

We can write the scale of the above model in other ways.

Scale 1 cm to 1 m

Scale 1 cm:1 m

Scale 1:100

When a scale is written using only numbers, the units are the same for measures in the model and measures of the actual object.

Example 1

What is the height of the actual fingerprint?
Scale 2:1

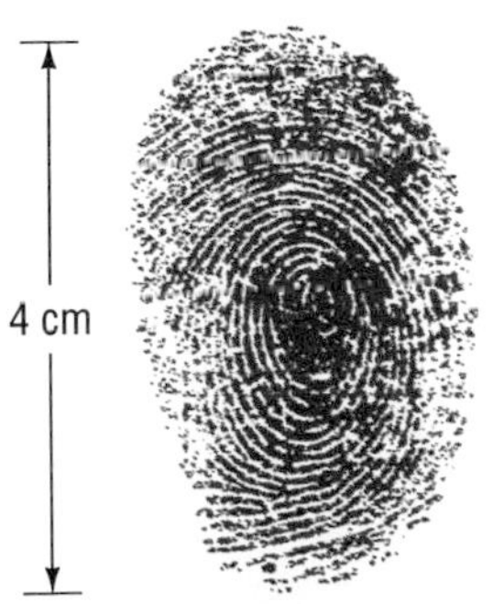

Solution

The height of the diagram is 4 cm. Let the actual height of the fingerprint be x.

So, $\frac{2}{1} = \frac{4}{x}$

$2x = 4$

$x = 2$

The actual height of the fingerprint is 2 cm.

CONTINUED

Example 2

Use the floor plan to determine

a) the actual length of the living room.

b) the length of the line needed to represent a 2-m closet wall.

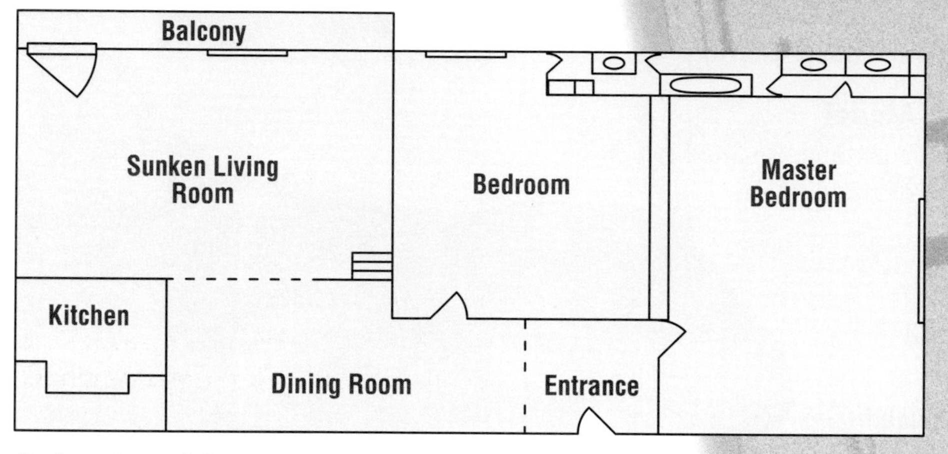

Scale 1 cm:0.8 m

Solution

a) The length of the model of the living room is 5 cm. Let the actual length of the living room be x.

$$\frac{1}{0.8} = \frac{5}{x}$$

$$x = 0.8 \times 5$$

$$= 4$$

EST $1 \times 5 = 5$

The actual length of the living room is 4 m.

b) The actual length of the closet wall is 2 m. Let the length of the line needed to represent the closet wall be x.

$$\frac{1}{0.8} = \frac{x}{2}$$

$$2 = 0.8x$$

$$\frac{2}{0.8} = \frac{0.8x}{0.8}$$

$$2.5 = x$$

EST $2 \div 1 = 2$

The length of the line needed to represent the closet wall is 2.5 cm.

Practice

Write each scale in 3 other ways.

1. Scale 1 cm represents 75 cm.

2. Scale 1 cm:50 cm

3. Scale 1:1000

4. Scale 1 cm to 5 cm

Rewrite each ratio in simplest form.

5. 3:75 **6.** 10:1000 **7.** 3:900

8. 200:10 **9.** 700:350 **10.** 1000:100

Write each scale in simplest form, expressing each term in centimetres.

11. Scale 1 cm to 25 km

12. Scale 1 cm:100 m

13. Scale 150 cm represents 100 mm.

Problems and Applications

14. Use the actual size and the size of the drawing to determine each scale.

a) The world's largest lizard is the Komodo monitor, which has been measured at up to about 3 m long.

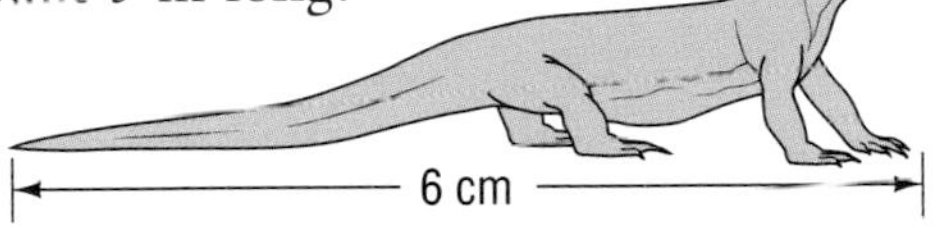

b) This cell is actually 0.34 mm across.

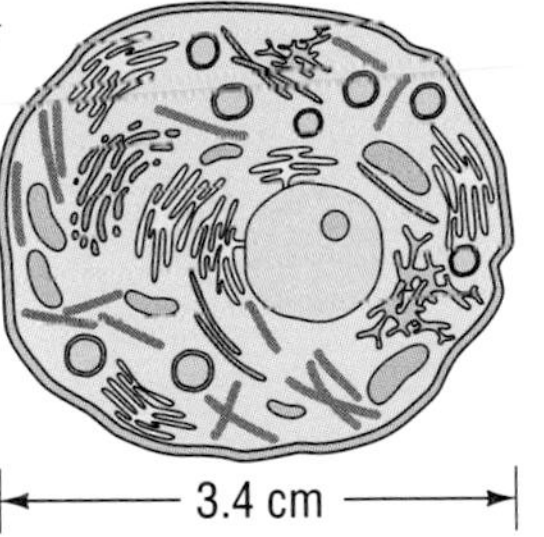

15. Use the scale shown for each drawing to determine the actual length.

a) the space shuttle *Discovery*, Scale 1:600

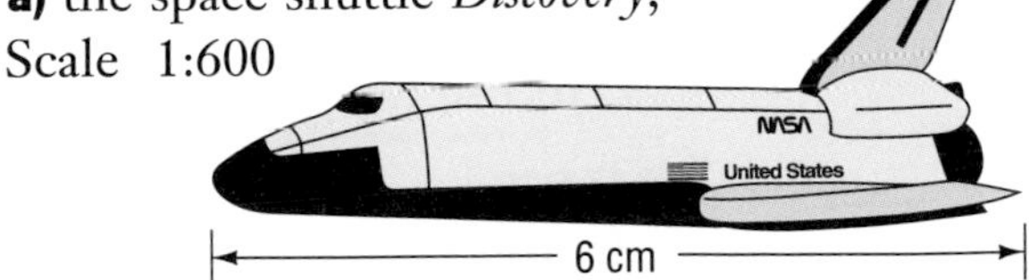

b) the smallest known starfish, Scale 5:1

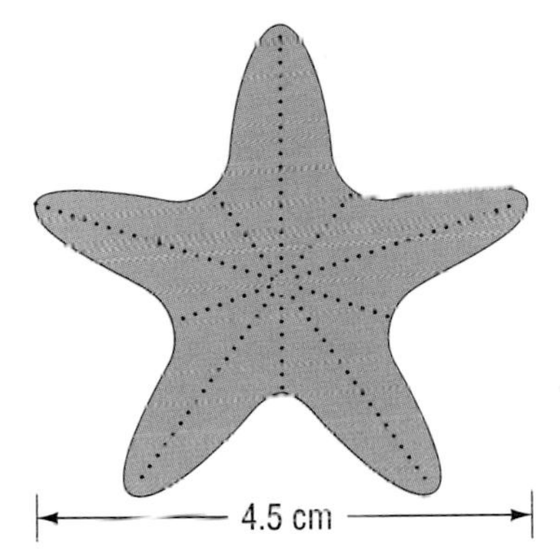

16. Many of the oldest Douglas fir trees are about 105 m high. What is the height of a diagram if the scale is 1:1000?

17. Mount Logan in the Yukon is Canada's highest mountain. If the height of a diagram is 5.9 cm, and the scale is 1:100 000, what is the actual height?

18. At about 2.6 mm long, the stirrup bone in the ear is the smallest bone in the human body. What is the length of a diagram if the scale is 20:1?

19. The length of a diagram of Euglena, a form of freshwater algae, is 2.8 cm. What is the actual length in millimetres if the scale is 100:1?

20. Measure the length and width of your desktop. Make a scale drawing and state the scale that you used.

21. A photograph taken through a microscope may be labelled as 2000 X.

a) Research the meaning of this notation.

b) How many centimetres wide is a 2000 X photograph of a cell that is 0.1 mm wide?

22. With a partner, design your ideal indoor/outdoor home recreation area. You might want to include such features as a dance floor and a swimming pool. Make a scale drawing and indicate the scale. Display your scale drawing and compare it with your classmates'.

9.6 Reading and Using Maps

Maps are examples of scale drawings that represent areas of the Earth. The scales represent the ratio of the distance on the map to the actual distance on the Earth.

Activity: Use the Map

The scale on this map of Canada can be reported in several ways.

Ratio Scale
1:42 000 000

Verbal Scale
1 cm represents 420 km

Graphic Scale

0 420 840 1260 1680 km

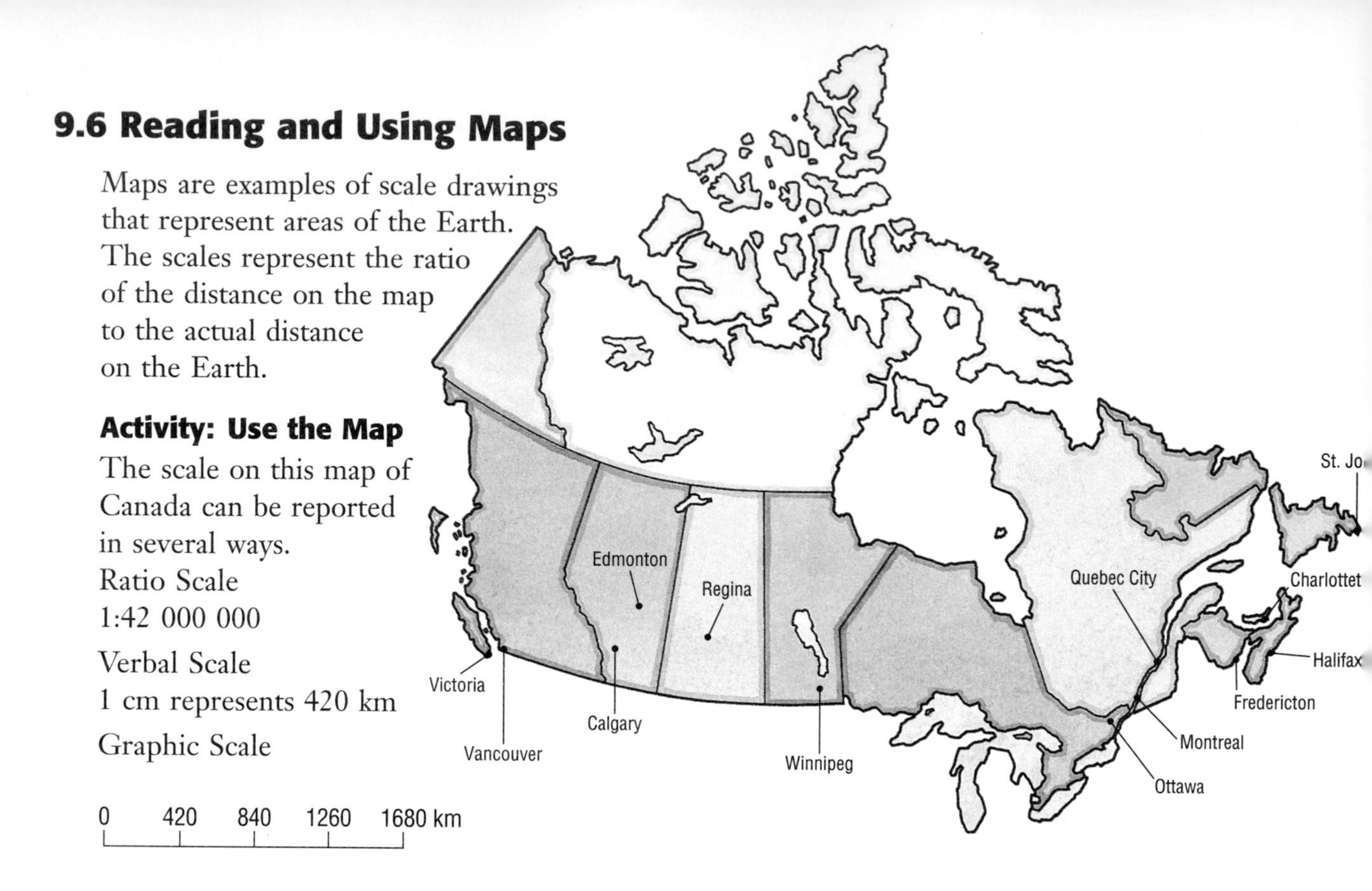

Inquire

1. Why is the name of each type of scale appropriate?

2. Explain how the 3 scales are equivalent.

3. What actual distance is represented by 2.4 cm on the map?

4. Measure the distance from Charlottetown to Calgary on the map to the nearest tenth of a centimetre.

5. Calculate the actual distance from Charlottetown to Calgary to the nearest hundred kilometres.

Example 1

Use the map to find the distance from Halifax to Vancouver to the nearest 100 km.

Solution

The distance from Halifax to Vancouver on the map is about 10.2 cm.
Set up a proportion.

$$\frac{1}{420} = \frac{10.2}{x}$$

$$x = 420 \times 10.2$$
$$= 4284$$

EST $400 \times 10 = 4000$

The distance from Halifax to Vancouver is about 4300 km.

Practice

Write the verbal scale showing centimetres to kilometres.

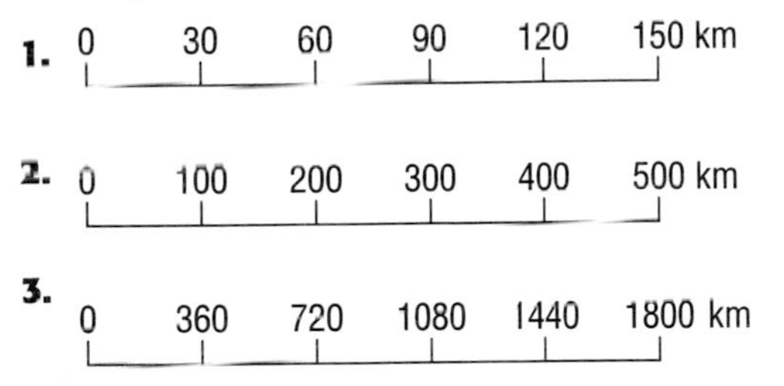

Write the ratio scale using centimetres for both terms.

4. 1 cm represents 300 km.

5. 1 cm represents 125 km.

6. 1 cm represents 50 km.

7. 1 cm represents 2 km.

Write each ratio as a verbal scale in centimetres to kilometres.

8. 1:78 000 000

9. 1:6 000 000

10. 1:4 500 000

11. 1:1 000 000

Use each scale and the distance on each map to find the distance on the Earth in kilometres.

12. 1 cm represents 100 km; 8 cm.

13. 1 cm represents 70 km; 6 cm.

14. 1 cm represents 15 km; 12.6 cm.

15. 1:100 000; 4 cm.

16. 1:250 000; 3.5 cm.

17. 1:1 000 000; 16.3 cm.

Use each scale and each distance on the Earth to find the distance on each map in centimetres.

18. 1 cm represents 10 km; 200 km.

19. 1 cm represents 100 km; 350 km.

20. 1 cm represents 75 km; 300 km.

21. 1:50 000 000; 200 km

22. 1:15 000 000; 450 km

23. 1:30 000 000; 3150 km

Problems and Applications

Use the map of Canada on the opposite page. Measure the distance between each of the following pairs of cities to the nearest tenth of a centimetre. Calculate the actual distance between each pair of cities.

24. Quebec City to Calgary

25. Edmonton to Montreal

26. Regina to Fredericton

27. Victoria to Winnipeg

28. Vancouver to St. John's

29. Ottawa to Fredericton

30. On a map of Ontario, the distance from Toronto to London is 5.5 cm. The actual distance is 165 km. Write the scale of the map in 3 ways.

31. The distance from Calgary to Edmonton is 248 km. How far apart are these cities on a map with a scale of 1:3 000 000? Round your answer to the nearest tenth of a centimetre.

32. By measuring distances in an atlas, find the distances between your nearest airport and 2 places you would like to visit.

33. Use an atlas to write a problem that requires the use of a scale. Have a classmate solve your problem.

PATTERN POWER

If the pattern continues, how many toothpicks make 9 triangles?

Percent

Activity ❶

1. Work with a classmate to list the uses of percent on these 2 pages.

2. List other uses of percent. Compare your list with your classmates'.

Friday, December 7
Mainly Cloudy
Probability of Precipitation 40%
Humidity 80%

may i help you?
half of me says yes. half says probably not.

36% of me says i'm making a mistake, 43% is wildly enthusiastic, and 21% is undecided.

70% of me believes i'm qualified, but 30% isn't sure i can pass the interview.

ah. you're here for the polling job.
i strongly agree.

Reprinted by permission of the cartoonist, Warren Clements

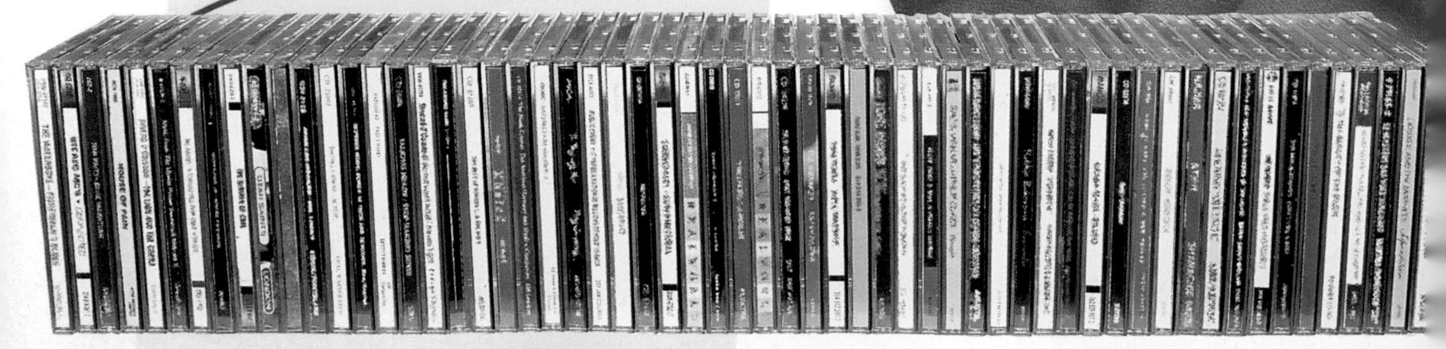

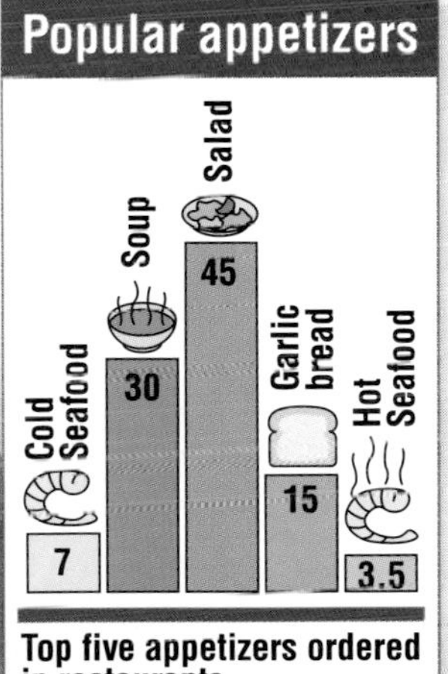

Blue Ridge has 45 runs.

25% Beginner
50% Intermediate
25% Advanced

Activity ❷

1. Collect some newspaper clippings in which percents appear.

2. Sort the articles into groups by use and give each group an appropriate name.

3. Prepare a display of the different groups you have created.

Player	Percent of Passes Completed
Tom Wilkinson	60.6
Tom Clements	60.3
Joe Paopao	56.9
Matt Dunigan	56.0
Ron Lancaster	54.3

Example 5

The ratio of endangered species of birds to all birds is 1:40. What percent of bird species are endangered?

Solution

Write the ratio as a decimal and multiply by 100%.

$$1:40 = \frac{1}{40}$$
$$= 0.025$$

C 1 ÷ 40 % 2.5

$0.025 \times 100\% = 2.5\%$

About 2.5% of bird species are endangered.

Example 6

Write as decimals.

a) 9% **b)** 12.5% **c)** 0.9%

Solution

Write the percent as a fraction and divide the numerator by the denominator.

a) $9\% = \frac{9}{100} = 0.09$ **b)** $12.5\% = \frac{12.5}{100} = 0.125$ **c)** $0.9\% = \frac{0.9}{100} = 0.009$

Practice

Write as percents.

1. 0.65 **2.** 0.83 **3.** 0.98
4. 0.02 **5.** 0.08 **6.** 0.6
7. 0.7 **8.** 0.752 **9.** 0.863
10. 0.005 **11.** 0.15 **12.** 0.49

Write as percents.

13. $\frac{1}{10}$ **14.** $\frac{3}{5}$ **15.** $\frac{3}{8}$
16. $\frac{1}{2}$ **17.** $\frac{9}{10}$ **18.** $\frac{7}{20}$
19. $\frac{13}{16}$ **20.** $\frac{11}{25}$ **21.** $\frac{36}{45}$

Write as percents to the nearest tenth of a percent.

22. $\frac{1}{3}$ **23.** $\frac{5}{11}$ **24.** $\frac{7}{9}$ **25.** $\frac{7}{24}$

Write the ratios as percents.

26. 1:2 **27.** 4:5 **28.** 9:20
29. 1:8 **30.** 9:16 **31.** 2:25

Write the percents as decimals.

32. 19% **33.** 35% **34.** 90%
35. 1% **36.** 66.6% **37.** 37.5%
38. 4.5% **39.** 0.8% **40.** 0.55%

Write each percent as a fraction in lowest terms.

41. 30% **42.** 24% **43.** 52% **44.** 75%
45. 88% **46.** 5% **47.** 2% **48.** 64%

Copy and complete the table.

	Fraction	Decimal	Percent
49.	$\frac{7}{10}$		
50.		0.03	
51.			17%

Calculate.

52. What percent is 25 of 125?

53. What percent is 90 of 200?

54. What percent is 10 of 40?

55. What percent is 45 of 50?

56. What percent is 6 of 24?

Problems and Applications

57. Four out of every 10 Canadians read at least 10 books a year. What percent of Canadians read at least 10 books a year?

58. There are 83 national parks in North America, and 34 of them are in Canada. What percent of the national parks are in Canada? Express your answer to the nearest tenth.

59. The table shows 4 of the longest Canadian rivers.

River	Length (km)
Mackenzie	4241
Yukon	3185
St. Lawrence	3058
Nelson	2575

Express each of the following percents to the nearest tenth.

a) What is the length of the Nelson as a percent of the length of the St. Lawrence?

b) What is the length of the Nelson as a percent of the length of the Yukon?

c) What is the length of the Yukon as a percent of the length of the Mackenzie?

60. One kilogram of chicken that has been cooked without the skin contains about 0.07 kg of fat.

a) What is the percent of fat in the cooked chicken?

b) What percent of the cooked chicken is not fat?

61. When Silken Laumann rowed in the Olympic final, she finished third. There were 8 competitors in the final.

a) What percent of the competitors finished ahead of her?

b) What percent of the competitors finished behind her?

c) Explain why the sum of your answers to parts a) and b) is not 100%.

62. Work with a partner to arrange the digits to make a fraction in which the percent is as close as possible to the target given. Use your calculator.

	Digits	Target
a)	1, 3, 6, 7	50%
b)	4, 3, 6, 7	65%
c)	0, 2, 8, 9	2%
d)	1, 2, 4, 0	4%

PATTERN POWER

Arrange the following dominoes in a rectangle that is not square, so that each side of the rectangle contains the same number of dots.

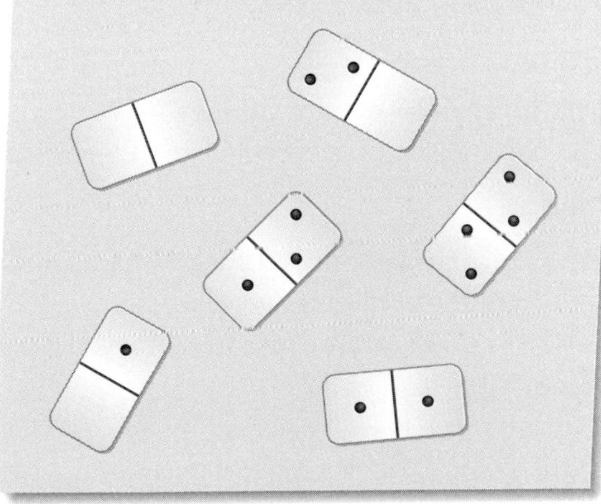

9.9 Using Percents

A species becomes extinct when the last animal or plant of that species dies. The black rhinoceros is in danger of extinction. In 1979, there were about 15 000 black rhinoceroses. Now there are about $\frac{1}{5}$ of that number.

$$\begin{aligned}\frac{1}{5} \text{ of } 15\ 000 &= \frac{1}{5} \times 15\ 000 \\ &= 0.2 \times 15\ 000 \\ &= 3000\end{aligned}$$

Changes in animal populations are often recorded in percents, not fractions.

Activity: Interpret the Data

In 1900, there were about 40 000 tigers in India. By 1972, the number of tigers had decreased to 4% of the number in 1900. Now that hunting tigers has been banned, the number of tigers is 10% of the number in 1900.

Inquire

1. What is 4% of 40 000?

2. What is 10% of 40 000?

3. What is the increase in tiger population since 1972?

4. In 1979, there were about 1 300 000 African elephants. Now there are 46% of that number. How many are there now?

5. Write a rule for finding the percent of a number.

Example 1

Each Canadian produces approximately 12 kg of garbage per week. Only 15% of this amount is recycled. What mass of each Canadian's garbage is recycled each week?

Solution

Each Canadian produces 12 kg of garbage.
Of this amount, 15% is recycled.
Find 15% of 12 to find the mass recycled each week.

$$\begin{aligned}15\% \text{ of } 12 &= 0.15 \times 12 \\ &= 1.8\end{aligned}$$

C . 15 × 12 = 1.8

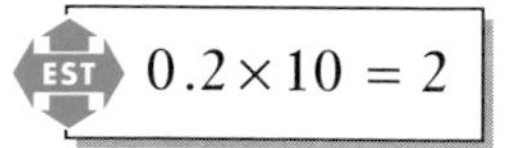

So, 1.8 kg of each Canadian's garbage is recycled each week.

Example 2

Canadians are being urged to conserve water. A family of 4 uses about 225 L of water per day. If a leaking tap is dripping at a rate of 1 drop/s, water use goes up to about 250 L per day. What percent of the 250 L of water is wasted because of the leaking tap?

Solution

The leaking tap wastes 250 L – 225 L or 25 L. Find what percent 25 L is of 250 L.

$$\frac{25}{250} \times 100\% = 0.1 \times 100\%$$
$$= 10\%$$

So, 10% of the 250 L is wasted because of the leaking tap.

Example 3

There are 20 000 species of butterflies. This number is 80% of the number of species of orchids. How many species of orchids are there?

Solution

80% of the number of species of orchids is 20 000

1% of the number of species of orchids is $\frac{20\ 000}{80}$

100% of the number of species of orchids is

$$\frac{20\ 000}{80} \times 100 = 25\ 000$$

There are 25 000 species of orchids.

Practice

Calculate.

1. What is 40% of 80?

2. What is 25% of 120?

3. What is 68% of 200?

4. What is 22% of 45?

5. What is 8% of $75.00?

6. What is 5% of $9.50 to the nearest cent?

7. What is 16% of $50.00 to the nearest cent?

Calculate.

8. What percent of 150 is 75?

9. What percent of 300 is 81?

10. What percent of 600 is 60?

11. What percent of 30 is 6?

12. What percent of 45 is 18?

Determine the number.

13. 50% of a number is 20

14. 25% of a number is 5

15. 20% of a number is 30

16. 10% of a number is 70

17. 2% of a number is 8

Problems and Applications

18. Jan scored 24 baskets out of 30 during a basketball free-throw contest. What percent of Jan's shots were baskets?

CONTINUED

19. On a written driver's test, you need to score 80% to pass. There are 60 questions on the test. How many questions must you answer correctly to pass the test?

20. In a student's council election, 80% of the students voted. If there are 290 students in the school, how many students voted?

21. The maximum speed of a lion has been recorded at 80 km/h. The maximum recorded speed of an elk is 72 km/h. What percent of 80 is 72?

22. The top-selling Canadian single was "Tears Are Not Enough," which sold approximately 300 000 copies. The song "Someday" by Glass Tiger sold approximately $16\frac{2}{3}$% of this amount. How many records of "Someday" were sold?

23. Canadian teams won 7 of the first 15 women's world curling championships. What percent of the championships did they win?

24. In 1830, there were about 60 000 000 bison. Today, there are about 0.15% of that number. About how many bison are there?

25. In 1800, the population of the world was 1 billion people. That is 10% of the projected population in the year 2100. What is the projected population for 2100?

26. Marie Curie, who studied radioactivity, was one of the few people to win 2 Nobel Prizes. She lived for 23 years in Poland and 43 years in France. For what percent of her life was her home in France? Round your answer to the nearest percent.

27. There are about 600 000 words in the English language.
a) People use about 5000 of them when talking. What percent of the words do they use?
b) People use about 10 000 words when writing. What percent of the words do they use?

28. Solve these problems, then compare your solutions with a classmate's. Suggest other ways to solve the problems.
a) The sides of one square are 20% as long as the sides of another square. What percent of the area of the larger square is the area of the smaller square?
b) The sides of one cube are 50% as long as the sides of another cube. What percent of the volume of the larger cube is the volume of the smaller cube?

29. Selling tapes, CDs, and records is a big business. The table shows who buys them.

Age	Percent Bought
10 to 19	24%
20 to 39	57%
40+	19%

Research the total sales of tapes, CDs, and records in Canada in dollars last year. Calculate how much was spent by each age group.

LOGIC POWER

Draw a chessboard in your notebook. Put a marker on each of 8 different squares so that no 2 markers are in the same row, column, or diagonal.

9.10 Percents and Business

Robert Bateman is a world-famous wildlife artist. In his paintings, he tries to portray an animal living in its natural habitat.

Activity: Solve the Problem

The original price of a Bateman print in a store was $600. To promote the sale of the print, the art dealer offered it at a discount of 20%. You want to know what you would pay for the print.

Inquire

1. What is 20% of $600?

2. What is the sale price of the print?

3. To find what you would pay, you now have to add taxes to the sale price. What is the current rate of the Goods and Services Tax (GST)? Calculate the GST owing on the sale price of the print.

4. What is the rate of Provincial Sales Tax (PST) in your province? Is the PST calculated on just the sale price or the sale price plus the GST? Calculate the PST owing on the print.

5. What is the total cost of the print?

Example 1

The cost price of a CD for a retailer was $13.50. She marked it up by 30%. What was the selling price before taxes?

Solution

Selling Price = Cost Price + Markup

Markup = 30% of $13.50
= 0.3 × $13.50
= $4.05

EST: 0.3 × 10 = 3

Selling Price = $13.50 + $4.05
= $17.55

The selling price before taxes was $17.55.

Example 2

Sandra works at a clothing store. She earns $100 a week plus a commission of 15% of all her sales. In 1 week, she sold $9500 of merchandise. What did she earn that week?

Solution

Commission = 15% of $9500
= 0.15 × 9500
= $1425

EST: 0.2 × 10 000 = 2000

Earnings = Commission + $100
= $1425 + $100
= $1525

Sandra earned $1525.

CONTINUED

Practice

Calculate to the nearest cent.

1. 1% of \$5.60

2. 1% of \$7.95

3. 2% of \$65.00

4. 2% of \$129.05

Each price is discounted by the percent shown. Find the amount of the discount and the selling price before taxes.

5. 10% of \$35

6. 15% of \$10

7. 20% of \$55.60

8. 25% of \$80.76

9. 50% of \$5.85

10. 10% of \$49.99

Each price is marked up by the percent shown. Find the amount of the markup and the selling price before taxes.

11. 10% of \$5.50

12. 15% of \$6.00

13. 25% of \$8.80

14. 20% of \$75.00

15. 30% of \$60.00

16. $33\frac{1}{3}$% of \$9.75

Problems and Applications

17. A merchant can purchase a T-shirt for \$18.90. The price charged to the customer includes a markup of 40%. What price does the customer pay for the T-shirt?

18. A skate company's total sales were \$122 130. The company's profit before taxes was \$18 630. What percent profit did the company make?

19. Jennifer bought a tennis racquet which regularly cost \$64 before tax. During a sale, the discount was 20%. What was the amount of the discount?

20. A portable compact disc player costs \$199.99 less a 35% discount.

a) How much is the discount?

b) What is the sale price?

21. A desk phone costs \$45 less a \$9 rebate. What percent discount is the rebate?

22. A video cassette of your favourite musical group costs \$39.95 plus PST and GST.

a) Find the PST in your province.

b) Find the GST.

23. The selling price of a windsurfer is \$399 plus PST and GST. The PST is 8% of the selling price. What is the total cost of the windsurfer?

24. A pair of basketball shoes sells for \$99.95 plus PST and GST. If the PST is 7% of the selling price, what is the total cost of the shoes.

25. An antique car dealer sold a car for \$12 000 that had cost \$15 000.

a) How much did the dealer lose on the car?

b) What was the percent loss on the car?

c) What other costs should be considered in determining the total loss on the car?

26. Two clock radios are on sale. The models have the same features. One sells for \$65.50 less 15%. The other sells for \$72.50 less 20%. Which is the better buy?

27. Linda sells cars on commission. She earns 2.5% of all sales. In 1 week, she sold 3 cars for a total of \$53 760. How much did she earn?

28. You have used a percent discount to find the amount of a discount. Subtracting this amount from the price gave the discount price.

a) Suggest another method for finding the discount price. Write an example to show how it works. Compare your method with a classmate's.

b) Can you modify your method to calculate a markup? Explain.

29. Write a question that involves buying or selling. Have a classmate solve your problem.

9.11 Percents Greater Than 100%

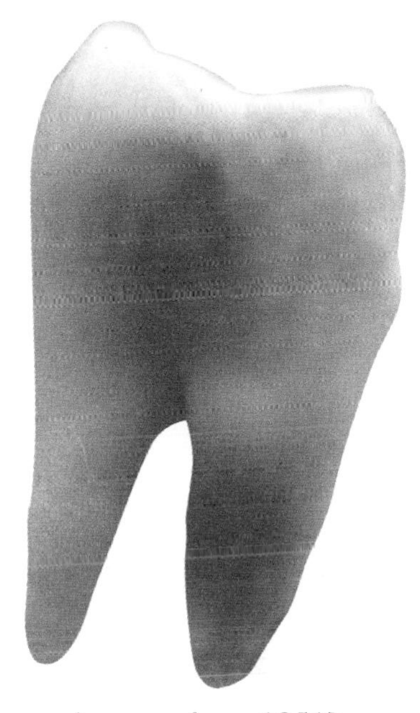

A child with a full set of milk teeth has 20 teeth. An adult with a full set of permanent teeth has 32 teeth. The percent increase in the number of teeth from childhood to adulthood is given by

$$\begin{aligned}\text{Percent increase} &= \frac{\text{increase}}{\text{original number}} \times 100\% \\ &= \frac{12}{20} \times 100\% \\ &= 0.6 \times 100\% \\ &= 60\%\end{aligned}$$

Activity: Study the Information

In 1951, it cost $1.00 to go to a movie, and the cost of an LP record was $4.50.

Inquire

1. What is the average cost of a movie today?

2. What is the increase in cost since 1951?

3. Calculate the percent increase.

4. The CD is the modern equivalent of the LP record. What is the average cost of a CD today?

5. What is the increase over the cost of an LP in 1951?

6. What is the percent increase?

7. Why do you think there has been such a big difference in the percent increases?

Example 1

Write as decimals.
a) 200%
b) 150%
c) 325%

Solution

a) $200\% = \frac{200}{100} = 2.0$

b) $150\% = \frac{150}{100} = 1.5$

c) $325\% = \frac{325}{100} = 3.25$

Example 2

Calculate the following.
a) 150% of 64
b) 220% of 85

Solution

a) 150% of 64 $= 1.5 \times 64 = 96$

C 150 % × 64 =   96.

b) 220% of 85 $= 2.2 \times 85 = 187$

C 220 % × 85 =   187.

Example 3

If you can jump 2 m on Earth, you can jump 12 m on the moon. What percent of the height of a jump on Earth is the height of a jump on the moon?

Solution

$\frac{12}{2} \times 100\% = 6 \times 100\% = 600\%$

C 12 ÷ 2 %   600.

The height of a jump on the moon is 600% of the height of a jump on Earth.

CONTINUED

Practice

Copy and complete the table.

	Percent	Fraction	Decimal
1.	50%		
2.		$\frac{1}{4}$	
3.	120%		
4.		$\frac{11}{10}$	
5.	300%		
6.		$\frac{3}{2}$	
7.			1.6
8.	105%		
9.			2

Write as decimals.

10. 200% **11.** 400% **12.** 150%

13. 500% **14.** 110% **15.** 115%

16. 125% **17.** 101% **18.** 250%

Write as percents.

19. $\frac{6}{5}$ **20.** $\frac{7}{4}$ **21.** $\frac{5}{2}$

22. 3 **23.** 1.5 **24.** 2.6

Calculate.

25. 50% of 80 **26.** 200% of 400

27. 300% of 60 **28.** 175% of 1200

29. 106% of 4000 **30.** 120% of 250

Problems and Applications

31. A magazine sells for $6.00. Last year it sold for $5.00.

a) What is the percent increase over last year's price?

b) What percent of the old price is the new price?

32. A box of cereal was enlarged to contain 540 g. The original box contained 400 g.

a) What is the percent increase over the original box?

b) What percent of the old size is the new size?

33. The Earth is approximately 150 000 000 km from the sun. Neptune is approximately 4 500 000 000 km from the sun. What percent is Neptune's distance from the sun of the Earth's distance from the sun?

34. The Yukon Territory covers about 5% of the area of Canada. Express Canada's area as a percent of the area of the Yukon Territory.

35. A newspaper recently reported that the most fuel-efficient car required 5.4 L of gas for every 100 km. The most gas required by a car was reported as 22.7 L for every 100 km. What percent of the gas required for the most fuel-efficient car is the gas required for the least fuel-efficient car? Express your answer to the nearest percent.

36. American bison are the first animals to be saved from extinction by breeding them in captivity. In 1830, there were about 60 000 000 bison. By 1894, there were 100. Now there are about 90 000.

a) What percent of the bison population in 1830 is the bison population today?

b) What percent of the bison population in 1894 is the bison population today?

37. The Mackenzie River and the Columbia River are 2 of the longest rivers in Canada. The length of the Mackenzie River is 212% of the length of the Columbia River. The Columbia River is 2000 km long. How long is the Mackenzie River?

38. What is wrong with each statement?

a) The price increased by 200%, which means it doubled.

b) A dollar is worth 400% more than a quarter.

c) If the new price is 125% of the old price, the old price was 75% of the new price.

39. Write a problem that involves a percent greater than 100%. Have a classmate solve your problem.

9.12 Simple Interest

Interest is money that is paid for the use of money over a period of time. If you deposit money in a bank, the bank will pay you interest for the use of that money. If you borrow money, the bank will charge you interest for the use of that money. The sum of money you deposit or borrow is called the **principal**. The total of the principal and the interest is called the **amount**.

Activity: Complete the Table

Copy and complete the table for the given bank deposits. The length of time of each deposit is expressed in years.

Principal ($)	Interest Rate (per year)	Time (years)	Interest ($)	Amount ($)
1000	5	2	$1000 \times 0.05 \times 2 = 100$	1100
2000	6	3		
3000	7	4		
2000	4	2		
1500	7.5	3		

Inquire

1. Calculate the amount if $7000 is invested at 6% per year for 6 months.

2. Calculate the amount if $800 is invested at 5% per year for 90 days.

3. Which interest rate is usually higher, the rate you receive for money you deposit or the rate you are charged for money you borrow? Explain.

The formula to determine **simple interest** is $I = Prt$

where I represents the interest paid for 1 year (per annum),
P represents the principal or the money invested or borrowed,
r represents the rate per year (per annum),
t represents the time in years.

Example 1

Find the simple interest for 3 years for $200 invested at 8% per year.

Solution

$P = 200$ $\quad r = 8\% = 0.08$ $\quad t = 3$

$I = Prt$
$= 200 \times 0.08 \times 3$ $\quad$ C 200 × 8 % × 3 = 48.
$= 48$

The interest is $48.

CONTINUED

Example 2

Find the simple interest on \$150 borrowed for 73 days (d) at 16.75% per year.

Solution

$P = 150 \quad r = 16.75\% \quad t = \frac{73}{365}$

$\qquad\qquad\quad = 0.1675$

73 d is $\frac{73}{365}$ year

$I = Prt$

$= 150 \times 0.1675 \times \frac{73}{365}$

C 150 × 16 . 75 % × 73 ÷ 365 = 5.025

$= 5.03$

The interest is \$5.03.

Practice

Express the following percents as decimals.

1. 9% **2.** 12% **3.** 6%

4. 14% **5.** $8\frac{1}{2}\%$ **6.** $11\frac{1}{4}\%$

7. $14\frac{3}{4}\%$ **8.** $20\frac{1}{2}\%$ **9.** $6\frac{9}{10}\%$

Express as fractions of a year.

10. 73 days **11.** 20 weeks, 6 d

12. 219 d **13.** 292 d

14. 30 d **15.** 5 weeks

16. 60 d **17.** 90 d

Problems and Applications

18. Michel received a scholarship of \$250 for his high marks in school. He plans to deposit the money for 9 months in an account that pays 8% interest per year.
a) How much interest will Michel earn in 9 months?
b) How much money will he have altogether in 9 months?

19. Hania borrowed \$135 for 60 d, at 13% interest per year.
a) Calculate the simple interest she will owe.
b) What will be the total amount that Hania will have to repay after 60 d?

20. The student council invested \$6000 of the money collected as student activity fees. The council invested the first \$4000 at 12% for 90 d and the remainder at 14% for 30 d. Calculate the total interest earned.

21. Find the total interest earned when \$3000 is invested for 90 d at 10%, and then the total amount is reinvested for 60 d at 12.5%.

22. Greg wants to use the interest from his savings to buy a cassette player for \$225. He invests \$1500 at 10% simple interest per year. How long will he have to wait to buy the cassette player?

23. Raisa earns \$100 a year in interest from her savings. They are invested at 8% simple interest per year. Calculate how much she has in her savings.

24. a) Consult the financial section of a newspaper to find the bank or trust company that pays the best rate of interest.
b) For what period of time is interest usually quoted?
c) Explain the difference between savings accounts and guaranteed investment certificates (GICs).

Compound Interest

Activity ❶ Compounding Mosquitoes

A mosquito population increased by 10% per day. There were 10 000 mosquitoes on Monday. On Tuesday there were 10% more. On Wednesday there were 10% more than there were on Tuesday. Copy and complete the table to find out how many mosquitoes there were on Friday.

Day	Increase	Number of Mosquitoes
Monday		10 000
Tuesday	10% of 10 000 = 1000	11 000
Wednesday	10% of 11 000 =	
Thursday	10% of	
Friday	10% of	

Activity ❷ Compounding Money

When the interest you earn on a deposit is compounded annually, the amount at the end of one year becomes the principal for the next year. Copy and complete the table for $4000 invested at 5% per annum compounded annually.

Year	Principal	Interest	Amount
1	4000	$4000 \times 0.05 = 200$	4200
2	4200	$4200 \times 0.05 =$	
3			
4			
5			

Activity ❸ Comparing Simple and Compound Interest

1. Calculate the simple interest earned if $10 000 is invested at 4% per annum for 5 years.

2. Calculate the compound interest earned if $10 000 is invested at 4% per annum for 5 years.

3. What is the difference in the amounts of interest earned?

4. Why is there a difference?

What is Happening to Black Bears?

Activity ❶

Black bears come in many colours, blond, cinnamon, light brown, dark brown, and black. There are even white black bears found on islands off the coast of British Columbia.

About 75% of the black bear's diet is vegetarian. Black bears eat for 5 to 7 months and then hibernate for the rest of the year. Adult males have masses between 120 kg and 280 kg. Adult females have masses between 45 kg and 180 kg. Black bears are between 1.5 m and 1.8 m long, and they can run at speeds of up to 40 km/h.

The table shows the 1990 population of black bears in Canada by province or territory and the number killed in each province or territory that year.

Province/Territory	Population Estimate	Number Killed
Alberta	48 000	2800
British Columbia	63 000	3500–4000
Manitoba	Unknown	1500–1700
New Brunswick	Unknown	1000
Newfoundland	6000	100
Northwest Territories	Unknown	Unknown
Nova Scotia	3000	500
Ontario	65 000–75 000	8700
Prince Edward Island	0	0
Quebec	60 000	2000
Saskatchewan	30 000	1600
Yukon	10 000	100

1. Estimate the total black bear population in Canada.

2. Estimate the total number of black bears killed in a year.

3. Estimate the percent of black bears killed in Canada in one year.

4. What percent of the black bear population in Canada is in British Columbia? Ontario? Alberta?

5. The black bear population in North America is 460 000. About what percent of the bears are in Canada?

Activity ❷

The map shows where black bears once lived in North America and where they live now.

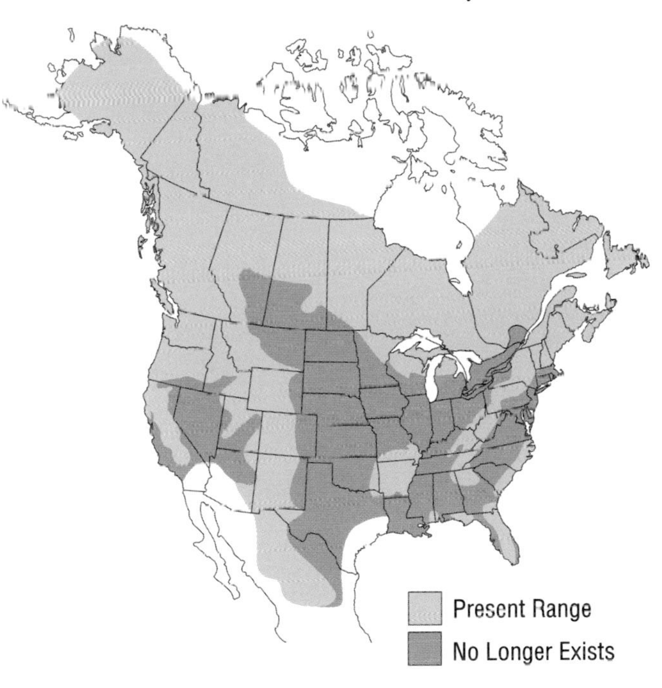

1. Estimate the percent of its range the black bear has lost in Canada.

2. Estimate the percent of its range the black bear has lost in the United States.

3. Estimate the percent of its range the black bear has lost in North America.

4. Which American states that once had black bears no longer have them?

Activity ❸

1. The black bear population in North America is declining. List some reasons for the decline.

2. List some things that could be done to save the black bears in Canada.

Review

Write each ratio in simplest form.

1. 4:6 **2.** 35:28 **3.** 60:80

Express as a ratio in simplest form.

4. 3 days to 18 h **5.** 4 L to 60 mL

Solve for x.

6. $\frac{x}{5} = \frac{27}{45}$ **7.** $\frac{12}{9} = \frac{16}{x}$

8. $10:6 = x:24$ **9.** $4:x = 20:15$

Express each rate in the stated units.

10. 720 km in 9 h (kilometres per hour)

11. $52.95 for 3 CDs (dollars per CD)

Write each percent as a decimal.

12. 79% **13.** 20% **14.** 8%

Write each percent as a fraction in lowest terms.

15. 50% **16.** 40% **17.** 5%

Write as percents.

18. $\frac{4}{5}$ **19.** $\frac{115}{46}$ **20.** 0.43

21. 2:8 **22.** 6.4 **23.** 0.008

Calculate.

24. $\frac{1}{2}$% of 78 **25.** $12\frac{1}{2}$% of 64

26. 275% of 80 **27.** 50% of $17.50

28. Some professional baseball teams have 10 pitchers, 3 catchers, 6 outfielders, 6 infielders. Express each ratio in lowest terms.

a) pitchers to catchers
b) infielders to catchers
c) outfielders to infielders

29. If a 40-m rope is cut into 2 pieces in the ratio 3:5, how long is each piece?

30. A punch recipe calls for orange juice and ginger ale in the ratio 3:5. How much orange juice should you mix with 6 L of ginger ale?

31. Renata invested $8000 in a new business. Her partner, Ben, invested $7000. Their profit at the end of the first year was $3000. How much did Renata receive?

32. Sylvie Daigle of Canada set a world record for speed skating by winning a 500-m race in 46.72 s. What was her average speed to the nearest tenth of a metre per second?

33. Christine took 45 min to finish 3 homework questions. At this rate, how long would it take her to finish 5 questions?

34. Maria can swim 3300 m in 1 h. John can swim 1980 m in 40 min. Who swims faster?

35. If 350 mL of shampoo is $4.59 and 150 mL is $2.25, which is the better buy?

36. Measure the line in each drawing, then state the scale.

a) The highest altitude of a flowering plant is 6400 m.

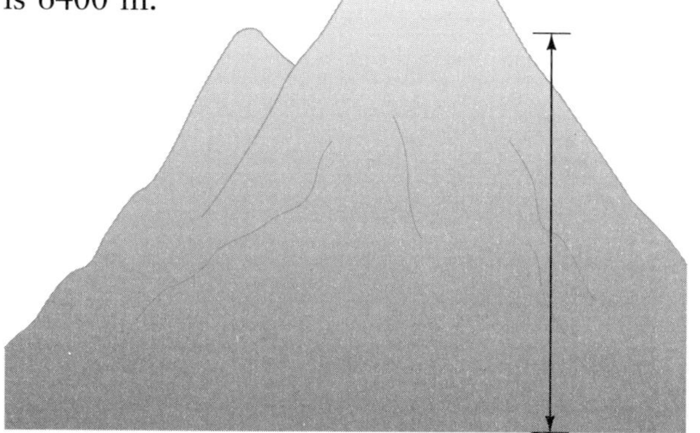

b) A common flea can grow to 3 mm in length.

37. Write each map scale in 2 other ways.

a) 1: 4 500 000
b) 1 cm represents 100 km.
c) 0 150 300 450 600 750 km

38. If 80% of a number is 240, find the number.

39. Use the map of Canada on page 330. Measure the distances between the following pairs of cities to the nearest tenth of a centimetre. Calculate each actual distance to the nearest hundred kilometres.

a) Quebec City and St. John's

b) Edmonton and Charlottetown

40. a) What percent of 65 is 39?

b) What percent of 190 is 95?

c) What percent of 25 is 5?

d) What percent of 100 is $\frac{1}{4}$?

41. For a collection of 5 nickels and 3 quarters, express as a percent of the total value the value of

a) 5 nickels **b)** 1 quarter

c) 3 nickels and 2 quarters

42. A sports bag was on sale at 30% off the original price of $45.00.

a) How much was the discount?

b) Find the sale price including PST and GST. Use the PST rate for your province.

43. A dealer bought a book for $20 and sold it for $30. What was the markup? the percent markup?

44. If 24 of 60 people chose cycling as their favourite physical activity, what percent did not choose cycling?

45. A recent Statistics Canada survey showed that 59% of the 8 552 290 households in Canada owned a video cassette recorder (VCR). How many households owned VCRs? Round your answer to the nearest thousand.

46. The Concorde has a wingspan of about 25.56 m. The overall length is about 241% of the wingspan. What is the length of the Concorde to the nearest tenth of a metre?

47. Sarah sold 2 cars for a total of $60 450. Her rate of commission is 2.5%. How much did she earn?

48. Simple interest for a savings account is calculated on the minimum balance for the previous 6 months. The minimum balance over one 6-month period was $75.80, and the interest rate was 8.5% per annum. Calculate the interest earned to the nearest cent.

Group Decision Making

Researching Community Service Careers

1. Brainstorm as a class the possible careers to investigate. They might include careers in the police force, fire department, library, social services, post office, or parks and recreation. As a class, choose 6 careers. Assign each career to 2 pairs of students.

1 2 | 3 4 | 5 6 | 7 8 | 9 10 | 11 12

Pairs

1 2 | 3 4 | 5 6 | 7 8 | 9 10 | 11 12

2. In your pair, list the questions you want answered, then research the answers. Make sure that you research how math is used.

3. Exchange questions with the pair who researched the same career as you. In your pair, research any of the other pair's questions you did not already answer.

4. Pool information with the pair who researched the same career as you. Work as a group of 4 to prepare a report.

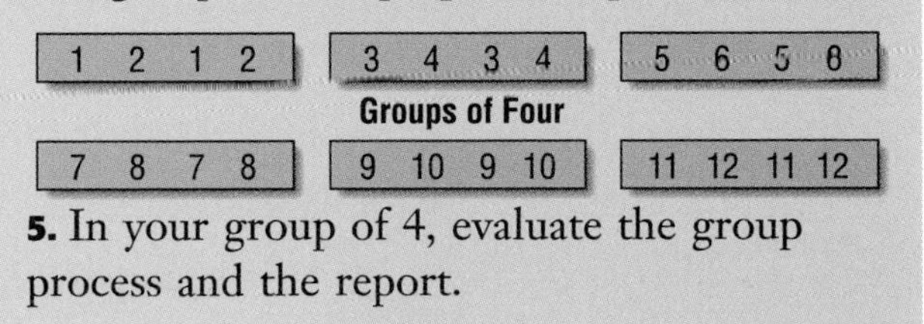

5. In your group of 4, evaluate the group process and the report.

Chapter Check

Express as ratios in simplest form.

1. 6:9 **2.** 4 g raisins:12 g peanuts

Express as a unit rate.

3. $1.32 for 4 cans of soda

4. 12 km in 9 min

Solve for x.

5. $x:5 = 39:13$ **6.** $2:x = 7:21$

7. $4:3 = x:15$ **8.** $1.8:3 = 1.5:x$

Write as a decimal and as a fraction in lowest terms.

9. 85% **10.** 6% **11.** 0.5%

Find.

12. 50% of 350 **13.** 220% of 25

14. The highest reported amount of caffeine in a 355-mL can of cola is 45 mg. To the nearest milligram, how much caffeine is present in 500 mL of this cola?

15. An alloy called coinage metal is used to make "silver" coins. Coinage metal contains copper and nickel in the ratio 3:1, by mass. How much copper is present for every 25 g of nickel?

16. A music store's stock of recorded music includes new releases and top ten hits in the ratio 3:5. How many top ten hits are there in a shipment of 80 items?

17. Canada's Marilyn Bell was the first person to swim across Lake Ontario. She completed the 51.5-km trip in 21 h. How fast did she swim? Round your answer to the nearest tenth of a kilometre per hour.

18. Twenty-four 355-mL cans of juice cost $4.99, and a 2-L container of juice costs $1.49. Which is the better buy?

19. Use a scale of 1:300 to draw a line that represents 7.5 m.

20. Find the actual length of the blue whale.

Scale 2 cm represents 10 m

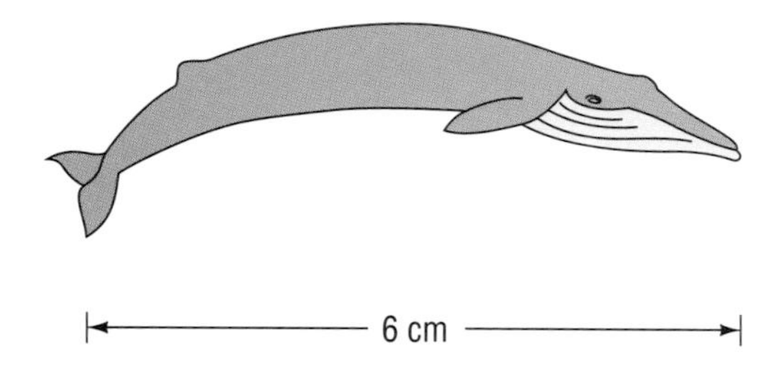

21. The distance from London, England, to Dublin, Ireland, is about 15.5 cm on a map with a scale of 1:3 000 000. What is the approximate distance between these cities?

22. The new price of a car was $10 900. Its old price was $900 less. What percent of the old price is the new price?

23. On a driving test, Chris answered 80% of the 20 questions correctly. How many questions did Chris get right?

24. A bicycle was on sale for 35% off. What was the sale price before tax if the original price was $375?

25. The cost of producing a manual is $9.50. If the markup is 20%, what is the selling price before taxes?

26. Find the cost of a portable colour TV that lists for $320. Include the PST for your province and the GST.

27. Calculate the simple interest on $300 invested for 3 years at 11% per year.

28. One week, Carl sold $8000 worth of paintings. He earned $200 plus a commission of 5% of sales. What did he earn?

Using the Strategies

1. Find the smallest number that, when divided by each of the integers 2, 3, 4, 5, 6, and 7, always gives a remainder that is 1 less than the divisor.

2. Place 10 markers in 5 rows of 4 markers each.

3. If it costs \$1.20 to cut a log into 4 equal segments, how much does it cost to cut it into 8 equal segments?

4. Use six 9s and the order of operations to write a number that equals 100.

5. Beth walks at a constant rate of 10 min/km and jogs at a constant rate of 5 min/km. Her home is on the corner of a block that is 1 km square. One day, Beth began her warm up by walking from her home to A. There, she stopped to talk with a friend for 5 min. She then walked to B and waited for 2 min for a light to change. After jogging to C, she realized that she had dropped her keys at B and jogged back to get them. At B, she spent 13 min finding and retrieving her keys, which had been handed in at a corner store. She jogged back to C and then on to D. To cool down, she walked home from D.

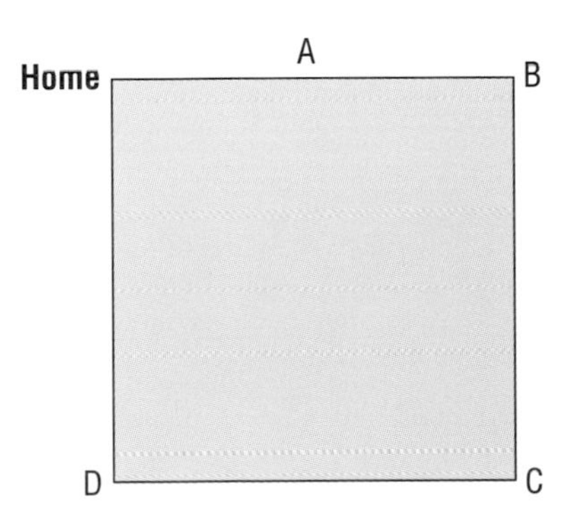

a) Use the diagram and the information from the story to complete the chart.

Location	Time from Start (min)	Total Distance Travelled (km)
Home	0	0
A	5	$\frac{1}{2}$
	10	$\frac{1}{2}$
B	15	1
B	17	
C	22	
B		3
	40	3
C		
	50	

b) Use the chart to draw a graph of total distance travelled versus time from the start.

c) The graph below shows the information for another trip on another day. Construct a chart like the chart in part a). Then, create a story that fits the information. Compare your story with your classmates'.

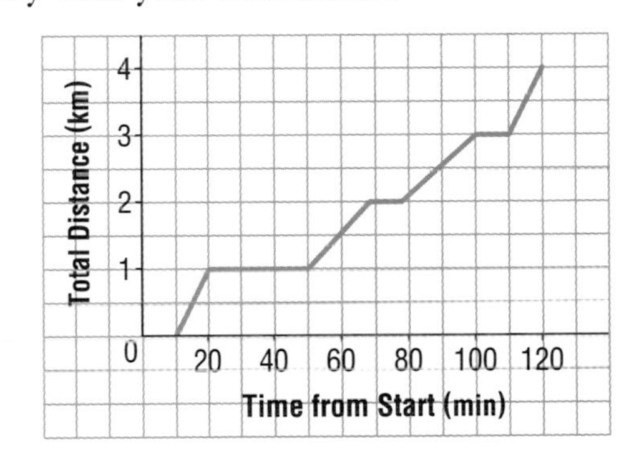

DATA BANK

1. Write the ratio of the speed of the strongest strong breeze to the speed of the strongest moderate breeze. Express your answer in lowest terms.

2. What percent of the total annual production of the world's top 10 gold producers does Canada produce?

ESP
RFSTARS

CHAPTER 10

Statistics and Probability

The graph shows the world's projected population growth.

Suppose you hold a concert in Nova Scotia in the year 2000. If each person needs an area of 1 m^2, could the whole world attend your concert? Could the whole world attend in the year 2060?

Estimate the first year in which the whole world could not attend the concert. What assumptions have you made? Are your assumptions reasonable?

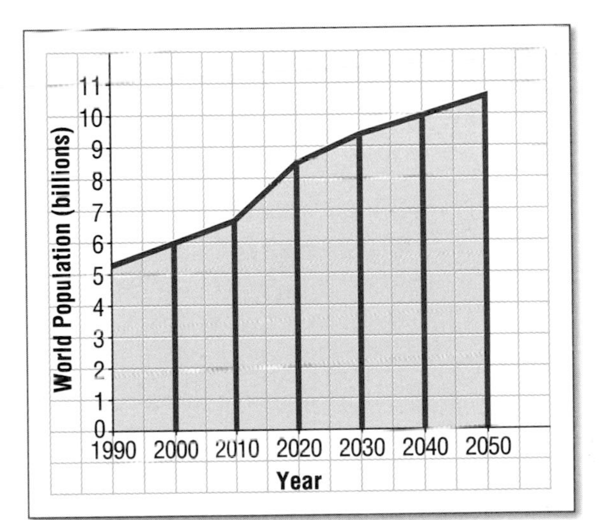

Canadian Weather Facts

The table shows the percent chance of different types of weather in some Canadian cities.

Type of Weather	City: Vancouver	Edmonton	Regina	Winnipeg	Toronto	Montreal	Halifax	St. John's
January Thaw	100	96	84	54	98	98	100	100
10-cm Snowfall in January	38	17	3	27	49	81	72	84
Temperature of –20°C or Lower in Winter	0	100	100	100	89	100	76	23
April Gale	8	0	35	19	27	12	10	68
May Snowfall	0	58	43	53	9	21	44	79
June Frost	0	21	43	30	0	0	0	60
June Fog	23	41	58	50	85	69	100	100
At Least 10 Wet Days in July	21	96	51	72	40	77	68	84
Temperature of 35°C or more in Summer	0	0	64	43	32	9	0	0
September Frost	2	92	97	87	40	30	8	14
December Thunderstorm	7	0	0	7	21	11	13	11
Hail During the Year	69	83	89	82	64	68	36	25

Activity ❶

Use the table to answer these questions.

1. Which cities have the same chance of a January thaw?

2. Which city has the highest chance of a temperature of 35°C or more in summer?

3. Which city has the lowest chance of an April gale?

4. Is the chance of a December thunderstorm higher in Toronto or in Montreal?

5. Which 2 cities always get a January thaw and a June fog?

6. How many times higher is the chance of

a) a June frost in St. John's than a June frost in Winnipeg?

b) a September frost in Edmonton than a September frost in Vancouver?

c) a 10-cm snowfall in January in St. John's than a 10-cm snowfall in January in Regina?

Activity ❷

Use the data in the table to write 3 more problems. Have a classmate solve your problems.

Warm Up

Write each fraction in lowest terms.

1. $\frac{4}{6}$
2. $\frac{8}{10}$
3. $\frac{10}{15}$
4. $\frac{25}{30}$
5. $\frac{18}{24}$
6. $\frac{12}{36}$

Write each fraction as a percent.

7. $\frac{3}{5}$
8. $\frac{5}{20}$
9. $\frac{8}{24}$
10. $\frac{15}{30}$
11. $\frac{90}{360}$
12. $\frac{245}{400}$

Calculate the number of degrees given by each percent of 1 full circle (360°).

13. 5%
14. 12%
15. 25%
16. 30%
17. 60%
18. 16.7%

Draw each angle.

19. 25°
20. 45°
21. 60°
22. 150°

Calculate.

23. 40% of 360°
24. 60% of of 360°
25. 20% of 360°
26. 1% of 360°
27. 90% of 360°
28. 22% of of 360°

29. The table gives the ages of students at a camp.

Age	8	9	10	11
Number of Campers	36	45	27	12

What percent of the campers are in each age group?

Mental Math

Calculate each product. Look for easy combinations.

1. $5 \times 2 \times 8$
2. $25 \times 4 \times 7$
3. $7 \times 5 \times 2$
4. $12 \times 4 \times 25$
5. $4 \times 6 \times 25$
6. $5 \times 18 \times 2$

Find each sum. Think quarters.

7. \$12.75 + \$8.25
8. \$3.25 + \$4.50
9. \$2.75 + \$4.50
10. \$5.00 + \$2.50 + \$3.50
11. \$6.25 + \$2.75 + \$5.50
12. \$8.50 + \$3.75 + \$2.75

Find each sum. Look for sums of 50.

13. 21 + 34 + 16 + 19
14. 27 + 35 + 23 + 10
15. 42 + 19 + 6 + 31

Estimate the total bill.

16. \$0.69 + \$1.89 + \$1.49 + \$0.87
17. \$0.87 + \$2.69 + \$1.34 + \$2.45
18. \$2.59 + \$1.89 + \$5.99 + \$3.49
19. \$2.29 + \$1.89 + \$1.49 + \$4.69

Estimate each product.

20. 6.02×4.07
21. 3.1×12.6
22. 42.8×9.8
23. 312×4.9

Estimate each sum.

24. $3\frac{1}{3} + 4\frac{1}{4}$
25. $8\frac{2}{3} + 16\frac{1}{4}$
26. $5\frac{1}{3} + 2\frac{1}{2}$
27. $12\frac{3}{4} + 3\frac{3}{8}$

10.1 Collecting and Recording Data

Statistics is the science of collecting, organizing, and interpreting data.

Activity: Use the Survey

In a Gallup poll conducted by telephone, 1026 Canadians of 18 years of age or older were asked this question.

Would you say that your standard of living is higher, lower or the same as the standard of living of your parents?

The results were as follows.

Response	Higher	Lower	Same	Don't Know
Percent	50	23	25	2

How many people gave each response?

Inquire

1. The results of the survey were rounded to the nearest percent. Explain why.
2. Why did the survey include only people who were 18 years of age or older?
3. Why were about 1000 people surveyed and not
   a) 100 people? **b)** 5000 people?
   c) all the adults in Canada?

All the people eligible for a survey make up the **population.** In the above survey, the population included all Canadians 18 years of age or older. In a **census**, data are collected from every member of a population. When a population is very large, a **sample** of the population may be surveyed. In the above survey, the sample included 1026 people.

If you look up the results of a Gallup poll, you are using a **secondary data-gathering method**. This means that you are using someone else's findings. Other secondary methods include the use of books and databases. If you gather your own information, you are using a **primary data-gathering method**. You might do this by carrying out your own telephone survey or by interviewing people in person.

As data are collected, they must be recorded. Polling companies generally use computers to record and organize data. When people record by hand, they may initially leave data in an unorganized form or they may record them in a table.

A **frequency distribution table** lists each item once. Each piece of data is recorded in the tally column. Then, the tallies are counted and the results recorded in the frequency column.

Example 1

Sara was undecided about running for president of the Student Council. Her friend, Miguel, conducted a survey. He asked 30 students which candidate they would vote for. He recorded these results.

Henri	Sara	Sara	Henri	Henri
Sara	Monique	Gino	Sara	Henri
Gino	Gino	Henri	Sara	Gino
Henri	Sara	Monique	Sara	Sara
Gino	Sara	Henri	Monique	Monique
Gino	Monique	Monique	Sara	Gino

a) Record the results in a frequency distribution table.
b) Who was most likely to be elected?

Solution

a)

Chosen Candidates		
Candidate	Tally	Frequency
Henri	$\cancel{||||}$ ||	7
Sara	$\cancel{||||}$ $\cancel{||||}$	10
Gino	$\cancel{||||}$ ||	7
Monique	$\cancel{||||}$ |	6
Total		30

b) Sara was most likely to be elected.

We often use a sample to make a prediction about a population. For the result to be reliable, we must use a **representative sample**. To predict who will win a student election, we select the representative sample from among the students. We do not include teachers, because they are unable to vote.

Example 2

In a school of 520 students, a survey of 20 students showed their favourite ways of listening to music. If the survey used a representative sample, what percent of the student population preferred to listen to tapes?

Favourite Music Sources	
Music Source	Number of Students
Tapes	7
FM Radio	5
CDs	3
Records	1
AM Radio	4
Total	20

Solution

Of the 20 students in the sample, 7 preferred tapes.

$$\frac{7}{20} = \frac{7 \times 5}{20 \times 5}$$
$$= \frac{35}{100}$$
$$= 35\%$$

Assume that this percent was true for the whole population, and find 35% of 520.

$520 \times 0.35 = 182$

EST $500 \times 0.4 = 200$

So, 182 of the 500 students preferred to listen to tapes.

CONTINUED

It is important that a representative sample be random and unbiased. In a **random sample**, each member of the population has an equal chance of being chosen. In an **unbiased sample**, all groups in the population are fairly represented.

Example 3

How can you randomly select 6 members from a class of 20 for a special project?

Solution

List all the students and assign each student a number from 1 to 20.

01 Paul	06 Katlin	11 Sam	16 Pierre
02 Karen	07 Jeffrey	12 Katherine	17 Misha
03 Samir	08 Milton	13 Ross	18 Tom
04 Kirsten	09 Soo Ping	14 Michela	19 Ken
05 Namath	10 Peter	15 Gus	20 Ayla

Use a random number table. Start anywhere in the table and move systematically in one direction. Circle the appropriate digits and select the matching students. If a digit repeats, skip it and select the next unused digit. In this case, we will consider the first 2 digits in each set of 3. We will start at the top left and move from left to right along the rows.

Random Number Table

786	582	625	313	338	(05)8	584	669	896	640	(13)7	536	460
(14)8	243	(12)8	(11)7	710	222	826	240	721	302	469	337	734
(09)8	110	867	778	541	402	544	193	541	485	572	303	810

The 6 numbers chosen are 05, 13, 14, 12, 11, and 09.
The 6 students are Namath, Ross, Michela, Katherine, Sam, and Soo Ping.

Example 4

There are 175 students in Grade 7, 175 in Grade 8, and 150 in Grade 9. How many students from each grade should an unbiased sample of 20 students contain?

Solution

The sample should contain the same percent of students from each grade as there are in the total population of the school.

Grades 7 and 8: $\frac{175}{500} = \frac{35}{100}$
$= 35\%$

Grade 9: $\frac{150}{500} = \frac{30}{100}$
$= 30\%$

Thus, 35% of the sample should come from Grade 7, 35% from Grade 8, and 30% from Grade 9.

The sample should include 7 students from Grade 7, 7 from Grade 8, and 6 from grade 9.

In this example, the sample took account of the different groups within the population. When we do this, we use a **stratified sample**.

Practice

Would each of the following likely come from a sample or a census? Explain.

1. The student enrolment at Silver Springs High School is 574.

2. Two-fifths of teenage students prefer sugarless gum.

3. A Canadian family spends $55/year on books.

4. Of last year's 490 subscribers, 452 will buy season tickets to the theatre this year.

5. There are 25 000 fish in the lake.

State whether each of the following is a primary or secondary source of data.

6. newspaper or magazine

7. computer database

8. reading an opinion poll

9. conducting an interview

10. measuring the ozone layer

11. performing a door-to-door survey

12. almanac

13. testing a drug in a laboratory

14. encyclopedia

Problems and Applications

15. What method would you use to find each of the following? Explain your choice.

a) the world's 10 biggest countries

b) the populations of the provincial capitals

c) your classmates' favourite foods

d) the most popular radio station in your home town

16. There are 30 girls and 20 boys in the ski club. You want to survey a stratified sample of 10 members to find out what trips to plan.

a) How many boys should you survey? Explain.

b) Are there other ways in which you could stratify the sample? Explain.

17. In a survey, people were asked to estimate the percent of the world's rainforests that have been destroyed. These are the results.

10	20	40	80	20	40
20	30	50	20	40	40
20	40	30	60	30	10
20	40	40	10	20	30
40	60	30	50	30	50

a) Complete a frequency distribution table for these data.

b) How many people were surveyed?

c) The percent with the greatest frequency is actually the correct value. What percent of rainforests have been destroyed?

d) Is the most frequent estimate necessarily the correct value? Explain.

18. The editor of the school newsletter wants to publish information on students' favourite TV and movie stars. Could a survey of each of the following groups be biased? Explain.

a) drama club students **b)** female students

c) 10 students at random

19. Suppose you were asked to recommend graduation activities for your school by surveying a sample of students. Would you stratify the sample or would you select randomly from the whole school? Explain.

20. What methods could you use to choose a random sample of 5 people from your class? Compare your methods with your classmates'.

LOGIC POWER

The 3-by-3-by-3 cube is made from 1-by-1-by-1 cubes. Three rows of the smaller cubes have been removed. How many smaller cubes are left?

10.2 Reading and Drawing Bar Graphs

Bar graphs are used to make comparisons. The bars can be horizontal or vertical.

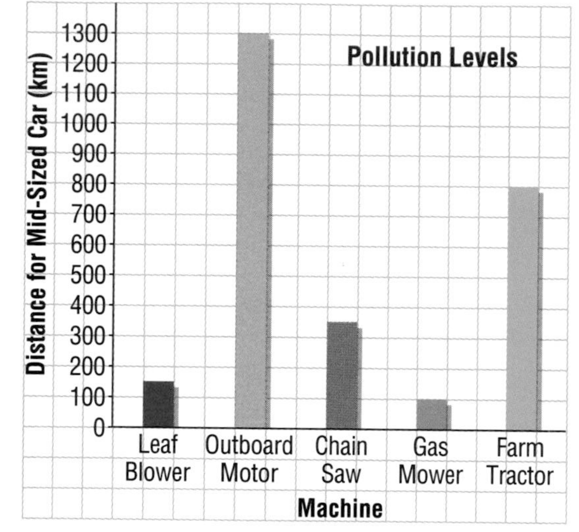

Activity: Study the Graph

Machines with gasoline-powered motors are a major cause of air pollution. The bar graph shows approximately how many kilometres a mid-sized car must be driven to produce as much pollution as each of these machines does in one hour.

Inquire

1. List the machines in decreasing order of the pollution they produce.

2. Why is the bar graph more effective than a list of numbers for showing the amounts of pollution?

3. About how many times more pollution does a farm tractor produce than a gas mower?

Double or multiple bar graphs are often used to compare groups of data.

Example 1

The table gives the percents of people in different age groups who snore and talk in their sleep. Display the data on a double bar graph.

	Frequency (%)	
Age	**Snoring**	**Talking**
15–24	17	29
25–34	30	23
35–44	42	15
45+	47	9

Solution

Draw the horizontal axis, then draw the vertical axis using an appropriate scale. In this case, the range of numbers is 38, from a high of 47 to a low of 9. Choose a uniform width for the bars. Plot the bars from the given values.

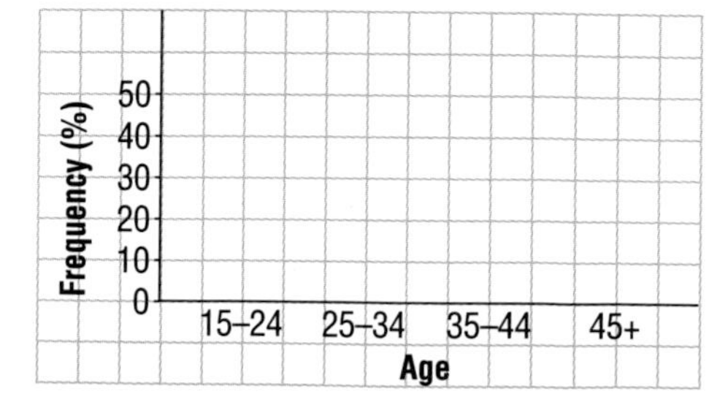

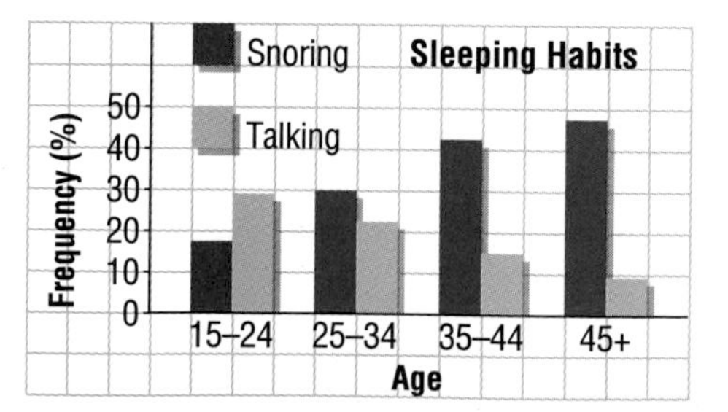

Problems and Applications

1. The bar graph shows the percent of car owners who do various tasks on their cars.

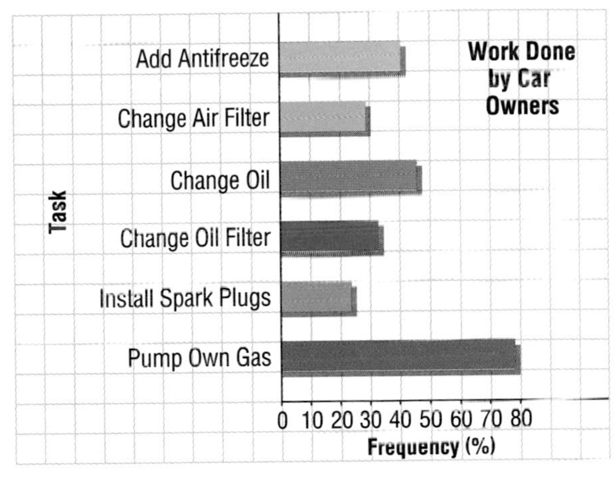

a) What is the only task that most car owners do for themselves?
b) About how many times more car owners change the oil than install spark plugs?

2. The double bar graph shows the greatest and least average snowfalls per year recorded at weather stations in different provinces.

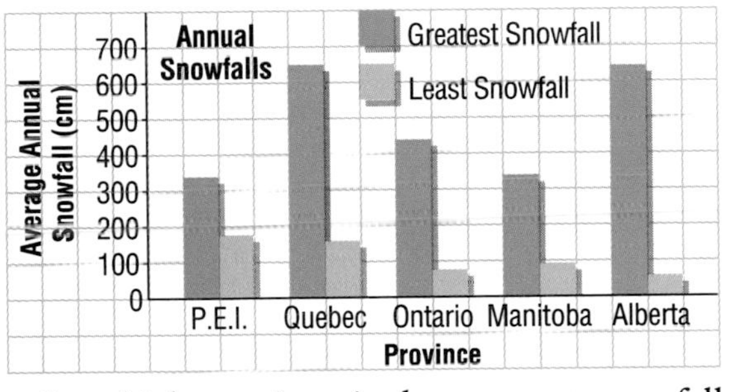

a) In which province is the greatest snowfall about the same as the greatest snowfall in Alberta?
b) Which province has the biggest difference between the greatest and least snowfalls?
c) In which province is the greatest snowfall closest to 4 times the least snowfall?

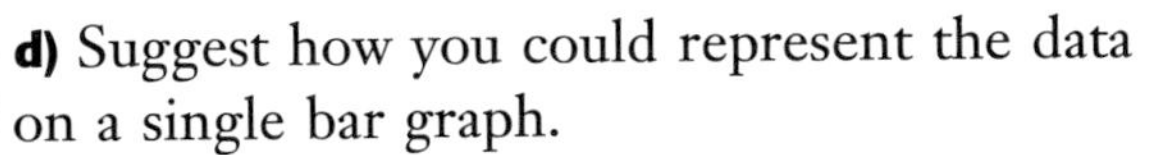

d) Suggest how you could represent the data on a single bar graph.

3. The table shows the incubation periods for the eggs of different birds. Display the data on a bar graph.

Bird	Incubation Period (days)
Emperor Penguin	63
Falcon	28
Ostrich	42
Royal Albatross	79
Swan	30
Wren	16

4. The table shows the percent of the world's land and the percent of the world's population on each continent.

Continent	Percent of Land	Percent of Population
Asia	30	60
Africa	20	12
N. America	16	8
S. America	12	5
Antarctica	9	0
Europe	7	14
Australia	5	0.3

a) Draw a double bar graph.
b) Which continents have a greater percent of the world's population than they have of the world's land?
c) Which are the 2 most crowded continents? Explain.
d) Which is the least crowded continent? Explain.

5. The table gives the heavy metal pollution in milligrams per litre for some rivers.

River	Heavy Metal Pollution (mg/L)
Danube	79
Meuse	11
Mississippi	12
Rhine	32
St. Lawrence	2
Sakayra	73
Seine	70
Thames	19
Tornionjoki	12

a) Draw a bar graph.
b) Why is this bar graph difficult to draw?

6. Research the winners of the Stanley Cup and draw a bar graph to show how many times each city has won it.

10.3 Reading and Drawing Broken-Line Graphs

A broken-line graph shows how something changes.

Activity: Interpret the Graph

The broken-line graph shows how the world's annual fresh water consumption changed over 50 years.

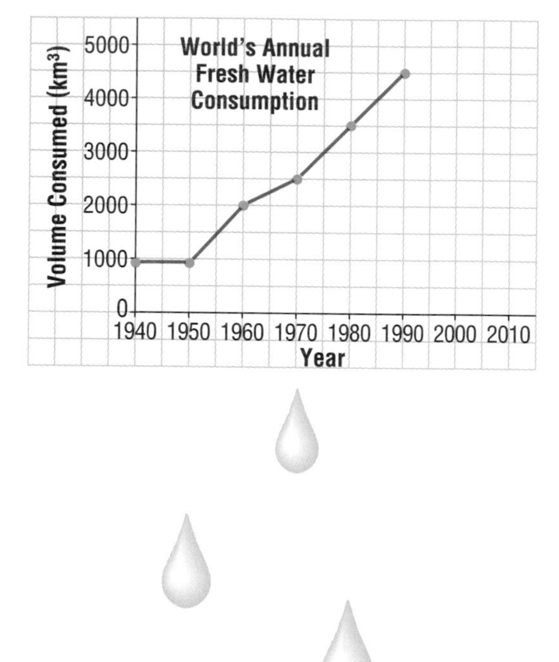

Inquire

1. About how much fresh water was used in 1940? 1960? 1980? 1990?

2. About how many times more fresh water is used now than in 1940?

3. a) What will the world's fresh water consumption be in the year 2000? the year 2010?
b) State the assumptions you made.

4. It is estimated that there are 39 000 km^3 of fresh water available for use in the world each year. Why is there concern about the use of fresh water?

5. Could you use a bar graph to represent the water consumption data? Explain.

Example 1

The table shows changes in the land speed record. Plot the data on a broken-line graph.

Year	Land Speed Record (km/h)
1900	60
1910	200
1920	240
1930	350
1940	600
1950	650
1960	750
1970	1000
1980	1000
1990	1000

Solution

Draw and label the axes. Choose a scale that allows you to plot all the data.

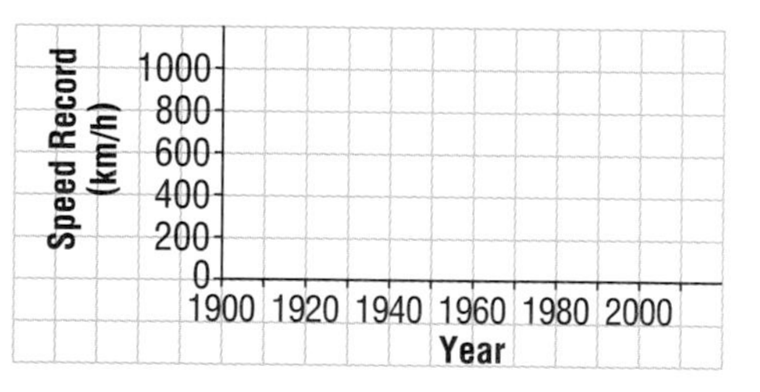

Plot the data and join the points with straight lines. Give the graph a title.

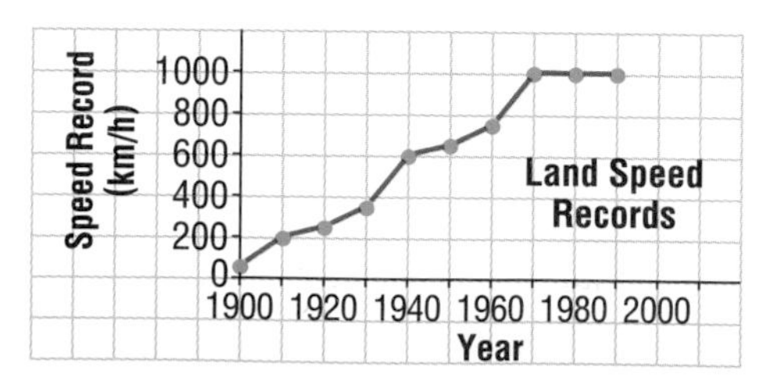

Problems and Applications

1. The broken-line graph shows Winnipeg's average daily temperature in degrees Celsius for each month of a year.

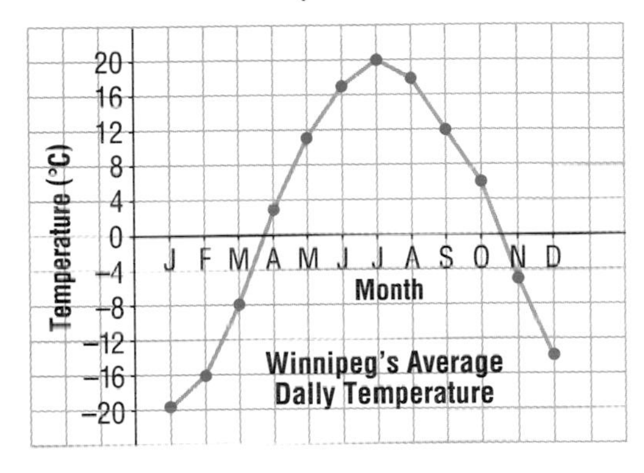

a) What is the difference in the average temperatures for the hottest month and the coldest month?

b) Between which 2 months does the temperature climb fastest?

c) Between which 2 months does the temperature drop fastest?

2. The diagram shows 2 broken-line graphs on the same set of axes. The graphs show the winning throws in the women's discus and the women's javelin at the Olympics.

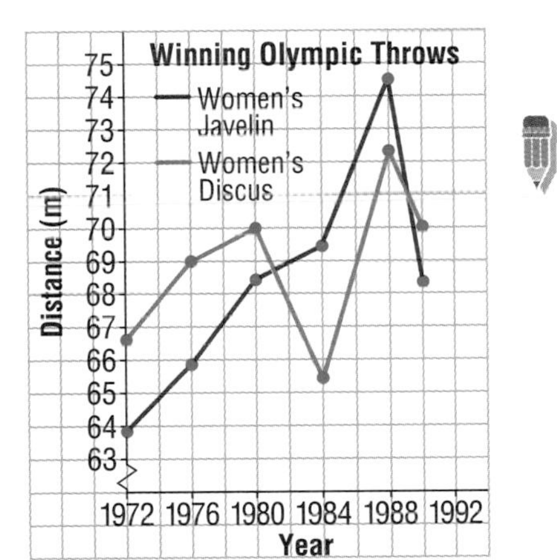

a) In which years was the javelin thrown further than the discus?

b) What was the difference in the winning distances in 1976?

c) What does the jagged line on the vertical axis mean?

3. These are the winners of the Blue Riband Award for the fastest Atlantic crossing by a liner.

Year	Winner	Speed (km/h)
1840	*Britannia*	19.6
1863	*Scotia*	25.9
1882	*Alaska*	31.9
1897	*Kaiser Wilhelm der Grosse*	41.5
1909	*Mauretania*	48.0
1929	*Bremen*	51.7
1938	*Queen Mary*	58.7
1952	*United States*	66.0

a) Display the data on a broken-line graph.

b) Which liner crossed at about half the speed of the *Bremen*?

4. The table shows changes in the percent of the Canadian population in different age groups.

	Age		
Year	Under 19	20 – 44	Over 44
1966	42	32	26
1971	39	34	27
1976	36	36	28
1981	32	39	29
1986	29	41	30
1991	27	41	32

a) Represent the data by plotting 3 broken-line graphs on the same set of axes.

b) Describe the trend in the average age.

5. a) Research the population of your province in every tenth year from 1911 to 1991. Plot the data on a broken-line graph.

b) In which 10-year period was there the biggest change in population?

WORD POWER

The word WRITER has a silent W. How many other letters of the alphabet can you find in words where they are silent? Compare your findings with a classmate's.

10.4 Reading and Drawing Circle Graphs

Circle graphs are used to show how something is divided.

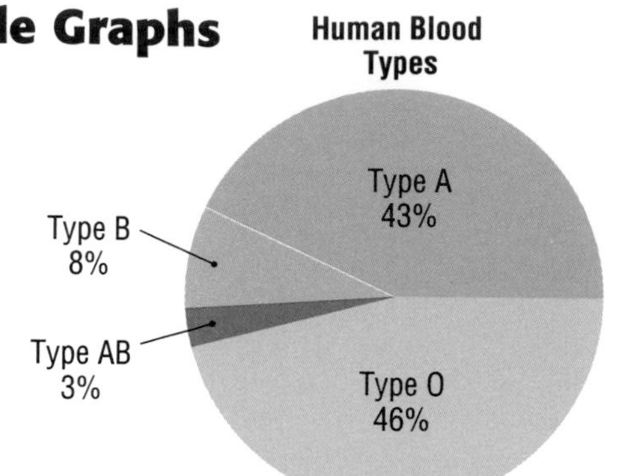

Activity: Analyze the Graph

Examine this circle graph. Then, describe its meaning.

Inquire

1. a) Find the total of the percents on the graph.

b) Would you expect the same total for another circle graph? Explain.

2. a) Measure the number of degrees in the sector that represents Type B.

b) What percent of a circle is this number of degrees?

c) Compare your answer to part b) with the percent shown for Type B on the graph.

3. How many times more people have Type O blood than have Type B blood?

4. In a sample of 5000 Canadians, how many do you expect to have Type A blood?

Example 1

Canada is a major producer of natural gas. The table shows the approximate volumes of natural gas produced in different provinces. Display the data on a circle graph.

Province	Natural Gas Production (millions of cubic metres)
Saskatchewan	6000
Alberta	83 000
British Columbia	10 000
Other	1000
Total	100 000

Solution

Write each volume as a percent of the total.

Saskatchewan $\frac{6000}{100\ 000} \times 100\% = 6\%$

Alberta $\frac{83\ 000}{100\ 000} \times 100\% = 83\%$

British Columbia $\frac{10\ 000}{100\ 000} \times 100\% = 10\%$

Other $\frac{1000}{100\ 000} \times 100\% = 1\%$

Calculate this percent of a circle.

$0.06 \times 360° = 21.6°$

$0.83 \times 360° = 298.8°$

$0.10 \times 360° = 36°$

$0.01 \times 360° = 3.6°$

Draw a circle and measure each angle with a protractor. Approximate where necessary. Label each sector with a name and a percent. Give the graph a title.

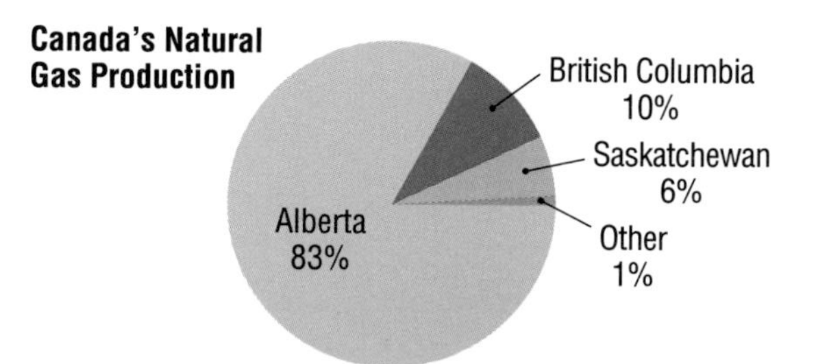

Problems and Applications

1. The circle graph shows data on the suitability of land for farming.

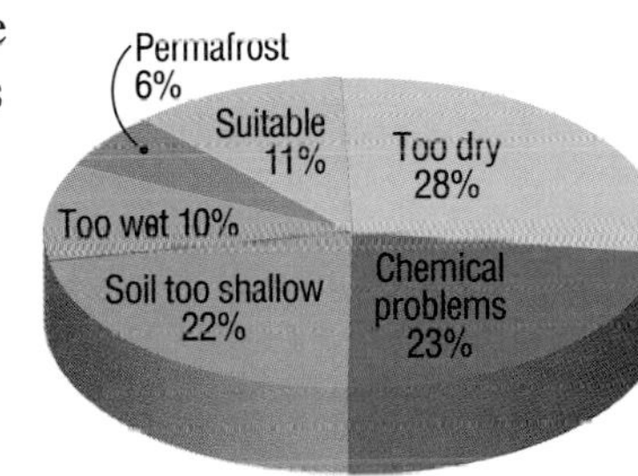

Suitability of Land for Farming

a) Which 2 categories account for over half the land?
b) For every 500 ha of land, how many hectares are suitable for farming?

2. The circle graph shows the world's uranium deposits.

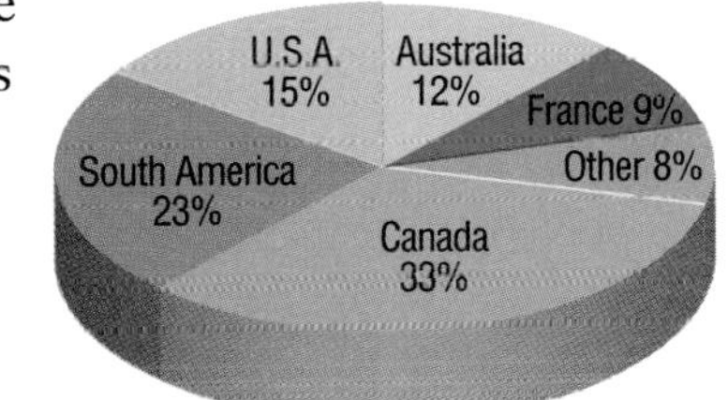

World's Uranium Deposits

a) Rank the deposits in decreasing order.
b) Calculate the angle in the sector that represents Canada's uranium.
c) If you measured this angle with a protractor, why might you not get the same answer as you calculated?

3. The table shows the composition of municipal waste. Draw, label, and title a circle graph for these data.

Type of Waste	Percent
Paper	40
Yard Waste	18
Food Waste	12
Glass	8
Plastic	7
Steel	7
Other	8

4. The table shows the number of species of different types of birds that breed in Canada. Use the data to draw a circle graph.

Type of Bird	Breeding Species in Canada
Ducks	36
Birds of Prey	19
Shorebirds	71
Owls	14
Perching Birds	180
Other	80

5. Would you represent each of the following on a circle graph, a broken-line graph, or a bar graph? Explain each choice and compare it with your classmates'.
a) the heights of the world's 10 tallest buildings
b) how you spend a school day
c) your shoe size at different ages

6. a) Find out your classmates' blood types. Record the data in a frequency chart and draw a circle graph.
b) Compare your graph with the graph of blood types on the opposite page.

PATTERN POWER

Write each sum as a power of 2.

$2^1 + 2^1 = \blacksquare$
$2^2 + 2^2 = \blacksquare$
$2^3 + 2^3 = \blacksquare$
$2^4 + 2^4 = \blacksquare$

a) Describe the pattern.
b) Explain why it works.
c) Use the pattern to predict $2^8 + 2^8$. Use a calculator to check your answer.
d) Can you write a similar pattern for powers with another base?

10.5 Reading and Drawing Pictographs

Pictographs display data visually with pictures or symbols.

Activity: Interpret the Pictograph

The pictograph shows the number of telephones per 100 people in different countries. List the numbers shown in the pictograph.

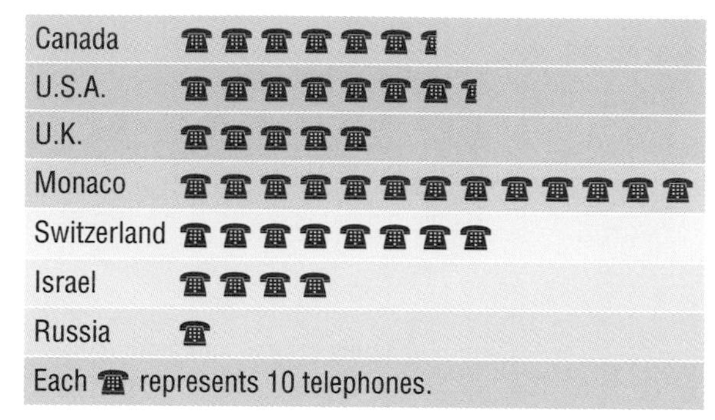

Inquire

1. How many telephones are there per 100 people in
a) Switzerland? **b)** Canada? **c)** U.S.A.?

2. Could you use each of the following graphs to represent the data shown in the pictograph? Explain.
a) circle graph **b)** bar graph **c)** broken-line graph

Problems and Applications

1. The pictograph represents the populations of the countries with the most people.

Brazil	🚹🚹
China	🚹🚹🚹🚹🚹🚹🚹🚹🚹🚹🚹🚹
India	🚹🚹🚹🚹🚹🚹🚹🚹🚹
Indonesia	🚹🚹
Russia	🚹🚹
United States	🚹🚹🚹
Each 🚹 represents 100 000 000 people.	

a) Which 3 countries have a total population that about equals China's?
b) The population of Canada is about 27 000 000. Which country has about 10 times as many people as Canada? about 30 times as many people as Canada?

2. The table gives the top speeds of 5 fish. Display the information on a pictograph.

Fish	Top Speed (km/h)
Barracuda	43
Mackerel	33
Sailfish	110
Shark	69
Swordfish	90

3. The table shows the approximate circulations of 6 magazines in Canada.

Magazine	Circulation
Canadian Geographic	250 000
Chatelaine	890 000
Equinox	170 000
L'Actualité	225 000
Maclean's	565 000
Westworld	825 000

a) Display the data on a pictograph.
b) What are the advantages and disadvantages of using a bar graph instead of a pictograph to display these data?
c) Could you represent these data on a circle graph? a broken-line graph? Explain. Compare your answers with a classmate's.

LOGIC POWER

The circle and the triangle intersect at 2 points. Draw diagrams to show a circle and a triangle intersecting at 3, 4, 5, and 6 points.

10.6 Mean, Median, Mode, and Range

In previous sections, graphs were used to describe and display data. Statisticians are also interested in finding a number to describe the middle or centre of a set of data.

Activity: Examine the Table

It is dangerous for aircraft to fly through thunderstorms, so air traffic controllers keep a careful watch on thunderstorm activity. The table shows the days of thunderstorms per year at airports in some Canadian cities.

Airport	Days of Thunderstorms
Edmonton	27
Halifax	9
Montreal	25
Ottawa	24
Regina	23
St. John's	3
Toronto	27
Vancouver	6
Winnipeg	27

Inquire

1. Rank the data in ascending order, that is, from lowest to highest.

2. a) What is the middle number in the new order?

b) What is the number that occurs most often?

c) Take the average of the numbers in the list.

3. The numbers from question 2 are 3 different ways to describe the centre of a set of data. Compare how effective they are at showing the centre of the data in the table.

4. What is the difference between the highest and lowest values in the data?

The 3 numbers you determined in question 2 above, are known as **measures of central tendency**.

The **mean**, or arithmetic average, is calculated by finding the sum of the data and dividing by the number of pieces of data. It is the most sensitive to changes in the data, because each piece of data helps determine it.

The **median** is the middle value when the data are arranged in numerical order. If the number of pieces of data is even, there are 2 middle values. The median is the arithmetic average of these 2 values. The median is especially useful when there are unusual data that distort the mean.

The **mode** represents popularity. It is the number that occurs most frequently. A set of data may have no mode or more than one mode.

The **range** is not a measure of central tendency. It tells us how spread out the data are. The range is the difference between the highest and lowest values. It does not give any information about the other values.

The mean, median, mode, and range help you analyze data and determine how they are grouped.

CONTINUED

Activity: Use the Data

The golf coach asked Carla and Alicia to play 9 holes of golf to decide who would make the golf team. The score card shows the results. Alicia hit her ball into the lake on the fifth hole.

Hole	1	2	3	4	5	6	7	8	9	Total
Carla	6	4	3	7	7	6	3	4	6	46
Alicia	4	4	3	4	15	5	4	3	6	48

Inquire

1. Who won the match?
2. Who won more holes?
3. Calculate the mean score for each player. Round each answer to the nearest tenth.
4. Who had the better mean score?
5. Calculate the median score for each player.
6. Who had the better median score?
7. Determine the mode of each player's scores.
8. Who had the better mode?
9. Who should make the golf team? Explain.

Example

Frank wrote 6 tests when he took his flying course. His results were as follows.

65, 66, 100, 63, 64, 63

a) Find the mean, median, and mode of the data.

b) Which measure of central tendency best represents Frank's knowledge of flying? Explain.

Solution

a) Mean $= \frac{65 + 66 + 100 + 63 + 64 + 63}{6}$

$= \frac{421}{6}$

$\doteq 70.2$

EST $70 + 70 + 100 + 60 + 60 + 60 = 420$
$420 \div 6 = 70$

The mean is 70.2 to the nearest tenth.

To find the median, arrange the numbers in order.

63, 63, 64, 65, 66, 100

There are 2 middle numbers, 64 and 65.

The median is $\frac{64 + 65}{2}$ or 64.5.

The mode is 63.

So, the mean is about 70.2, the median is 64.5, and the mode is 63.

b) The median value, 64.5, probably best represents Frank's knowledge of flying. The mode is at the low end of the data, and the mean is distorted by Frank's one perfect score.

Practice

State the measure of central tendency that best describes each of the following.

1. the most requested song on a radio station

2. the test marks 12, 80, 82, 84, 87

3. the most popular baseball cap size

4. the mass of an adult elephant

5. the typical salary paid by a company

6. your bowling ability

Find the mean, median, mode, and range for each set of data.

7. 21, 25, 27, 26, 25

8. 13, 21, 16, 25, 18, 28, 32, 31

9. 8, 16, 28, 41, 16, 11, 8

10. 80, 40, 35, 62, 11, 80

11. 3800, 2700, 1650, 1120, 1360, 4500

Problems and Applications

12. The list gives the shots on goal for the Hornets in their last 10 games.

18 18 25 19 27
18 25 19 22 27

a) Find the mean, mode, and median.

b) Explain how the following calculates the mean.

$$\frac{3 \times 18 + 2 \times 19 + 22 + 2 \times 25 + 2 \times 27}{10}$$

13. Describe the effect of each change on the mean of this set of numbers.

10 11 12 13 14 15

a) Increase each number by 2.

b) Decrease each number by 2.

c) Double each number.

14. These are Tanya's bowling scores.

145 145 168 170 174 182

a) Does the mode give an accurate indication of how well she bowls? Why?

b) Does the median give a more accurate indication of how well she bowls? Why?

15. The table shows the salaries of all the employees in a small company.

Position	Number	Annual Salary ($)
President	1	100 000
Vice-President	1	60 000
Senior Staff	4	40 000
Junior Staff	2	30 000

a) List all 8 salaries in ascending or descending order.

b) Find the mean, median, and mode.

c) Which measure of central tendency is closest to the centre of the data? Explain.

d) Drop the highest salary. Find the mean, median, and mode of the remaining 7 salaries.

e) Drop the lowest salary. Find the mean, median, and mode of the remaining 7 salaries.

f) Which measure of central tendency is most affected by the extreme (highest and lowest) salary values?

16. Is each statement always true, sometimes true, or never true? Explain.

a) If a list of numbers has a mode, it is one of the numbers in the list.

b) The median of a list of whole numbers is a whole number.

c) The mean of a list of numbers is one of the numbers in the list.

d) The mean, median, and mode of a list of numbers are not equal.

17. Write the following sets of numbers in ascending order. Compare your answers with a classmate's. Decide whether different answers are possible in each case.

a) 5 numbers with a mean of 15, a median of 12, and a mode of 11

b) 4 numbers with a mean of 12.5, a median of 11.5, and a mode of 10

c) 6 numbers with a mean of 20, a median of 22.5, and no mode

d) 5 numbers whose mean, median, mode, and range all equal 16

10.7 Stem-and-Leaf Plots

Activity: Interpret the Data

Babe Ruth was one of the greatest home run hitters. He hit these numbers of home runs in 15 seasons with the New York Yankees.

Babe Ruth's Home Runs in 15 Seasons

54	59	35	41	46
25	47	60	54	46
49	46	41	34	22

Organize the data by listing the tens digits, called **stems**. Beside them, list the units digits, called **leaves**.

2	5 2
3	5 4
4	1 6 7 6 9 6 1
5	4 9 4
6	0

The same data are shown below in a type of table called a **stem-and-leaf plot**. Describe how the data are organized in this plot.

Babe Ruth's Home Runs in 15 Seasons	
2	2 5
3	4 5
4	1 1 6 6 6 7 9
5	4 4 9
6	0

Inquire

1. From which table can you read the median most easily? Explain.

2. What was Babe Ruth's median number of home runs?

3. What other type of representation does the stem-and-leaf plot resemble? Why?

4. What was the mode of Babe Ruth's numbers of home runs?

5. What was Babe Ruth's mean number of home runs?

6. Which measure of central tendency best represents these data? Explain.

Back-to-back stem-and-leaf plots can be used to compare data from 2 groups.

Example

a) Construct a back-to-back stem-and-leaf plot for these data.

Hours Spent Watching Television Per Week	
Female	**Male**
13, 16, 17, 18, 19	11, 18
21, 27	21, 26, 26
35, 38	31, 34, 34, 36
40, 42	41

b) How many people were surveyed?

c) Was the mean value higher for males or females?

Solution

a) The stem is the 10s digit in each number. The leaf is the 1s digit in each number.

b) There were 21 people surveyed, 11 females and 10 males.

c) The 10 males watched 278 h of television for a mean of 27.8 h.
The 11 females watched 286 h of television for a mean of 26 h.
So, the mean value was higher for males.

Hours Spent Watching Television Per Week		
Female		Male
9 8 7 6 3	1	1 8
7 1	2	1 6 6
8 5	3	1 4 4 6
2 0	4	1

Problems and Applications

1. Weather stations in each of Canada's provinces and territories have recorded the greatest number of consecutive days with the highest temperature above 32°C. The data are shown in the stem-and-leaf plot.

Stem	Leaf
0	3 4 4 5 7 8
1	1 6 6
2	2 2
3	1

a) How many weather stations were considered?
b) The greatest number of consecutive days with the highest temperature above 32°C occurred in British Columbia in 1971. What was the number of days?
c) Do you agree that most of Canada's provinces and territories have had at least 10 consecutive days with the highest temperature above 32°C? Explain.
d) What are the 3 modes in the data?

2. The list shows the peak wind speeds, in kilometres per hour, recorded at 16 Canadian airports.

109	129	127	148	153	129	124	135
161	177	146	132	177	193	106	106

a) Construct a stem-and-leaf plot.
b) What is the range of the data?
c) What is the median peak wind speed?

3. a) How many more 14-year-olds than 18-year-olds are represented in this back-to-back plot?

Weekly Television Viewing (h/week)		
14-Year-Olds		18-Year-Olds
9 7	0	7 7 9 9
9 8 8 4 4 2 1	1	2 3 4 4 4 8 8 9 9
8 8 6 6 4 3 1	2	1 2 3 4 8
8 8 5 5 5	3	6
2 2	4	1
6	5	0

b) What is the range of hours of television watched by 14-year-olds?
c) Describe the difference in viewing habits of the 2 age groups.
d) What might contribute to the difference in part c)?

4. The following data show the points scored by the Celtics girls basketball team in home and away games.

Home Games	62, 68, 73, 43, 51, 55, 53, 78, 64, 66, 67
Away Games	42, 50, 62, 51, 53, 58, 63, 64, 71, 57, 43

a) Construct a back-to-back stem-and-leaf plot to show the comparison.
b) What was the median number of points in each category?

10.8 Box-and-Whisker Plots

Kim Campbell was 46 years old when she became Canada's nineteenth Prime Minister.

The mean age of Canada's first 19 Prime Ministers when they began their terms in office was about 55 years. Their ages in ascending order were as follows.

39	45	46	46	47	47	48	51	52	54	55	57	60	61	65	66	70	70	74

Inquire

1. What is the lowest value in the data?

2. What is the highest value in the data?

3. What is the median value?

4. Consider the 9 values above the median.
What is the median of these 9 values?

5. Consider the 9 values below the median.
What is the median of these 9 values?

The number you found in question 4 is known as the **upper quartile**. The number you found in question 5 is known as the **lower quartile**. The 5 numbers you found in questions 1–5 are known as a **five-number summary**.

Example 1

a) Use the above data to construct a box-and-whisker plot.
b) About what percent of the values lie in the box? in each whisker?

Solution

a) The 5-number summary is 39, 47, 54, 65, 74. Plot these 5 numbers on a number line. Then, draw vertical line segments through the median and the upper and lower quartiles. Draw a box between the upper and lower quartiles. Draw horizontal segments to the highest and lowest values from the ends of the box. These segments are the whiskers.

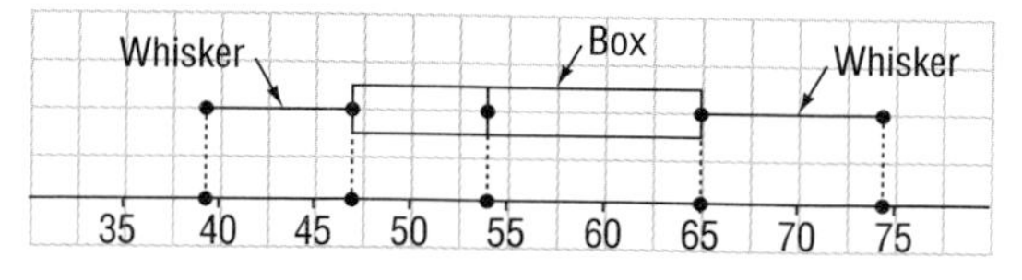

b) About 50% of the numbers lie in the box. Each whisker contains about 25% of the numbers.

Example 2

Mariko scored 75 on a math test. The results, in ascending order, for the whole class were as shown.

45	48	52	52	58	64	64	64	65	67
67	70	72	75	75	75	80	82	86	90

a) In what percentile was Mariko's score? **b)** What did this percentile show?

Solution

a) The percentile of Mariko's score or her percentile rank is given by the expression

$$\frac{b + \frac{1}{2}e}{n} \times 100$$

where b represents the number of scores below hers, e represents the number of scores that equalled hers, and n represents the total number of scores.

In this case, $b = 13$, $e = 2$, and $n = 20$.

Mariko's percentile rank is

$$\frac{13 + \frac{1}{2}(2)}{20} \times 100 = \frac{14}{20} \times 100$$
$$= 0.7 \times 100$$
$$= 70$$

So, Mariko's score was in the 70th percentile.

b) Her percentile rank of 70 showed that Mariko scored better than about 70% of the students, and that about 30% of the students scored better than Mariko.

Problems and Applications

1. The number of hours some bands spend practising a new song before they record is shown below.

a) What fraction of the bands spend more than 30 h practising?
b) Within what range of hours do one-half of the bands practise?
c) What fraction of the bands spend less than 45 h practising?

2. a) Draw a box-and-whisker plot to represent the following test scores.

40	47	50	50	50	54	56	56
60	60	62	62	63	65	70	70
72	76	77	80	85	85	95	

b) Find the percentile rank of a score of 60.

3. a) The median of a set of data is also the 50th percentile. Explain why.
b) What percentiles represent the upper and lower quartiles? Explain.

4. The plots show the math marks for 3 students.

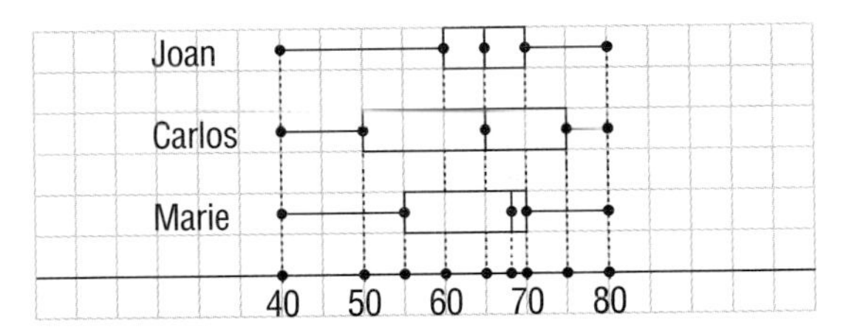

a) One-half of Joan's test scores lie between which 2 values?
b) Which student's test scores are most inconsistent? Explain.
c) What percent of Marie's test scores are above 70%?

5. If you score 80% on a test are you at the 80th percentile? Explain.

6. Measure your pulse rate for 1 min. Plot the results for the whole class in beats per minute on
a) a stem-and-leaf plot
b) a box-and-whisker plot

Histograms

Activity 1

The table shows the number of space shuttle launches in the first 12 years of shuttle missions.

Years	1981–82	1983–84	1985–86	1987–88	1989–90	1991–92
Number of Launches	5	9	11	2	11	14

We can display the data on a type of graph called a **histogram**.

Number of Launches
14
13
12
11
10
9
8
7
6
5
4
3
2
1
0
1982 1984 1986 1988 1990 1992
Year

1. In what ways is the histogram like a bar graph?

2. In what ways is the histogram different from a bar graph?

3. Why are there no spaces between the bars of the histogram?

4. Can you tell from the histogram how many launches there were in 1985? Explain.

5. How many shuttle flights were there from the beginning of 1981 to the end of 1992?

Activity 2

A survey of the owners of 100 personal stereos produced the following data.

Age	0–9	10–19	20–29	30–39	40–49	50–59	60–69	70–79	80+
Frequency	19	26	17	13	7	6	5	4	3

Display the data on a histogram.

Misleading Statistics

Activity ❶

A graph can be used to misrepresent data.

1. These 2 graphs display the same data in different ways.

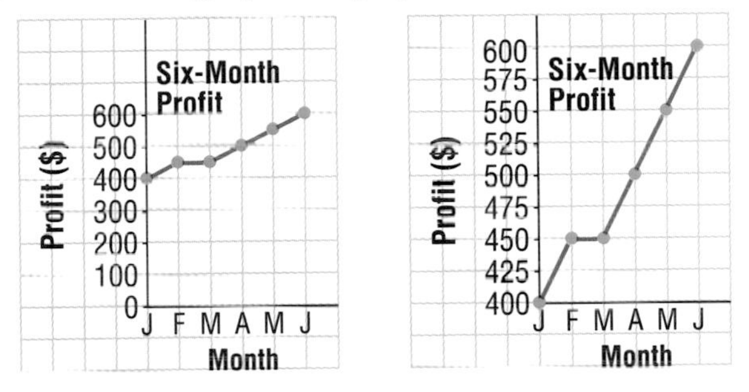

a) How are the graphs the same?
b) How are the graphs different?
c) What impression is the second graph intended to create?

2. Suppose your first math mark was 63 and your next 2 math marks were 70 and 71. Draw a graph to make your improvement look terrific.

Activity ❷

1. In a newspaper or magazine, find an example of a graph that misleads the reader. Write a report on how the graph does this.

2. Graphs do not give the only examples of the misuse of statistics. Statements that distort the meaning of data may create headlines in newspapers and lead stories on TV and radio. Find examples of such statements and report them to the class.

3. Find examples of advertisers who use before-and-after pictures to sell their products. Describe any examples that attempt to mislead.

4. How do car and athletic-shoe commercials try to convince you to buy the product? Describe any techniques that are misleading.

Activity ❸

Statistics are used to highlight the significance of events and to clarify their meaning. This headline

WORST STORM IN 100 YEARS

is more effective than this headline.

STORM CLOSES CITY

Find examples of the effective use of statistics in the media.

Millions lost in huge scam

Graphics Software Packages

Research has shown that some people learn better when information is presented to them visually. One way to present information visually is to use graphics software packages. There are 2 types of packages.

An *analytical* graphics package takes numerical data that are in rows and columns and displays them in bar graphs, broken-line graphs, and circle graphs. The graphical display is much easier to analyze than rows and columns of numbers. You can display analytical graphics on a monitor or print them out.

Presentation graphics packages are used to deliver a message by people who work in sales and marketing. Presentation graphics look more sophisticated than analytical graphics and use techniques a graphic artist might use. The graphics displays can be converted into overhead transparencies or slides. Some packages allow you to make animated graphics on a computer and run them on a VCR.

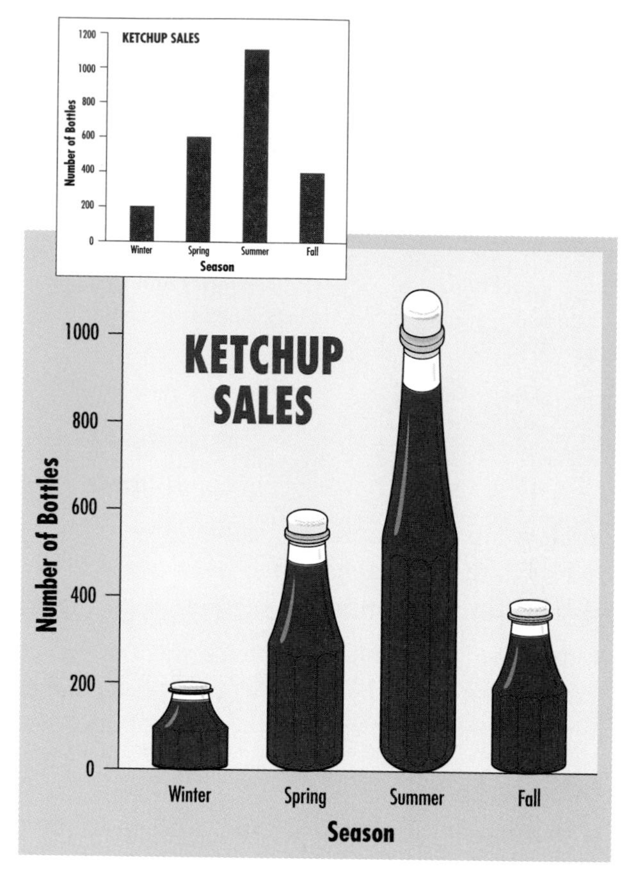

Activity ❶

1. If a computer with a graphics software package is available, try different graphs to display these survey data.

Kinds of Athletic Shoes People Buy	
Type of Shoe	**Frequency (%)**
Running	10
Golf	6
Tennis	11
Basketball	26
Aerobics	6
Walking	11
Cross-training	18
Other	12

2. Display the information on a bar graph in a creative way.

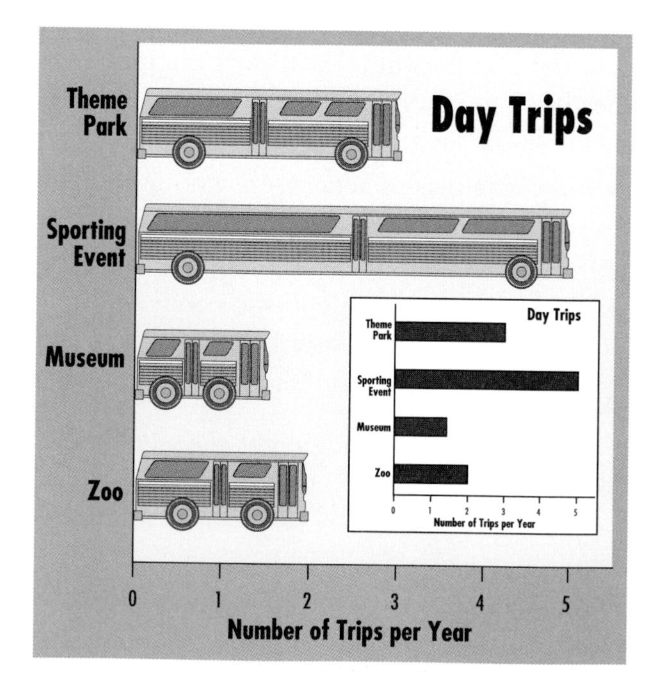

Activity 2

The following 3 graphs display the data from Activity 1. Which graph do you think best represents the data? Give reasons for your choice.

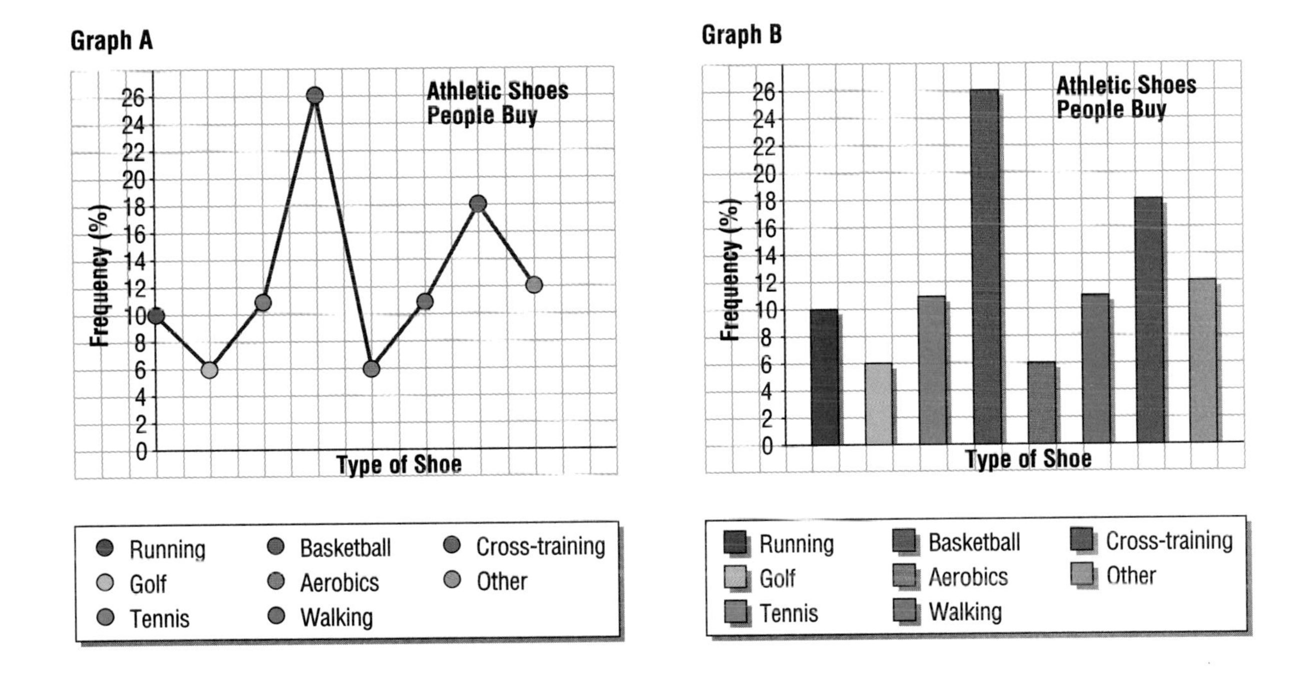

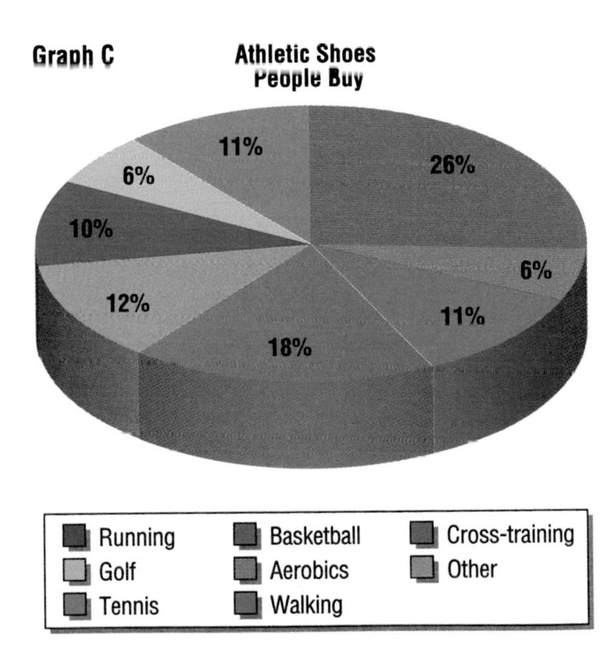

Activity 3

Suppose you had a presentation graphics package you could use to design animated graphics and run them on a VCR. How would you display the data from Activity 1?

10.9 Possible Outcomes

Activity: Compare the Methods

When red shorthorn cattle and white shorthorn cattle interbreed, some of their offspring have a mixture of red and white hair. One way to represent the possible offspring from red and white shorthorn cattle is with a **tree diagram**.

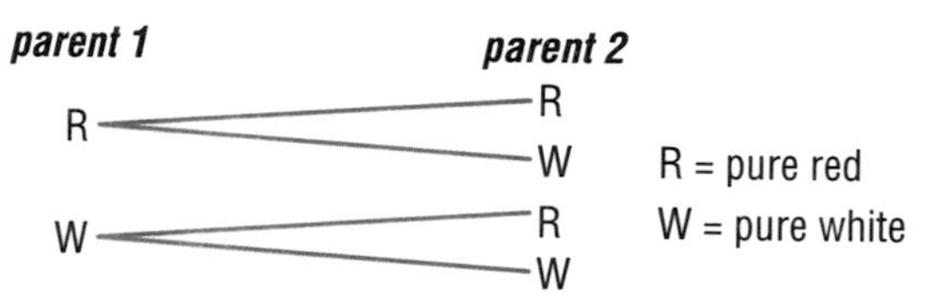

A second method is to use a table.

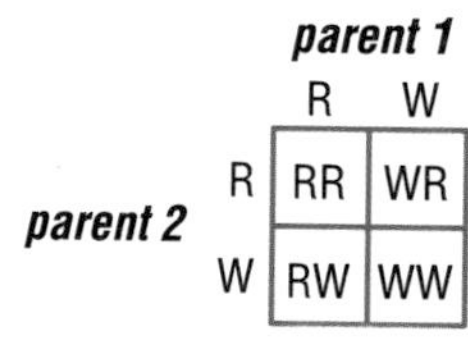

		parent 1	
		R	W
parent 2	R	RR	WR
	W	RW	WW

Both methods show that the 4 possible outcomes are RR, WR, RW, and WW. Offspring from 2 red cattle, RR, are red. Offspring from 2 white cattle, WW, are white. Offspring from 1 red parent and 1 white parent are represented by WR and RW. They have a mixture of red and white hair.

Inquire

1. Which method do you prefer for representing the possible outcomes in this case? Why?

2. When red and white shorthorn cattle interbreed, 50% of the offspring have a mixture of red and white hair. Explain how the tree diagram and the table show this.

3. What percent of the offspring from the red and white cattle are red? Explain.

The possible outcomes from an experiment are often referred to as the **sample space**. The sample space for the cattle offspring is RR, WR, RW, and WW.

Example 1

Spinners are used in some board games.

a) How many possible outcomes are there for each spin of this spinner?

b) What is the sample space for this spinner?

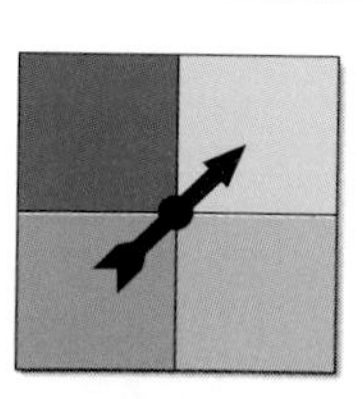

Solution

a) There are 4 possible outcomes for each spin.

b) The sample space is red, yellow, green, and blue.

Practice

1. When you roll a die, what are the possible outcomes?

2. When you toss a coin, what is the sample space?

3. a) If you choose a card from a standard deck of cards, how many possible outcomes are there if you are concerned only with the colour of the card?

b) If you are concerned only with the suit of the card, how many outcomes are there?

Problems and Applications

4. The tree diagrams show the possible outcomes for an experiment in which a coin is tossed and a die is rolled.

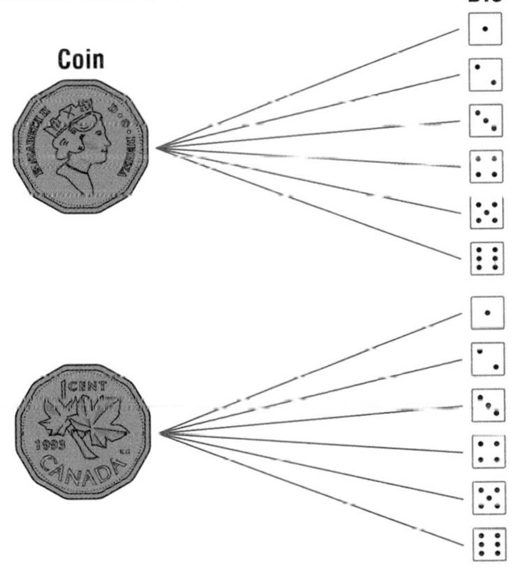

a) How many possible outcomes are there?

b) What is the sample space?

5. a) What are the possible outcomes for the numbered spinner? for the coloured spinner?

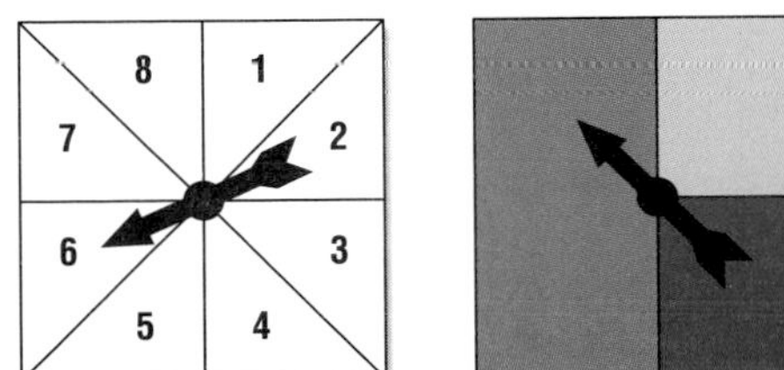

b) Draw a tree diagram to show the possible outcomes when both spinners are used at the same time.

c) How many possible outcomes are there?

d) Is "7, yellow" as likely an outcome as "7, blue"?

6. Draw a tree diagram that shows the sample space when a red die and a white die are rolled.

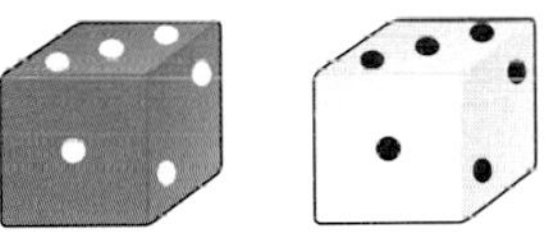

7. a) In some games, the sum of the values on 2 dice is important. Copy and complete the table to show possible sums on 2 dice.

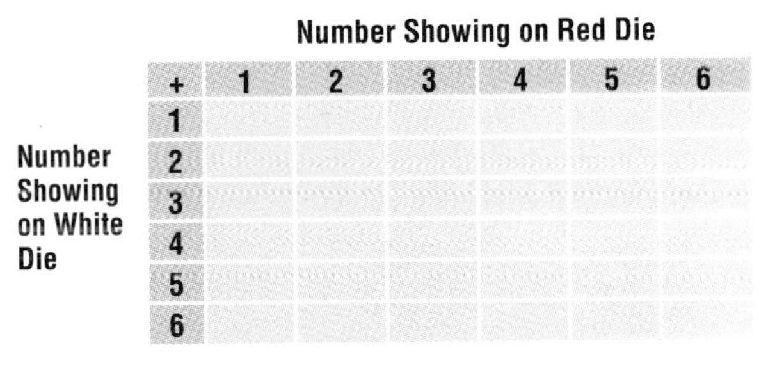

Number Showing on Red Die

Number Showing on White Die

+	1	2	3	4	5	6
1						
2						
3						
4						
5						
6						

b) What is the sample space?

c) In how many ways can a sum of 7 occur? List them.

8. If you toss a nickel, a dime, and a quarter at the same time, what is the sample space? Compare your answer with a classmate's.

LOGIC POWER

The diagram shows one triangle on a 9-point grid. How many other sizes of triangle can you draw on a 9-point grid?

10.10 The Probability Formula

Activity: Use the Pattern

When you roll a die, there are 6 possible outcomes, 1, 2, 3, 4, 5, and 6. Each outcome has an equal chance of happening or is **equally likely**. So, the chance or **probability** of rolling a 3, $P(3)$, is 1 out of 6.

$P(3) = \frac{1}{6}$ ← Number of ways to get a 3 / ← Total number of possible outcomes

Inquire

1. What is the probability of each of the following outcomes?
a) $P(6)$ **b)** P(even number) **c)** P(prime number)
d) P(number divisible by 3) **e)** P(number divisible by 7)
f) P(number less than 3)

2. a) How many possible outcomes does a toss of a coin have?
b) What is the probability of each outcome?

When all the outcomes are known and equally likely, the probability of a single outcome is given by the probability formula.

$$\text{Probability of an outcome} = \frac{\text{number of favourable outcomes}}{\text{total number of possible outcomes}}$$

Example 1

Find the probability of each of the following outcomes for this spinner. Express each answer as a percent.

a) a 4
b) an odd number
c) a 4 or a 7
d) a 9
e) a number from 1 to 8
f) the colour green

Solution

There are 8 equal sectors on the spinner.

a) There is one 4.
So, $P(4) = \frac{1}{8}$ or 12.5%

b) There are 4 odd numbers.
So, $P(\text{odd}) = \frac{4}{8}$ or $\frac{1}{2}$ or 50%

c) There is one 4 and one 7.
So, $P(4 \text{ or } 7) = \frac{2}{8}$ or $\frac{1}{4}$ or 25%

d) There is no 9.
So, $P(9) = \frac{0}{8}$ or 0 or 0%

e) There are 8 numbers from 1 to 8.
So, $P(1 \text{ to } 8) = \frac{8}{8}$ or 1 or 100%

f) There are 4 green sectors.
So, $P(\text{green}) = \frac{4}{8}$ or $\frac{1}{2}$

You can see from Example 1 that an impossible outcome has a probability of 0, and a certain outcome has a probability of 1. Outcomes that are neither impossible nor certain have probabilities between 0 and 1 or between 0% and 100%.

Practice

1. What is the probability of spinning each of the following numbers with this spinner?

a) 1

b) a prime number

c) 1, 2, 3, or 4

d) 6

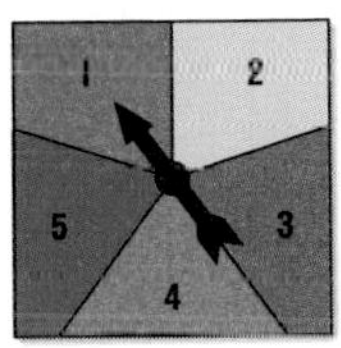

2. Which number or numbers have these probabilities for the spinner in question 1?

a) 0 **b)** $\frac{1}{5}$ **c)** 1

3. What is the probability of rolling each of these numbers with a die? Express each answer as a percent.

a) 2 **b)** an odd number

c) 4, 5, or 6 **d)** a number from 1 to 6

Problems and Applications

4. a) On this spinner, what is P(blue)?

b) What is P(red)?

c) How many times more likely is the spinner to land on blue than on red?

5. The Junior Hockey League is raising money by raffling a car. There are 20 000 tickets. You buy 2 tickets.

a) What is your probability of winning if all the tickets are sold?

b) What is your probability of winning if only half the tickets are sold and the draw is made from the stubs of the sold tickets?

6. There are 4 green, 15 red, 6 yellow, and 5 black marbles in a bag. You remove 1 marble without looking. State the probability that it is

a) red **b)** black **c)** purple

d) green, yellow, red, or black

7. Each letter of the word IMPOSSIBLE is on a different card. All the cards are the same size. The cards are placed face down and shuffled. State the probability that you will randomly draw each of the following letters. Write each answer as a decimal.

a) I **b)** N **c)** S **d)** L

8. State the probability of drawing these cards from a standard deck of cards.

a) the 2 of clubs **b)** a black card

c) a heart **d)** a red jack

9. If you add the probabilities of all the possible outcomes for a spinner or die, what is the result? Explain.

10. If the probability of spinning a 1 on a spinner is 0.2, what is the probability of not spinning a 1?

11. A driving instructor says that the probability of her students passing their driving test on the first try is 0.80.

a) If the instructor now has 25 students, how many do you expect to pass on the first try?

b) If you take lessons from this instructor, what is your chance of passing on the first try? Explain.

12. a) Draw a spinner that gives the following probabilities.

$P(\text{red}) = \frac{1}{8}$ $P(\text{blue}) = \frac{3}{8}$

$P(\text{green}) = \frac{1}{2}$ $P(\text{yellow}) = 0$

b) Describe your method and compare it with a classmate's.

c) Predict the outcomes from spinning the spinner 1200 times.

13. Design your own spinner and write a problem to be answered with it. Have a classmate solve your problem.

10.11 Independent Events

Activity: Conduct an Experiment

Work with a partner and toss a penny and a nickel at the same time. Repeat another 19 times. Record the results of the 20 trials and calculate the probability of each outcome. Collect data from the rest of the class and calculate each probability using all the data.

Inquire

1. How many different outcomes are possible when 2 coins are tossed at the same time?

2. What is the sample space when 2 coins are tossed at the same time?

3. What is the probability of throwing

a) 2 heads? **b)** 2 tails?

c) a head and a tail? **d)** 2 heads or 2 tails?

4. Write each probability from question 3 as a percent.

When 2 coins are tossed simultaneously, the outcome for one coin has no effect on the outcome for the other. The events are said to be **independent** of each other.

Example

A coin is tossed and a die is rolled at the same time. What is the probability of getting a head and a 6?

Solution

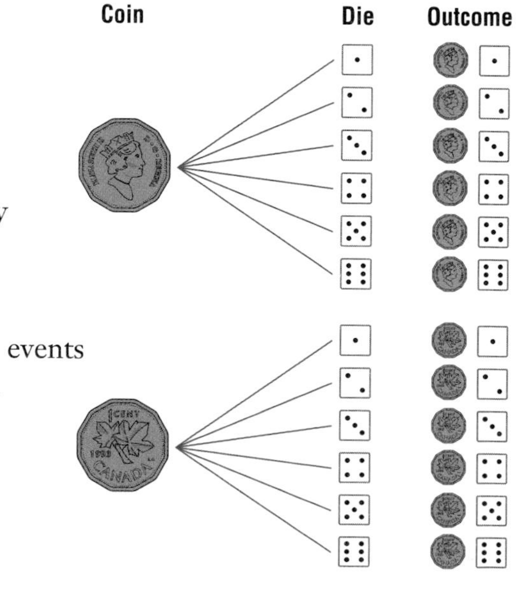

Construct a tree diagram to show the sample space. There is only 1 favourable outcome out of 12 equally likely outcomes.

$P(\text{H6}) = \frac{1}{12}$

Another way to find the probabilities of independent events is to use the probabilities of the individual outcomes.

For a toss of a coin, $P(\text{H}) = \frac{1}{2}$

For a roll of a die, $P(6) = \frac{1}{6}$

$P(\text{H6})$ is $\frac{1}{2}$ of $\frac{1}{6}$

$= \frac{1}{2} \times \frac{1}{6}$

$= \frac{1}{12}$

Problems and Applications

1. This spinner and die are used for an experiment.

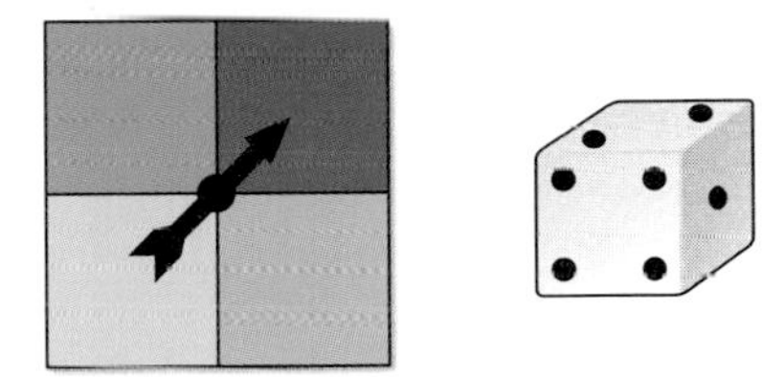

What is the probability of each of the following?

a) spinning red
b) rolling a 5
c) spinning red and rolling a 5
d) rolling an odd number and spinning green

2. A red die and a green die are rolled. What are the probabilities of these outcomes?

a) The red die shows a 6 and the green die shows a 1.
b) Both dice show 6s.
c) The red die shows an even number and the green die shows an odd number.

3. A bag contains 3 red and 2 blue balls. What is the probability of drawing each of the following if each ball is replaced after it is drawn?

a) a red ball, then a blue ball
b) 2 red balls

4. For the following spinners, which game would you rather play? Why?

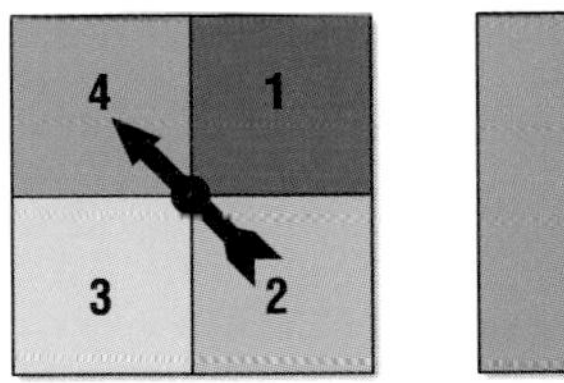

a) If you play on the numbered spinner, you win if you spin a 1.
b) You win on the second spinner if, in 2 turns, you spin red then blue.

5. a) What is the theoretical probability of a child being female for any birth in a family?
b) What is the theoretical probability in a family with 2 children that both are girls?
c) What is the theoretical probability in a family with 3 children that all 3 are boys?

6. The following activity is played by picking a card from a bag, replacing the card, and then spinning the spinner.

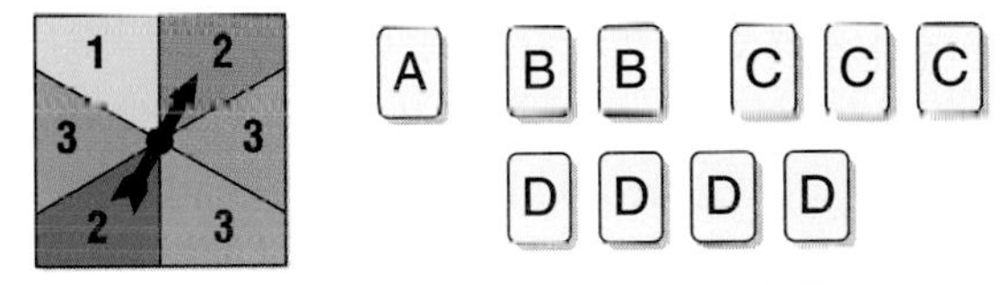

What are the probabilities of the following outcomes?

a) D then 3 **b)** C then 2 **c)** A then 3
d) D then 2 **e)** B then 1 **f)** A then 1

7. If you toss a coin and it shows a head, are your chances greater than one-half of tossing a tail on the next toss? Explain.

8. Design a spinner so that, when you toss a coin then spin the spinner, the probability of "a head then red" is

a) $\frac{1}{10}$ **b)** $\frac{1}{6}$ **c)** 0

9. Design an experiment that involves finding the probability of independent events. Have a classmate perform your experiment.

LOGIC POWER

Lin drove from her home to her office at 90 km/h. At the end of the day, there was more traffic, and Lin drove home from her office along the same highway at 60 km/h. What was her average speed for the 2 journeys?

10.12 Dependent Events

Activity: Perform an Experiment

Place 2 green and 2 red linking cubes in a bag. If you remove 1 cube, place it back in the bag, then remove another cube, there are 4 possible outcomes. These are 2 reds, 2 greens, red then green, and green then red. The probability of each outcome is $P = \frac{1}{2} \times \frac{1}{2}$ or $\frac{1}{4}$

Perform the experiment, but *do not* return the first cube to the bag before you remove the second. Carry out 30 trials and record the outcomes. Calculate the probability of each outcome.

Calculate each probability using data from the whole class.

Inquire

1. Predict the probability of picking a red cube first.

2. If you pick a red cube first, what cubes remain in the bag?

3. If you pick a red cube first and do not return it to the bag, predict the probability of the second cube you pick being
a) red **b)** green

4. Multiply probabilities you found in questions 1 and 3 to predict the probability of drawing
a) red then green **b)** green then red **c)** 2 reds **d)** 2 greens

5. Compare your predictions from question 4 with the results of the experiment.

If the outcome of one event influences the outcome of another, the events are **dependent**.

Example

A bag holds 10 white marbles and 15 red marbles. What is the probability of drawing a white marble then a red marble if you do not return the first one to the bag?

Solution

At first, there are 10 + 15 or 25 marbles.
$P(\text{white}) = \frac{10}{25}$ or $\frac{2}{5}$

Of the remaining 24 marbles, 15 are red.
$P(\text{red}) = \frac{15}{24}$ or $\frac{5}{8}$

$$\begin{aligned} P(\text{white then red}) &= \frac{2}{5} \times \frac{5}{8} \\ &= \frac{10}{40} \\ &= \frac{1}{4} \end{aligned}$$

The probability of drawing a white marble then a red marble is $\frac{1}{4}$.

Problems and Applications

1. The 6 cards show the letters in the word CANADA.

If you choose randomly, what is the probability of choosing

a) an A?

b) another A if the first is not replaced?

c) 2 As in a row if the first is not replaced?

2. There are 3 quarters and 2 dimes in a bag. What is the probability of picking in 2 picks

a) 2 quarters if you replace the coin you picked first?

b) 2 quarters if you do not replace the coin you picked first?

c) 2 dimes if you replace the coin you picked first?

d) 2 dimes if you do not replace the coin you picked first?

e) a quarter then a dime if you replace the coin you picked first?

f) a quarter then a dime if you do not replace the coin you picked first?

g) a dime then a quarter if you replace the coin you picked first?

h) a dime then a quarter if you do not replace the coin you picked first?

3. Each card shows 1 letter from the word BANANA.

a) If 2 cards are chosen one after the other with replacement of the first, what is the probability of drawing 2 As? 2 Ns? 2 consonants?

b) If 2 cards are chosen one after the other without replacement of the first, what is the probability of drawing 2 As? 2 Ns? 2 consonants?

4. Each card shows 1 letter from the word EXPERIMENT.

What are the probabilities of picking the following pairs of cards one after the other without replacing the first?

a) 2 Es **b)** 2 vowels **c)** 2 consonants

d) a vowel then a consonant

e) a consonant then an E

f) an E then a T

5. A bag contains 5 red, 3 blue, and 2 green marbles. What are the probabilities of drawing the following without replacement?

a) a red marble then a blue marble

b) a green marble then a blue marble

c) 2 red marbles **d)** 2 green marbles

6. A bag contains 10 cards, each carrying the name of a different Canadian province. What are the probabilities of drawing the following pairs without replacement?

a) 2 names beginning with N

b) a name beginning with a vowel then a name beginning with N

c) a name beginning with P then a name beginning with N

d) 2 names ending in A

e) 2 names ending in vowels

f) a name beginning with an S then a name ending in A

7. What are the probabilities of drawing the following cards, one after the other, from a shuffled deck?

a) 2 red cards with replacement

b) 2 red cards without replacement

c) 2 clubs with replacement

d) 2 clubs without replacement

e) 2 aces with replacement

f) 2 aces without replacement

8. There are equal numbers of red marbles and blue marbles in a bag. Is the probability of choosing 2 red marbles in 2 picks higher with or without replacement? Explain.

9. Write a problem that involves dependent events. Have classmate solve your problem.

Experimental Probability

For some events, such as tossing a coin or rolling a die, you can determine the probability of an outcome mathematically, without doing the experiment. For other events, you must determine the probability of an outcome by experiment.

Activity ❶ The Paper Cup

When you throw a paper cup, there are 3 ways it can land: on its side, on its top, and on its bottom.

1. Estimate the probability of a tossed cup landing in each of the 3 positions.

2. Toss a paper cup 25 times and record your results in a table, like the one shown.

Outcome	Tally
Side	
Top	
Bottom	

3. Combine your results with your classmates'.

4. Use the class results to find the experimental probability of a cup landing in each of the 3 positions.

5. Compare the experimental results with your estimates.

Activity ❷ Dropping Cubes

Set up an experiment to determine the probability of dropping a plastic cube into a container, such as a paper bag. To perform the experiment, stand with the container behind you at your feet. Hold the cube over your shoulder and face forward. Each group in the class should use the same size of container.

1. Estimate the probability of your group dropping the cube into the container.

2. Have each member of the group try the experiment 10 times. Combine the results.

3. Use the results to find the experimental probability of your group dropping the cube into the container.

4. Compare the probability for your group with the probabilities found by other groups.

Activity ❸ Thumbtacks

When you roll a thumbtack, there are 2 ways it can stop: point down or point up.

1. Estimate the probability of a rolled thumbtack stopping in each position.

2. Roll 10 thumbtacks 10 times and record your results.

3. Combine your results with the results of your classmates.

4. Use the class results to find the experimental probability of a rolled thumbtack stopping in each position.

5. Compare the experimental results with your estimates.

Activity ❹ Other Experimental Probabilities

A sports commentator knows the probability of a certain baseball player getting a hit because of the number of "experiments" the hitter has conducted. List 5 other events for which the probability is found by experiment or on the basis of actual happenings.

Stylometry

Did Sir Francis Bacon write some of the plays attributed to Shakespeare? This is one of the questions a stylometer tries to answer. **Stylometry** is the science of measuring written words. It is used to show that one particular person has written something.

Authors, like burglars, leave "fingerprints." The "fingerprints" of an author are verbal. From year to year, a certain author uses roughly the same proportion of 5-letter words in written pieces. The same is true for words of any other length. But the proportion of 5-letter words will likely differ from one author to another. When analyzing writing, the first task of a stylometer is to graph how someone writes.

The following excerpt is from a piece by Canadian humorist Stephen Leacock (1869–1944).

"I've been reading some very interesting statistics," he was saying to the other thinker.

"Ah, statistics!" said the other; "wonderful things, sir, statistics; very fond of them myself."

"I find, for instance," the first man went on, "that a drop of water is filled with little...with little...I forget just what you call them...little — er — things, every cubic inch containing — er — containing...let me see..."

"Say a million," said the other thinker, encouragingly.

"Yes, a million, or possibly a billion...but at any rate, ever so many of them."

"Is it possible?" said the other. "But really, you know, there are wonderful things in the world. Now, coal...take coal...."

"Very good," said his friend, "let us take coal," settling back in his seat with the air of an intellect about to feed itself.

"Do you know that every ton of coal burnt in an engine will drag a train of cars as long as...I forget the exact length, but say a train of cars of such and such a length, and weighing, say so much...from...from...hum! for the moment the exact distance escapes me...drag it from..."

"From here to the moon," suggested the other.

"Ah, very likely; yes, from here to the moon. Wonderful, isn't it?"

"But the most stupendous calculation of all, sir, is in regard to the distance from the earth to the sun. Positively, sir, a cannon-ball — er — fired at the sun..."

"Fired at the sun," nodded the other, approvingly, as if he had often seen it done.

"And travelling at the rate of...of..."

"Of three cents a mile," hinted the listener.

"No, no, you misunderstand me — but travelling at a fearful rate, simply fearful, sir, would take a hundred million — no, a hundred billion — in short would take a scandalously long time in getting there — "

Activity ❶

1. To draw the graph of how Stephen Leacock wrote, copy and complete the following table for the excerpt.

Word Length in Letters	Frequency	
	Number of Words	Percent of Total
1		
2		
3		
4		
5		
6		
7		
8		
9		
10		
11		
12		
13+		

2. Graph the percent of total values versus the word length values.

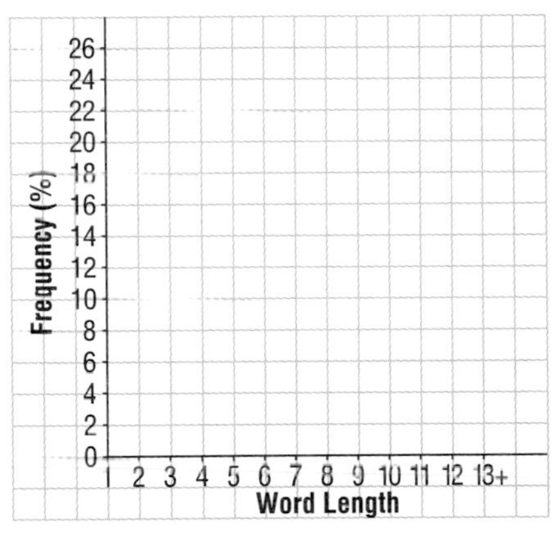

Activity ❷

1. Construct the graph for a newspaper article. Do not count any proper names.

2. Compare the graph with the graph you drew for Stephen Leacock.

Activity ❸

1. Construct the graph for something you have written. Do not count any proper names.

2. Compare your graph with Stephen Leacock's.

3. Compare your graph with your classmates'.

4. Why might the graphs for Stephen Leacock's writing and your writing not be as accurate as they could be?

Review

1. The table shows the world's 8 longest tunnels for road traffic.

Tunnel	Length (km)
Brooklyn-Battery (U.S.A.)	3.4
Great St. Bernard (Switzerland–Italy)	5.5
Lincoln (U.S.A.)	4.0
Mt. Blanc (France–Italy)	12.1
Mt. Ena (Japan)	8.5
Mount Royal (Canada)	5.1
Queensway Road (U.K.)	3.5
St. Gotthard (Switzerland)	16.4

a) Represent the data on a bar graph.
b) Which tunnel is 1.5 times longer than the Brooklyn-Battery Tunnel?

2. The ranges of some aircraft are shown in the table. Represent the data on a pictograph.

Aircraft	Range (km)
L–1011	8850
B–767	9500
B–727	4340
B–747	6600
DC–9	2400

3. The table gives the energy content, in kilojoules (kJ), of 15 mL servings of certain food items.

	Energy Content (kJ)	
Product	Regular	Light
Salad Dressing	334	120
Mayonnaise	418	334
Jam	226	167
Catsup	67	50

Display the data on a double bar graph.

4. The following speeds, in kilometres per hour, were measured for cars passing through a 50 km/h zone.

44, 50, 52, 63, 46, 50, 54, 56, 59, 63, 56, 39, 44, 45, 51

a) Display the data on a stem-and-leaf plot.
b) What was the median speed?

5. The circle graph shows the population distribution in the Atlantic Provinces.

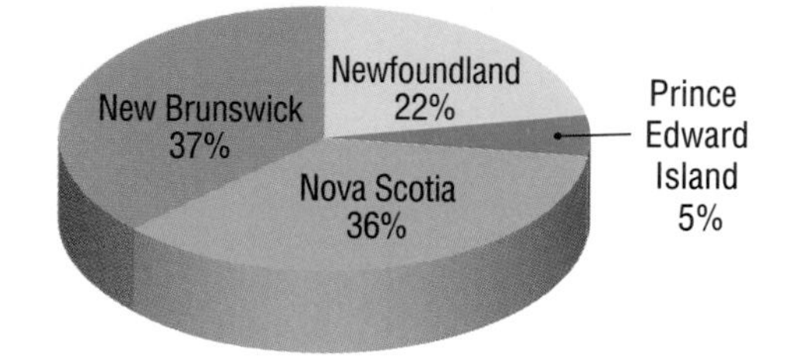

a) Which 2 provinces account for nearly three-quarters of the population of the Atlantic Provinces?
b) The total population of the Atlantic provinces is about 2 500 000. About how many people live in Nova Scotia?

6. The Ski Club includes 68 Grade 9 students, 77 Grade 10 students, 41 Grade 11 students, and 25 grade 12 students. The executive of the Club is to include 10 students. If each grade is to be represented fairly, how many members of the executive should be chosen from each grade?

7. The table gives the average daily high temperatures, in degrees Celsius, in Montreal and Alice Springs for each month of one year.

	Temperature (°C)	
Month	Montreal	Alice Springs
Jan	–3	38
Feb	0	33
Mar	9	30
Apr	11	26
May	19	19
Jun	22	20
Jul	25	21
Aug	22	24
Sep	18	31
Oct	12	33
Nov	0	35
Dec	–2	36

a) Display the data as 2 broken-line graphs on the same set of axes.
b) Is the summer warmer in Montreal or Alice Springs? Explain.

8. These are Stephan's marks on 9 sailing tests.

35, 86, 88, 37, 90, 41, 12, 89, 37

a) Determine the mean, median, and mode of the data.

b) Which of the 3 measures best represents Stephan's sailing ability?

9. Draw a box-and-whisker plot to represent the following driving test scores.

85, 86, 60, 55, 73, 75, 81, 83, 84, 98, 56, 64, 68, 66, 88, 90, 92

10. Tara scored 78 on a swimming test. The scores of the other students in the class were as follows.

56, 57, 66, 68, 70, 72,
74, 78, 80, 82, 84, 88

In what percentile was Tara's score?

11. Use a tree diagram to find the number of possible outcomes when you toss a dime and spin the spinner.

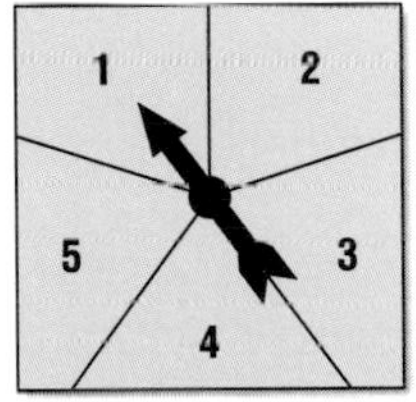

12. Find each probability.

a) P(H)

b) P(H or I)

c) P(blue)

d) P(K)

e) P(yellow)

f) P(blue or green)

g) P(G, H, or I)

h) P(G, H, I, J, K, or L)

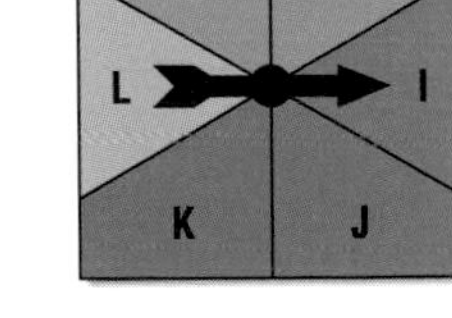

13. A bag contains 4 red cubes and 6 blue cubes. What is the probability of picking 2 red cubes in a row

a) if you replace the first cube you pick?

b) if you do not replace the first cube you pick?

Group Decision Making

Writing A Story

1. In your home group, write the names of 6 different objects on 6 strips of paper, with 1 object per strip. Each object must have at least 1 number on it. Be creative. The objects might include such things as a Dick Tracy watch, an airline schedule, an American dollar bill, a 1970 calendar, and a credit card.

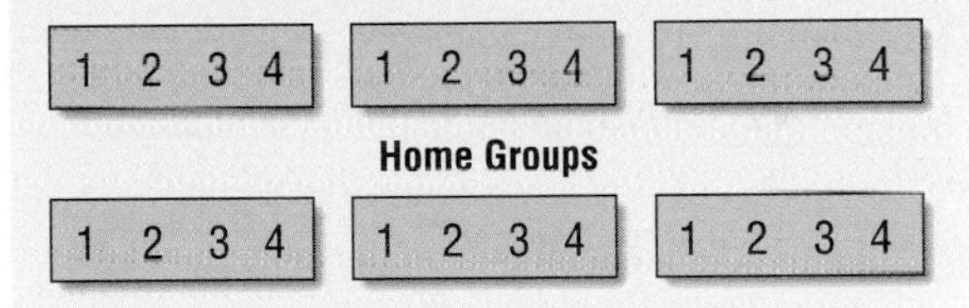

2. Put your group's strips of paper, and the strips from the other groups, into a container at the front of the class. Have a representative from each group select 4 strips of paper from the container without looking. Do not return the selected strips to the container.

3. In your home group, write a story in which each of the 4 objects chosen by your group plays an important role. In your story, make use of the numbers on the objects as much as possible.

4. Decide how to present your story to the class. You may want to have someone read it, or you may want to act it out or use some other method.

5. Meet as a class to present your stories. Before you start, announce the 4 objects you had to work with.

6. As a class, evaluate the stories for their entertainment value and creativity. Remember to consider the objects each group had to work with.

Chapter Check

1. The circle graph shows the percent, by mass, of different chemical elements in the human body.

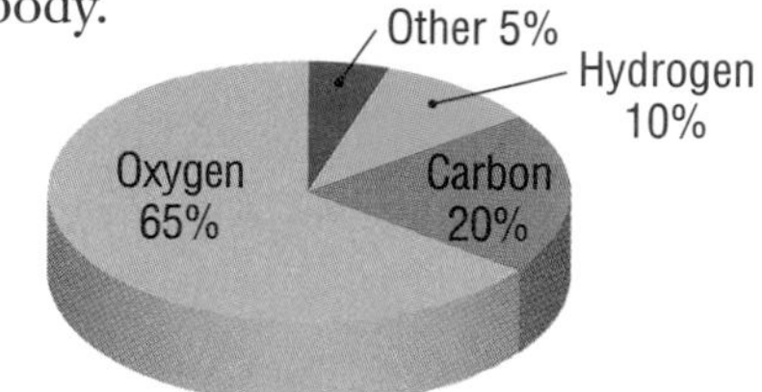

a) For which element is the mass twice the mass of hydrogen?
b) For which element is the mass 6.5 times the mass of hydrogen?

2. The table includes the seating capacities of several aircraft. Display the data on a graph.

Aircraft	Seating Capacity
L–1011	212
B–767	179
B–727	136
B–747	382
DC–9	100

3. Find the mean, median, and mode of these test scores.
64, 85, 86, 70, 72, 80, 83, 72, 65, 80

4. Lucia got 62 on a music test. The marks of the other students were as follows.
50, 56, 58, 60, 62, 62, 64, 68, 70, 78, 80
In what percentile was Lucia's mark?

5. These are the speeds, in kilometres per hour, of 15 go-carts after the first lap of a race.
30, 35, 41, 40, 39, 29, 28, 36,
44, 40, 33, 29, 37, 42, 46
Display the data on
a) a stem-and-leaf plot
b) a box-and-whisker plot

6. Use a tree diagram to find the number of possible outcomes when you roll a die and spin the spinner.

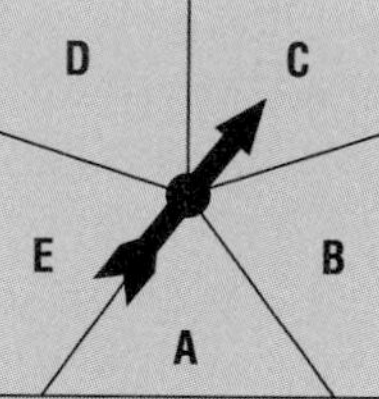

7. Find each probability.

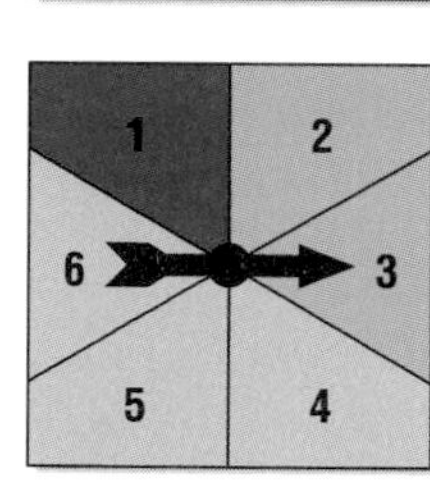

a) $P(1)$
b) $P(\text{yellow})$
c) $P(1 \text{ or } 2)$
d) $P(\text{white})$
e) $P(\text{red, green, or yellow})$ **f)** $P(5 \text{ or } 6)$

8. A box contains 6 white marbles and 4 green marbles. What is the probability of picking 2 white marbles in a row if you
a) replace the first marble you pick?
b) do not replace the first marble you pick?

Using the Strategies

1. Copy the diagram. Place the numbers from 1 to 10 in the circles so that the numbers on each side of the pentagon total 14. The 5 has been placed for you.

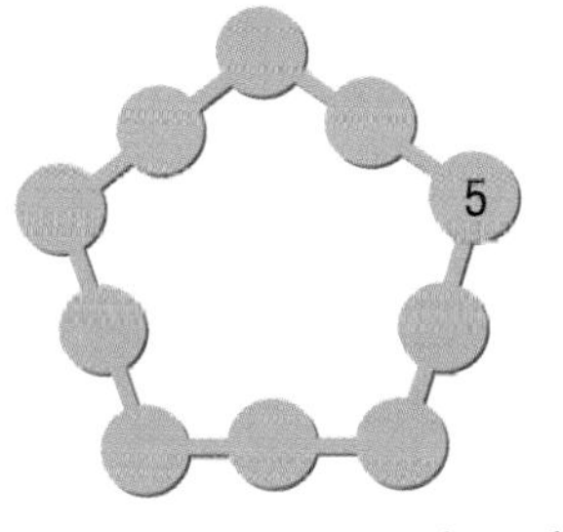

2. A block of cheese, covered with wax, measures 12 cm by 12 cm by 10 cm. The block is cut into 2-cm cubes.

a) How many cubes are there?
b) How many cubes have wax on 3 faces?
c) How many cubes have wax on 2 faces?
d) How many cubes have no wax on them?

3. The diagram shows how 20 sheep have been placed in 8 pens so that there are 6 sheep in each row of 3 pens.

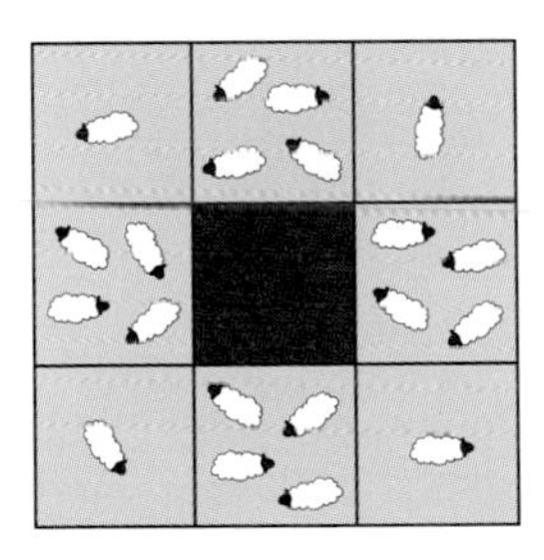

Rearrange the sheep so that there are 7 sheep in each row of 3 pens.

4. You are standing in line at the cafeteria. You are seventh from the front and eighth from the end. How many people are in the line?

5. The number 9 has 3 different factors, 1, 3, and 9. Find all the positive integers less than 50 that have an odd number of different factors.

6. A team gets 2 points for a win, 1 point for a tie, and no points for a loss. The Bears have played 28 games. They have 27 points and 7 losses. How many wins do they have?

7. The perimeter of the figure is 12 units.

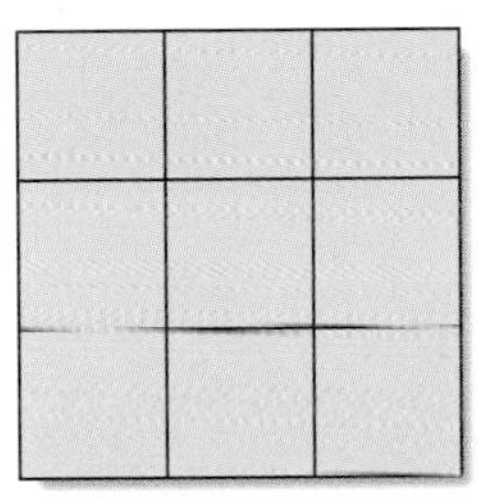

Sketch your answers to the following.

a) Remove 1 square and keep the perimeter the same.
b) Remove 2 squares and keep the perimeter the same.
c) Remove 1 square and increase the perimeter by 2.
d) Remove 2 squares and increase the perimeter by 2.
e) Remove 2 squares and increase the perimeter by 4.

8. Sketch a graph of the time needed to decorate the gym for a graduation dance versus the number of people who volunteer to help.

9. If 2 months in a row have a Friday the 13th, what months are they? Explain.

DATA BANK

1. A plane left Calgary at 16:00 and flew to Montreal at a speed of 700 km/h. At what time did it land in Montreal?

2. When you travel from east to west across the International Date Line, do you lose a day or gain a day?

CHAPTER 11

Perimeter, Area, and Volume

The average number of people per square metre on a Tokyo train filled to capacity is officially 4. In fact, the average number of people per square metre on Tokyo trains in rush hour is 10. Can 10 people in your class stand on 1 m^2 of floor?

Estimate the area of your classroom floor in square metres. About how many people are there per square metre in your classroom?

If your classroom were as crowded as a Tokyo train in rush hour, about how many people would be in the room?

Tangrams

A tangram is a 7-piece puzzle that comes from China. The origin of the word "tangram" is itself interesting. The ending of the word, *gram*, refers to something drawn, such as a diagram. The beginning of the word, *tan*, dates back to the Tang dynasty, A.D. 618-906, the greatest dynasty in Chinese history. The 7 tangram pieces are shown assembled as a square.

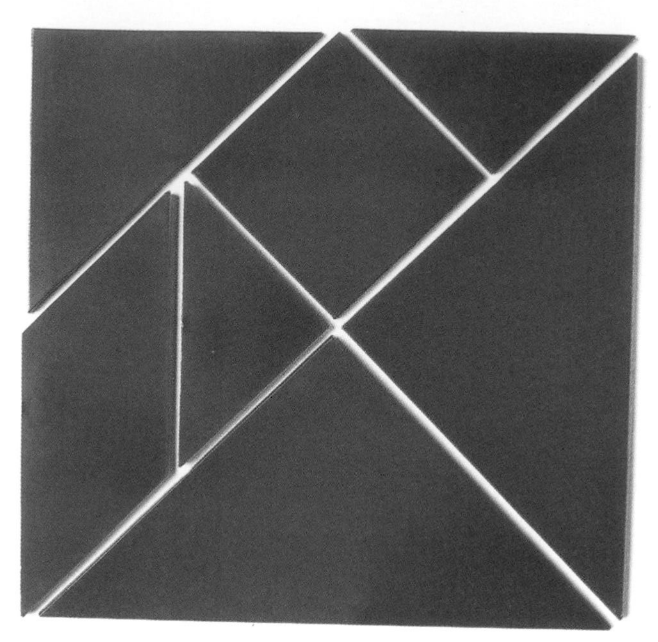

Activity ❶ Convex and Concave Figures

Polygons are closed geometric figures with line segments as sides. You are already familiar with triangles (3 sides), quadrilaterals (4 sides), pentagons (5 sides), and hexagons (6 sides).

Polygons can be convex or concave. If you wrap a piece of string around the perimeter of a convex polygon and pull the string tight, the string makes contact with every point of the polygon. On the other hand, if you wrap a piece of string around a **concave polygon**, the string does not make contact with every point of the polygon. A concave polygon bends in at some point or points, leaving gaps.

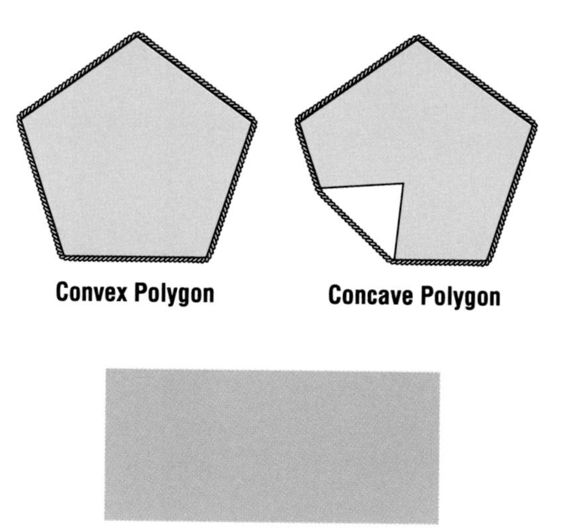

Using all 7 tangram pieces, it is possible to construct 6 different convex quadrilaterals without holes in the middle. One of them is the square shown at the top of this page. The others are 1 rectangle, 1 parallelogram, and 3 trapezoids. Use your 7 tangram pieces to construct these 5 quadrilaterals. Sketch your solutions on grid paper.

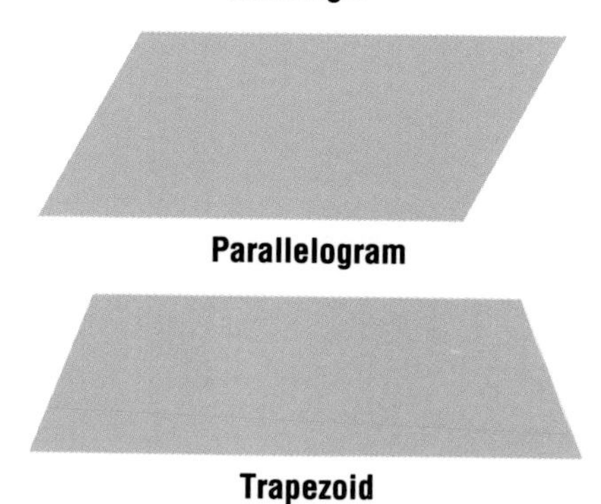

Activity ❷ Quadrilaterals

1. From a tangram, select the 4 triangles shown. Use the 4 triangles to make the following quadrilaterals. Sketch your solutions on grid paper.

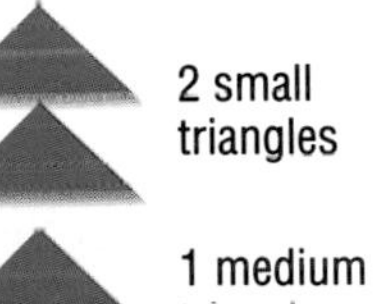

a) parallelogram **b)** square

c) rectangle **d)** trapezoid

2. Are any of these quadrilaterals regular polygons? Explain.

Activity ❸ More Convex Figures

1. There are 7 other convex polygons without holes in the middle that can be constructed. Use the 7 tangram pieces to construct these polygons.

a) 1 triangle **b)** 3 pentagons **c)** 3 hexagons

2. Are any of these polygons regular polygons? Explain.

Activity ❹ Area with Tangrams

Let the area of the smallest tangram piece be 1 square unit. Calculate the areas of the other tangram pieces, and record the areas in a chart.

Mental Math

Calculate.

1. 3.14×10 **2.** 3.14×100 **3.** 3.14×1000 **4.** $3.14 \times 10\ 000$

5. $3.14 \div 10$ **6.** $3.14 \div 100$ **7.** $3.14 \div 1000$ **8.** $3.14 \div 10\ 000$

Calculate.

9. $5 \times 25 \div 1000$ **10.** 12×8 **11.** 14×7 **12.** 2.5×6

13. $8 \times 5 \times 3$ **14.** $10 \times 7 \times 9$ **15.** $4(7 + 6)$ **16.** $15.5 \div 5$

Estimate.

17. $(3.14)(1.2)$ **18.** $(3.14)(1.8)$ **19.** $(3.14)(5.8)$ **20.** $(3.14)(0.9)$

Estimate, then calculate.

21. $\frac{1}{2}(6.2 + 4.8)$ **22.** $\frac{1}{2}(2.2 + 5.8)$ **23.** $\frac{1}{2}(3.9 + 8.1)$ **24.** $\frac{1}{2}(10.2 + 9.8)$

Estimate.

25. $(3.14)(2.2)^2$ **26.** $(3.14)(3.3)^2$ **27.** $(3.14)(5.4)^2$ **28.** $(3.14)(4.9)^2$

Estimate.

29. $(3.14)(2.1)(3.2)$ **30.** $(3.14)(5.2)(2.4)$ **31.** $(3.14)(5.6)(5.1)$ **32.** $(3.14)(7.2)(10.4)$

Estimate.

33. $(3.14)(2.4)^3$ **34.** $(3.14)(4.3)^3$ **35.** $(3.14)(9.8)^3$ **36.** $(3.14)(5.1)^3$

11.1 Perimeter

Activity: Explore the Shapes

Pentominoes are made up of 5 identical squares joined at their edges. A number of different pentominoes can be formed by rearranging the squares. Two pentominoes are shown. The third shape is not a pentomino.

Draw as many different pentominoes as possible on centimetre grid paper.

Pentomino

Pentomino

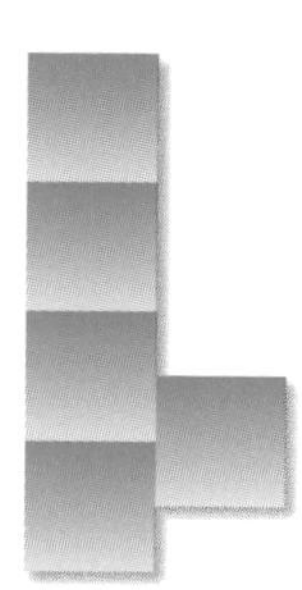

Inquire

1. Is this a pentomino? Explain.

2. How many different pentominoes are possible? Compare your answer with your classmates'.
3. Find the perimeter of each pentomino.
4. Which shape has the smallest perimeter?

Example

The Toronto-Dominion Bank Tower is one of Canada's tallest buildings, with an approximate height of 168 m. Its base is a rectangle that measures approximately 60 m by 36 m. What is the perimeter of the base of the Toronto-Dominion Bank Tower?

Solution

The formula for the perimeter of a rectangle is $P = 2(l + w)$. The base is a 60 m by 36 m rectangle.

$$\begin{aligned} P &= 2(l + w) \\ &= 2(60 + 36) \\ &= 2(96) \\ &= 192 \end{aligned}$$

EST: $2(100) = 200$

The perimeter of the base of the Toronto-Dominion Bank Tower is 192 m.

Practice

Find the perimeter of each figure. The side of each square is 1 unit.

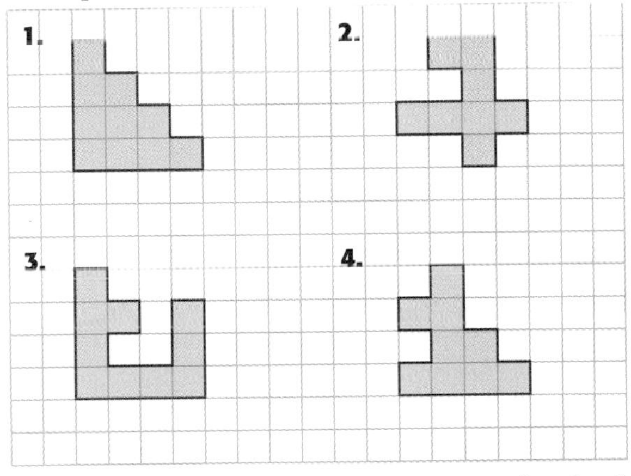

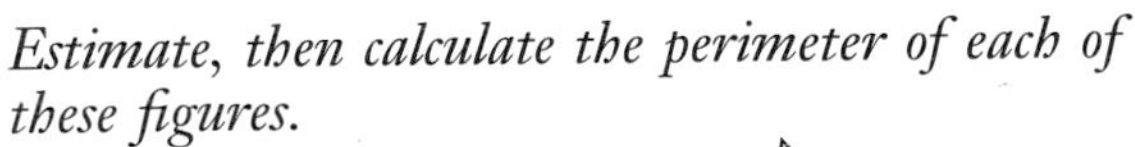

Estimate, then calculate the perimeter of each of these figures.

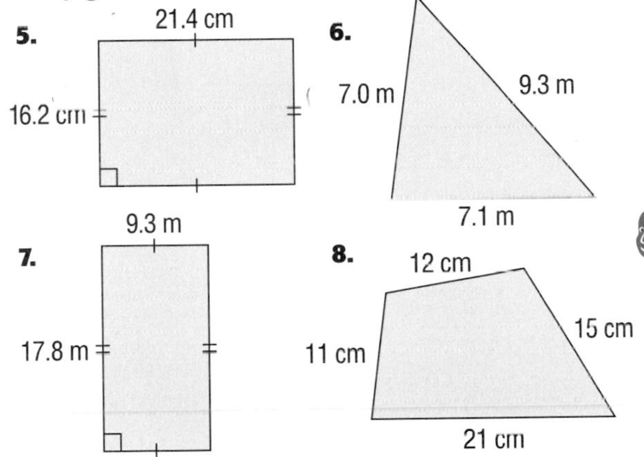

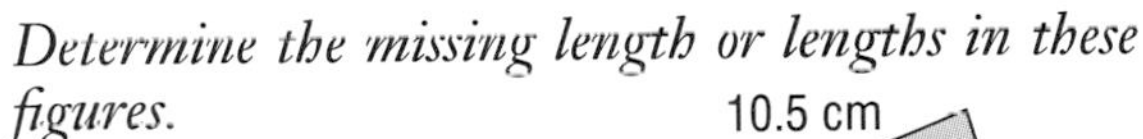

Determine the missing length or lengths in these figures.

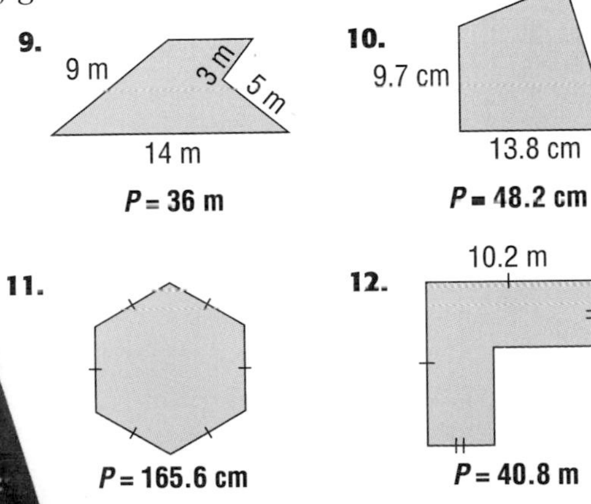

Problems and Applications

13. For which of the following do you need to know the perimeter of the object?

a) the amount of paint required to paint your bedroom

b) the number of metres of paving stone needed to edge the garden

c) the amount of time you need to walk around the outside of the school if you walk at 2 m/s

14. a) Find the perimeter of the ceiling of a room that is 5.2 m in length and 3.2 m in width.

b) A wallpaper border costs $10.99/m. How much will it cost to border this room?

15. The Pentagon, in Arlington, Virginia, is the world's largest office building. It is a regular pentagon of perimeter 1405 m. What is the length of each side?

16. A square and a rectangle have a side in common. The side of the square is 5 cm, and the perimeter of the rectangle is 50 cm. What are the dimensions of the rectangle?

17. Hexominoes are made up of 6 squares joined at their edges. Use grid paper and work with a classmate to find the hexomino with the smallest perimeter.

LOGIC POWER

Assume that there are no cubes missing from the back of this stack.

How many cubes are in the stack?

11.2 Circumference of a Circle

Activity: Discover the Relationship

Use string and a ruler to measure and record the circumference and the diameter of 4 different circular objects. Then, use a calculator to complete the following table.

Circular Object	Circumference (C)	Diameter (d)	$\frac{C}{d}$	Circular Object	Circumference (C)	Diameter (d)	$\frac{C}{d}$
1.				3.			
2.				4.			

Inquire

1. Round the values you found for $\frac{C}{d}$ to the nearest tenth. Compare your answers with your classmates'.

2. Press the pi (π) key on your calculator, and round that value to the nearest tenth. Compare the result with the values you found in question 1.

The perimeter of a circle is called the **circumference**. The distance across a circle, through its centre, is its **diameter**. The diameter of a circle is twice the length of its radius. In all circles, the ratio of the circumference to the diameter is a constant called π.

$$\pi = \frac{C}{d}$$
$$= 3.141\ 592\ 653\ 589\ 793...$$

Pi (π) is an irrational number. It is a nonrepeating, nonterminating decimal. The number of decimal places used depends on the desired accuracy. In this book, use $\pi = 3.14$.

Example

The Chubb Crater in northern Quebec is circular, with a radius of about 1.5 km. What is the circumference of the crater to the nearest tenth of a kilometre?

Solution

$C = 2\pi r$ or πd
$= 2\pi(1.5)$
$= 2(3.14)(1.5)$
$= 9.42$

EST $(2)(3)(2) = 12$

C 2 × π × 1 . 5 = 9.4247780

The circumference of the Chubb Crater is about 9.4 km.

Practice

Find the radius of each circle.

1. $d = 10$ cm **2.** $d = 15$ m

3. $d = 70$ cm **4.** $d = 7$ m

Find the diameter of each circle.

5. $r = 2$ cm **6.** $r = 6.8$ m

7. $r = 10.2$ cm **8.** $r = 1$ m

Estimate, then calculate the circumference of each circle to the nearest centimetre or metre.

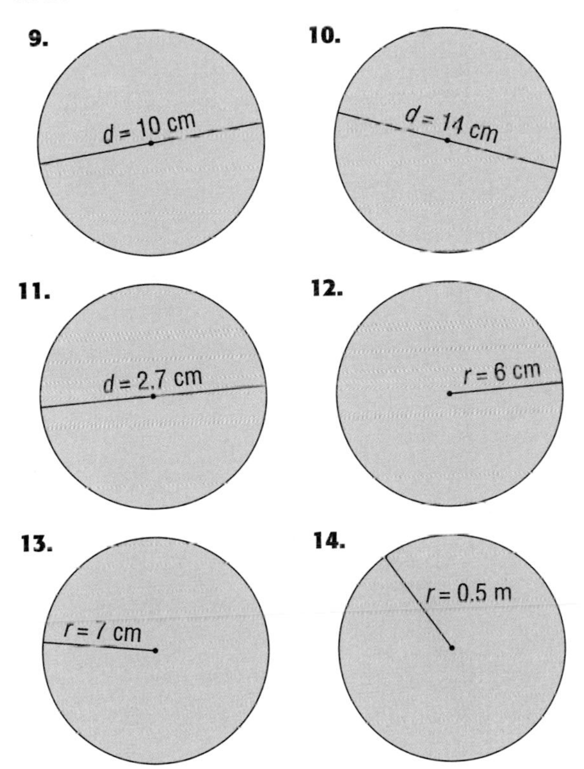

Find the circumference of each circle to the nearest tenth of a unit.

15. $r = 11$ cm **16.** $d = 11.9$ m

17. $d = 10$ cm **18.** $r = 12$ m

Find the diameter of each circle to the nearest unit.

19. $C = 3.14$ cm **20.** $C = 9.42$ m

21. $C = 27$ cm **22.** $C = 7$ m

Problems and Applications

23. a) How would you describe the shape of this window?

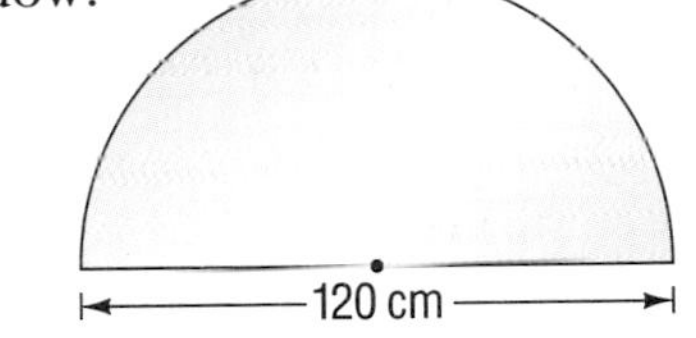

b) Write a formula to calculate its perimeter.

c) Calculate the perimeter of the window.

24. Calculate the perimeter of this mirror.

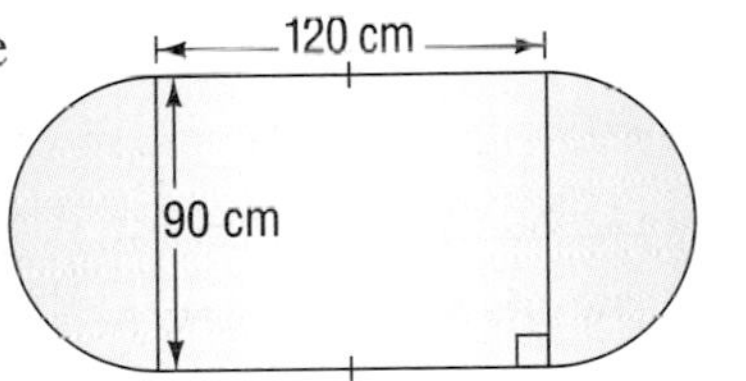

25. a) Calculate the circumference of each of the 3 small circles within the large one to the nearest tenth of a centimetre.

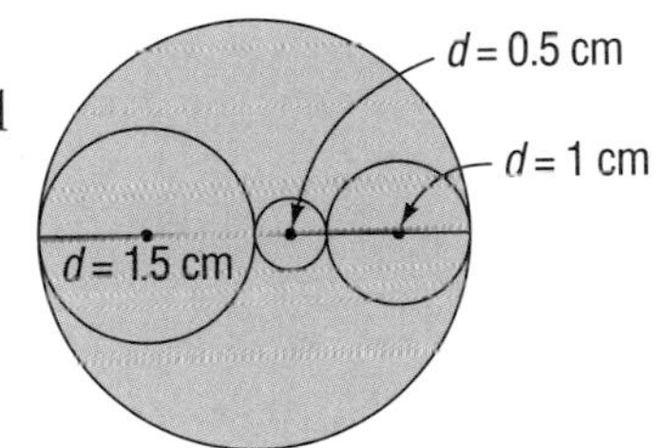

b) What is the diameter of the large, outer circle?

c) Calculate the circumference of the large, outer circle to the nearest tenth of a centimetre.

26. a) The diameter of the Earth at the equator is approximately 12 750 km. Work with a partner and use $\pi = 3.14$ to calculate the circumference of the Earth at the equator to the nearest thousand kilometres.

b) The closest the moon comes to the Earth is about 357 000 km. How many planet Earths, placed side-by-side, would it take to reach the moon from the Earth?

27. Write a problem that requires the calculation of the circumference of a circle. Have a classmate solve your problem.

11.3 Areas of Rectangles and Squares

Activity: Complete the Table

On 1-cm grid paper, draw as many different rectangles as possible that have the same area as the rectangle shown. Then, copy and complete the table.

A

Rectangle	Area (cm^2)	Length (cm)	Width (cm)
A	36	9	4
B			

Inquire

1. Is a square a rectangle? Explain.
2. How many different rectangles can be formed? Compare your answer with other classmates'.
3. What is the formula for the area of a rectangle? of a square?
4. Describe how the factors of 36 are related to the dimensions of each rectangle that you formed.

Example

The Canadian flag has 2 red rectangles, with a square in the middle. Calculate the area of 1 red rectangle.

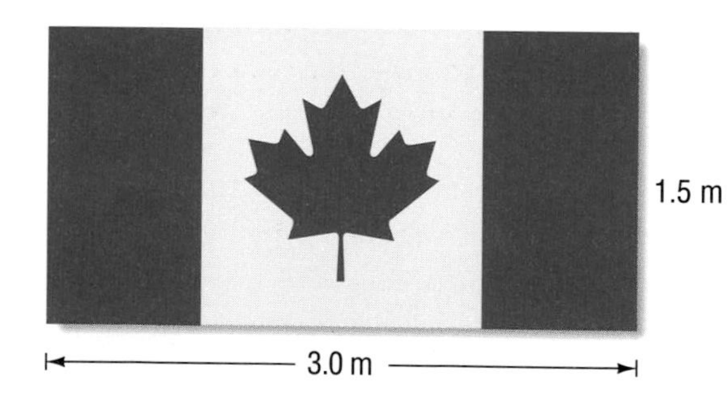

Solution

The formula for the area of a rectangle is $A = lw$. The flag is a 3.0 m by 1.5 m rectangle.

$$\begin{aligned} A &= lw \\ &= (3.0)(1.5) \\ &= 4.5 \end{aligned}$$

The formula for the area of a square is $A = s^2$. The side length of the white square is 1.5 m.

$$\begin{aligned} A &= s^2 \\ &= (1.5)^2 \\ &= 2.25 \end{aligned}$$

$$\begin{aligned} \text{Area of 2 red rectangles} &= 4.5 - 2.25 \\ &= 2.25 \end{aligned}$$

$$\begin{aligned} \text{Area of 1 red rectangle} &= \frac{2.25}{2} \\ &= 1.125 \end{aligned}$$

The area of 1 red rectangle is 1.125 m^2.

Practice

Estimate, then calculate the area of each figure by counting squares. The area of each square is one square unit.

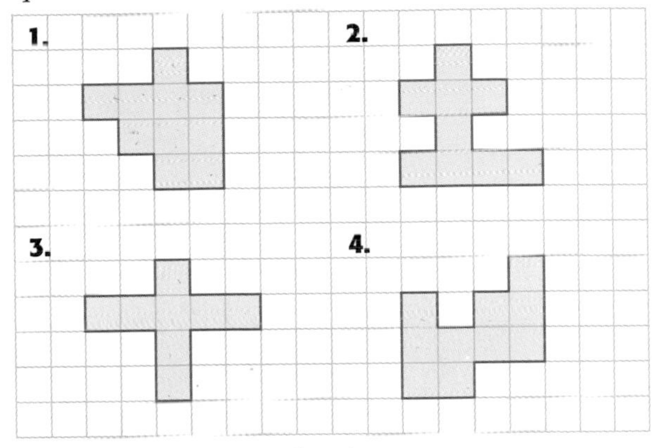

Determine the area of each square.

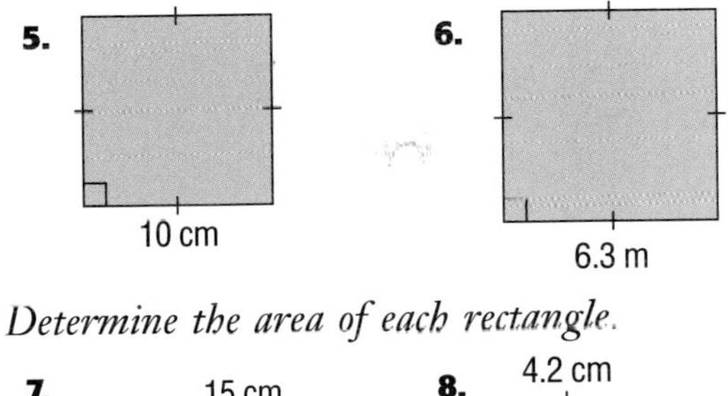

Determine the area of each rectangle.

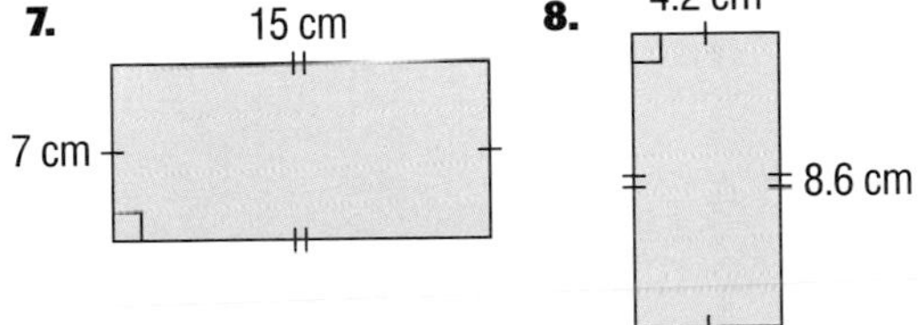

Problems and Applications

Calculate the area of each patio.

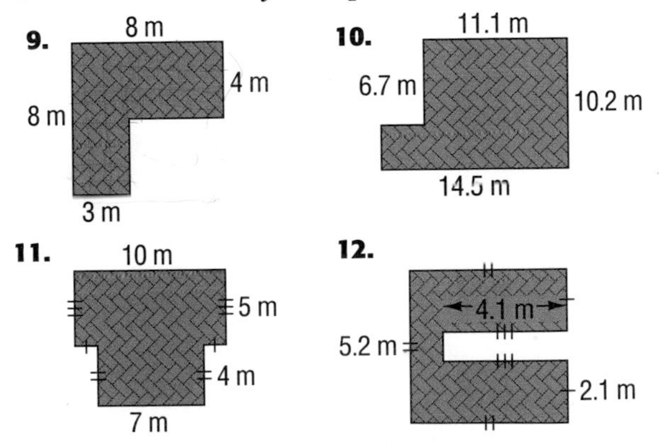

13. Calculate the area of a page in this book.

14. a) Calculate the area of the stonework around this rectangular garden.

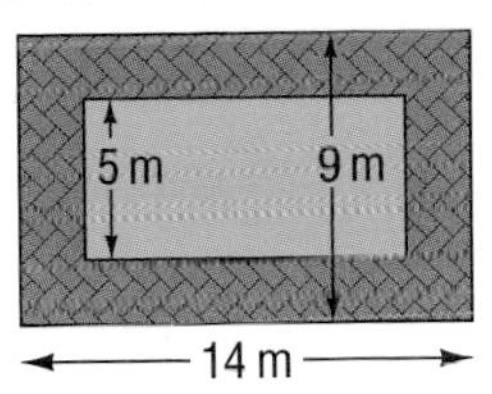

b) What is the total cost of landscape fabric needed to cover the garden if the fabric costs $0.99/m²? Allow an extra 10% for overlapping.

15. Find the side of a square with each area.

a) 100 m² **b)** 225 m²

16. The Taj Mahal in India is built on a square piece of land that has an area of 9101.16 m². What is the length of each side of the square?

17. A rectangular lot measures 66 m by 107 m. A sidewalk surrounds the lot. The sidewalk is 2 m wide. What is the area of the sidewalk?

18. You want to plant a 16-m² rectangular garden and build the shortest possible fence around it. What shape should you make the garden?

19. If a rectangle has a perimeter of 200 cm, what whole number dimensions give the greatest area?

20. Write a problem that requires the calculation of the area of a rectangle or square. Have a classmate solve your problem.

LOGIC POWER

Use toothpicks to construct the figure shown. Then, move 3 toothpicks so that only 5 squares remain.

11.4 Areas of Parallelograms, Triangles, and Trapezoids

Some of the most distinctive architecture is composed of parallelograms, triangles, and trapezoids.
This photograph of the ceiling in the Hockey Hall of Fame is an example.

Activity: Determine the Formula

Copy the parallelogram onto grid paper. Cut a right triangle from the parallelogram along the height, h. Then, fit the right triangle onto the opposite end of the remainder of the parallelogram so that the two pieces form a rectangle.

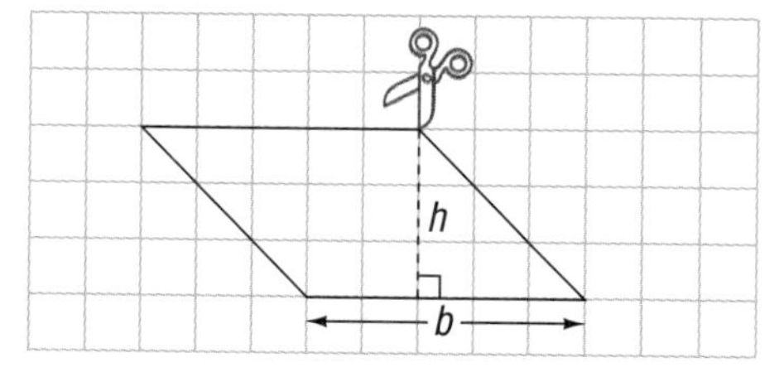

1. What is the area of the rectangle in square units?
2. Write a formula for the area of the rectangle in terms of the variables b and h.
3. What is the area of the parallelogram in square units?
4. Write a formula for the area of the parallelogram in terms of b and h.

Example 1

Many Canadian heritage quilts are made from simple geometric shapes such as identical parallelograms. If the side of each square on the grid is 5 cm, what is the area of each parallelogram in this section of a quilt?

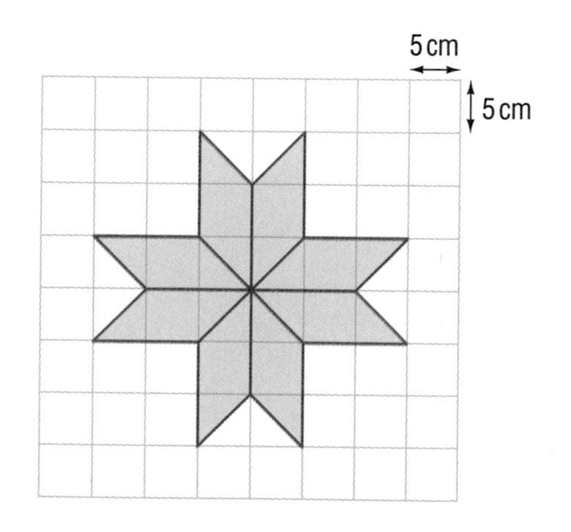

Solution

The formula for the area of a parallelogram is $A = bh$. The base of each parallelogram in this quilt is 10 cm and each height is 5 cm.

$$\begin{aligned} A &= bh \\ &= (10)(5) \\ &= 50 \end{aligned}$$

The area of each parallelogram in the quilt is 50 cm^2.

Activity: Determine the Formula

Copy this triangle and trapezoid onto grid paper.

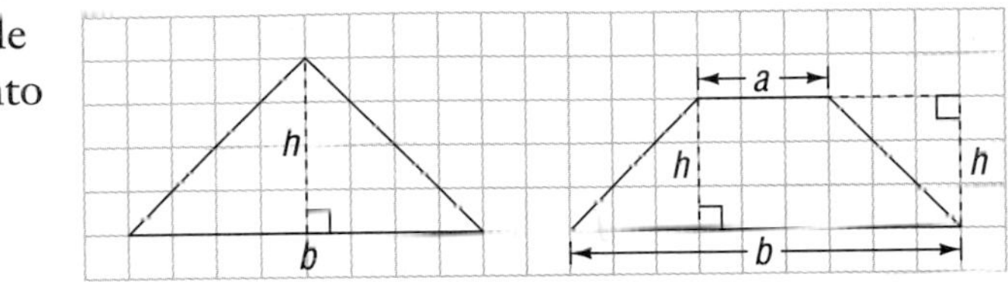

Inquire

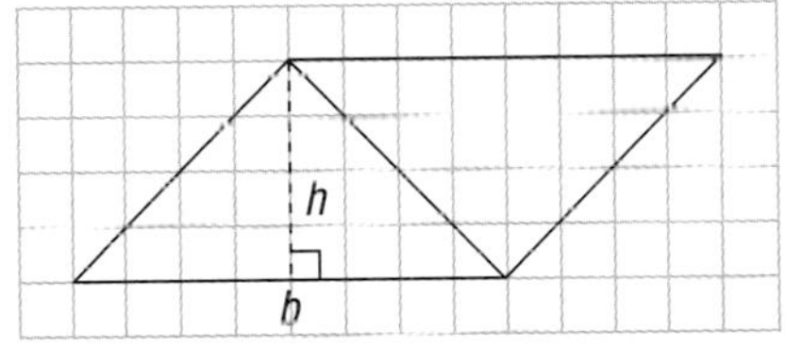

1. Along one side of the triangle, draw another identical triangle to form a parallelogram similar to the one shown.

2. Write a formula for the area of the parallelogram in terms of b and h.

3. How is the area of the triangle related to the area of the parallelogram?

4. Write a formula for the area of the triangle in terms of b and h.

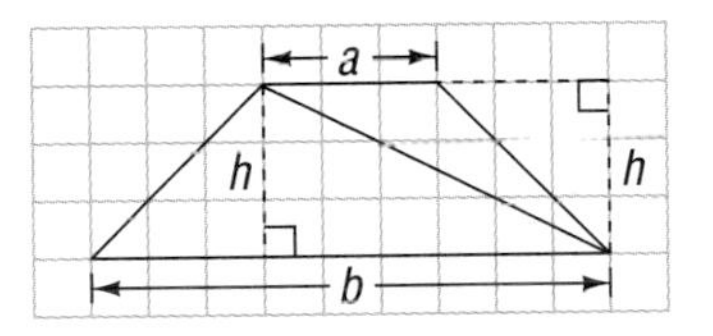

5. In the trapezoid, draw a diagonal to form 2 triangles as shown. Then, write the area of each triangle in terms of its base and height.

6. What is the sum of the areas of both triangles in terms of a, b, and h?

7. Factor the sum you found in question 6 and, write a formula for the area of a trapezoid in terms of a, b, and h.

Example 2

Calculate the area of each figure to the nearest tenth of a square centimetre.

a)

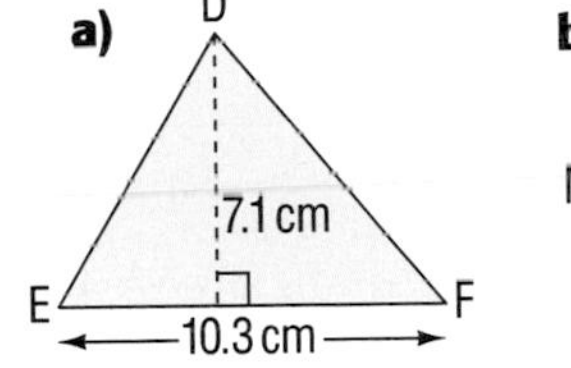

b)

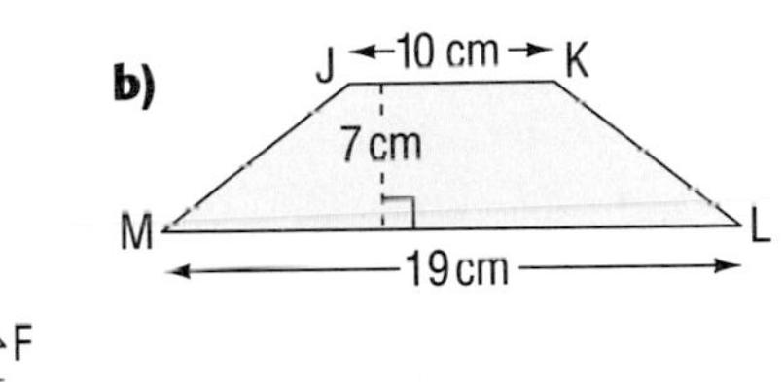

Solution

a) The formula for the **area of a triangle** is $A = \frac{1}{2}bh$.

$$\begin{aligned} A &= \frac{1}{2}bh \\ &= \frac{1}{2}(10.3)(7.1) \\ &= \frac{1}{2}(73.13) \\ &= 36.565 \end{aligned}$$

EST $(0.5)(10)(7) = 35$

The area of ΔDEF is about 36.6 cm^2.

b) The formula for the **area of a trapezoid** is $A = \frac{1}{2}(a + b)h$.

$$\begin{aligned} A &= \frac{1}{2}(a + b)h \\ &= \frac{1}{2}(10 + 19)(7) \\ &= \frac{1}{2}(29)(7) \\ &= \frac{1}{2}(203) \\ &= 101.5 \end{aligned}$$

EST $(0.5)(30)(7) = 105$

The area of trapezoid JKLM is 101.5 cm^2.

CONTINUED

Practice

Calculate the area of each figure.

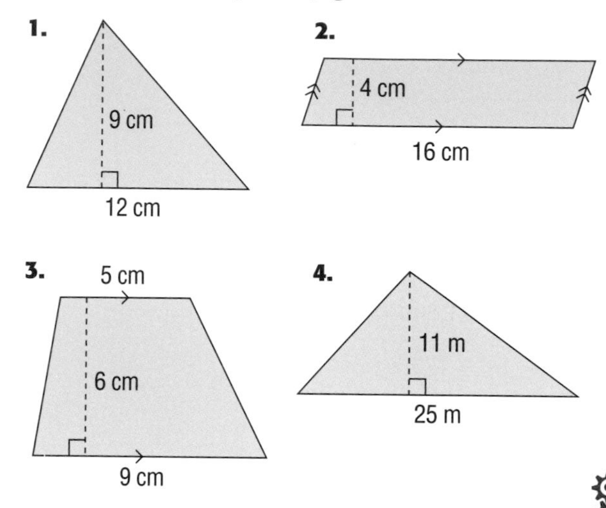

Estimate each area. Then, calculate it to the nearest square centimetre or square metre.

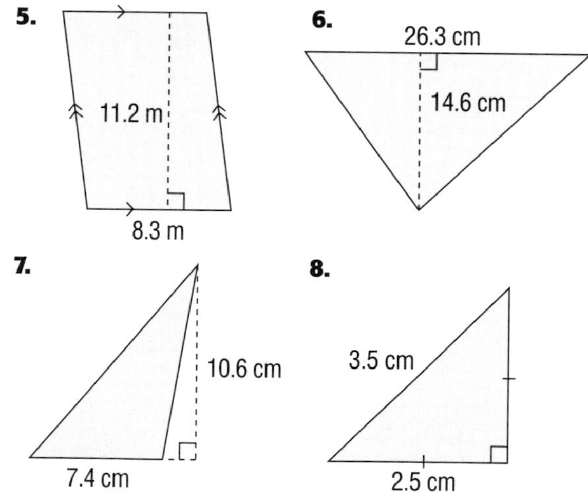

Problems and Applications

9. Calculate the area of the shaded region in each diagram.

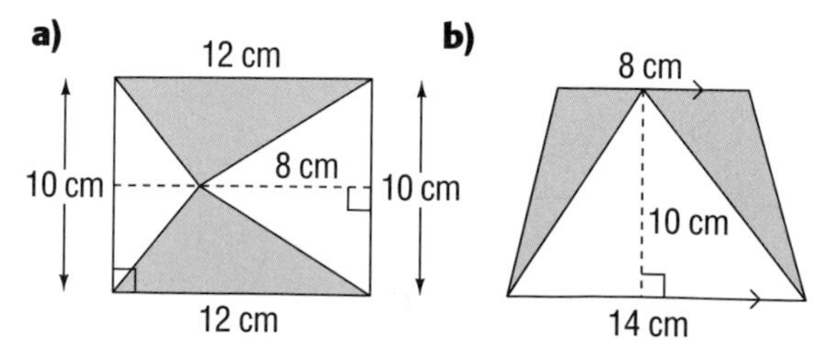

10. The area of a parallelogram is 8400 cm^2. Its height is 60 cm. What length is its base?

11. Many Canadian cities use the Blue Box for recycling. The sides, the front, and the back of the Blue Box are trapezoids. What is the area of a side and a front or back?

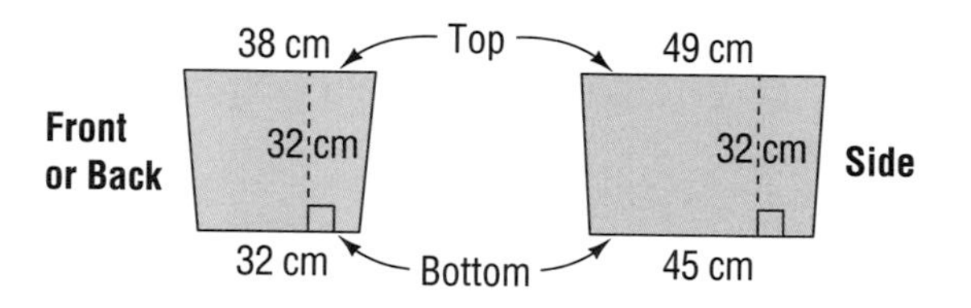

12. The area of a triangle is 14.4 cm^2, and the length of its base is 6.4 cm. What is its height?

13. Without calculating any areas, decide how the areas of the triangles, parallelogram, and trapezoid compare. Explain.

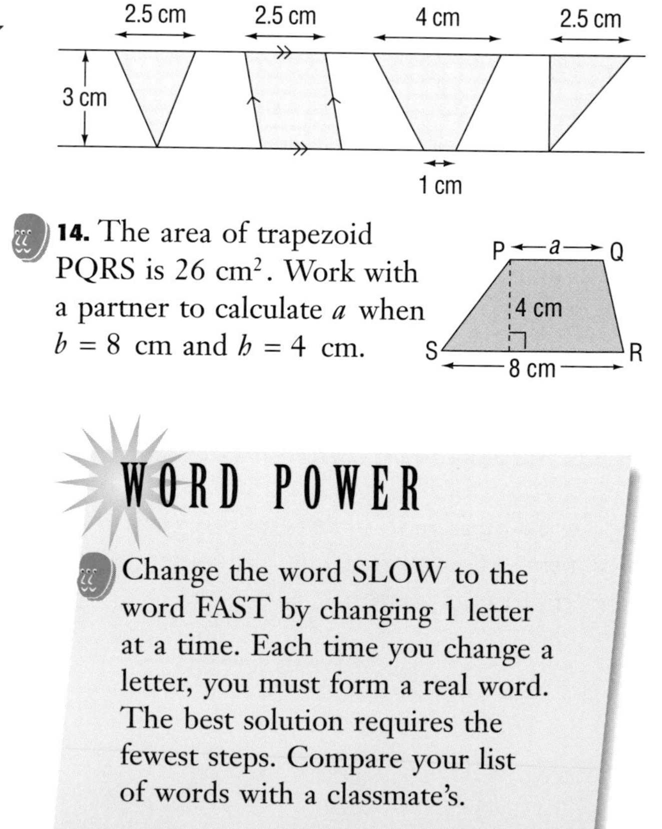

14. The area of trapezoid PQRS is 26 cm^2. Work with a partner to calculate a when $b = 8$ cm and $h = 4$ cm.

WORD POWER

Change the word SLOW to the word FAST by changing 1 letter at a time. Each time you change a letter, you must form a real word. The best solution requires the fewest steps. Compare your list of words with a classmate's.

Areas of Irregular Figures

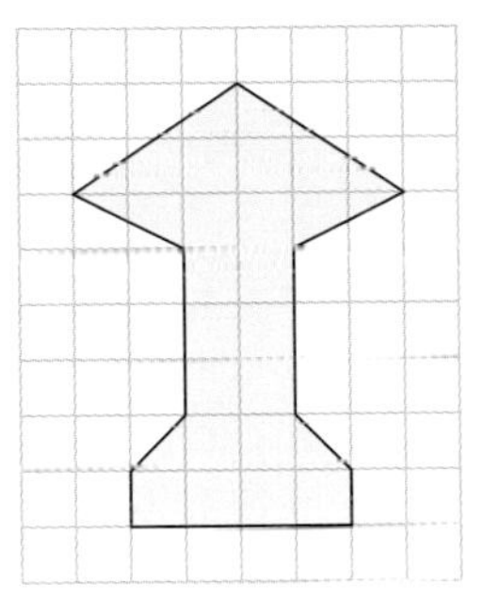

Look at the figure of the arrow. It has been drawn on a grid with each square equal to 1 square unit.

The area of the arrow can be estimated by counting the number of whole and part squares that it covers.

The number of whole squares is 16. The number of part squares is 14.

$$\begin{aligned}\text{Total Area} &= (\text{Number of whole squares}) + \tfrac{1}{2}(\text{Number of part squares})\\ &= 16 + 0.5(14)\\ &= 16 + 7\\ &= 23\end{aligned}$$

The total area of the arrow is about 23 square units.

Why do you think you calculate one-half of the part squares?

Activity

1. Trace your hand on a sheet of 1-cm grid paper.

2. Estimate the area of your hand by counting whole squares and part squares.

3. Measure your arm length in centimetres.

4. Collect a class set of arm lengths in centimetres and hand areas in square centimetres.

5. Graph the data on a grid with arm length along the horizontal axis and hand area along the vertical axis. Can you make a general conclusion about arm length and hand area from the data? If you can, what is it?

11.5 Area of a Circle

Activity: Develop a Formula

This circle has been divided into twelve 30° wedges. Copy the circle and cut it into 2 semicircles along the diameter, as shown. Cut along each wedge from the centre almost to the edge of the semicircle. Then, open each semicircle and place the 2 sections together to approximate the shape of a parallelogram.

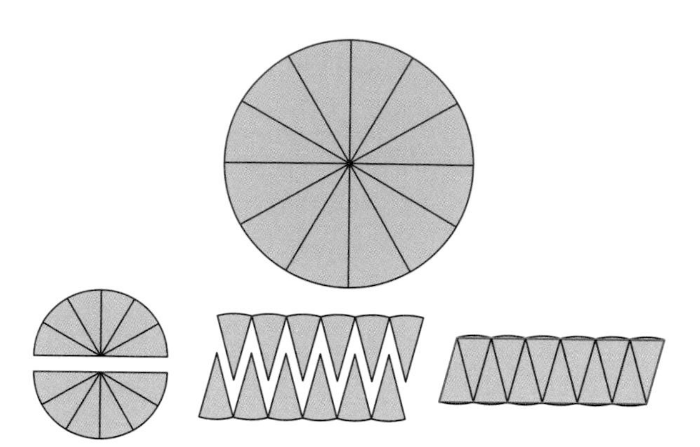

Inquire

1. How is the radius, r, of the circle related to the height of the parallelogram.
2. What is the expression for the distance around a circle in terms of r?
3. How much of the circumference of the circle is the base of the parallelogram?
4. What is the expression for the base of the parallelogram in terms of r?
5. What is the expression for the area of the parallelogram?
6. Is the area of the parallelogram also the area of the circle? Why?
7. What is the expression for the area of the circle?

Example

This ancient medicine wheel, typical of those made by the native peoples of Alberta and Saskatchewan, is in the shape of a circle. One spoke of the wheel measures 5 m. What is the area of the medicine wheel to the nearest tenth of a square metre?

Solution

A spoke of the wheel is the radius of the circle. So, $r = 5$ m. The formula for the **area of a circle** is $A = \pi r^2$.

$$\begin{aligned} A &= \pi r^2 \\ &= (3.14)(5)^2 \\ &= (3.14)(25) \\ &= 78.5 \end{aligned}$$

EST $(3)(25) = 75$

C π × 5 x² = 78.539816

The area of the medicine wheel is about 78.5 m^2.

Practice

Calculate the area of each circle to the nearest square centimetre.

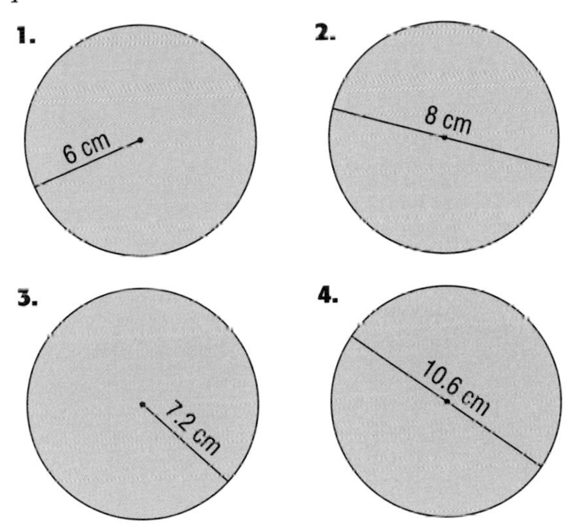

Estimate, then measure each radius. Then, calculate the area of the circle with each radius to the nearest tenth of a square centimetre.

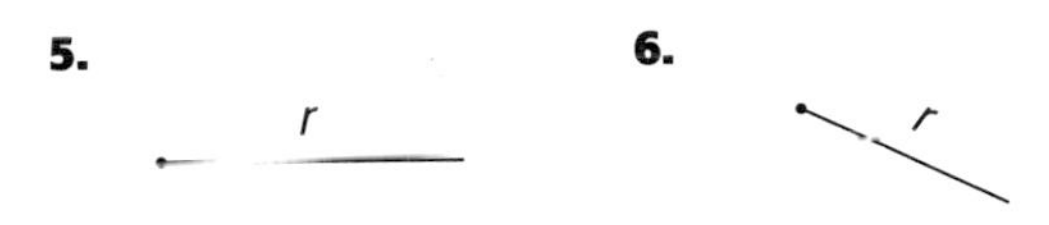

Problems and Applications

7. Calculate each area to the nearest square unit.

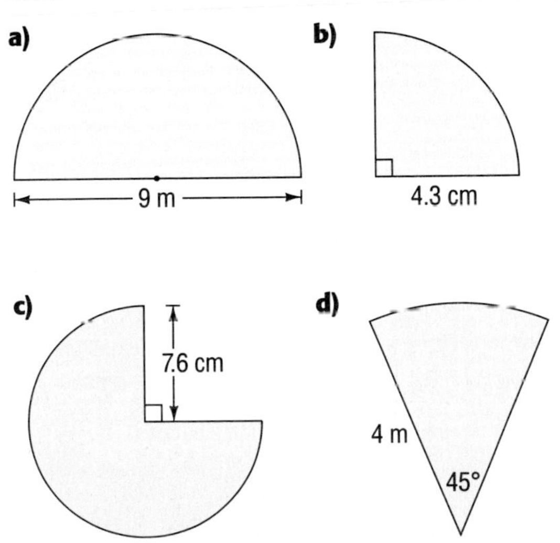

8. a) Calculate the area of the glass in this window to the nearest tenth of a square metre.

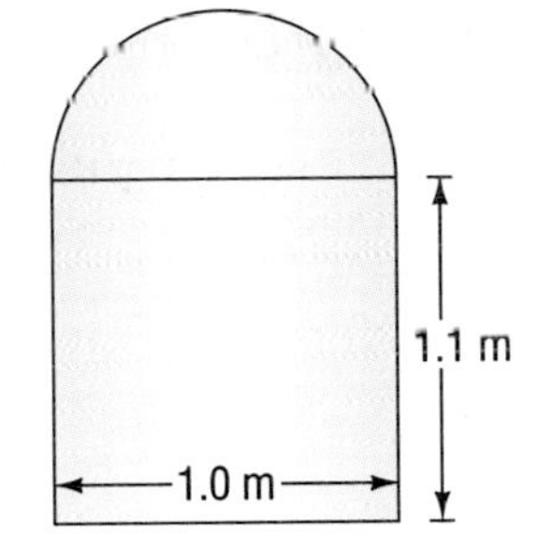

b) A protective storm window for this window costs $\$19.99/m^2$. What is the cost of the storm window?

9. Suneel tied his dog to the corner of a shed. The shed is 6 m by 3 m. The rope is 5 m long. Calculate the area within the dog's reach to the nearest square metre.

10. A rectangular yard is watered by 2 sprinklers as shown.

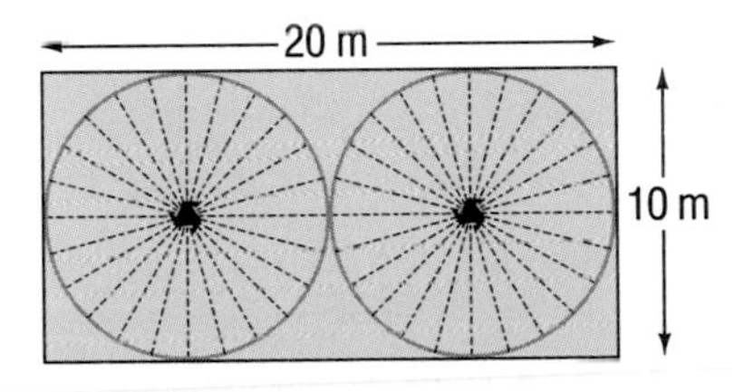

Find each of the following areas to the nearest square metre.

a) the area watered by the sprinklers

b) the area that is not watered

11. The area of the circle is 153.86 m^2. Work with a partner to calculate the area of the shaded region. Use $\pi = 3.14$.

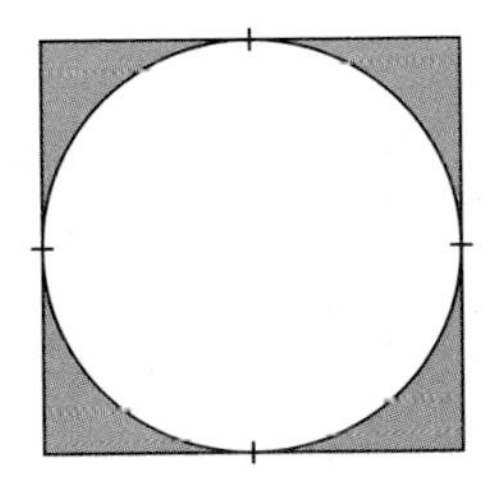

12. Write a problem that requires the calculation of the area of a circle. Have a classmate solve your problem.

Modelling Polyhedrons—Prisms

Polyhedrons are 3-dimensional objects in which the faces are polygons. A **prism** is a polyhedron with 2 parallel congruent bases that are in the shape of a polygon. The other faces are parallelograms.

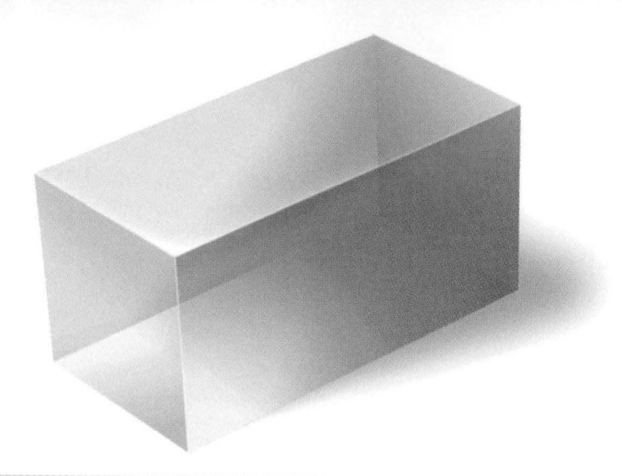

Activity ❶

1. Name the types of faces contained in the net of prism 1.

2. What prism is formed when the net of prism 1 is cut out and folded along the dotted lines?

3. Build a similar shape using 2-cm interlocking cubes.

4. What is different about the interiors of the paper prism and the one built from cubes?

5. What names do we use to describe these 2 different types of models? Explain.

6. On grid paper, draw a 3-dimensional copy of this prism.

7. Name a real-world example of this geometric shape.

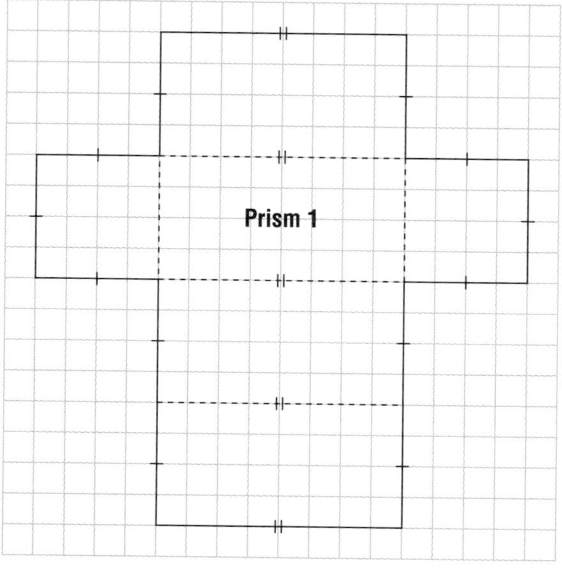

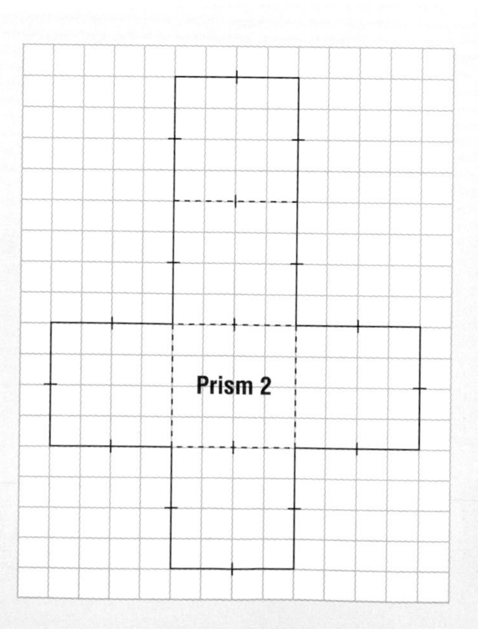

Activity ❷

1. Name the types of faces contained in the net of prism 2.

2. What prism is formed when the net of prism 2 is cut out and folded along the dotted lines?

3. Build a similar shape using 2-cm interlocking cubes.

4. Count the number of faces to calculate how much area is exposed.

5. How many cubes represent the amount of space occupied by this prism?

6. On grid paper, draw a 3-dimensional copy of this prism.

Activity ❸

1. What type of prism is formed when the net of prism 3 is cut out and folded along the dotted lines?

2. Name the polygons that make up this prism.

3. What plane figure is its base?

4. Draw and label a 3-dimensional copy of this prism.

5. Name a real-world example of this prism.

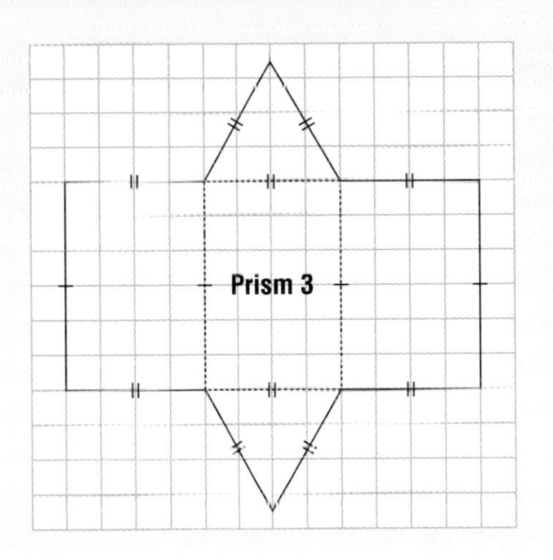

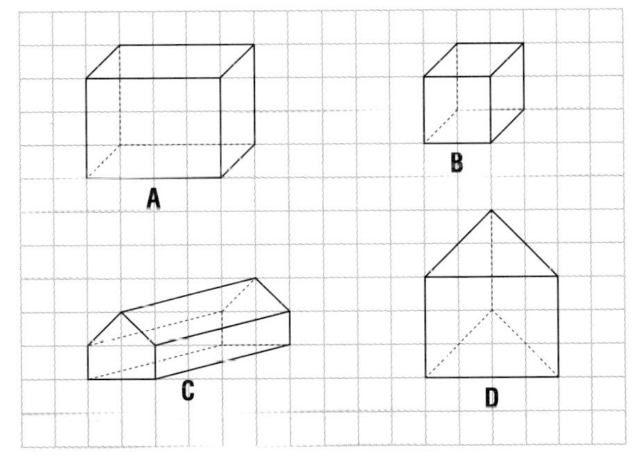

Activity ❹

1. Which of these prisms did you not consider in Activities 1, 2, or 3?

2. Name the types of faces and the number of each type contained in this prism.

Activity ❺

1. Which of the patterns on the right could be folded to form a cube?

2. There are 11 different patterns of 6 squares that can be folded to form a cube. Sketch the patterns on grid paper.

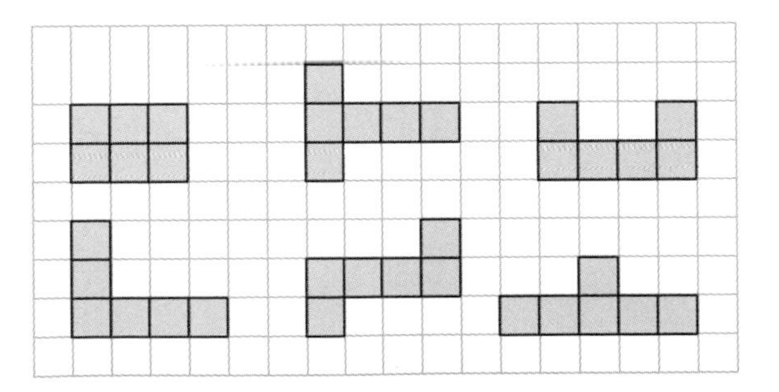

11.6 Surface Area and Volume of a Prism

Activity: Solve the Problem

Mark has to wrap a gift for his friend. The gift box measures 40 cm by 20 cm by 20 cm. Mark has 16 000 cm^2 of wrapping paper. Mark wants to know if he has enough paper.

Inquire

1. What type of prism is the box?
2. What types of figures are the faces of the box?
3. Calculate the area of each face.
4. What is the sum of all the areas?
5. Does Mark have enough paper? What assumptions did you make?

Example 1

Fort York was built next to Lake Ontario during the War of 1812. The 2.1 m high parapet that surrounds it forms a triangle with sides of 309 m, 153 m, and 324 m, as shown. The shape of the Fort resembles a triangular prism, but without a top. Find the interior surface area of the Fort to the nearest square metre.

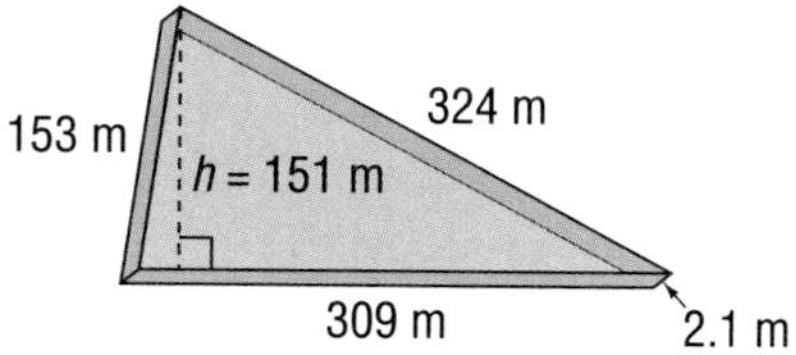

Solution

The surface area of a prism is the sum of the areas of the faces. In this case, the faces are the triangular base of the Fort and the 3 rectangular walls. The area of the triangular base is given by

$$A = \frac{1}{2}bh$$
$$= \frac{1}{2}(309 \times 151)$$
$$= \frac{1}{2}(46\ 659)$$
$$= 23\ 329.5$$

EST $\frac{1}{2}(300)(150) = 22\ 500$

C 0 . 5 × 309 × 151 = 23329.5

The area of the 3 rectangular faces is

$$2.1(309) + 2.1(153) + 2.1(324)$$
$$= 648.9 + 321.3 + 680.4$$
$$= 1650.6$$

EST $2(300)+2(200)+2(300) = 1600$

C 2 . 1 × (309 + 153 + 324) = 1650.6

The surface area is

$$23\ 329.5 + 1650.6 = 24\ 980.1$$

EST $23\ 000 + 2000 = 25\ 000$

The interior surface area of Fort York is 24 980 m^2 to the nearest square metre.

Activity: Investigate Volume

Build this rectangular prism with interlocking cubes.

Inquire

1. How many cubes represent the amount of space occupied by the prism?

2. Let the side of 1 cube be 1 unit of length,

a) What is the height of the prism?

b) What is the area of the base of the prism in square units?

3. Multiply the area of the base by the height, and compare the result with the amount of space you calculated in question 1.

4. The amount of space occupied by a prism is called the **volume**. Write a formula to calculate the volume of this prism. Use the variables V, B, and h, where B is the area of the base and h is the height.

Example 2

Calculate each volume.

a)

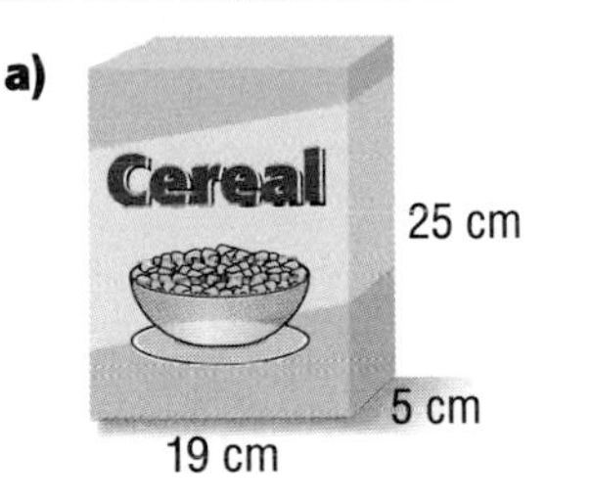

b)

Solution

The volume of a prism is the area of its base times its height.

a) The cereal box is a rectangular prism.

$V = B \times h$

$= [(19)(5)](25)$

$= 2375$

EST $(20)(5)(25) = 2500$

The volume of the box is 2375 cm^3.

b) The cheese box is a triangular prism. The base and height of the triangle are 10 cm each.

$V = B \times h$

$= [\frac{1}{2}(10)(10)](2)$

$= (50)(2)$

$= 100$

The volume of the box is 100 cm^3.

Example 3

A **composite solid** is made up of 2 or more prisms joined together. Find the volume of this composite solid.

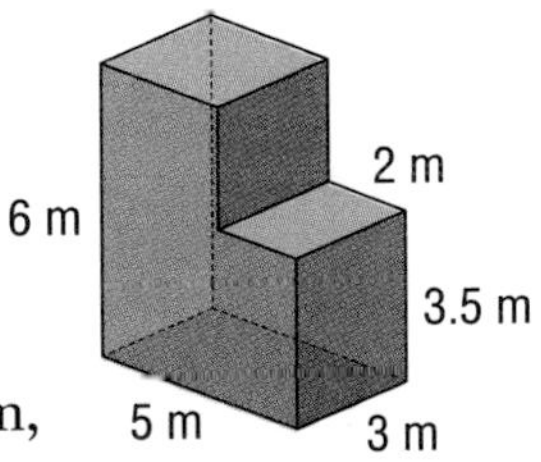

Solution

The solid is composed of a smaller prism, where l = 3 m, w = 2 m, and h = 3.5 m and a larger prism, where l = 3 m, w = 3 m, and h = 6 m. The volume of the smaller prism is (3)(2)(3.5) or 21 cm^3. The volume of the larger prism is (3)(3)(6) or 54 cm^3.

Total volume = 21 + 54

= 75

The volume of the composite solid is 75 cm^3.

CONTINUED

Practice

Name each prism.

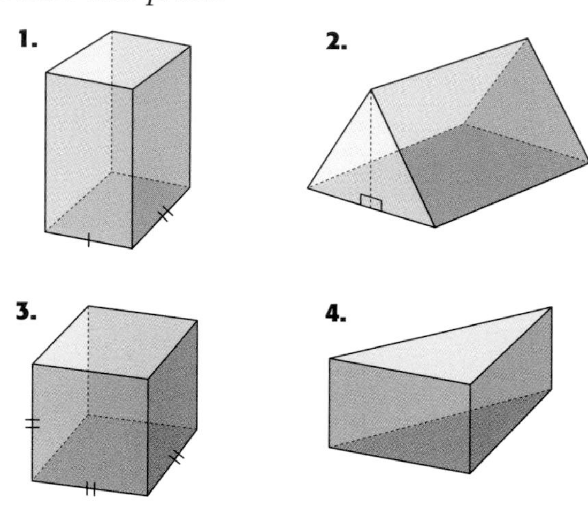

Which prism can be formed from each net?

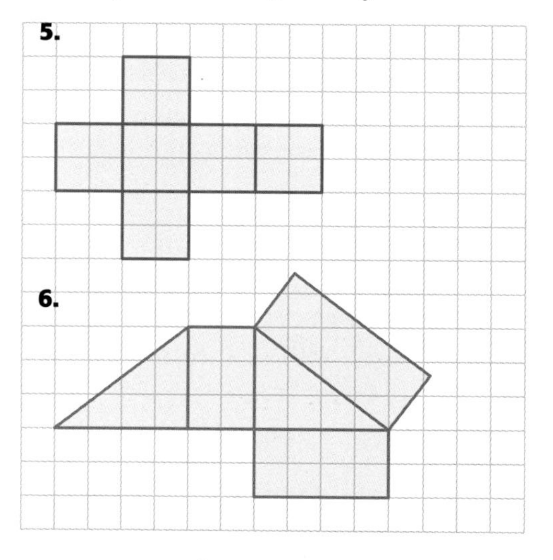

Estimate the surface area of each prism. Then, calculate it to the nearest square centimetre or square metre.

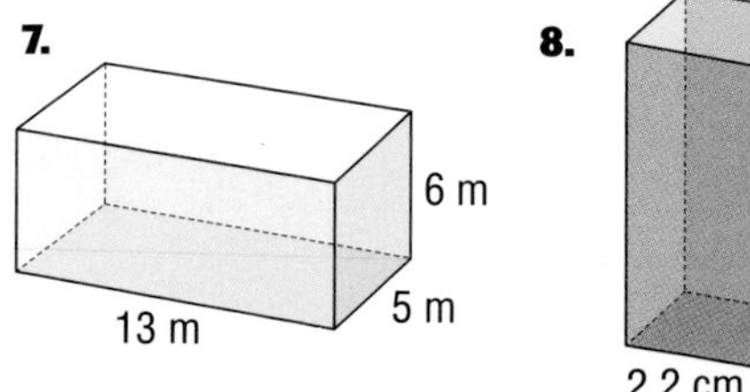

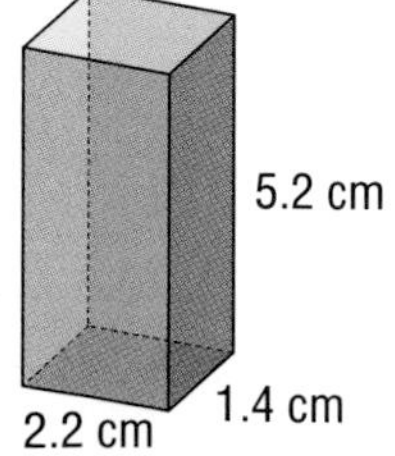

Calculate each surface area.

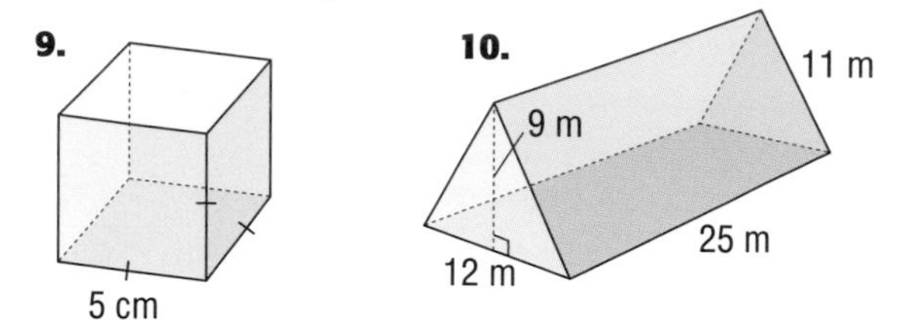

Estimate, then calculate the volume of each rectangular prism.

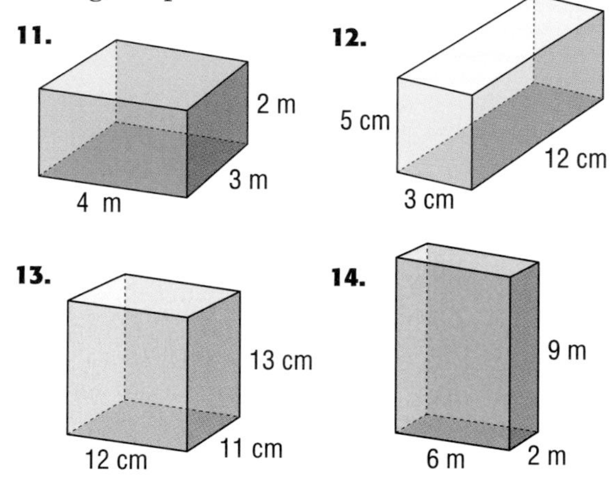

Calculate the surface area and volume of each prism to the nearest square or cubic unit.

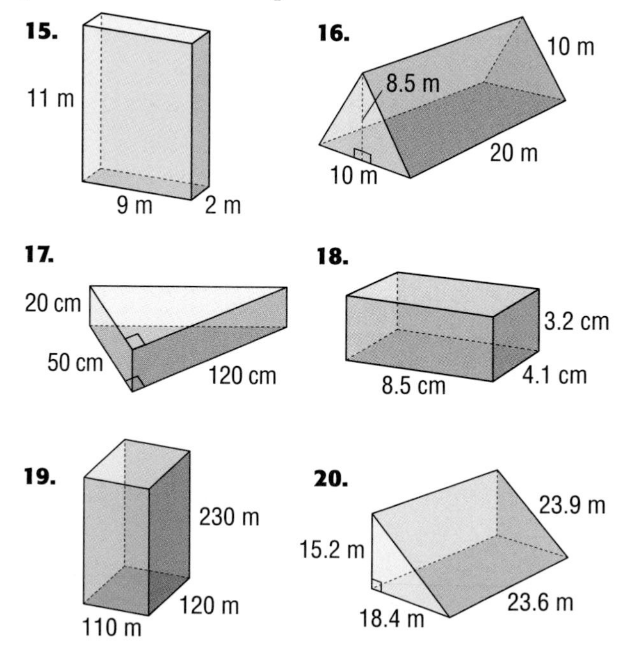

Problems and Applications

21. A covered garbage bin is to be built so that it measures 1.5 m by 1.2 m by 1.0 m.

a) How much plywood will it take to build the garbage bin?

b) How many cubic metres of garbage will it hold?

22. a) Calculate the surface area of this room.

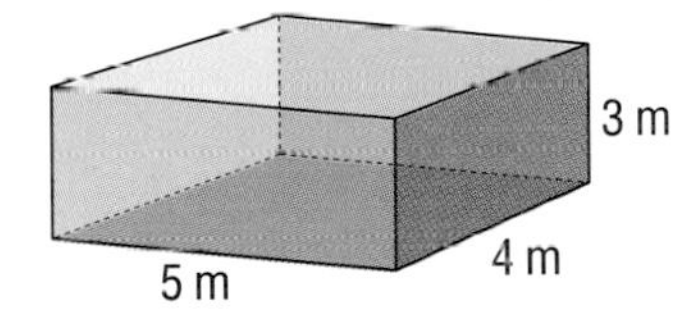

b) One 4-L can of paint will cover 36 m^2. If you want to give the ceiling and walls of the room 2 coats of paint, how many 4-L cans will you need? What assumptions have you made?

23. a) The dimensions of the base of a composter are 1 m by 1 m. Its height is 0.65 m. It is a prism with a top, a bottom, and 4 sides. Calculate its surface area.

b) The cost of material to build this composter is \$9.98/m^2. What is the total cost of the material?

c) If a town has set aside \$1 250 000 for the materials to build these composters, how many composters can be built?

24. The surface area of a cube is 216 cm^2. What are the dimensions of this cube?

25. a) A prism has a height of 10 cm. Find its surface area if the dimensions of the base are 8 cm by 2 cm.

b) Draw and label a diagram of the prism on dot paper or centimetre grid paper.

c) What is the name of the prism?

26. How many cubic metres of air does the tent contain?

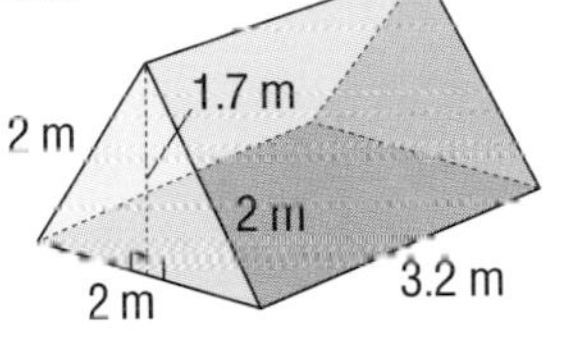

27. Canada's Anik E1 is a domestic communications satellite. Launched in September, 1991, it is a rectangular prism with l = 23 m, w = 8.5 m, and h = 4.3 m. Calculate the surface area and volume of Anik E1 to the nearest square or cubic unit.

28. The diagram shows the side view of a pool.

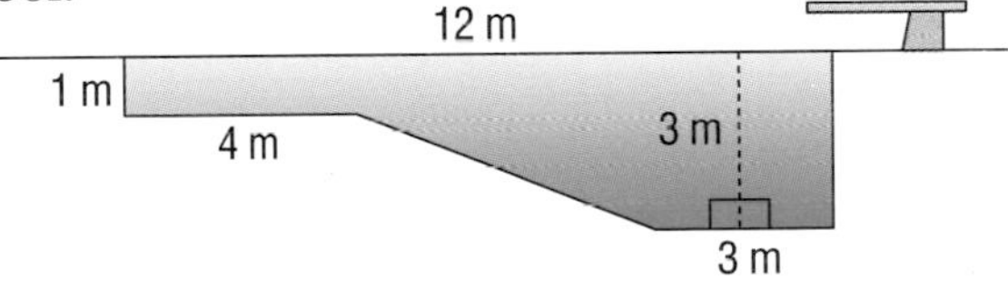

a) The pool is 5 m wide. Calculate its volume.

b) A pump can drain water from the pool at 0.3 m^3/min. How long does it take to drain the pool?

29. A garden storage shed is to be built in the shape of a rectangular prism before the roof is added. The volume of the shed before the roof is put on is 24 m^3. What are the most appropriate dimensions for the rectangular prism?

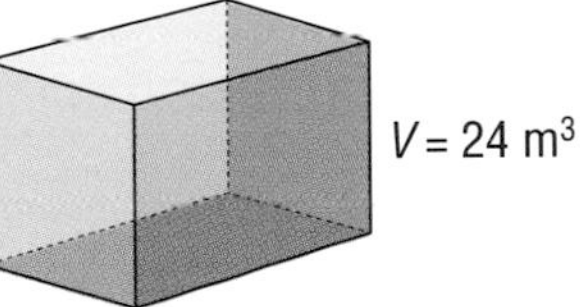

30. Work with a partner to calculate the surface area and volume of the interior of your classroom.

31. Write a problem that requires the calculation of the surface area and volume of a prism. Have another classmate solve your problem.

Volumes from a Spreadsheet

A spreadsheet can be used to calculate the volume of a rectangular prism. The dimensions of the rectangular prism shown here are 20 cm by 10 cm by 8 cm. Examine the spreadsheet.

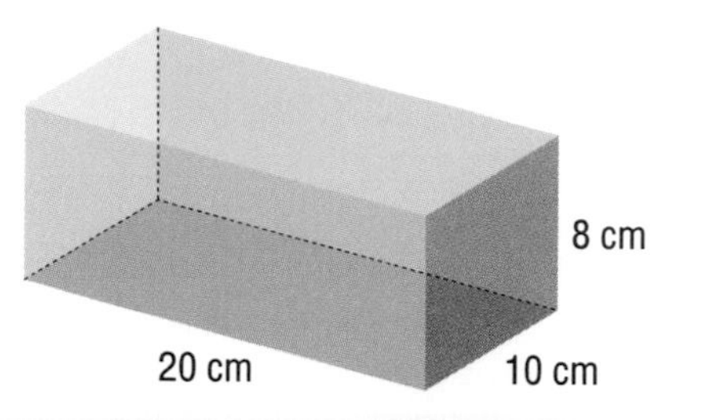

	A	B	C	D	E
1	Volume of a Rectangular Prism				
2					
3	Length =	20			
4					
5	Width =	10			
6					
7	Height =	8			
8					
9	Volume =	1600			
10	L x W x H				
11					
12					
13	B9 Calculates B3*B5*B7				
14					

The length is entered into cell B3, the width into cell B5, and the height into cell B7. The formula $V = l \times w \times h$ is entered into cell B9 as +B3*B5*B7. The + sign at the beginning is used to indicate that a formula is being entered. After the formula is entered, the resulting number appears in cell B9.

Activity ❶

1. Set up the spreadsheet on your computer, and use it to determine the volumes of the following rectangular prisms.

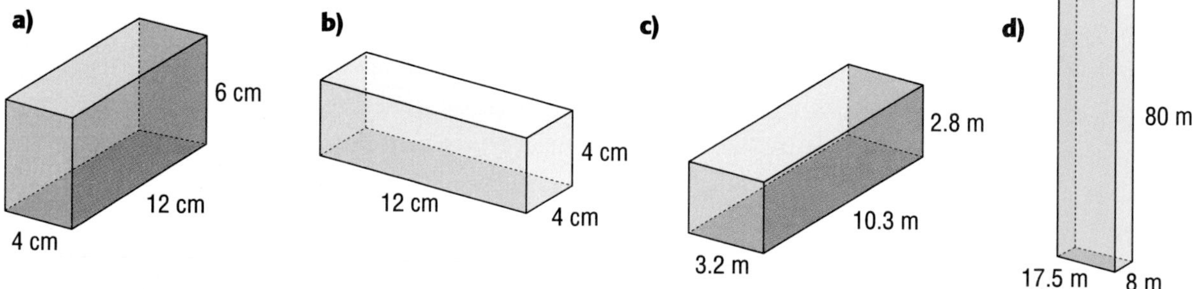

2. Modify the program to calculate the volume of a triangular prism. Check that the program works correctly.

A spreadsheet can also be used to see how changes in dimensions affect the volume of a rectangular prism.

Activity ❷

Use this spreadsheet to compare volumes when the length, width, and height of a rectangular prism are changed.

	A	B	C	D	E
1	Volume of a Rectangular Prism				
2	as Length, Width, and				
3	Height Change.				
4					
5		Original	New		
6		Prism	Prism		
7					
8	Length =	20	10		
9					
10	Width =	10	5		
11					
12	Height =	8	4		
13					
14	Volume =	1600	200		
15	L x W x H				
16					
17					
18	B14 Calculates B8*B10*B12				
19	C14 Calculates C8*C10*C12				

We can use a rectangular prism with dimensions 20 cm by 10 cm by 8 cm as the original prism and calculate the new volume as we make changes in the dimensions.

1. Double, triple, and quadruple the length of the original prism. Calculate each new volume.
2. What happens to the volume when the length is doubled, tripled, and quadrupled?
3. Double, triple, and quadruple both the length and width of the original prism, and calculate each new volume.
4. What happens to the volume when both the length and width are doubled, tripled, and quadrupled?
5. What do you think will happen to the volume if all 3 of the original dimensions are doubled, tripled, and quadrupled? Relate your answer to the previous results and explain.

11.7 Volume, Capacity, and Mass

The world's sixth largest reservoir is the Daniel Johnson Reservoir at Manicouagan, Quebec. The volume of water in this reservoir when it is full is 141 852 000 000 m^3. The greatest volume a container, such as a reservoir, can hold is known as the container's **capacity**. The capacity of a container is expressed in the units litres and millilitres. Large capacities may be expressed in kilolitres.

$$1000 \text{ mL} = 1 \text{ L}$$
$$1000 \text{ L} = 1 \text{ kL}$$

The **mass** of an object is a measure of the quantity of matter it is made up of. Mass is measured in milligrams, grams, kilograms, and tonnes.

$$1000 \text{ mg} = 1 \text{ g}$$
$$1000 \text{ g} = 1 \text{ kg}$$
$$1000 \text{ kg} = 1 \text{ t}$$

Activity: Interpret the Diagrams

The diagrams of 3 different containers filled with water illustrate a special relationship between the volume and mass of water. Note that the diagrams are not drawn to scale.

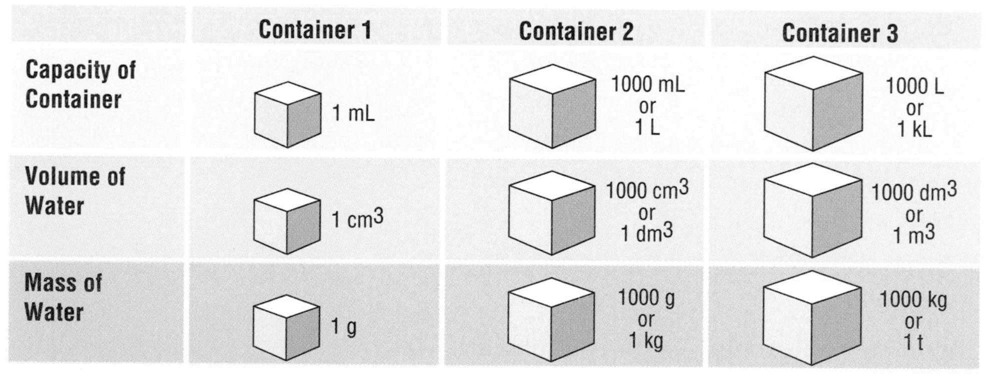

	Container 1	Container 2	Container 3
Capacity of Container	1 mL	1000 mL or 1 L	1000 L or 1 kL
Volume of Water	1 cm^3	1000 cm^3 or 1 dm^3	1000 dm^3 or 1 m^3
Mass of Water	1 g	1000 g or 1 kg	1000 kg or 1 t

Inquire

1. a) Which container has a capacity of 1 kL?
b) What is the volume of water in this container?
c) What is the mass of water in this container?

2. a) Which container has a capacity of 1 L?
b) What is the volume of water in this container?
c) What is the mass of water in this container?

3. State the capacity of the Daniel Johnson Reservoir in kilolitres.

4. State the mass of water in the Daniel Johnson Reservoir in tonnes.

5. What mass of water could a 100-mL container hold?

6. State the capacity of a container that holds 5 kg of water.

Practice

Express each volume in cubic centimetres.

1. 5 dm^3

2. 0.032 m^3

3. 0.4 dm^3

4. 10 m^3

Express each volume in cubic metres.

5. 5000 dm^3

6. 45 000 cm^3

7. 1 cm^3

8. 25 200 dm^3

Express each capacity in litres.

9. 275 mL

10. 15 mL

11. 1.5 kL

12. 0.1 kL

Express each capacity in millilitres.

13. 7 L

14. 0.25 L

15. 1 kL

16. 0.045 L

Express each mass in grams.

17. 15 kg

18. 4500 mg

19. 325 mg

20. 0.35 kg

Express each mass in kilograms.

21. 28 000 g

22. 15 t

23. 540 g

24. 100 mg

State the volume, in cubic centimetres, of each mass of water.

25. 10 g

26. 6.1 kg

27. 3.3 mg

28. 0.1 g

Problems and Applications

29. State the capacity, in kilolitres, of the smallest container needed to hold 50 t of water.

30. Food products are sold in different units of measurement.

a) List 3 food products sold in units of mass.

b) List 3 food products sold in units of capacity.

31. About how many 150-mL servings of grape juice can you pour from a 1-L container?

32. A can of frozen orange juice holds 355 mL of concentrate. If you add 3 times this volume of water, how many litres of orange juice do you have?

33. A trench on a building site is 75% full of water. The trench measures 3 m long by 1 m wide by 2 m deep.

a) What is the volume of water in the trench?

b) What is the capacity of the trench?

c) How long does it take a pump that draws water at 0.6 m^3/min to empty the trench?

34. Two containers have the same outside dimensions but different capacities. Explain how this situation is possible.

35. A can with a capacity of 4 L holds 4 dm^3 of gasoline. Can you state the mass of the gasoline? Explain.

36. a) What is the volume of milk in a 1-L carton?

b) Is the capacity of the carton exactly 1 L?

c) Would it be better to label the carton as 1 L of milk or 1 dm^3 of milk? Explain.

37. In your own words, distinguish the meanings of the terms "capacity" and "volume." Compare your answer with a classmate's.

LOGIC POWER

You can use 12 counters to make a square with 4 counters on each side.

How can you use 12 counters to make a square with 5 counters on each side?

Different Views

Knowing how to represent 3-dimensional objects in 2 dimensions is an important skill. Taking 2-dimensional drawings of an object and building the 3-dimensional figure is also useful.

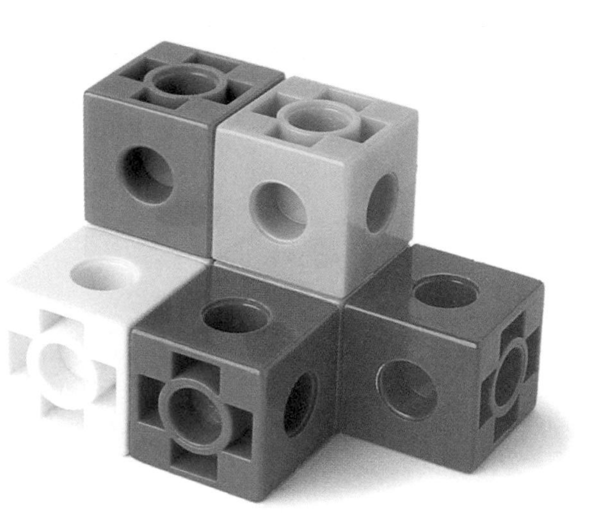

Activity ❶ Front, Side, and Top Views

This 3-dimensional shape was built using 7 cubes. The front view, right-side view, top view, and base design are shown. The base design is the top view with numbers giving the number of cubes in each position.

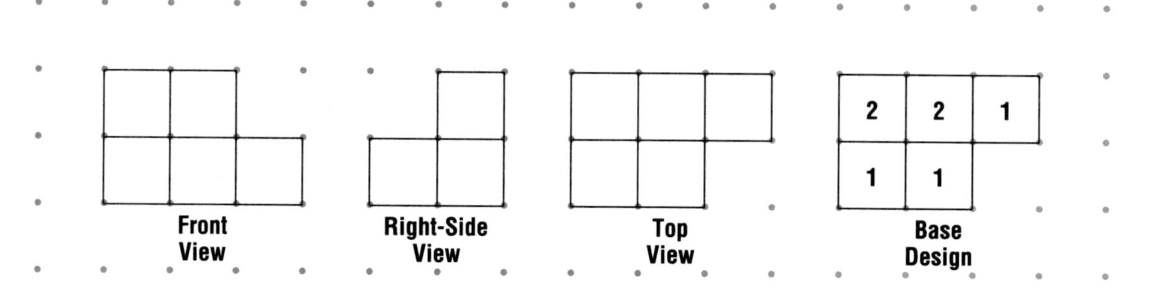

On grid paper, sketch the front view, right-side view, top view, and base design for each of the following shapes.

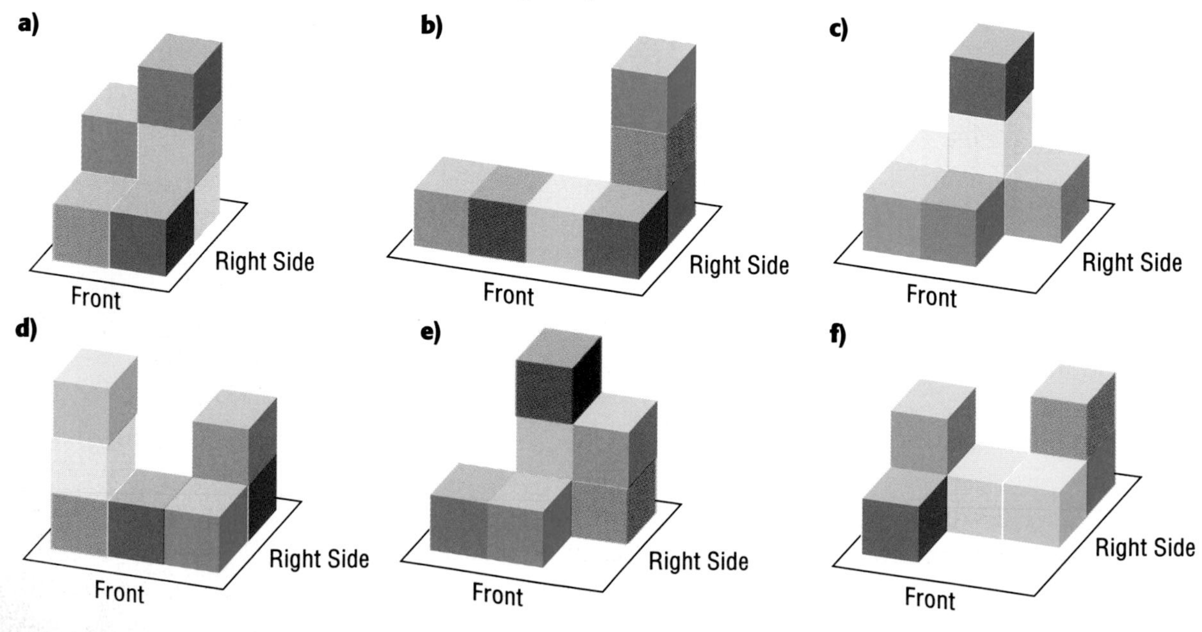

Activity ❷ Build the Model

Use these 2-dimensional drawings of 3 views to build the 3-dimensional shape. Draw the base design on grid paper.

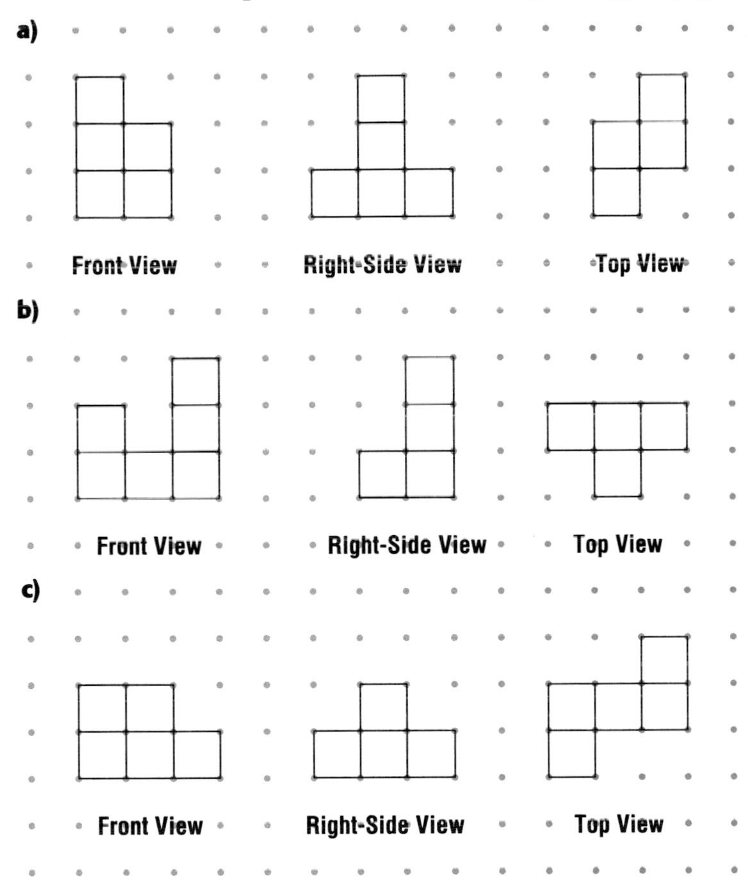

Activity ❸ Build the Model

The front and top views of models built with 7 cubes are shown.

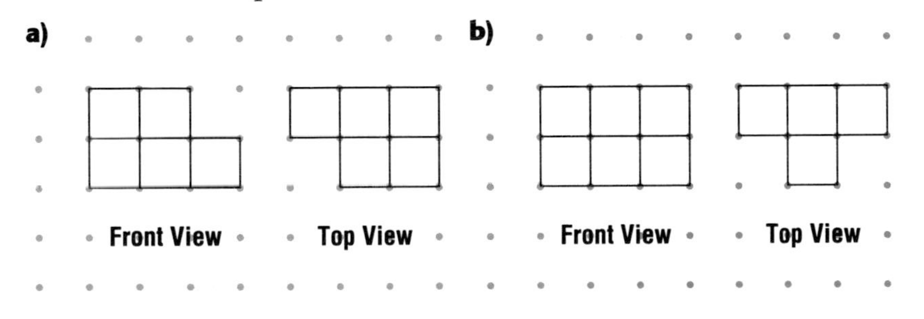

Build the model. On grid paper, draw the right-side view and the base design. Is there more than 1 solution? Compare with a classmate's solution.

11.8 Surface Area and Volume of a Cylinder and a Cone

Nearly perfect cylindrical shapes can be found in nature. An example is this part of the trunk of one of Canada's biggest trees. It is a Douglas fir that grows near Port Renfrew on Vancouver Island.

Activity: Develop a Formula

Join the shorter edges of a sheet of paper to form a cylinder. Do not crease the paper.

Inquire

1. a) Name the shapes of the top and bottom of your cylinder.
b) Write a formula for the combined area of these shapes.

2. a) What name is given to the distance around the top of the cylinder?
b) Write a formula for this distance in terms of r.

3. a) What was the original shape of the sheet of paper?
b) How does the length of this shape compare with the distance around the top of the cylinder?
c) What is the length of this shape in terms of r?

4. How does the height, h, of the cylinder compare with the width of the sheet of paper?

5. Write the formula for the area of the sheet of paper in terms of r and h.

6. Use the results of questions 1 b) and 5 to write a formula for the surface area of a cylinder that has a top and a bottom.

Example 1

The cylindrical can of juice has a diameter of 6 cm and a height of 13 cm. Find the surface area to the nearest tenth of a square centimetre.

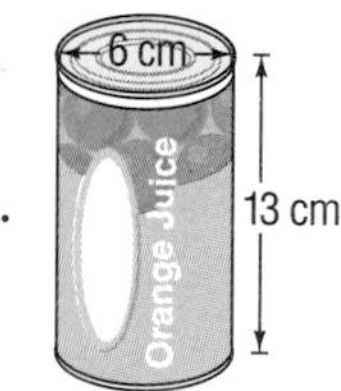

Solution

The area of the curved surface is the area of a rectangle. The top and bottom are circles. The formula for the **surface area of a closed cylinder**, a cylinder with a top and bottom, is $2\pi r^2 + 2\pi rh$.

$$\begin{aligned}\text{Surface Area} &= 2\pi r^2 + 2\pi rh \\ &= 2(3.14)(3)^2 + 2(3.14)(3)(13) \\ &= 56.52 + 244.92 \\ &= 301.44\end{aligned}$$

EST
$$\begin{aligned}&(2)(3)(10) + (2)(3)(3)(10) \\ &= 60 + 180 \\ &= 240\end{aligned}$$

The surface area of the juice can is 301.4 cm^2 to the nearest tenth of a square centimetre.

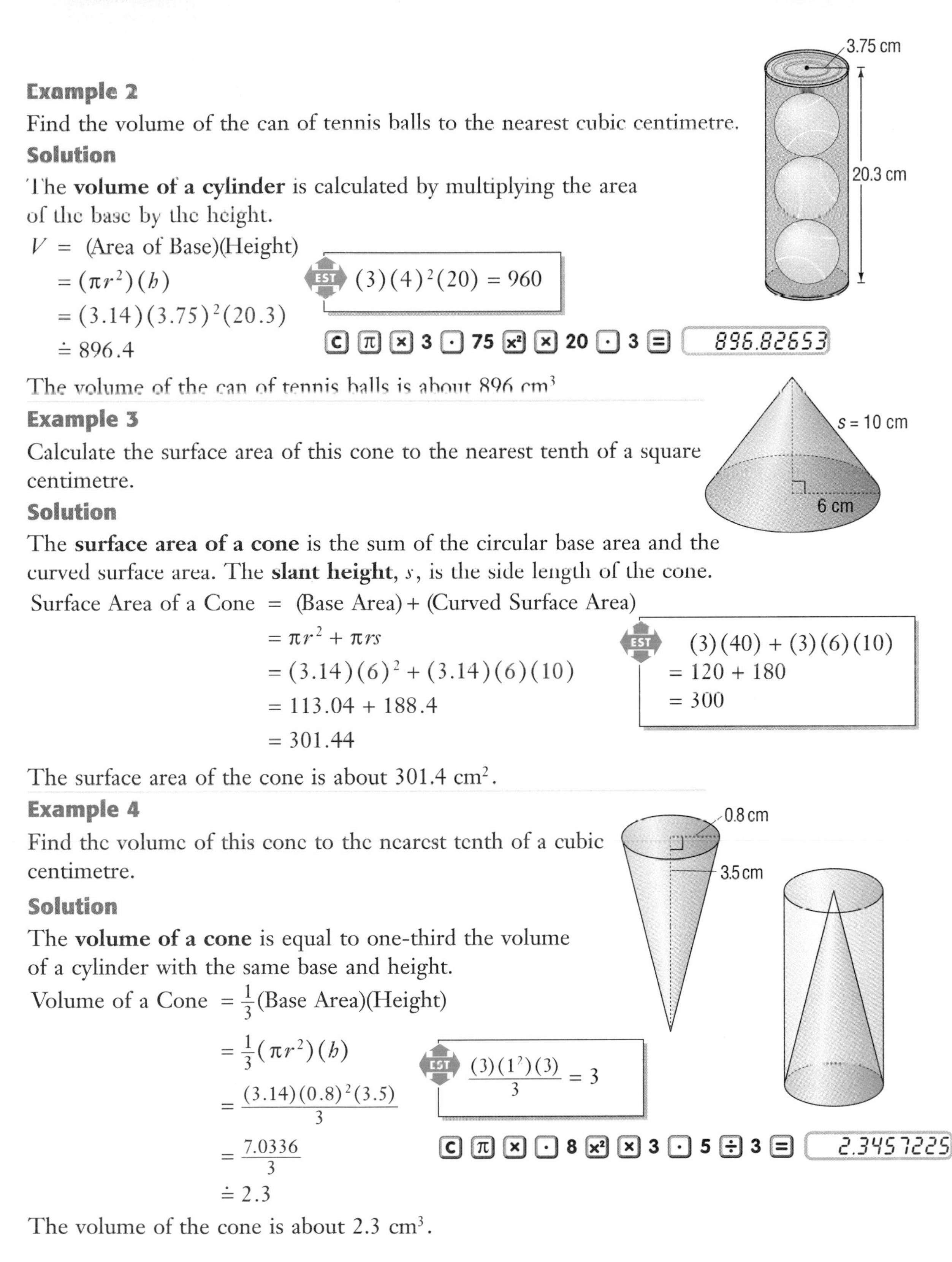

Example 2

Find the volume of the can of tennis balls to the nearest cubic centimetre.

Solution

The **volume of a cylinder** is calculated by multiplying the area of the base by the height.

V = (Area of Base)(Height)

$= (\pi r^2)(h)$

$= (3.14)(3.75)^2(20.3)$

$\doteq 896.4$

EST: $(3)(4)^2(20) = 960$

C π × 3 . 75 x² × 20 . 3 = 896.82653

The volume of the can of tennis balls is about 896 cm^3.

Example 3

Calculate the surface area of this cone to the nearest tenth of a square centimetre.

Solution

The **surface area of a cone** is the sum of the circular base area and the curved surface area. The **slant height**, s, is the side length of the cone.

Surface Area of a Cone = (Base Area) + (Curved Surface Area)

$= \pi r^2 + \pi rs$

$= (3.14)(6)^2 + (3.14)(6)(10)$

$= 113.04 + 188.4$

$= 301.44$

EST: $(3)(40) + (3)(6)(10)$
$= 120 + 180$
$= 300$

The surface area of the cone is about 301.4 cm^2.

Example 4

Find the volume of this cone to the nearest tenth of a cubic centimetre.

Solution

The **volume of a cone** is equal to one-third the volume of a cylinder with the same base and height.

Volume of a Cone $= \frac{1}{3}$(Base Area)(Height)

$= \frac{1}{3}(\pi r^2)(h)$

$= \frac{(3.14)(0.8)^2(3.5)}{3}$

$= \frac{7.0336}{3}$

$\doteq 2.3$

EST: $\frac{(3)(1^2)(3)}{3} = 3$

C π × . 8 x² × 3 . 5 ÷ 3 = 2.3457225

The volume of the cone is about 2.3 cm^3.

CONTINUED

Practice

Calculate each surface area to the nearest square centimetre.

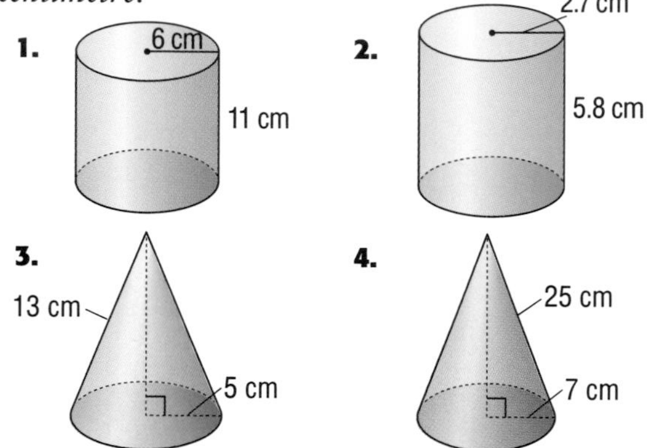

Estimate, then calculate each volume.

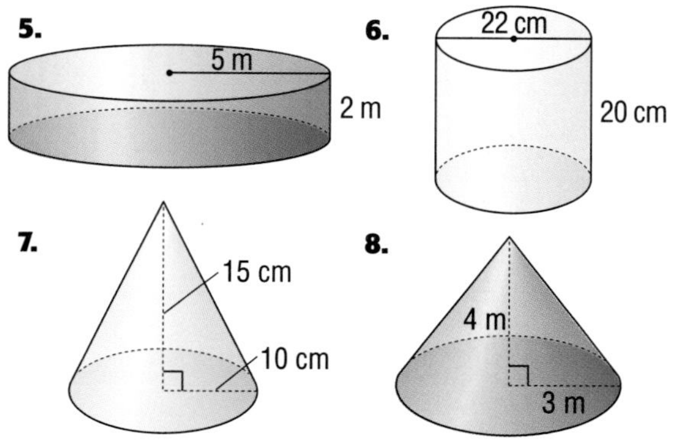

Calculate the surface area and volume of the following.

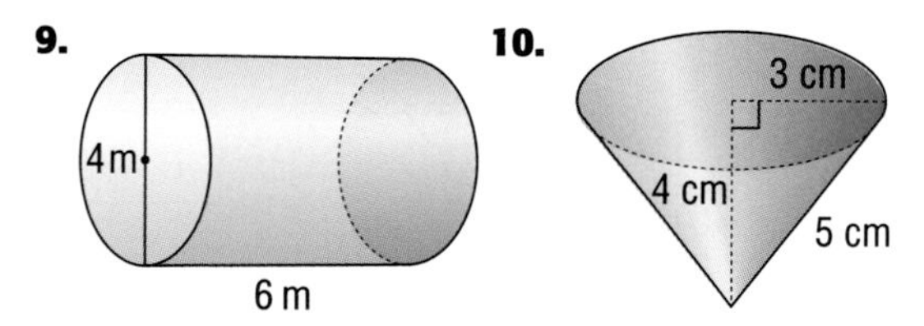

Problems and Applications

11. A paper cup at a water dispenser has a conical shape. The radius of the cup is 3 cm, and its height is 6 cm.

a) Find the slant height of the cup to the nearest tenth of a centimetre.

b) Calculate how much paper, to the nearest square centimetre, is required to make the cup.

12. The Canadarm used on the space shuttle is a cylinder that is 15.2 m long with a diameter of 38 cm. What is its surface area to the nearest tenth of a square unit?

13. A cone-shaped container has a diameter of 20 cm and a height of 20 cm. How many litres of water will the cone hold to the nearest litre?

14. A hobby club runs remote-controlled boats in a tank that is shaped as shown.

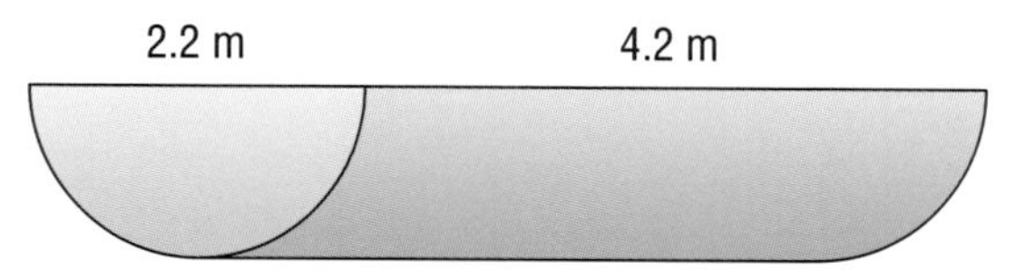

What volume of water can this tank hold to the nearest cubic metre?

15. Paper towels are sold in packages of 2 rolls. Each roll is a cylinder with a height of 30 cm and an outer diameter of 12 cm.

a) Design a shipping carton that will contain 12 packages of paper towels.

b) The inner diameter of each roll is 4 cm. How much wasted space is there in the carton?

16. Write a problem that requires the calculation of the surface area and volume of a cone. Have a classmate solve your problem.

NUMBER POWER

Place the digits from 1 to 9 in the circles so that the consecutive sums of the four numbers on each side differ by 1.

Modelling Polyhedrons—Pyramids

A **pyramid** is a polyhedron with one base and the same number of triangular faces as there are sides on the base.

Activity ❶

1. What polygon forms the base of each of the following pyramids?

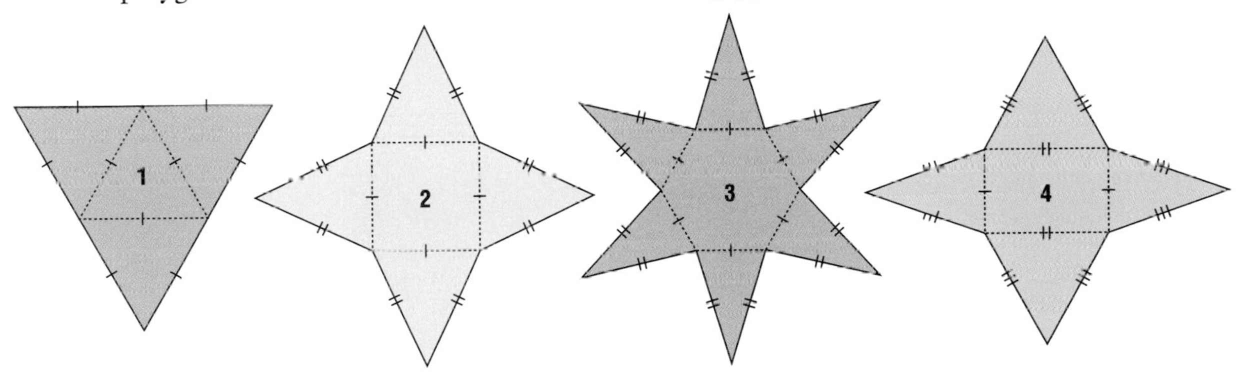

2. Name the pyramids that would be formed if you were to cut out the nets and fold along the dotted lines.

3. Draw and label a 3-dimensional copy of each pyramid on dot or grid paper.

4. A *tetrahedron* is a pyramid with 4 identical faces. Which of the above pyramids is a tetrahedron?

Activity ❷

1. a) Which one of these pyramids cannot be built from a net in Activity 1?

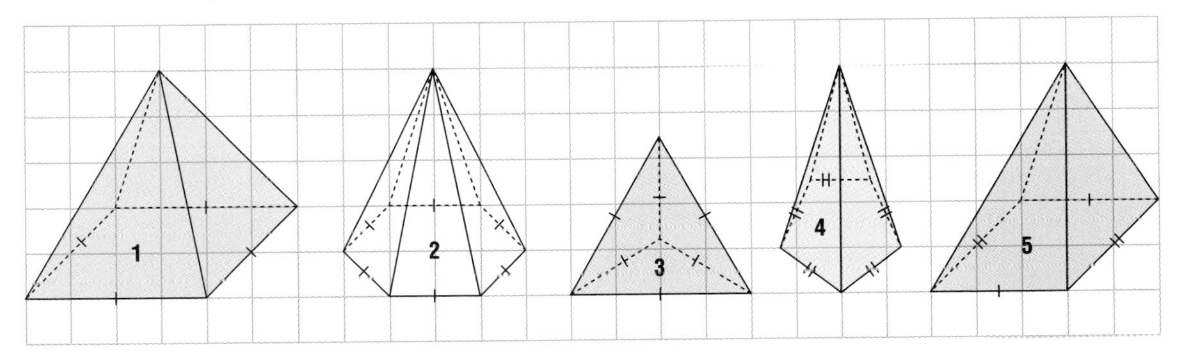

b) Name the types of faces and the number of each on the pyramid you identified in question 1.

2. Draw the net of this pyramid on dot or grid paper.

11.9 Surface Area and Volume of a Pyramid and a Sphere

The Great Pyramid of Khufu is shown in this photograph. It was built by an ancient Egyptian civilization around 2500 B.C. In spite of its age, its sophisticated design continues to amaze people.

Inquire

1. Name the different geometric figures the Great Pyramid is made up of.

2. What is the name of this type of pyramid?

3. The area of the square base of Khufu is 53 058 m^2. What is the length of each side of the base to the nearest metre?

4. The **slant height** of the pyramid, or the height of each triangular face, is 187 m. What is the surface area of each face?

5. What is the total amount of exposed surface area on this pyramid?

Example 1

Calculate the surface area of this square-based pyramid.

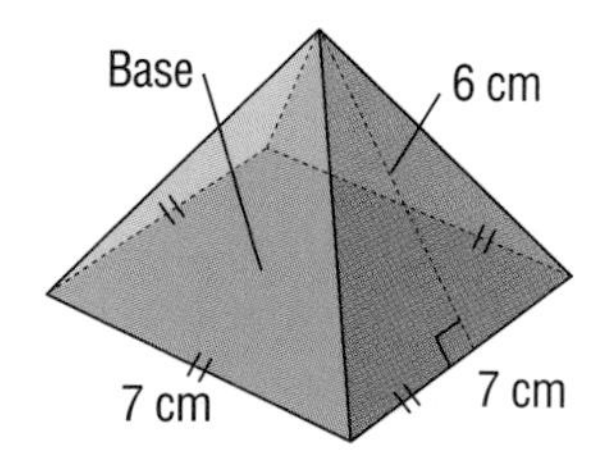

Solution

The **surface area of a pyramid** is the sum of the area of the base and the areas of the triangular faces.

The base of the pyramid is a square.

$$\begin{aligned} \text{Base Area} &= (7)(7) \\ &= 49 \end{aligned}$$

The 4 triangular faces are identical.

$$\begin{aligned} \text{Face Area} &= 4[\tfrac{1}{2}(7)(6)] \\ &= 4(21) \\ &= 84 \end{aligned}$$

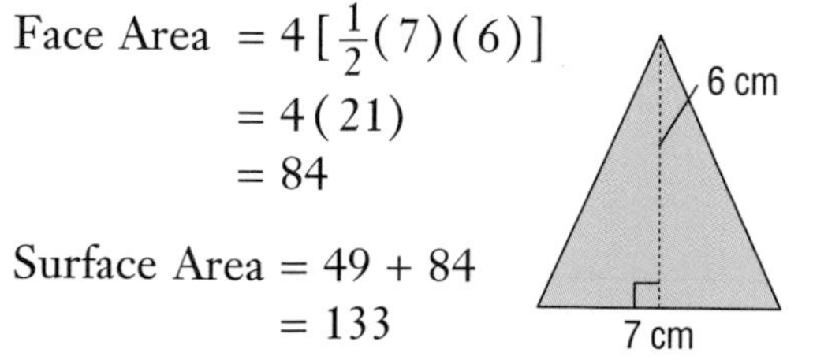

$$\begin{aligned} \text{Surface Area} &= 49 + 84 \\ &= 133 \end{aligned}$$

The surface area of the pyramid is 133 cm^2.

Example 2

Calculate the volume of this pyramid.

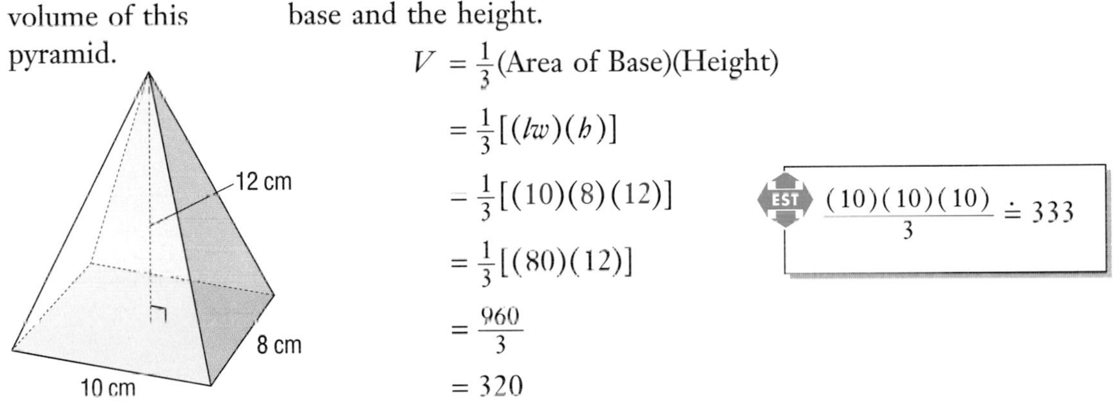

Solution

The **volume of a pyramid** is one-third the product of the area of the base and the height.

$$\begin{aligned} V &= \tfrac{1}{3}(\text{Area of Base})(\text{Height}) \\ &= \tfrac{1}{3}[(lw)(h)] \\ &= \tfrac{1}{3}[(10)(8)(12)] \\ &= \tfrac{1}{3}[(80)(12)] \\ &= \frac{960}{3} \\ &= 320 \end{aligned}$$

EST $\frac{(10)(10)(10)}{3} \doteq 333$

The volume of the pyramid is 320 cm^3.

A **sphere** is like a ball. The radius of a sphere is a line segment joining the centre and a point on the surface of the sphere.

Example 3

The radius of the Cinesphere at Ontario Place in Toronto is 19 m. Calculate the sphere's surface area and volume to the nearest square or cubic unit.

Solution

The formula for the **surface area of a sphere** is $4\pi r^2$.
The radius of the Cinesphere is 19 m.

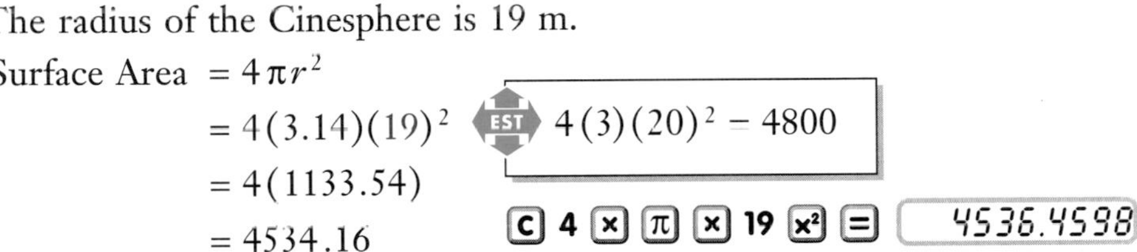

$$\begin{aligned} \text{Surface Area} &= 4\pi r^2 \\ &= 4(3.14)(19)^2 \\ &= 4(1133.54) \\ &= 4534.16 \end{aligned}$$

EST $4(3)(20)^2 = 4800$

C 4 × π × 19 x² = 4536.4598

The surface area of the Cinesphere is about 4534 m^2.

The formula for the **volume of a sphere** is $\frac{4}{3}\pi r^3$.

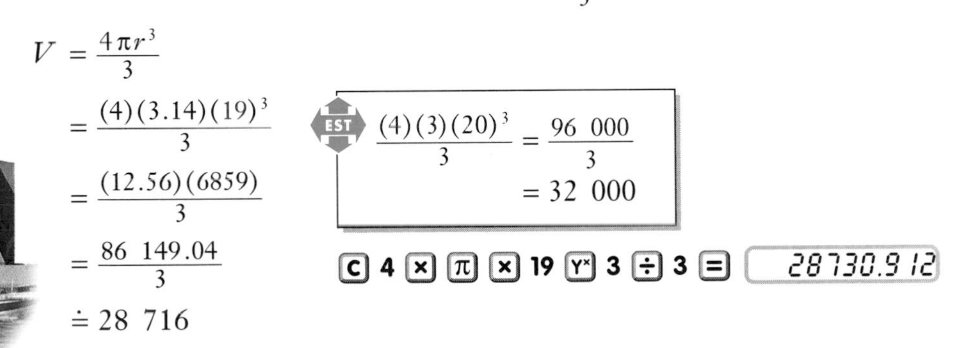

$$\begin{aligned} V &= \frac{4\pi r^3}{3} \\ &= \frac{(4)(3.14)(19)^3}{3} \\ &= \frac{(12.56)(6859)}{3} \\ &= \frac{86\ 149.04}{3} \\ &\doteq 28\ 716 \end{aligned}$$

EST $\frac{(4)(3)(20)^3}{3} = \frac{96\ 000}{3} = 32\ 000$

C 4 × π × 19 Yˣ 3 ÷ 3 = 28730.912

The volume of the Cinesphere is about 28 716 m^3.

CONTINUED

Practice

Calculate the surface area.

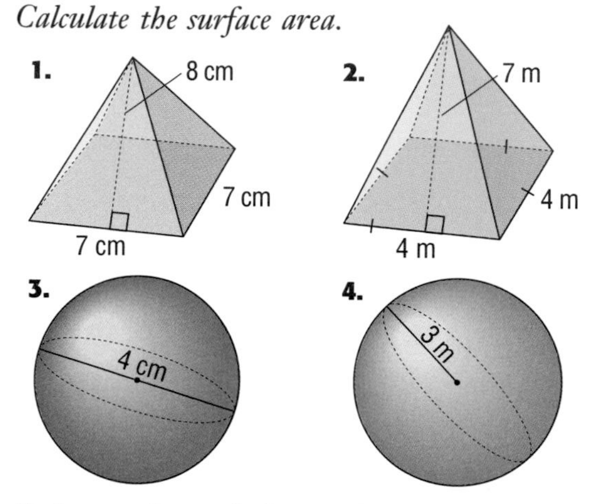

Estimate, then calculate each volume to the nearest cubic unit.

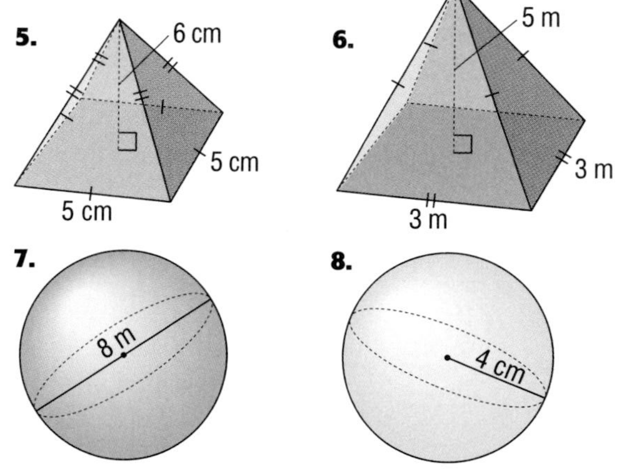

Problems and Applications

9. The rectangular base of a pyramid measures 10 cm × 9 cm. The height of the pyramid is 12 cm. Calculate its volume.

10. A golf ball is in the shape of a sphere with radius 20 mm. What is its surface area?

11. The length of one side of the base of the Great Pyramid is 230 m, and the height of the pyramid is 146 m. What is its volume?

12. The diameter of the Earth is 12 756 km.

a) Calculate its surface area to the nearest square kilometre.

b) What assumptions have you made?

13. The surface area of Canada is 9 970 610 km^2. Use your answer from question 12 to estimate the fraction of the Earth's surface that is covered by Canada.

14. The slant height of a pyramid is about 120 m and each side of its base measures about 180 m. Calculate the surface area to the nearest thousand square metres.

15. If an orange of diameter 6 cm costs 25¢, and an orange of diameter 9 cm costs 50¢, which is the better value? Explain your answer, and state any assumptions that you made.

LOGIC POWER

The cubes are identical. Each face has a ○, □, ■, or ● on it.

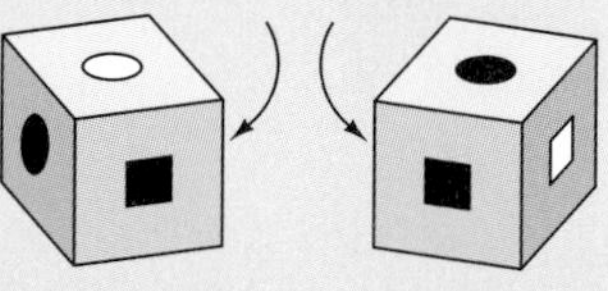

What symbol is on each of the faces indicated by the arrows?

Programming Formulas in BASIC

Activity ❶ Cylinders and Cones

1. Work with a partner to examine the following computer programs, and describe what each line of the program does or asks you to do.

a)
```
NEW
10 PRINT"SURFACE AREA"
20 PRINT"OF A CYLINDER"
30 INPUT"RADIUS IS";R
40 INPUT"HEIGHT IS";H
50 A=2*3.14*R^2+2*3.14*R*H
60 PRINT"THE SURFACE AREA IS";A
70 END
```

b)
```
NEW
10 PRINT"VOLUME OF A CYLINDER"
20 INPUT"RADIUS IS";R
30 INPUT"HEIGHT IS";H
40 V=3.14*R^2*H
50 PRINT"VOLUME IS";V
60 END
```

c)
```
NEW
10 PRINT"SURFACE AREA OF A CONE"
20 INPUT"RADIUS IS";R
30 INPUT"SLANT HEIGHT IS";S
40 A=3.14*R*(R+S)
50 PRINT"THE SURFACE AREA IS";A
60 END
```

d)
```
NEW
10 PRINT"VOLUME OF A CONE"
20 INPUT"RADIUS OF BASE IS";R
30 INPUT"HEIGHT IS";H
40 V=3.14*R^2*H/3
50 PRINT"VOLUME IS";V
60 END
```

Activity ❷ Spheres

1. What does each line of the following programs do or have you do?

a)
```
NEW
10 PRINT"SURFACE AREA"
20 PRINT"OF A SPHERE"
30 INPUT"RADIUS IS";R
40 A=4*3.14*R^2
50 PRINT"SURFACE AREA IS";A
60 END
```

b)
```
NEW
10 PRINT"VOLUME OF A SPHERE"
20 INPUT"RADIUS OF SPHERE IS";R
30 V=4*3.14*R^3/3
40 PRINT"VOLUME IS";V
50 END
```

2. Estimate the surface area and volume for each of these spheres.

a) $r = 3$ cm **b)** $d = 4$ cm

c) $r = 6$ m **d)** $d = 3$ m

3. Use the values from question 2 in the above programs to verify your answers.

4. List the advantages of using a computer program to perform these types of calculations. Compare your list with a classmate's.

5. Using the above programs as models, write a program that calculates the volume of a square-based pyramid.

Vanishing Points and Perspective

Many artists from the 14th to 16th century, a period known as the Renaissance, studied mathematics and engineering. The artists used geometry to help them show the three-dimensional world on a two-dimensional surface. They used the concepts of perspective and vanishing points to help create great works of art.

Activity ❶ One Vanishing Point

1. a) To sketch a cube using 1 vanishing point, draw a square for the front face of the cube.
b) Draw a horizontal line parallel to the horizontal edges of the square. This line is called the **eye level.**
c) Mark a vanishing point above the top edge of the square on the eye level line.

2. Draw a line from each corner of the square to the vanishing point. These lines are called **vanishing lines** .

3. Complete the sketch of the cube using the vanishing lines as edges of the cube.

4. Erase the eye level line and any unnecessary parts of the vanishing lines. The viewer sees the cube from above and in front.

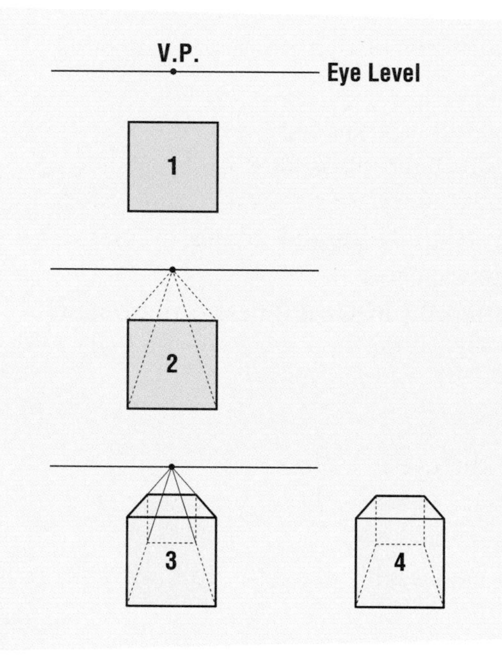

Activity ❷ Perspectives I

1. Sketch a cube where the eye level line is below the cube and the vanishing point is to the right of the cube. How does the viewer see the cube?

2. Sketch a cube where the eye level line is through the centre of the cube and the vanishing point is to the right of the cube. How does the viewer see the cube?

3. Sketch a cube where the eye level line is below the cube and the vanishing point is directly below the cube. How does the viewer see the cube?

Activity ❸ Two Vanishing Points

1. a) To sketch a cube using 2 vanishing points, draw the front edge of the cube.
b) Draw an eye level line above the front edge of the cube, and mark 2 vanishing points on the line.

2. Draw vanishing lines from each end of the front edge of the cube to the vanishing points.

3. a) To form the remaining front vertical edges, draw vertical line segments, parallel to the front edge. Draw these lines to intersect the vanishing lines.
b) Draw more vanishing lines to these new edges to determine the back edges of the cube.

4. Erase the eye level line and any unnecessary parts of the vanishing lines.

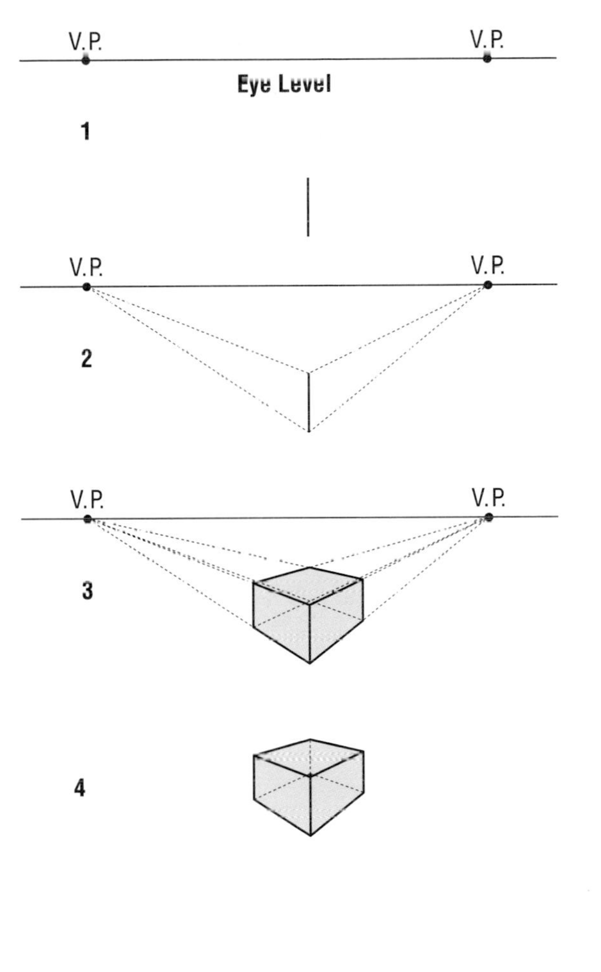

Activity ❹ Perspectives II

1. Use 2 vanishing points to sketch a cube with the eye level line above the cube.

2. Use 2 vanishing points to sketch a cube with the eye level line passing through the cube.

Activity ❺ Sketching Letters in Perspective

1. Sketch the letter M using 1 vanishing point.

2. Sketch the letter T using 2 vanishing points.

3. Sketch your initials using 1 or 2 vanishing points.

Review

Calculate the perimeter.

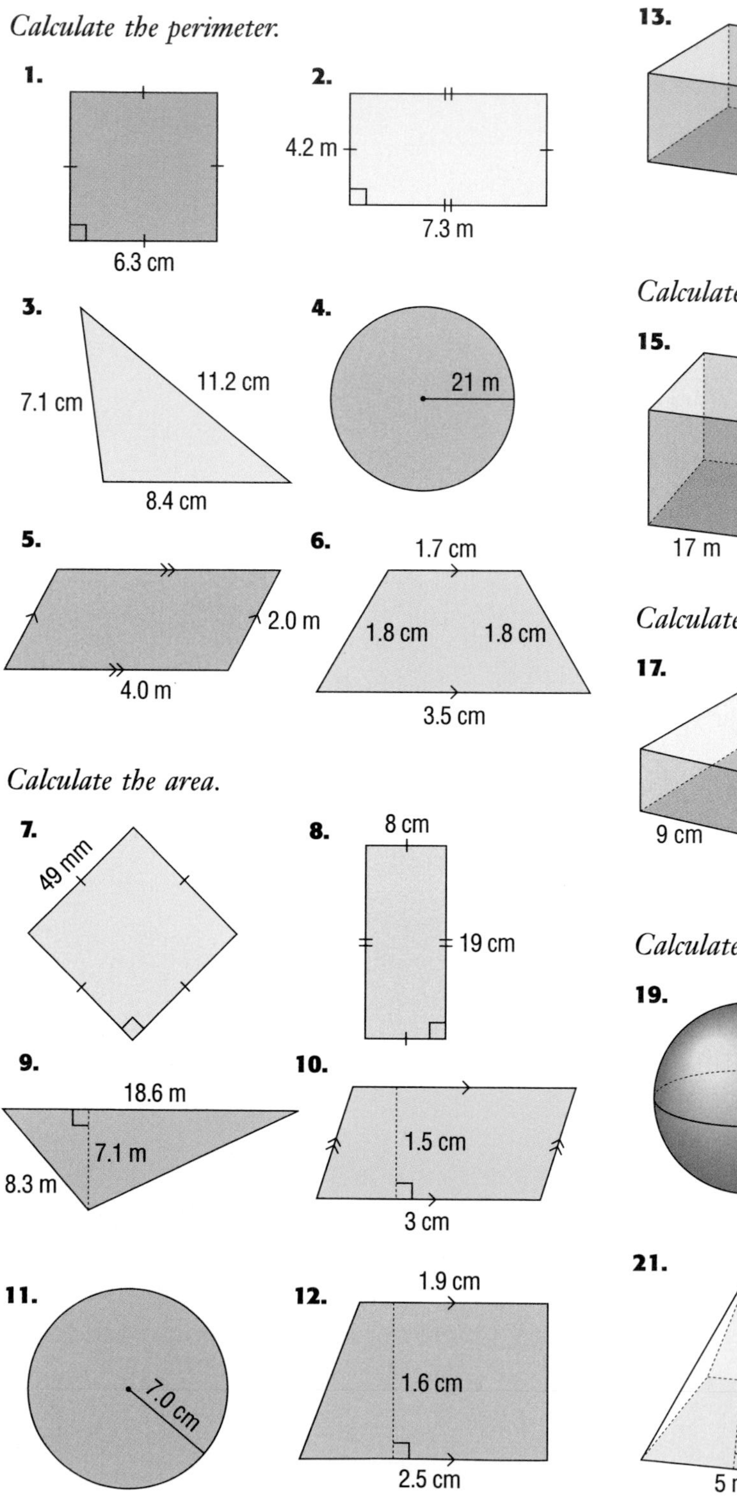

Name each prism.

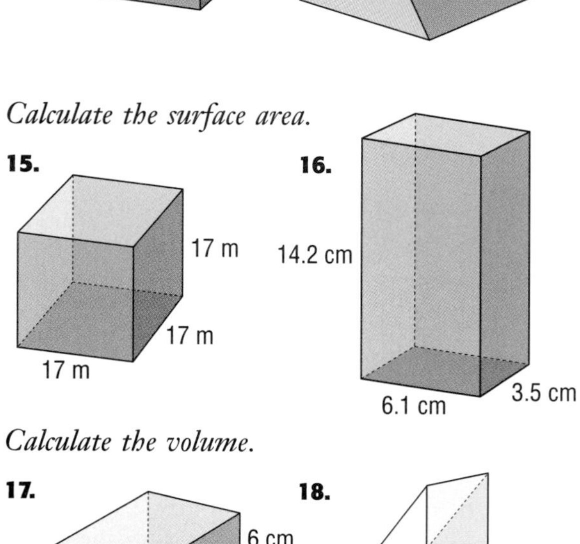

Calculate the volume.

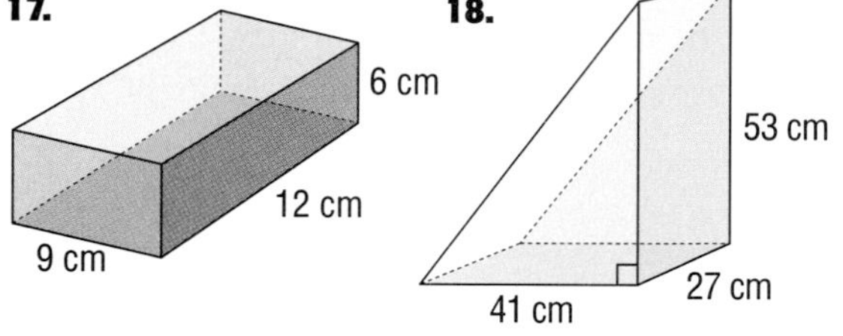

Calculate the surface area.

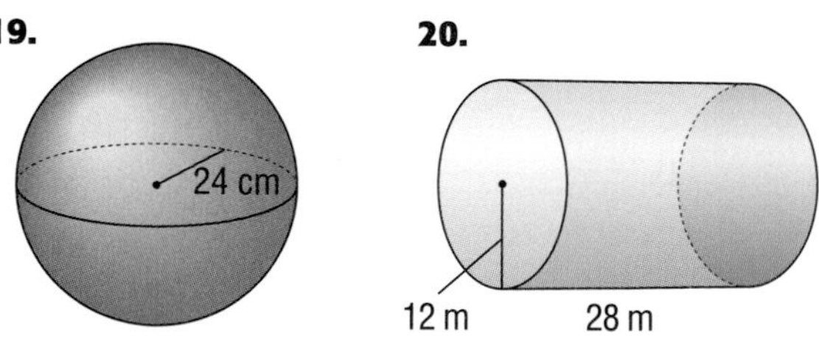

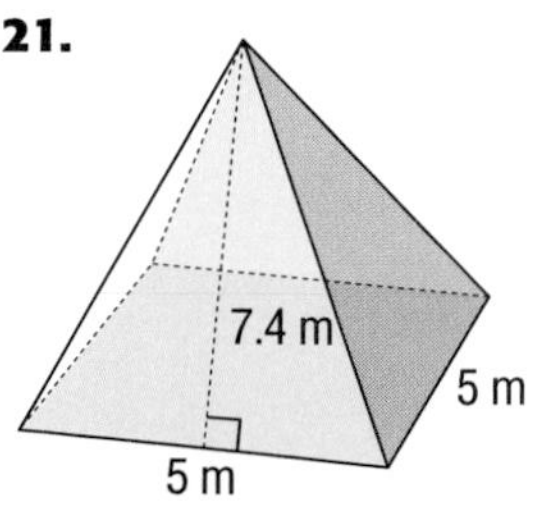

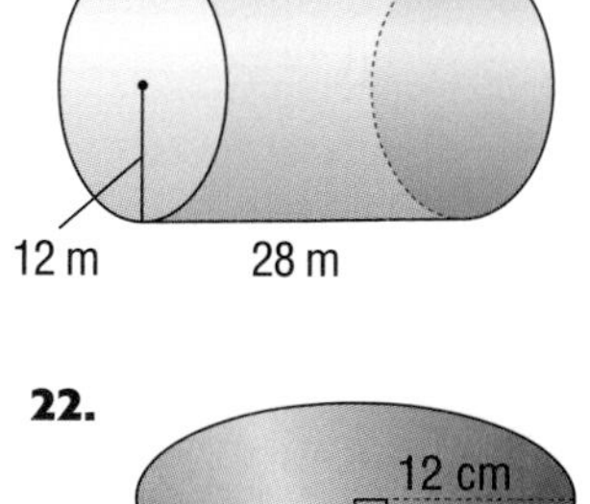

Calculate the volume to the nearest unit.

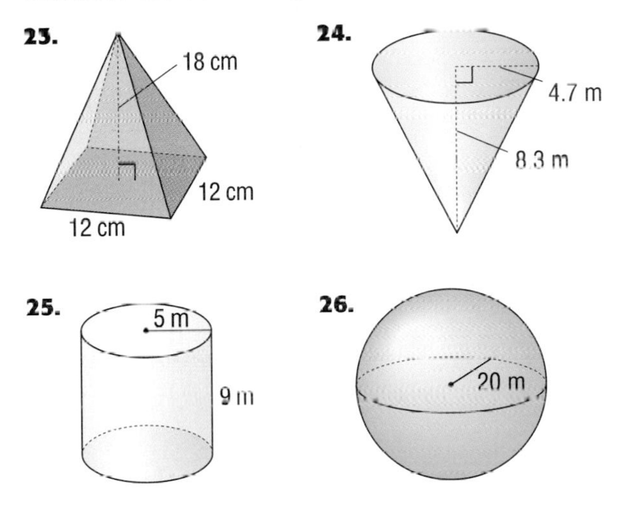

27. What polyhedron can be formed using these faces? How many of each face do you need?

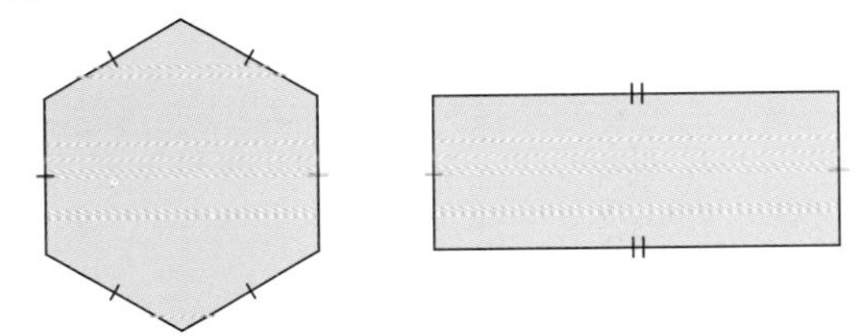

28. A container has the shape of a rectangular prism. Its inside dimensions are 12 cm by 10 cm by 15 cm.

a) Find the volume of water, in dm^3, the container holds.

b) Find the mass of water, in kg, the container holds.

c) Find the capacity, in L, of the container.

29. A 4 m by 4 m square is removed from the centre of a 15 m by 30 m rectangle. What is the area of the remaining part?

30. A pizza with a diameter of 45 cm has a circular ring of tomato sauce with a radius of 20 cm. Find the area of the bare crust around the edge of the pizza to the nearest square centimetre.

Group Decision Making

Making Math Games for Grades 1–8

You are going to make a math game for students in one of the grades from 1 to 8.

1. Decide as a class the grade or grades different groups will make games for. You might select the same grade or choose different grades. If you choose different grades, decide which grade each group will have.

2.

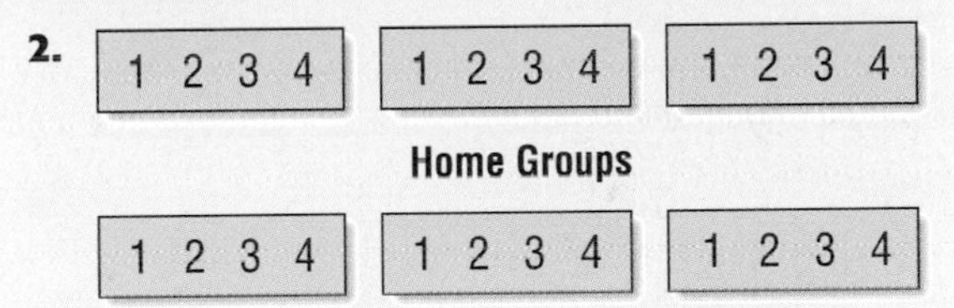

As a group, research the topics taught in your chosen grade and the level of difficulty required. Textbooks are good sources of information. Ask teachers what games are already available to students at their grade level. This information may give you some ideas about the kind of game you want to make.

3. As a group, brainstorm the types of games you could make. Decide on the type of game and on the math you want to use.

4. Make the game and test it with your classmates. Ask for suggestions for improvement. Refine your game and test it with the students it was designed for. Ask the students how they like it. Have their teacher evaluate the game.

5. In your group, prepare a report on your game for the class. Describe how well students liked the game and what you might do differently the next time.

Chapter Check

Calculate the perimeter and area.

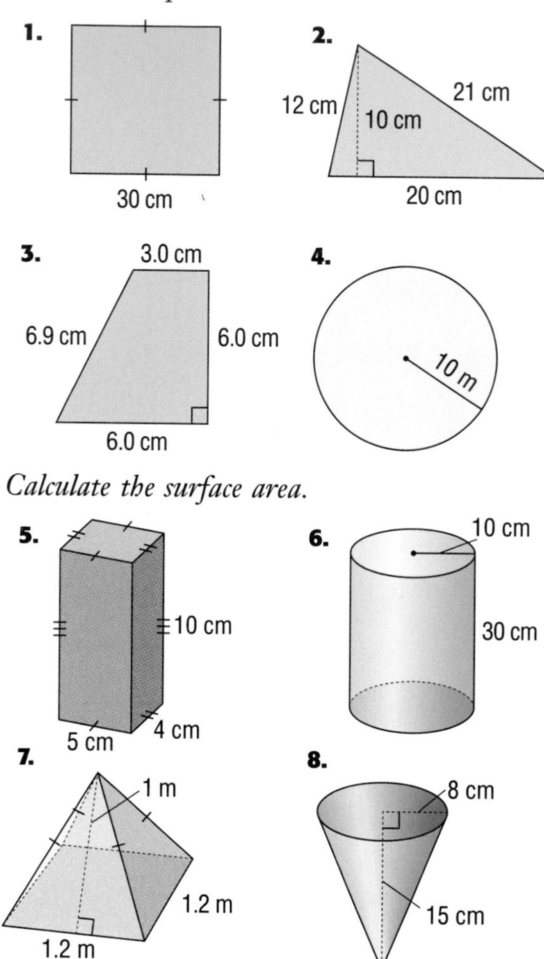

Calculate the volume to the nearest unit.

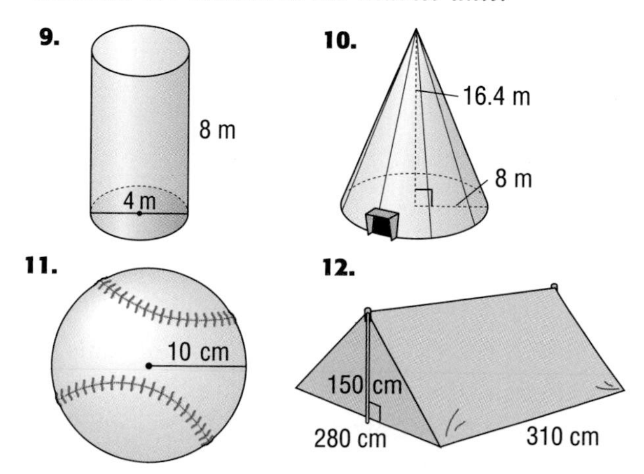

What single 3-dimensional object can you make using both figures? How many of each face do you need?

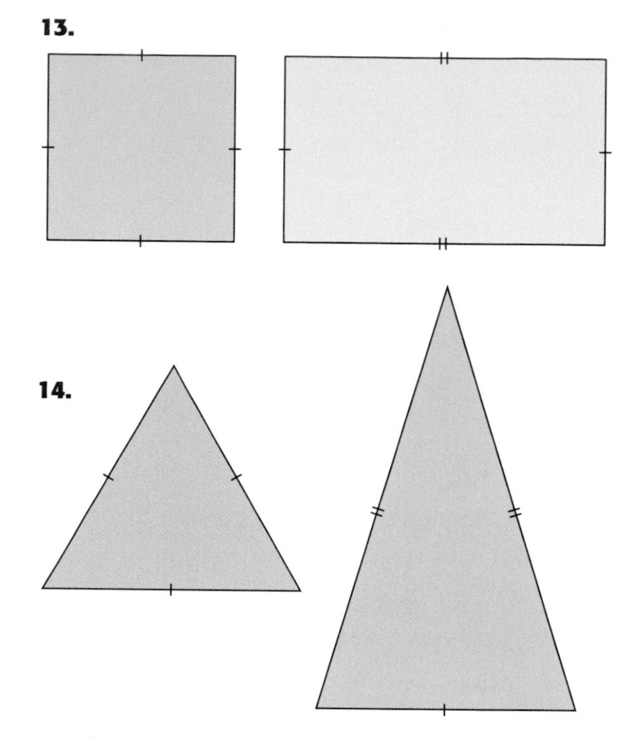

15. Circular patio stones are placed as shown to make a walkway that is 7.2 m long. How many patio stones are used?

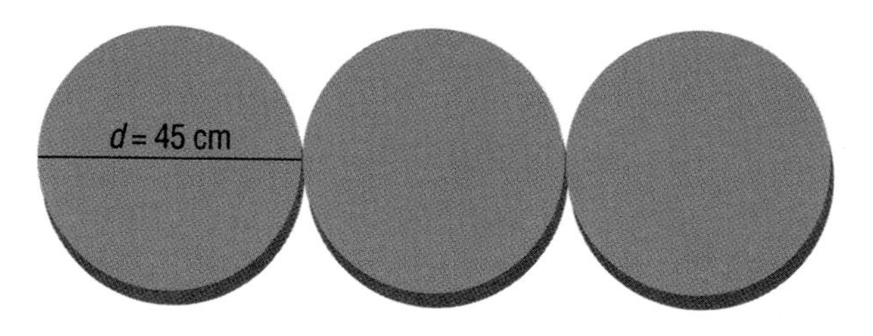

16. A juice box measures 10 cm × 6 cm × 3 cm.

a) What is its surface area?

b) How many millilitres of juice can the juice box hold?

17. A block is a rectangular prism of volume 600 cm^3. What is the height of the block if its length is 20 cm and its width is 10 cm?

Using the Strategies

1. How many squares are in this diagram?

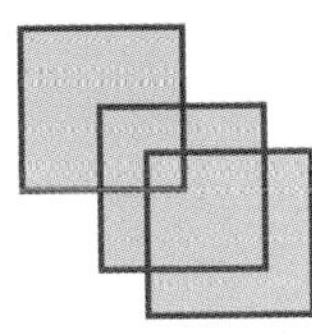

2. The numbers in the large squares are found by adding the numbers in the small squares. What are the numbers in the small squares?

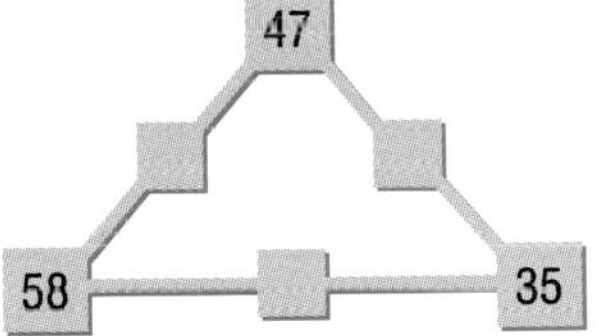

3. The Beach bus leaves every 20 min and the Sand bus leaves every 45 min. If they leave together at noon, when is the next time that they will leave together?

4. The amount of a purchase is $12.43. How can the exact amount be paid without using a $10.00 bill, but using the smallest number of bills and coins?

5. This is a 4 by 4 square.

One way to separate it into 2 congruent shapes, made up of smaller squares, is shown.

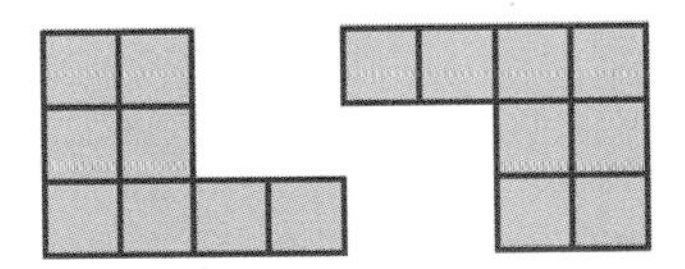

Find at least 5 other ways.

6. Six consecutive even numbers are written on a piece of paper. If the sum of the first 3 numbers is 60, what is the sum of the last 3?

7. Which of these pizzas is the best buy? Why?

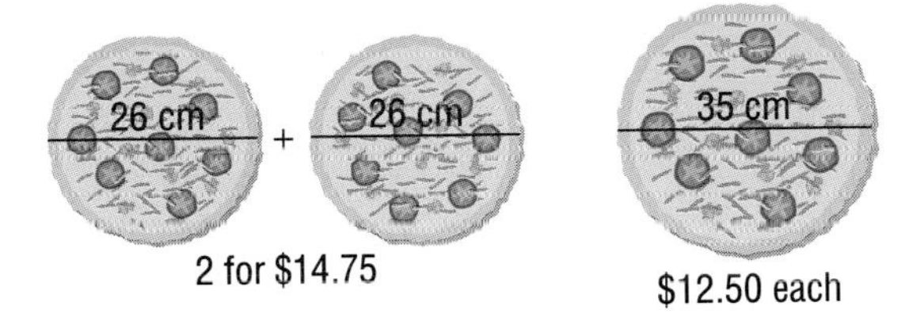

8. Six water glasses are in a row. The first 3 are full, and the last 3 are empty. How can 1 glass be moved so that full glasses and empty glasses alternate?

9. What is the minimum number of mass comparisons an inspector needs to make to find 1 counterfeit coin in a collection of 40 coins. Assume that the counterfeit coin is lighter than the others.

10. You fill a glass half full of water from the tap. Then, you add enough ice cubes to the water to fill the glass. Sketch a graph of temperature versus time from the moment you add the ice to the water until the moment when the ice has all melted.

1. A bathtub can hold 142 L of water up to the overflow. How many bathtubs can the world's largest reservoir fill when it is full?

2. Which 2 provinces are closest to each other in surface area? Which province is larger and by how much?

CHAPTER 12

Geometry

The construction of the Alexander Graham Bell Museum in Baddeck, Nova Scotia, uses a common geometric shape. Name the shape.

The photographs below show an experiment in which a ball is supported by a strip of Bristol board. Explain how the result of this experiment is related to the construction of the museum.

List 5 other examples of how the shape you identified above is used for strength.

Alexander Graham Bell Museum

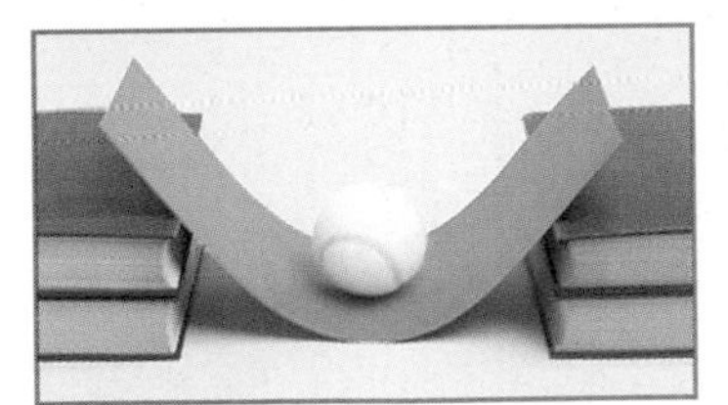

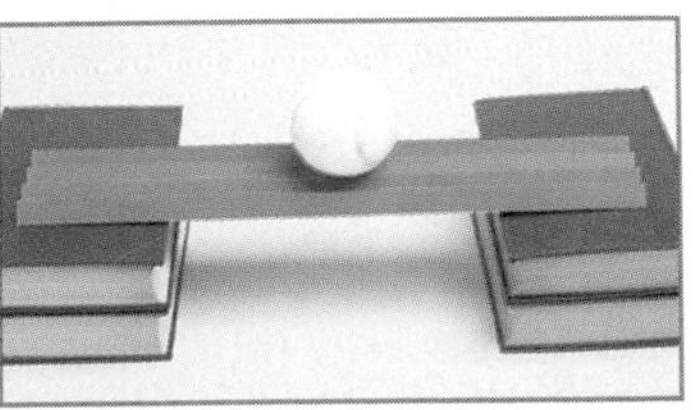

Seeing Shapes

Some of the information that we receive comes from pictures that can create illusions.

The spoon looks broken because light rays bend as they pass from one material into another. For example, they bend when they pass from air into water.

Activity ❶ Length

1. Horizontal and vertical lines can create an illusion involving length. In the diagram, does the horizontal or the vertical line seem longer?

2. Measure the lines and compare their lengths.

3. Draw a similar diagram in which the horizontal and vertical lines appear to be equal in length. Then, measure them to find the difference in their lengths.

Activity ❷ Perspective

1. Do the sides of the triangle appear to be straight or bent?

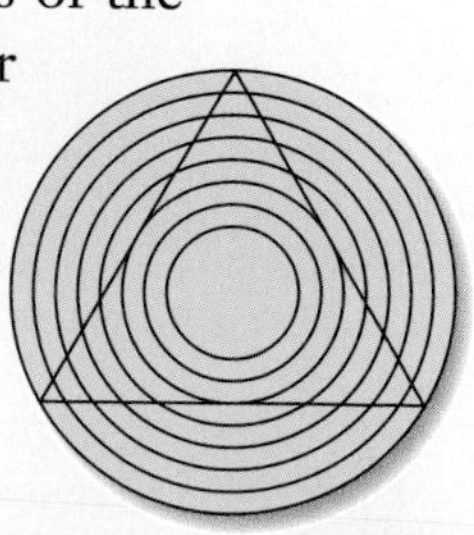

2. Describe the effect the circles have on the sides of the triangle.

Activity ❸ Reversing

1. Focus on this box until you see it change position.

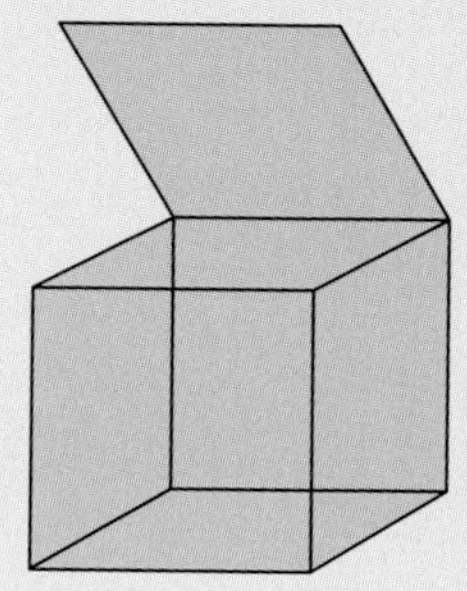

2. Describe the second position as you perceive it. Compare your description with your classmates'.

Activity ❹ Impossible Figures

1. Focus on the figure and describe what you see.

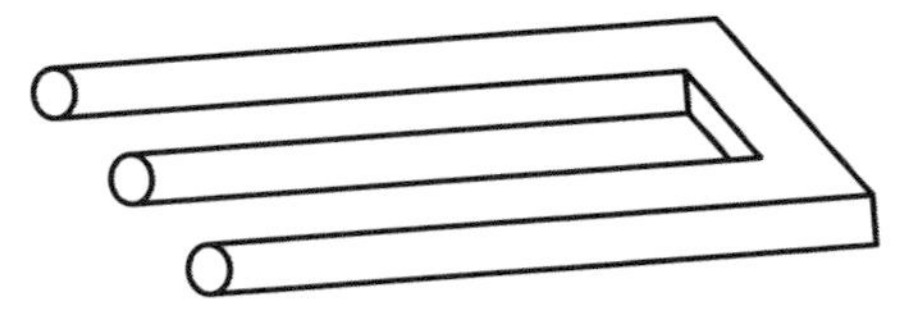

2. Is this figure possible in the real world? Why?

Activity 5

1. Which line segment is longer?

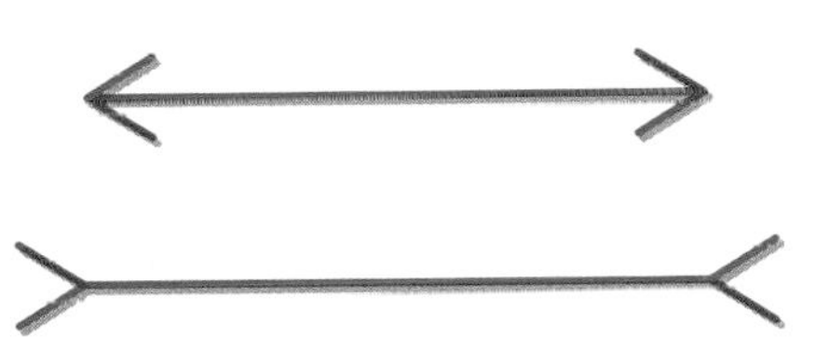

2. Are the horizontal lines straight or bent?

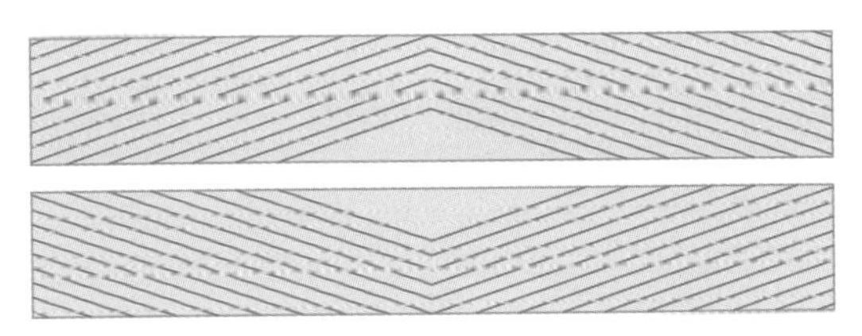

3. Describe the different ways you see each figure.

a) **b)**

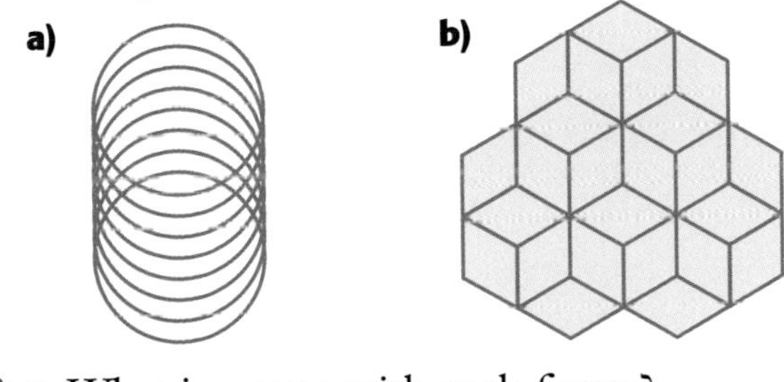

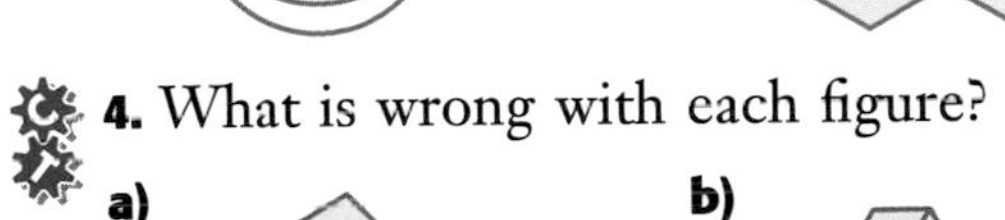

4. What is wrong with each figure?

a) **b)**

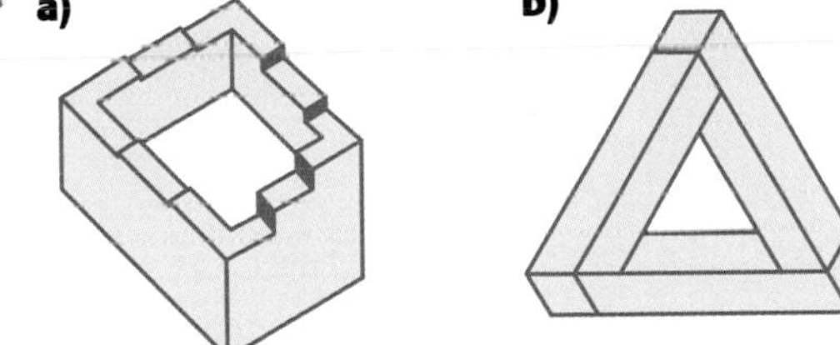

5. Are these 3 girls all the same height?

Mental Math

Calculate the perimeter of each figure.

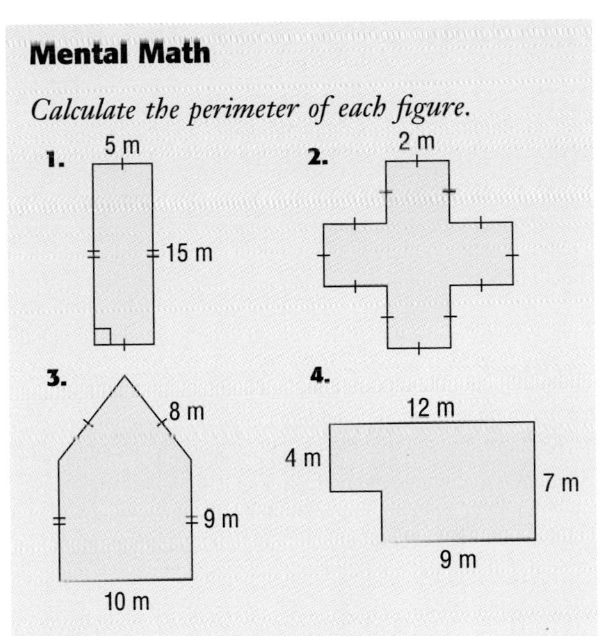

Calculate the area of each figure.

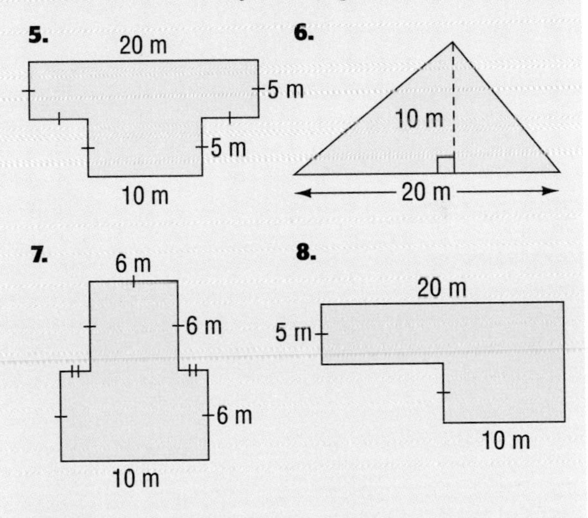

Calculate.

9. $37 + 23 + 20 + 10$

10. $52 - 12 + 30 - 20$

11. $84 + 16 - 25 + 5$

12. $61 - 31 + 47 - 7$

13. $92 - 12 - 35 - 15$

14. $73 + 17 - 30 - 20$

15. $56 + 44 - 80 + 20$

12.1 Points, Lines, Rays, and Angles

Activity: Study the Photographs

Inquire

1. Name all the different geometric shapes you see in the first photograph of the Bluenose II.

2. List as many examples of line segments and angles as you can in the remainder of the photographs.

3. List 5 different careers in which you would use lines, angles, or other geometric shapes.

The word *geometry* comes from the 2 Greek words, *geo*, meaning earth and *metron*, meaning measure. Geometry uses special words to describe the world around you.

The 2 basic concepts in geometry are the point and the line.

Most television screens are covered by more than 300 000 red, green, and blue dots. They are so close together that your eye does not see the dots but rather the whole picture.

The dots on a television screen suggest the simplest figure in geometry, the **point**. A point itself has no size, but it is represented by a dot that does have size.

Points A and B are shown here. They are usually named by capital letters. • A • B

The idea of tightly packed points, such as those in colour television images, gives us the concept of a **line**. The following chart suggests that a line is made up of an unlimited number of closely packed points.

• •	two points
• • • •	four points
• • • • • • • •	eight points
•••••••••••••••	more points
••••••••••••••••••••	more points
─────────	many more points
←──────→	line

Lines can be named in different ways. If specific points are not indicated, then we label the line with a lower case letter, often l. If specific points are indicated, such as A and B, then we write line AB (symbol $\overleftrightarrow{AB}$) or line BA (symbol $\overleftrightarrow{BA}$). The arrowheads mean that the line extends in both directions without ending. A drawing of a line has some thickness. However, the line itself has no thickness.

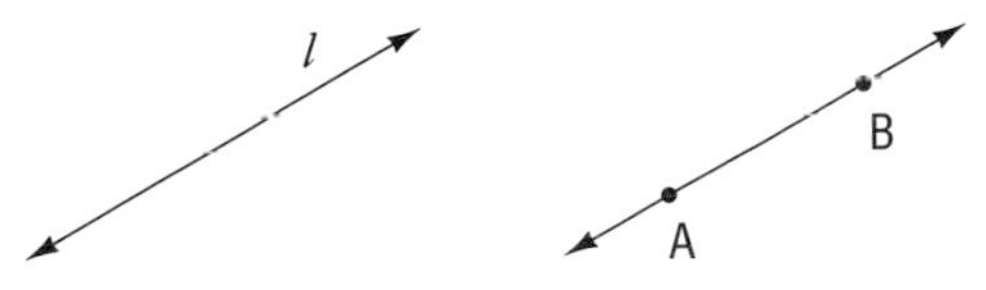

CONTINUED

A **ray** is a part of a line. A ray contains only 1 endpoint. A **line segment** is also a part of a line. A line segment is clearly defined by 2 endpoints. Line segment AB and ray AB are shown below.

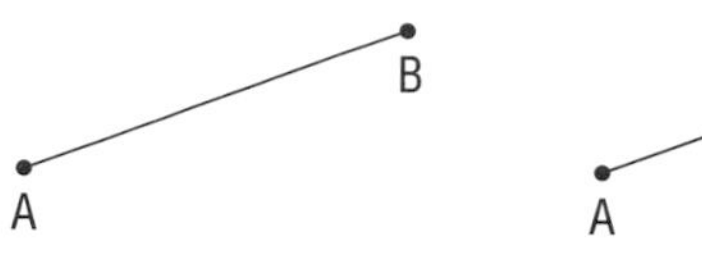

The following chart summarizes the diagrams, words, and symbols for the basic facts needed to study geometry.

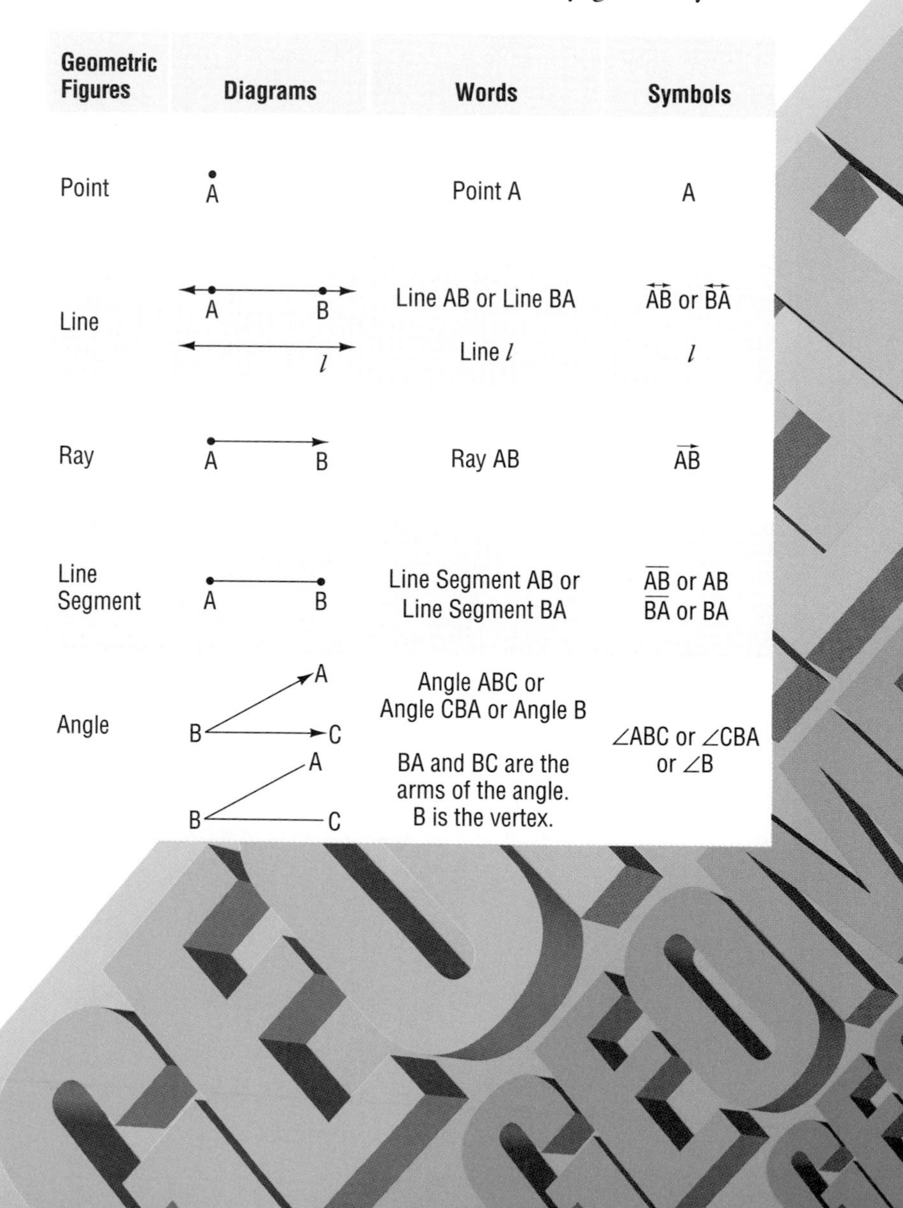

Geometric Figures	Diagrams	Words	Symbols
Point	A	Point A	A
Line	A B	Line AB or Line BA	$\overleftrightarrow{AB}$ or $\overleftrightarrow{BA}$
	l	Line l	l
Ray	A B	Ray AB	$\overrightarrow{AB}$
Line Segment	A B	Line Segment AB or Line Segment BA	$\overline{AB}$ or AB $\overline{BA}$ or BA
Angle	A B C	Angle ABC or Angle CBA or Angle B	∠ABC or ∠CBA or ∠B
	A B C	BA and BC are the arms of the angle. B is the vertex.	

Practice

State what each symbol means in words.

1. $\overleftrightarrow{CD}$ **2.** $\overrightarrow{ST}$ **3.** $\angle DEF$

4. $\overline{RT}$ **5.** $\angle M$ **6.** $\angle CAR$

7. List 5 points and 5 lines in the diagram.

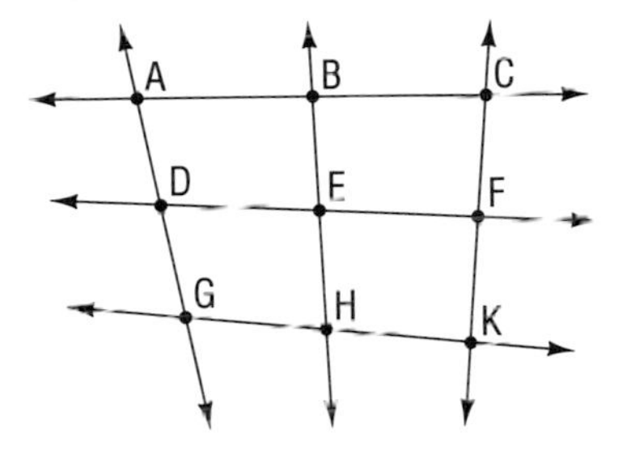

8. Name 5 line segments in this diagram.

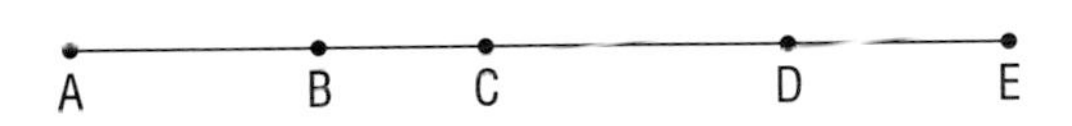

9. Name 8 angles in this diagram.

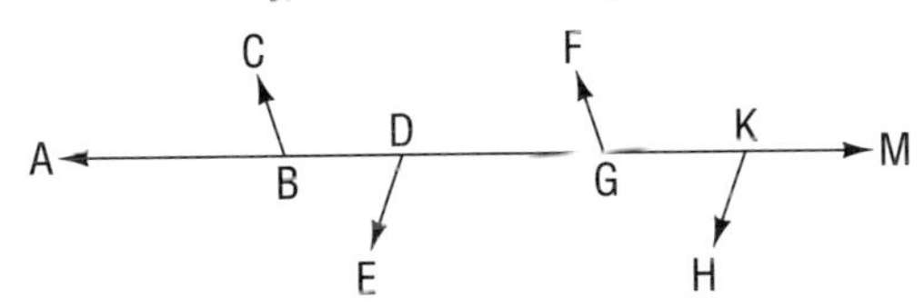

Name the following from the figure shown below.

10. three points **11.** three lines

12. six angles **13.** three rays

14. three line segments

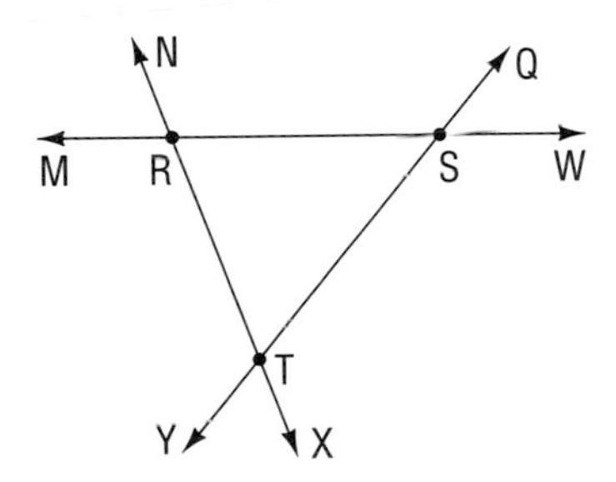

Draw a labelled diagram for each of the following.

15. angle RST **16.** line PQ

17. ray BT **18.** line segment DF

Problems and Applications

19. In how many different ways can 2 lines intersect? Illustrate your answer.

20. In how many different ways can 3 lines intersect? Draw diagrams to illustrate your answer.

21. How many line segments are in this diagram? Name them.

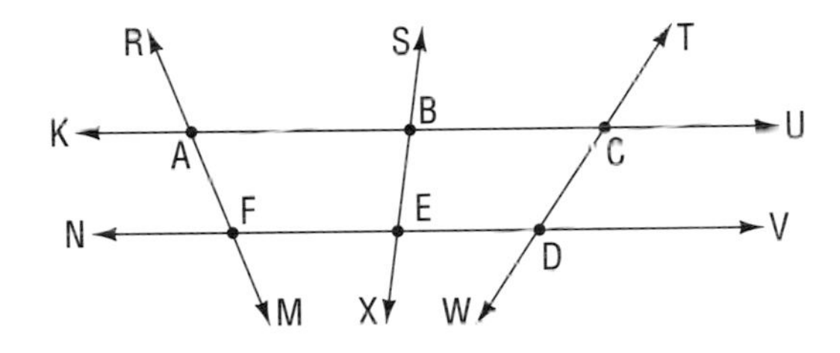

22. a) How many different line segments make up a cube?

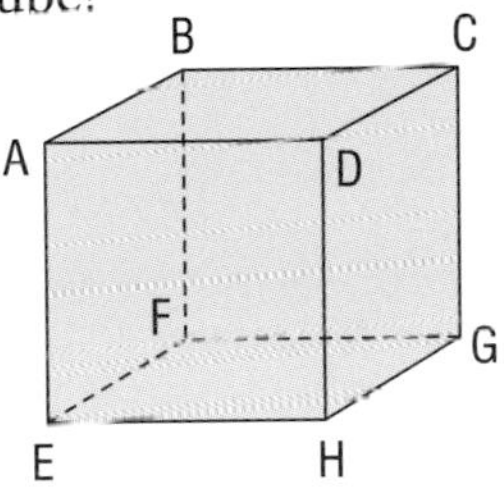

b) How many angles does the cube contain? Name them.

23. Write a problem that involves naming line segments and angles. Have a classmate solve your problem.

LOGIC POWER

Nim is a game for 2 players. Arrange 9 toothpicks as shown.

Take turns removing 1, 2, or 3 toothpicks. The player who must take the last toothpick loses.

12.2 Angles and Intersecting Lines

One of the most fascinating uses for angles and intersecting lines can be seen in suspension bridges. The photograph shows the Lion's Gate Bridge in British Columbia.

Angles are classified according to their size in degrees.

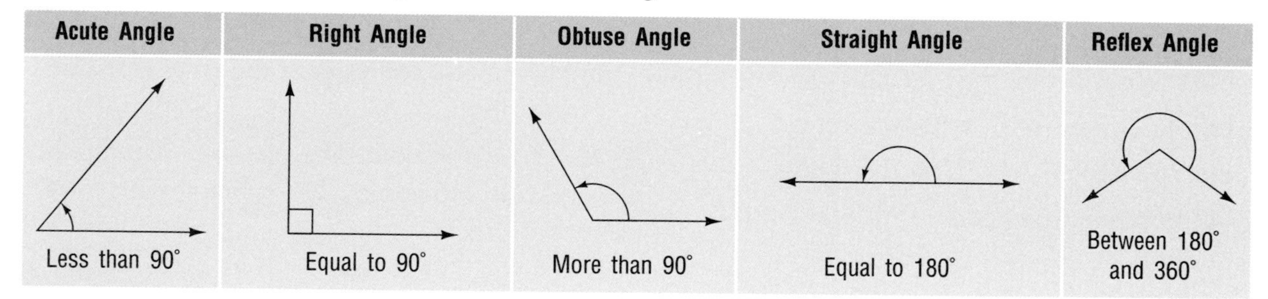

Acute Angle
Less than 90°
Right Angle
Equal to 90°
Obtuse Angle
More than 90°
Straight Angle
Equal to 180°
Reflex Angle
Between 180° and 360°

Some pairs of angles have special names.

Supplementary angles add to 180°.

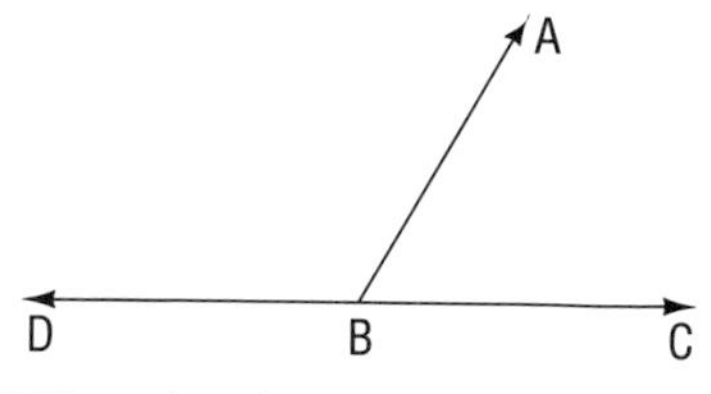

∠ABC and ∠ABD are supplementary because $\angle ABC + \angle ABD = 180^\circ$.

Complementary angles add to 90°.

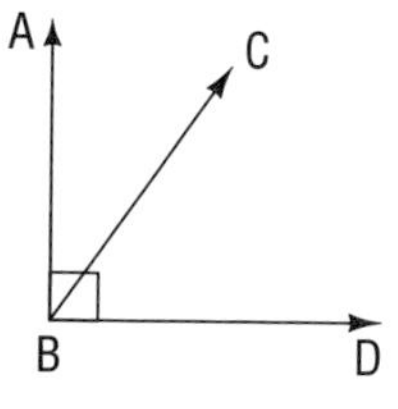

∠ABC and ∠CBD are complementary because $\angle ABC + \angle CBD = 90^\circ$.

Activity: Discover the Relationship

When 2 lines intersect, 2 pairs of **opposite angles** are formed. Draw 2 intersecting lines and label them as shown. Measure the angles marked a, b, c, and d.

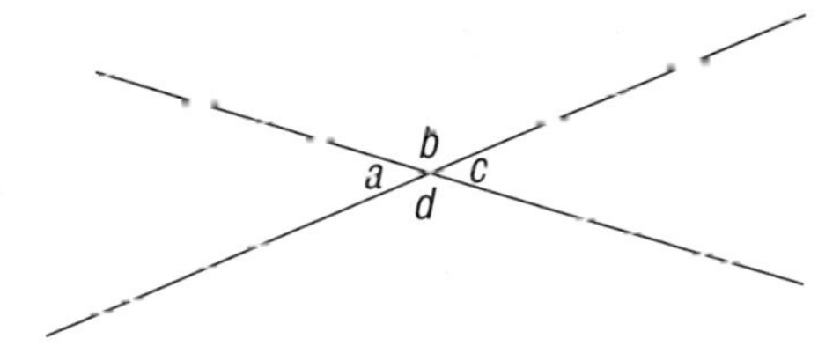

Inquire

1. How are a and c related?

2. How are b and d related?

3. Repeat the activity for 2 other intersecting lines.

4. Write a statement that describes how opposite angles are related when two lines intersect.

5. Describe where intersecting lines occur in the Lion's Gate Bridge.

Example

a) Find the measures of $\angle$CAD and $\angle$CAE.
b) Name 1 of each of the following angles: acute, obtuse, right, straight.
c) Name 1 pair of complementary and 1 pair of supplementary angles.

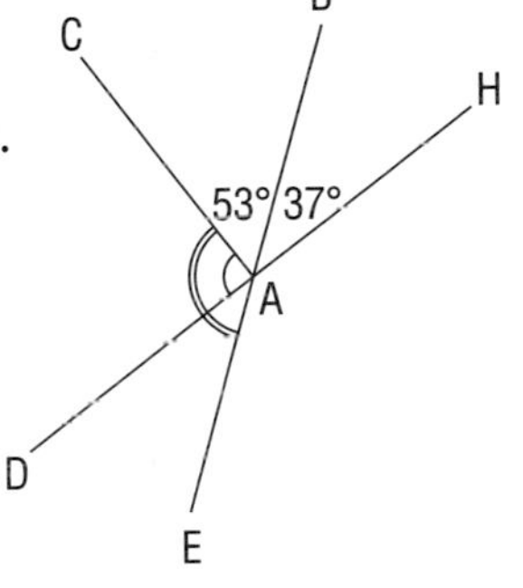

Solution

a) $\angle$DAH is a straight angle.

$$\angle CAD + \angle BAC + \angle BAH = 180° \text{ (straight angle)}$$
$$\angle CAD + 53° + 37° = 180°$$
$$\angle CAD + 90° = 180°$$
$$\angle CAD = 90°$$

The measure of $\angle$CAD is 90°.

$\angle$DAE and $\angle$BAH are opposite angles.

$$\angle DAE = \angle BAH$$
$$\angle DAE = 37°$$
$$\angle CAE = \angle CAD + \angle DAE$$
$$= 90° + 37°$$
$$= 127°$$

The measure of $\angle$CAE is 127°.

b) $\angle DAE = 37°$. Since $37° < 90°$, $\angle$DAE is acute.
$\angle CAE = 127°$. Since $127° > 90°$, $\angle$CAE is obtuse.
$\angle$CAD is a right angle. It measures 90°.
$\angle$DAH is a straight angle. It measures 180°.

c) $\angle$BAH and $\angle$BAC are a pair of complementary angles. Their sum is 90°.
$\angle$CAE and $\angle$BAC are a pair of supplementary angles. Their sum is 180°.

CONTINUED

Practice

Determine the size of each of the angles drawn on the protractor.

1. ∠BAC **2.** ∠BAD **3.** ∠BAE

4. ∠BAF **5.** ∠HAJ **6.** ∠GAJ

7. ∠IAJ **8.** ∠FAJ **9.** ∠DAJ

10. ∠BAH **11.** ∠BAI **12.** ∠BAJ

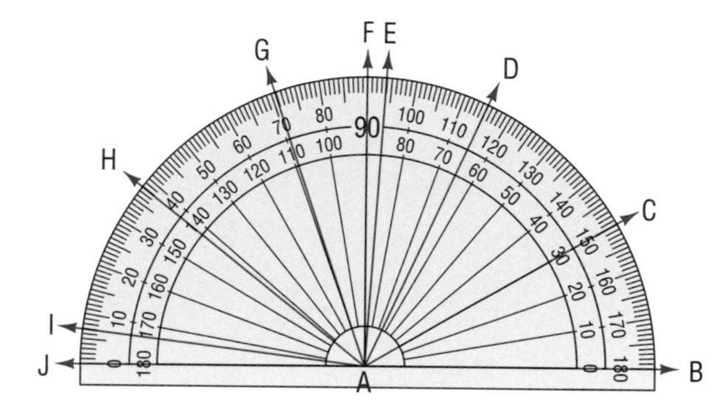

13. Classify the angles in 1–12 as acute, obtuse, right, or straight.

Use a protractor to construct these angles.

14. ∠ABC = 35° **15.** ∠RST = 75°

16. ∠DEF = 21° **17.** ∠LMN = 83°

18. ∠XYZ = 90° **19.** ∠GHI = 145°

20. ∠PQR = 156° **21.** ∠CDE = 180°

Estimate the size of each angle. Then, classify each angle.

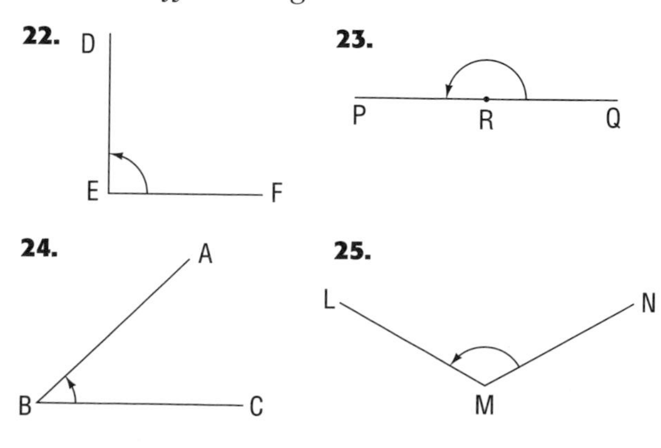

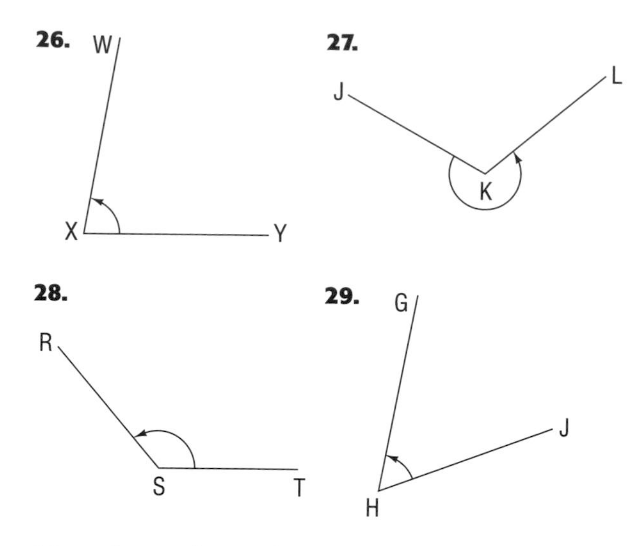

Use only a ruler to draw each angle. Check your angles with a protractor.

30. 45° **31.** 10° **32.** 70° **33.** 80°

34. 60° **35.** 30° **36.** 90° **37.** 150°

Write the complement of each angle.

38. 30° **39.** 42° **40.** 75° **41.** 9°

42. 89° **43.** 66° **44.** 15° **45.** 1°

Write the supplement of each angle.

46. 40° **47.** 70° **48.** 85° **49.** 150°

50. 125° **51.** 8° **52.** 110° **53.** 179°

Find the measure of each unknown angle.

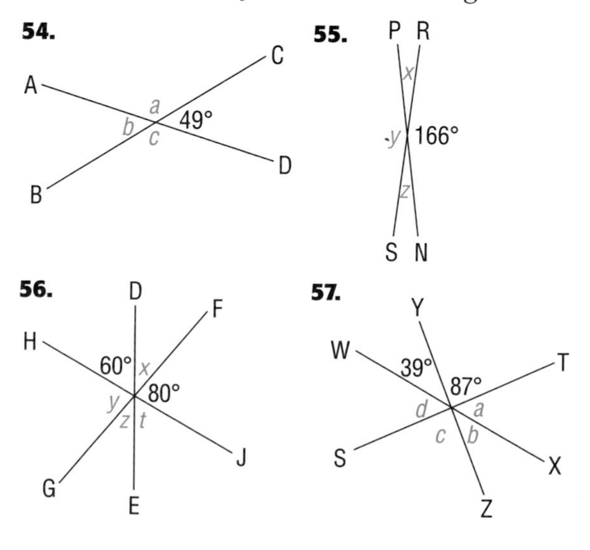

Find the measure of each unknown angle.

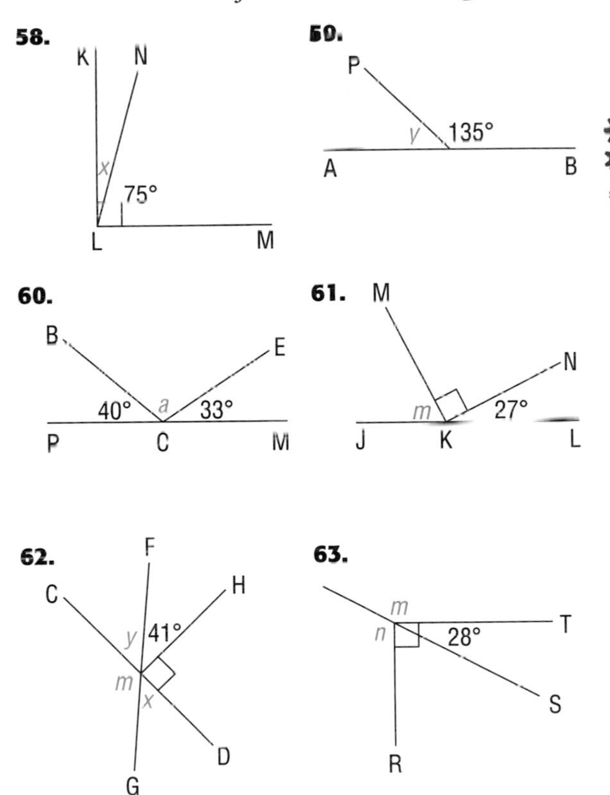

Problems and Applications

Describe how you would use a protractor to construct each of the following angles. Construct each angle.

64. $\angle UVW = 256°$ **65.** $\angle EFG = 270°$

66. $\angle JKL = 300°$ **67.** $\angle TUV = 330°$

68. Two lines intersect so that there are no 90° angles. What can be said about the angles?

69. If one angle measures 108°, describe how you would calculate its supplement.

70. $\angle PQR$ measures 89°. What is the measure of reflex angle PQR?

71. Through how many degrees does the minute-hand of a clock turn in these lengths of time?

a) 15 min **b)** 30 min **c)** 45 min

72. a) Draw and label a diagram where AB and CD intersect at point E and $\angle AED = 38°$. Name the other angles and calculate their measures.

b) Explain why this conclusion is true.

$$\angle AED + \angle BED = 180°$$
$$\angle BEC + \angle BED = 180°$$
$$\text{Thus, } \angle AED = \angle BEC$$

73. The Ferris wheel has 12 cars. The wheel turns counterclockwise. Through how many degrees does the wheel turn to achieve the following car positions?

a) the blue car moves to the bottom

b) the green car moves to the bottom

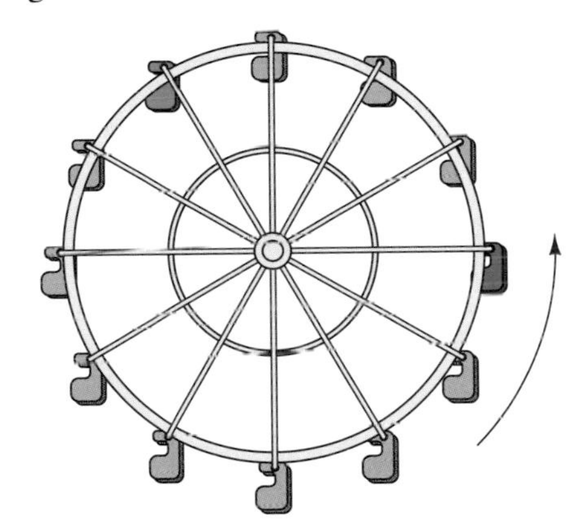

74. Write a problem that involves angles and intersecting lines. Have a classmate solve your problem.

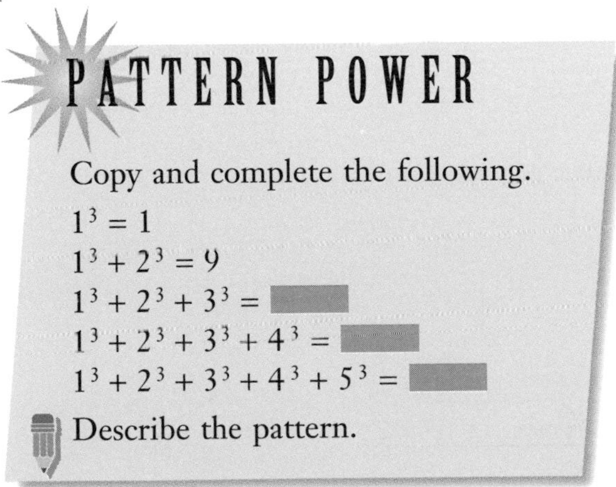

PATTERN POWER

Copy and complete the following.

$1^3 = 1$

$1^3 + 2^3 = 9$

$1^3 + 2^3 + 3^3 =$ ▬

$1^3 + 2^3 + 3^3 + 4^3 =$ ▬

$1^3 + 2^3 + 3^3 + 4^3 + 5^3 =$ ▬

Describe the pattern.

12.3 Angles and Triangles

Triangles are seen in many types of construction. How many different types of triangles do you see in this photograph of the Canadian Mint in Winnipeg?

Activity: Conduct an Experiment

Use an index card or piece of cardboard to cut out 4 thin strips 3 cm, 4 cm, 5 cm, and 8 cm long. Use the strips to try to make triangles with the following sides.

a) 3 cm, 4 cm, 5 cm **b)** 3 cm, 4 cm, 8 cm

c) 4 cm, 5 cm, 8 cm **d)** 3 cm, 5 cm, 8 cm

Inquire

1. Which set or sets of 3 strips can form a triangle?

2. Which set or sets of 3 strips cannot form a triangle? Why not?

3. Predict which of the following sets of 3 strips will form a triangle.

a) 5 cm, 7 cm, 9 cm **b)** 4 cm, 6 cm, 11 cm

c) 5 cm, 6 cm, 10 cm **d)** 3 cm, 4 cm, 7 cm

4. Copy and complete this statement: In a triangle, the sum of the lengths of any 2 sides must be ______ than the length of the third side.

Triangles can be classified by the lengths of their sides.

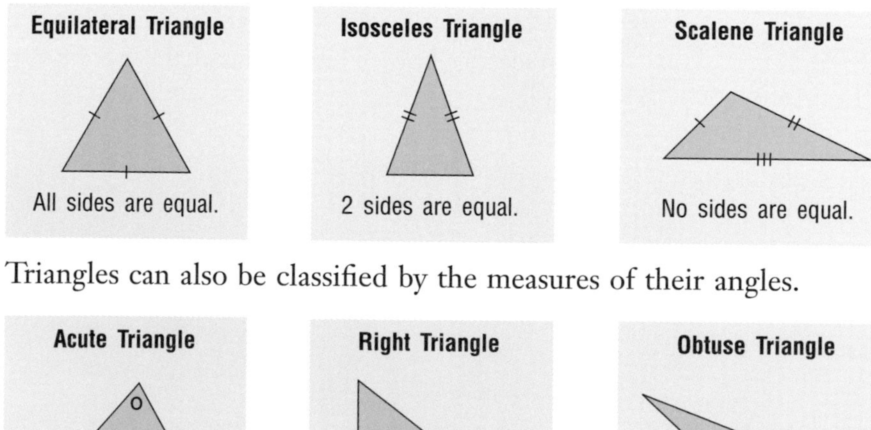

Triangles can also be classified by the measures of their angles.

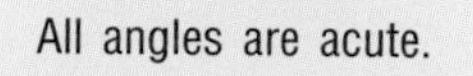

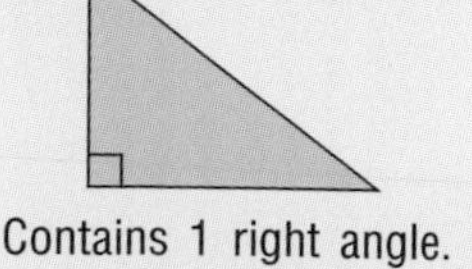

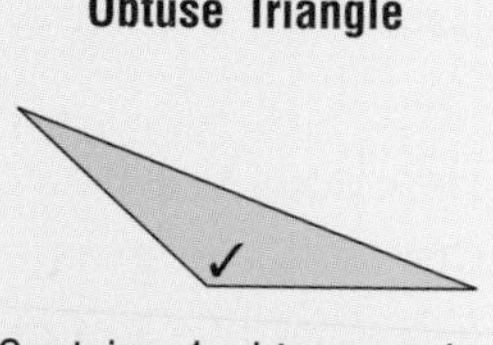

Activity: Discover the Relationship

Trace these triangles into your notebook. Then, measure and record the lengths of the 3 sides and the sizes of the 3 angles in each triangle. (If necessary, extend the sides of the triangles in order to accurately measure the size of each angle.)

a)

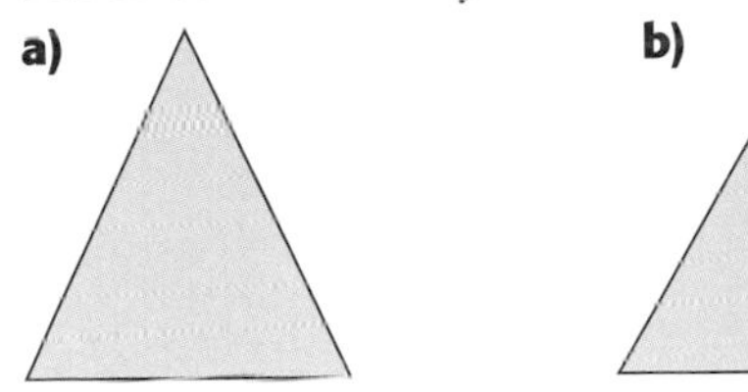

b)

c)

Inquire

1. Use the lengths of the sides to classify each triangle as scalene, isosceles, or equilateral.

2. What is the sum of the angles in each of these triangles? Is this sum true for all triangles? Compare your findings with your classmates'.

3. Describe another relationship between the angles of the following types of triangles.

a) scalene **b)** isosceles **c)** equilateral

4. Is it possible for a triangle to have two 90° angles? Explain.

5. Is it possible for a triangle to have more than 1 obtuse angle? Explain.

Example

Name each type of triangle and find the missing values.

a)

b)

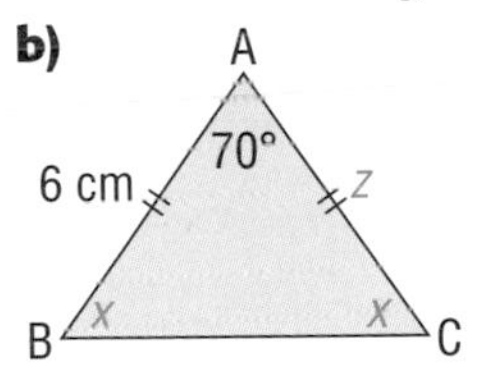

c)

Solution

a) ΔPQR has 3 equal sides and 3 equal angles. It is an equilateral triangle.

$$3x = 180°$$
$$x = 60°$$

b) ΔABC has 2 equal sides and 2 equal angles. It is an isosceles triangle.

$$2x = 180° - 70°$$
$$2x = 110°$$
$$x = 55°$$
$$AB = AC$$
$$z = 6 \text{ cm}$$

c) ΔDEF has no equal parts. It is a scalene triangle.

$$x + 73° + 56° = 180°$$
$$x + 129° = 180°$$
$$x = 51°$$

CONTINUED

Practice

Which of the following lengths will form a triangle?

1. 5 m, 6 m, 7 m **2.** 4 m, 5 m, 9 m

3. 2 m, 3 m, 4 m **4.** 2 m, 3 m, 6 m

Classify these triangles by angle measures.

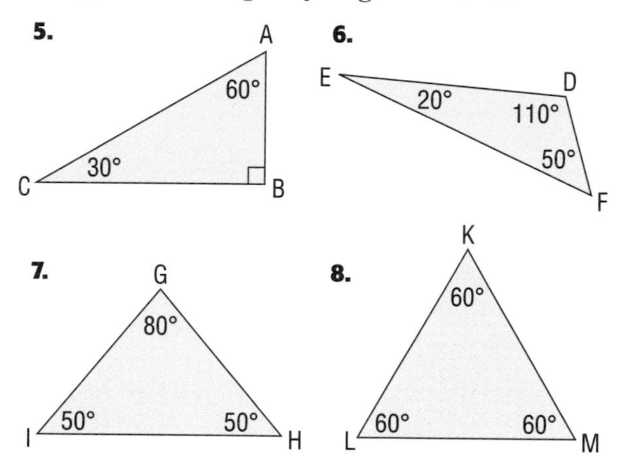

Classify these triangles by side lengths.

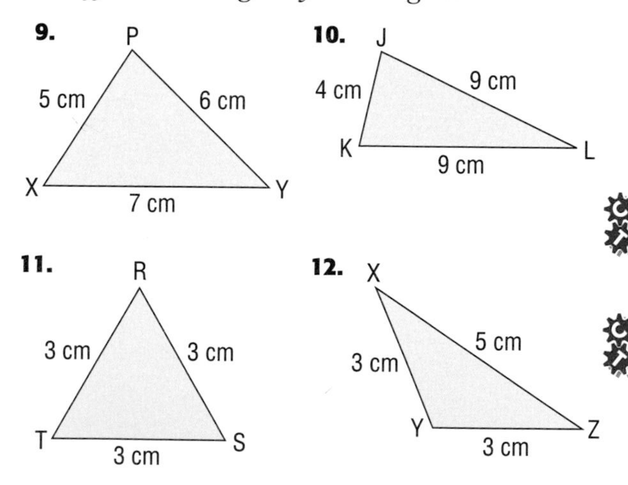

Find the missing measures.

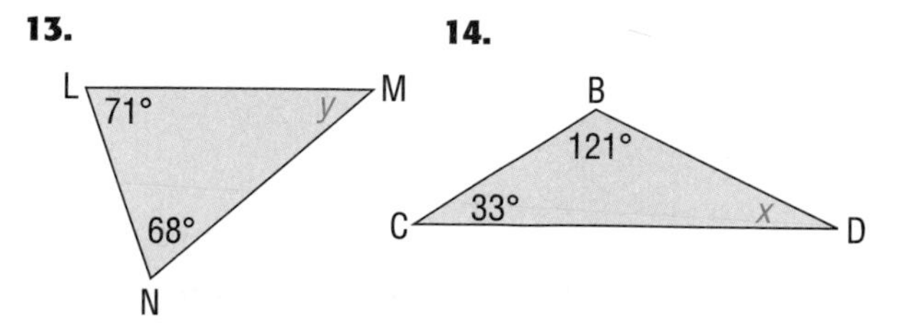

Find the missing values.

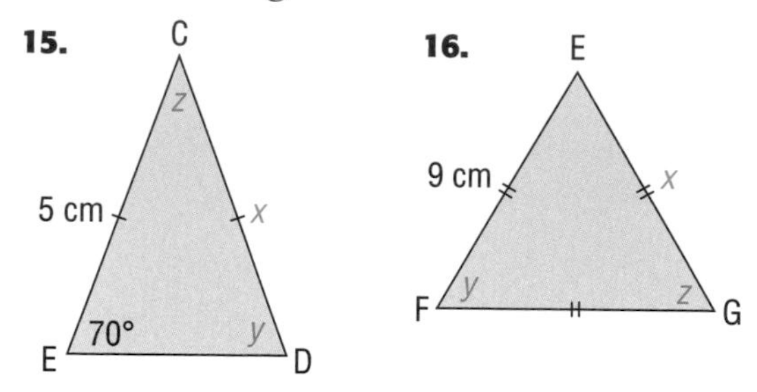

Problems and Applications

17. An **exterior angle** of a triangle is formed by extending 1 of its sides. There are 2 interior and opposite angles for every exterior angle. In ΔABC, $\angle ACD$ is an exterior angle. Draw ΔABC. Cut out the figure. Then, tear off $\angle A$ and $\angle B$ and arrange them as shown.

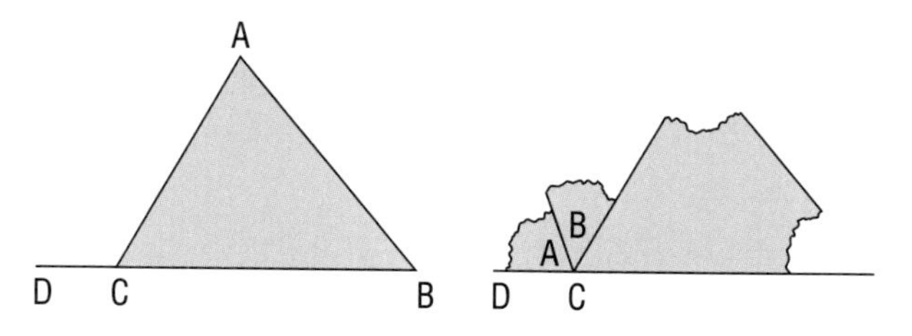

Write a statement about the relationship between an exterior angle and its 2 interior and opposite angles.

18. One angle of an isosceles triangle has a measure of 40°. What are the possible measures of the other angles?

19. One angle of an isosceles triangle has a measure of 140°. What are the possible measures of the other angles?

Find the measure of each unknown angle.

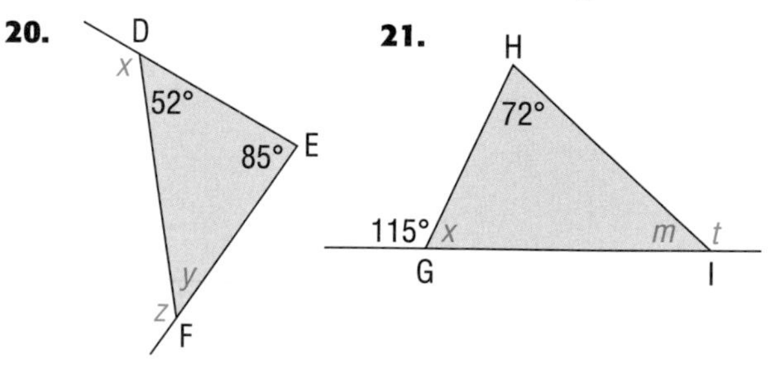

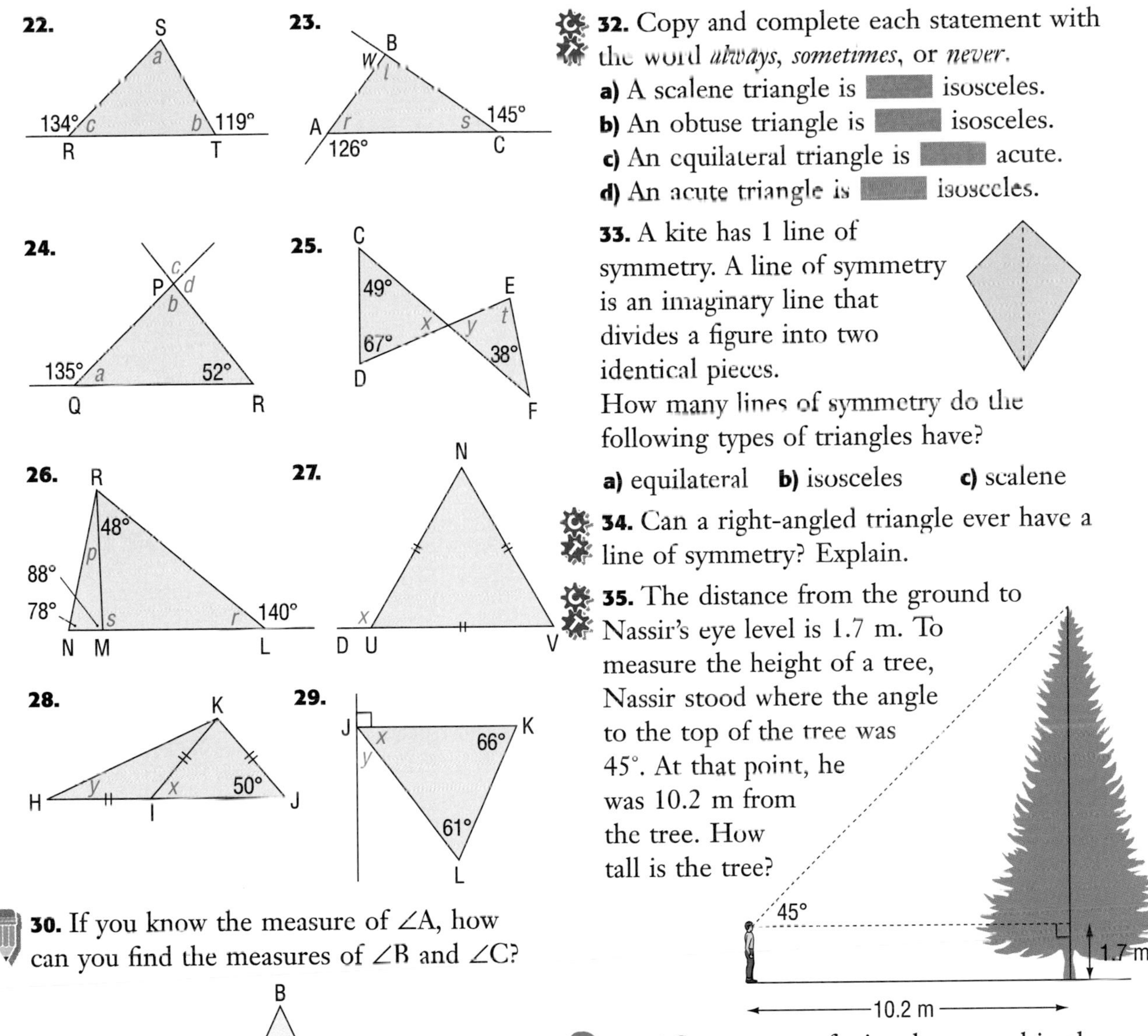

30. If you know the measure of ∠A, how can you find the measures of ∠B and ∠C?

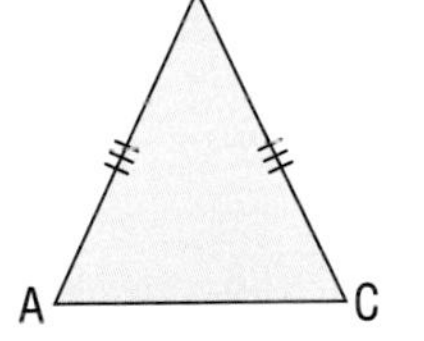

31. If you know the measure of ∠D, how can you find the measure of ∠A?

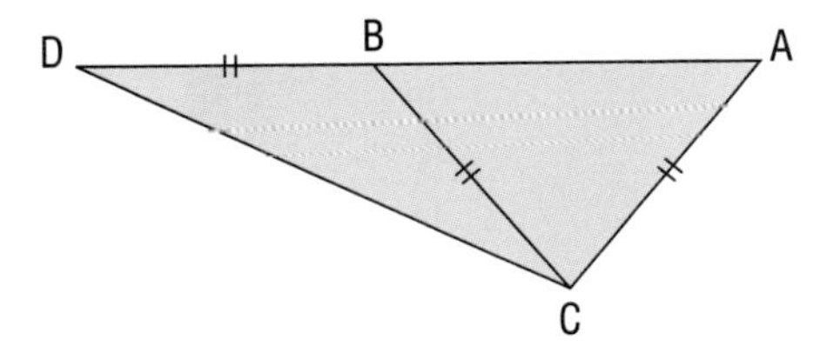

32. Copy and complete each statement with the word *always*, *sometimes*, or *never*.

a) A scalene triangle is ▆▆▆ isosceles.

b) An obtuse triangle is ▆▆▆ isosceles.

c) An equilateral triangle is ▆▆▆ acute.

d) An acute triangle is ▆▆▆ isosceles.

33. A kite has 1 line of symmetry. A line of symmetry is an imaginary line that divides a figure into two identical pieces.
How many lines of symmetry do the following types of triangles have?

a) equilateral **b)** isosceles **c)** scalene

34. Can a right-angled triangle ever have a line of symmetry? Explain.

35. The distance from the ground to Nassir's eye level is 1.7 m. To measure the height of a tree, Nassir stood where the angle to the top of the tree was 45°. At that point, he was 10.2 m from the tree. How tall is the tree?

36. a) Some types of triangles named in the table exist, while others do not. Work with a classmate to draw the types of triangles that exist.

	Scalene	Isosceles	Equilateral
Acute	A	B	C
Right	D	E	F
Obtuse	G	H	I

b) If a triangle does not exist, explain why not. You may wish to use a diagram.

12.4 Angles and Parallel Lines

A **transversal** is a line that crosses or intersects 2 or more lines, each at a different point. Some pairs of angles formed by a transversal have special names.

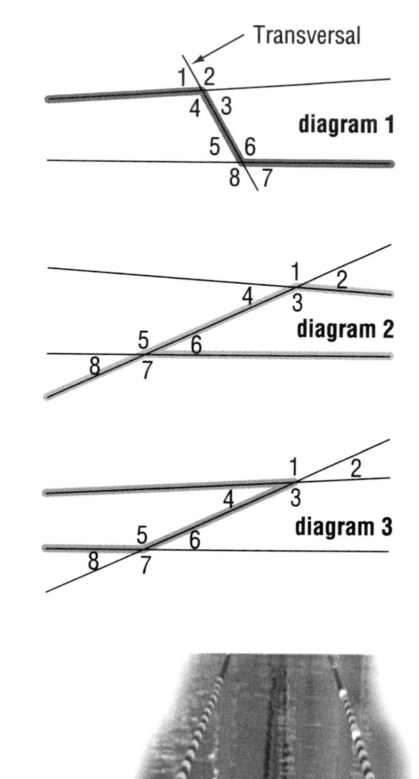

In diagram 1, $\angle 4$ and $\angle 6$ are **alternate interior angles**, or simply **alternate angles**. Another pair of alternate angles is $\angle 3$ and $\angle 5$. Alternate angles form a Z pattern or a reversed Ƨ pattern.

In diagram 2, $\angle 3$ and $\angle 7$ are a pair of **corresponding angles**. Other pairs of corresponding angles are $\angle 2$ and $\angle 6$, $\angle 4$ and $\angle 8$, and $\angle 1$ and $\angle 5$. Corresponding angles form an F pattern, or an ꟻ, Ⅎ, or ⅎ pattern.

In diagram 3, $\angle 4$ and $\angle 5$ are a pair of **co-interior angles**. Co-interior angles are on the same side of the transversal. Another pair of co-interior angles is $\angle 3$ and $\angle 6$. Co-interior angles form a ⊏ pattern or a reversed ⊐ pattern.

Canada's Mark Tewksbury set a new Olympic Record of 53.98 s in 1992 for the 100-m backstroke. Swimmers like Mark compete in lanes separated by parallel dividers. In geometry, we often use **parallel lines**. Parallel lines are lines in the same plane that do not intersect.

Activity: Discover the Relationship

Use a ruler to draw 2 parallel lines and a transversal. Label the angles as shown. Measure the 8 angles formed.

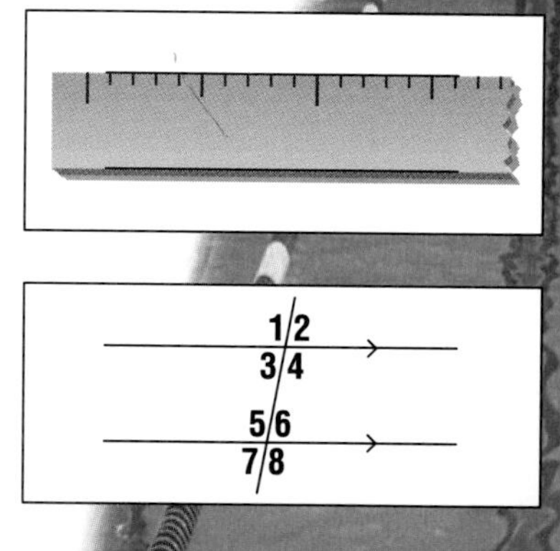

Inquire

1. How are the pairs of alternate angles related?

2. How are the pairs of corresponding angles related?

3. How are the pairs of co-interior angles related?

4. Repeat the activity for 2 more parallel lines and another transversal.

5. Write a statement to describe the relationship between each of the following pairs of angles formed when a transversal intersects 2 parallel lines.

a) alternate angles

b) corresponding angles

c) co-interior angles

Example 1

Find the measures of the unknown angles.

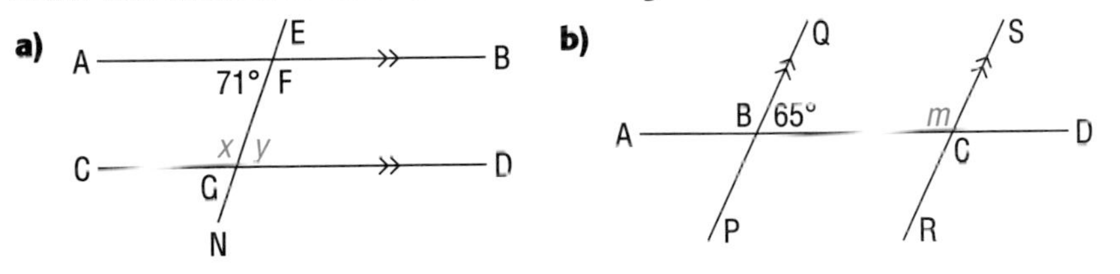

Solution

a) Since AB || CD, alternate angles are equal.

|| means "is parallel to"

$$\angle DGF = \angle AFG \text{ (alternate angles)}$$

$$y = 71^\circ$$

$$\angle DGF + \angle CGF = 180^\circ \text{ (straight angle)}$$

$$71^\circ + x = 180^\circ$$

$$x = 109^\circ$$

The measures are $x = 109^\circ$ and $y = 71^\circ$.

b) Since PQ || RS, co-interior angles are supplementary.

$$\angle BCS + \angle CBQ = 180^\circ \text{ (co-interior angles)}$$

$$m + 65^\circ = 180^\circ$$

$$m = 115^\circ$$

The measure of the unknown angle is $m = 115^\circ$.

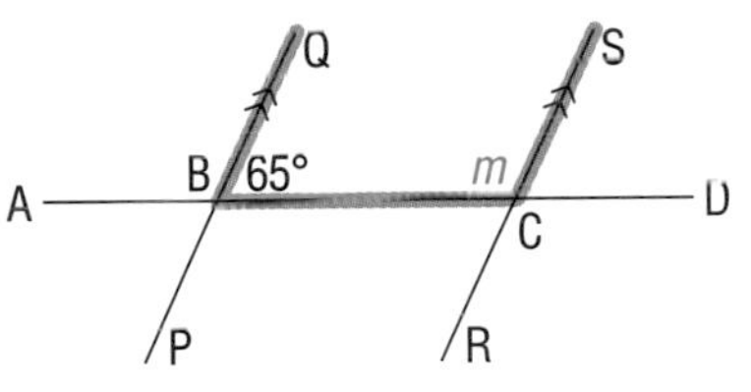

Example 2

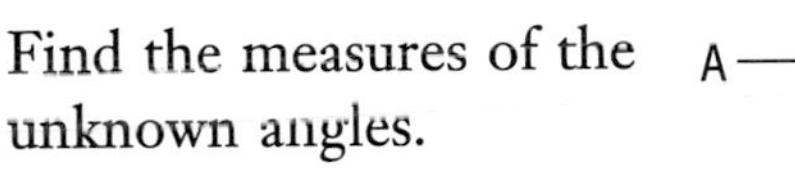

Find the measures of the unknown angles.

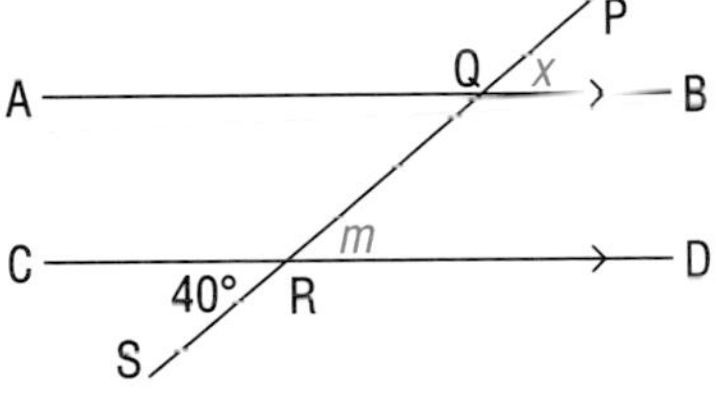

Solution

$$\angle DRQ = \angle CRS \text{ (opposite angles)}$$

$$m = 40^\circ$$

Since AB || CD, corresponding angles are equal.

$$\angle BQP = \angle DRQ \text{ (corresponding angles)}$$

$$x = 40^\circ$$

The measures are $m = 40^\circ$ and $x = 40^\circ$.

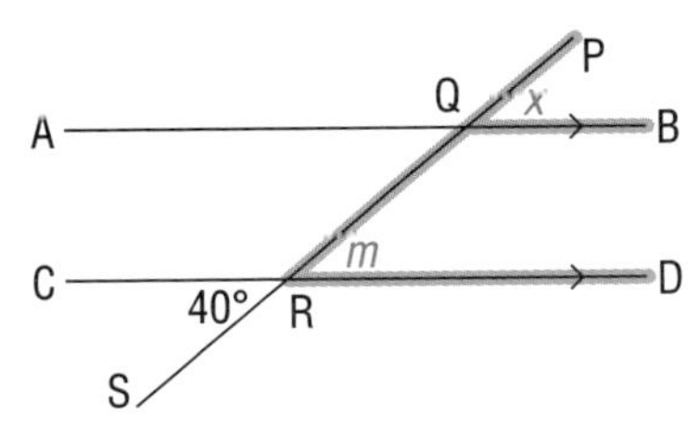

CONTINUED

Example 3

What are the measures of m and n in this diagram?

Solution

Since GK || PB, corresponding angles are equal.

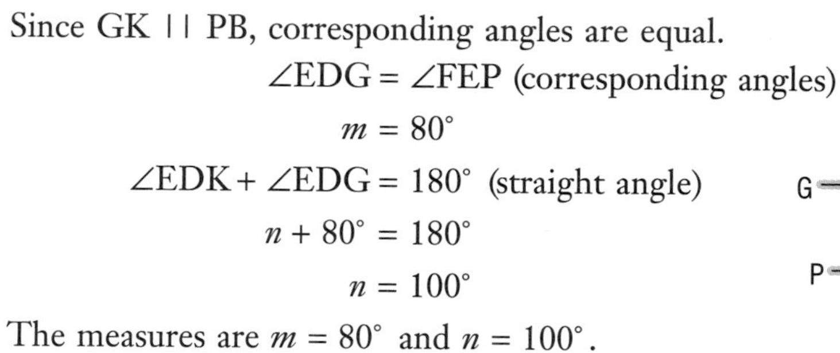

$$\angle EDG = \angle FEP \text{ (corresponding angles)}$$
$$m = 80^\circ$$
$$\angle EDK + \angle EDG = 180^\circ \text{ (straight angle)}$$
$$n + 80^\circ = 180^\circ$$
$$n = 100^\circ$$

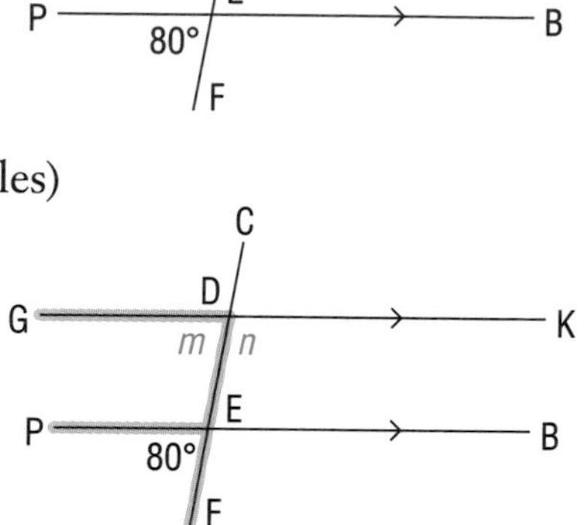

The measures are $m = 80^\circ$ and $n = 100^\circ$.

Example 4

Find the measures of the unknown angles.

a)

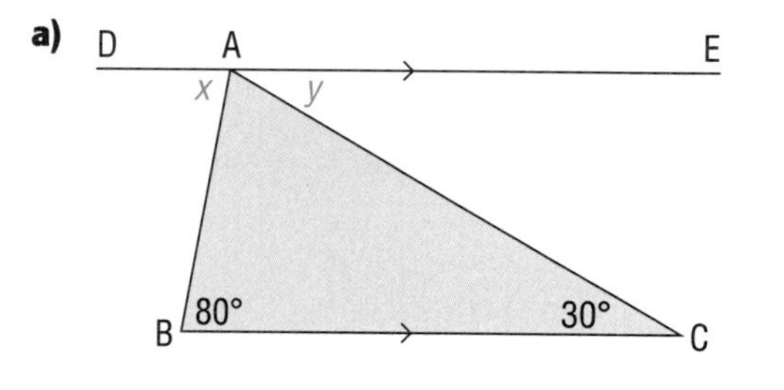

b)

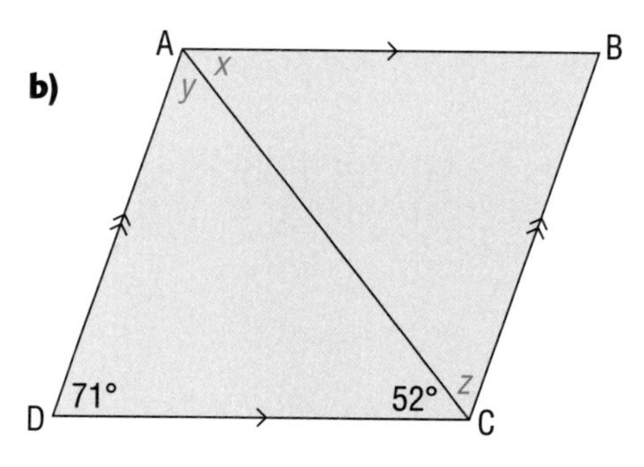

Solution

a) Since DE || BC, alternate angles are equal.

$\angle BAD = \angle ABC$ (alternate angles)

$x = 80^\circ$

$\angle CAE = \angle ACB$ (alternate angles)

$y = 30^\circ$

The measures are $x = 80^\circ$ and $y = 30^\circ$.

b) Since AB || DC, alternate angles are equal.

$\angle BAC = \angle ACD$ (alternate angles)

$x = 52^\circ$

In ΔADC, the sum of the angles is 180°.

$y + 71^\circ + 52^\circ = 180^\circ$

$y + 123^\circ = 180^\circ$

$y = 57^\circ$

Since AD || BC, alternate angles are equal.

$\angle ACB = \angle CAD$ (alternate angles)

$z = 57^\circ$

The measures are $x = 52^\circ$, $y = 57^\circ$, and $z = 57^\circ$.

Practice

Find the missing measures. Give reasons for your answers.

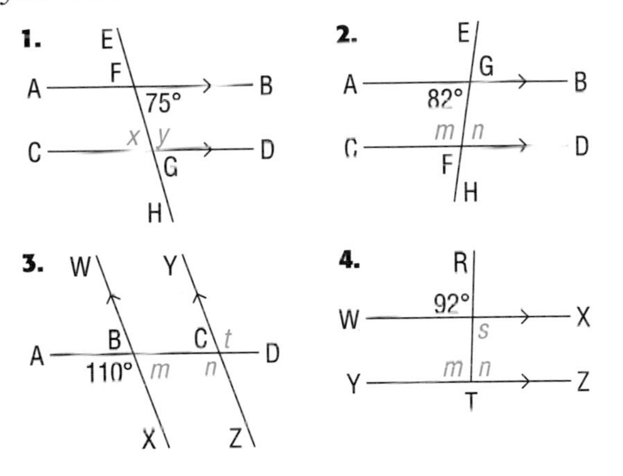

Find the measure of each unknown angle.

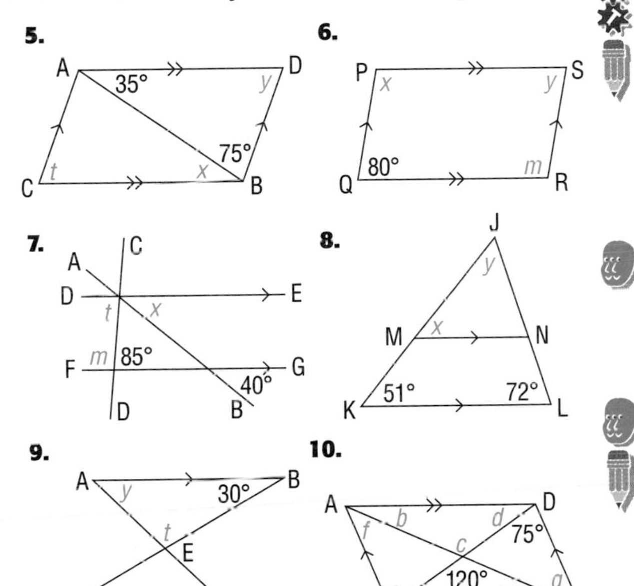

Problems and Applications

11. List all the angles in this diagram. Then, calculate each angle's measure.

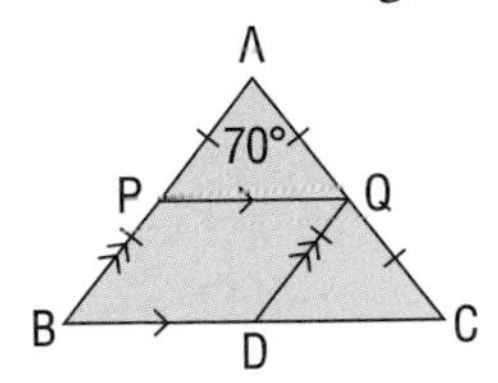

12. One way to show the sum of the interior angles of a triangle is to draw a line through 1 vertex of the triangle parallel to the opposite side. How does this show the following?
$\angle BAC + \angle B + \angle C = 180°$

13. Places on Earth are located using parallels of latitude and meridians of longitude.

a) Research the meaning of a parallel of latitude.

b) Are parallels of latitude really parallel lines? Explain why or why not.

14. Lines that do not lie in the same plane are called **skew lines.** Lines *m* and *n* in this diagram are skew lines. How are skew lines and parallel lines different? How are they the same?

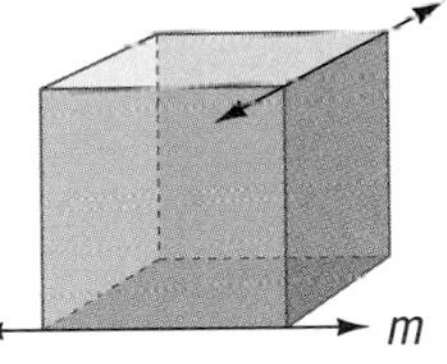

15. Parallel dividers create lanes for swimming competitions. Work with a partner to list other examples of the use of parallel lines in sports.

16. Write a problem that involves angles and parallel lines. Have a classmate solve your problem.

NUMBER POWER

When groups of 2, 3, 4, 5, or 6 people were formed to play a game at a party, there was always 1 person left over. What was the smallest number of people that could have been at the party?

12.5 Polygons

A **polygon** is a closed figure formed by joining 3 or more line segments at their endpoints. Polygons of many different types appear in nature. What type of polygon is most prominent on this black rat snake, a reptile that is native to Ontario?

Polygons are classified according to their number of sides.

Polygon	Number of Sides	Polygon	Number of Sides	Polygon	Number of Sides
Triangle	3	Hexagon	6	Nonagon	9
Quadrilateral	4	Heptagon	7	Decagon	10
Pentagon	5	Octagon	8	Dodecagon	12

Each point where 2 sides meet is called a **vertex** of the polygon. A **diagonal** is the line segment that joins any 2 nonconsecutive vertices of a polygon.

Activity: Discover the Relationship

The sum of the interior angles in a triangle is 180°. We can use this fact to find the sum of the interior angles of any polygon. To do this, separate the polygon into triangles using diagonals. Draw all the diagonals from 1 vertex.

Study this chart. Then, copy and complete the chart for 4-to 10-sided polygons.

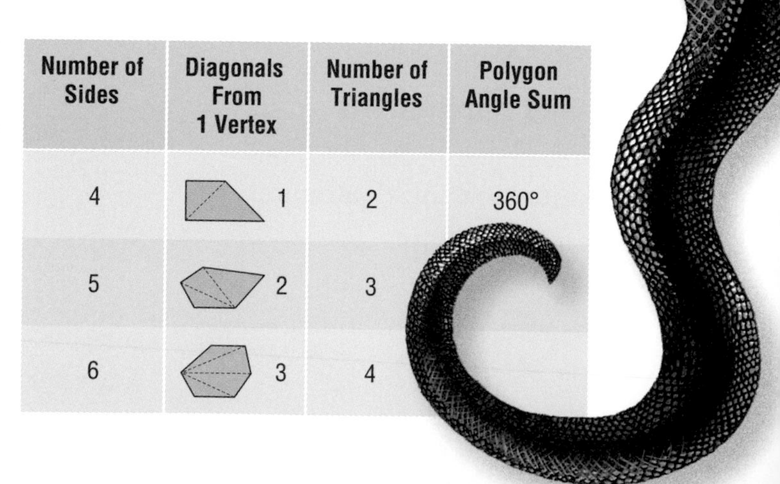

Number of Sides	Diagonals From 1 Vertex	Number of Triangles	Polygon Angle Sum
4	1	2	360°
5	2	3	
6	3	4	

Inquire

1. How are the number of sides and the number of triangles related?
2. How many triangles would be formed in a 12-sided polygon?
3. What is the sum of the interior angles in a 12-sided polygon?
4. Write a rule to determine the sum of the interior angles for any polygon.

Example 1

Find the measure of the unknown angle in this pentagon.

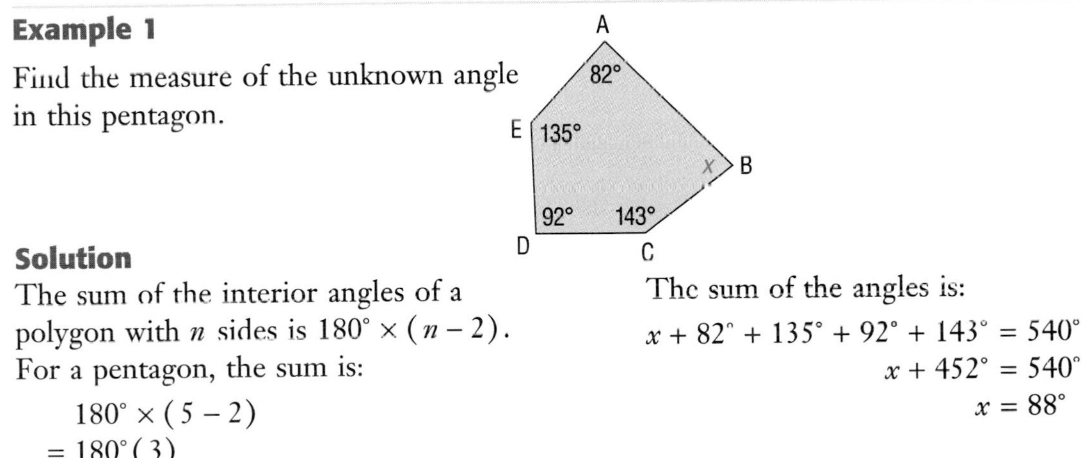

Solution

The sum of the interior angles of a polygon with n sides is $180° \times (n - 2)$.
For a pentagon, the sum is:

$$
\begin{aligned}
&180° \times (5 - 2) \\
&= 180°(3) \\
&= 540°
\end{aligned}
$$

The sum of the angles is:

$$
\begin{aligned}
x + 82° + 135° + 92° + 143° &= 540° \\
x + 452° &= 540° \\
x &= 88°
\end{aligned}
$$

The measure of the unknown angle is $x = 88°$.

Polygons can be *equilateral* (all sides equal) and *equiangular* (all angles equal). When a polygon is both equilateral and equiangular, it is called a **regular polygon**.

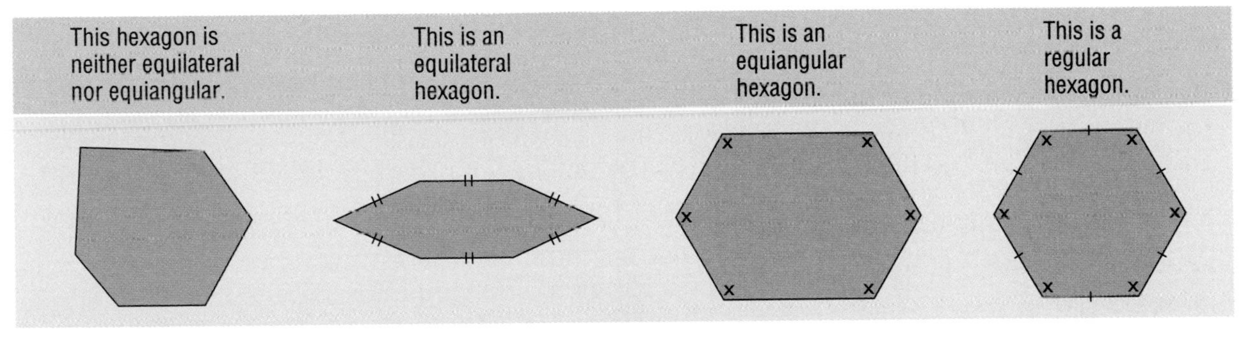

This hexagon is neither equilateral nor equiangular.
This is an equilateral hexagon.
This is an equiangular hexagon.
This is a regular hexagon.

Example 2

Find the measure of each angle in a regular octagon.

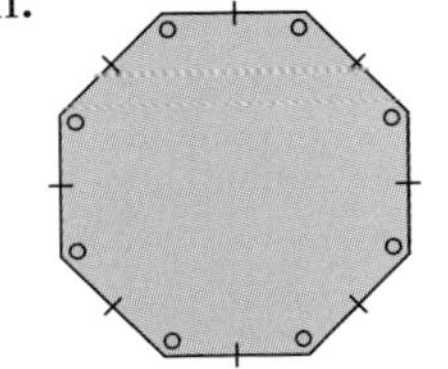

Solution

A **regular octagon** has 8 equal sides and 8 equal angles. The sum of the interior angles is:

$$
\begin{aligned}
&180° \times (n - 2) \\
&= 180(6) \\
&= 1080°
\end{aligned}
$$

Each angle in a regular octagon is $\frac{(1080)}{8}$ or $135°$.

CONTINUED

Practice

Why are these figures not polygons?

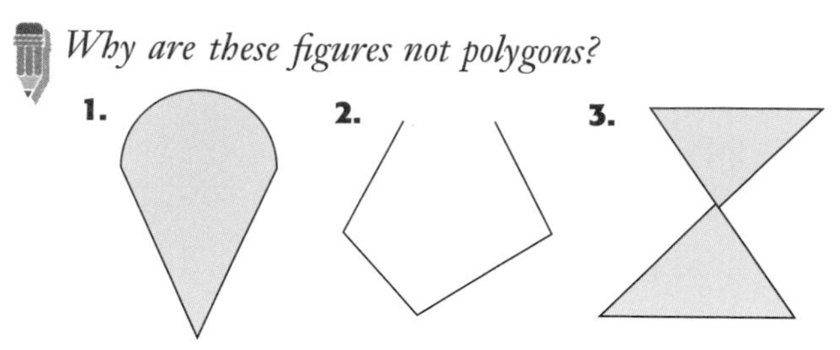

Calculate the measure of the unknown angle in each figure.

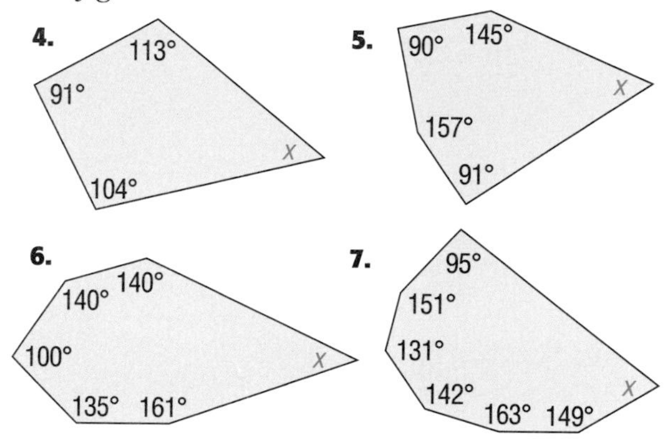

Calculate the measure of each angle for these regular polygons.

8. heptagon **9.** decagon

10. nonagon **11.** dodecagon

Problems and Applications

12. Calculate the sum of the interior angles of an 11-sided polygon.

13. Sketch each of the following figures.

a) a quadrilateral that is equiangular but not equilateral

b) a quadrilateral that is equilateral but not equiangular

14. Research the shape of home plate on a baseball diamond. What is the measure of each angle on home plate?

15. A honeycomb is made up of interlocking regular hexagons. What is the measure of each interior angle of these hexagons?

16. What is the measure of each interior angle of these polygons?

a) highway stop sign

b) Sudbury's Big Nickel

17. The 5 exterior angles of a pentagon are shown.

The sum of the 5 straight angles at each vertex is $5 \times 180°$ or $900°$. The sum of the interior angles is $180° \times (5 - 2)$ or $540°$. Therefore, the sum of the 5 exterior angles is $900° - 540°$ or $360°$.

Use this procedure to calculate the sum of the exterior angles of these polygons.

a) triangle **b)** hexagon **c)** heptagon

d) octagon **e)** decagon **f)** dodecagon

18. Use your findings in question 17 to write a statement about the sum of the exterior angles of a polygon.

19. Work with a classmate to calculate the size of each exterior angle for the following regular polygons.

a) triangle **b)** quadrilateral

c) pentagon **d)** hexagon

e) octagon **f)** decagon

20. Work with a classmate. Use your research skills to list reptiles and other animals on which you can see shapes of polygons.

Congruent Figures

Congruent figures have the same size and shape.

Activity 1

1. Find the pairs of congruent figures. You may want to trace the figures and slide, flip, or turn the tracings.

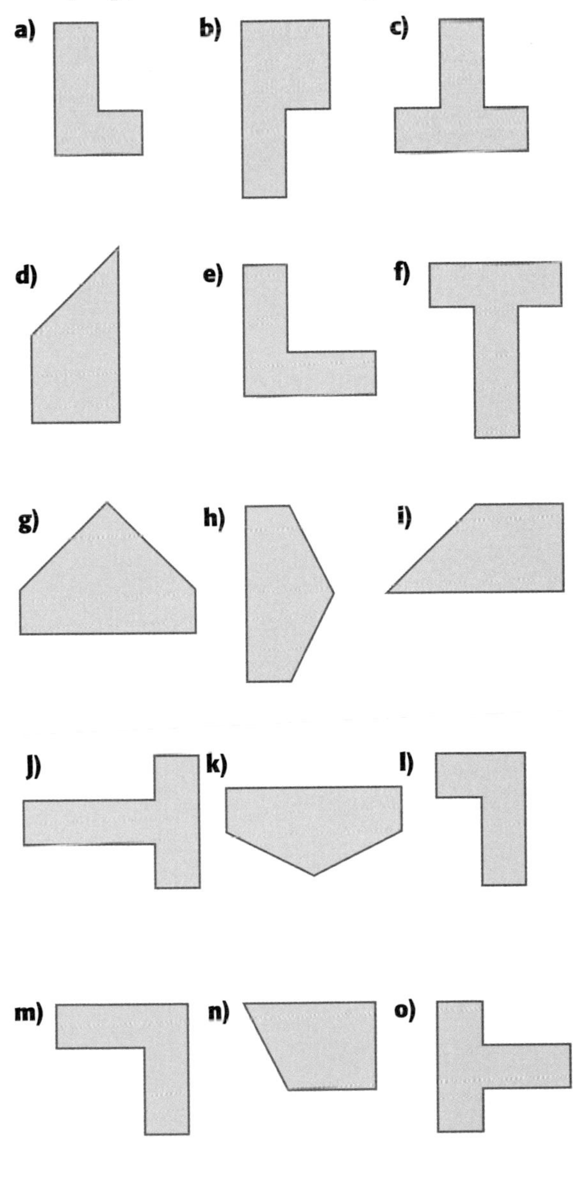

Activity 2

1. Copy the given triangle and line segment onto grid paper. Locate a point D to make ΔDEF the same size as ΔABC. There may be more than one place to put D.

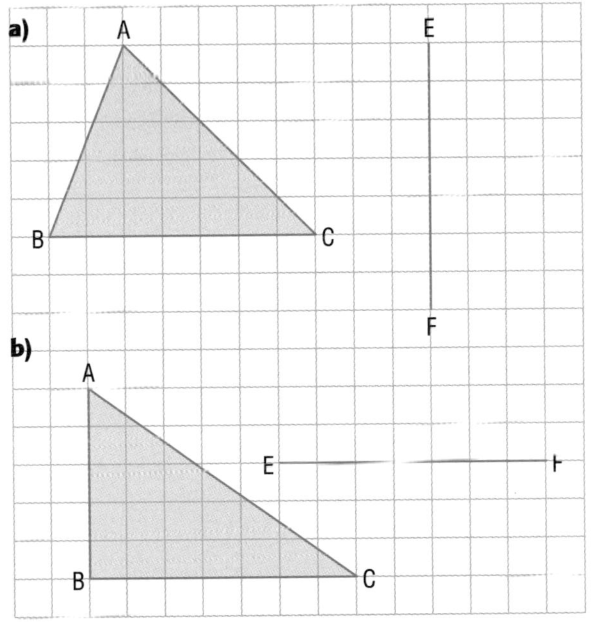

2. In the following diagram, locate 2 points, R and S, so that quadrilateral PQRS is the same size as quadrilateral ABCD.

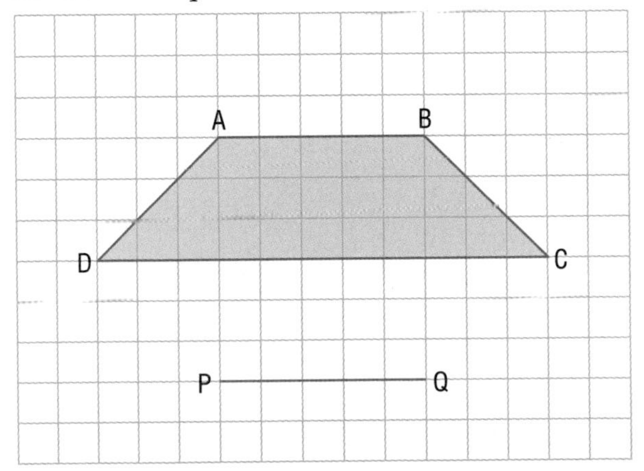

Congruent Triangles

Congruent triangles are triangles that have the same size and shape.

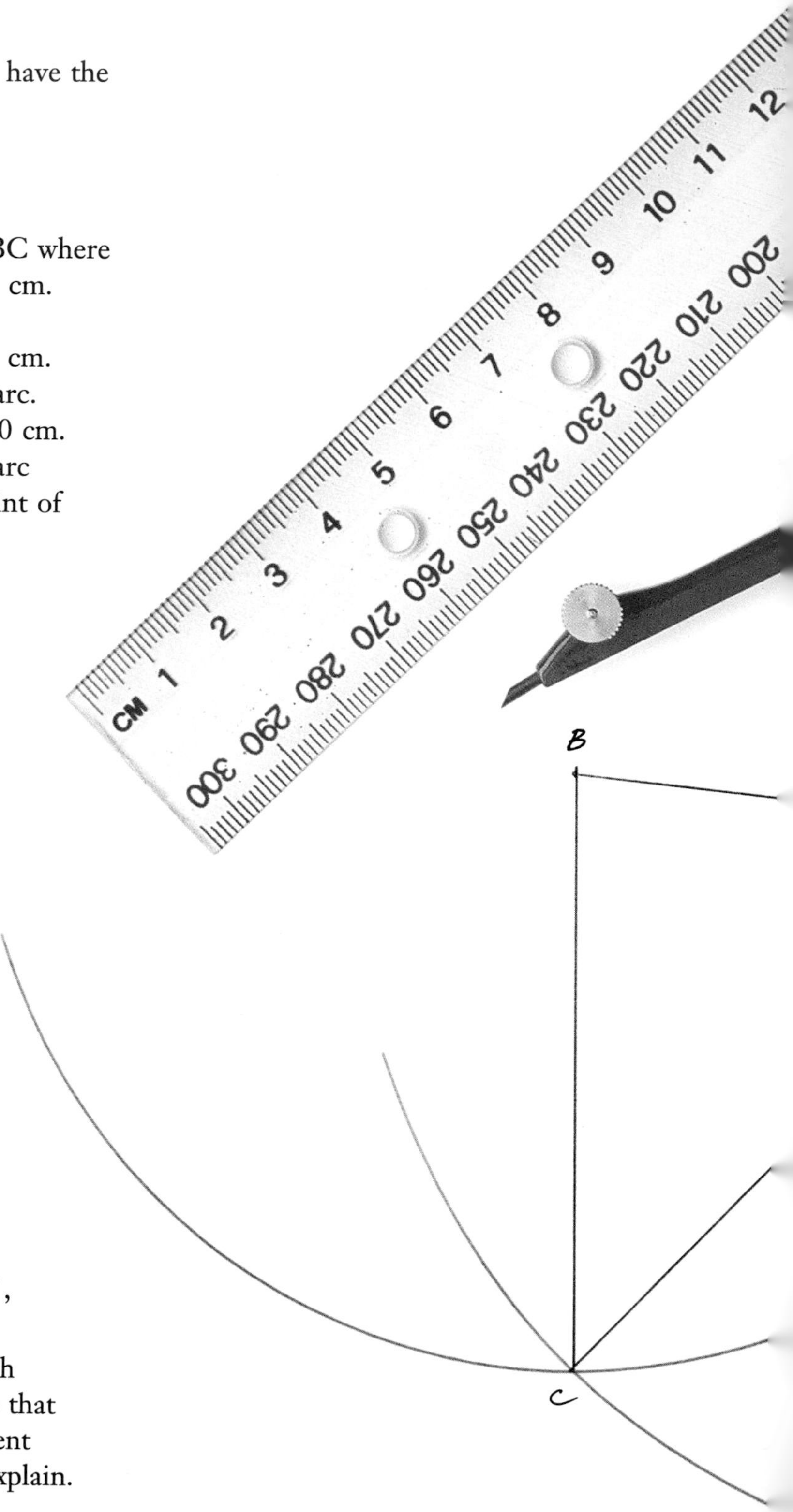

Activity 1

1. Follow these steps to construct △ABC where AB = 7 cm, BC = 8 cm, and AC = 10 cm.

a) Draw line segment AB = 7 cm.

b) Set your compasses to a radius of 8 cm. Then, use B as a centre and draw an arc.

c) Set your compasses to a radius of 10 cm. Then, use A as a centre and draw an arc to intersect the first arc. Label the point of intersection C.

d) Draw BC and AC.

2. Is it possible to construct △DEF where DE = 7 cm, EF = 8 cm, and DF = 10 cm so that the size and shape of △DEF are different from the size and shape of △ABC? Explain.

Activity 2

1. Construct △GHI so that GH = 5 cm, and GI = 7 cm.

2. Is it possible to construct △JKL with JK = 5 cm, and JL = 7 cm so that the size and shape of △JKL are different from the size and shape of △GHI? Explain.

Activity 3

1. Construct △MNO so that ∠M = 35°, ∠N = 65°, and ∠O = 80°.

2. Is it possible to construct △PQR with ∠P = 35°, ∠Q = 65°, and ∠R = 80° so that the size and shape of △PQR are different from the size and shape of △MNO? Explain.

Activity ❹

1. Construct ΔRST so that $\angle R = 70°$, $RS = 6$ cm, and $RT = 7$ cm.

2. Is it possible to construct ΔXYZ with $\angle X = 70°$, $XY = 6$ cm, and $XZ = 7$ cm so that the size and shape of ΔXYZ are different from the size and shape of ΔRST? Explain.

Activity ❺

1. Construct ΔABC so that $\angle A = 40°$, $AB = 7$ cm, and $\angle B = 55°$.

2. Is it possible to construct ΔDEF with $\angle D = 40°$, $DE = 7$ cm, and $\angle E = 55°$ so that the size and shape of ΔDEF are different from the size and shape of ΔABC? Explain.

Activity ❻

1. Construct ΔPQR so that $\angle P = 35°$ and $PQ = 7$ cm.

2. Is it possible to construct ΔXYZ with $\angle X = 35°$ and $XY = 7$ cm, so that the size and shape of ΔXYZ are different from the size and shape of ΔPQR? Explain.

Activity ❼

1. Use the results of the first 6 activities to decide which sets of 3 facts need to be given so that only 1 triangle can be constructed.

12.6 Congruent Triangles

Many beautiful and unusual structures are built using congruent triangles. One such structure is shown in this photograph of Katimavik, built for Expo 67 in Montreal.

Activity: Use the Diagrams

Each of these 4 triangles is positioned differently but they all have the same size and shape. Copy or trace them into your notebook.

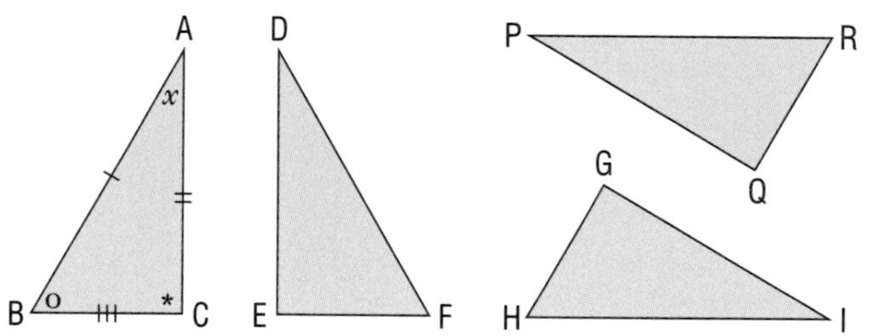

Inquire

1. Mark AB with a single tick as shown in the diagram above. Then, in the other 3 triangles, use a single tick to mark the sides that are the same length as AB.

2. Mark AC with 2 ticks. Then, in the other 3 triangles, use two ticks to mark the sides that are the same length as AC.

3. Use 3 ticks to mark each of the remaining equal sides.

4. Mark ∠A with an x. Then, in the other 3 triangles, use an x to mark the angles that have the same measure as ∠A.

5. Mark ∠B with an o. Then, in the other 3 triangles, use an o to mark the angles that have the same measure as ∠B.

6. Use an asterisk, *, to mark the remaining equal angles.

A triangle has 6 parts, 3 angles and 3 sides. Two triangles are **congruent** if you can match up their vertices so that all pairs of corresponding angles and corresponding sides are equal.
In ΔABC and ΔDEF, the following corresponding parts are equal.

∠A = ∠D AB = DE
∠B = ∠E BC = EF
∠C = ∠F AC = DF

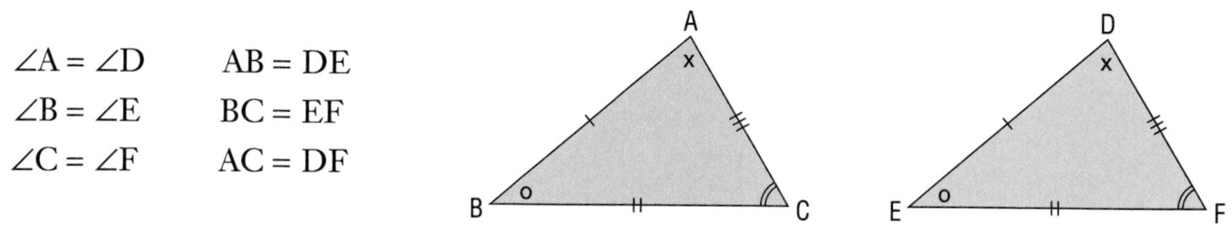

We say ΔABC ≅ ΔDEF. The symbol for congruence, ≅, is read "is congruent to." Note that the vertices of each triangle are listed in the same order as their corresponding angles.

You do not need to know that all 6 parts of 1 triangle are equal to the 6 corresponding parts of another triangle to prove congruency. Knowing that 3 parts of one triangle are equal to 3 corresponding parts of another triangle may be sufficient to say that the triangles are congruent.

The chart gives you 3 ways to state that 2 triangles are congruent.

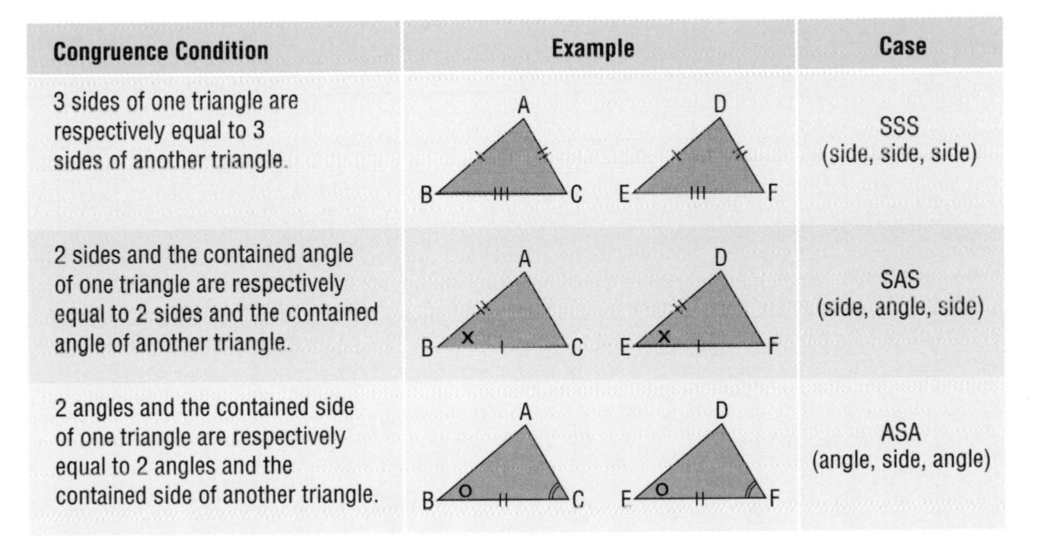

Congruence Condition	Example	Case
3 sides of one triangle are respectively equal to 3 sides of another triangle.		SSS (side, side, side)
2 sides and the contained angle of one triangle are respectively equal to 2 sides and the contained angle of another triangle.		SAS (side, angle, side)
2 angles and the contained side of one triangle are respectively equal to 2 angles and the contained side of another triangle.		ASA (angle, side, angle)

Example 1

a) State why $\Delta RST \cong \Delta LMN$.

b) List the other equal sides and angles.

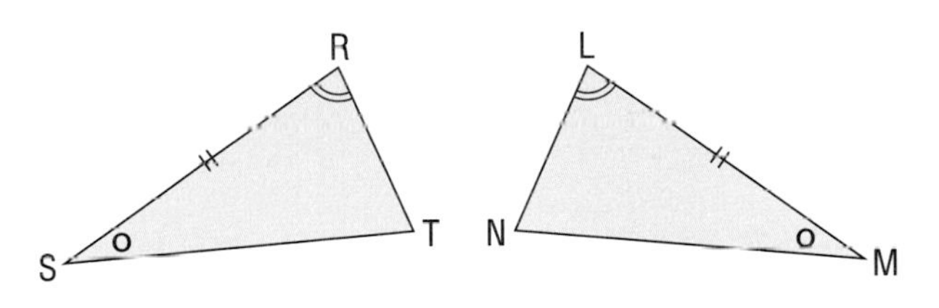

Solution

a) In ΔRST and ΔLMN,

$\angle R = \angle L$
$RS = LM$
$\angle S = \angle M$

Therefore, $\Delta RST \cong \Delta LMN$ **(ASA)**

b) $RT = LN$
$ST = MN$
$\angle T = \angle N$

Example 2

a) State why $\Delta PQR \cong \Delta BCA$.

b) List the other equal parts.

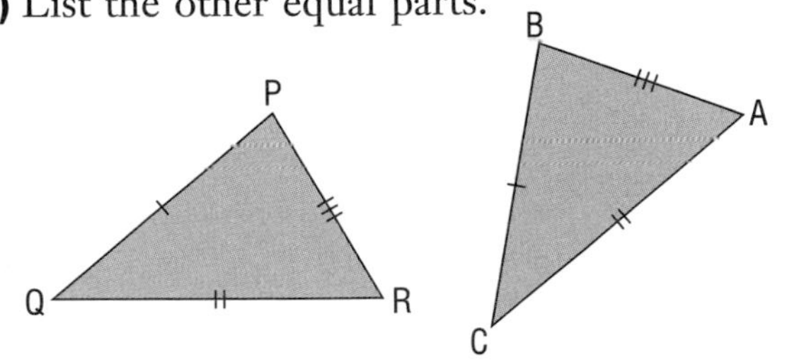

Solution

a) In ΔPQR and ΔBCA,

$PQ = BC$
$QR = CA$
$PR = BA$

Therefore, $\Delta PQR \cong \Delta BCA$ **(SSS)**

b) $\angle P = \angle B$
$\angle Q = \angle C$
$\angle R = \angle A$

CONTINUED

Example 3

a) State why $\Delta PQR \cong \Delta STR$.

b) List the other corresponding equal parts.

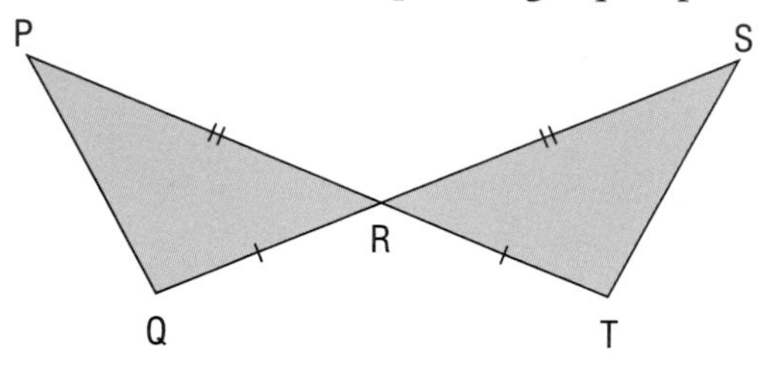

Solution

a) In ΔPQR and ΔSTR,

$PR = SR$

$\angle PRQ = \angle SRT$ (opposite angles)

$QR = TR$

Therefore, $\Delta PQR \cong \Delta STR$ **(SAS)**

b) $PQ = ST$

$\angle P = \angle S$

$\angle Q = \angle T$

Practice

Name the equal sides and angles for these pairs of congruent triangles.

1.

2.

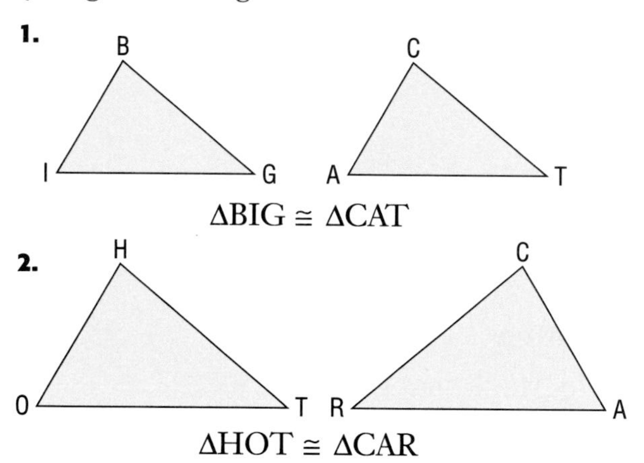

$\Delta BIG \cong \Delta CAT$

$\Delta HOT \cong \Delta CAR$

Is each pair of triangles congruent in questions 3-9? If they are, what case did you use?

3.

4.

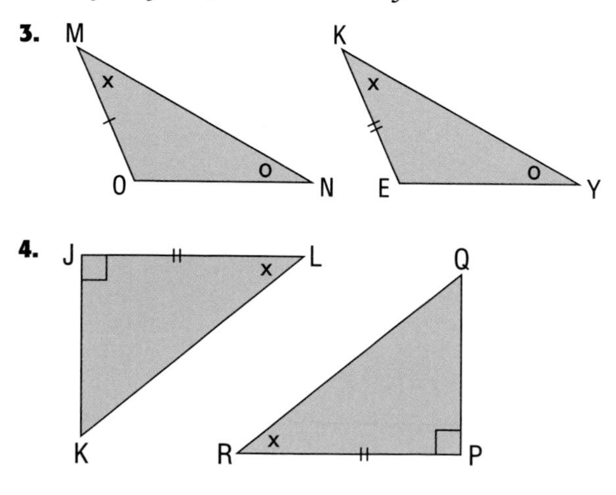

5.

6.

7.

8.

9.

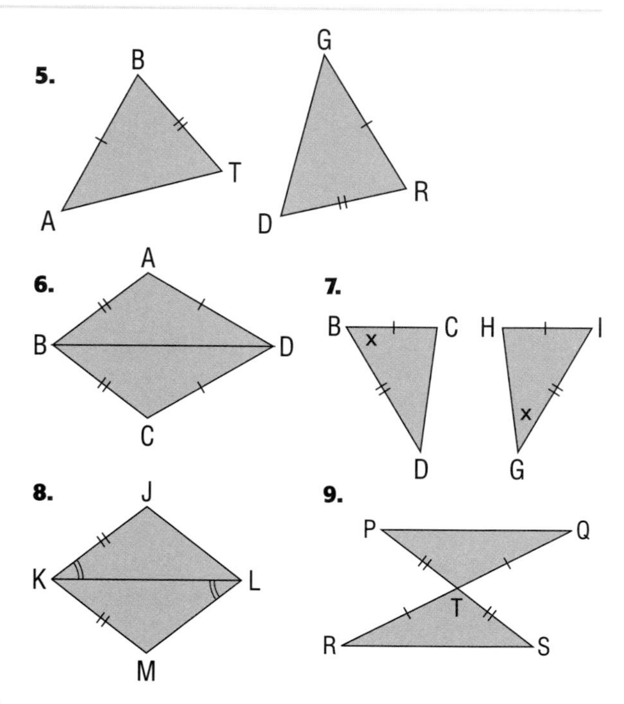

What is the fewest number of other parts that must be equal before you can state that the following pairs of triangles are congruent? Explain.

10.

11.

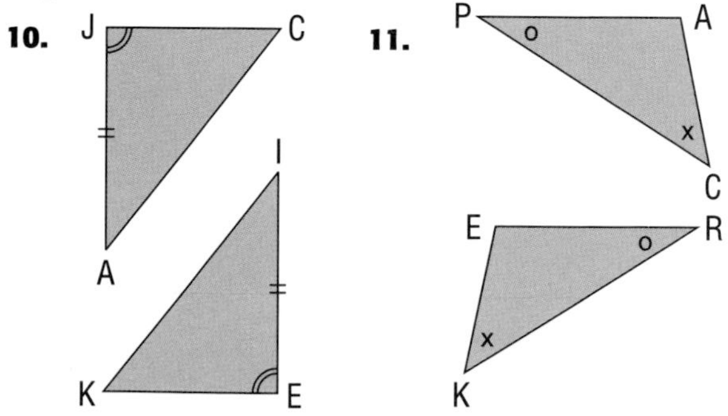

Are these pairs of triangles congruent? If they are, give the case and list all the corresponding equal parts.

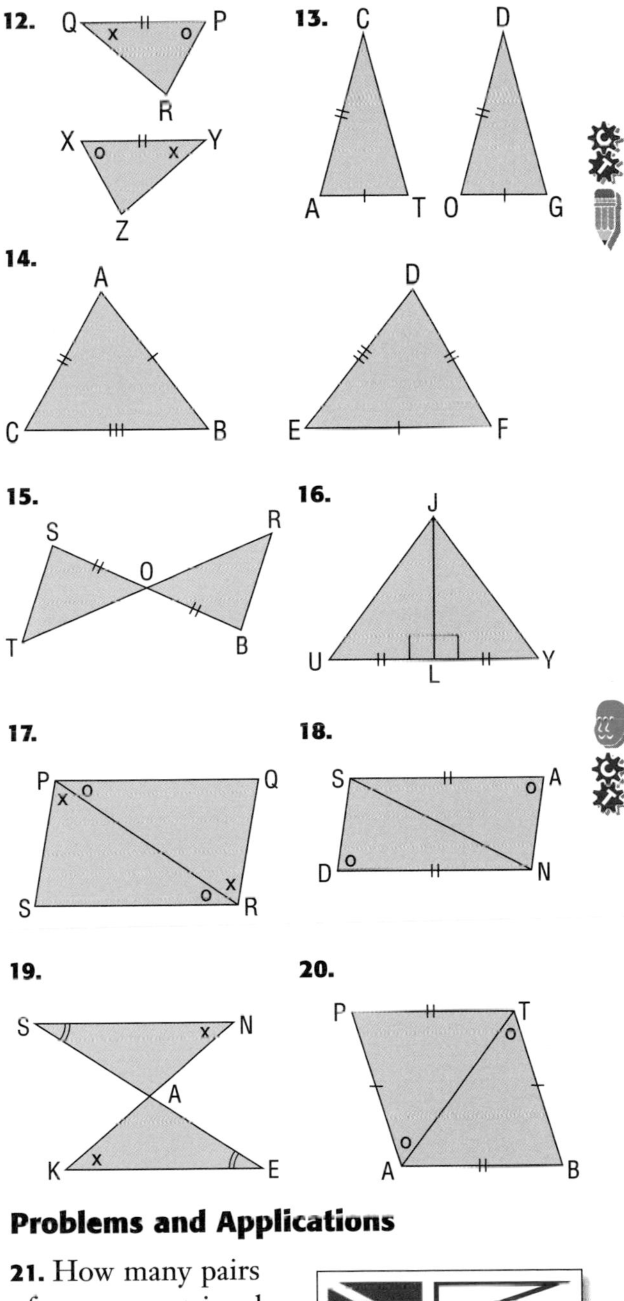

Problems and Applications

21. How many pairs of congruent triangles are there on the flag of Newfoundland and Labrador?

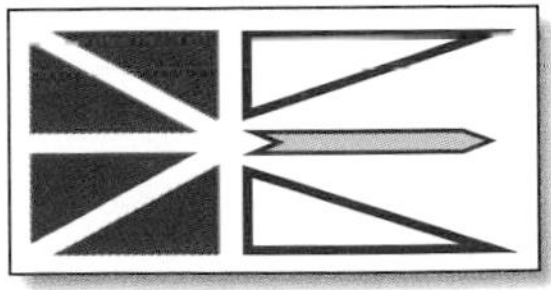

22. Do the gables on these 2 roofs form congruent triangles? Explain.

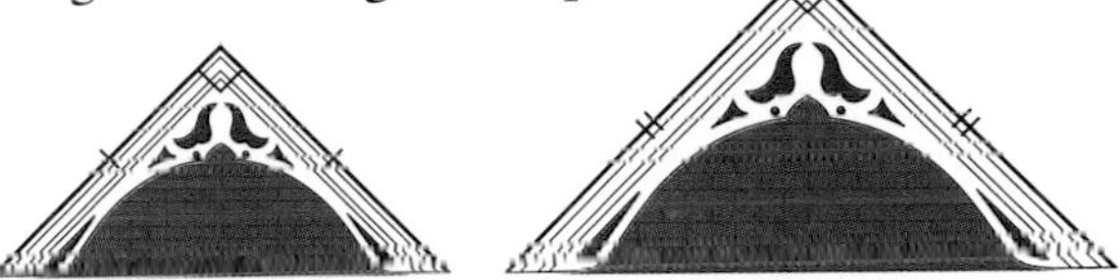

23. a) Triangles are used in construction because they hold their shape. To change the shape of this triangle, you have to change the length of at least 1 of the sides. Why?

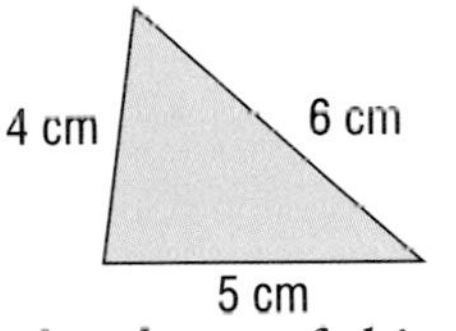

b) To change the shape of this figure, you do not have to change the length of a side. Why not?

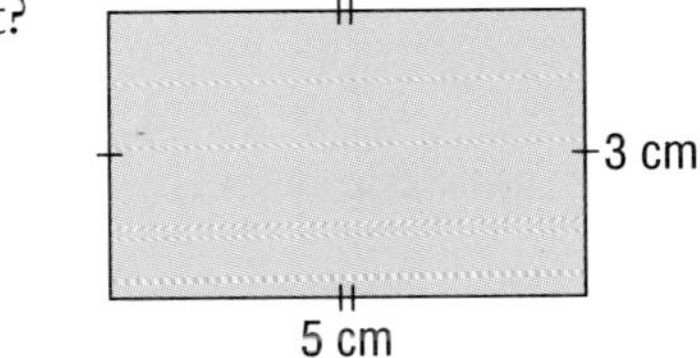

24. Work with a classmate to decide which of the following are always congruent, sometimes congruent, or never congruent. Illustrate your answer with diagrams.

a) 2 triangles with the same perimeter

b) 2 rectangles with the same area

c) 2 squares with the same perimeter

d) 2 rectangles with the same perimeter

WORD POWER

a) Louisiana is 1 of 6 American states in which the name contains the consecutive letters IS. What are the other 5 states?

b) Which 2 Canadian provinces also contain the consecutive letters IS?

Investigating Right Triangles

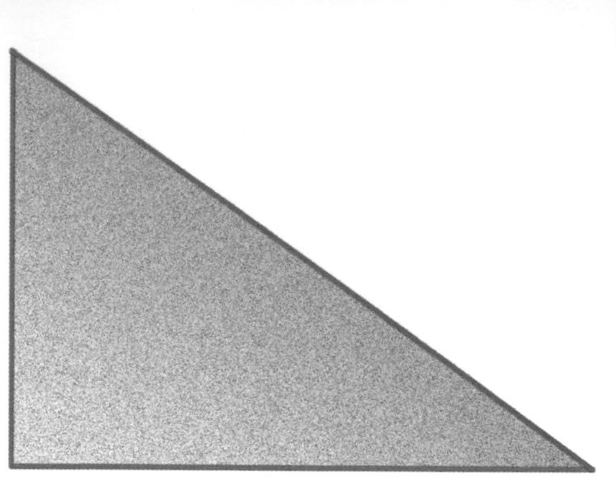

A right triangle has one right angle. The side opposite the right angle is called the **hypotenuse**. The other two sides are called the **legs**.

Activity ❶ Areas of Squares on a Grid

The area of a square drawn on a grid can be calculated by dividing the square into right triangles and, in some cases, a smaller square. Determine each area by counting squares on the grid. Compare your method with your classmates'.

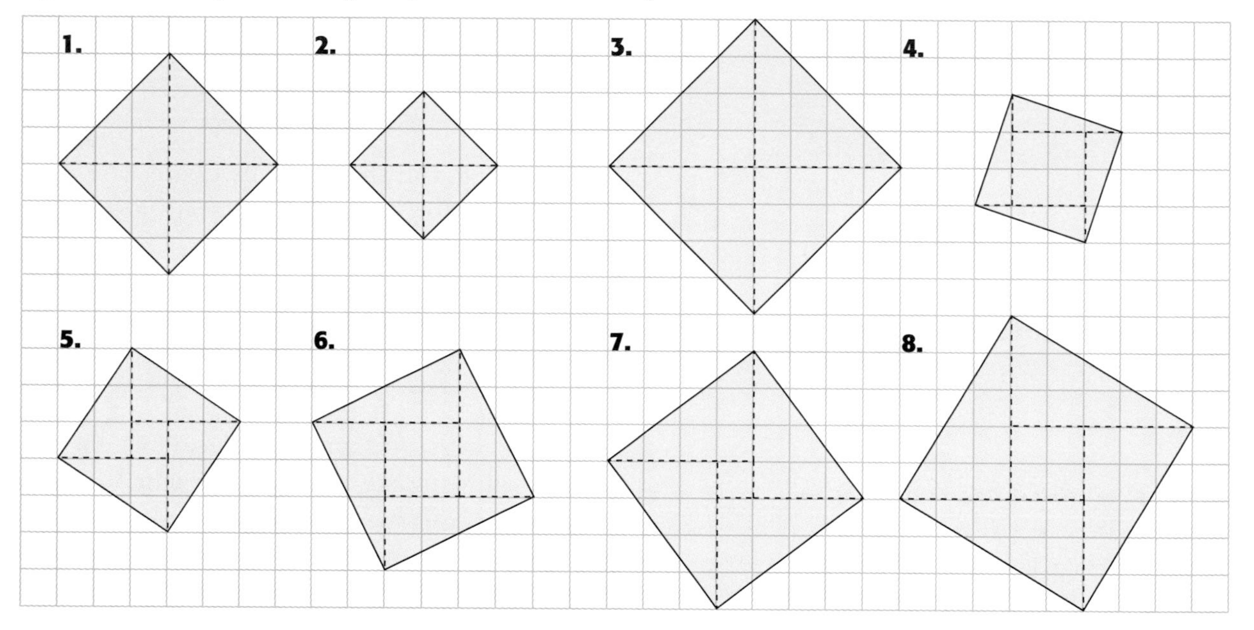

Activity ❷ More Squares on a Grid

Copy the squares onto grid paper. Divide each square into right triangles and a smaller square. Determine the area of each square by counting squares on the grid.

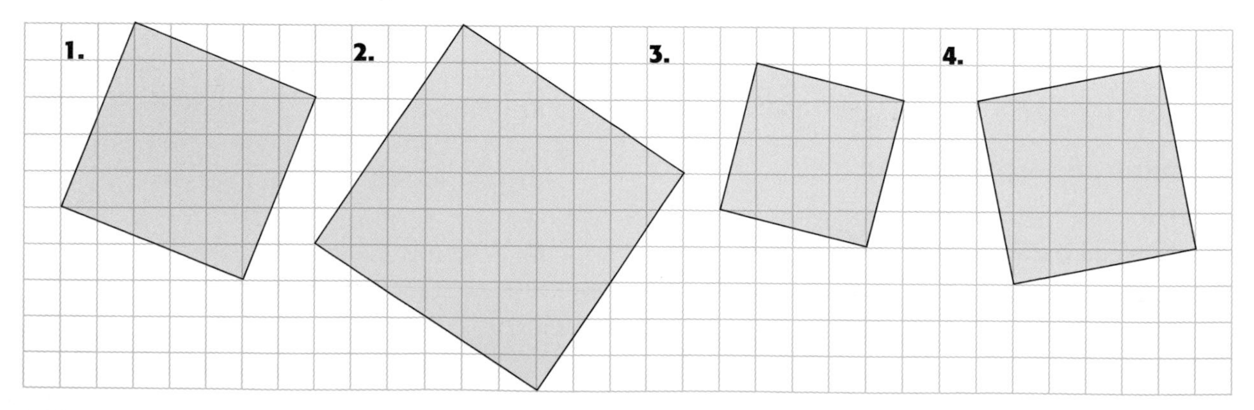

Activity ❸ Squares on the Sides of a Right Triangle

The diagrams show squares on the sides of each triangle. Copy the table. Count squares on the grid to help complete the table.

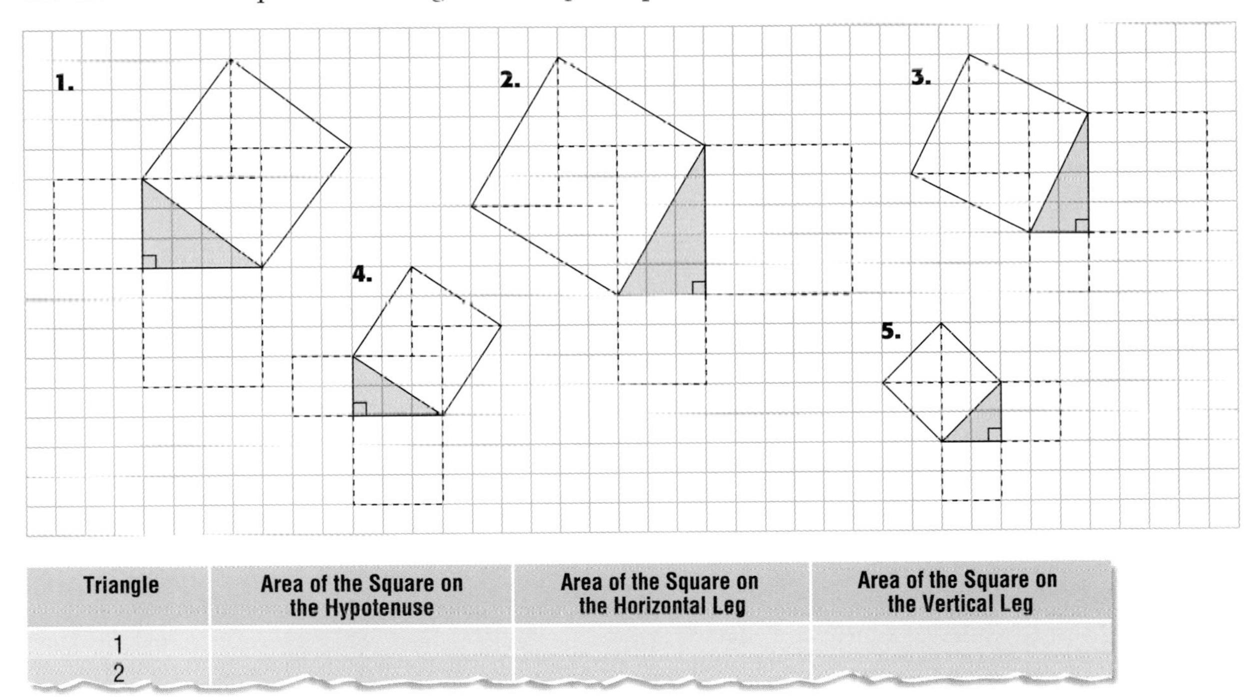

Triangle	Area of the Square on the Hypotenuse	Area of the Square on the Horizontal Leg	Area of the Square on the Vertical Leg
1			
2			

Activity ❹ More Right Triangles

Use the technique from Activity 3 to draw squares on the sides of these right triangles. Copy and complete the table.

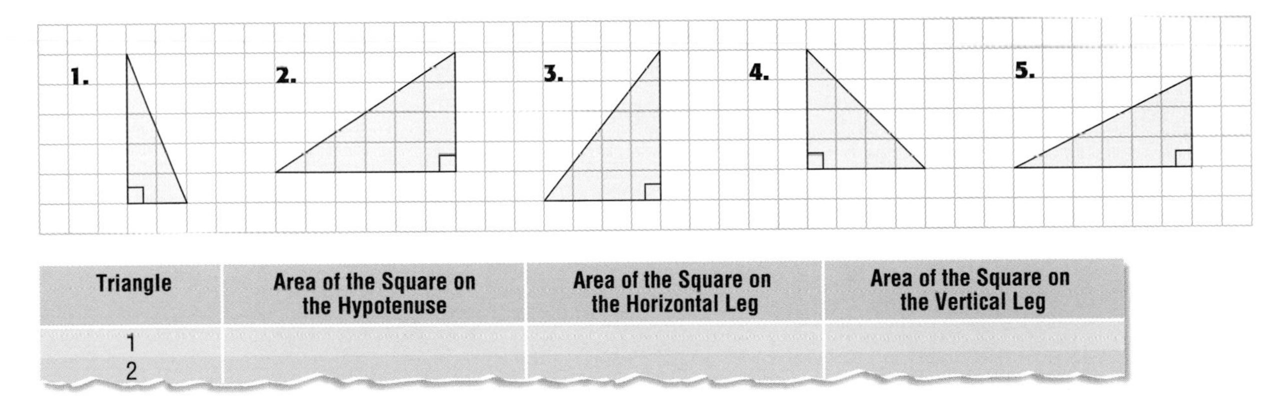

Triangle	Area of the Square on the Hypotenuse	Area of the Square on the Horizontal Leg	Area of the Square on the Vertical Leg
1			
2			

Activity ❺ The Pythagorean Theorem

Describe the relationship between the areas of the squares on the legs of a right triangle and the area of the square on the hypotenuse.

12.7 The Pythagorean Theorem

A baseball diamond is a square 27.5 m on each side. The catcher often throws the ball from home plate to second base to stop an opponent from stealing a base. To find how far the ball is thrown, you could make a scale drawing, or measure the distance, or use the Pythagorean Theorem.

Activity: Discover the Relationship

Draw 3 right triangles on grid paper. Number the triangles 1, 2, and 3. Label the sides a, b, and c, as shown.

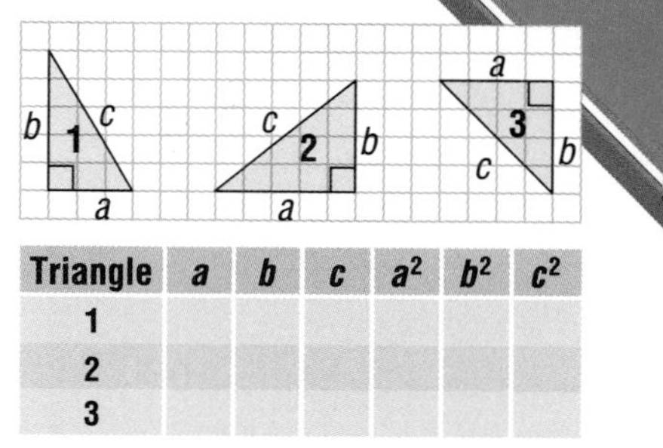

Copy the table. Measure sides to the nearest tenth of a centimetre. Complete the table. Use a calculator for the last 3 columns.

Triangle	a	b	c	a^2	b^2	c^2
1						
2						
3						

Inquire

1. a) For each triangle, calculate the sum of a^2 and b^2.

b) How is the sum $a^2 + b^2$ related to c^2?

2. Use your conclusion from question 2 to calculate the length of the hypotenuse in each of the following triangles.

a) **b)** **c)**

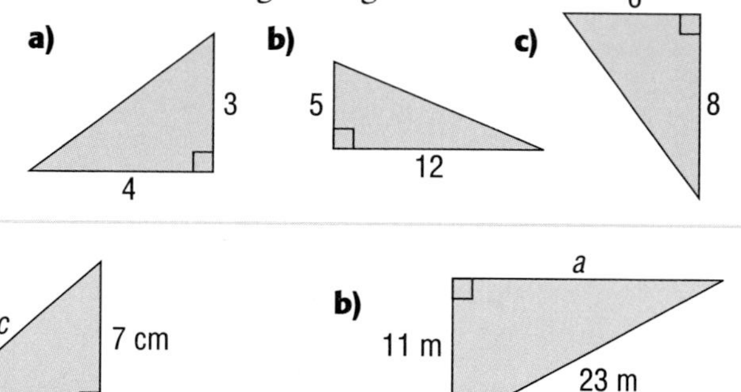

3. Find the distance from home plate to second base, to the nearest tenth of a metre.

Example

Calculate the length of the unknown side, to the nearest tenth of a unit.

a) **b)**

11 m

23 m

a

Solution

The Pythagorean Theorem states that in any right triangle, if c is the length of the hypotenuse and a and b are the lengths of the legs, then $c^2 = a^2 + b^2$.

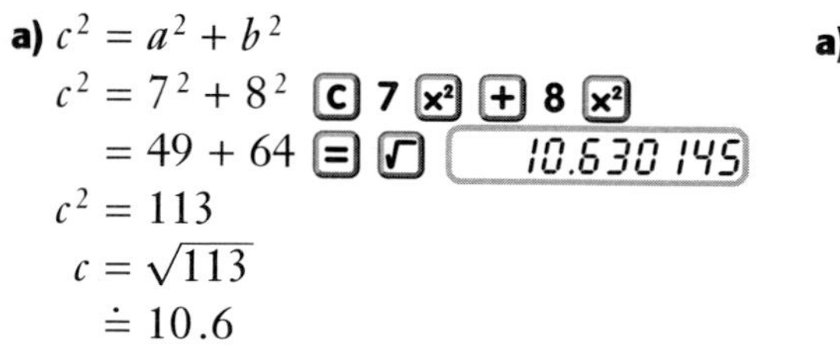

a) $c^2 = a^2 + b^2$

$c^2 = 7^2 + 8^2$

$= 49 + 64$

$c^2 = 113$

$c = \sqrt{113}$

$\doteq 10.6$

The length is about 10.6 cm.

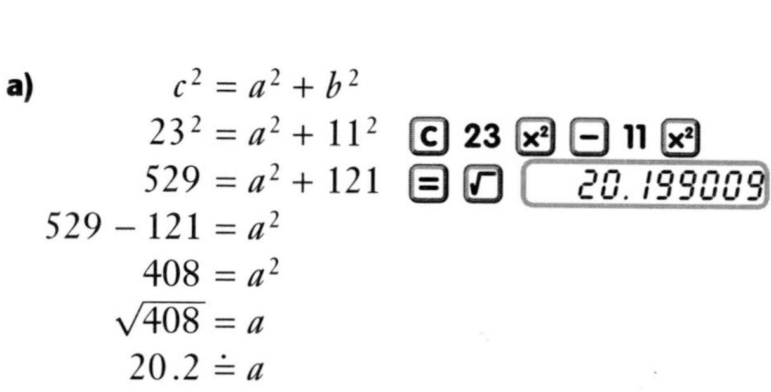

a) $c^2 = a^2 + b^2$

$23^2 = a^2 + 11^2$

$529 = a^2 + 121$

$529 - 121 = a^2$

$408 = a^2$

$\sqrt{408} = a$

$20.2 \doteq a$

The length is about 20.2 m

Practice

State the Pythagorean relationship for each triangle.

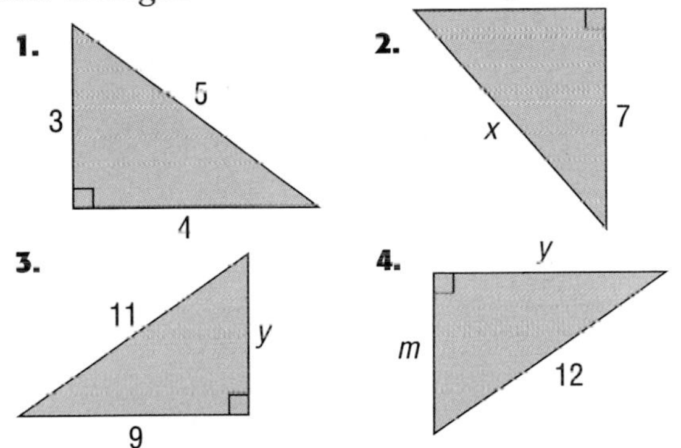

Calculate the length of the unknown side.

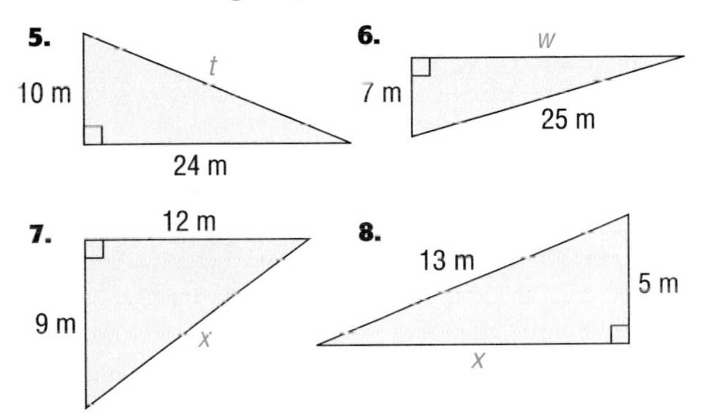

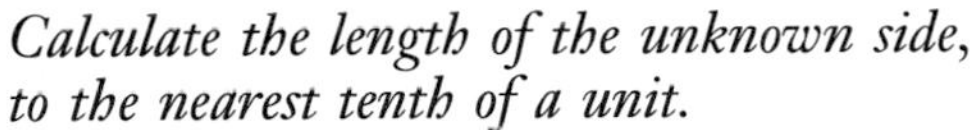

Calculate the length of the unknown side, to the nearest tenth of a unit.

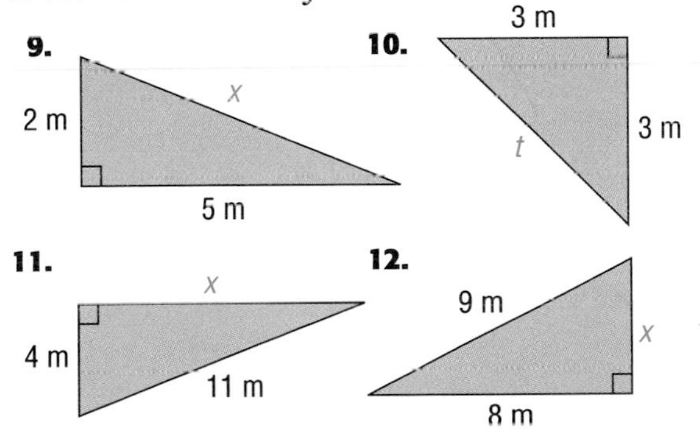

Problems and Applications

13. The foot of a 6-m ladder is placed 2 m from the base of a building. How far up the building wall does the ladder reach, to the nearest tenth of a metre?

14. If the two diagonals of a field are the same length, the field has square corners. How long is each diagonal in a field measuring 80 m by 70 m with square corners?

15. A radio tower is 10 m high. Four wires are attached from the top of the tower to points on the ground 15 m from the base of the tower. What is the total length of the wires, to the nearest tenth of a metre?

16. To find the length of a lake, a surveyor puts stakes at each end of the lake and then locates a stake at C so that $\angle ACB = 90°$. How long is the lake, to the nearest tenth of a metre?

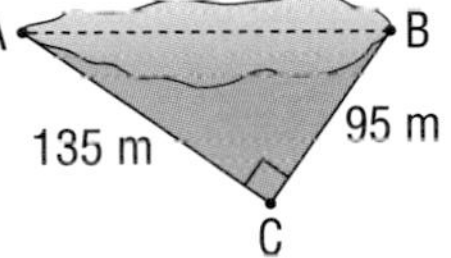

17. Calculate x, to the nearest tenth of a metre.

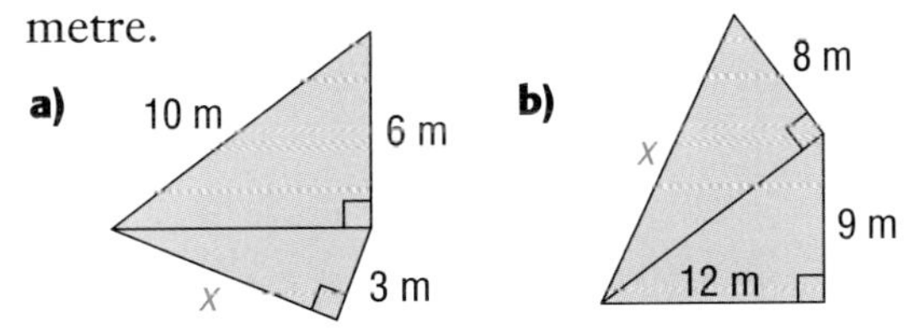

18. a) What is the longest walking stick that can be placed in the bottom of a trunk of length 80 cm and width 60 cm?

b) What is the longest walking stick that can be placed in a trunk that is 80 cm long, 60 cm wide, and 20 cm high? Answer to the nearest centimetre.

19. A Pythagorean Triple is a set of 3 whole numbers that can be the lengths of the sides of a right triangle. The numbers 3, 4, and 5 make up the simplest Pythagorean Triple. With a classmate, list the other Pythagorean Triples in which no number is more than 50.

20. Write a problem that can be solved with the Pythagorean Theorem. Have a classmate solve your problem.

12.8 Angles in a Circle

This clock tower is part of the Parliament Buildings in Ottawa. If we extended the hands of the clock to intersect the circle, they would define an angle in the circle. The following diagrams show some parts of a circle and the angles in a circle.

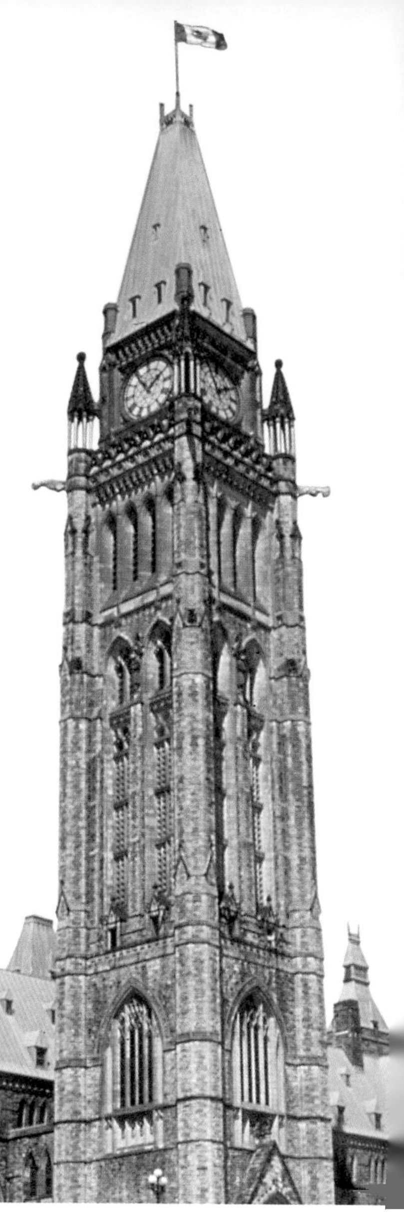

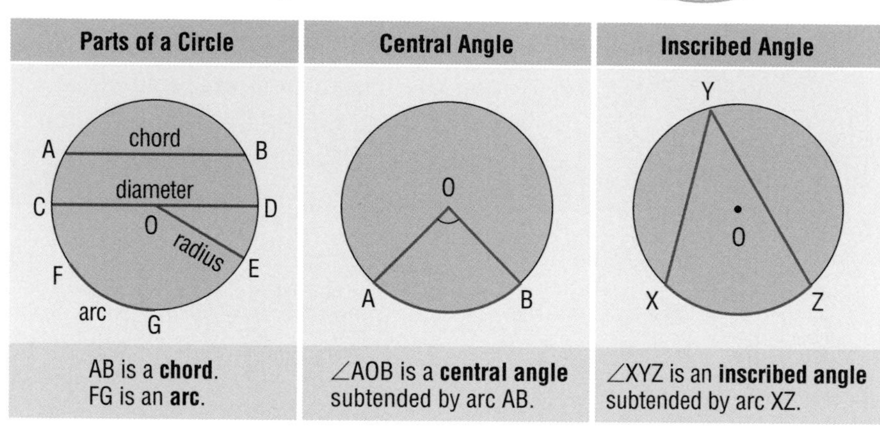

AB is a **chord**.
FG is an **arc**.

∠AOB is a **central angle** subtended by arc AB.

∠XYZ is an **inscribed angle** subtended by arc XZ.

A **central angle** has its vertex at the centre of the circle. The arms of the angle are radii of the circle. An **inscribed angle** has its vertex on the circle. The arms of the angle are chords of the circle.

Activity: Discover the Relationship

Draw a circle with centre O and radius 4 cm.
Draw an inscribed angle and a central angle on arc AB.
Measure the inscribed angle ∠ACB.
Measure the central angle ∠AOB.

Inquire

1. How is the size of ∠ACB related to the size of ∠AOB?

2. Repeat the activity using 2 other circles of different radii.

3. Write a statement about the relationship between the size of a central angle and the size of an inscribed angle subtended by the same arc.

4. Repeat the activity for inscribed angles and central angles subtended by the same arc that overlap as shown in the diagram.

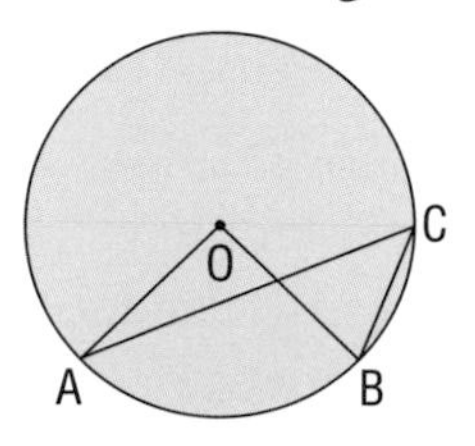

5. Are the overlapping angles related in the same way as you found in question 3?

Activity: Discover the Relationship

Draw a circle with centre O and radius 4 cm.
Draw 2 inscribed angles on arc AB.
Measure inscribed ∠ACB and inscribed ∠ADB.

Inquire

1. How is the size of ∠ACB related to the size of ∠ADB?

2. Repeat the activity using 2 other circles of different radii.

3. Write a statement about the relationship between the size of 2 inscribed angles subtended by the same arc.

Activity: Discover the Relationship

Draw a circle with radius 5 cm. Draw a diameter AB as shown in the diagram on the right. Construct an inscribed angle, ∠ACB on the diameter. Measure ∠ACB.

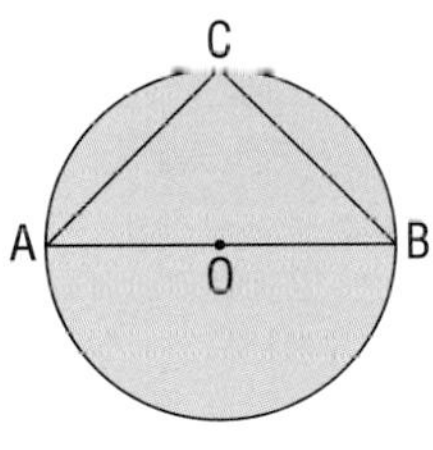

Inquire

1. What is the measure of ∠ACB?

2. Repeat the activity using 2 different circles.

3. Write a statement about the size of an inscribed angle on a diameter.

Example

Find the value of each unknown angle.

a)

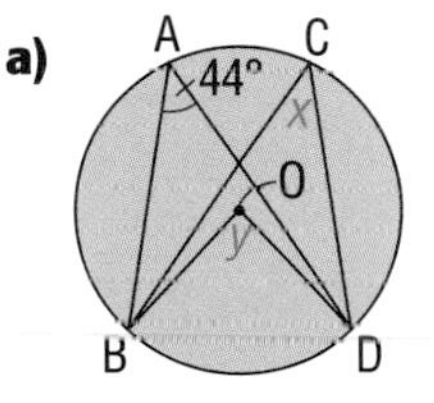

b)

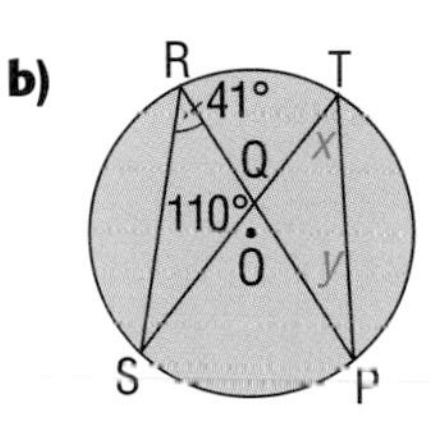

Solution

a) ∠A and ∠C are subtended by the same arc BD.

$$\angle C = \angle A$$
$$x = 44^\circ$$

∠BOD is a central angle subtended by the same arc as inscribed ∠A.

$$\angle BOD = 2 \times \angle A$$
$$= 2(44^\circ)$$
$$y = 88^\circ$$

b) ∠R and ∠T are subtended by the same arc SP.

$$\angle T = \angle R$$
$$x = 41^\circ$$

In ΔQRS,

$$\angle R + \angle RQS + \angle S = 180^\circ$$
$$41^\circ + 110^\circ + \angle S = 180^\circ$$
$$151^\circ + \angle S = 180^\circ$$
$$\angle S = 29^\circ$$

∠S and ∠P are subtended by the same arc RT.

$$\angle P = \angle S$$
$$y = 29^\circ$$

CONTINUED

Practice

In each diagram, name the inscribed angles and the central angles subtended by arc AB.

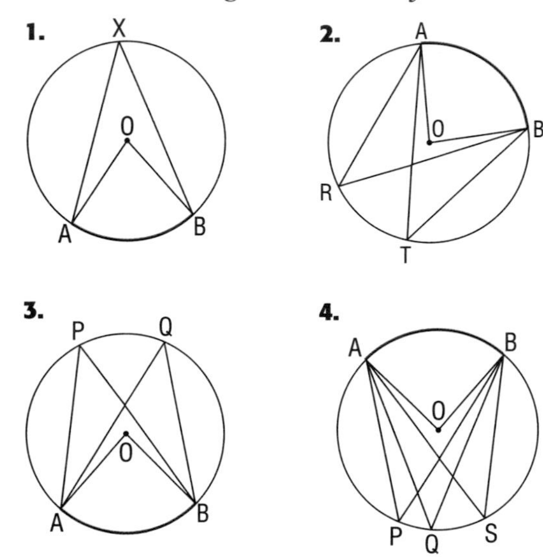

Find the measure of each unknown angle.

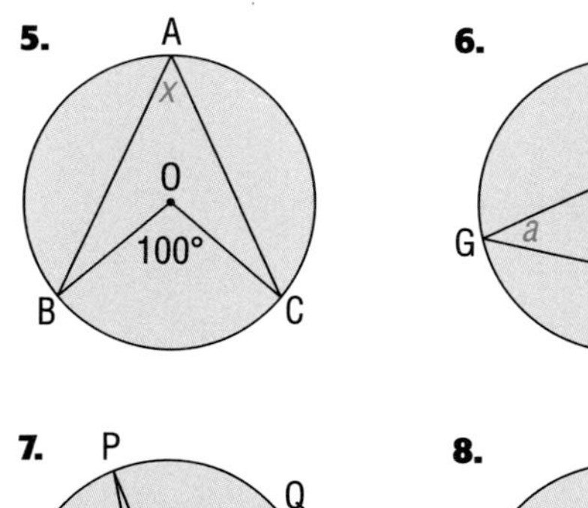

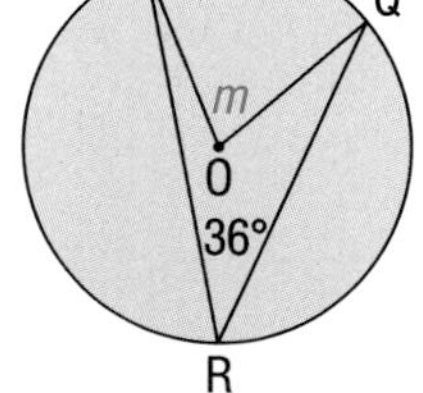

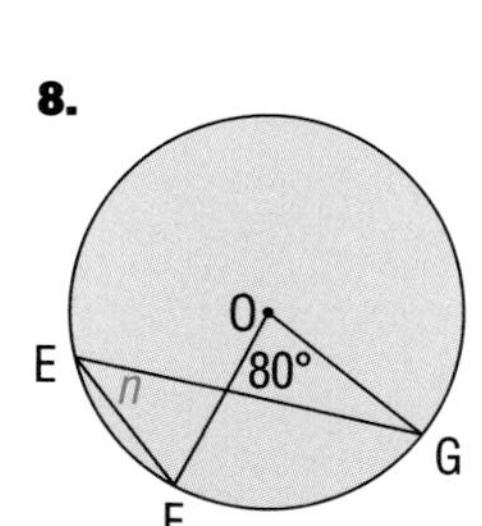

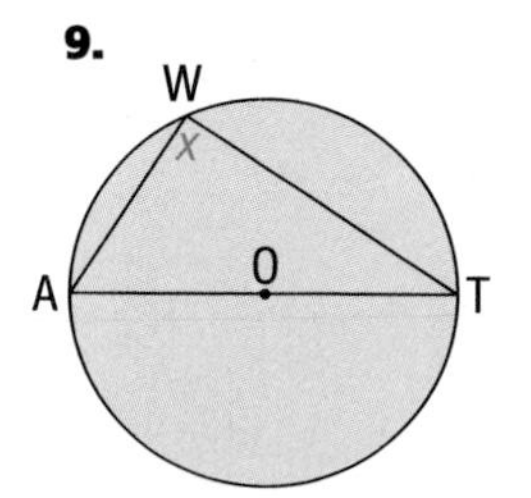

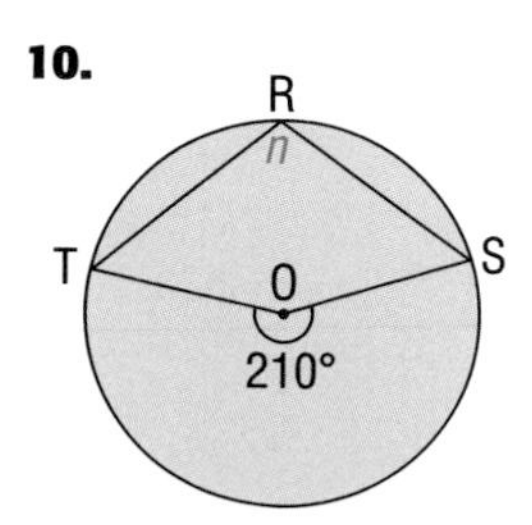

Find the measure of each unknown angle.

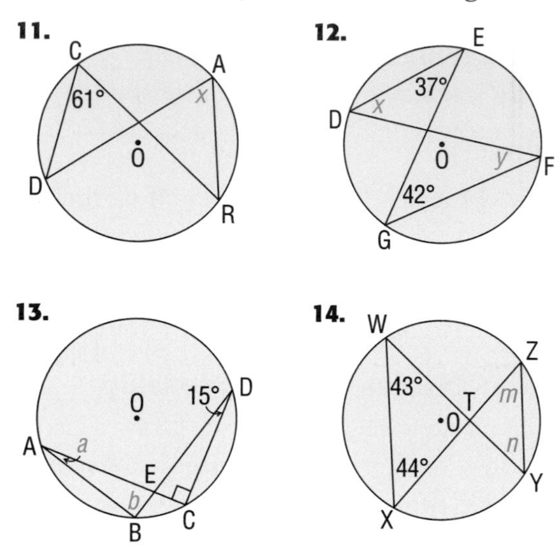

Find the measure of each unknown angle.

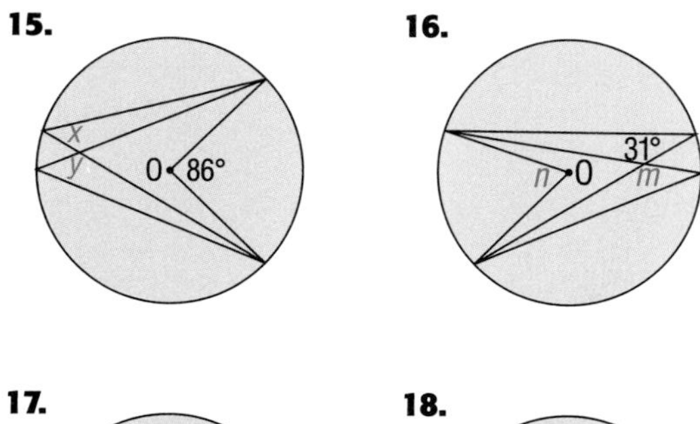

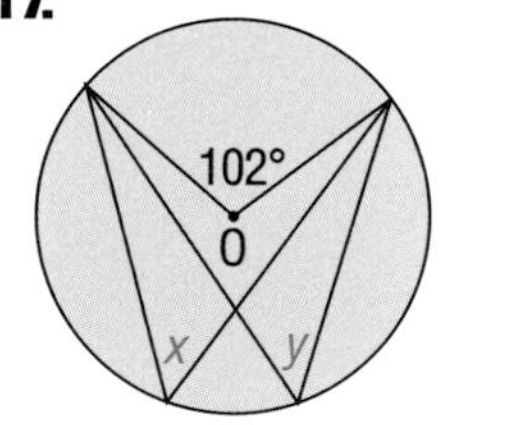

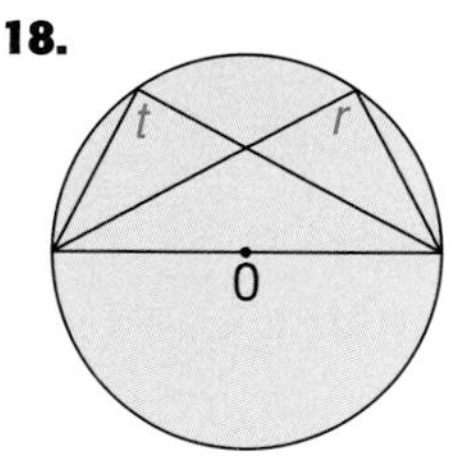

Find the measure of each unknown angle.

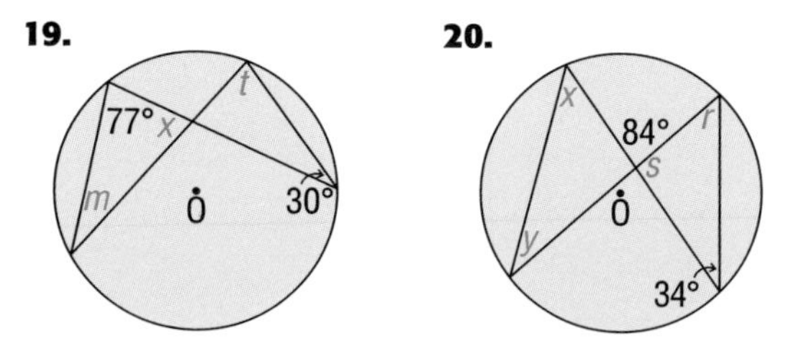

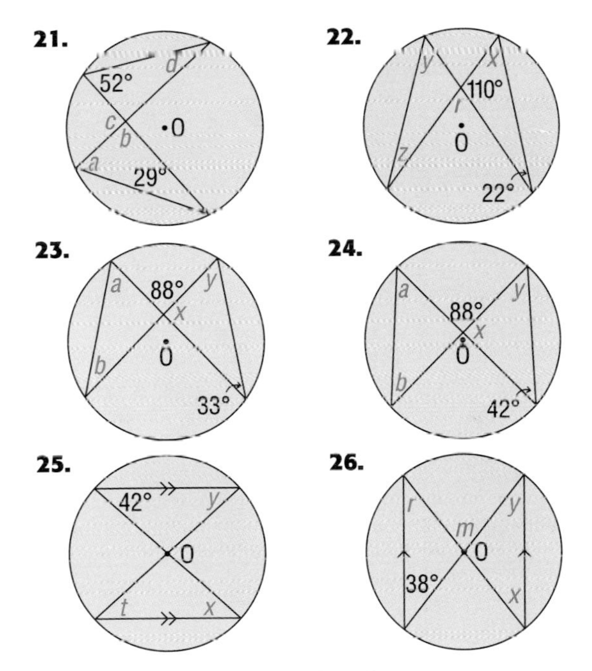

Problems and Applications

27. Copy this diagram. Then, draw and label an angle that is twice as large as ∠XYZ.

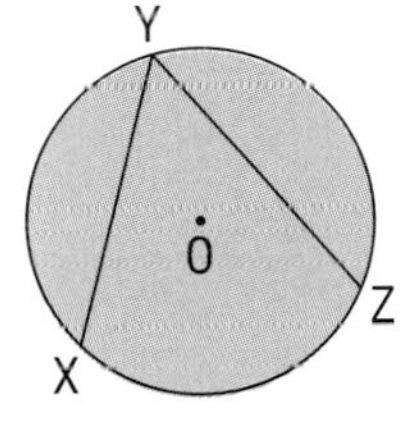

28. Describe how an angle the same size as ∠PQR can be drawn on this diagram.

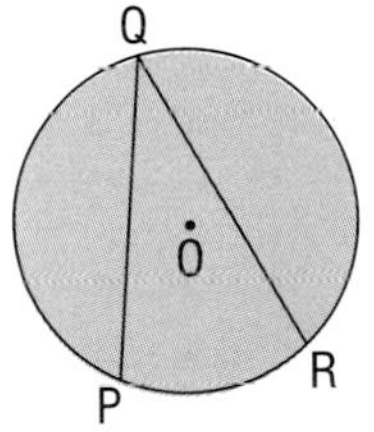

29. A cooking timer will run for up to 60 min. Through what central angle does the arm turn if the timer runs for 25 min? 40 min?

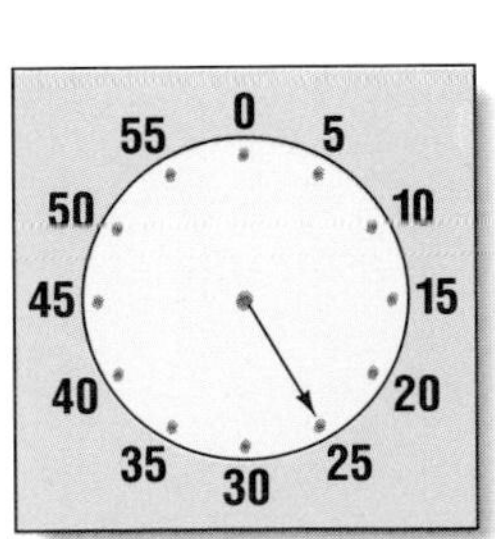

30. a) A round apple pie is cut into 8 equal slices through the centre. What is the measure of the central angle of each slice?
b) Another round pie is cut through the centre so that the central angle of each piece is 60°. How many pieces of this pie could you eat and leave half of the pie uneaten?

31. a) Find the lengths of chords AB and CD.

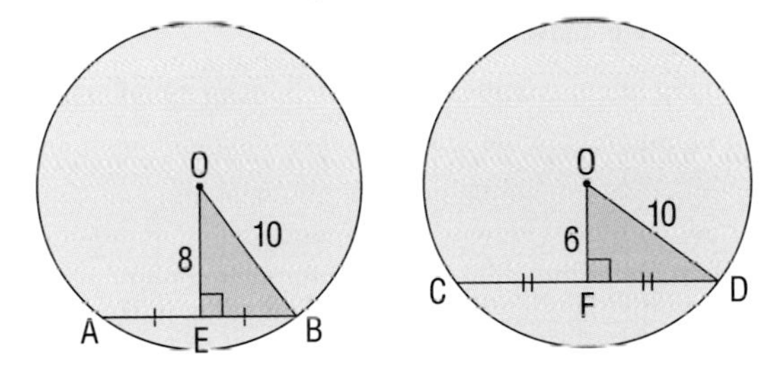

b) Write a statement that describes the relationship between the length of a chord and its distance from the centre of the circle.

32. If the circumference of a circle is 60 cm and the central angle ∠AOC is 60°, how long is arc AC?

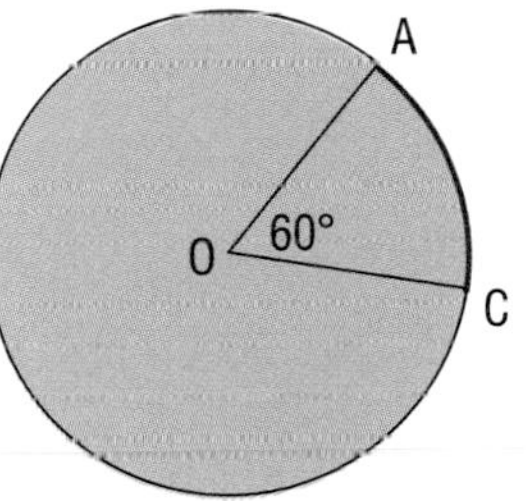

33. Write a problem that involves angles in a circle. Have a classmate solve your problem.

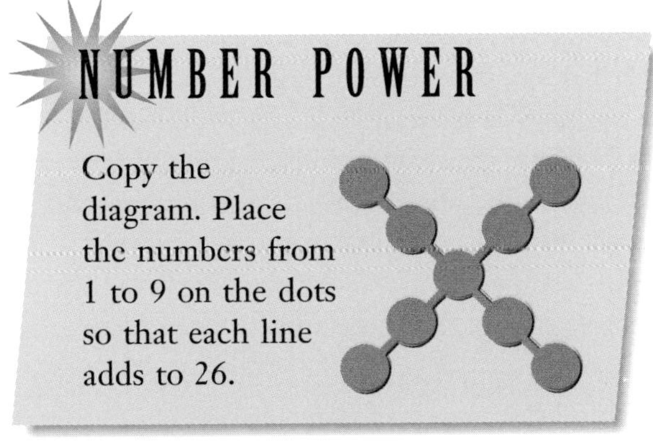

Copy the diagram. Place the numbers from 1 to 9 on the dots so that each line adds to 26.

Fractal Geometry

One of the newest branches of mathematics is **fractal geometry**. The word "fractal" comes from the Latin verb *frangere* which means to break. The study of fractals has led to the development of new computer graphics programs and may give scientists a way to describe complex things in nature.

Activity ❶ The H Fractal

The diagrams show the first 3 stages of an H Fractal with a reduction factor of $\frac{1}{2}$.

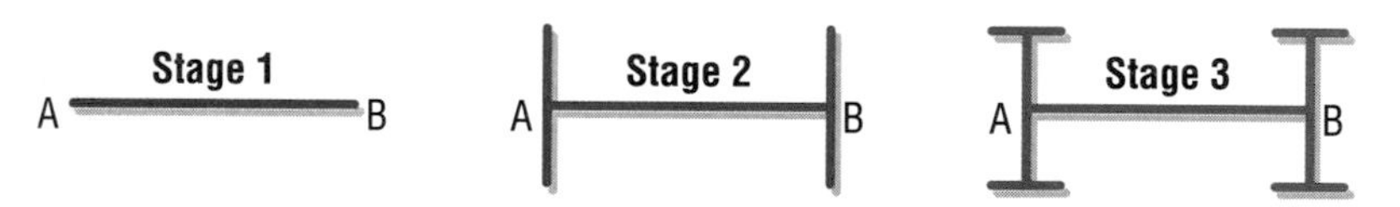

In stage 1, begin with a horizontal line segment AB of length 1 unit. In stage 2, add 2 vertical line segments, each of length $\frac{1}{2}$, perpendicular to AB at A and B. In stage 3, add 4 horizontal line segments, each of length $\frac{1}{4}$, perpendicular to the 2 vertical line segments. This H fractal has a reduction factor of $\frac{1}{2}$ because each new set of line segments is half as long as the previous set.

1. On grid paper, draw the first 4 stages of an H fractal with a reduction factor of $\frac{1}{2}$.

2. On grid paper, draw the first 4 stages of an H fractal with a reduction factor of $\frac{3}{4}$.

Activity ❷ The Fractal Tree

The diagrams show the first 3 stages of a fractal tree with a reduction factor of 0.7.

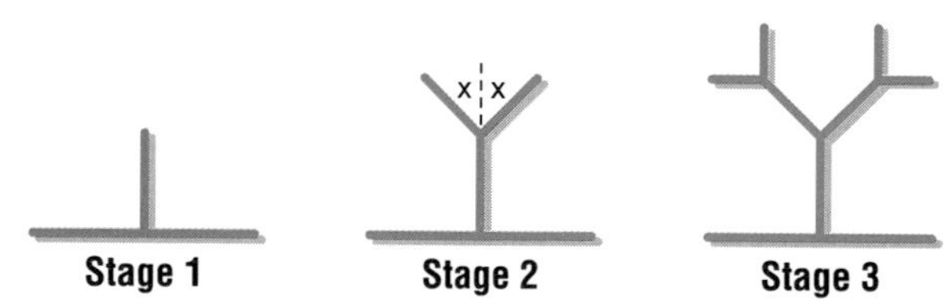

To produce the second stage from the first, 2 perpendicular branches are added to the end of the previous branch. Notice that if we continued the previous branch, it would bisect the right angle formed by the new branches. We can construct a third stage from the second in the same way.

Draw the first 4 stages of a fractal tree using a reduction factor of your choice.

Activity ❸ The Binary Tree

The diagrams show the first 3 stages of a binary tree with a reduction factor of $\frac{1}{2}$.

Begin with a vertical branch of length 1. Each horizontal branch is twice the length of the vertical branch above it. Each vertical branch is $\frac{1}{2}$ the length of the vertical branch above it.

Draw the first 6 stages of a binary tree using a reduction factor of your choice.

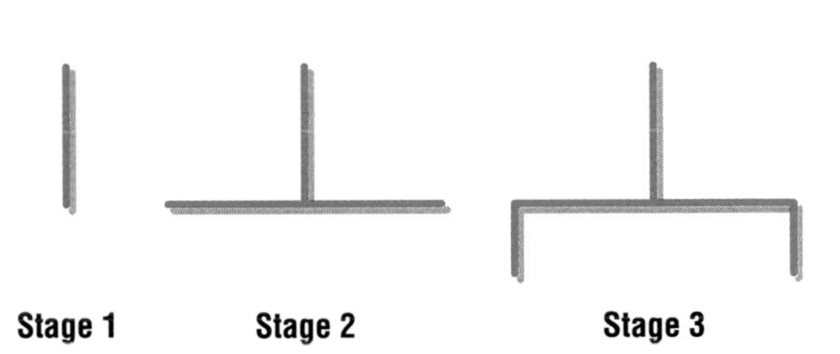

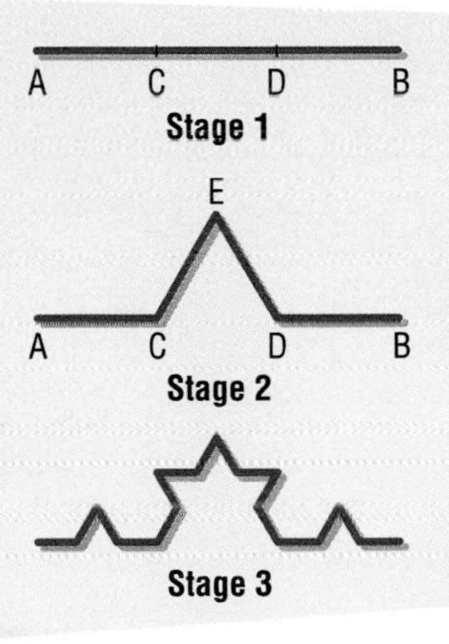

Activity ❹ The "Snowflake" Fractal

This is the most famous fractal. The first 3 stages are shown.

Begin with a line segment AB and divide it into 3 equal parts. In stage 2, an equilateral triangle CED is constructed on the middle line segment. The middle segment CD is then removed. The process can be repeated as many times as you wish.

Construct a fractal curve by drawing squares instead of triangles.

Activity ❺ Sierpinski's Sieve

The diagrams show the first 3 stages of Sierpinski's Sieve.

Begin with an equilateral triangle. Divide the triangle into 4 smaller congruent equilateral triangles, and remove the inner triangle. At each stage, a triangle is removed from the centre of each triangular region.

Draw the first 4 stages of Sierpinski's Sieve.

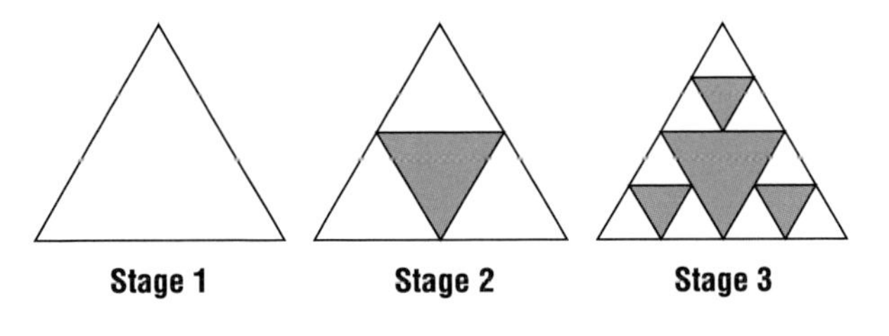

Activity ❻ Uses of Fractals

1. Use your research skills to find out how computer graphics programs based on fractal geometry are used in movies such as *Star Trek II*.

2. Can you find any other applications of fractals? If so, describe them.

12.9 Cyclic Quadrilaterals

Who says you cannot fit a square peg into a round hole? Of course you can if the square peg is small enough. The largest square peg that fits into a round hole just touches the side of the hole. A drawing of one view of the arrangement looks like the diagram on the right. In this drawing, the vertices of the square appear to lie on the circle. The square is an example of a **cyclic quadrilateral**, which is a quadrilateral inscribed in a circle. The vertices of a cyclic quadrilateral lie on the circumference of the circle.

Activity: Discover the Relationship

Draw a circle with radius 5 cm.
Draw cyclic quadrilateral ABCD.
Measure $\angle A$ and $\angle C$, two opposite angles.
Measure $\angle B$ and $\angle D$, two opposite angles.

Inquire

1. How are $\angle A$ and $\angle C$ related?

2. How are $\angle B$ and $\angle D$ related?

3. Repeat the activity using 2 other circles of different radii.

4. Write a statement describing the relationship between the opposite angles of a quadrilateral inscribed in a circle.

Example 1

Determine the measures of $\angle E$ and $\angle H$.

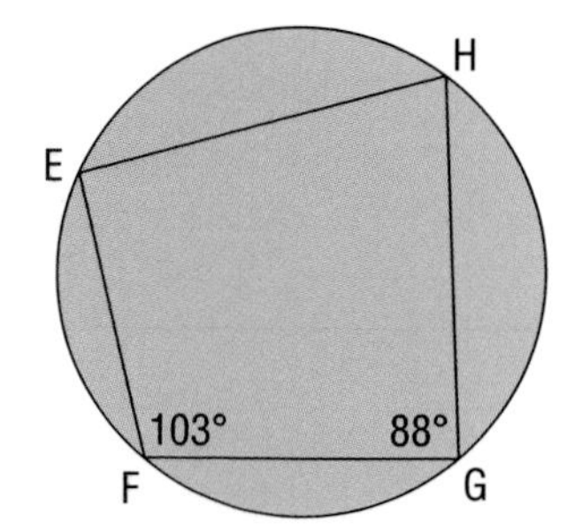

Solution

Since quadrilateral EFGH is inscribed in a circle, the opposite angles add to 180°.

$$\angle E + \angle G = 180^\circ \qquad \angle H + \angle F = 180^\circ$$
$$\angle E + 88^\circ = 180^\circ \qquad \angle H + 103^\circ = 180^\circ$$
$$\angle E = 92^\circ \qquad \angle H = 77^\circ$$

Example 2

Determine the measure of ∠DCB.

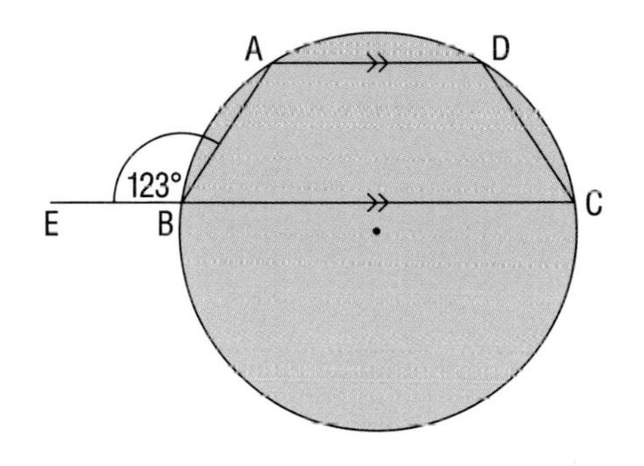

Solution

Since AD || EC,

∠ABE = ∠BAD (alternate angles)

∠BAD = 123°

Since quadrilateral ABCD is inscribed in a circle,

$$\angle BAD + \angle BCD = 180^\circ$$
$$123^\circ + \angle BCD = 180^\circ$$
$$\angle BCD = 57^\circ$$

Practice

Find the measure of each unknown angle.

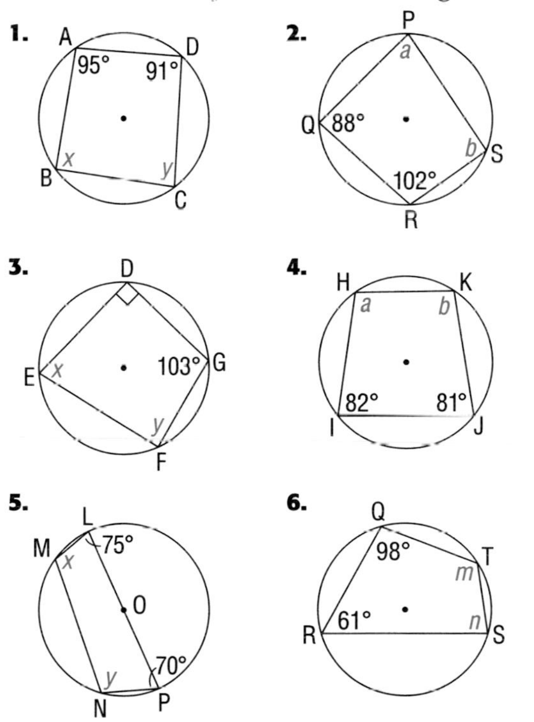

Find the measure of each unknown angle.

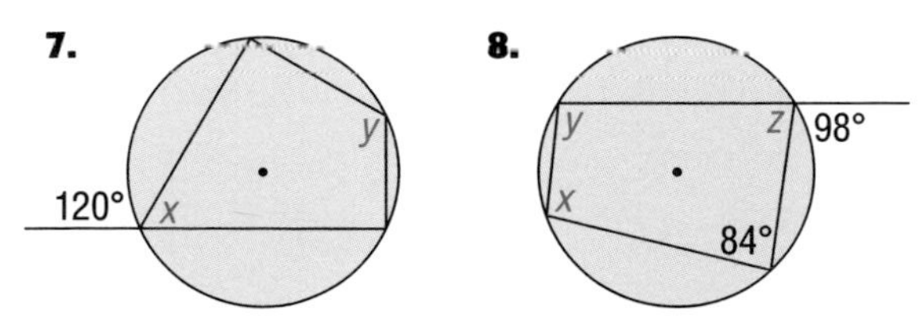

9.

104°
a
d
96°
b
c

10.

m
114°
111°
y
x

Find the measure of each unknown.

11.

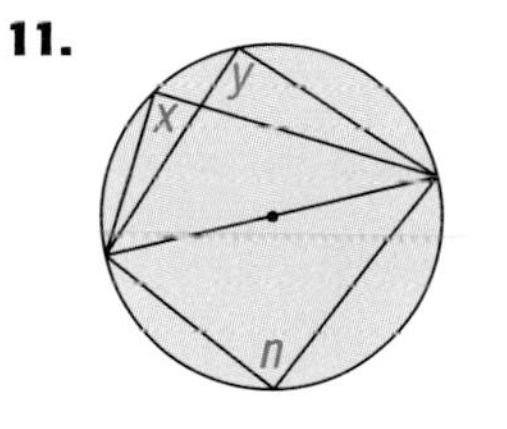

12.

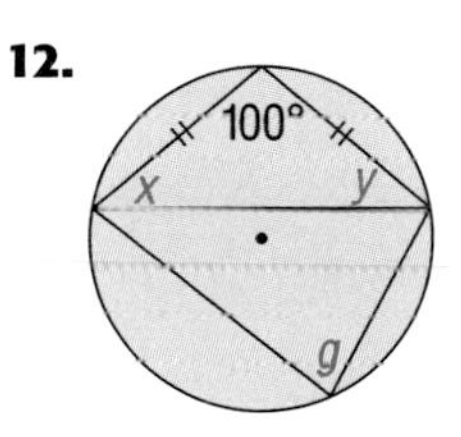

13.

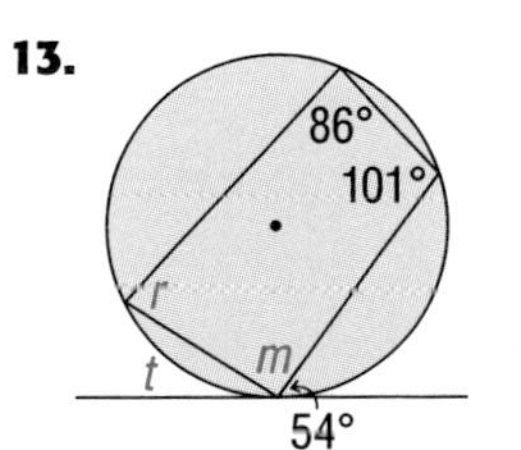

14.

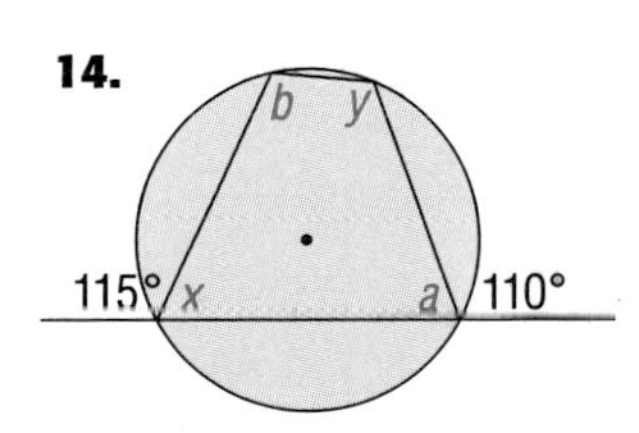

CONTINUED

Problems and Applications

Find the measure of each unknown angle.

15.

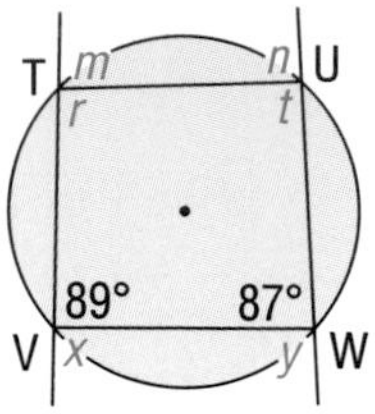

16.

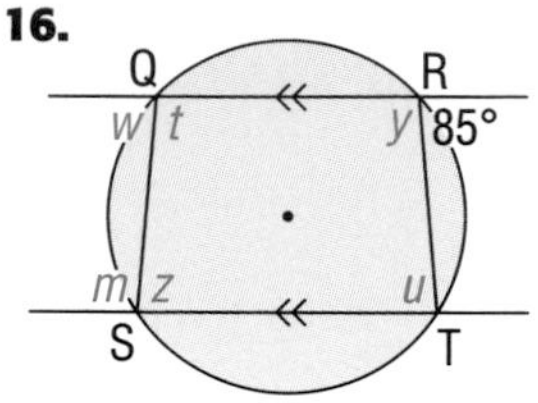

17.

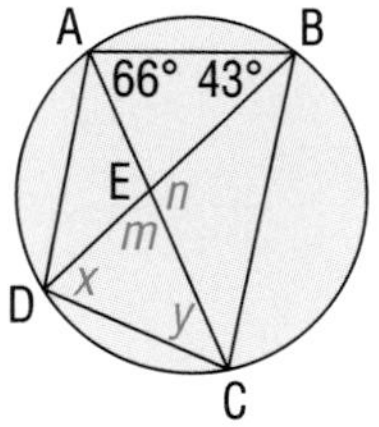

18.

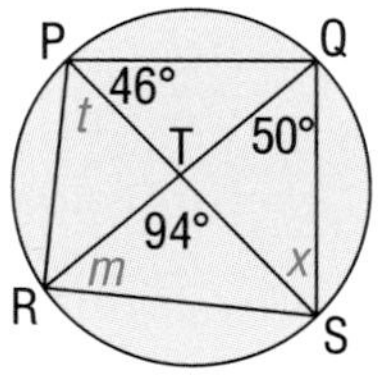

19.

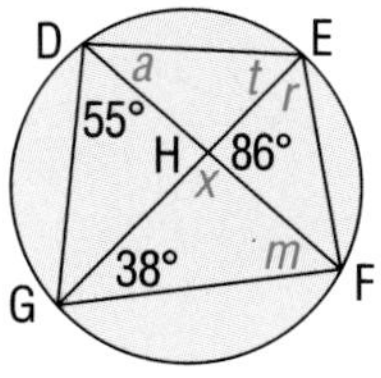

20.

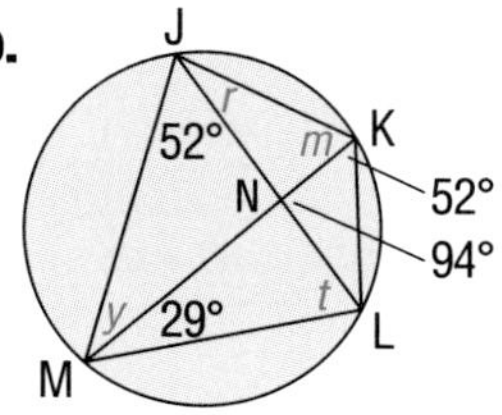

21. In the diagram, if x is known, how can y be found?

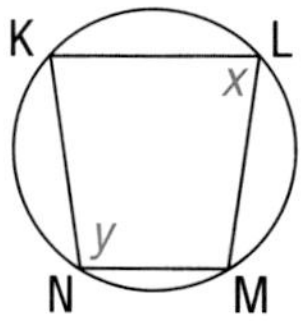

22. What are the measures of ∠B and ∠C in cyclic quadrilateral ABCD?

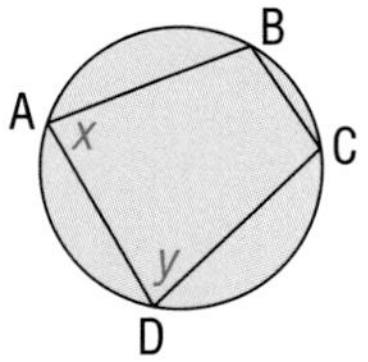

23. What other name can be given to the opposite angles in a cyclic quadrilateral? Why?

24. a) What is the sum of the interior angles in cyclic quadrilateral QRST?

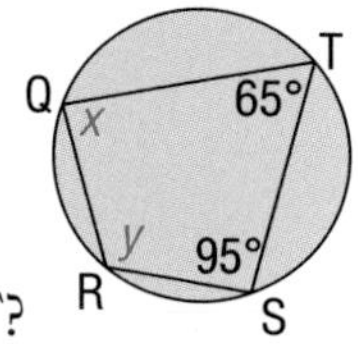

b) Is this angle sum true for all cyclic quadrilaterals? Why?

25. a) What is the measure of each angle in this cyclic quadrilateral?

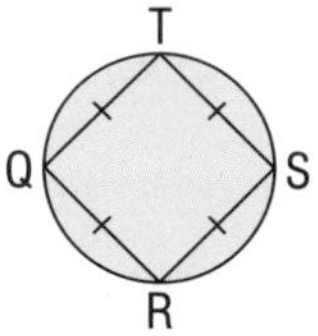

b) What is another name for the figure QRST?

26. How can the values of x and y be calculated in quadrilateral WXYZ?

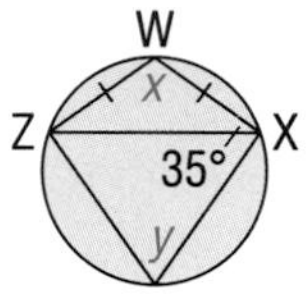

27. a) PQRS is a quadrilateral inscribed in a circle. Draw a similar diagram in your notebook.

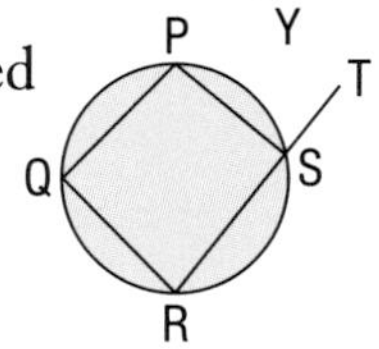

∠PST is an exterior angle of the quadrilateral. Measure ∠PST and ∠Q. How are the 2 angles related?

b) Repeat part a) for two other circles and inscribed quadrilaterals.

c) Write a statement about the size of an exterior angle of a quadrilateral inscribed in a circle.

28. a) Use the conclusion you made in question 27 to find the value of the unknowns in this cyclic quadrilateral.

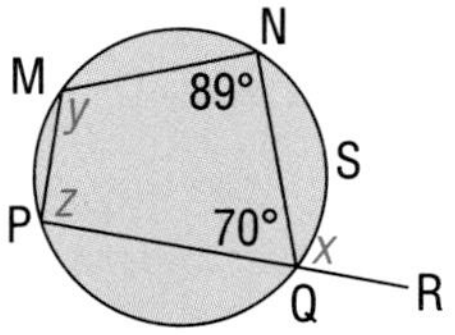

b) How could you find the unknowns without using your conclusion from question 27?

LOGIC POWER

How many cubes are in this figure? What assumptions have you made?

Geometric Constructions I

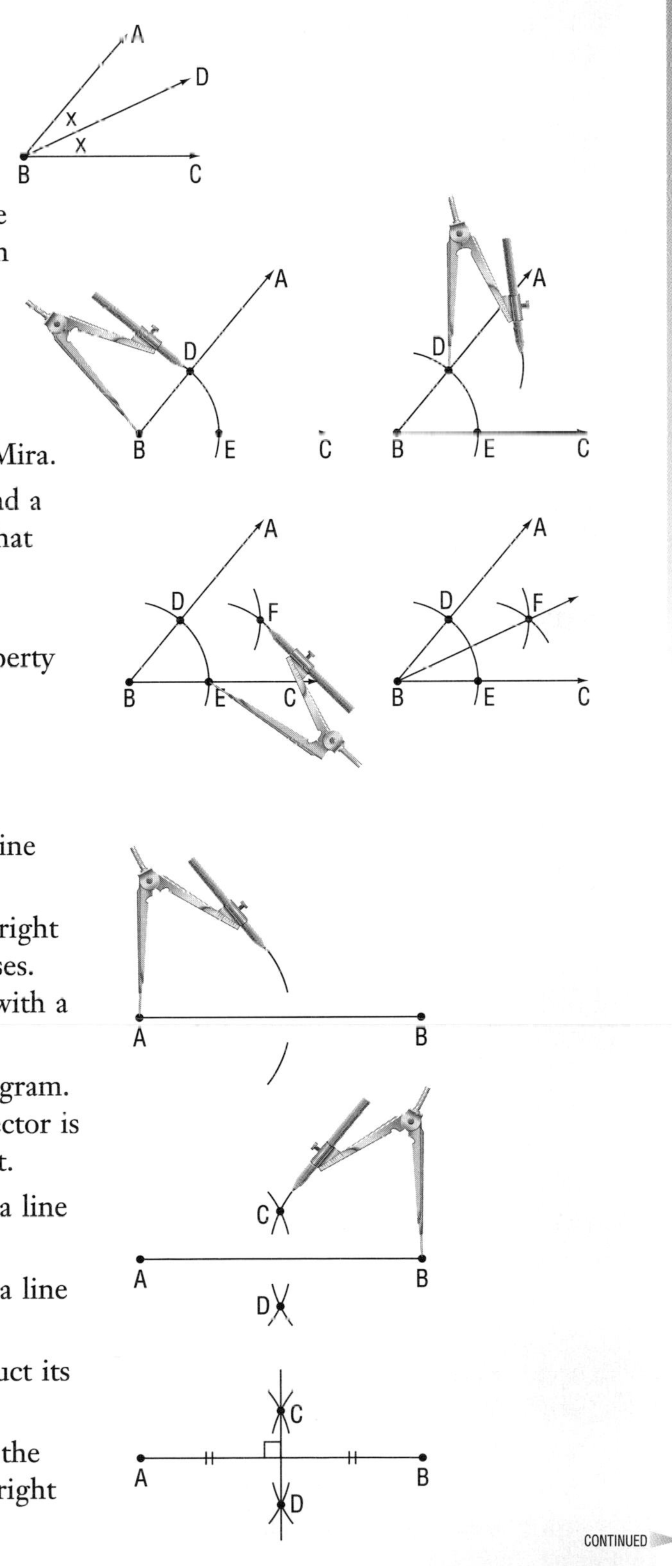

Activity ❶ Angle Bisector

The **bisector** of an angle divides the angle into 2 equal parts.

1. The following steps show how to bisect an angle using ruler and compasses. Repeat the construction in your notebook. Work with a partner to write a description of the steps.

2. Describe how to bisect an angle using paper folding.

3. Describe how an angle can be bisected using a Mira.

4. Draw a 60°-angle, a 120°-angle, a right angle, and a straight angle. Then, bisect each of the angles. What is the measure of each new angle?

5. Every point on the bisector of an angle is equidistant from the arms of the angle. Use a property of congruent triangles to explain why this is true.

Activity ❷ Right Bisector

The **right bisector** of a line segment divides the line segment into 2 equal parts at right angles.

1. The following steps show how to construct the right bisector of a line segment using ruler and compasses. Repeat the construction in your notebook. Work with a partner to write a description of the steps.

2. Join AC and BC or AD and BD in your last diagram. Explain why every point on the perpendicular bisector is equidistant from the endpoints of the line segment.

3. Describe how to construct the right bisector of a line segment by paper folding.

4. Describe how to construct the right bisector of a line segment using a Mira.

5. Draw a line segment at least 8 cm long. Construct its right bisector.

6. Draw a line segment at least 8 cm long. Divide the line segment into 4 equal parts by constructing 3 right bisectors.

CONTINUED

Activity ❸ Perpendicular Lines I

Perpendicular lines meet at 90°. There are several ways to construct a perpendicular to a line from a point not on the line.

1. The following steps show how to construct a perpendicular to a line from a point not on the line. Repeat the construction in your notebook. Work with a partner to write a description of the steps.

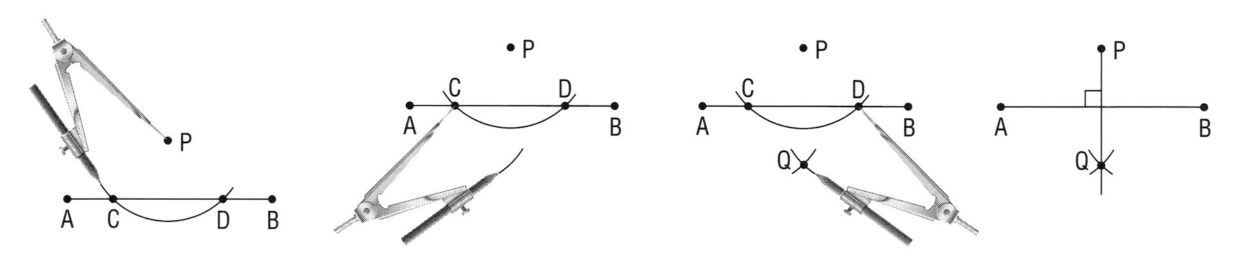

2. Describe how to draw a perpendicular to a line from a point not on the line using paper folding.

3. Describe how to draw a perpendicular to a line from a point not on the line using a Mira.

4. Copy these diagrams and construct perpendiculars from P to AB.

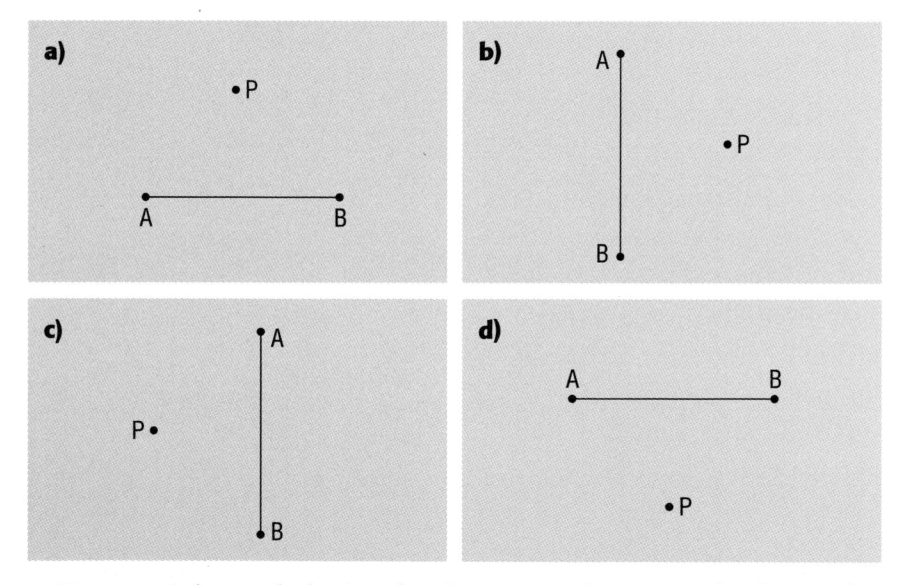

5. Draw a right-angled triangle. Construct the perpendicular to the hypotenuse from the opposite vertex.

6. Draw an acute-angled triangle. Construct a perpendicular from each vertex to the opposite side.

7. Draw an obtuse-angled triangle. Construct the perpendicular from each vertex to the opposite side. Extend the sides if necessary.

Activity 4 Perpendicular Lines II

Constructing a perpendicular to a line at a point on the line is the same construction as bisecting a straight angle.

1. The following steps show how to construct a perpendicular to a line at a point on the line. Repeat the construction in your notebook. Work with a partner to write a description of the steps.

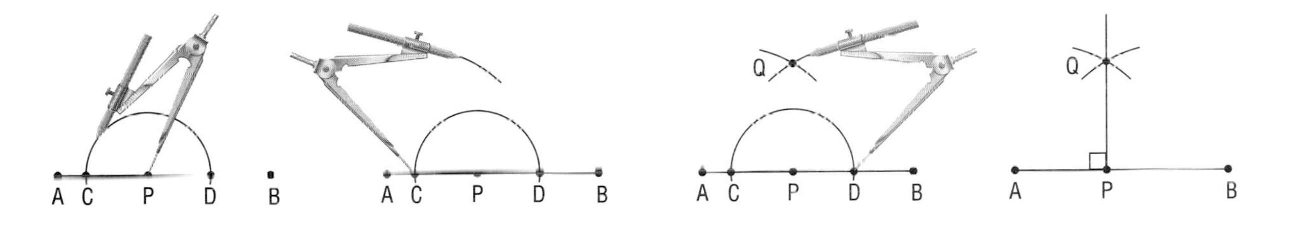

2. Describe how to draw a perpendicular to a line at a point on the line using paper folding.

3. Describe how a perpendicular to a line at a point on the line can be drawn with a Mira.

4. Copy the diagrams and construct perpendiculars to AB at P.

a) **b)** **c)**

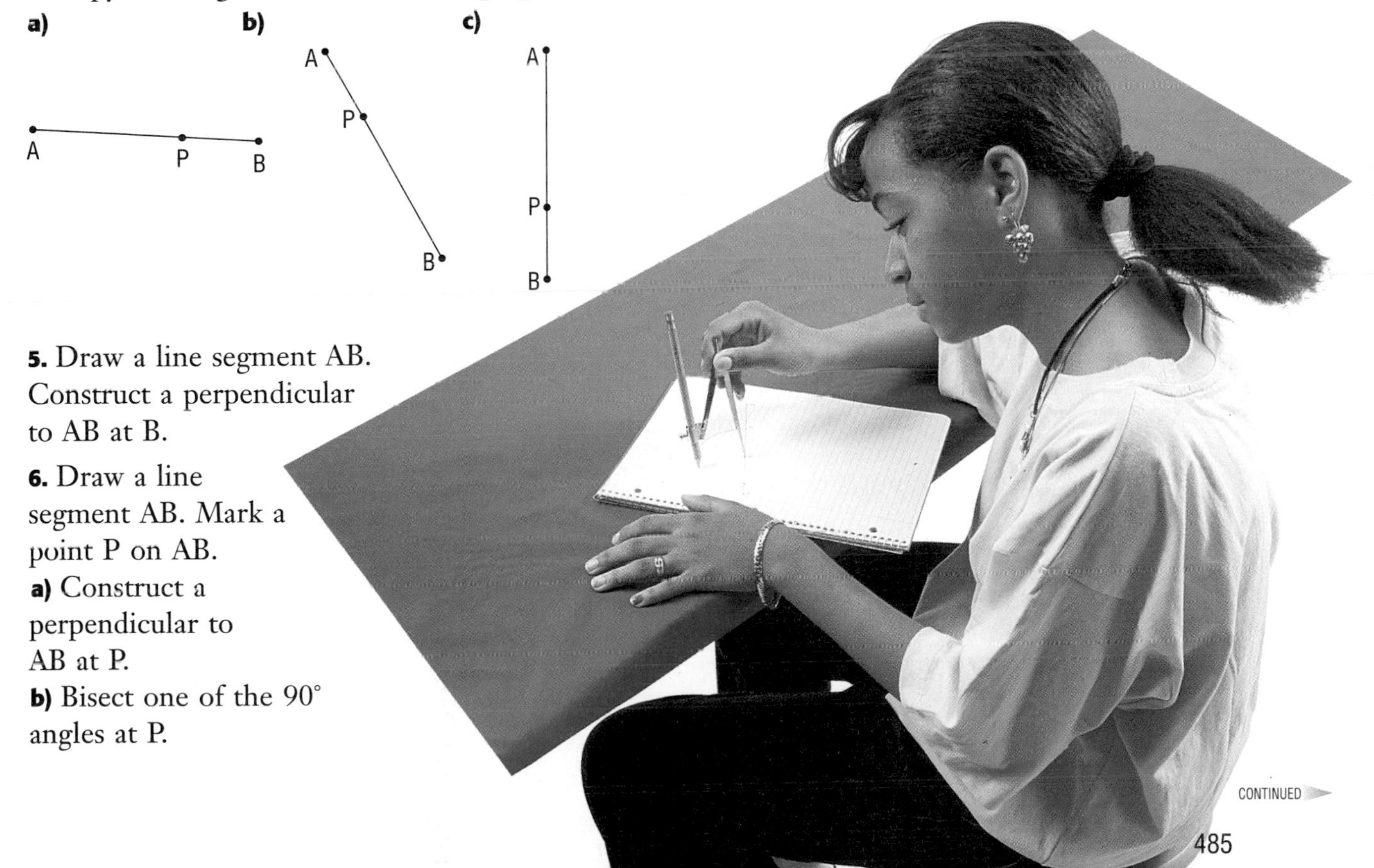

5. Draw a line segment AB. Construct a perpendicular to AB at B.

6. Draw a line segment AB. Mark a point P on AB.

a) Construct a perpendicular to AB at P.

b) Bisect one of the 90° angles at P.

CONTINUED

Activity 5 Parallel Lines

Here are 3 ways to construct a line parallel to a given line through a point not on the line.

1. The following steps show how to construct a line parallel to a given line through a point not on the line using ruler and compasses. Repeat the construction in your notebook. Work with a partner to write a description of the steps.

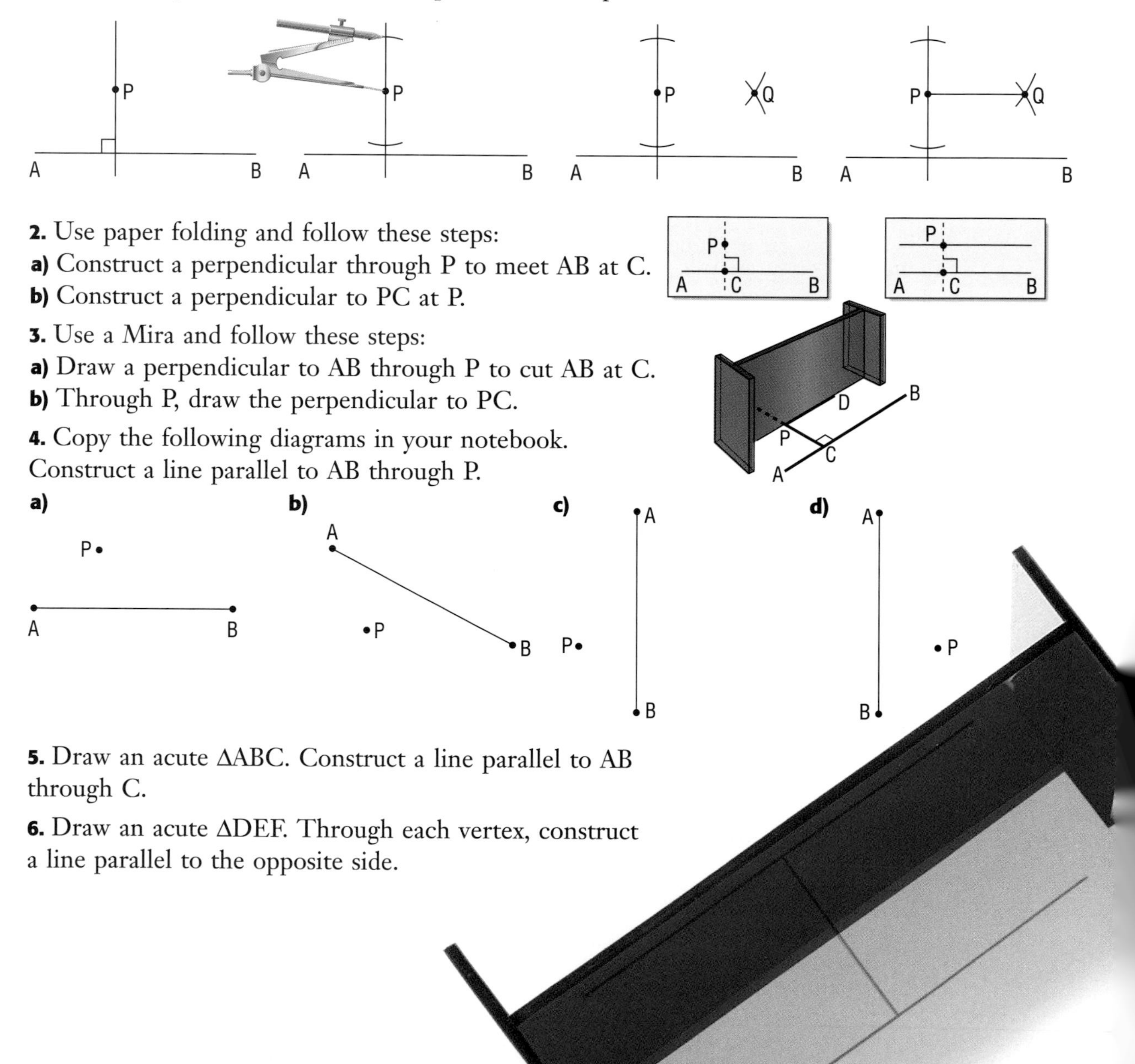

2. Use paper folding and follow these steps:

a) Construct a perpendicular through P to meet AB at C.

b) Construct a perpendicular to PC at P.

3. Use a Mira and follow these steps:

a) Draw a perpendicular to AB through P to cut AB at C.

b) Through P, draw the perpendicular to PC.

4. Copy the following diagrams in your notebook. Construct a line parallel to AB through P.

a) **b)** **c)** **d)**

5. Draw an acute ΔABC. Construct a line parallel to AB through C.

6. Draw an acute ΔDEF. Through each vertex, construct a line parallel to the opposite side.

Geometric Constructions II

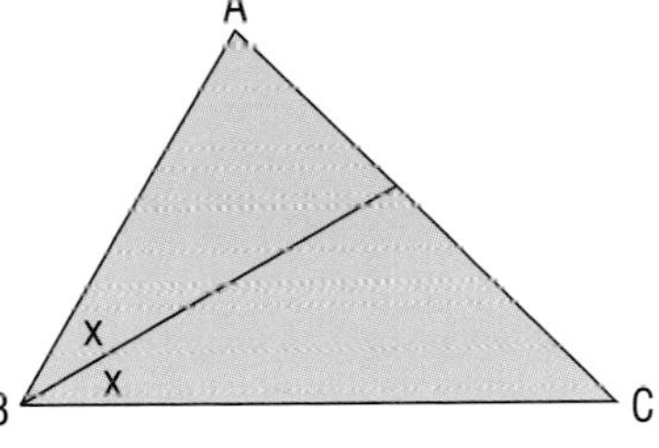

Activity ❶ The Incircle of a Triangle

1. Draw an acute ΔABC.

2. Construct the bisector of each angle. What do you notice?

3. Let D be the point of intersection of the angle bisectors. Use D as a centre and draw a circle that touches all 3 sides of the triangle. This circle is the **incircle** of the triangle. D is the **incentre**.

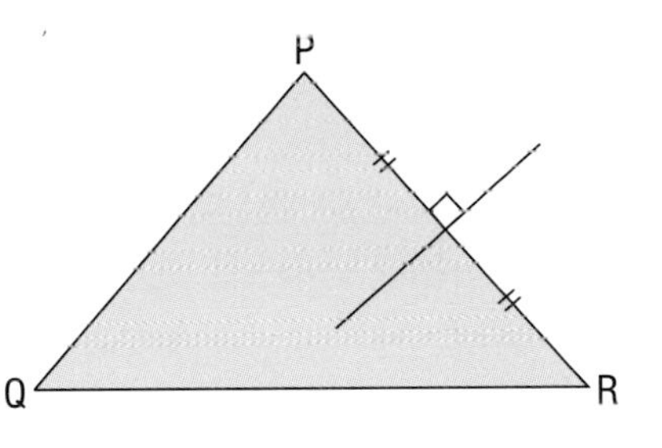

Activity ❷ The Circumcircle of a Triangle

1. Draw an acute ΔPQR.

2. Construct the right bisector of each side of the triangle. What do you notice?

3. Let D be the point of intersection of the right bisectors. Use D as a centre and draw a circle that passes through each vertex of the triangle. This circle is the **circumcircle** of the triangle.

4. Draw an obtuse triangle and repeat steps 2 and 3 above to draw the circumcircle of this triangle.

5. Describe how you can draw a circle through any three points that do not lie on the same line.

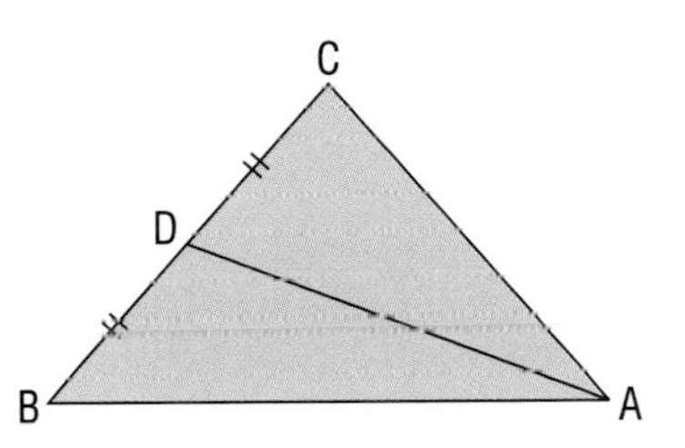

Activity ❸ The Medians of a Triangle

1. Draw an acute ΔABC.

2. Construct the right bisector of BC and use D to label the midpoint of BC.

3. Join AD. AD is called a **median** of ΔABC.

4. Construct the medians from B to AC and from C to AB.

5. What do you notice about the three medians of the triangle?

6. Repeat the above steps for an obtuse triangle.

7. The point at which the 3 medians of a triangle meet is called the **centre of gravity**. Research the meaning of this term.

Activity ❹ Polygon Construction

1. Use a ruler and a protractor or compasses, or use a computer to construct a square with each side 4 cm long.

2. Use a method of your choice to draw a regular pentagon, a regular hexagon, and a regular octagon, each having sides 3 cm long.

3. Compare your methods of construction with your classmates'.

Sight

Even though a horse's eyes are twice as large as a human's, the horse cannot see an area directly in front of it. This area is called the **blind spot**. When horses, such as Canada's Big Ben, compete in jumping competitions they actually jump the barriers blind.

Peripheral vision in humans and animals is the range or angle through which vision is possible. The angle of peripheral vision for a horse is 170° on one side and 170° on the other side. A horse cannot see an area within an angle of 10° directly in front of it or 10° directly behind it.

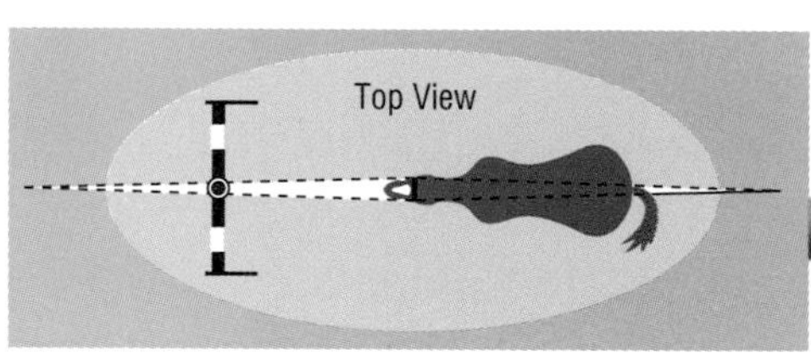

Activity ❶ The Blind Spot

1. To locate your blind spot, mark 2 black dots 5 cm apart on a blank piece of paper as shown.

•   •

2. Hold the paper in front of you with your arm extended.

3. Cover your left eye with your hand and look at the left dot with your right eye.

4. Continue looking at the left dot and move the paper toward you slowly until the right dot disappears. The point at which this happens is called the blind spot for your right eye.

5. Have a classmate measure the distance from your eye to the paper.

6. Repeat the experiment to find the blind spot for your left eye by covering your right eye and looking at the right dot.

7. Are the blind spot distances the same for both eyes?

Activity ❷ Peripheral Vision

1. To determine your angle of peripheral vision, you will need a blank index card and an index card with a black dot drawn on it that has a diameter of 1 cm.

2. Use chalk to draw a circle with a radius of 1 m on a large piece of paper placed on the floor.

3. Stand on the centre of the circle and have 1 classmate stand in front of you at the edge of the circle holding the card with the black dot on it. Have another classmate hold the blank card beside the card with the black dot on it. The blank card should be to your left of the card with the dot on it.

4. Stare at the black dot while the classmate with the blank card moves slowly to your left, along the edge of the circle. Say, "Stop" when you can no longer see the blank card.

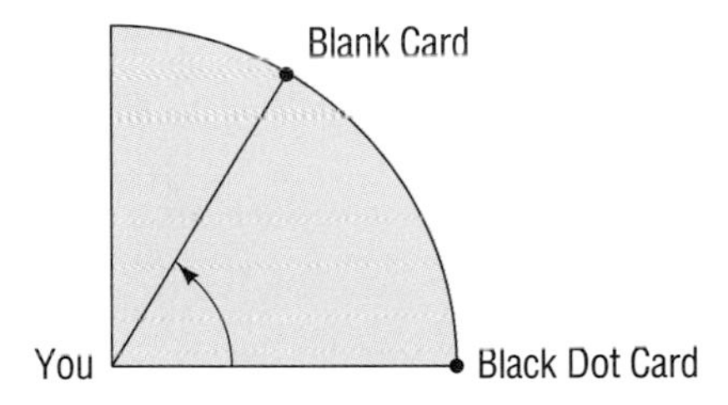

5. Have another classmate draw a line from you to the person holding the card with the dot and a line from you to the person holding the blank card.

6. Measure and record the angle between the 2 lines drawn in step 5.

7. Repeat the procedure with the blank card moving to your right.

8. Add the 2 angles you have found and record them. The sum of these 2 angles is your angle of peripheral vision.

9. Find the average angle of peripheral vision for your class.

Review

Name the following in the diagram shown below.

1. five points
2. five lines
3. five rays
4. eight angles
5. five segments

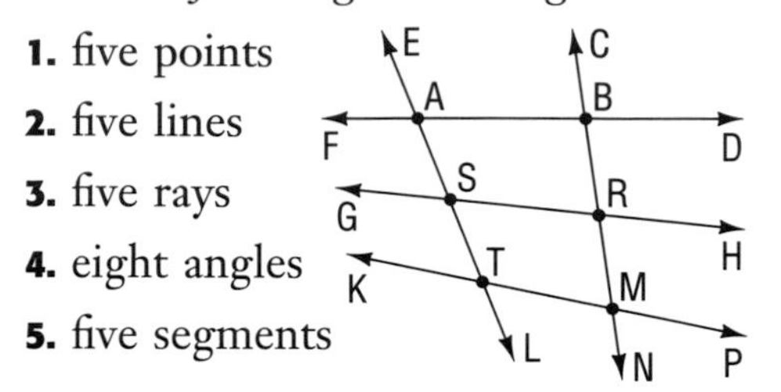

Use a protractor to draw these angles. Then, classify them.

6. $\angle HAT = 45°$ **7.** $\angle DOG = 90°$

8. $\angle ANT = 17°$ **9.** $\angle FUN = 180°$

10. $\angle CUP = 165°$ **11.** $\angle CAT = 235°$

Find the measure of each unknown angle.

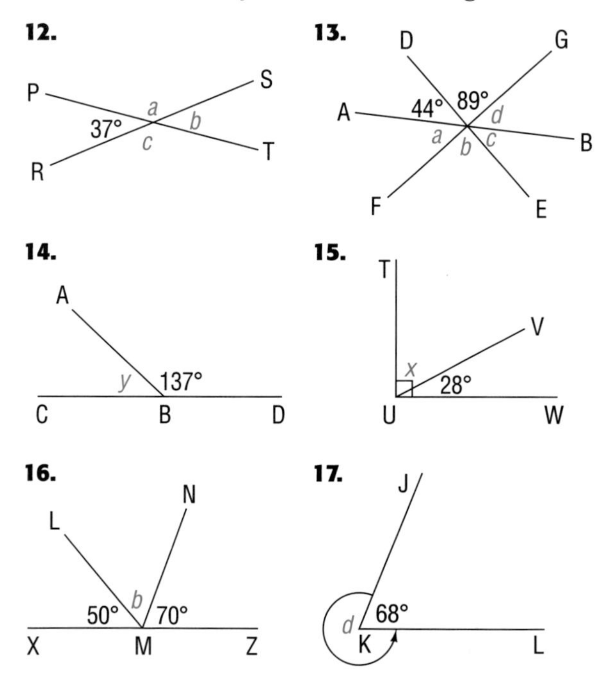

Find the measure of each unknown angle.

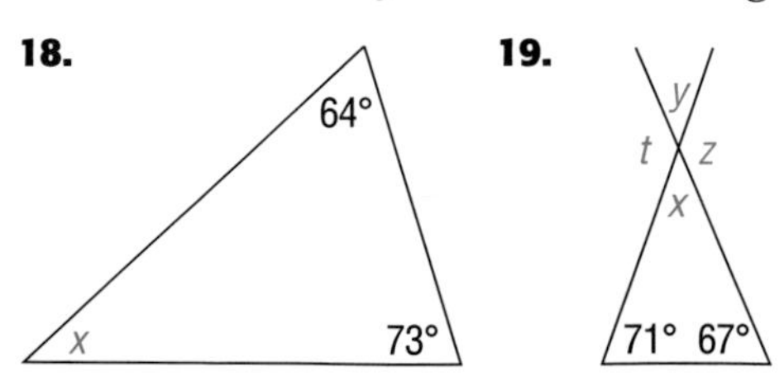

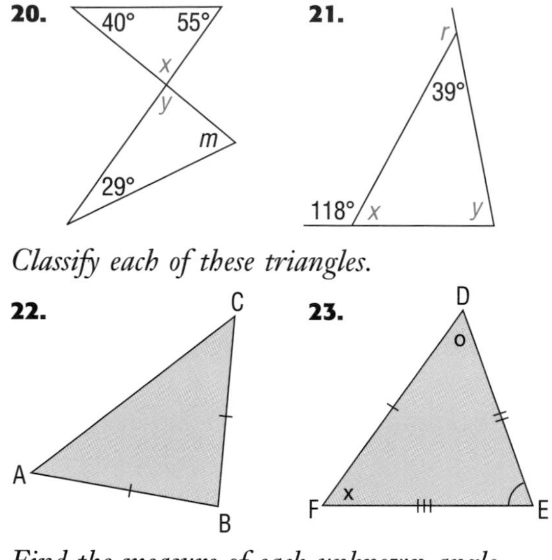

Classify each of these triangles.

Find the measure of each unknown angle.

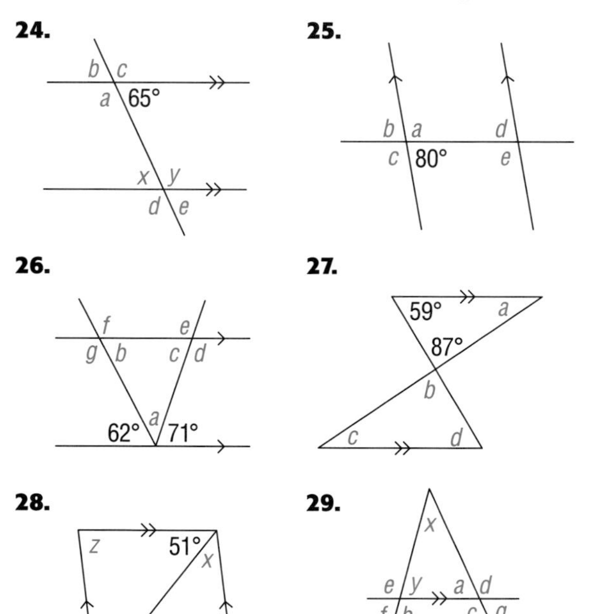

30. Write the formula to calculate the sum of the interior angles of a polygon with n sides.

Calculate the sum of the interior angles of polygons with these numbers of sides.

31. five **32.** seven **33.** twelve

34. fifteen **35.** twenty **36.** thirty

Which pairs of triangles are congruent? Explain.

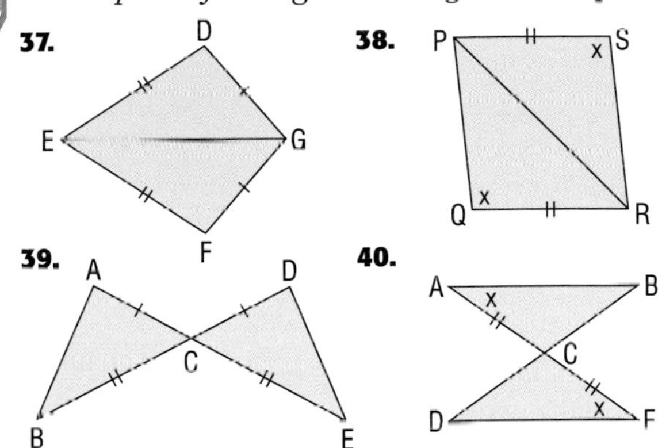

Calculate the length of the unknown side.

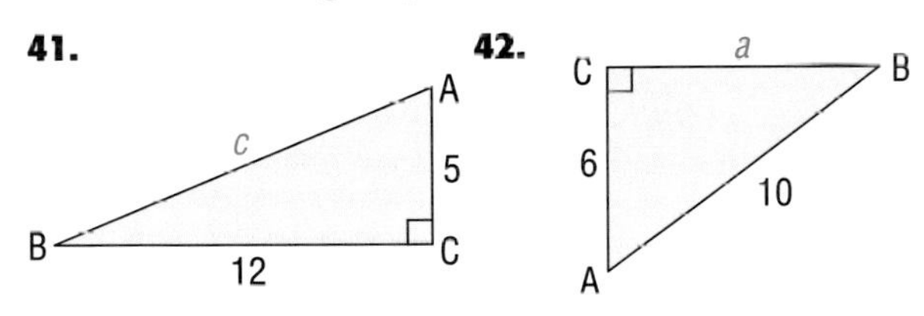

Calculate x, to the nearest tenth of a metre.

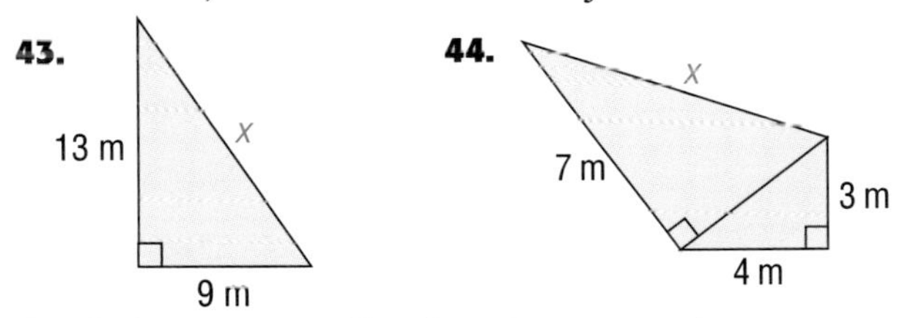

Find the measure of each unknown angle.

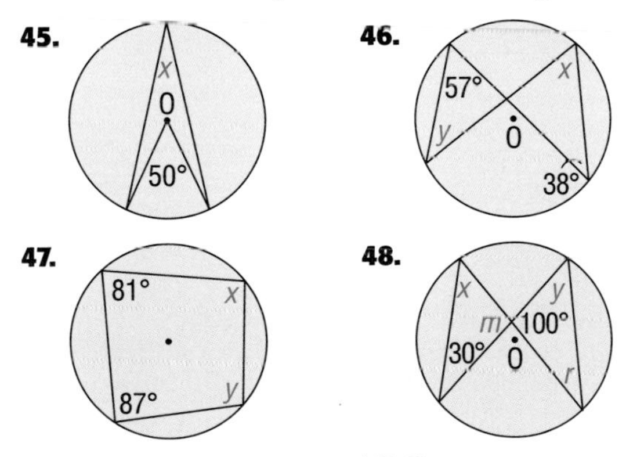

49. a) Draw an acute ΔABC.

b) Construct the bisector of $\angle ABC$.

c) Construct the perpendicular from A to BC.

d) Construct the right bisector of AB.

Group Decision Making

Researching Construction Careers

1. Meet as a class to choose 6 careers in the construction industry. Your choices might include a plumber, an electrician, a crane operator, an architect, or a building contractor.

2. Go to home groups. Assign one career to each home group member.

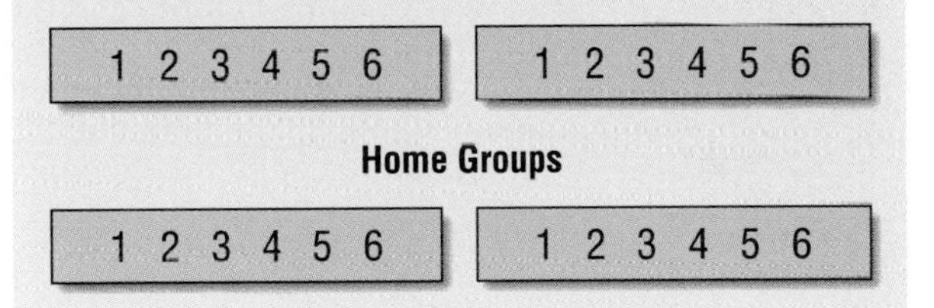

3. Form an expert group of 4 with students who have the same career as you to research. Decide on the questions you want to answer about your career. One of the questions should be: "How is math used in this career?" Research your assigned career in your expert group.

4. Return to your home group and share your findings about the career you researched.

5. In your home group, prepare a report on the 6 careers.

6. Meet as a class to discuss your group work. Decide what went well and what you would do differently next time.

Chapter Check

Find the value of each unknown angle.

1. **2.**

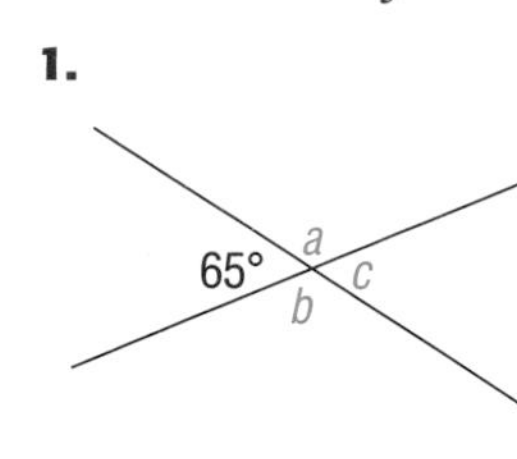

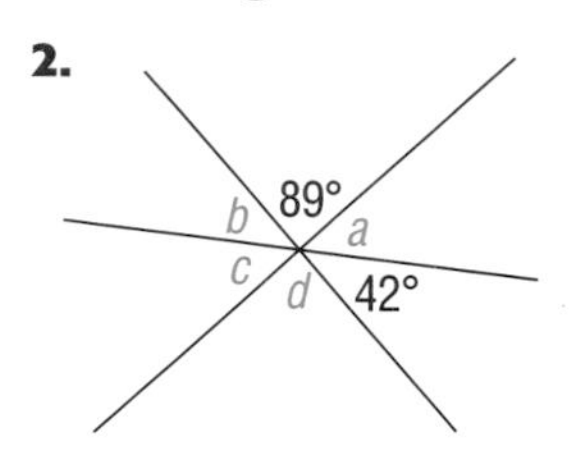

3. **4.**

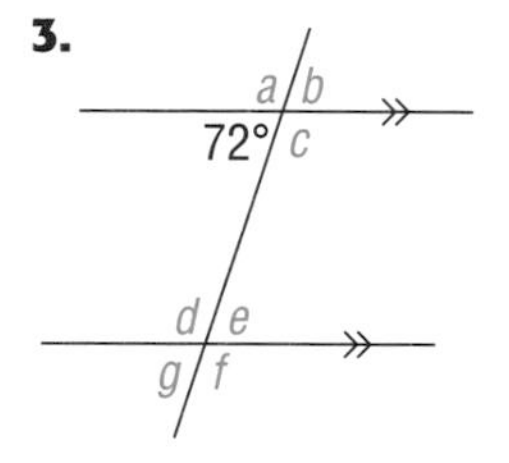

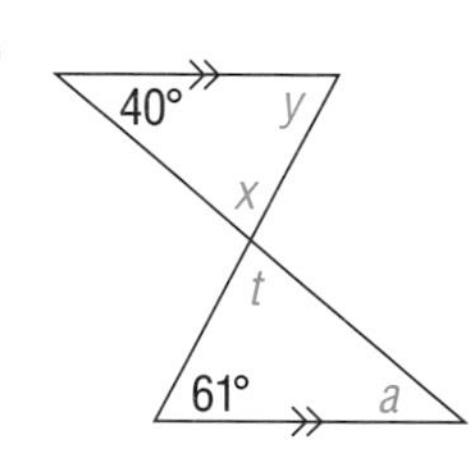

Name these polygons.

5. **6.** **7.** **8.**

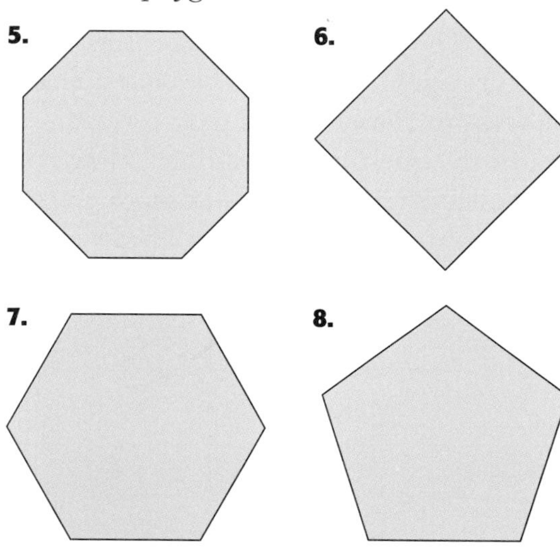

Which of these triangles are congruent? Explain.

9. **10.**

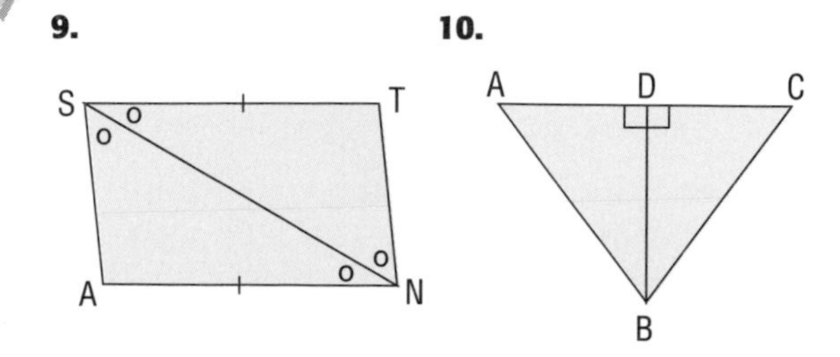

11. **12.**

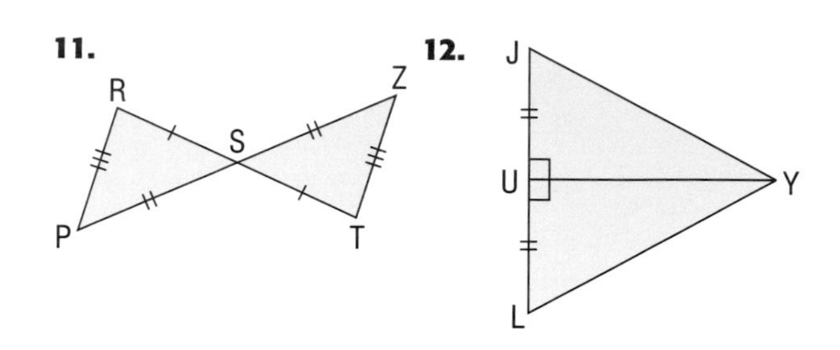

Find the value of each unknown angle.

13. **14.** **15.** **16.**

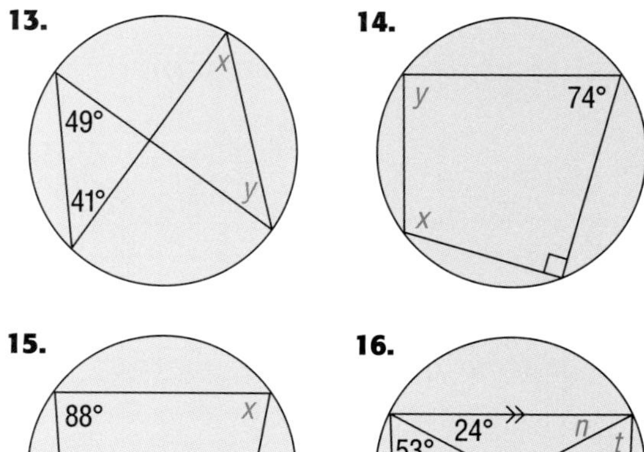

Calculate the length of the unknown side, to the nearest tenth of a unit.

17. **18.**

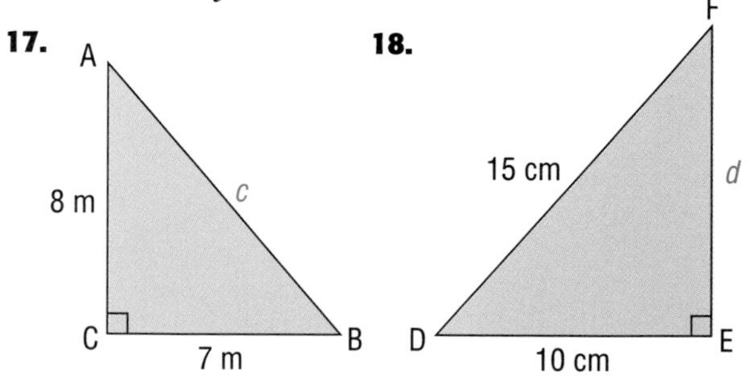

Using the Strategies

1. What is the sum of these numbers?
a) first 3 odd numbers
b) first 4 odd numbers
c) first 5 odd numbers
d) first 6 odd numbers
e) first 25 odd numbers
f) first 105 odd numbers

2. If a year has 2 months in a row with Friday the 13th, what months are they?

3. The area of the large square is 64 cm^2.

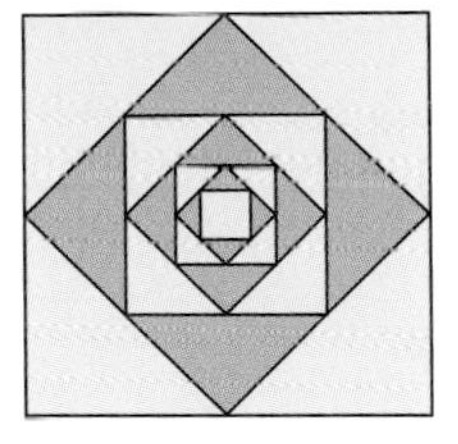

Each smaller square is formed by joining the midpoints of the sides of the next larger square. What is the area of the smallest square?

4. How many pieces of string each 8.5 cm long can be cut from a spool of string 400 cm long? What assumptions have you made?

5. The number 2601 is a 4-digit number that is a perfect square because $51^2 = 2601$. What is the smallest 4-digit number that is a perfect square and that has all even digits?

6. The average of two numbers is 21. When a third number is included, the average of the three numbers is 23. What is the third number?

7. Justine, Chris, and Meelang plan to travel together in a van. They will all sit in the front seat. In how many different arrangements can they sit in each of the following situations?
a) all three can drive
b) only Chris and Meelang can drive
c) only Justine can drive

8. List the different ways you can make change for a dollar using only quarters and nickels.

9. Starting at the letter A, how many different pathways can you follow to spell ANGLE?

10. A square piece of paper is folded in half as shown. The perimeter of each new rectangle formed is 24 cm. What is the perimeter of the original square?

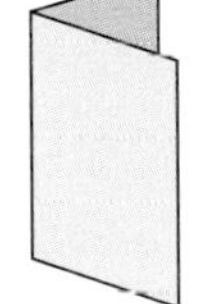

11. The ship will leave the harbour at 08:30. You have to be on board 20 min before departure. It takes 25 min to drive to the ship from your hotel. You must allow 15 min to check out of the hotel. It will take you 20 min to pack. You need half an hour to eat breakfast and at least 45 min to shower and dress. For what time should you place your wake-up call?

1. How many times will Mercury orbit the sun in the time it takes Jupiter to orbit the sun once?

2. Use the DATA BANK to make up your own problem. Have a classmate solve your problem. Check your classmate's solution.

Chapter 9

Write 2 equivalent ratios for each of the following.

1. 12 to 10

2. 3:4

Solve using a proportion.

3. Find the cost of 5 oranges if 12 sell for $2.28.

4. Find the distance travelled in 5 h if the speed is 80 km/h.

5. Find the actual length of a room if the length of a diagram measures 3 cm and the scale is 1 cm represents 5 m.

6. Which rate is better?
a) 60 m in 10 s or 120 m in 24 s
b) $7.08 for twelve 175-mL containers of yogurt or three 175-mL containers of yogurt for $1.95

7. Use the following grid to complete the table.

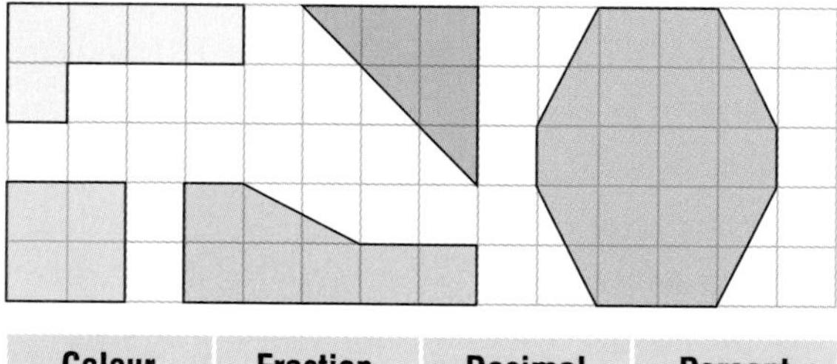

Colour	Fraction	Decimal	Percent
Green			
Yellow			
Purple			
Orange			
Pink			
White			

8. Find the selling price.
a) $125.90 jacket at 40% off
b) $380 television with a 30% markup

9. A video cassette costs $6.95 plus PST and GST. Find the total cost of the video cassette to the nearest cent in your province.

Chapter 10

1. The table gives the approximate distribution of blood in the body during relaxation.

Brain	13%
Liver	27%
Kidneys	17%
Muscles	14%
Rest of Body	29%

Display this information on a circle graph.

2. The table gives the passenger capacities of several aircraft.

Aircraft	Capacity
B-727	136
L-1011	212
B-767	179
B-747	256
DC-9	100

Display this information on a bar graph.

3. The table gives the average ocean temperatures at different latitudes.

Latitude	Temp	Latitude	Temp
50°N	7°C	50°S	7°C
40°N	16°C	40°S	9°C
30°N	22°C	30°S	16°C
20°N	25°C	20°S	19°C
10°N	26°C	10°S	23°C
0°	24°C		

Display this information on a broken-line graph.

4. The following are the written test scores for 11 playground supervisor candidates.
81, 79, 88, 94, 80, 79, 86, 82, 91, 79, 81
Find the mean, median, and mode.

5. A bag of marbles contains 5 red marbles, 4 blue marbles, and 2 white marbles. If you pick 1 marble from the bag, find the probability that it is

a) red **b)** white **c)** blue
d) green **e)** red or blue

Chapter 11

Calculate the perimeter and area of each figure.

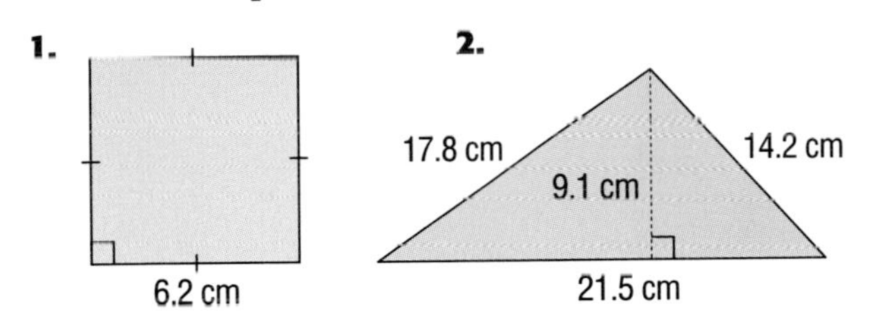

Calculate the surface area and volume of each figure.

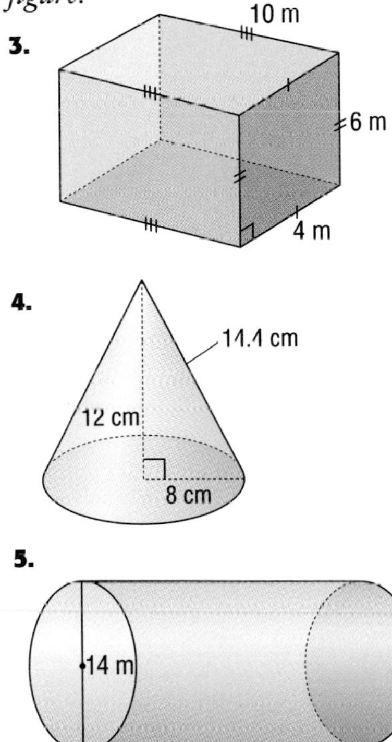

6. Calculate the volume of the sphere.

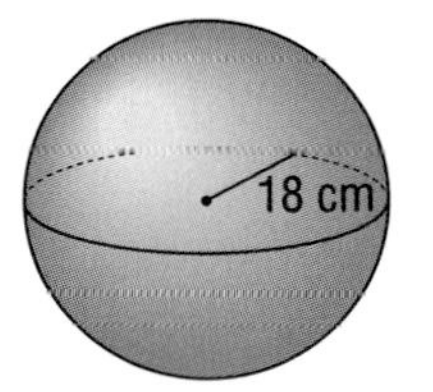

7. How many boxes 1 m by 1 m by 0.5 m will fit into a storage container 3 m by 4 m by 10 m?

Chapter 12

Find the measure of each unknown angle.

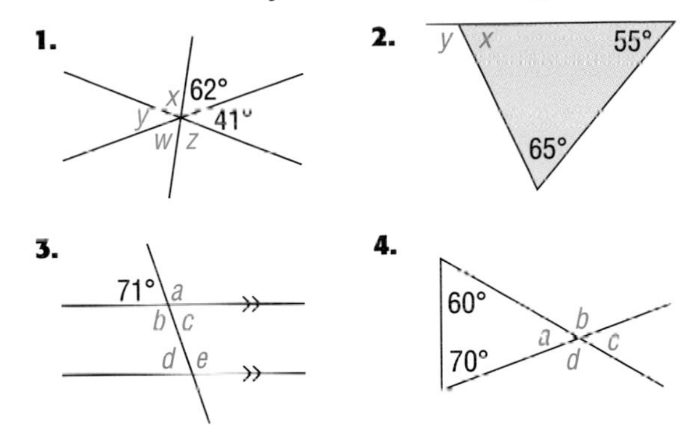

Which pairs of triangles are congruent? Explain.

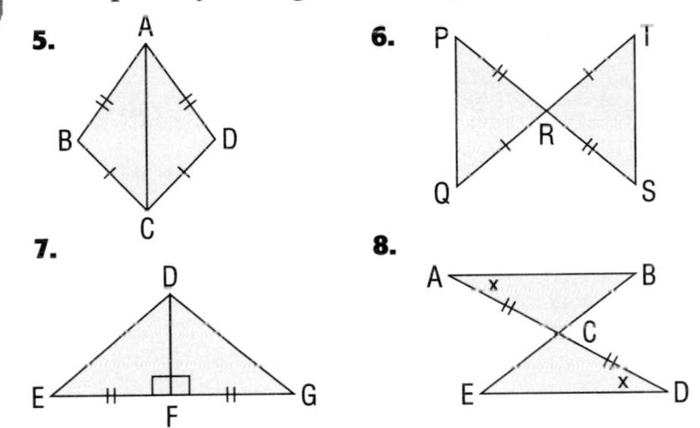

Calculate x to the nearest tenth of a metre.

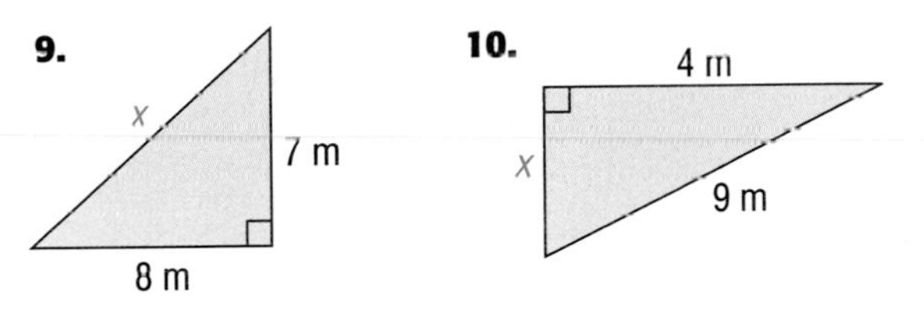

Find the measure of each unknown angle.

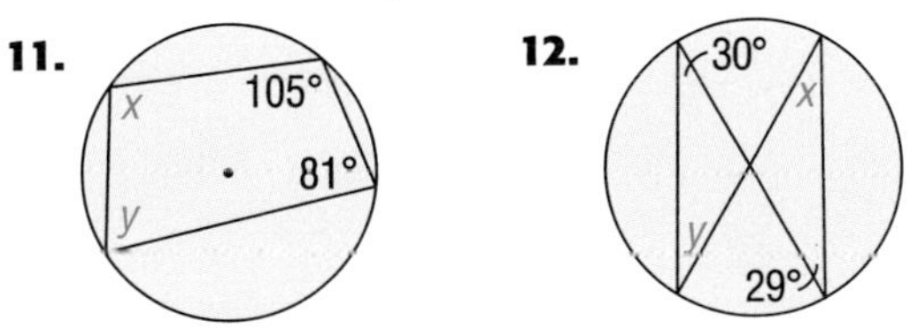

13. Draw acute ΔPQR.

a) Construct the right bisector of QR.

b) Bisect $\angle P$.

c) Construct the perpendicular from R to PQ.

T. NITTO

CHAPTER 13

Coordinate Geometry

A location's antipodal point is the place on Earth most distant from it. The South Pole is the antipodal point to the North Pole.

On a globe, the antipodal point is found by imagining a line drawn from the location, through the centre of the globe, and out the other side.

What is the antipodal point to where you live?

Activity: A Grid Game

1. Draw 2 grids and mark each of them from 0 to 8 along the horizontal and vertical axes as shown.

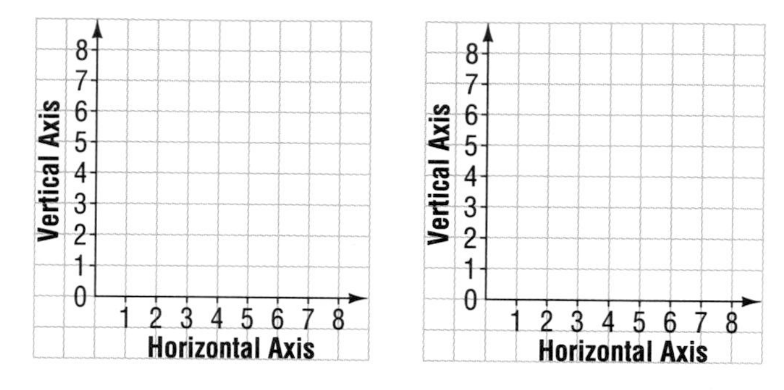

2. Use 1 of the 2 grids and draw 4 spaceships on it, either as horizontal or vertical straight lines. Begin the game with spaceships drawn through the following numbers of points on the grid.

a) Starbase
(5 points on the grid)
b) Starship
(4 points on the grid)
c) Rescue Ship
(3 points on the grid)
d) Transporter
(2 points on the grid)

3. Flip a coin to decide who will begin the game.

4. Call out pairs of numbers such as (4, 5), where 4 is the value along the horizontal axis and 5 is the value along the vertical axis. Your opponent should tell you if that pair successfully locates a spaceship or not. For example, the location of the Rescue Ship shown in the diagram includes (4, 3), (4, 4), and (4, 5), in any order. If you successfully locate a point on a spaceship, you take another turn. Once all of the points of a spaceship have been called, the spaceship is removed from the game. If you miss 1 of the points, it is your opponent's turn. Keep track of your successes and misses on the second grid. The winner is the first player to remove all of the opponent's fleet.

Warm Up

Copy and complete these tables of values.

1.

x	$x + 7$
13	
5	
−2	
−5	

2.

x	$x - 4$
5	
2	
−3	
−5	

3.

b	$0.8b$
−0.5	
−0.25	
0.25	
0.5	

4.

n	$n \div 3$
12	
6	
−18	
−30	

Evaluate each expression for $m = 15$.

5. $m - 13$ **6.** $5m$ **7.** $2m \div 30$

Evaluate each expression for $k = -6$.

8. $30 \div k$ **9.** $k + 5$ **10.** $6 - k$

Find the value of each expression after substituting $-2, -1, 0, 1,$ *and* 2 *for the variable.*

11. $5 + 3x$ **12.** $n \div 2$ **13.** $-2p$

Copy and complete each table.

14.

y	$y - 5$
	4
	6
	11
	14
	17

15.

x	$x + 9$
	10
	8
	−3
	−6
	−9

16.

m	$3m$
	−15
	−18
	−24
	−36
	−42

17.

n	$n \div 2$
	12
	9
	−1
	−3
	−9

Mental Math

Solve for x.

1. $x + 1 = 4$ **2.** $x - 15 = 9$

3. $x + 3 = -9$ **4.** $x - 2 = -5$

5. $x + 7 = -2$ **6.** $x - 9 = 15$

7. $x + \frac{1}{2} = -8$ **8.** $x - 3 = 12$

Study the patterns. State how y is related to x for each table.

9.

x	2	1	0	−1
y	2	1	0	−1

10.

x	1	2	3	4
y	2	4	6	8

11.

x	2	3	4	5
y	−6	−9	−12	−15

12.

x	2	1	0	−1
y	3	2	1	0

13.

x	2	1	0	−1
y	7	6	5	4

14.

x	2	1	0	−1
y	−1	−2	−3	−4

15.

x	4	3	2	1
y	2	1.5	1	0.5

16.

x	2	3	4	5
y	−2	−3	−4	−5

17.

x	2	3	4	5
y	4	9	16	25

18.

x	1	2	3	4
y	−1	−8	−27	−64

13.1 Relations as Ordered Pairs

In 1992, Canada's Marnie McBean and Kathleen Heddle won Olympic gold medals in a pairs rowing event. For athletes, there is a relationship between how hard they train and how successful they are.

There are many examples of how things relate or depend on each other. For example:

- The cost of a concert ticket depends on the popularity of the group.
- The distance travelled over time in a car depends on the speed of the car.
- The number of people in a mall depends on the time of day.

Mathematics has been described as the study of patterns or how numbers relate to each other.

Activity: Look for a Pattern

One cube has a surface area of 6 square units.

Two cubes, attached as shown, have a surface area of 10 square units.

Three cubes can be attached in the same way.

Copy and complete the table.

Number of Cubes	Surface Area
1	6 square units
2	10 square units
.	
.	
7	

Inquire

1. What is the surface area of 3 cubes? 4 cubes? 7 cubes?

2. What is the pattern in the surface area column?

3. Write the results from the table as a set of ordered pairs. For 1 cube, the ordered pair is (1, 6), where the first number is the number of cubes and the second number is the surface area.

4. What is the surface area of 8 cubes? 9 cubes?

5. Describe in words the relationship between the number of cubes and the surface area.

6. Let n represent the number of cubes and write an equation in the form $S = \blacksquare$ to find the surface area.

7. Use your equation to calculate the surface area for 20 cubes; 50 cubes.

8. For the cubes and surface area, what does the ordered pair (15, 62) mean? Why does the ordered pair (10, 10) have no meaning?

A **relation** is a set of ordered pairs. A relation can be described in other ways. Some of them are:

- by a table of values
- in words
- by an equation

Many relations in mathematics are expressed as formulas with 2 or more variables. Here are some of the familiar ones.

$A = l \times w$ $P = 2(l + w)$ $A = B \times h$

Example 1

Use the equation $x + y = 5$.

a) Describe the relation in words.

b) Complete a table of values for $x = 2, 1, 0, -1, -2$.

c) Write the relation as a set of ordered pairs.

Solution

a) The sum of the x-and y-values equals five.

b) Calculate the y-values.

$x + y = 5$
$2 + 3 = 5$
$1 + 4 = 5$
$0 + 5 = 5$
$-1 + 6 = 5$
$-2 + 7 = 5$

Complete the table of values.

x	y
2	3
1	4
0	5
–1	6
–2	7

c) The set of ordered pairs is (2, 3), (1, 4), (0, 5), (–1, 6), (–2, 7).

Example 2

Use the equation $y = 2x + 1$.

a) Describe the relation in words.

b) Complete a table of values for $x = 2, 1, 0, -1, -2$.

c) Write the relation as a set of ordered pairs.

Solution

a) The value of y is two times the value of x, plus one.

b) Calculate the y-values.

$2x + 1$
$2(2) + 1$
$2(1) + 1$
$2(0) + 1$
$2(-1) + 1$
$2(-2) + 1$

Complete the table of values.

x	y
2	5
1	3
0	1
–1	–1
–2	-3

c) The set of ordered pairs is (2, 5), (1, 3), (0, 1), (–1, –1), (–2, –3).

CONTINUED

Practice

1. Describe the relationship between the following.

a) the length of a tree's shadow on a sunny day and the time of day

b) the number of greeting cards sold by a store and the month of the year

c) the number of stars you can see on a clear night and where you live

2. A baseball commentator uses ordered pairs to describe the ball and strike count for a batter. What are the possible ordered pairs? Is there any relationship between the 2 numbers?

Use each of the following equations.

a) *Describe the relation in words.*

b) *Copy and complete the table of values.*

c) *Write the relation as a set of ordered pairs.*

3. $x + y = 4$

x	y
2	
1	
0	
–1	
–2	

4. $x - y = 1$

x	y
3	
4	
5	
6	
7	

5. $m + n = -1$

m	n
2	
1	
0	
–1	
–2	

6. $s - t = 3$

s	t
4	
3	
2	
1	
0	

7. For the equation $x + y = 8$, find the missing value in each ordered pair.

a) (1, ■) **b)** (6, ■)
c) (–5, ■) **d)** (■, 2)
e) (■, –2) **f)** (■, –5)

8. For the equation $x - y = 2$, find the missing value in each ordered pair.

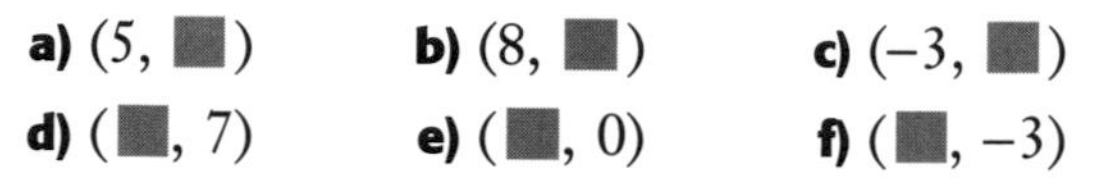

a) (5, ■) **b)** (8, ■) **c)** (–3, ■)
d) (■, 7) **e)** (■, 0) **f)** (■, –3)

Use each of the following equations.

a) *Describe the relation in words.*

b) *Copy and complete the table of values.*

c) *Write the relation as a set of ordered pairs.*

9. $y = 4x + 2$

x	y
2	
1	
0	
–1	
–2	

10. $y = 3x - 4$

x	y
2	
1	
0	
–1	
–2	

11. $y = -2x + 7$

x	y
2	
1	
0	
–1	
–2	

12. $y = -x - 2$

x	y
–4	
–3	
–2	
–1	
0	

Find 5 ordered pairs for each relation.

13. $x + y = 11$ **14.** $x - y = 3$

15. $y = 4x + 5$ **16.** $y = 2x - 7$

Write an equation that describes each relation.

17.

x	y
1	8
2	7
3	6
4	5

18.

x	y
1	3
2	6
3	9
4	12

19.

x	y
4	2
3	1
2	0
1	–1
0	–2

20.

x	y
2	–4
1	–2
0	0
–1	2
–2	4

Problems and Applications

21. a) Copy and complete the table for a train travelling at 80 km/h.

Time (h)	Distance (km)
1	
2	
3	
4	

b) How far would the train travel in 5 h? 6 h?

c) Describe the pattern in the distance column.

d) Write the results from the table as a set of ordered pairs.

e) Describe in words the relationship between the time and the distance travelled.

f) Let t represent the number of hours travelled. Write an equation in the form $D = $ ▬ to find the distance travelled.

g) Use your equation to determine how far the train will travel in 13 h; 21 h.

22. When you fold a piece of paper once you divide it into 2 regions. If you fold it again, perpendicular to the last fold, you get 4 regions.

a) Copy and complete the table.

Number of Folds	Number of Regions
1	2
2	4
3	
4	
5	
6	

b) What is the pattern in the number of regions column?

c) How many regions would there be after 7 folds? 8 folds?

d) Write the results from the table as a set of ordered pairs.

e) Describe in words the relationship between the number of folds and the number of regions.

f) Let n represent the number of folds. Write an equation in the form $R = $ ▬ to find the number of regions.

g) Use your equation to calculate the number of regions after 10 folds; 12 folds.

23. A diagonal is a straight line that joins 2 vertices of a polygon. A quadrilateral has 2 diagonals.

a) Work with a partner. Copy and complete this table.

Name of Polygon	Number of Sides	Number of Diagonals
Quadrilateral	4	2
Pentagon	5	
Hexagon	6	
Heptagon	7	
Octagon	8	
Nonagon	9	
Decagon	10	
Hendecagon	11	

b) What is the pattern in the number of diagonals?

c) How many diagonals does a 12-sided polygon have? 13-sided?

d) Write the results from the table as a set of ordered pairs.

e) Describe in words the relationship between the number of sides and the number of diagonals.

f) Let s represent the number of sides. Write an equation in the form $D = $ ▬ to find the number of diagonals.

g) Use your equation to find the number of diagonals for a polygon with 20 sides; 50 sides; 100 sides.

13.2 Graphing Ordered Pairs

Air traffic controllers at a busy airport, such as Mirabel in Montreal, must quickly and easily identify an approaching aircraft's position.

To do this, air traffic controllers observe the lighted, electronic dots on their radar screens. Each aircraft is represented by a dot that is labelled with the aircraft's flight number, altitude, and ground speed.

Activity: Analyze the Number Lines

1. Is an approaching aircraft's distance along the ground from the airport a horizontal or a vertical dimension?

2. Is an approaching aircraft's height above the ground a horizontal or a vertical dimension?

3. a) On which of the following number lines would you plot the aircraft's distance along the ground from the airport in kilometres?
b) On which of the number lines would you plot the aircraft's height above the ground in metres?

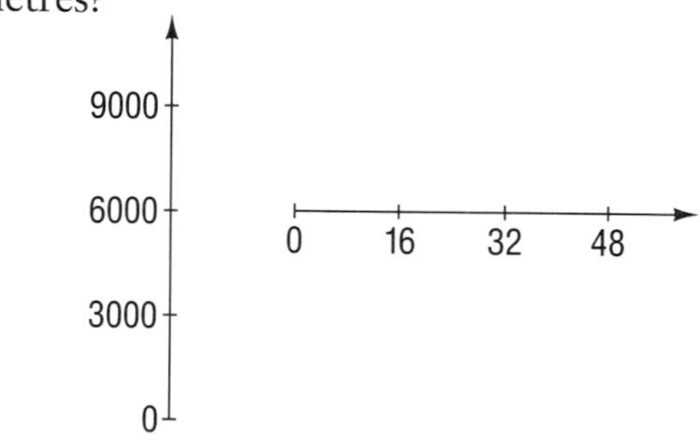

4. Can either number line show both distance along the ground and height above the ground at the same time?

5. Describe how the 2 number lines can be put together so that both sets of values can be plotted at once.

6. What is the advantage of the system you described in question 5?

René Descartes, a 17th century mathematician, developed a system for graphing ordered pairs on a grid. This system is called the **Cartesian Coordinate System.**

In this system, ordered pairs are graphed on a grid made up of 2 perpendicular number lines. The lines meet at a point called the **origin**. The horizontal number line is called the ***x*-axis.** The vertical number line is called the ***y*-axis.** The x- and y-axes divide the plane into 4 quadrants.

Example 1

What ordered pairs describe the points shown on this grid?

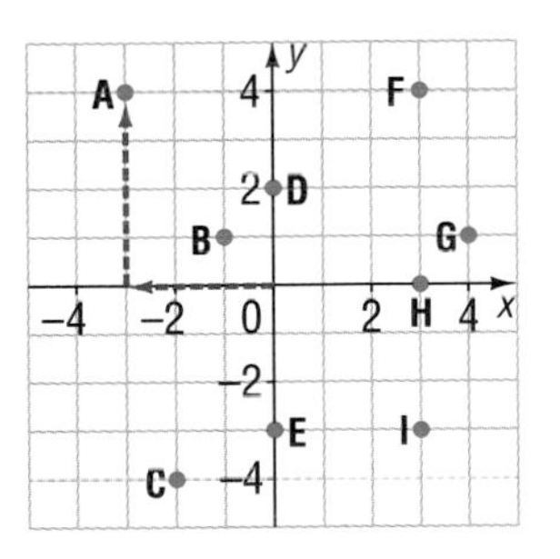

Solution

First, read along the x-axis to identify the x-value. This is called the ***x*-coordinate.**

Then, read along the y-axis to identify the matching y-value. This is called the ***y*-coordinate.**

For point A, the x-coordinate is -3 and the y-coordinate is 4. The ordered pair is $(-3, 4)$.

The ordered pairs that describe the points are:

A(−3, 4), B(−1, 1), C(−2, −4), D(0, 2), E(0, −3), F(3, 4), G(4, 1), H(3, 0), and I(3, −3)

Example 2

a) Plot the points A(−2, 3), B(2, 3), C(2, −3), and D(−2, −3) on a grid.

b) Join AB, BC, CD, and DA.

c) Identify the polygon and its dimensions.

Solution

a)

b)

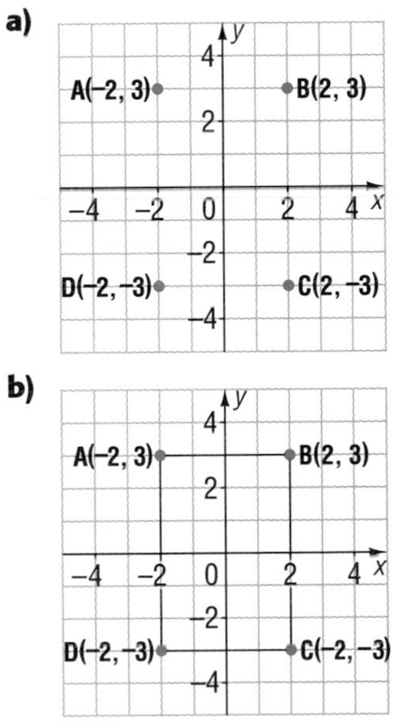

c) The polygon is a rectangle, where $l = 6$ units and $w = 4$ units.

CONTINUED

Practice

Name the point on the following grid given by these coordinates.

1. (3, 1) **2.** (−5, 1) **3.** (2, 3)

4. (4, −1) **5.** (0, −2) **6.** (5, 2)

7. (−3, 0) **8.** (1, −5) **9.** (−6, −4)

10. (−2, 2) **11.** (3, 5) **12.** (6, −3)

13. (−1, −4) **14.** (0, 0) **15.** (−4, −1)

16. Write the coordinates of each point shown on the following grid.

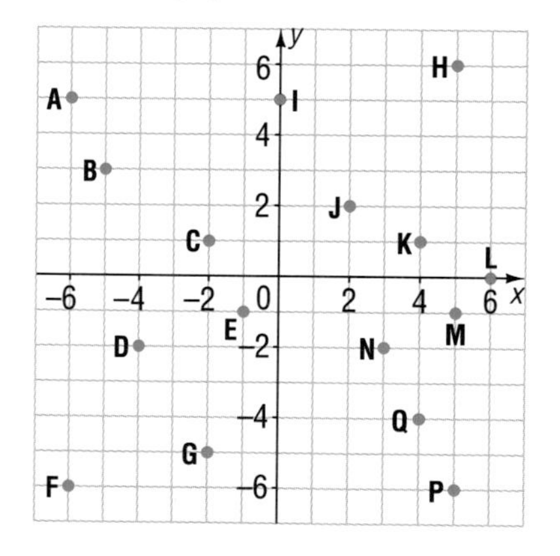

Plot each point on a grid.

17. A(3, 5) **18.** B(5, 3) **19.** C(−3, 5)

20. D(−5, 3) **21.** E(3, −5) **22.** F(5, −3)

23. G(0, 4) **24.** H(4, 0) **25.** I(−4, 0)

26. J(0, −4) **27.** K(1, 1) **28.** L(−1, 1)

29. M(1, −1) **30.** N(−1, −1) **31.** P(0, 0)

Problems and Applications

32. Plot and join these sets of points in order. Join the last point to the first point to form a closed figure. Identify each figure. Calculate the area of each figure in square units.

a) A(3, 1), B(3, 5), Z(8, 4)

b) D(1, 1), E(1, −2), F(−5, −2), G(−5, 1)

c) J(6, 0), K(9, 1), L(9, −4), M(6, −5)

d) Q(−3, 1), R(−3, −1), S(3, −1), T(3, 1).

e) G(−3, 0), H(0, −4), I(5, −4), J(8, 0), N(5, 4), T(0, 4)

33. Plot the points A(−2, 3) and B(5, −1). Find the coordinates of a third point that will form a right triangle with these 2 points. There is more than 1 solution.

34. Plot the points D(5, 1) and E(3, −3) on a grid. Find the coordinates of a third point, F, so that ΔDEF is isosceles and DE = DF. How many solutions are there?

35. a) Plot the points A(−4, 1), W(−3, −4), and Z(5, 0) on a grid.

b) Find all the possible points that form a parallelogram with these 3 points.

c) How many solutions are there?

36. Decide if each statement is always true, sometimes true, or never true. Explain.

a) A point with 1 positive coordinate and 1 negative coordinate lies in the 4th quadrant.

b) A point whose x- and y-coordinates are equal lies in the 1st or 3rd quadrant.

c) The points on a vertical line have the same x-coordinate.

37. Make up a problem where the solution requires that points be plotted on a grid. Have a classmate solve your problem.

13.3 Graphing Relations

The following ad was used to help raise money to save rainforests.

"In the time it takes you to read this, about six hectares of rainforest will be lost forever. Please help to save our trees."

Activity: Draw the Graph

Assume it takes 10 s to the read the above ad. Copy and complete this table of values.

Time (s)	Area of Rainforest Lost (ha)
0	
10	
20	
30	
40	
50	

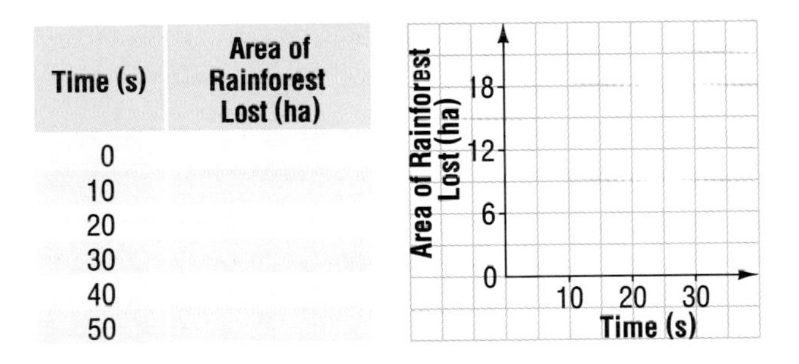

Area of Rainforest Lost (ha)
18
12
6
0
10
20
30
Time (s)

Inquire

1. Copy the grid and plot the ordered pairs for this relation. Join the points.

2. Is the line a straight line?

3. Extend the line to $t = 60$ s. How much rainforest is lost in this amount of time?

4. Use the graph to estimate about how much time has elapsed if the number of hectares lost is 10.

5. Use the graph to estimate the number of hectares of rainforest lost in 25 s.

6. Work with a partner. Use the variables A, r, and t to write an equation that relates the time elapsed and the number of hectares of rainforest lost. A is the number of hectares of rainforest lost, r is the rate of rainforest loss, and t is the time in seconds.

7. Describe how you can check that your equation is correct.

"IN THE TIME IT TAKES YOU TO READ THIS, ABOUT SIX HECTARES OF RAINFOREST WILL BE LOST FOREVER!"
Please help to save our trees.

CONTINUED

Example 1

The time that elapses before you hear a thunderclap after you see a flash of lightning depends on your distance from the lightning. With every kilometre, the time that elapses increases by 3 s.

a) Make a table of values for distances from 0 km to 4 km.
b) Graph this relation. State what type of relation it is.
c) How much time elapses before you hear the thunderclap at a distance of 3.5 km from the lightning?
d) How far away are you if you hear the thunderclap in 1.5 s?

Solution

a) The table of values that satisfies this relation is:

Time (s)	Distance (km)
0	0
3	1
6	2
9	3
12	4

b)

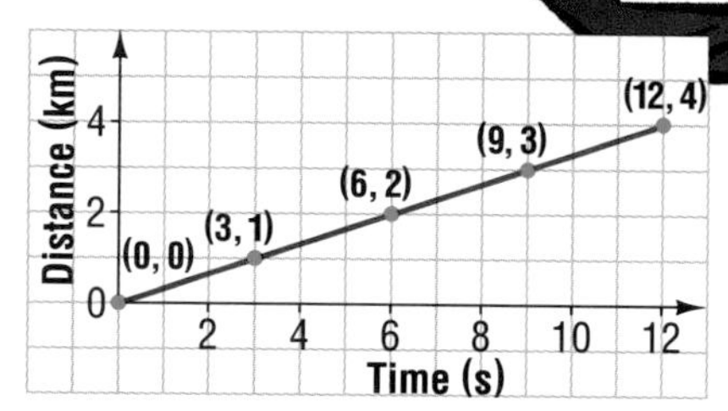

The points lie in a straight line. The relation is a **linear relation.**

c) From the graph, the time that elapses before you hear the thunderclap at 3.5 km is about 10.5 s.
d) From the graph, at 1.5 s you are about 0.5 km away from the lightning.

Example 2

The temperature of the air varies with the altitude.

a) Use the values in the table to graph the relation between air temperature and altitude.
b) Mount Logan is 5959 m high. Estimate the temperature at the peak of Mount Logan when the temperature at the base is 16°C.

Altitude (m)	Temperature (°C)
0	16
2000	12
4000	8
6000	4
8000	0

Solution

a) Plot the 2 endpoints of the data and join them with a straight line.
b) Mount Logan is 5959 m high. The temperature at its peak is approximately 4°C when the temperature at the base is 16°C.

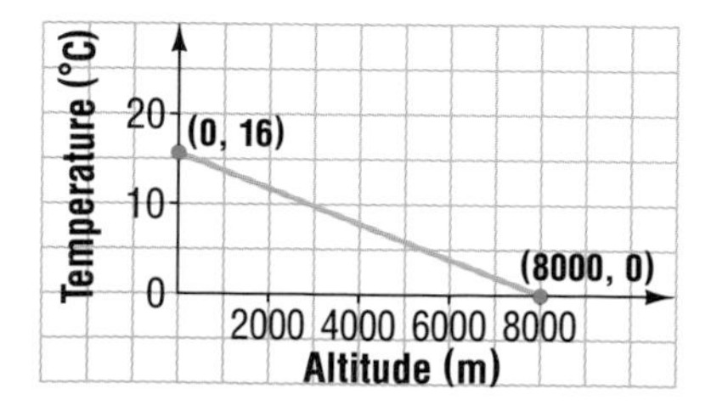

Practice

Name the coordinates of the points for each linear relation.

1.

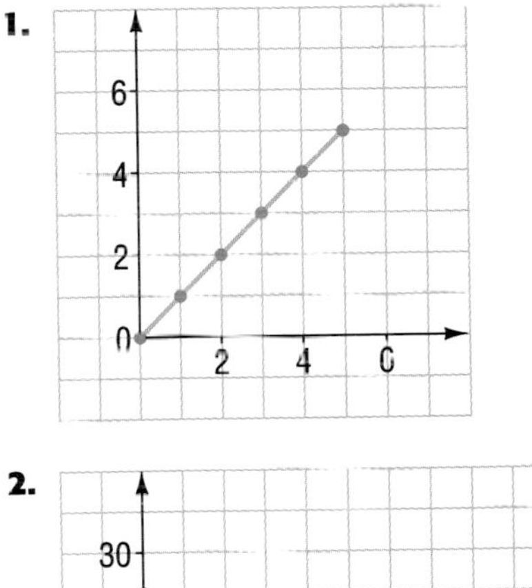

2.

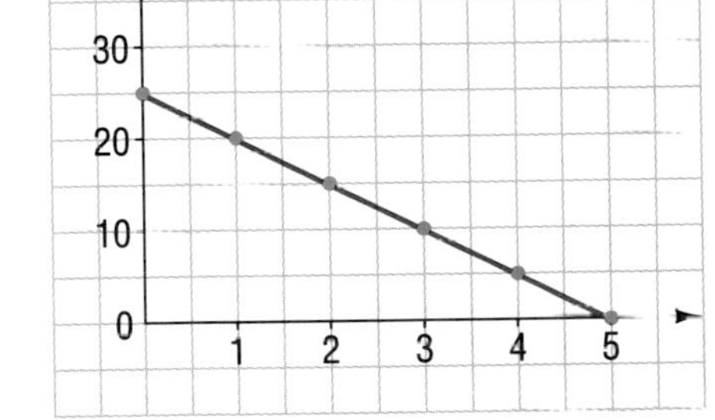

Copy and complete these tables of values. Graph each relation.

3.

Time (h)	Distance (km)
0	0
1	90
2	180
3	
4	
5	

4.

Time (h)	Cost ($)
0	0
1	6.00
2	12.00
3	
4	
5	

5.

Time (h)	Distance (km)
0	0
1	30
2	60
3	
4	
5	

6.

Time (h)	Cost ($)
0	30
1	80
2	130
3	
4	
5	

Problems and Applications

7. This table of values shows the effect of temperature on the volume of a gas.

Temperature (°C)	Volume (mL)
10	849
15	864
20	879
25	894
30	909

a) Graph the relation.
b) By how many millilitres does the volume of the gas change per degree?
c) What is the volume of the gas at 21°C? at 29°C?
d) Determine the volume of the gas when the temperature is 0°C.

8. Tina repairs video games. She charges $35.00 to assess the problem and $15.00 for each half hour of labour needed to repair the game.
a) Complete a table of values that shows her fees for up to and including 3 h.
b) Graph the relation.
c) Does this graph pass through the origin? Explain.

9. Operator-assisted telephone calls through the International Maritime Satellite Organization are expensive. The satellite serves ships, oil rigs, and ports of maritime member nations. The initial 3-min period or part of this period costs $36.00. Each additional minute or part of a minute costs $12.00.

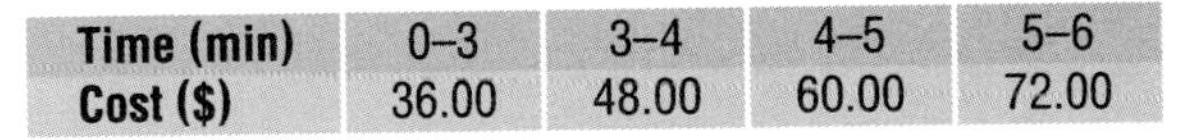

Time (min)	0–3	3–4	4–5	5–6
Cost ($)	36.00	48.00	60.00	72.00

a) Graph this relation.
b) How much is a call that lasts 30 s?
c) How much is a call that lasts 3 min and 1 s?

10. Write a problem that involves graphing a relation. Have a classmate solve your problem.

Street Geography and Geometry

In mathematics and other subjects, you are sometimes asked to find the distance between 2 places. This usually means the straight-line distance.

However, the walking distance or driving distance between 2 places in a city or town is sometimes measured in blocks. Buildings prevent you from travelling in a straight line from place to place. For example, the block distance from P to Q in the diagram is 8.

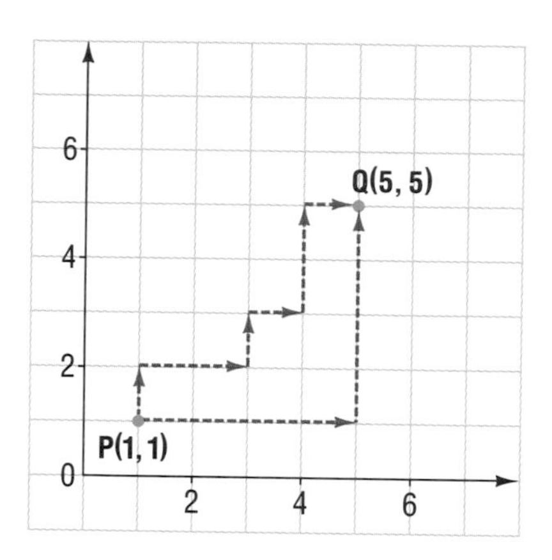

Activity ❶

The points U(6, 6) and L(3, 2) are plotted on a grid. Suppose the point U represents a university, and the point L represents a library.

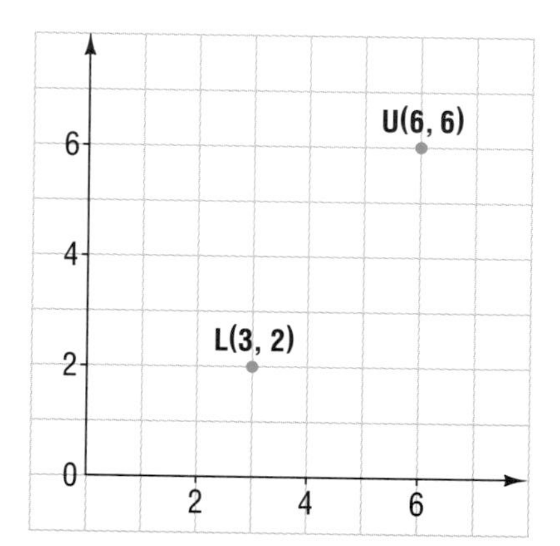

1. Kerry wants to find an apartment on a corner so that the combined walking distance from her apartment to the university and from her apartment to the library is as short as possible. Copy the grid and locate the corners where Kerry could have an apartment.

2. What is the shortest total number of blocks Kerry has to walk?

3. a) How many corners does Kerry have to choose from for her apartment?
b) What are the coordinates of these corners?

4. Is there any corner on which Kerry can locate her apartment so that the distance from her apartment to the university and the distance from her apartment to the library are the same?

5. Is there a relationship between the coordinates of U and L and the shortest total distance Kerry has to walk? If so, describe the relationship.

6. Suppose Kerry decides that she would like to take an apartment so that the sum of the distances from the apartment to the university and from the apartment to the library is 9 blocks. Can she find such a corner location?

Activity ❷

Gino is looking for an apartment. The college is located at C(1, 2), the library at L(5, 6), and the grocery store at S(6, 1).

1. Plot these locations on a grid.

2. Find the corners on which Gino could locate his apartment so that the total walking distance from his apartment to the college, his apartment to the library, and his apartment to the store is as short as possible.

3. a) How many locations for his apartment have the shortest total walking distance?

b) What are the coordinates of these locations?

4. How many locations have the second shortest total walking distance?

Activity ❸

The college is located at C(2, 9), the library at L(3, 1), the elementary school at E(8, 7), and the high school at H(6, 3).

1. Plot these locations on a grid.

2. Where should the city locate a bus stop so that the total walking distance from the bus stop to the college, the bus stop to the library, the bus stop to the elementary school, and the bus stop to the high school is as short as possible?

3. How many possible locations are there?

13.4 Equations and Linear Relations

Canada's national breed of horse, called the *little iron horse*, originated from the stables of King Louis XIV of France in 1665. A gift of a pair of horses arrived in New France at Stadacona, the fort and fur-trade centre that later became Quebec City.

In spite of its small size, the little iron horse had speed, strength, and endurance. In the 19th century, one of these horses covered the 250-km distance from Quebec City to Montreal faster than an overnight steamship could cover the same distance.

Activity: Draw the Graph

The equation $d = 20t$, calculates the distance in kilometres travelled by the little iron horse in a time of t hours. Copy and complete the table of values.

Time (h)	Distance (km)
0	
5	
10	
15	

Distance (km)
300
200
100
0
5 10 15
Time (h)

Inquire

1. Plot the ordered pairs on a grid similar to the one shown.

2. How many kilometres did the horse travel each hour?

3. What was the speed of the horse?

4. Approximately how long did it take the horse to travel the 250 km from Quebec City to Montreal?

5. Approximately what distance did the horse travel in 2.5 h?

6. Approximately how long did the horse take to travel 130 km?

7. Use the graph to determine how far the horse travelled in 7 h.

Example 1

Plot the points given in these tables of values and write an equation to describe each relation.

a)

x	y
–2	–1
–1	0
0	1
1	2
2	3

b)

x	y
–2	–4
–1	–2
0	0
1	2
2	4

Solution

a) Each y-value is 1 more than each x-value.
The equation is $y = x + 1$.

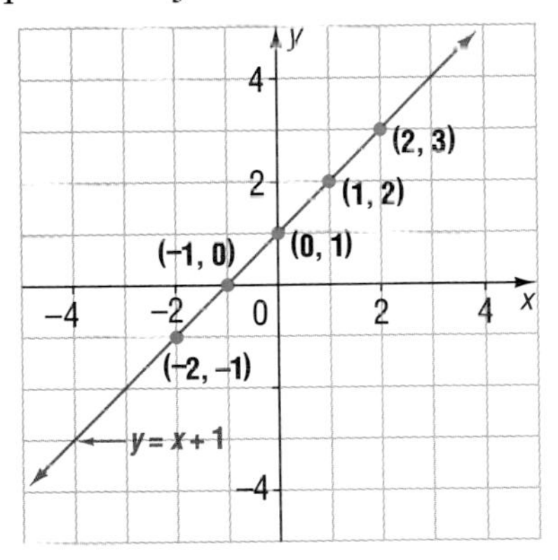

b) Each y-value is double each x-value.
The equation is $y = 2x$.

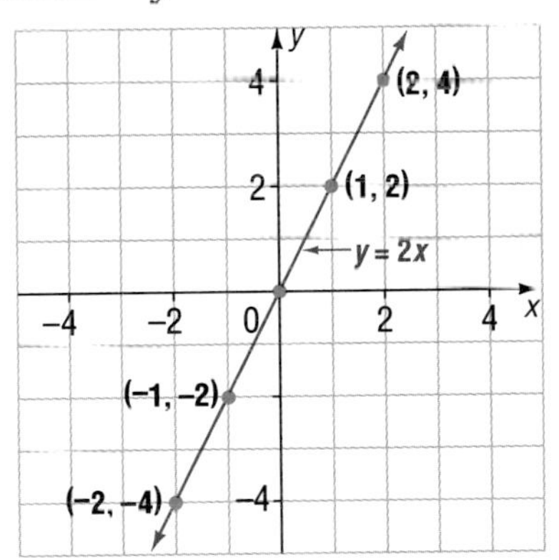

At the point where a graph crosses the x-axis, the coordinates are $(x, 0)$.
At the point where a graph crosses the y-axis, the coordinates are $(0, y)$.

Example 2

a) Find the points where the graph of $2x + 3y = 6$ crosses the x- and y-axes.

b) Graph the equation.

Solution

a) When $y = 0$,

$$2x + 3(0) = 6$$
$$2x = 6$$
$$x = 3$$

When $x = 0$,

$$2(0) + 3y = 6$$
$$3y = 6$$
$$y = 2$$

The points are (3, 0) and (0, 2).

b) We can use these 2 points to draw the graph of the equation $2x + 3y = 6$.

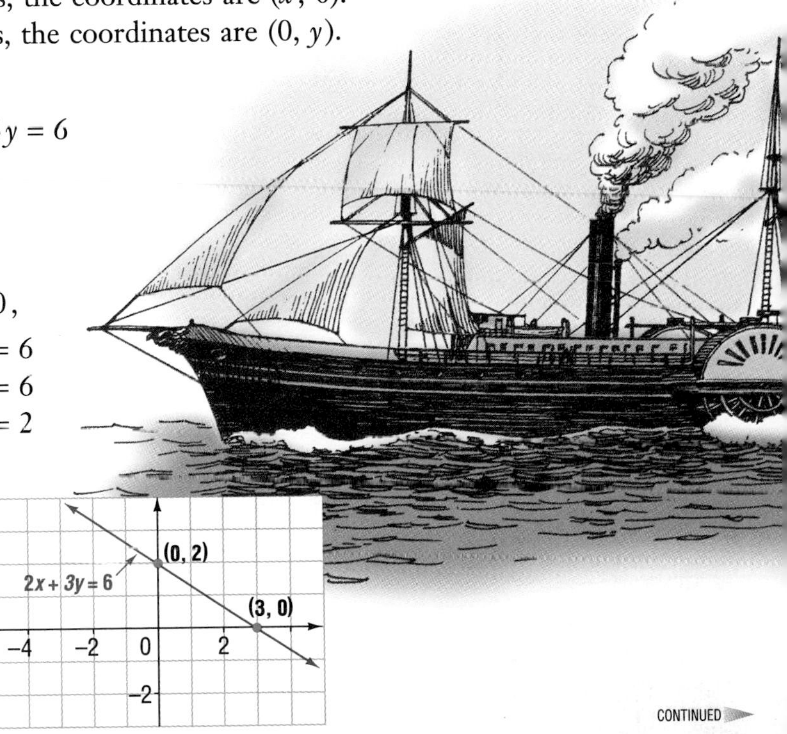

CONTINUED

Practice

Copy and complete each table of values.

1. $y = x - 1$

x	y
−2	
−1	
0	
1	
2	

2. $y = x + 2$

x	y
−2	
−1	
0	
1	
2	

3. $x + y = 4$

x	y
−2	
−1	
0	
1	
2	

4. $x - y = 2$

x	y
−2	
−1	
0	
1	
2	

5. Given the table of values, write an equation for each relation.

a)

x	y
−2	0
−1	1
0	2
1	3
2	4

b)

x	y
−2	7
−1	6
0	5
1	4
2	3

Construct a table of values and graph each equation.

6. $y = x$

7. $y = x - 2$

8. $y = 3x$

9. $y = x + 3$

10. $y = 2x + 3$

11. $y = 3 - 2x$

12. $x + y = 7$

13. $x - y = 5$

Graph each equation by finding the points where the line crosses the axes.

14. $x + y = 1$

15. $x - y = 5$

16. $x - y = -2$

17. $2x + 3y = 12$

18. $3x - 4y = 12$

19. $3x + 5y = -15$

Problems and Applications

20. Let x equal the number of sides for each figure. Let y equal the number of diagonals from 1 vertex in each figure.

a) Complete a table of values for the first 10 diagrams.

b) Graph the relation.

c) Write an equation for the relation.

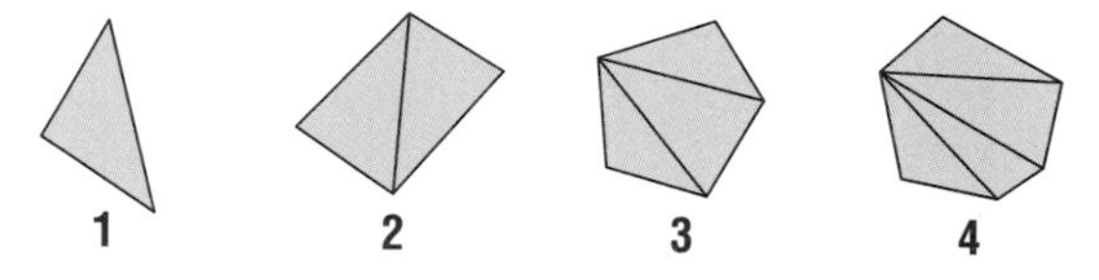

21. The equation $c = 1.30k + 2.20$ gives the cost to hire a taxi, where c is the cost in dollars and k is the distance travelled in kilometres.

a) Use $k = 0, 5, 10, 15, 20, 25$ to construct a table of values.

b) Draw a graph of cost versus distance travelled.

c) What is the meaning of the point where the line crosses the y-axis?

22. Let x equal the number of hexagons in each diagram and y equal the number of toothpicks in each diagram.

a) Make a table of values for the first 5 diagrams.

b) Graph the relation.

c) Write an equation for the relation.

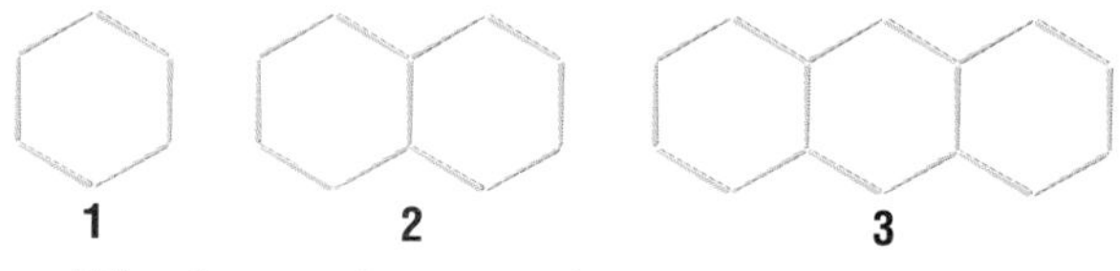

23. The base of a triangle is 6 cm. The table shows several different heights for this triangle. Work with a partner to do the following.

a) Copy and complete the table.

b) Graph the relation.

c) Write an equation that relates area and height.

Height (cm)	2	4	6	8
Area (cm²)				

The Graphing Calculator

A graphing calculator can be used to draw the graph of an equation. When you use a graphing calculator, there are 3 things you must remember to do.

- Express y in terms of x.
  All equations must be in the form $y =$ ▬.
- Set the range by entering the highest and the lowest x- and y-values.
- Set the scale (the units you want marked on the axes).

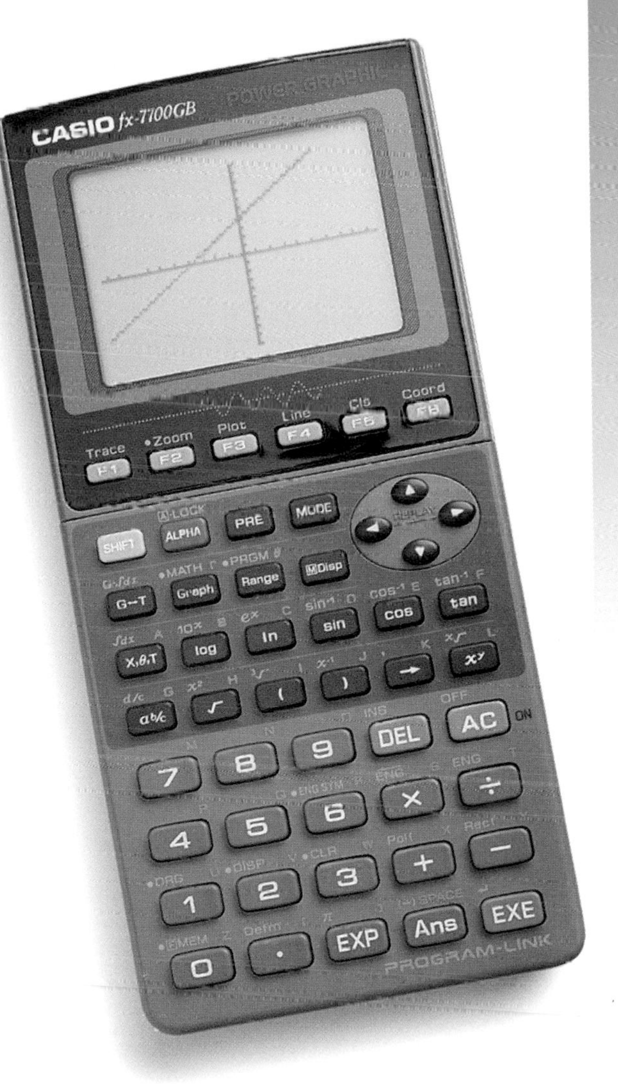

Activity ❶ Setting the Range

To graph $y = x + 4$,
first set the range.
RANGE
Xmin: -10
max: 10
scl: 1
Ymin: -10
max: 10
scl: 1

Graph the line.
GRAPH
Y = X + 4
EXE

1. Change the range so that x has values between 0 and 10 and the scale is 2. Draw the graph.

2. Change the range so that x and y have values between 0 and 10 and the scale is 5. Draw the graph.

3. Change the range so that x has values between 0 and 10, y has values between -10 and 0, and the scale is 1. Draw the graph. Explain the result.

Activity ❷ Graphing Lines

To graph the line $x + y = 4$, first rewrite the equation.

$$x + y = 4$$
$$x - x + y = 4 - x$$
$$y = 4 - x$$

Graph each of the following lines. Use a scale of your choice.

1. $x + y = 3$ **2.** $x + y = -2$ **3.** $2x + y = 4$
4. $x - y = 2$ **5.** $x + y = 0$ **6.** $x + 2y = 2$

Activity ❸ Intersecting Lines

1. Use your graphing calculator to find the point where the lines $y = x + 2$ and $y = 6 - x$ cross.

2. Compare your method with a classmate's.

Activity 1 The Line of Best Fit

The table shows selected winning heights in the men's Olympic high jump from 1912 to 1992.

Olympic Year	Winning Country	Jump Height (m)
1912	United States	1.93
1932	Canada	1.97
1952	United States	2.04
1972	USSR	2.23
1992	Cuba	2.34

1. Copy the following grid and plot the ordered pairs from the table.

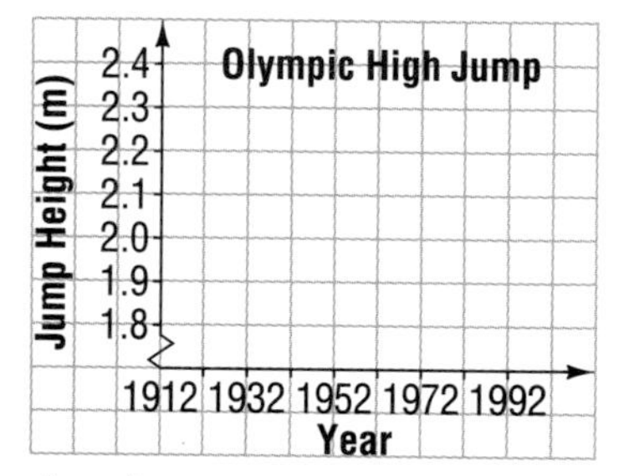

2. Do the data represent a linear relation? Explain.

3. What is the general trend in the data for the jump height over time?

4. Draw a straight line that is as close as possible to all of the points plotted on the grid. Compare the line you drew with a classmate's.

5. Statisticians call this line the **line of best fit**. Why?

6. Use your line of best fit to predict the winning jump height in the year 2012.

Activity ❷ Interpolation

When you **interpolate**, you estimate values that lie within a range of given values.

As you dive deeper into a lake, the amount of sunlight that penetrates the water decreases. The following table shows the percent of sunlight that penetrates water to different depths.

Water Depth (m)	Sunlight Penetration (%)	Water Depth (m)	Sunlight Penetration (%)
0	100	50	30
10	80	70	20
30	50	100	10

1. Plot the ordered pairs on a grid similar to the one shown.

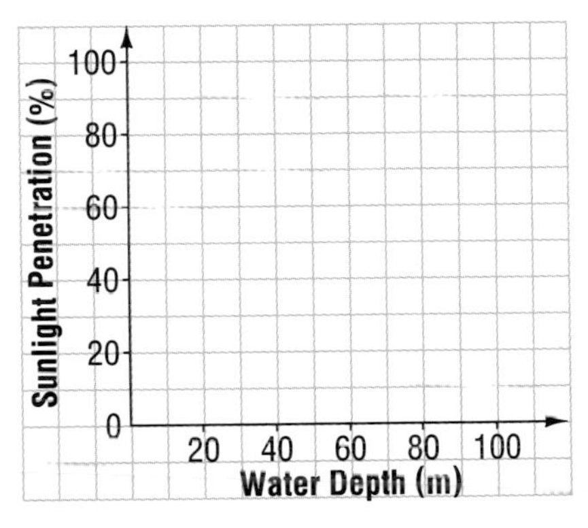

2. Explain why this relation is called a nonlinear relation.

3. Join the points you plotted to form a smooth curve.

4. What percent of sunlight penetrates to a depth of 30 m? 70 m?

5. About how deep is the water when 70% of the sunlight penetrates to that depth?

6. About what percent of sunlight penetrates water to
a depth of 90 m?

7. Which of your answers to the above questions were interpolated? Explain.

Activity ❸ Extrapolation

When you **extrapolate**, you extend a line to estimate points that lie outside a given range of values.

1. Plot the data on a grid.

Olympic Year	Men's 110-m Hurdles	Winning Time (s)
1980	Munkelt (E. Germany)	13.39
1984	Kingdom (U.S.)	13.20
1988	Kingdom (U.S.)	12.98
1992	McKoy (Canada)	13.12

2. Draw a line of best fit.

3. If the general trend were to continue, what might be the winning time for the 110-m hurdles in the year 2020?

4. Compare your answer with your classmates'.

13.5 Intersecting Lines

Manitoba is a province of weather extremes. The intersection of Portage Avenue and Main Street in Winnipeg is one of the coldest intersections in Canada.

The intersection of Portage and Main is an example of 2 intersecting streets. In mathematics, we often use intersecting lines.

Activity: Draw the Graphs

Two companies rent backhoes on 2 different daily plans.
Company A is $200.00 for the backhoe plus $20/h.
Company B is $60/h.
Copy and complete the table of values for each company's daily rental cost. Graph both lines on the same set of axes.

Company A							
Time (h)	1	2	3	4	5	6	7
Cost ($)							

Company B							
Time (h)	1	2	3	4	5	6	7
Cost ($)							

Inquire

1. Which plan is cheaper if you rent the backhoe for 3 h? 7 h?
2. What does the point where the 2 graphs cross each other mean?
3. How many hours' rental gives the same cost for both plans?

The point where the 2 lines meet is called the **point of intersection**. This point is common to both lines.

Example

Graph the lines $y = 2x + 1$ and $y = x + 3$ on the same set of axes and name the point of intersection.

Solution

Make a table of values for each line.

$y = 2x + 1$	
x	y
−2	−3
−1	−1
0	1
1	3
2	5

$y = x + 3$	
x	y
−2	1
−1	2
0	3
1	4
2	5

Graph the lines.

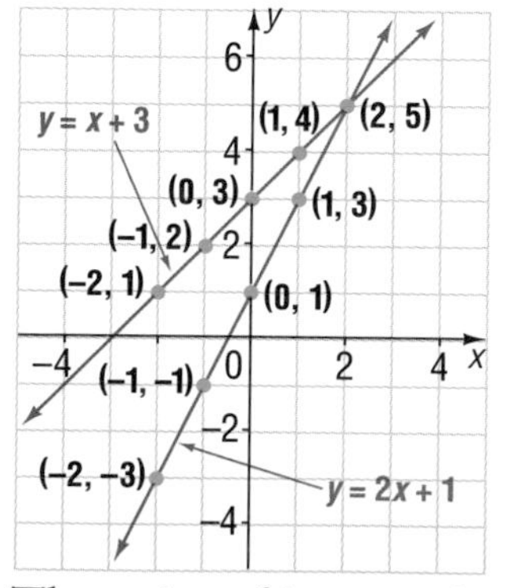

The point of intersection is (2, 5).

Practice

Name the point of intersection for each pair of lines.

1. **2.**

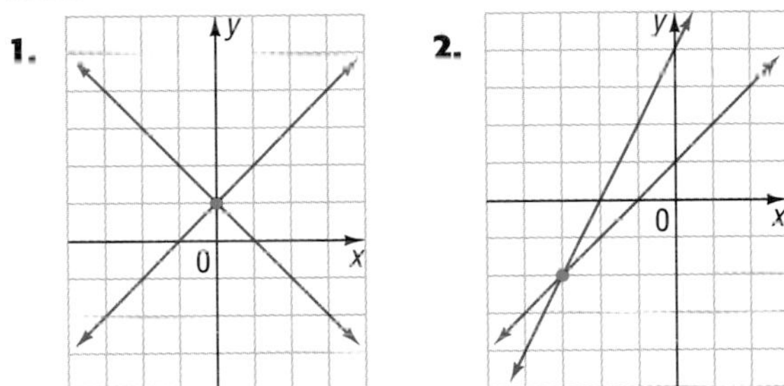

What is the point of intersection for the following pairs of lines?

3.

x	y
−2	−2
−1	−1
0	0
1	1
2	2

x	y
−2	2
−1	1
0	0
1	−1
2	−2

4.

x	y
2	1
1	0
0	−1
−1	−2
−2	−3

x	y
2	4
1	2
0	0
−1	−2
−2	−4

5. Write an equation for the line represented by each table of values in questions 3 and 4.

Make tables of values for each pair of lines. Find the point of intersection.

6. $y = 2x$ and $y = x + 3$

7. $y = x - 1$ and $y = 3 - x$

8. $y = x + 3$ and $y = 3 - 2x$

9. $y = x + 4$ and $y = 3x$

10. $y = x + 8$ and $y = -3x$

11. $y = x + 2$ and $y = 2x - 3$

12. $y = 2x - 2$ and $y = -2x$

Problems and Applications

13. a) Graph the lines $y = 2x - 4$ and $y = x - 4$ on the same set of axes.
b) What figure is formed by these lines and the x axis?

14. Two companies rent trucks. Company A charges $60.00 for the truck plus $0.20/km. Company B charges $0.50/km.
a) Write an equation for each company's rental cost.
b) Complete a table of values for each company. Use the following numbers of kilometres in your tables: 1, 50, 100, 150, 200, 250, 300.
c) Graph both equations on the same set of axes.
d) What is the point of intersection? What does it mean?

15. At 12:30, Kenji left town driving at 80 km/h. At 13:00, Yvette left town along the same highway driving at 100 km/h.
a) Construct a table of distance and time values for each.
b) Plot both graphs on the same grid.
c) At what time did Yvette catch up with Kenji?
d) How far had they travelled?

16. a) Graph all 3 lines on the same grid.

$y = -2x$ $\quad$ $y = 2x - 4$ $\quad$ $y = -x - 1$

b) These 3 lines are called **concurrent**. Explain what concurrent means.

17. Write a problem that creates 4 concurrent lines. Have a classmate solve your problem.

WORD POWER

How many real words of four letters or more can you form using the letters in the word CARTESIAN? Compare your list of words with a classmate's.

13.6 Slope of a Straight Line

Thousands of people throng to the mountains of the Okanagan Highland in British Columbia each winter for some of North America's best skiing.

Big White is the highest mountain in the region, with a summit elevation of 2319 m. There are long, but different, runs for the average skier, the beginner, and the expert skier. How are the 3 types of runs different?

Activity: Discover the Relationship

Copy these line segments onto centimetre grid paper.

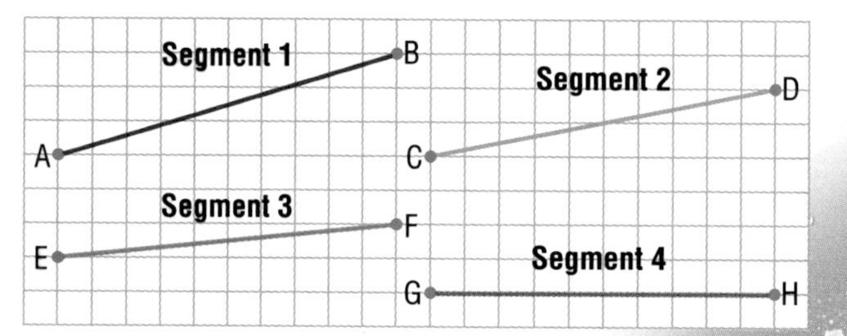

Inquire

1. Rank the 4 line segments in order of steepness, from least steep to most steep.

2. Describe how you decided on your rankings in question 1.

3. Copy and complete the following table for the above line segments.

Line Segment	1	2	3	4
Vertical Height (cm)				
Horizontal Length (cm)				

4. In fraction form, write each segment's dimensions as a ratio of vertical height to horizontal length.

5. Write each of the ratios that you found in question 4 as a decimal.

6. On mountain roads, drivers are often warned of steep-hill conditions by road signs like the one shown on the right. If the sign indicates a slope of 10%, it means that, for every 100 m along the horizontal, the vertical change is 10 m. Which of the above line segments has a 10% slope?

Example 1

Determine the slope of

a) line segment JK **b)** line segment YZ

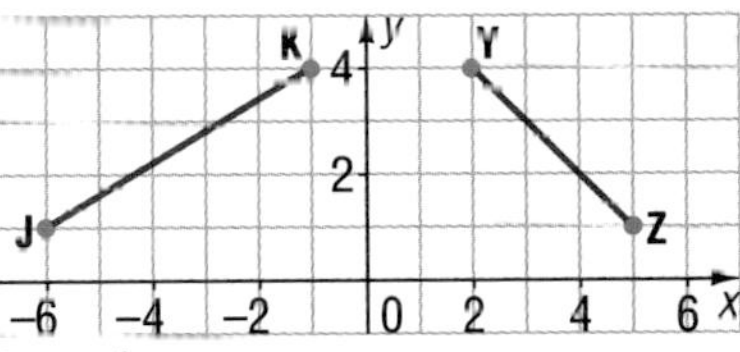

Solution

Slope is the ratio $\frac{\text{rise}}{\text{run}}$. We can find the slope by counting the number of units for the rise and for the run.

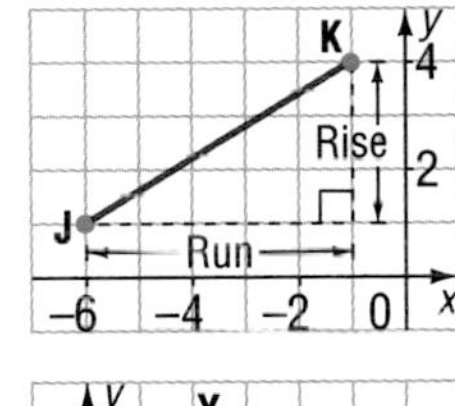

a) For JK, the height increases as we move from left to right along the x-axis. The rise is 3 units from J to K. The run is 5 units. The slope of JK is positive and equals $\frac{3}{5}$.

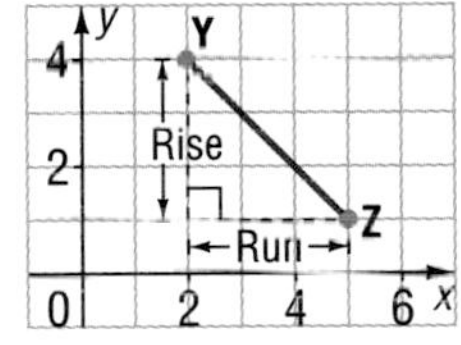

b) For YZ, the height decreases as we move from left to right along the x-axis. The rise is -3 units from Y to Z. The run is 3 units. The slope of YZ is negative and equals $\frac{-3}{3}$ or -1.

Note that a line segment with a **positive slope** rises from left to right and a line segment with a **negative slope** rises from right to left.

If you know the coordinates of 2 points on a line, you can calculate the slope using the following formula.

$$\text{slope} = \frac{\text{rise}}{\text{run}} = \frac{\text{difference in the } y\text{-coordinates}}{\text{difference in the } x\text{-coordinates}}$$

Example 2

Calculate the slope of

a) AB **b)** CD **c)** EF

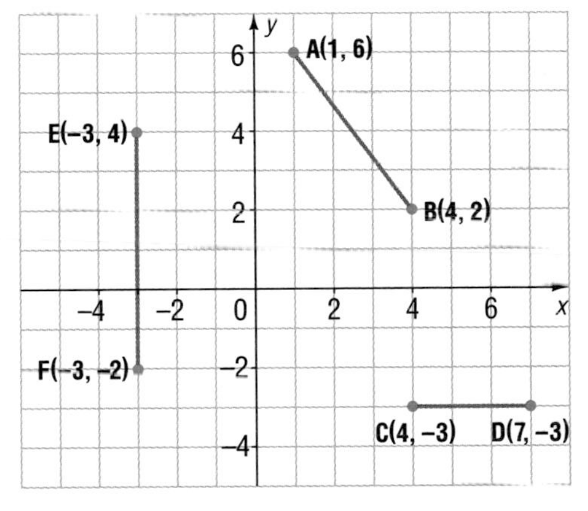

Solution

a) The 2 points are A(1, 6) and B(4, 2).

$$\begin{aligned} \text{slope} &= \frac{\text{difference in the } y\text{-coordinates}}{\text{difference in the } x\text{-coordinates}} \\ &= \frac{6-2}{1-4} \\ &= \frac{4}{-3} \\ &= -\frac{4}{3} \end{aligned}$$

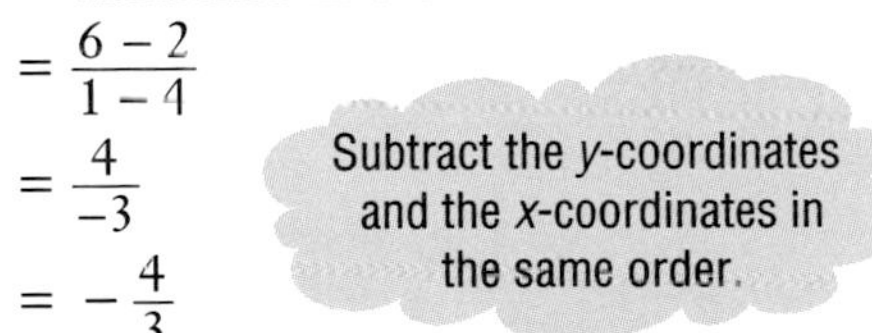

b) The 2 points are C(4, −3) and D(7, −3).

$$\begin{aligned} \text{slope} &= \frac{\text{difference in the } y\text{-coordinates}}{\text{difference in the } x\text{-coordinates}} \\ &= \frac{-3-(-3)}{4-7} \\ &= \frac{0}{-3} \\ &= 0 \end{aligned}$$

A horizontal line has no slope.

c) The 2 points are E(−3, 4) and F(−3, −2).

$$\begin{aligned} \text{slope} &= \frac{\text{difference in the } y\text{-coordinates}}{\text{difference in the } x\text{-coordinates}} \\ &= \frac{4-(-2)}{-3-(-3)} \\ &= \frac{6}{0} \end{aligned}$$

Since $\frac{6}{0}$ is undefined, the slope of a vertical line is not defined by a real number.

CONTINUED

Practice

1. State whether each slope is positive or negative.

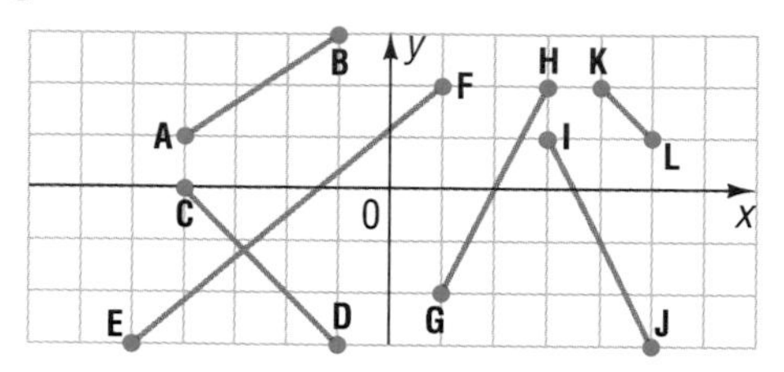

2. State the slope and the equation of each line segment.

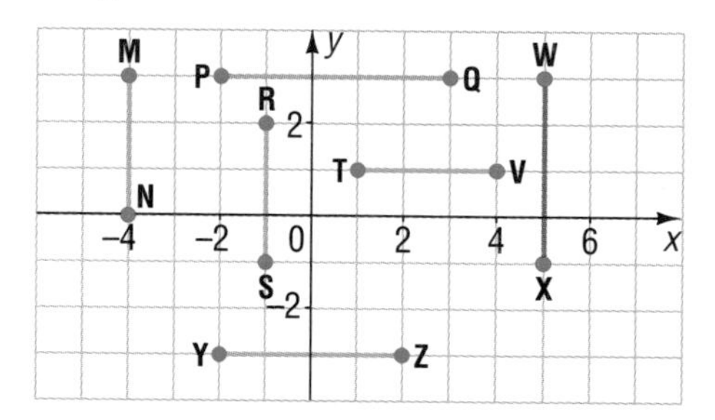

3. State the slope of each line segment.

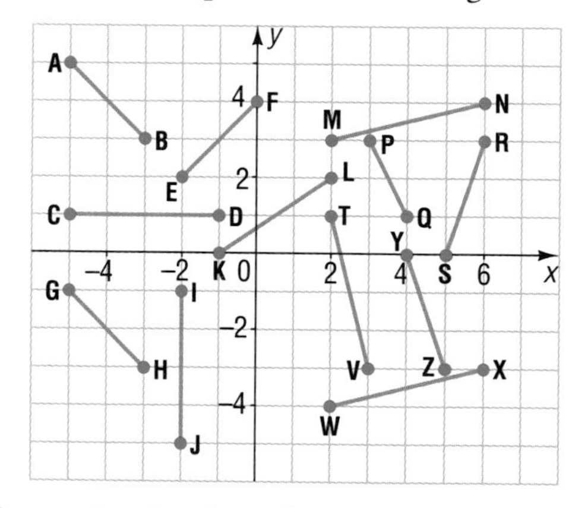

Determine the slope of the line passing through each pair of points.

4. A(5, 9), B(7, 8)   **5.** C(3, 4), D(7, 4)

6. M(9, 3), N(0, 1)   **7.** P(0, –5), Q(0, 4)

8. K(2, 5), L(0, 8)   **9.** A(–2, –3), B(3, 3)

State whether the slope is positive or negative for each pair of points.

10. C(3, 4), D(–2, –5)   **11.** P(–1, 3), Q(0, 2)

12. M(–6, 0), N(4, 5)   **13.** X(3, 5), Y(7, 2)

Problems and Applications

14. The graph shows how far Danzel cycled in 3 h.

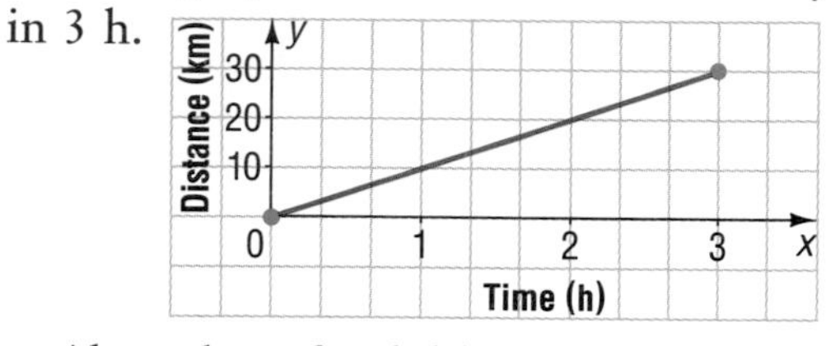

a) About how far did he travel in 2 h 45 min?

b) About how long did it take him to cycle 15 km?

c) What does the slope of this line tell you?

15. a) On the same grid, draw a line through A(1, 4) and B(–2, 2) and a line through P(–1, –1) and Q(2, 1).

b) Calculate the slope of each line in part a).

c) Use your answer to part b) to explain why these lines are parallel.

16. a) Plot the points P(–4, 2), Q(–1, –2), R(4, –2), and S(1, 2). Join PQ, QR, RS, and SP.

b) Draw diagonals PR and QS.

c) The diagonals PR and QS are perpendicular to each other. Find the slope of each diagonal. How are the slopes related?

d) Is this relationship true for any pair of perpendicular lines?

17. The lengths of 2 ski slopes are 625 m and 760 m. The horizontal distance from the start of the run to the end of the run for both slopes is 300 m.

a) What is the height of the higher ski slope to the nearest metre?

b) Which of the 2 ski slopes is steeper? Explain.

NUMBER POWER

Find the values of x and y that make the following true.

$$\frac{x^y}{y^x} = 1, x \neq y$$

Nonlinear Relations

Activity 1 Reading Graphs

Many people enjoy building and flying miniature rockets. The graph on the right represents the flight of a miniature rocket.

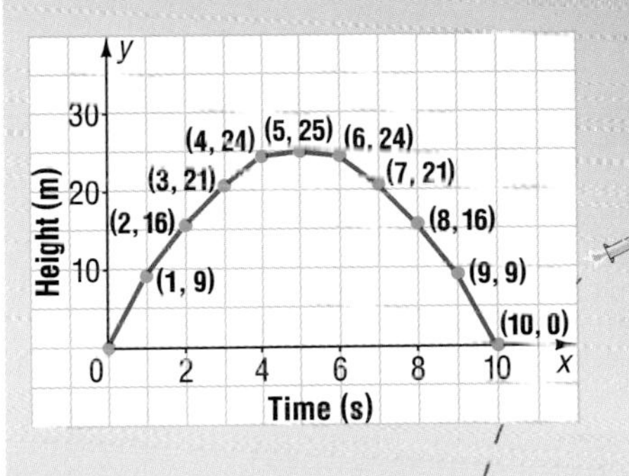

1. Is this relation a linear relation? Explain.
2. For how long is the rocket in upward flight?
3. How high is the rocket after 3 s? after 9 s?
4. At what times is the rocket 24 m high?
5. At what times in whole numbers of seconds after launch is the rocket at the same height?
6. What is the maximum height reached by the rocket?
7. After how many seconds does the rocket hit the ground?

Activity 2 Drawing Graphs

A **second degree relation** is one in which the highest power of any term is 2. For example, $y = x^2 - 3$ is a second degree relation.

1. Copy and complete the following tables of values for the perimeter and area of a square.

a)

Side Length	Perimeter
0	
1	
2	
3	
4	
5	
6	

b)

Side Length	Area
0	
1	
2	
3	
4	
5	
6	

2. Graph each of the relations in question 1 on a separate grid. Join the points of the nonlinear relation with a smooth curve.
3. Let x be the side length and y be the perimeter of the square in question 1a). Write an equation that describes the relation.
4. Is the relation in question 3 linear? Explain.
5. Let x be the side length and y be the area of the square in question 1b). Write an equation that describes the relation.
6. Is the relation in question 5 linear? Explain.

CONTINUED

Length

Area = Length^2

Activity ❸ Other Relations

A **third degree relation** is one in which the highest power of any term is 3. For example, $y = x^3 - 8$ is a third degree relation.

1. Use interlocking cubes to build cubes with side lengths of 1 unit, 2 units, 3 units, and 4 units.

2. Calculate the surface area of each cube in question 1.

3. Copy and complete the tables of values for the surface area and the volume of a cube.

a)

Side Length	Volume
0	
1	
2	
3	
4	

b)

Side Length	Total Surface Area
0	
1	
2	
3	
4	

4. Let x be the side length and y be the surface area of each cube in question 1. Write an equation that describes the relation.

5. Is the relation in question 4 a linear relation? Explain.

6. Let x be the side length and y be the volume of each cube in question 1. Write an equation that describes the relation.

7. Is the relation in question 6 a linear relation? Explain.

Activity ❹

1. Work with a classmate to find other graphs of non-linear relations.

2. List the examples that you find. Sketch the graphs.

3. Compare your examples with other classmates'.

The Length and Midpoint of a Line Segment

Activity 1

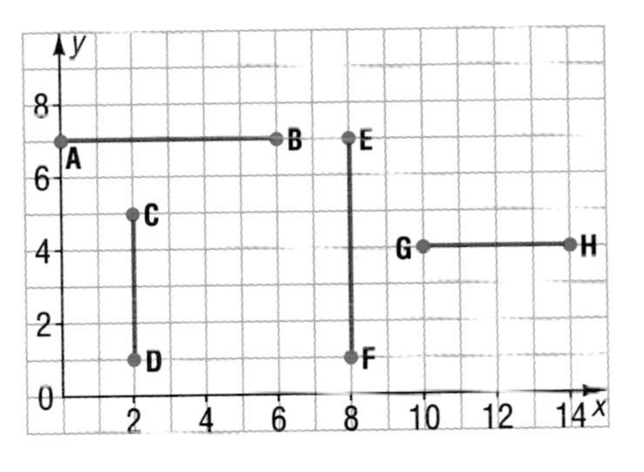

1. Find the length of each line segment by counting units.
2. Describe how these lengths can be found from the coordinates of the endpoints of the line segments.
3. Can the length of a line segment ever be a negative number? Explain.
4. Find the coordinates of the midpoint of each line segment.
5. Describe how the coordinates of the midpoint can be found from the coordinates of the endpoints.

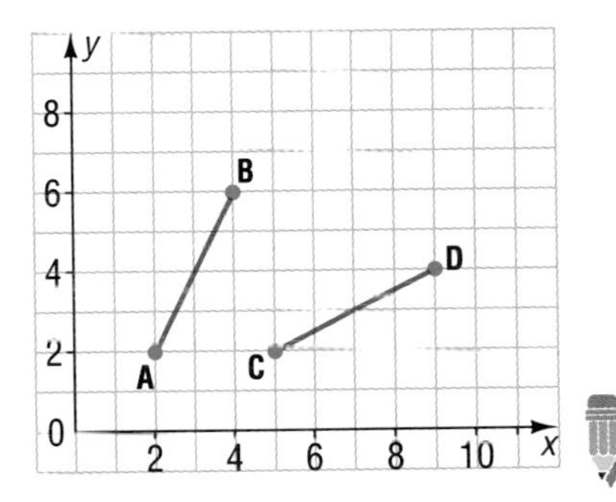

Activity 2

1. Copy the line segments shown onto grid paper.
2. Make 2 right triangles with each line segment as the hypotenuse of a triangle.
3. Write the coordinates of the 3 vertices for each triangle you drew in question 2.
4. Find the lengths of the horizontal and vertical line segments.
5. Describe how the lengths of the line segments AB and CD can be found using the Pythagorean Theorem.
6. Calculate the lengths of line segments AB and CD to the nearest tenth of a unit.
7. Find the coordinates of the midpoint of each line segment.
8. Describe how the coordinates of the midpoint can be found from the coordinates of the endpoints.

Activity 3

1. Plot the following points on a grid.

a) R(−3, 3) and S(0, 0) **b)** M(−4, −5) and N(−1, −2)

c) W(2, 1) and X(5, −1) **d)** J(1, −5) and K(7, −3)

2. Draw line segments RS, MN, WX, and JK and calculate their lengths to the nearest tenth of a unit.
3. Choose a pair of points. Have a classmate calculate the length of the line segment joining these 2 points. Check your classmate's answer.

Automobiles

Automobiles are designed to be visually appealing to potential buyers. They are also designed so that there will be less air resistance at high speeds.

Activity ❶

1. Draw 3 reference points on a blank piece of paper, as shown below.

2. Join your 3 reference points using curved lines.

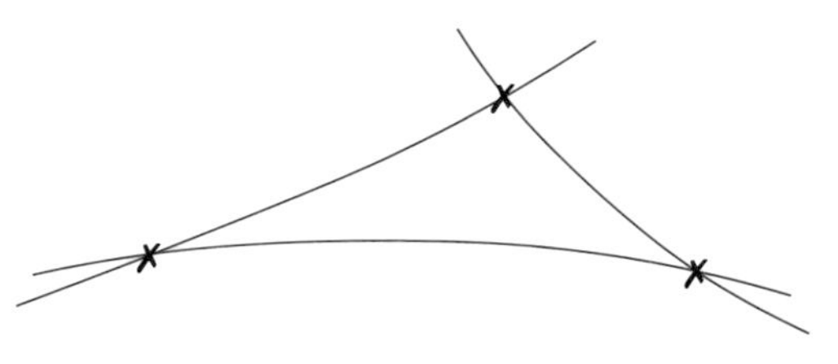

3. Draw the shape of an automobile within the curved lines.

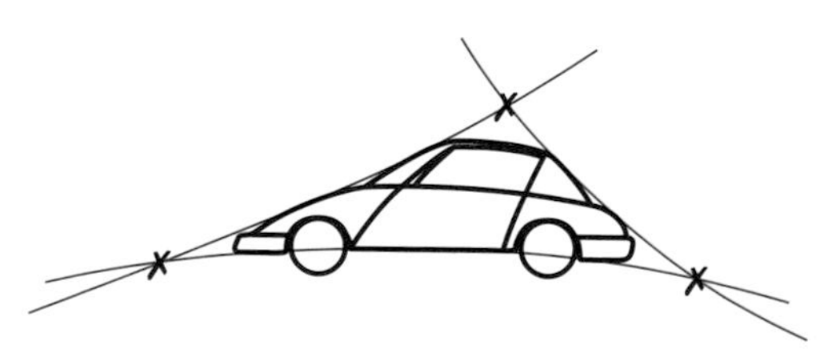

4. How would you locate the reference points to design a low sportscar? a family sedan?

5. Compare your answers to question 4 with your classmates'.

Activity ❷

1. Trace the automobile on the right onto a sheet of blank paper.

2. Draw its 3 reference points and the curved lines.

3. Trace the 3 reference points onto another sheet of blank paper and trade them for a classmate's.

4. Design a new automobile with your classmate's reference points.

5. How similar are your designs to each other's and to the original design?

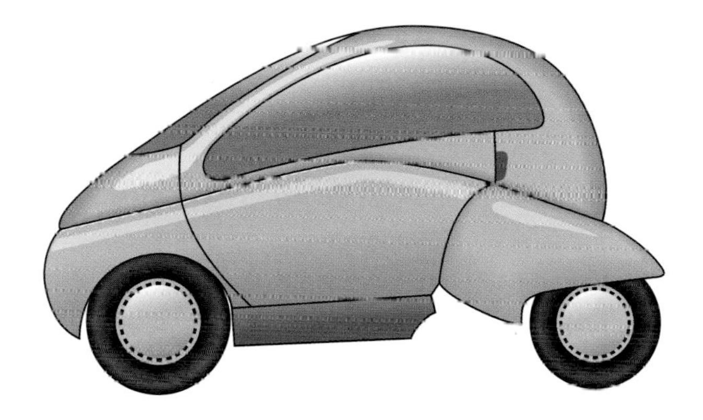

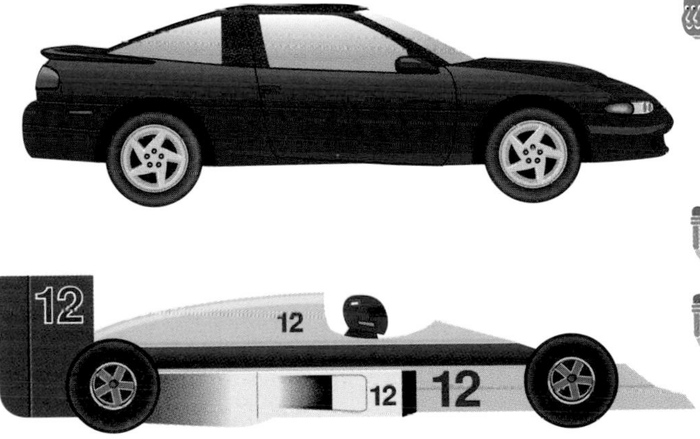

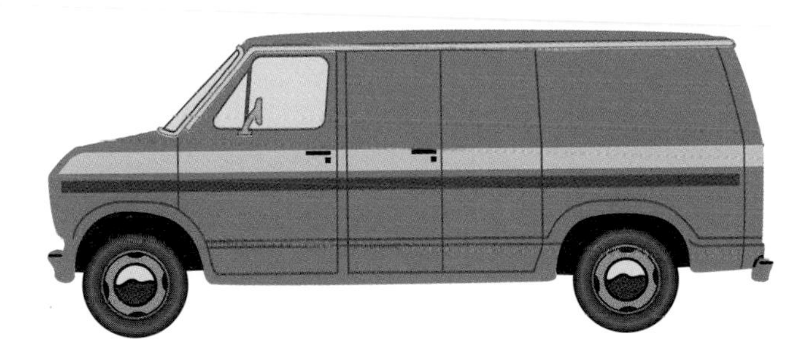

Activity ❸

1. Examine the pictures of the sportscar, racing car, and van shown on the left.

2. Rank these 3 vehicles in terms of their maximum speed, from lowest to highest.

3. Which of the vehicles is the most air resistant? Explain.

4. What shapes offer the least air resistance? Explain why this is so.

5. Trace each picture onto a blank sheet of paper.

6. Locate the 3 reference points for each type of vehicle.

7. Trade your 3 sets of reference points for a classmate's.

8. Design 3 new vehicles from your classmate's reference points.

9. Compare your designs with the original designs.

Activity ❹

1. Experiment with the design of an automobile using 4 and 5 reference points. Compare your results with your classmates'.

2. Research computer programs that are available to design automobiles.

Review

Copy and complete each table of values.

1. $y = 2 + x$

x	y
−1	
0	
1	

2. $x + y = 8$

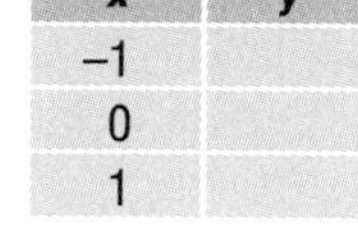

x	y
−1	
0	
1	

3. $x - y = 0$

x	y
−1	
0	
1	

4. $y = 2x + 5$

x	y
−1	
0	
1	

Make a table of values for each equation.
Use $x = -1, 0, 1$.

5. $y = 4x - 1$

6. $y = 1 - 2x$

7. $y = x$

8. $y = 2x$

9. $y = 5x - 1$

10. $y = -x$

11. Write the coordinates of each letter as an ordered pair.

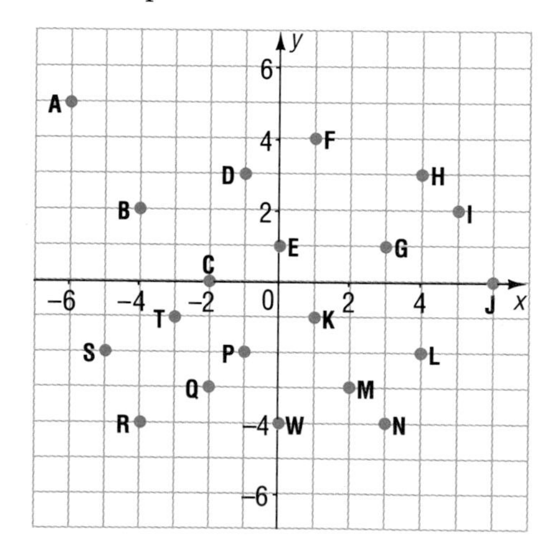

12. a) Plot the points A(−3, 6), B(2, 6), and C(−3, −2) on a grid.
b) Write the coordinates of the fourth point needed to form a rectangle.
c) What are the dimensions of this rectangle?

13. Plot and join each set of points in order. Join the last point to the first point in each set to form a closed figure. Name the figure.

a) P(6, 1), Q(2, −3), R(5, −2)

b) A(−5, 3), B(−4, −1), C(4, −1), D(3, 3)

c) P(0, −3), Q(−5, −3), R(−5, −2), S(0, −2)

14. a) Plot the points A(2, −3) and B(5, 0).
b) Find the coordinates of C so that $\triangle ABC$ is a right triangle.
c) How many different positions of C are possible?

15. Find 5 ordered pairs for each of the following equations.

a) $y = x + 2$

b) $y = x - 2$

c) $x + y = 6$

d) $x - y = 4$

16. Graph each equation in question 15.

17. Plot the points given in these tables of values. Write an equation that describes each relation.

a)

x	y
−1	1
0	2
1	3

b)

x	y
−1	−3
0	0
1	3

18. Graph each equation by finding the points where the line crosses the axes.

a) $x + y = 3$

b) $x - y = -1$

c) $2x + 3y = 6$

d) $2x - 5y = -10$

19. What is the point of intersection for each of the following pairs of relations?

a)

x	y
−2	−2
−1	−1
0	0
1	1

x	y
2	−2
1	−1
0	0
−1	1

b)

x	y
−3	−5
−2	−4
−1	−3
0	−2

x	y
−2	0
−1	−1
0	−2
1	−3

20. a) Make a table of values for the equations $y = 5 - x$ and $y = x + 7$.
Use $x = -2, -1, 0, 1, 2$.
b) Draw and label the graph of each equation on the same grid.
c) What is the point of intersection?

21. This graph shows the distance a train travels over time.

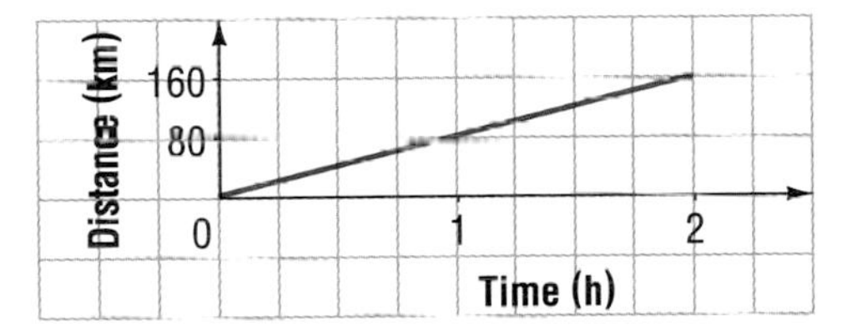

a) How far does the train travel in 1.25 h?
b) How long does it take the train to travel 20 km?

22. This graph shows the relation between the width of the image on a movie screen and the distance of the projector from the screen.

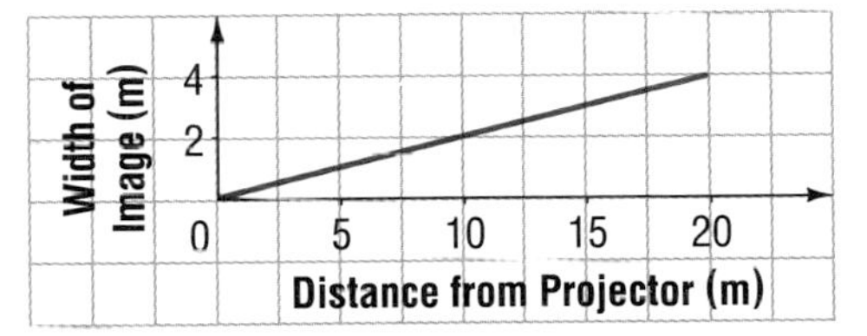

a) How wide is the image if the projector is 12.5 m from the screen?
b) How far is the projector from the screen if the width of the image is 3.5 m?

23. Determine the slope of the line passing through each pair of points.

a) A(4, 2), B(8, 6) **b)** P(5, 3), Q(1, 7)
c) M(1, 0), N(0, 0) **d)** C(0, 4), D(0, −5)
e) S(3, 4), T(2, 6) **f)** W(3, 7), X(5, 10)

24. Show that the line segment joining A(−2, 5) and B(1, 8) is parallel to the line segment joining C(6, 8) to D(9, 11).

25. a) Graph the equation $y = x^2$.
b) Is $y = x^2$ a linear equation? Explain.

Group Decision Making
Important Inventions

In this activity you will discuss and make a list of 20 of the most important inventions in the world.

1. Decide as a class on the difference between an invention and a discovery. As a class, list some categories of important inventions. The categories could include science, industry, transportation, medicine, and communications.

2. In your home group, list 20 important inventions.

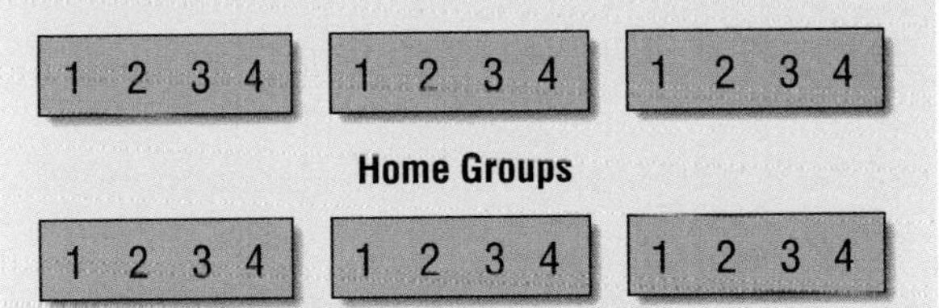

3. Discuss how the world would be different without these inventions. Use this discussion to order the inventions from most important to least important.

4. Make a presentation to the entire class explaining why your group chose each invention and why you ordered them the way you did.

5. As a class, list the 10 most important inventions. Decide which invention each home group will research.

6. In your home group, research the inventor of your invention. Discuss how the inventor might have used mathematics in the invention. Prepare a report of your findings. Present your report to the class. Be creative with your presentation.

7. As a class, evaluate the process.

Chapter Check

1. Copy and complete each table of values.

a) $y = 2 - x$

x	y
–2	
–1	
0	
1	

b) $x + y = 3$

x	y
2	
1	
0	
–1	

c) $y - x = 2$

x	y
–2	
–1	
0	
1	

d) $y = 3x + 1$

x	y
2	
1	
0	
–1	

2. Make a table of values for each equation. Use $x = -1, 0, 1$.

a) $y = 3x + 1$ **b)** $y = 1 - x$

c) $y = -7x$ **d)** $y = x - 1$

e) $y = 4x + 1$ **f)** $y = x + 3$

3. Write the coordinates of the letters as ordered pairs.

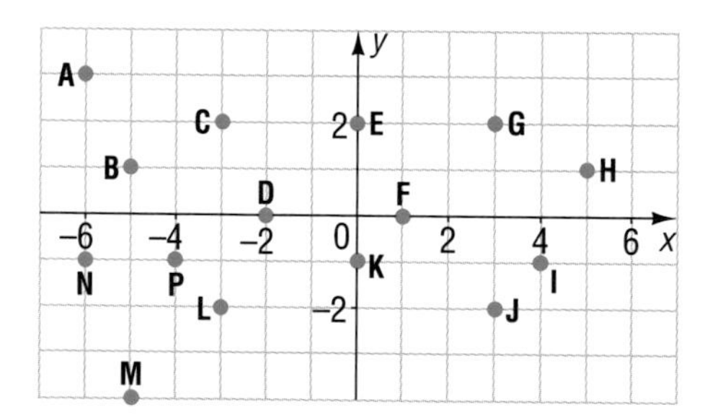

4. Plot the points A(–1, 4), B(1, 4), C(1, –4), and D(–1, –4). Join them in order. Join the last point to the first point to form a closed figure.

a) What figure have you drawn?

b) What are the figure's dimensions?

c) Which line segments have a slope of zero and which have a slope that is undefined?

5. Write 5 ordered pairs for the equation $y = 3x - 1$.

6. Write an equation that describes the relation in each table of values.

a)

x	y
1	4
0	3
–1	2

b)

x	y
1	–5
0	0
–1	5

7. Graph the equation $5x - 3y = -15$ by finding the points where the line crosses the x- and y-axes.

8. This graph shows how far Joan ran in 5 s.

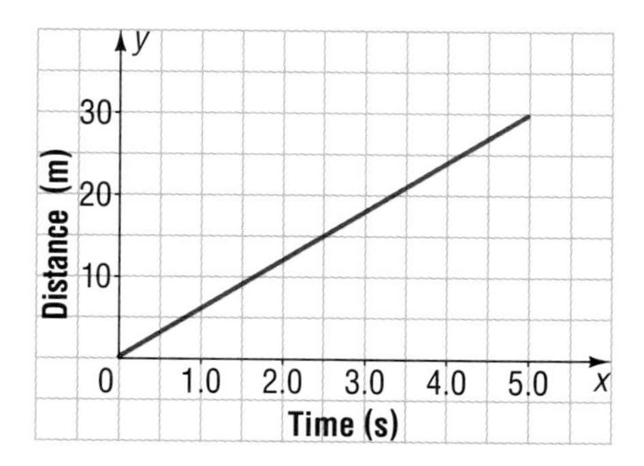

a) Approximately how far did Joan run in 1.75 s?

b) How long did it take her to run 7.5 m?

9. Graph the lines $x + y = 5$ and $x - y = 3$ on the same set of axes. What is the point of intersection?

10. a) Find the slope of the line passing through the points A(–2, 1) and B(3, 7).

b) Write 2 other points that lie on the same line.

11. a) Calculate the slope for 2 different pairs of points in this relation.

x	y
–2	3
–1	0
0	–1
1	0
2	3

b) Is this a linear relation? Explain.

Using the Strategies

1. Use any of the arithmetic operations and brackets, if necessary, to combine each set of numbers so that they equal the number in brackets.

a) 4, 5, 12 [8] **b)** 3, 7, 14 [10]

c) 4, 6, 7, 8 [15] **d)** 3, 5, 6, 11 [19]

2. The value of six factorial (6!) is

$$6 \times 5 \times 4 \times 3 \times 2 \times 1$$

Calculate.

a) 5! **b)** 4! **c)** 3! **d)** 2!

3. What number must be added to the numerator and denominator of the fraction $\frac{2}{11}$ to make it equal $\frac{1}{2}$?

4. This problem was part of the Canada/U.S. qualifying test for the First World Puzzle Team Championship held in 1992.

How many triangles with side length 1 unit, with side length 2 units, with side length 3 units, with side length 4 units, and with side length 5 units are in this figure?

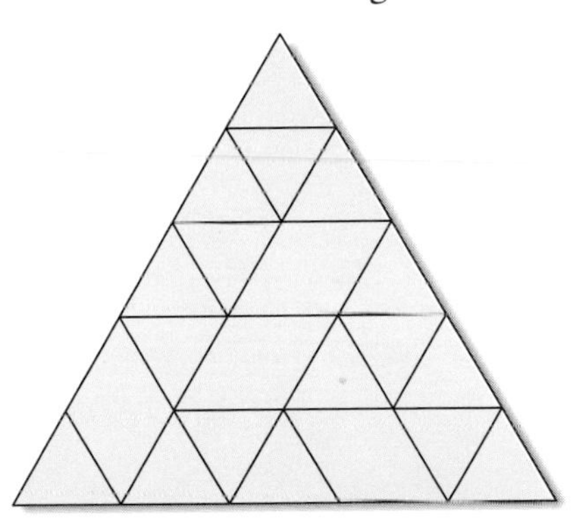

5. In a student council election, there are 2 candidates for president: Ali and Beth. There are 3 candidates for vice-president: Connie, Devo, and Eleanor. There are 3 candidates for treasurer: Franco, Gino, and Helen. One possible set of winners is Beth, Devo, and Helen. How many others are there?

6. Is it possible to guess any number from 1 to 1024 in 10 guesses or fewer if you are told on each guess that it is correct, too large, or too small? Explain.

7. You have been asked to buy the numbers for your baseball team's shirts. The shirts are numbered from 1 to 25. If each numeral costs $1.25, including tax, how much money will you need?

8. The square has been divided into 8 triangles. One way to shade 4 of the 8 triangles is shown. How many other different ways to shade 4 triangles are there?

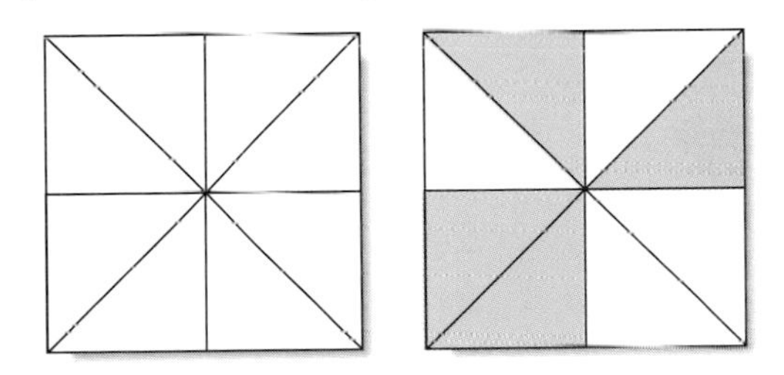

9. If 4 days after tomorrow is Thursday, what day is 3 days after yesterday?

10. The number 153 is equal to the sum of each of its digits cubed:

$$153 = 1^3 + 5^3 + 3^3$$

Find 1 other 3-digit number equal to the sum of each of its digits cubed.

1. One tonne of gold has a volume of about 0.052 m³.

a) What is the volume of Canada's annual gold production?

b) If Canada's annual gold production were made into a single cube, what would be the length of each side to the nearest 0.1 m?

CHAPTER 14

Transformations

Can you pass your body through a baseball card? You can if you change the card's shape with some scissors.

Fold the card lengthwise.

Starting at one end of the card, make the first cut from the fold, the second from the open side, and so on. Never cut all the way across from either side. Make as many cuts as possible. Begin the last cut from the fold.

Unfold the card and cut along the fold as shown, stopping before the ends.

Now stretch the card into a large circle and pass your body through it.

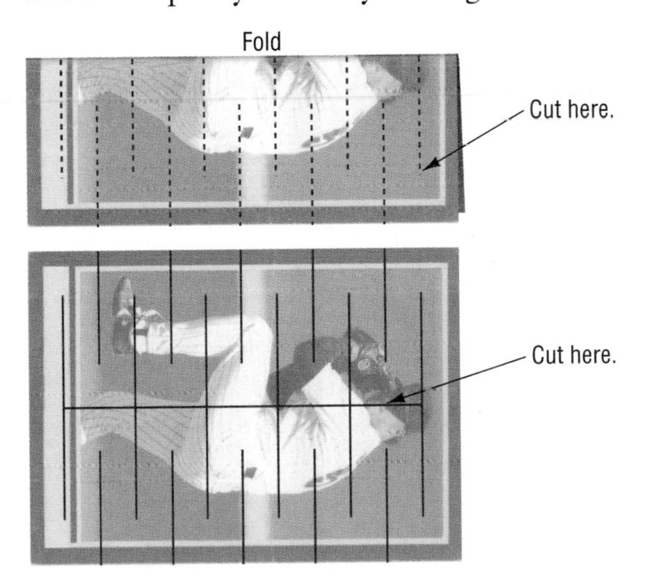

Activity ❶ Slides, Flips, and Turns

An elevator slides up and down in the elevator shaft. The elevator on the second floor can be called a slide image of the elevator on the ground floor.

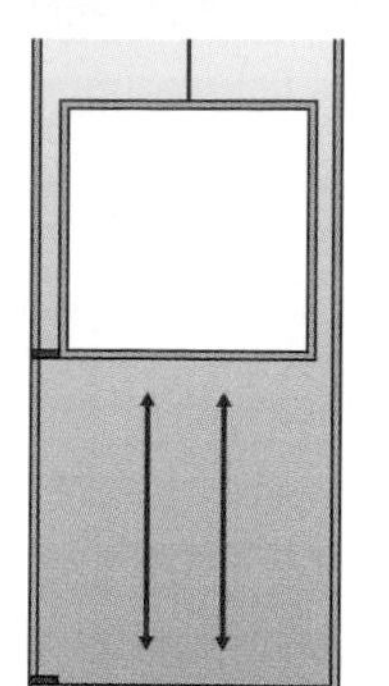

The reflection of a mountain in a still lake is a flip image of the mountain.

When a chair on a Ferris wheel moves from the bottom to the top, the chair at the top can be called a turn image of the chair at the bottom.

Find 3 more examples of each of the following and sketch the original figure and its image.

1. slides **2.** flips **3.** turns

Activity ❷ Patterns

1. Identify each pattern as a slide, flip, or turn.

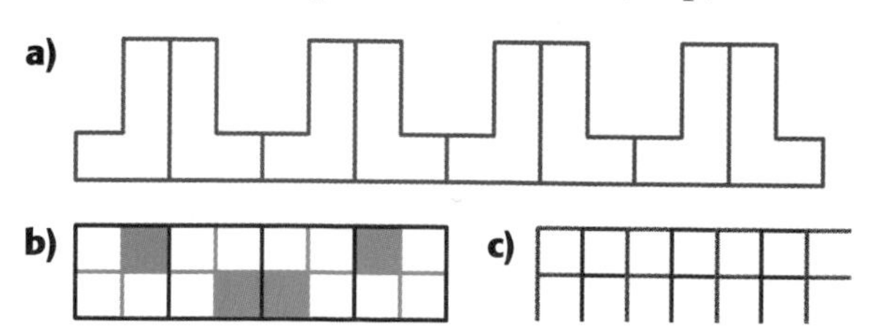

2. Draw a slide pattern, a flip pattern, and a turn pattern.

Activity ❸ The Square

Write the slide, flip, or turn that gives each image.

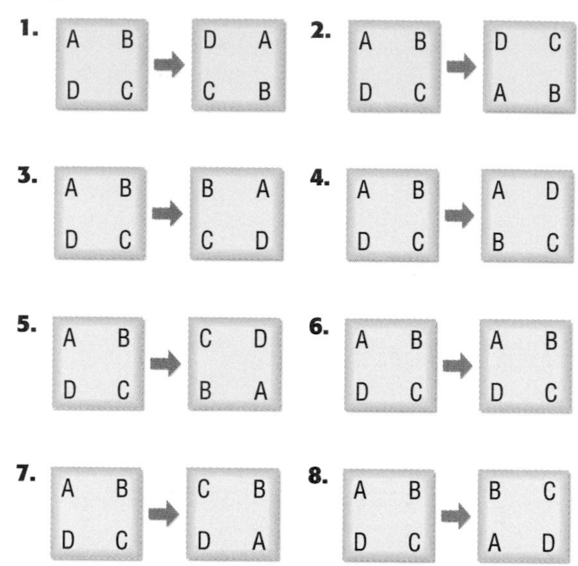

Activity ❹

Identify each figure as a slide, flip, or turn of the red figure.

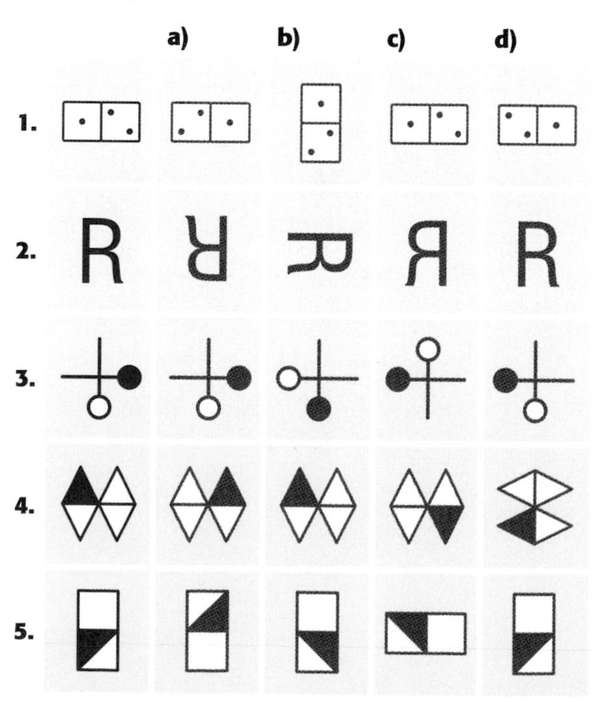

Activity ❺ Tile Patterns

In mathematics, tiling the plane means to cover an area with shapes so that there are no gaps and the shapes do not overlap.

Select 1 or more of these shapes and tile the plane to make an interesting design.

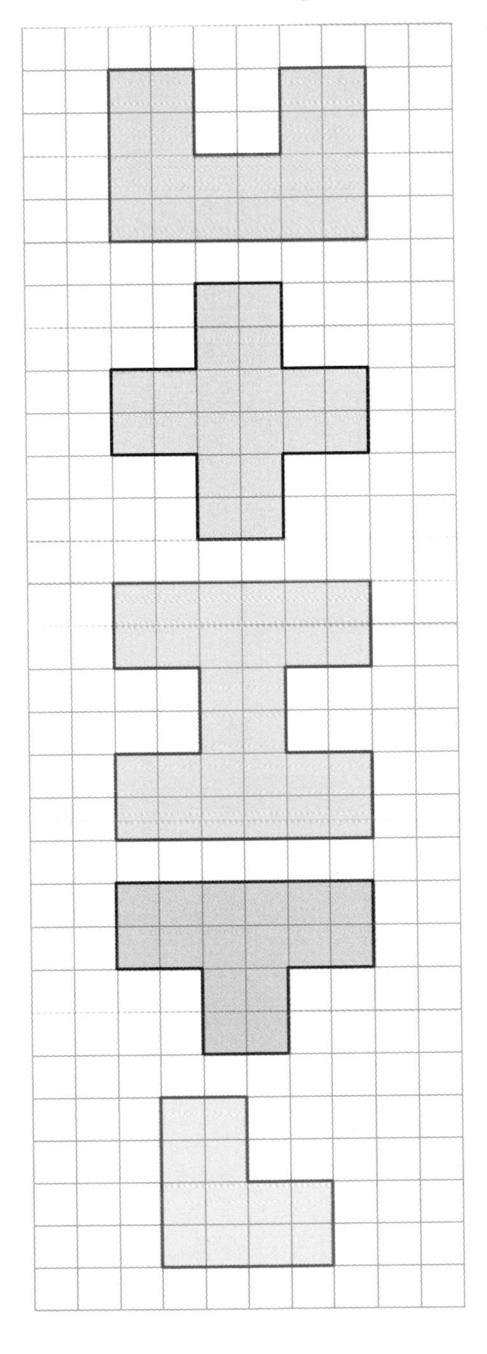

Mental Math

Calculate.

1. 220×0.1 **2.** 2000×0.01

3. $300 \div 100$ **4.** 4500×0.001

5. $0.2 \div 10$ **6.** $600 \div 0.01$

7. 0.003×1000 **8.** $0.01 \div 0.01$

Add.

9. \$3.99 + \$4.99 **10.** \$3.50 + \$4.25

11. \$3.98 + \$4.98 **12.** \$7.99 + \$2.95

13. \$5.90 + \$2.95 **14.** \$0.99 + \$2.45

15. \$1.95 + \$2.95 **16.** \$4.60 + \$3.60

Simplify.

17. $-7-(-3)$ **18.** $-12 \div (-6)$

19. $-4 \times (-5)$ **20.** $-6-4-2$

21. $-6+3-8$ **22.** $20 \div (-5)$

23. $9-8-7$ **24.** $5 \times (-2) \times (-1)$

Calculate.

25. 2 at \$2.99 **26.** 4 at \$2.98

27. 5 at \$0.99 **28.** 3 at \$19.98

29. 4 at \$3.95 **30.** 6 at \$4.98

31. 3 at \$2.95 **32.** 5 at \$1.98

Calculate.

33. $\frac{5}{6}-\frac{2}{3}$ **34.** $\frac{3}{4}+\frac{1}{8}$ **35.** $\frac{2}{5} \div \frac{2}{3}$

36. $\frac{1}{2} \times \frac{3}{5}$ **37.** $\frac{1}{3}$ of 72 **38.** $\frac{1}{4}$ of 84

Simplify.

39. 6^2-24+3 **40.** $100-8^2-6^2$

41. $500 \div 10+47$ **42.** $200-12^2$

43. $3^3+40+10$ **44.** $2^2+3^2+4^2$

45. $4 \times 50+23-1$ **46.** $7+20 \times 3+2$

47. $3^2 \times 100+47$ **48.** $29+29+29$

49. $19 \times 3-4$ **50.** 10^2+7^2-4

14.1 Translations

A **translation**, or slide, is a motion that is described by length and direction.

A helicopter can go straight up and down, or back and forth, or some combination of these two motions. If a helicopter goes down 30 m, then goes west 60 m, we say that the helicopter is "translated down 30 m," then "translated west 60 m." The description of each translation includes the direction and the distance.

Activity: Draw a Translation Image

Plot ΔABC with coordinates A(1, 1), B(5, 1), and C(5, 4). Translate each point 6 units to the right and 4 units up. Label the new points A′, B′, and C′. Join A′, B′, and C′ to form a triangle.

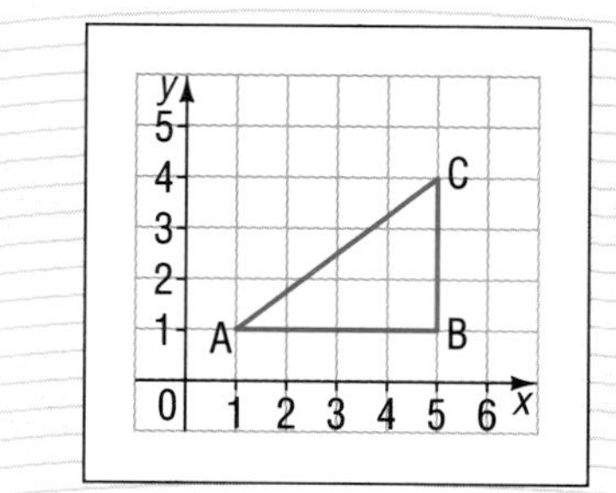

Inquire

1. What are the coordinates of the vertices of $\Delta A'B'C'$?

2. $\Delta A'B'C'$ is called the **translation image** of ΔABC. How do the lengths of the sides of $\Delta A'B'C'$ compare with the lengths of the sides of ΔABC?

3. Are the two triangles congruent? Explain.

4. What is true about the measures of the angles after a translation?

5. What number do you add to the x-coordinate of each vertex of ΔABC to give the x-coordinate of each vertex of $\Delta A'B'C'$?

6. What number do you add to the y-coordinate of each vertex of ΔABC to give the y-coordinate of each vertex of $\Delta A'B'C'$?

7. Describe a translation in your own words.

ΔDEF has been translated 5 units to the left and 3 units up (5L, 3U). $\Delta D'E'F'$ is the image of ΔDEF. Notice that D–E–F and D′–E′–F′ read in the same direction; in this case it is counterclockwise (ccw). We say that:

"ΔDEF and $\Delta D'E'F'$ have the same **sense**."

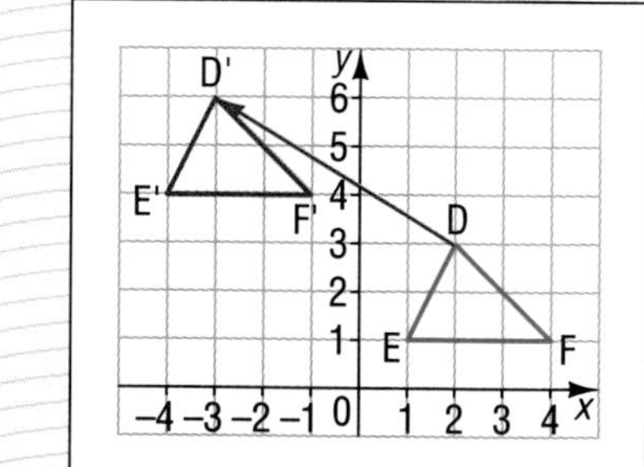

The translation arrow shows the distance the triangle is translated and the direction of the translation.

The translation can be described mathematically as the ordered pair [–5, 3] or as the following **mapping**.

$(x, y) \rightarrow (x - 5, y + 3)$

The lengths of line segments and the sizes of angles do not change in a translation. The original figure and its image have the same sense.

Example

ΔRST has vertices R(−3, −2), S(1, −3), and T(0, 4). Use the mapping $(x, y) \rightarrow (x - 2, y + 4)$ to draw the translation image of ΔRST.

Solution

Plot ΔRST and find the coordinates of the vertices of its image.

$$(x, y) \rightarrow (x - 2, y + 4)$$

For R: $(-3, -2) \rightarrow (-3 - 2, -2 + 4)$ so R′ is (−5, 2)
For S: $(1, -3) \rightarrow (1 - 2, -3 + 4)$ so S′ is (−1, 1)
For T: $(0, 4) \rightarrow (0 - 2, 4 + 4)$ so T′ is (−2, 8)

Plot R′, S′, and T′ and join them.
ΔR′S′T′ is the translation image of ΔRST.

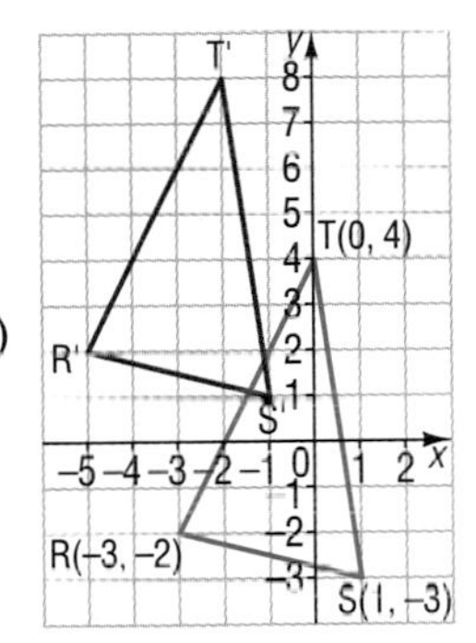

Practice

1. Which of the lettered figures are translation images of the green figure? Give reasons for your answers.

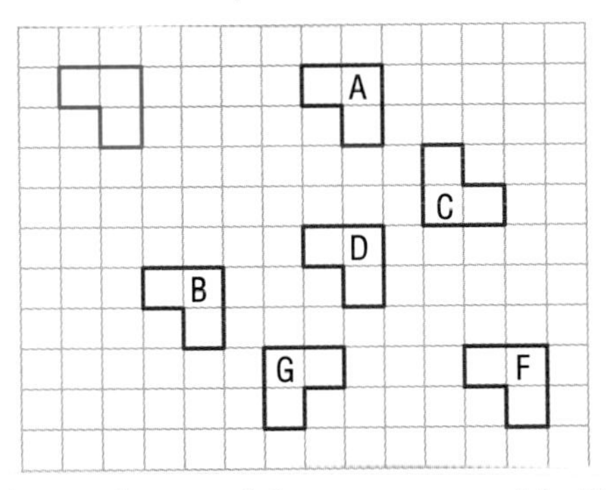

2. Plot each set of figures on a grid. Graph the image of each figure under the given translation.

a) 4 units right and 3 units up

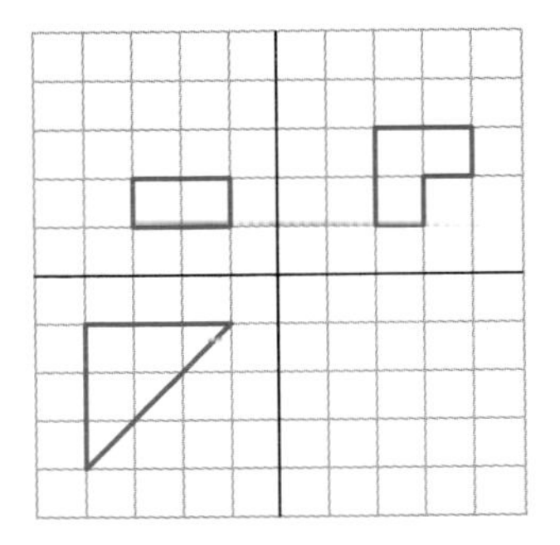

b) 2 units left and 5 units down

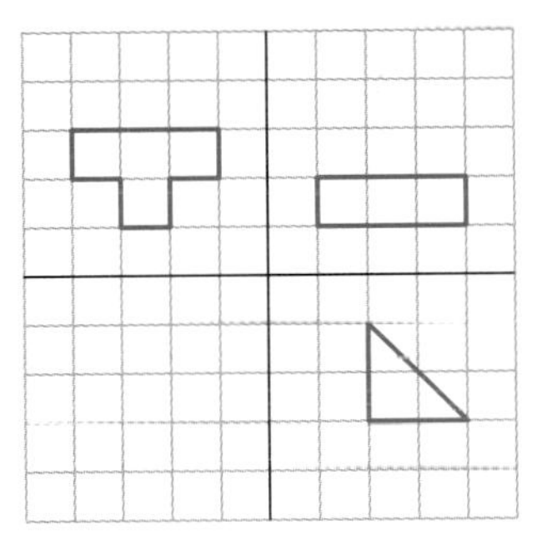

3. State the translation described by each arrow. Write your answers in the form $[x, y]$.

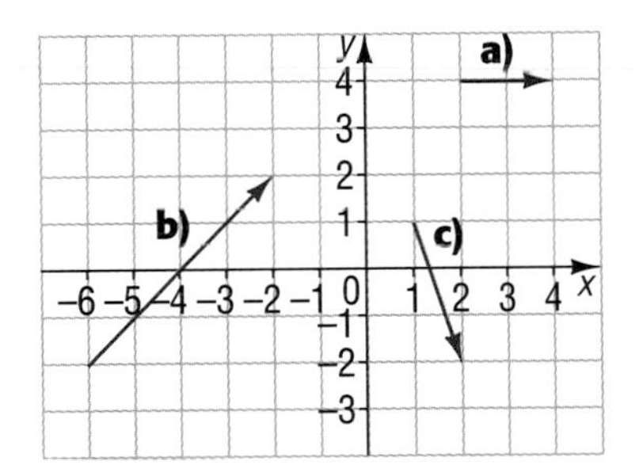

CONTINUED

Draw an arrow on grid paper to show each translation.

4. $[2,-3]$ **5.** $[-1,-5]$ **6.** $[1,-5]$ **7.** $[5,0]$
8. $[3,-2]$ **9.** $[-2,3]$ **10.** $[0,-5]$ **11.** $[5,-4]$

Name the translation that maps each green figure onto its purple image.

12. **13.**

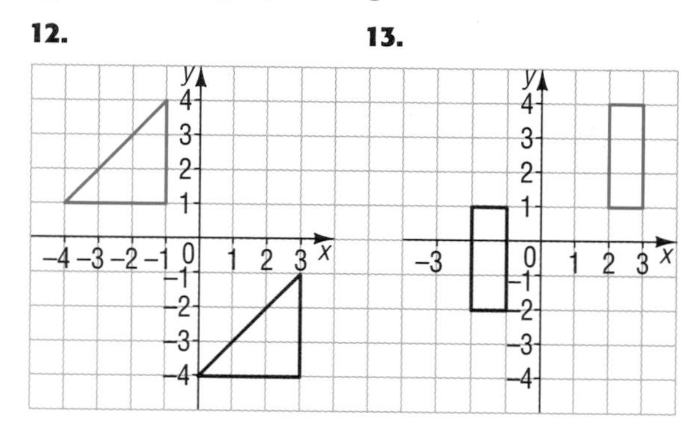

Describe each translation in words.

14. $(x,y) \rightarrow (x+3,y+2)$

15. $(x,y) \rightarrow (x-1,y+4)$

16. $(x,y) \rightarrow (x-2,y-3)$

17. $(x,y) \rightarrow (x+5,y-1)$

18. $(x,y) \rightarrow (x,y+6)$ **19.** $(x,y) \rightarrow (x-3,y)$

20. $[2,-3]$ **21.** $[-3,5]$ **22.** $[-4,2]$ **23.** $[1,-6]$

Find the coordinates of each point after the translation.

	Point	Translation
24.	(2, 3)	[3, 4]
25.	(–3, –1)	[2, –2]
26.	(5, –2)	[–1, 5]
27.	(–5, 7)	[–2, –6]

Express each translation as an ordered pair.

28. A(3, 5) to A′(7, 9)

29. C(2, 1) to C′(3, 0)

30. X(–2, –4) to X′(–3, –5)

31. M(0, –1) to M′(4, –6)

32. T(–1, –3) to T′(–6, 5)

33. P(0, 4) to P′(–1, –2)

Draw each triangle on grid paper. Then, draw the translation image.

34. A(3, 2), B(–1, 4), C(–3, –5)
$(x,y) \rightarrow (x-4,y+2)$

35. D(–2, 0), E(4, –1), F(2, –5)
$(x,y) \rightarrow (x+3,y-3)$

36. R(0, 0), S(–4, 0), T(–3, –5)
$(x,y) \rightarrow (x+2,y+3)$

Problems and Applications

37. The translation [2, 3] translates ΔABC so that the coordinates of the vertices of the image, ΔA′B′C′, are A′(3, 4), B′(0, 7), and C′(–2, –1). What are the coordinates of A, B, and C?

38. a) ΔRST has vertices R(3, 0), S(–1, 5), and T(–3, 1). Draw ΔRST on grid paper.
b) Determine the coordinates of the vertices of ΔR′S′T′ under the translation $(x,y) \rightarrow (x+4,y+2)$. Draw ΔR′S′T′.
c) Apply the translation $(x,y) \rightarrow (x+1,y-5)$ to ΔR′S′T′ and find the coordinates of the vertices of the image, ΔR″S″T″. Draw ΔR″S″T″.
d) Determine the translation that maps ΔRST onto ΔR″S″T″.

39. a) Draw the line $y = 2x + 1$ on grid paper.
b) Draw the image of the graph of the line after the translation $(x,y) \rightarrow (x+2,y+5)$.

40. What translations are needed to get a robot at point A to move an object from point B to point A?

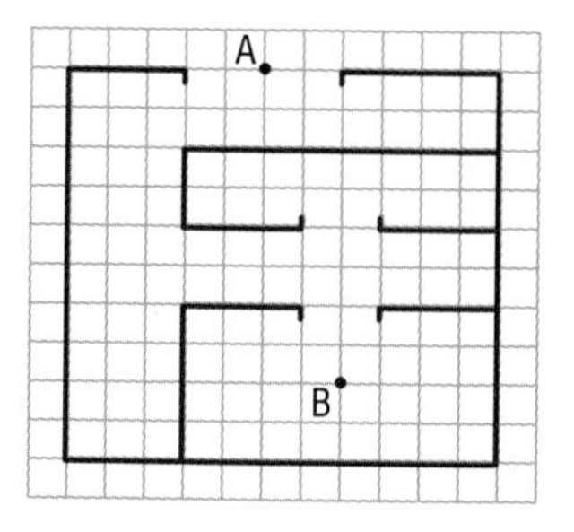

41. Write a problem like question 40. Have a classmate solve your problem.

14.2 Reflections

When you stand in front of a mirror, you see an image of yourself, your reflection. If you move closer to the mirror, the image moves closer. If you move back, the image moves back. The image is always the same distance from the mirror as you are. In mathematics, a **reflection** is a transformation in which a figure is reflected or flipped over a **mirror line** or **reflection line**.

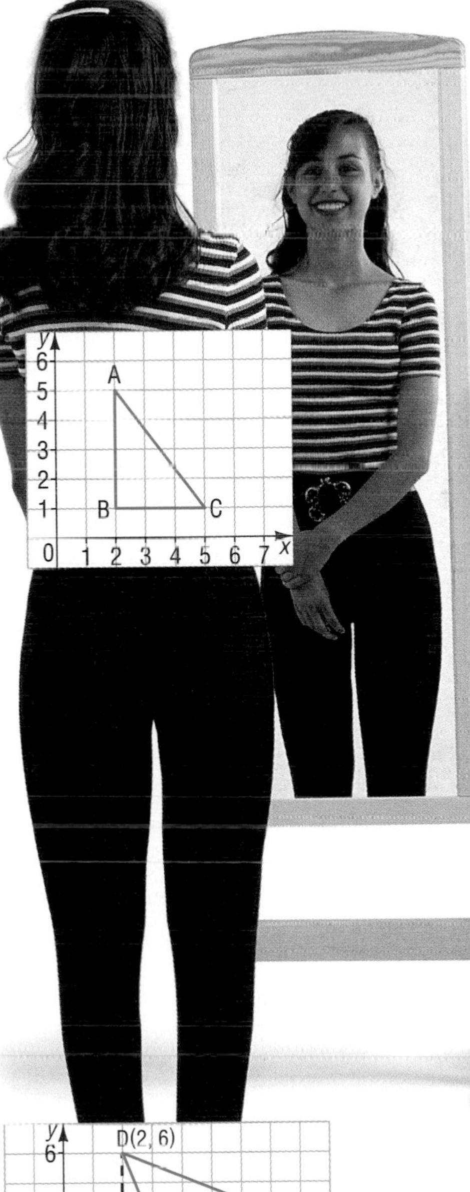

Activity: Draw a Reflection Image

Plot ΔABC with coordinates A(2, 5), B(2, 1), and C(5, 1). Reflect point A in the y-axis to locate point A′ so that A and A′ are the same perpendicular distance from the y-axis. Reflect points B and C in the y-axis to locate B′ and C′. Join A′, B′, and C′ to form a triangle.

Inquire

1. What are the coordinates of A′, B′, and C′?
2. $\Delta A'B'C'$ is the **reflection image** of ΔABC. How do the side lengths in $\Delta A'B'C'$ compare with the side lengths in ΔABC?
3. Are the two triangles congruent? Explain.
4. What is true about the measures of the angles after a reflection?
5. Do ΔABC and $\Delta A'B'C'$ have the same sense? Explain.
6. Describe a reflection in your own words.

Example

ΔDEF has vertices D(2, 6), E(4, 1), and F(7, 4). Draw the image of ΔDEF after a reflection in the x-axis.

Solution

Draw ΔDEF.
Locate D′ so that the perpendicular distance from D′ to the x-axis equals the perpendicular distance from D to the x-axis. The coordinates of D′ are (2, −6). Locate E′ and F′ in the same way.

D(2, 6) → D′(2, −6)
E(4, 1) → E′(4, −1)
F(7, 4) → F′(7, −4)

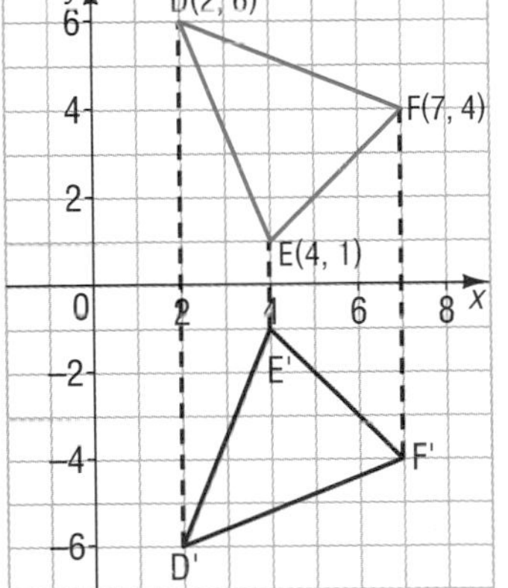

Join points D′, E′, and F′. $\Delta D'E'F'$ is the image of ΔDEF after a reflection in the x-axis.

The lengths of line segments and the sizes of angles do not change in a reflection. The sense of a reflection image is the reverse of the sense of the original figure.

CONTINUED

Practice

1. Which of the lettered figures are reflections of the green figure? Give reasons for your answer.

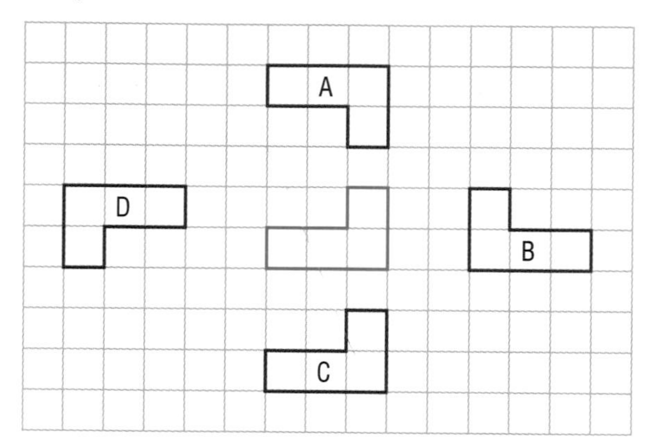

Trace each diagram. Use a Mira or paper folding to locate the reflection line.

2. **3.** **4.**

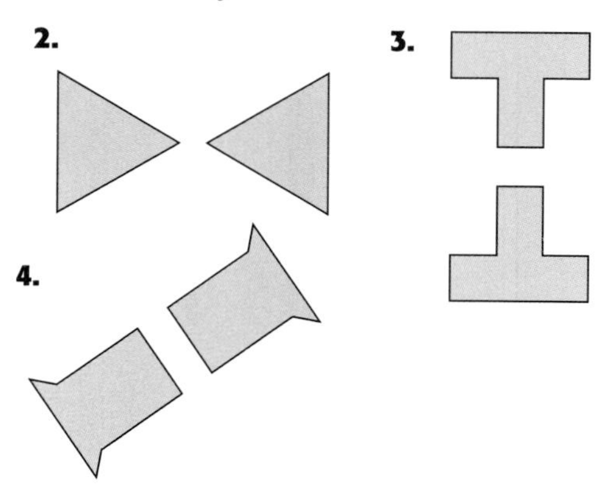

In each diagram, the reflection line is l. Trace each diagram and use a Mira or paper folding to draw the reflection image.

5. **6.**

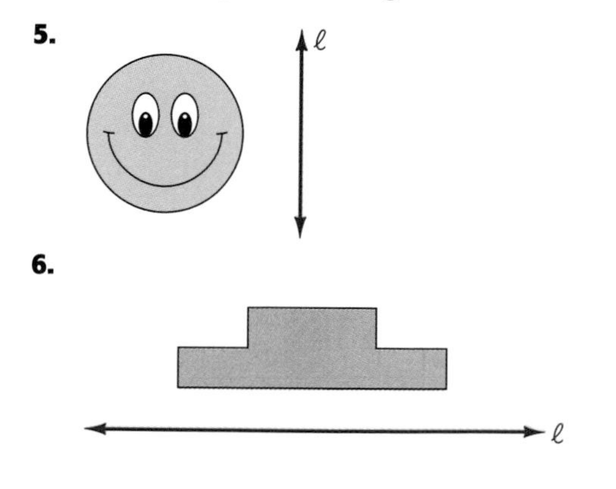

7.

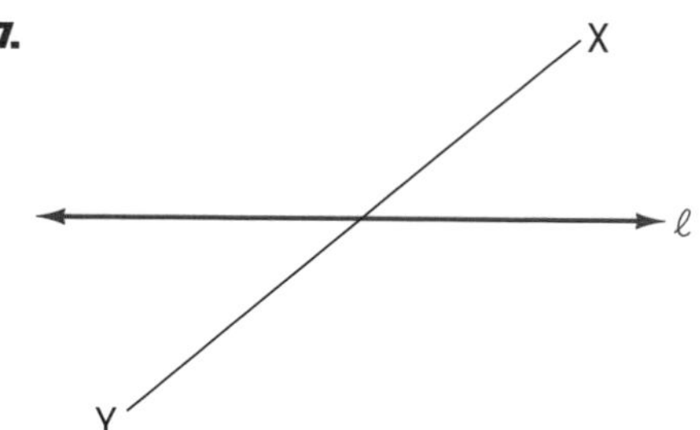

Copy each figure onto a grid and draw its reflection image in the x-axis.

8.

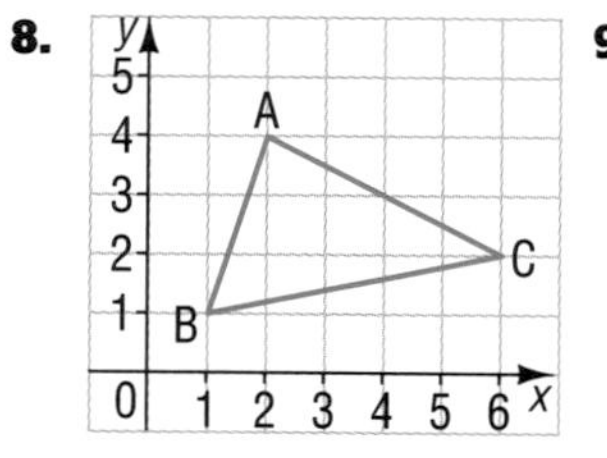

9.

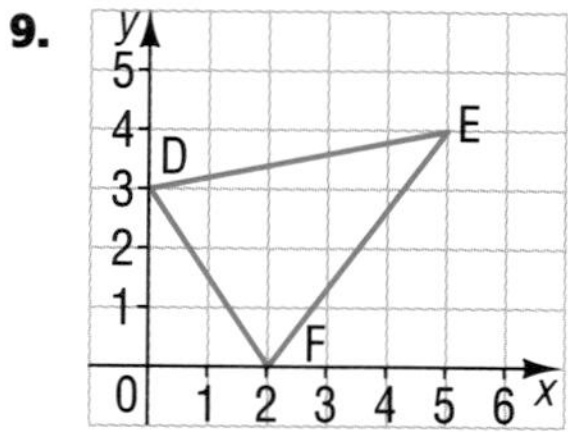

10. **11.**

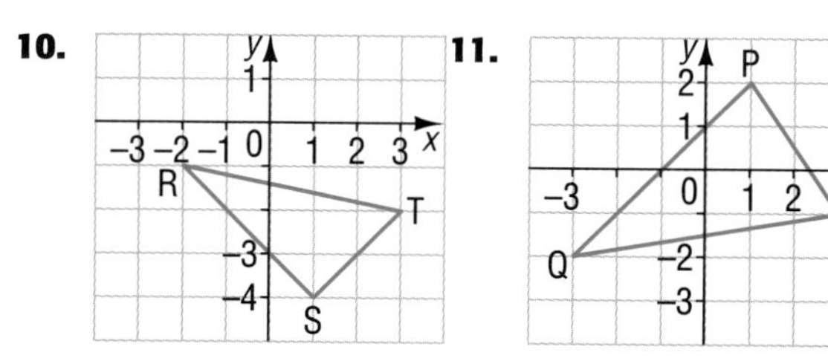

Draw each triangle on a grid. Draw the image after a reflection in the y-axis.

12.

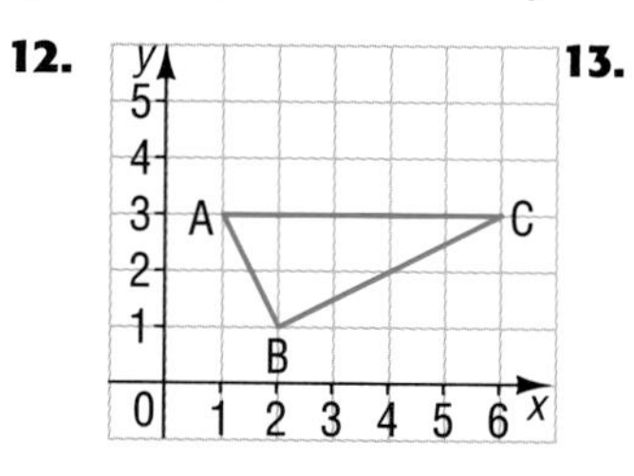

13.

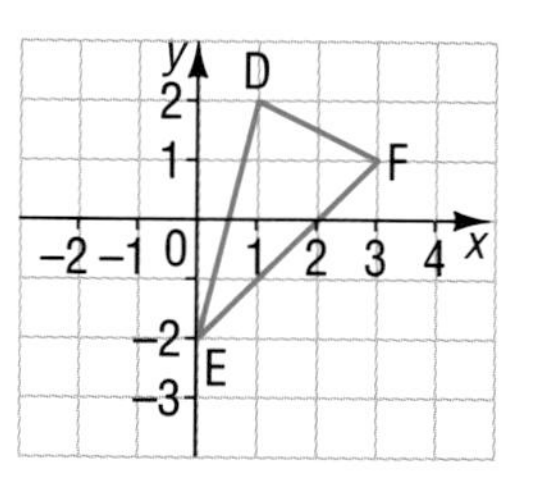

14.

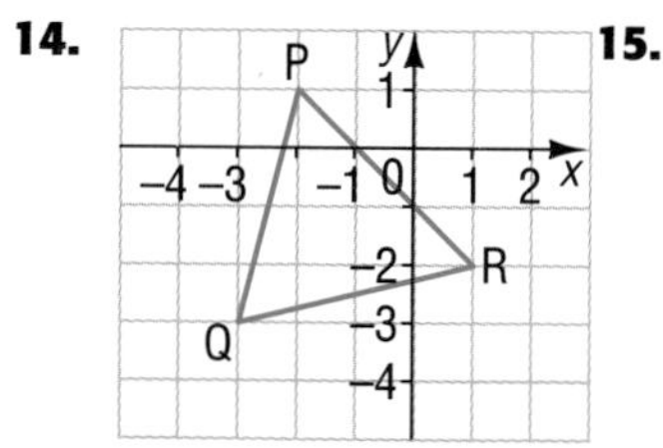

15.

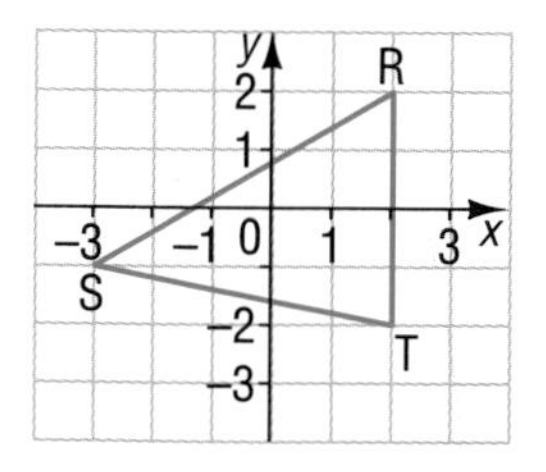

16. Find the coordinates of the image of each point after a reflection in each axis.

	Point	Reflection Line	
		x-axis	*y*-axis
a)	(1, 4)		
b)	(2, 3)		
c)	(–1, –2)		
d)	(–3, –2)		
e)	(–3, 2)		
f)	(4, 0)		

17. ΔABC has vertices A(1, 1), B(5, 2), and C(3, 6). Draw the image of ΔABC after a reflection in the y-axis.

18. ΔRST has vertices R(2, 5), S(–2, 4), and T(–1, –2). Draw the image of ΔRST after a reflection in the x-axis.

Problems and Applications

19. The letter M is a reflection of itself through the vertical red reflection line.

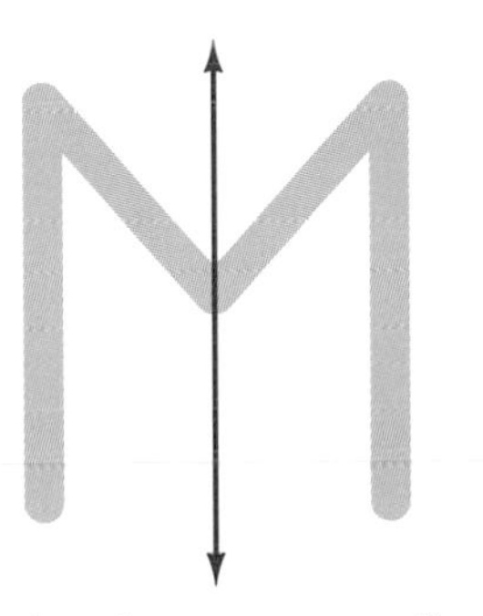

a) What other letters are reflections of themselves through vertical reflection lines?
b) What letters are reflections of themselves through horizontal reflection lines?
c) What letters are reflections of themselves through horizontal and vertical reflection lines?
d) Which of the ten digits are reflections of themselves?
e) Are you a reflection of yourself through a vertical reflection line? Explain.

Draw the triangles on a grid. Draw the image after a reflection in the given reflection line.

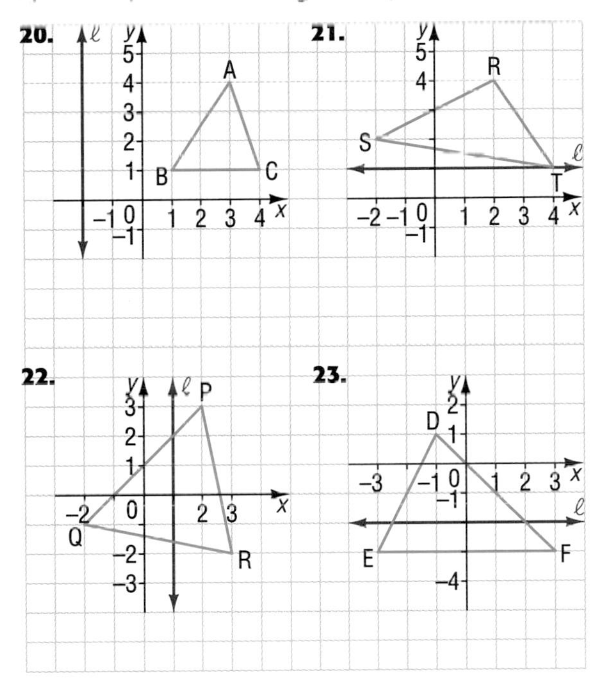

24. The coordinates of the vertices of ΔDEF are D(1, 5), E(–2, –2), and F(5, 1). Draw ΔDEF and its image after a reflection in the line $y = -1$.

25. a) ΔPQR has vertices P(–5, 6), Q(–7, 2), and R(–1, 1). Draw ΔPQR on grid paper.
b) Reflect ΔPQR in the y-axis and draw the image ΔP′Q′R′.
c) Reflect ΔP′Q′R′ in the x-axis and draw the image ΔP″Q″R″.

26. Which coordinate of a point does not change when the point is reflected
a) in the x-axis? **b)** in the y-axis?

27. a) Draw a graph of the line $x + y = 4$.
b) Reflect the line in the x-axis.
c) Reflect the line in the y-axis.

28. Why are the footprints you leave when you walk in the snow called a "glide reflection" of each other? Compare your answer with a classmate's.

14.3 Rotations

The hands of this antique clock turn or rotate about a point. This point is the centre of the clock face.

In mathematics, a **rotation** is a transformation in which a figure is turned or rotated about a point.

Activity: Draw a Rotation Image

Plot ΔABC with coordinates A(0, 0), B(3, 0), and C(3, 4). Use tracing paper to rotate ΔABC 90° counterclockwise (ccw) about the origin. Draw the **rotation image** of ΔABC and name it ΔA′B′C′.

Inquire

1. What are the coordinates of A′, B′, and C′?

2. How do the lengths of the sides of ΔABC compare with the lengths of the sides of ΔA′B′C′?

3. Are the two triangles congruent? Explain.

4. What is true about the measures of the angles after a rotation?

5. Do ΔABC and ΔA′B′C′ have the same sense? Explain.

6. A 90° rotation is called a $\frac{1}{4}$ turn.

a) What is a 180° rotation called?

b) What is a 270° rotation called?

7. Describe a rotation in your own words.

To rotate an object and find its image, you need to know the amount and direction of the rotation. You also need to know the **centre of rotation** or **turn centre**.

Example

ΔRST has vertices R(−5, 0), S(−1, 0), and T(−2, −3). Draw the image of ΔRST after a rotation of 180° clockwise (cw) about the origin.

Solution

Draw ΔRST. Find the rotation image of each vertex.

R(−5, 0) → R′(5, 0)
S(−1, 0) → S′(1, 0)
T(−2, −3) → T′(2, 3)
Join points R′, S′, and T′.
ΔR′S′T′ is the rotation image of ΔRST.

⊙ marks the turn centre

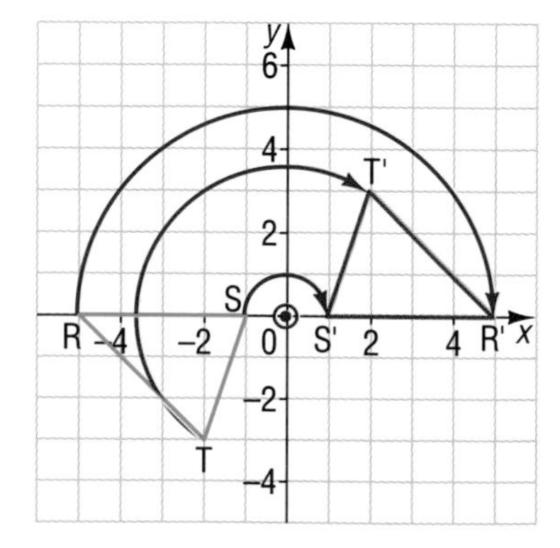

The lengths of line segments and the sizes of angles do not change in a rotation. The original figure and its image have the same sense.

Practice

The green flag has been rotated about the origin. The purple flag is the image. Name the rotation.

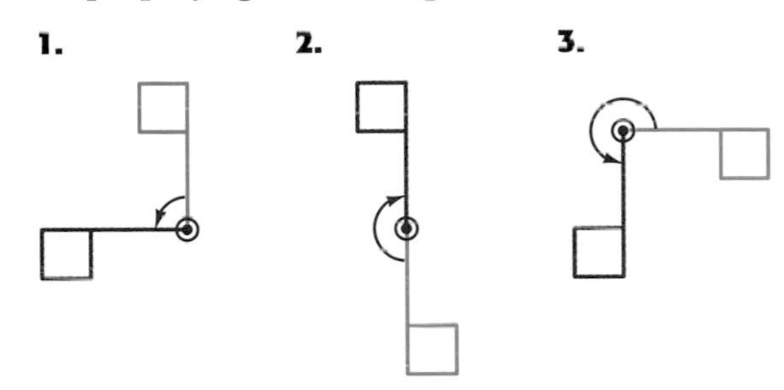

The green flag has been rotated about the origin. The purple flag is the image. Name 2 rotations for each case.

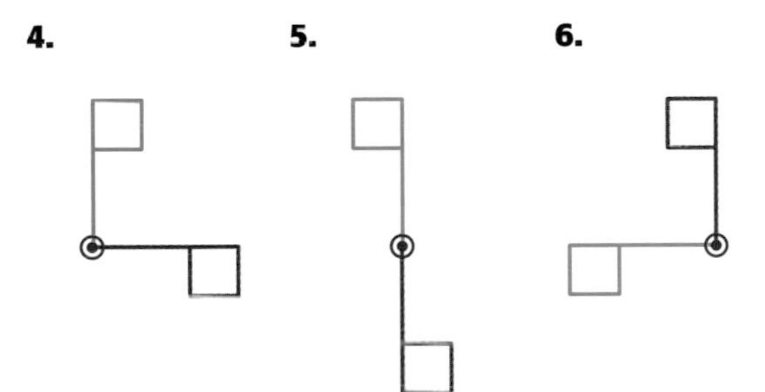

Copy each figure onto grid paper and draw the image after the given rotation about the red turn centre.

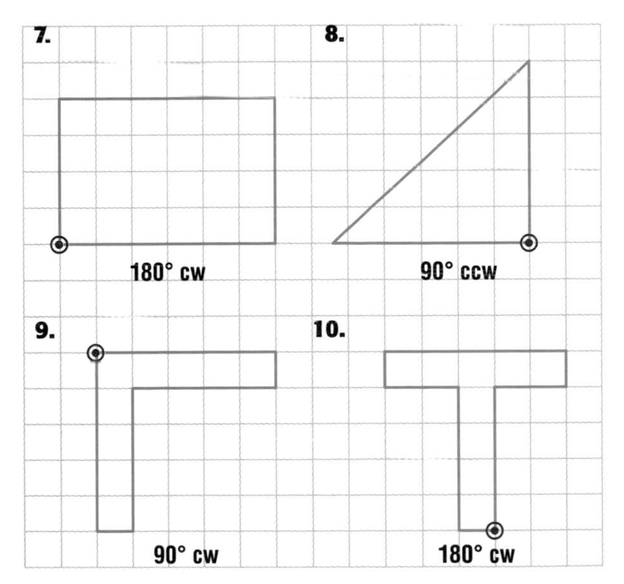

Draw the image of each figure after a 90° clockwise rotation about the origin.

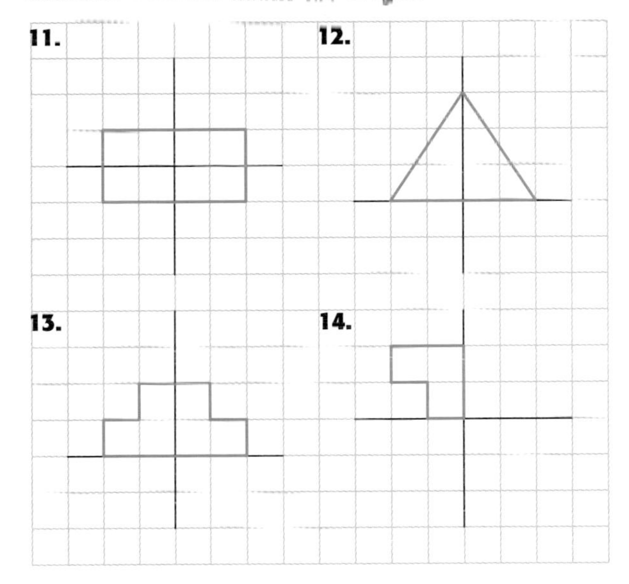

Draw the image of each figure after a 180° counterclockwise rotation about the origin.

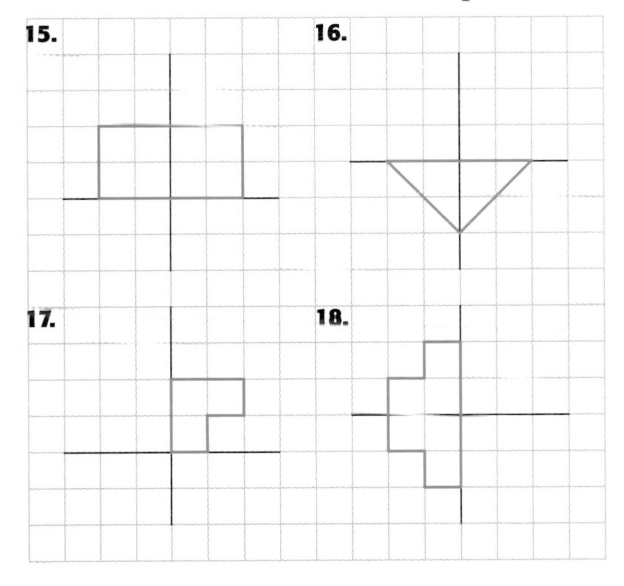

19. $\triangle DEF$ has vertices D(0, 0), E(–4, 2), and F(0, 6). Draw the image of $\triangle DEF$ after a 90° counterclockwise rotation about the origin.

20. $\triangle ABC$ has vertices A(1, 0), B(3, 4), and C(3, 0). Draw the image of $\triangle ABC$ after a 90° clockwise rotation about the origin.

CONTINUED

Copy each figure and find the image after the given rotation about the red turn centre.

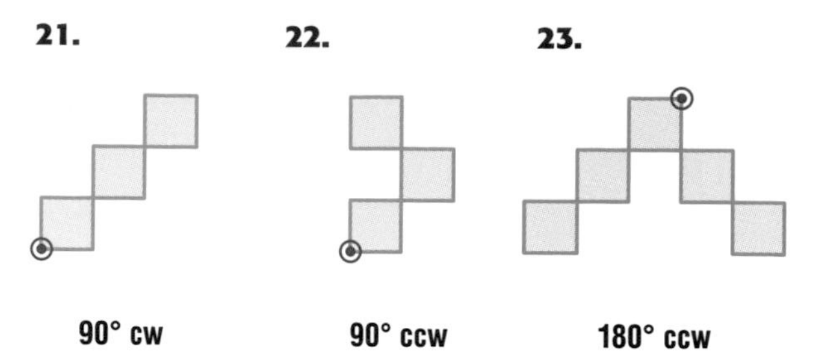

Copy and complete the table with the image of each rotation about the origin.

	Rotation			
	Point	90° CCW	180° CW	270° CCW
24.	(0, 6)			
25.	(–4, 0)			
26.	(0, –3)			
27.	(3, 4)			
28.	(–4, 3)			

Problems and Applications

29. Rectangle ABCD has vertices A(1, 0), B(1, 2), C(6, 2), and D(6, 0). Draw the image of the rectangle after a 180° counterclockwise rotation about the origin.

30. Parallelogram ABCD has vertices A(−1, 0), B(−2, 2), C(−6, 2), and D(−5, 0). Draw the image of the parallelogram after a 90° clockwise rotation about the origin.

31. Copy the figure onto grid paper and rotate it 90° clockwise about

a) point P **b)** the origin

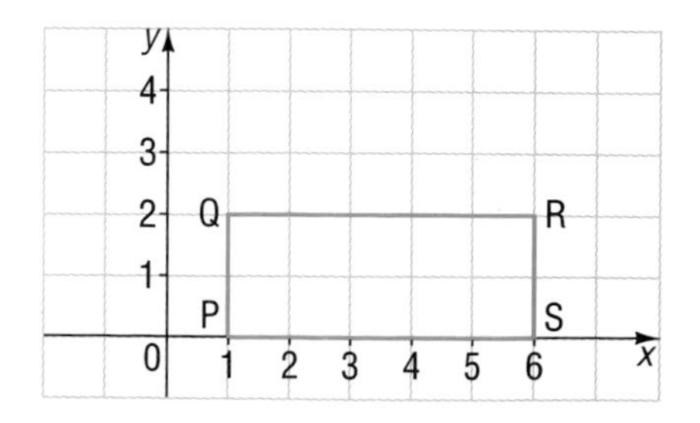

32. The parallelogram PQRS has vertices P(1, 0), Q(3, 3), R(8, 3), and S(6, 0). Draw the image of the parallelogram after a 90° clockwise rotation about S. What sort of figure is the image?

33. Rotate each letter by a $\frac{1}{2}$ turn clockwise about the indicated point. Which ones make readable letters when rotated?

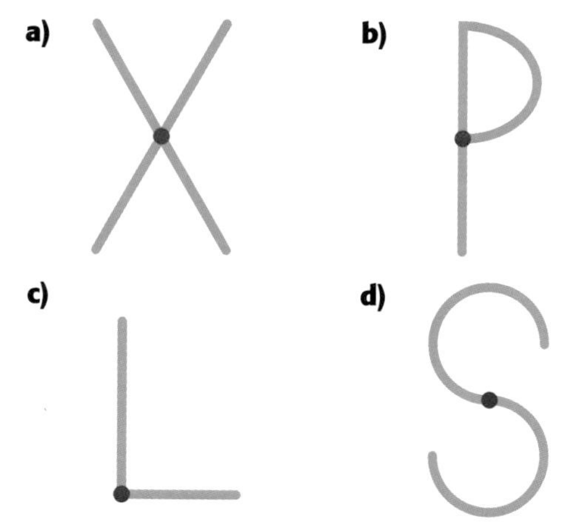

34. Plot the line $x + y = 2$ on a grid and rotate it 90° clockwise about the origin. At what angle do the line and its image intersect?

35. a) Plot the line $y = x - 3$ on a grid. Rotate the line 180° about the origin. How are the line and its image related?

b) Repeat part a) for the line $y = x$.

c) Explain why your findings in parts a) and b) are different.

36. a) Name another clockwise rotation that will give the same image of a figure as a 90° clockwise rotation about the origin.

b) Name a counterclockwise rotation that will give the same image of the figure as in part a).

37. The stars in the sky appear to rotate 360° about a point every day. What is that point?

38. List 3 examples of real objects that rotate about a turn centre. Compare your list with a classmate's. Choose objects that have not been mentioned in this section.

14.4 Dilatations

Lewis Carroll's *Alice's Adventures in Wonderland* and *Through the Looking Glass* are considered the most famous children's books written in English. In both books, the story is a dream in which Alice changes size and encounters fantastic creatures.

A transformation that changes the size of an object is called a **dilatation.** Dilatations are called **enlargements** or **reductions,** depending on the way in which the size is changed.

Activity 1 Draw a Dilatation Image

Plot rectangle ABCD with coordinates A(4, 2), B(−2, 2), C(−2, −2), and D(4, −2). Multiply the coordinates of A by 2 to give its image A′(8, 4). Multiply the coordinates of the other vertices by 2 to give B′, C′, and D′. Join A′, B′, C′, and D′ to make rectangle A′B′C′D′, which is the dilatation image of rectangle ABCD.

Inquire

1. How do the side lengths of rectangle A′B′C′D′ compare with the side lengths of rectangle ABCD?

2. How do the ratios $\frac{AB}{A'B'}$, $\frac{BC}{B'C'}$, $\frac{CD}{C'D'}$, and $\frac{AD}{A'D'}$ compare?

3. How do the angles of rectangle A′B′C′D′ compare with the angles of rectangle ABCD?

4. Calculate the area of each rectangle. How do the areas compare?

5. Does rectangle ABCD and its image, rectangle A′B′C′D′, have the same sense?

6. a) Draw a line from the origin, (0, 0), through A and extend it. Does it pass through A′?
b) Does a straight line pass through the origin, B, and B′? the origin, C, and C′? the origin, D, and D′?

7. We say that rectangle ABCD has been enlarged by a dilatation with **centre** (0, 0) and a **scale factor** of 2. Describe a dilatation of this type in your own words.

8. Repeat this activity by dividing each coordinate of rectangle ABCD by 2 to give the coordinates of the image rectangle A″B″C″D″. How do the lengths of the sides, the sizes of the angles, and the area compare with those of rectangle ABCD?

The dilatation with centre (0, 0) and scale factor k can be described mathematically as a mapping.

$(x, y) \rightarrow (kx, ky)$

When $k > 1$, the mapping gives an enlargement.

When $k < 1$, the mapping gives a reduction.

The image and the original figure are **similar**. They have the same shape, but not the same size.

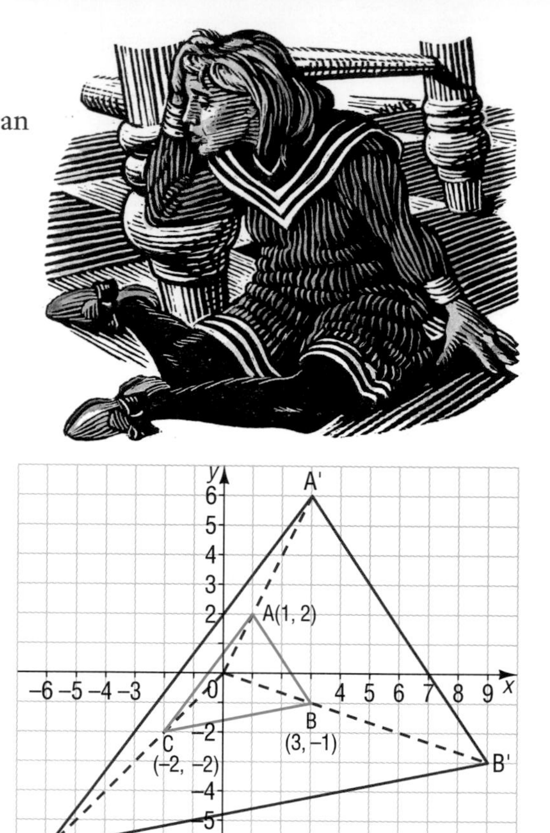

Example

ΔABC has vertices A(1, 2), B(3, −1), and C(−2, −2). Find the image of ΔABC under the mapping $(x, y) \rightarrow (3x, 3y)$.

Solution

Draw ΔABC.

For $(x, y) \rightarrow (3x, 3y)$

$A(1, 2) \rightarrow A'(3, 6)$

$B(3, -1) \rightarrow B'(9, -3)$

$C(-2, -2) \rightarrow C'(-6, -6)$

Plot A′, B′, and C′ and join them.

ΔA′B′C′ is the image of ΔABC.

Practice

A figure is shown with its image to the right. What is the scale factor?

1.

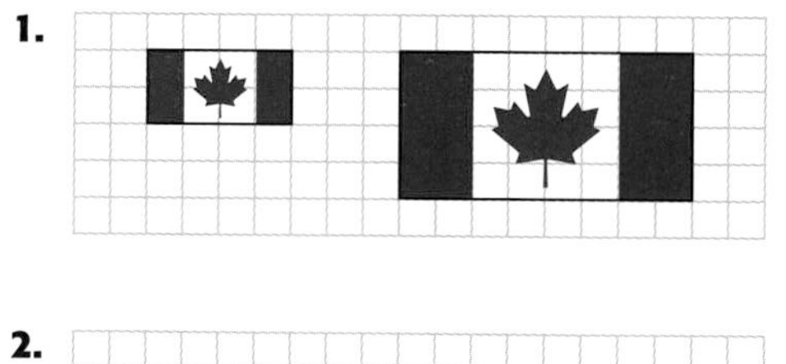

2.

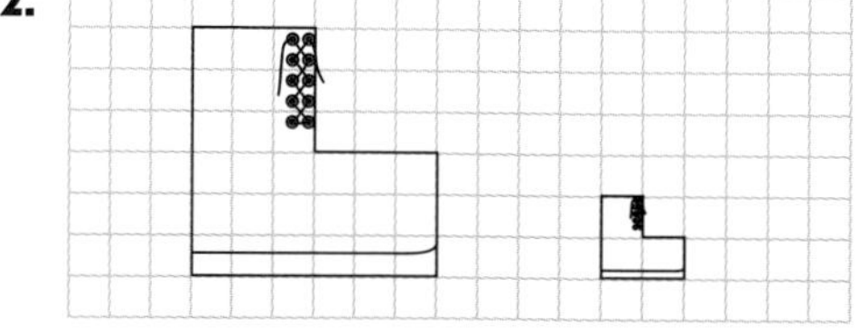

3. Copy the figure onto grid paper and enlarge it by a scale factor of 2.

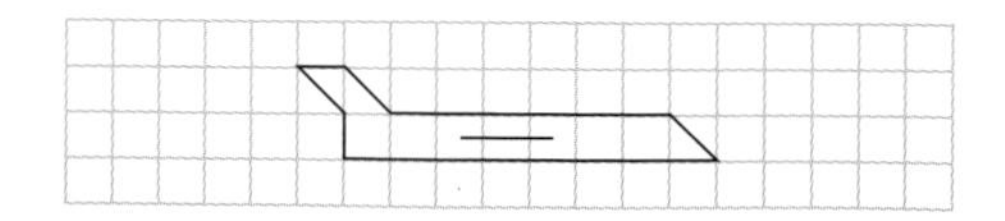

4. Copy the figure onto grid paper and reduce it by a scale factor of $\frac{1}{2}$.

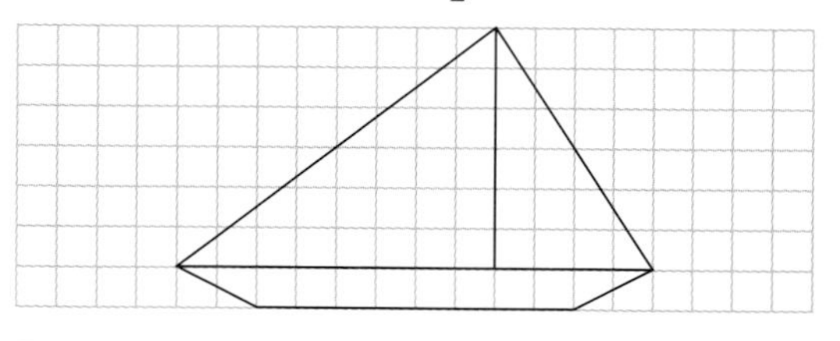

Draw the dilatation image of each line segment under the given mapping.

	Line Segment	Mapping
5.	A(3, 2), B(1, 4)	$(x, y) \rightarrow (2x, 2y)$
6.	C(6, 4), D(−2, 2)	$(x, y) \rightarrow (\frac{1}{2}x, \frac{1}{2}y)$
7.	E(−1, −1), F(1, 2)	$(x, y) \rightarrow (3x, 3y)$
8.	G(9, 3), H(−6, 0)	$(x, y) \rightarrow (\frac{1}{3}x, \frac{1}{3}y)$

9. ΔRST has vertices R(2, 3), S(−1, 4), and T(−3, −2). Find the image of ΔRST under the mapping $(x, y) \rightarrow (3x, 3y)$.

10. Quadrilateral DEFG has vertices D(6, 4), E(−2, 6), F(−4, −4), and G(4, −6). Find the image of quadrilateral DEFG under the mapping $(x, y) \rightarrow \left(\frac{1}{2}x, \frac{1}{2}y\right)$.

Problems and Applications

11. Copy this piece of an impossible chessboard onto grid paper and enlarge it by a scale factor of 2.

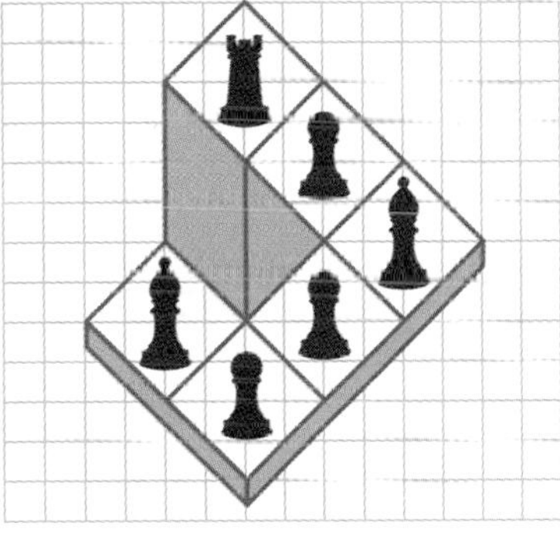

12. Copy this disappearing block puzzle onto grid paper and reduce it by a scale factor of $\frac{1}{2}$.

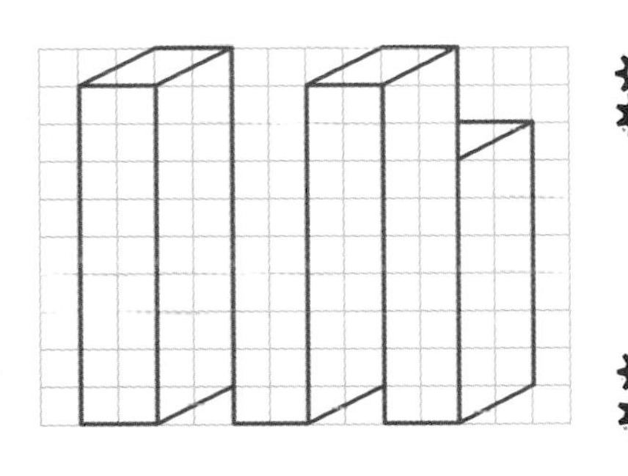

13. ΔPQR has vertices P(−2, 2), Q(−2, −2), and R(2, −2).

a) Draw ΔPQR on grid paper.

b) Calculate the area of ΔPQR.

c) Find the image of ΔPQR, ΔP′Q′R′, under the mapping $(x, y) \rightarrow (3x, 3y)$.

d) Calculate the area of ΔP′Q′R′.

e) Find the image of ΔPQR, ΔP″Q″R″, under the mapping $(x, y) \rightarrow \left(\frac{1}{2}x, \frac{1}{2}y\right)$.

f) Calculate the area of ΔP″Q″R″.

g) Write the following ratios.

area ΔP′Q′R′ : area ΔPQR

area ΔP″Q″R″ : area ΔPQR

h) How is each ratio related to the scale factor that produced the image?

14. a) State the types of dilatations you see in:

- a map of a province
- a movie screen
- a television screen
- a blueprint of a house

b) Find 2 other examples of enlargements or reductions in real-world objects. Estimate the scale factors involved.

In questions 15–17, copy the figure and its red image onto grid paper. Find the dilatation centre and state the scale factor.

15.

16.

17.

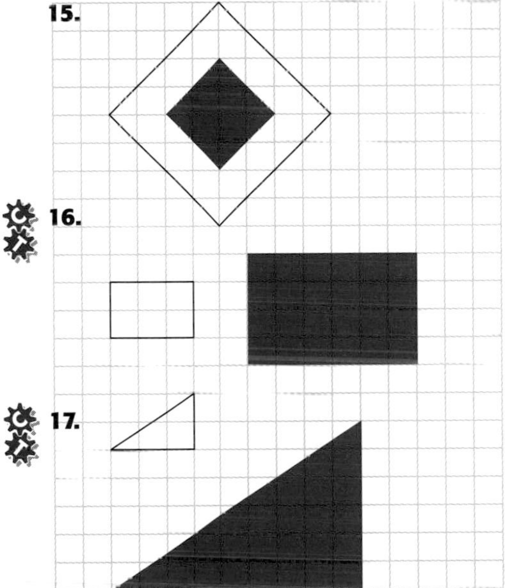

18. What would the image of a triangle look like under this mapping?
$(x, y) \rightarrow (1x, 1y)$

19. ΔABC has vertices A(4, 2), B(1, −2), and C(8, −3).

a) Draw ΔABC on grid paper.

b) Find the image of ΔABC for a dilatation with centre (4, 0) and a scale factor of 2.

20. Use a figure of your choice and investigate these mappings. Describe your findings

a) $(x, y) \rightarrow (2x, 4y)$

b) $(x, y) \rightarrow (-2x, -2y)$

Drawing with LOGO

In LOGO, the turtle draws on the screen.
Here are some LOGO commands.

Command	What the turtle does
FD	Moves forward
FD 30	Moves forward 30 units
BK	Moves backward
RT	Turns right a number of degrees
RT 90	Turns right 90°
LT	Turns left a number of degrees
PU	Does not draw a line (Pen Up)
PD	Draws a line (Pen Down)
DRAW	Returns to its home position (DRAW clears the screen)

Activity ❶ LOGO Commands

What does the turtle draw with the following commands?

1. **FD 50 RT 90**
   FD 50 RT 90
   FD 50 RT 90
   FD 50 RT 90
2. **RT 30 FD 60**
   RT 120 FD 60
   RT 120 FD 60
3. **FD 40 BK 40 RT 30**
   FD 40 BK 40 RT 30
   FD 40 BK 40 LT 90
   FD 40 BK 40 RT 30

Activity ❷ The REPEAT Command

In question 1 of Activity 1, the commands **FD 50 RT 90** were repeated 4 times. The **REPEAT** command could have been used instead.

REPEAT 4[FD 50 RT 90]

What does the turtle draw with the following commands?

1. **REPEAT 4[FD 60 RT 90]**
2. **REPEAT 4[FD 60 BK 20 RT 90]**
3. **REPEAT 2[FD 60 LT 90 FD 30 LT 90]**

Write LOGO commands to draw the following figures.

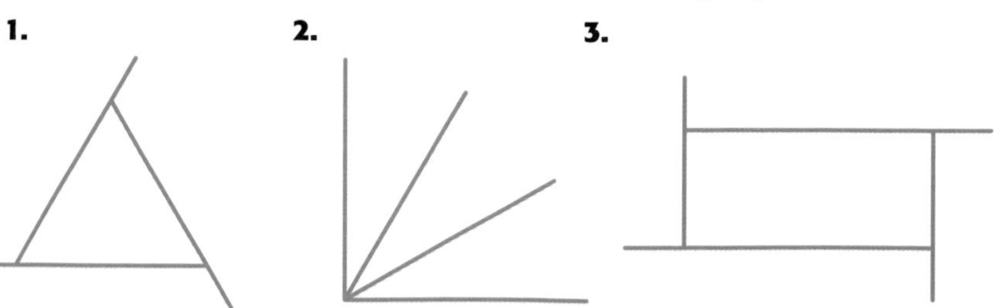

Activity ❸ Joining Figures

What does the turtle draw with the following commands?

```
REPEAT 2[FD 15 RT 90 FD 30 RT 90]
RT 90 FD 30 LT 90
REPEAT 2[FD 15 RT 90 FD 30 RT 90]
RT 90 FD 30 LT 90
REPEAT 2[FD 15 RT 90 FD 30 RT 90]
```

Write LOGO commands to draw the figures shown.

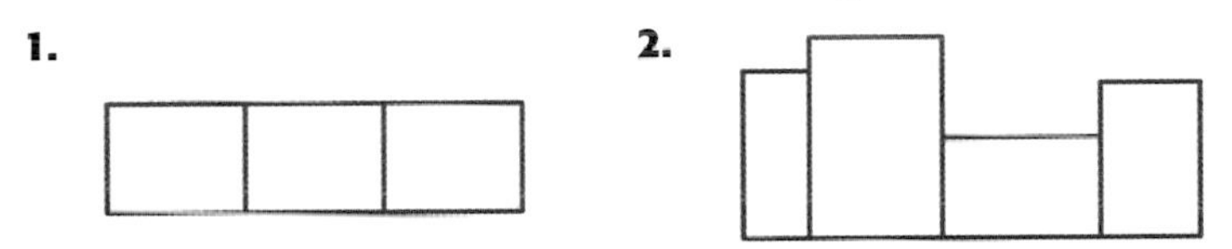

Activity ❹ Pen Up and Pen Down

What does the turtle draw with the following commands?

```
REPEAT 4[FD 40 RT 90]
PU RT 90 FD 70 LT 90 PD
REPEAT 4[FD 40 RT 90]
PU RT 90 FD 70 LT 90 PD
REPEAT 4[FD 40 RT 90]
```

Write LOGO commands to draw the patterns shown.

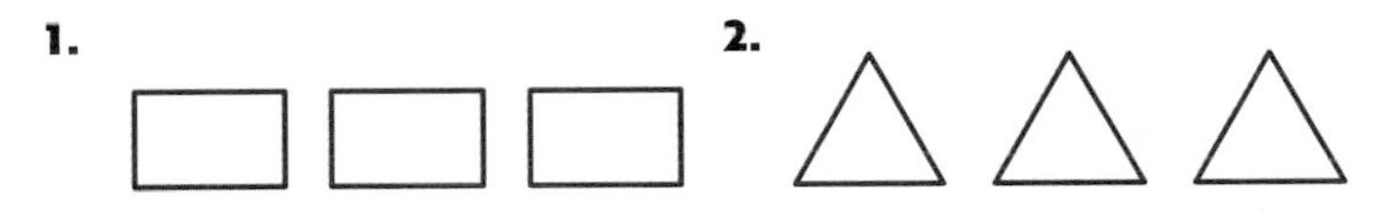

Activity ❺ Putting It Together

Write LOGO commands to draw the following patterns.

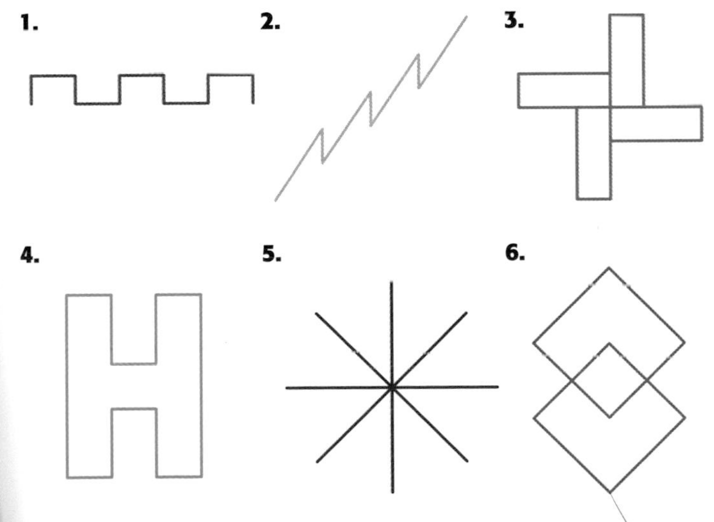

14.5 Similar Triangles

Before constructing a new plane, aeronautical engineers build and test a model. The model plane is the same shape as, or is similar to, the real plane.

Model Aircraft of 767 in old Air Canada livery.

Activity: Construct the Triangles

Draw ΔABC and ΔDEF on grid paper. Measure the angles and sides in each triangle and complete the table.

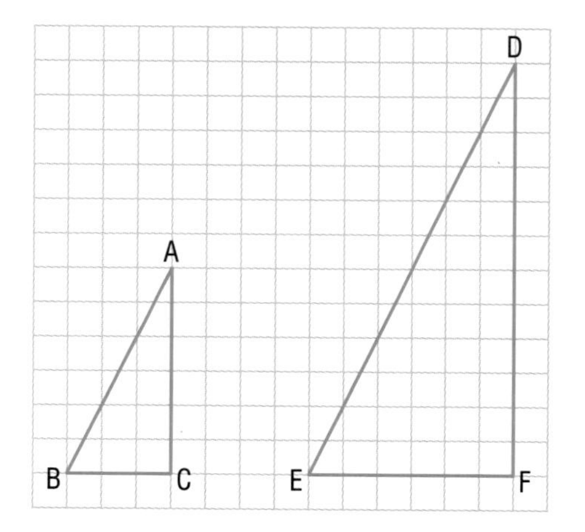

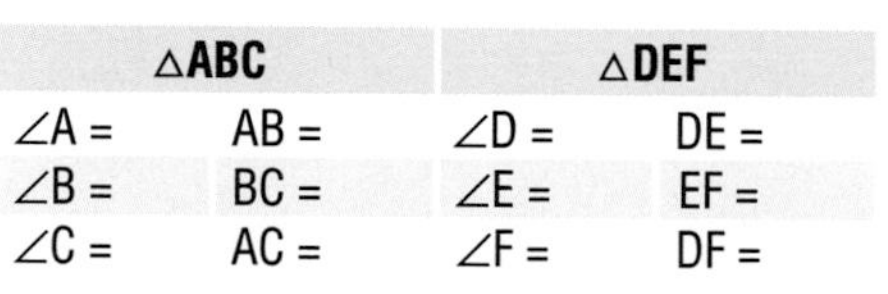

ΔABC		ΔDEF	
∠A =	AB =	∠D =	DE =
∠B =	BC =	∠E =	EF =
∠C =	AC =	∠F =	DF =

Inquire

1. How do the measures of the corresponding angles compare?

2. Find the ratios of the lengths of the corresponding sides, AB:DE, BC:EF, and AC:DF.

3. How do the ratios of the corresponding sides compare?

4. a) In what way are ΔABC and ΔDEF the same as congruent triangles?
b) In what way are they different from congruent triangles?

5. What is the ratio of the lengths of corresponding sides in congruent triangles?

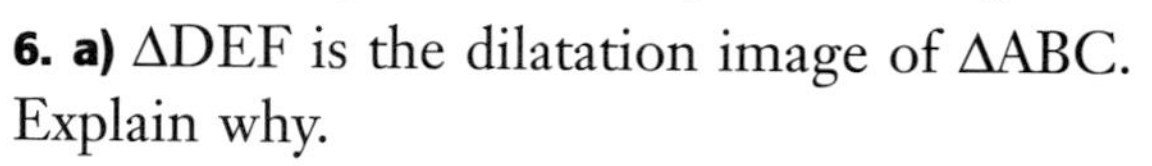

6. a) ΔDEF is the dilatation image of ΔABC. Explain why.
b) What is the scale factor?
c) Find the dilatation centre.

ΔPQR and ΔWXY are similar. This means that

- The corresponding angles are equal.

∠P = ∠W ∠Q = ∠X ∠R = ∠Y

- The ratios of the corresponding sides are equal.

$\frac{PQ}{WX} = \frac{QR}{XY} = \frac{PR}{WY}$ or $\frac{r}{y} = \frac{p}{w} = \frac{q}{x}$

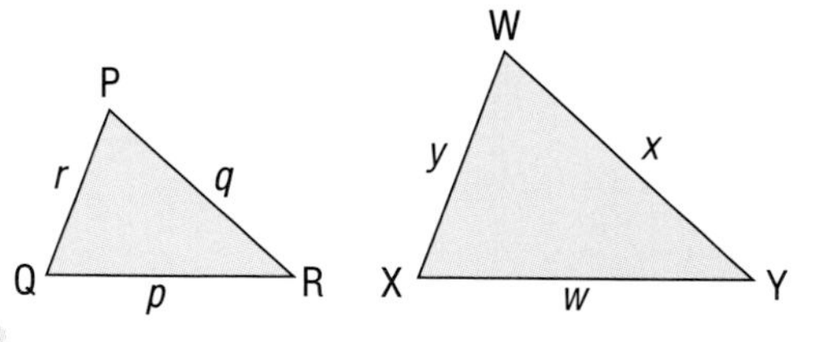

We write ΔPQR ~ ΔWXY.

~ means "is similar to"

You can state that 2 triangles, ΔABC and ΔDEF, are similar if you know that the corresponding pairs of angles are equal.

∠A = ∠D ∠B = ∠E ∠C = ∠F

You can state that 2 triangles, ΔABC and ΔDEF, are similar if you know that the ratios of the corresponding sides are equal.

$\frac{AB}{DE} = \frac{BC}{EF} = \frac{AC}{DF}$ or $\frac{c}{f} = \frac{a}{d} = \frac{b}{e}$

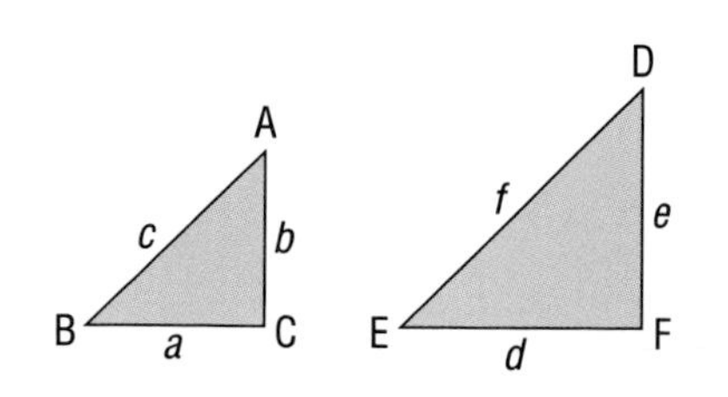

Example 1

$\triangle DEF \sim \triangle RST$. Find the values of r and e.

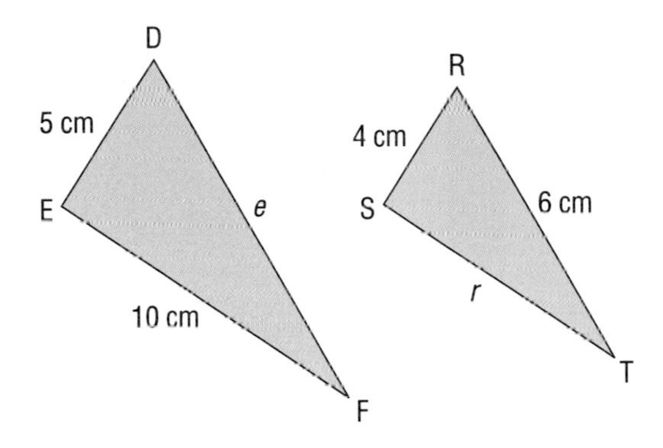

Solution

Since the triangles are similar, the ratios of the corresponding sides are equal.

$$\frac{DE}{RS} = \frac{EF}{ST} = \frac{DF}{RT}$$

Substitute. $\frac{5}{4} = \frac{10}{r} = \frac{e}{6}$

Take the ratios 2 at a time.

$$\frac{5}{4} = \frac{10}{r} \qquad \frac{5}{4} = \frac{e}{6}$$

$$5 \times r = 10 \times 4 \qquad 5 \times 6 = e \times 4$$

$$5r = 40 \qquad 30 = 4e$$

$$r = 8 \qquad 7.5 = e$$

So, r is 8 cm, and e is 7.5 cm.

Similar triangles can be used to find distances that are difficult to measure directly.

Example 2

The diagram shows how surveyors can lay out 2 triangles to find the width of the St. Lawrence River near Brockville. Use the triangles to calculate the width of the river, DE.

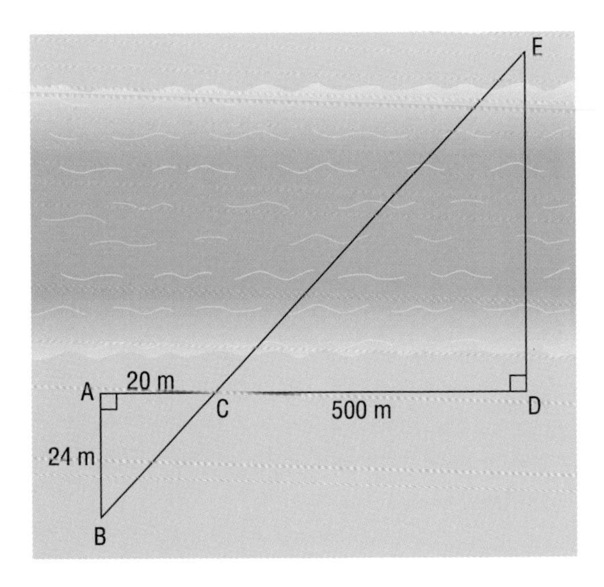

Solution

If $\triangle ABC$ is similar to $\triangle DEC$, you can write ratios of corresponding sides to find DE.

In $\triangle ABC$ and $\triangle DEC$

$\angle CAB = \angle CDE$ (both 90°)

$\angle ACB = \angle DCE$ (opposite angles)

If 2 angles in one triangle are equal to 2 angles in another triangle, the third angles in each triangle must be equal.

$\angle ABC = \angle DEC$

Since corresponding angles are equal, $\triangle ABC \sim \triangle DEC$.

Since the triangles are similar, the ratios of corresponding sides are equal.

$$\frac{AC}{DC} = \frac{AB}{DE}$$

$$\frac{20}{500} = \frac{24}{DE}$$

$$20 \times DE = 500 \times 24$$

$$20 \times DE = 12\ 000$$

$$DE = 600$$

The width of the river is 600 m.

CONTINUED

Practice

Use grid paper to make a figure similar to, but not congruent to, each of the following figures.

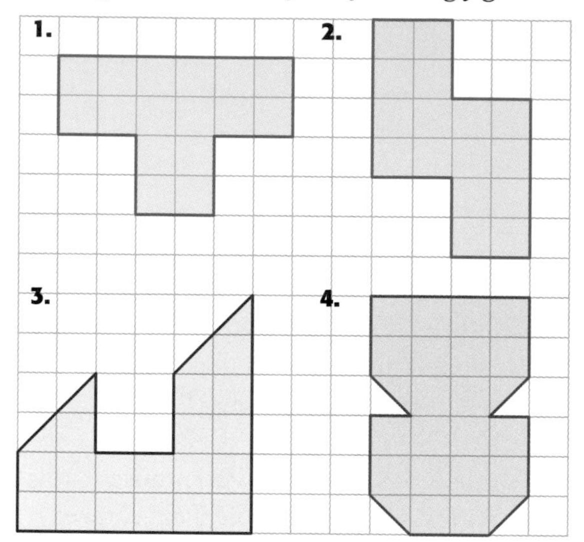

The triangles in each pair are similar. Find the unknown side lengths.

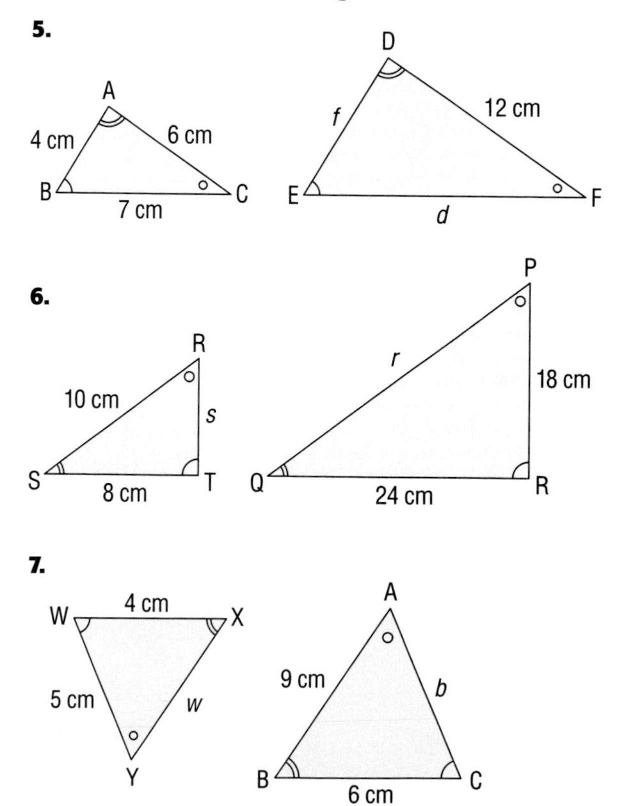

8.

D
8 cm
9 cm
E
F
10 cm
P
6 cm
r
R
Q
p

9.

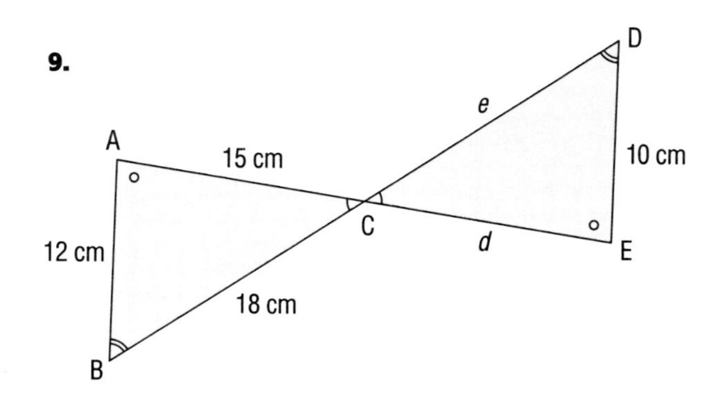

10. Find x.

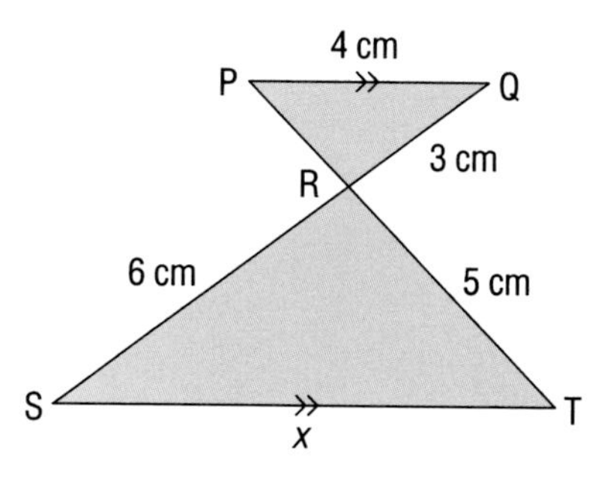

11. Find x.

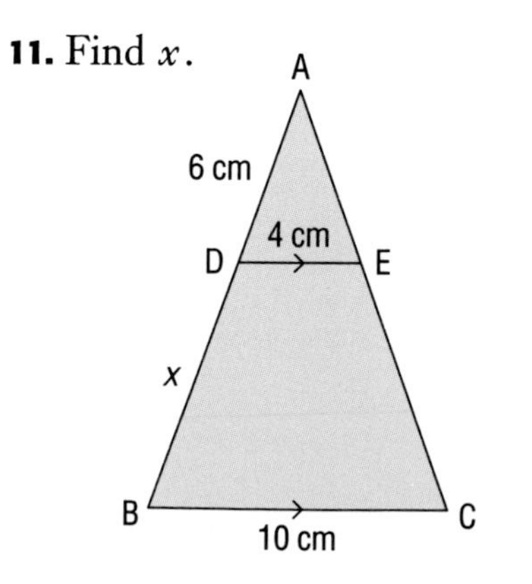

Problems and Applications

12. The 5-m flagpole casts a 4-m shadow at the same time of day as a building casts a 30-m shadow. How tall is the building?

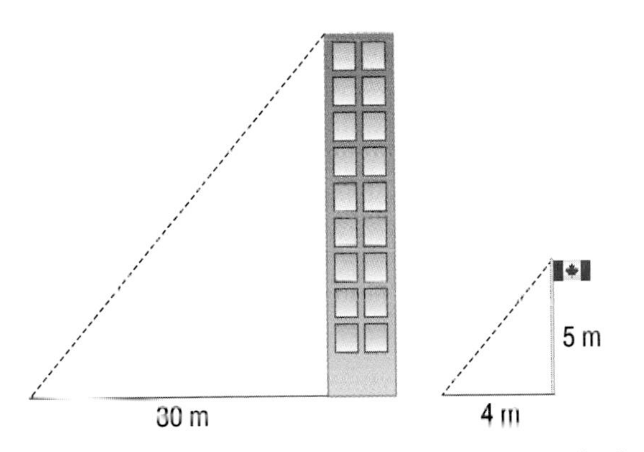

13. Surveyors have laid out triangles to find the length of the lake. Calculate this length.

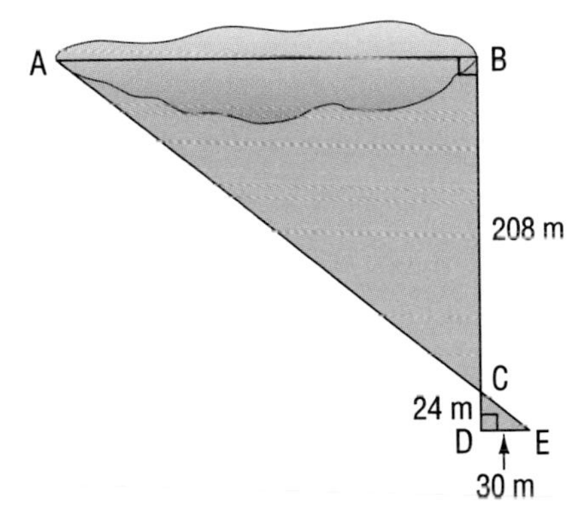

14. Surveyors have laid out these triangles to calculate the width of a canyon. Find this width to the nearest metre.

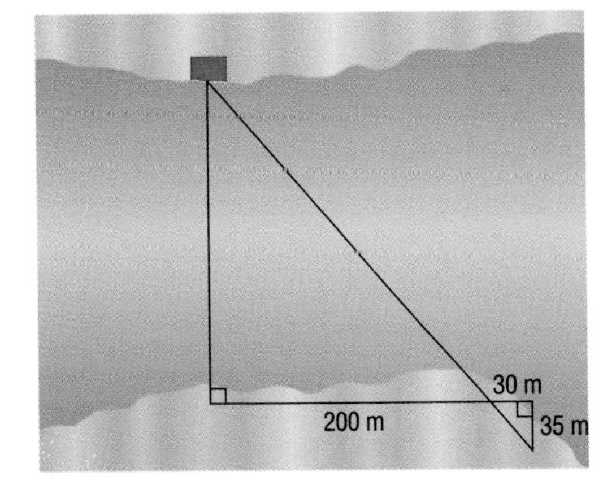

15. The two quadrilaterals are similar. Find the lengths of the unknown sides.

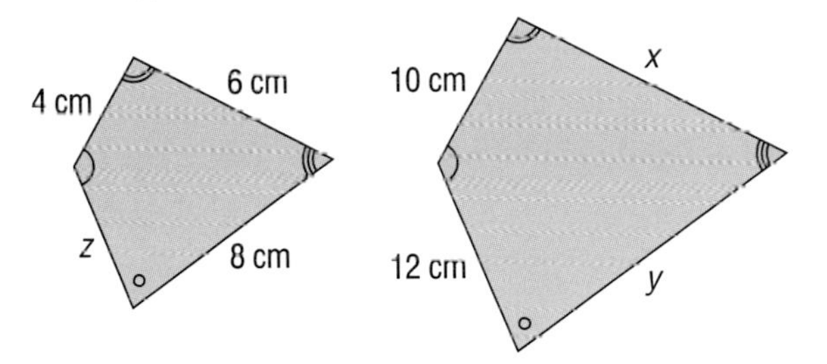

16. Two quadrilaterals are similar if the corresponding angles are equal and the ratios of the lengths of the corresponding sides are equal.

a) Are all squares similar? Explain.

b) Are all rectangles similar? Explain.

17. Decide on 3 outdoor objects whose heights you want to calculate. Your choices might include a street lamp, a flagpole, and the school. At the same time of day, find the length of the shadow cast by a metre-stick and the length of the shadow of each chosen object. Calculate the height of each object.

LOGIC POWER

Use your knowledge of geography to identify the province these towns are in. Then, check by finding this area on a map.

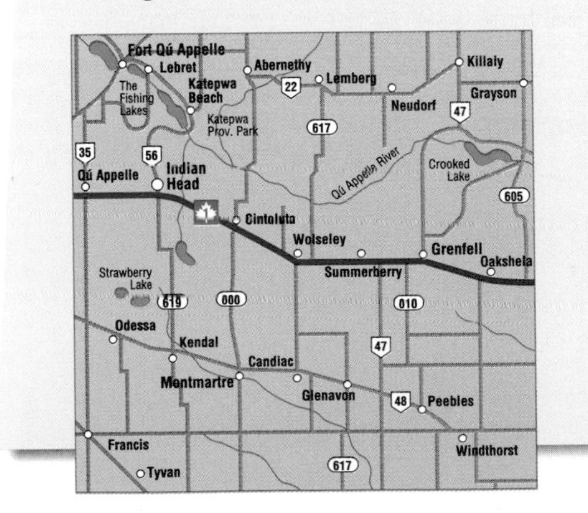

14.6 Symmetry

A **line of symmetry** is a mirror line that reflects an object onto itself. Line symmetry is also called **reflectional symmetry** or **mirror symmetry**. Insects, flowers, and many other natural objects have lines of symmetry.

Describe where the line of symmetry is on the ladybug.

Activity: Draw Lines of Symmetry

When water drops are sprinkled onto a dry, hot skillet, the drops dance across the skillet. The drops take many different shapes. A few of them are shown. Trace the water drops and draw all their lines of symmetry. You may want to use a Mira.

1 2 3 4 5

Inquire

1. How many lines of symmetry does each water drop have?
2. How many lines of symmetry does a square have?
3. How many lines of symmetry does an isosceles triangle have? an equilateral triangle?
4. List some objects in the classroom that have several lines of symmetry.

A figure that can be mapped onto itself with a turn of less than one complete rotation has **rotational symmetry** or **turn symmetry**. The number of times the figure matches with itself in a turn of 360° is the **order** of rotational symmetry. Rotational symmetry of order 2 is also called **point symmetry**.

Example 1

Determine the order of turn symmetry of water drop 2.

Solution

Trace the water drop and put the tracing on top of the original. Put a pencil point or pen point on the centre and turn the tracing. Count the number of times the tracing matches the original in a 360° turn.
The tracing matches 3 times.
The water drop has turn symmetry of order 3.

Practice

How many lines of symmetry does each figure have?

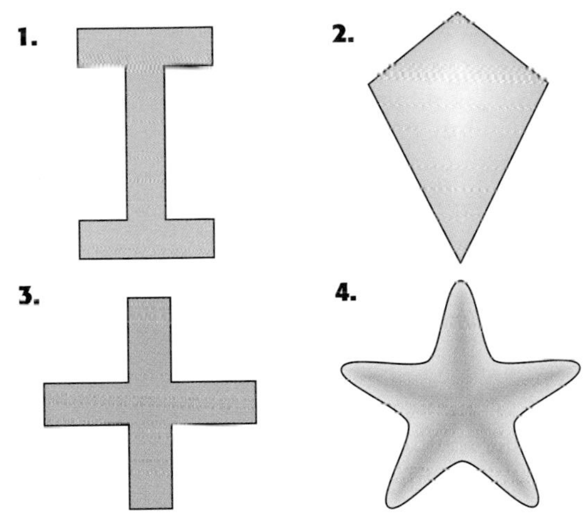

What is the order of turn symmetry for each figure?

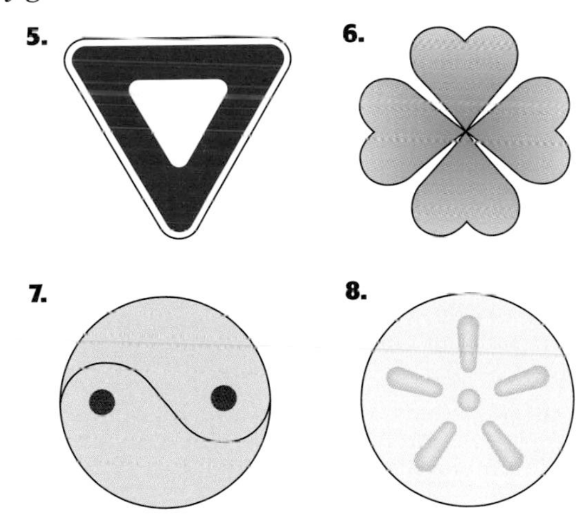

Problems and Applications

9. Print the capital letters of the alphabet.

a) Which letters have a vertical line of symmetry?

b) Which letters have a horizontal line of symmetry?

c) Which letters have both a vertical and a horizontal line of symmetry?

d) Which of the letters have point symmetry?

Trace each figure. Then, add enough parts so that it is symmetric about the red line.

10. **11.**

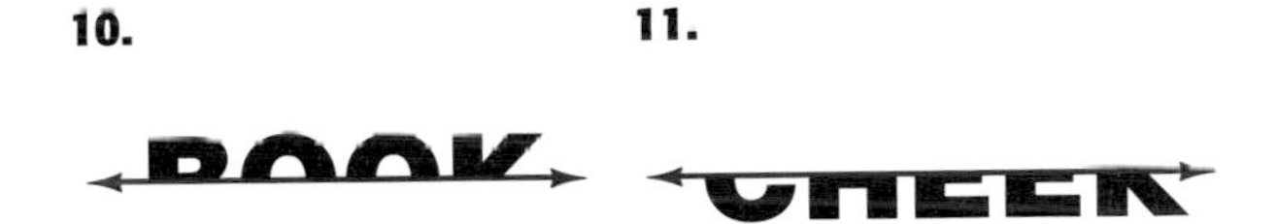

Trace each figure. Then, add parts to each of the other 3 quadrants so that the figure has

a) *1 line of symmetry*

b) *2 lines of symmetry*

c) *no lines of symmetry*

d) *rotational symmetry of order 4*

e) *rotational symmetry of order 2*

12. **13.**

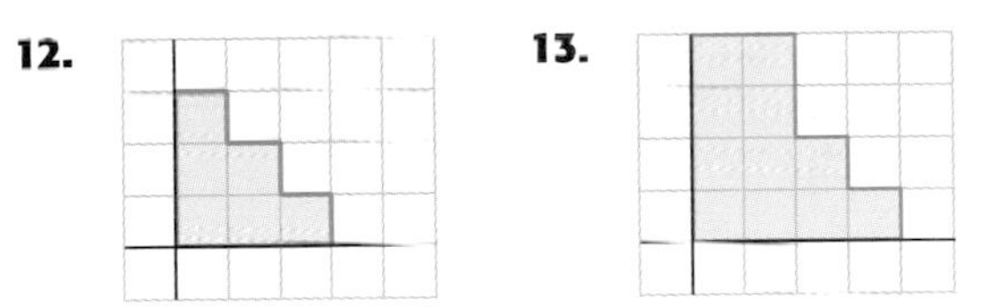

14. A double six set of dominoes uses seven patterns from "blank" to six. The set starts with the double blank domino and goes to the double six domino.

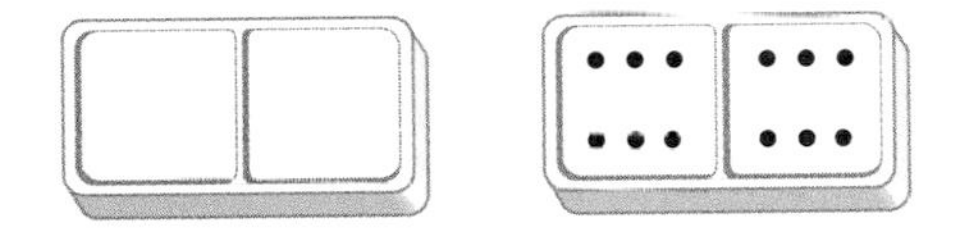

In a double six domino set, how many dominoes have

a) no lines of symmetry?

b) 1 line of symmetry?

c) 2 lines of symmetry?

d) rotational symmetry of order 2?

15. Many company logos have lines of symmetry and rotational symmetry. Find examples of company logos with symmetry. Copy them into your notebook and record the type of symmetry they have. Why do logo designers use symmetry?

16. Work with a classmate to find flags of the world that have lines of symmetry. Sketch the flags.

Distortions on a Grid

When you stand in front of a mirror in a fun house, your image is distorted. When you distort a figure in mathematics, you may stretch, shrink, and turn it in many directions.

Activity 1

1. Draw the arrow on grid paper.

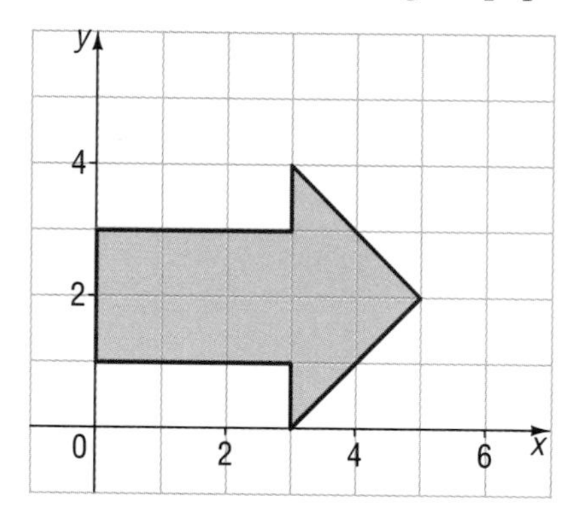

Find the image of the arrow under the following mappings.

a) $(x, y) \to (2x, y)$

b) $(x, y) \to (x, 2y)$

Describe how each image has been distorted from the original arrow.

2. A kite ABCD has vertices A(0, 0), B(2, 4), C(0, 6), and D(–2, 4). Draw the kite on grid paper, then draw the images of the kite under the following mappings.

a) $(x, y) \to (3x, y)$

b) $(x, y) \to (x, 2y)$

c) $(x, y) \to \left(x, \frac{y}{2}\right)$

d) $(x, y) \to (2x, 2y)$

Does each distortion give an image that is also a kite?

3. Draw a square with vertices (3, 3), (–3, 3), (–3, –3), and (3, –3). If you draw the square under each of the mappings you used for the kite in question 2, is the image always a square?

Activity 2

Draw the T on grid paper.

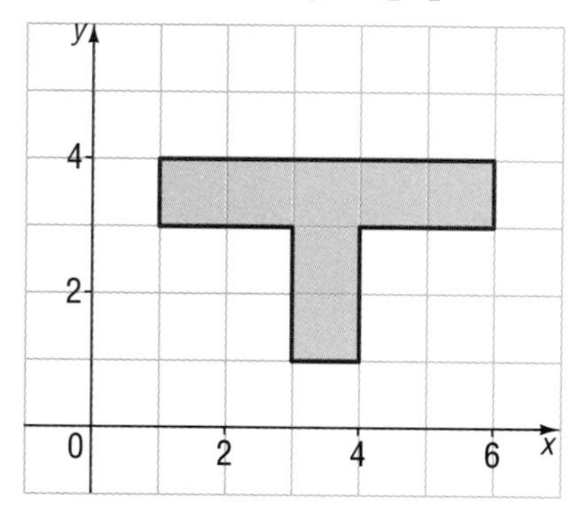

Draw the images of the T under the following mappings.

a) $(x, y) \to (-x, 2y)$ **b)** $(x, y) \to (2x, -y)$

Describe how each image has been distorted from the original T-shape.

Activity 3

Draw the figure on grid paper.

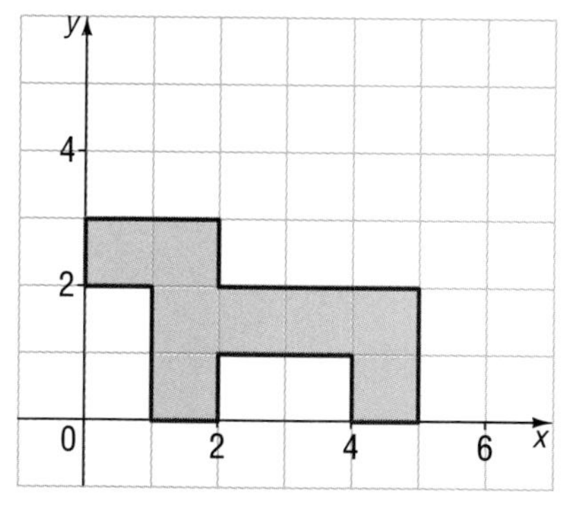

Draw the images of the figure under the following mappings.

a) $(x, y) \to \left(\frac{x}{2}, 3y\right)$ **b)** $(x, y) \to \left(3x, \frac{y}{2}\right)$

Describe how each image has been distorted from the original figure.

Activity 4

The rectangle ABCD has been transformed under the mapping $(x, y) \rightarrow \left(2x, \frac{x+y}{2}\right)$.

$$(x, y) \rightarrow \left(2x, \frac{x+y}{2}\right)$$

$A(0,0) \rightarrow A'(0,0)$
$B(4,0) \rightarrow B'(8,2)$
$C(4,3) \rightarrow C'(8,3.5)$
$D(0,3) \rightarrow D'(0,1.5)$

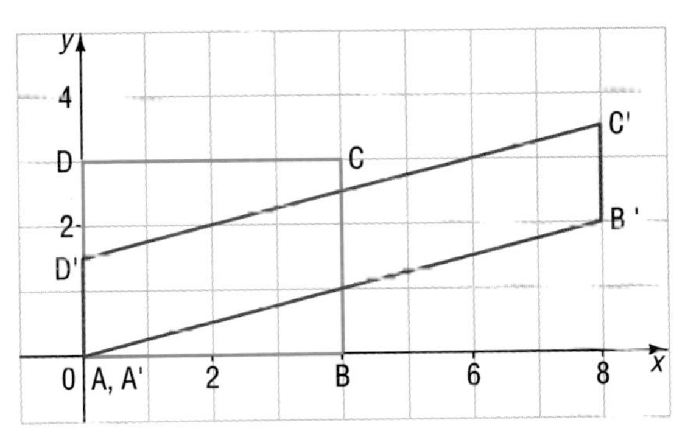

1. Draw the following figure onto grid paper.

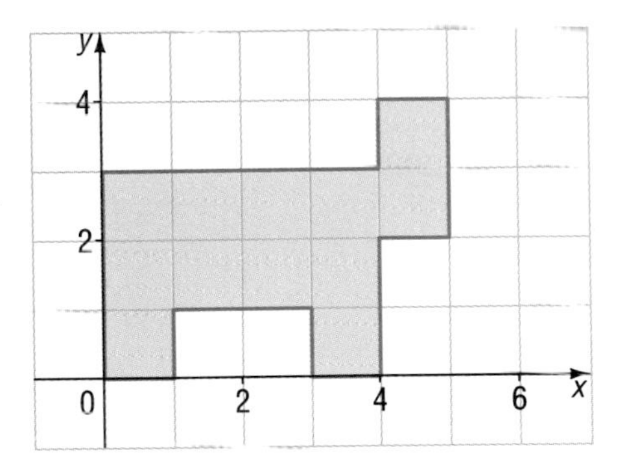

Draw the image of the figure under this mapping.

$$(x, y) \rightarrow \left(2x, \frac{x+y}{2}\right)$$

2. Draw the image of the same figure under this mapping.

$$(x, y) \rightarrow \left(\frac{x+y}{2}, 2y\right)$$

3. Redraw the figure away from the axes and apply the mappings from steps 1 and 2 of this activity to the figure.

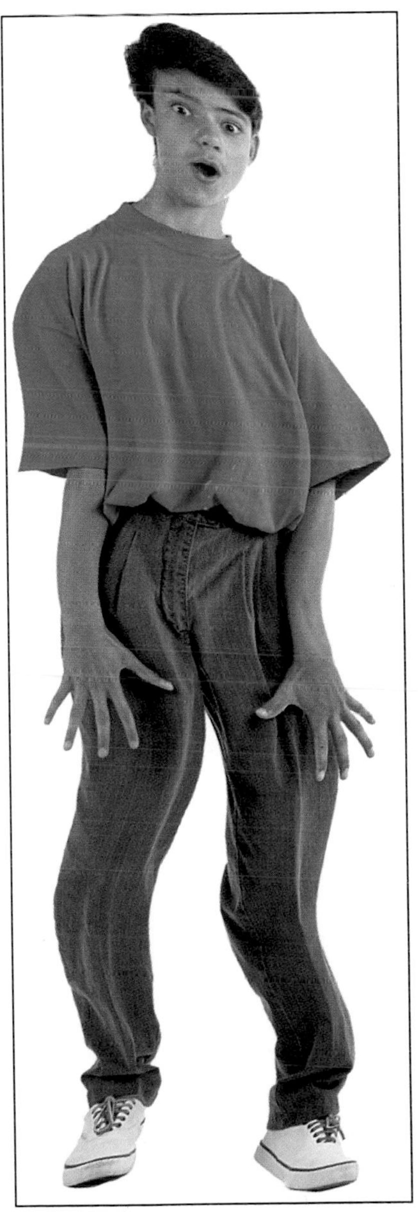

Activity 5

1. With any combinations of the mappings you have used, write your own mapping that will distort a figure in an unusual way.

2. Have a classmate use your mapping to distort a figure.

Escher Drawings

Islamic artists use tessellations in much of their work. A **tessellation** or tiling is an arrangement of shapes that completely covers the plane without overlapping or leaving gaps. The Dutch artist M.C. Escher became fascinated with tile patterns he found in Spain. Escher used translations, rotations, and reflections to make tessellations of animals and humans. This Escher tessellation is based on translations in parallelograms.

Activity ❶ Geometric Figures that Tessellate

Only 3 regular polygons will tile the plane: an equilateral triangle, a square, and a regular hexagon. However, many irregular figures will tile the plane.

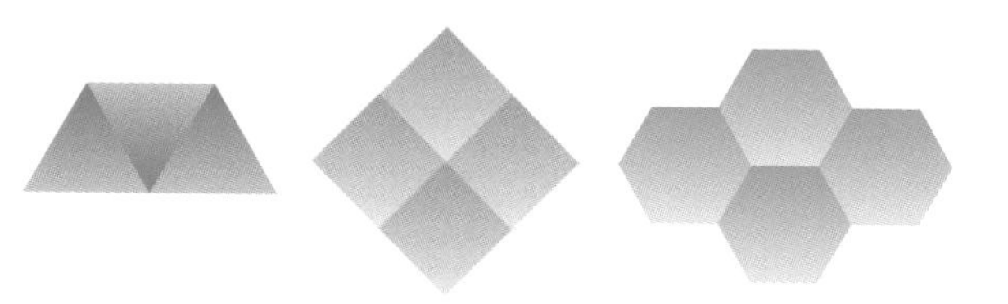

1. Use grid paper to show how an irregular triangle tiles the plane.

2. Draw an example of an irregular quadrilateral that tiles the plane.

3. The diagram shows one example of an irregular hexagon that tiles the plane. Draw another example.

Activity ❷ Tessellations Using Translations

1. You can alter the opposite sides of a square and use translations to tile the plane.

a) Make an alteration to one side.

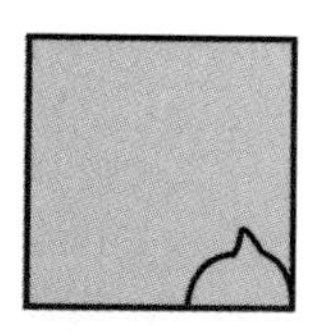

b) Translate the alteration to the opposite side.

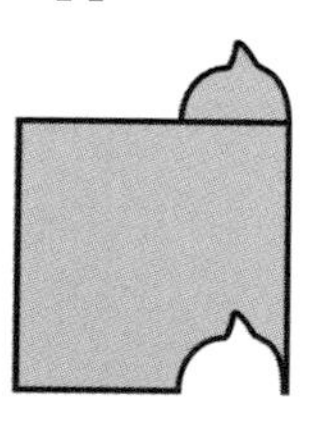

c) Alter the other sides in a similar way.

d) Use the new figure to tile the plane.

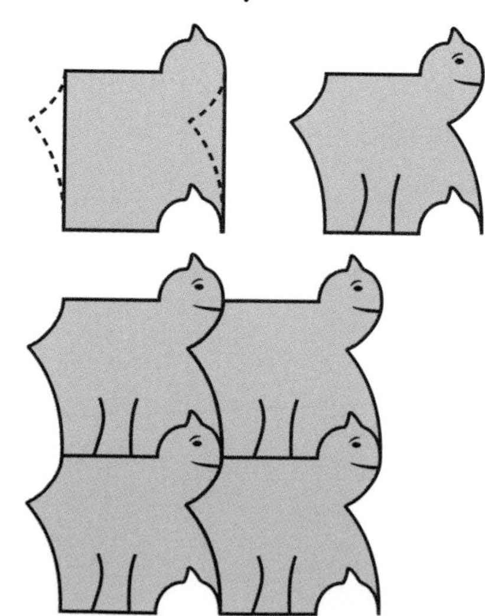

2. Alter the opposite sides of a parallelogram to tile the plane by translation.

3. Alter the opposite sides of a hexagon to tile the plane by translation.

Activity 3 Tessellations Using Rotations

This Escher tessellation is based on rotations in regular hexagons.

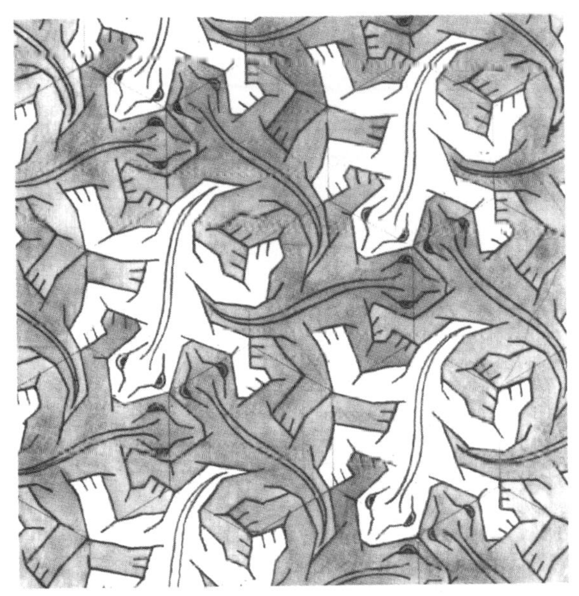

1. You can alter the sides of a square and use rotations to tile the plane.

a) Make an alteration on one side.

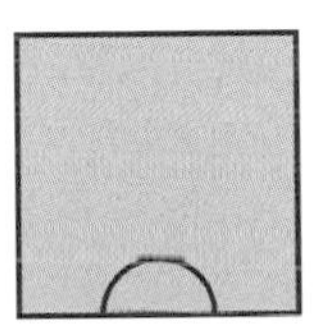

b) Rotate the alteration.

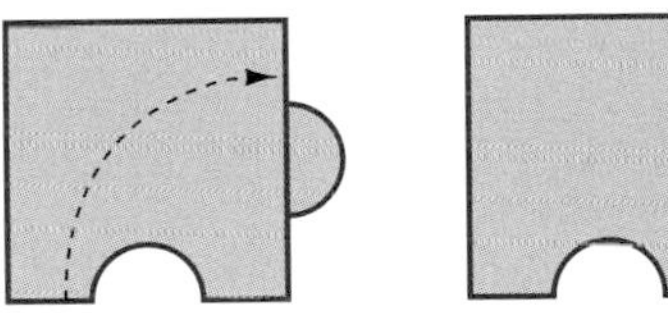

c) Use the new figure to tile the plane.

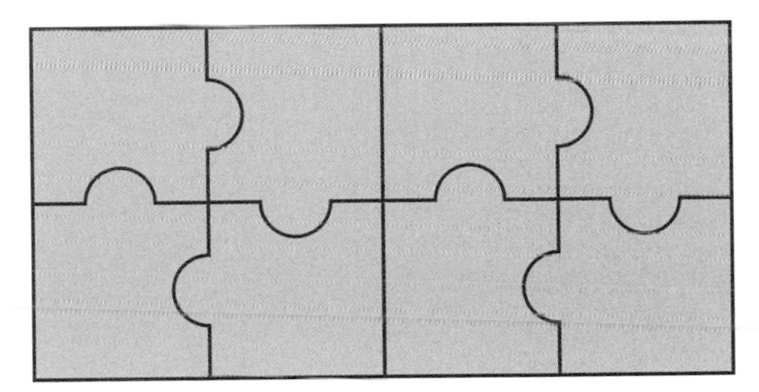

2. Alter the sides of a rhombus to form a figure to tile the plane by rotation.

3. Alter the sides of a triangle to form a figure to tile the plane by rotation.

Activity 4 Impossible Figures

Escher also used impossible geometric figures, such as the one to the right, in his art.

1. What are impossible figures?

2. Research other examples of Escher's art that include these figures.

Review

State each translation arrow as an ordered pair.

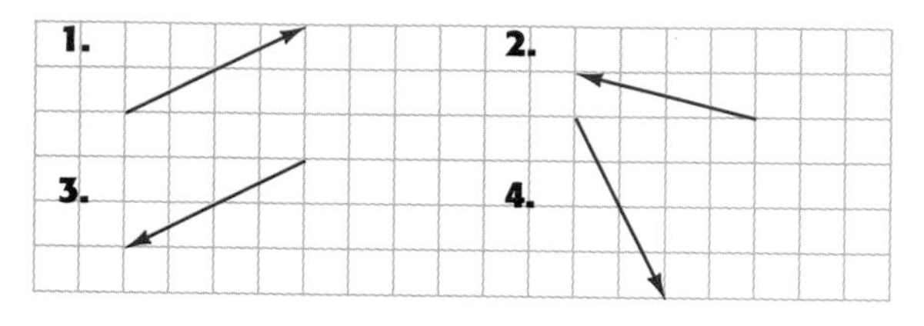

Describe each translation in words.

5. $(x, y) \rightarrow (x + 2, y + 3)$

6. $(x, y) \rightarrow (x - 3, y - 1)$

7. $[-4, 5]$ **8.** $[0, 6]$

Express each translation as an ordered pair.

9. $A(5, 4) \rightarrow A'(6, 2)$

10. $B(3, -1) \rightarrow B'(-1, 5)$

11. $C(0, -3) \rightarrow C'(2, -6)$

12. ΔABC has vertices A(3, 2), B(4, −1), and C(−2, 1). Find the coordinates of the image of ΔABC under this mapping.
$(x, y) \rightarrow (x + 4, y + 3)$

13. ΔRIK has vertices R(−2, 3), I(3, 0), and K(2, −3). Find the image of ΔRIK under the mapping $(x, y) \rightarrow (x - 3, y - 2)$.

Find the coordinates of the image after a reflection in each axis.

	Point	x-axis	y-axis
14.	(4, 5)		
15.	(5, −2)		
16.	(−3, 6)		
17.	(−3, −2)		

Copy each figure onto a grid and draw its reflection image in the x-axis.

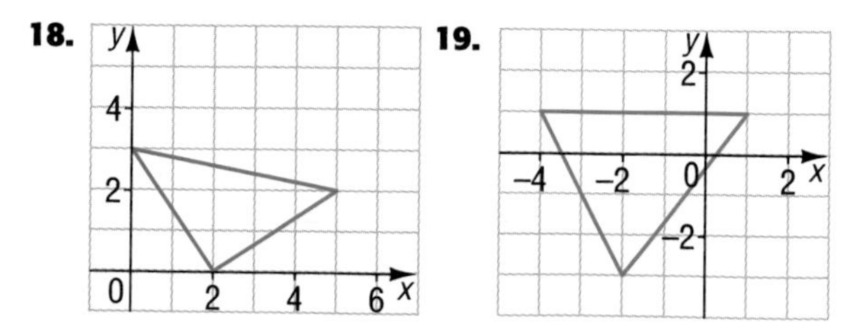

Copy each figure onto a grid and draw its reflection image in the y-axis.

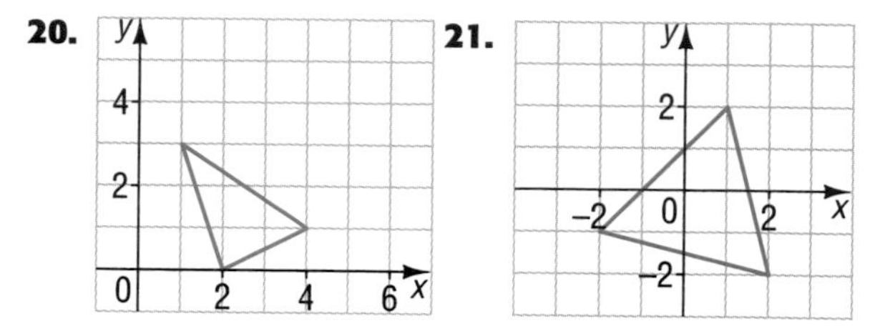

22. ΔDEF has vertices D(2, 3), E(5, 1), and F(4, 6). Find the coordinates of the image of ΔDEF after a reflection in the x-axis.

23. ΔRST has vertices R(3, −1), S(−3, 2), and T(−4, −3). Find the coordinates of the image of ΔRST after a reflection in the y-axis.

Draw the image of each figure after a 90° counterclockwise rotation about the origin.

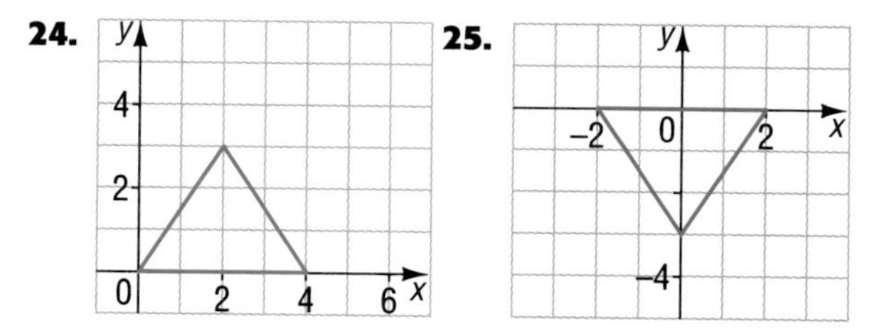

Draw the image of each figure after a 180° clockwise rotation about the origin.

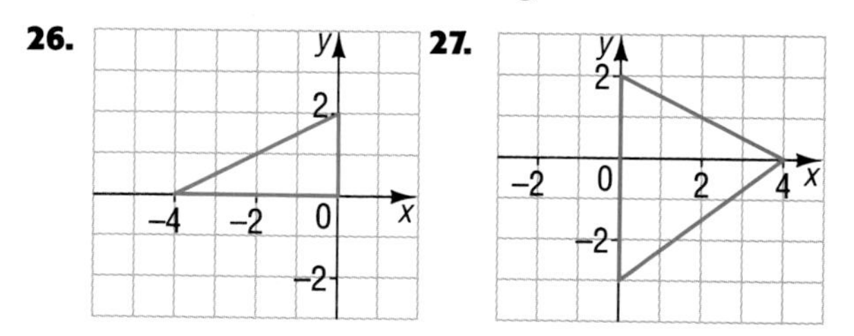

28. ΔDEF has vertices D(3, 0), E(3, −4), and F(0, −4). Draw the image of ΔDEF after a 180° counterclockwise rotation about the origin.

29. ΔABC has vertices A(3, 0), B(3, 5), and C(6, 0). Draw the image of ΔABC after a 90° clockwise rotation about the origin.

Draw the dilatation image of each line segment under the given mapping.

	Line Segment	Mapping
30.	A(2, 3), B(5, 0)	$(x, y) \rightarrow (2x, 2y)$
31.	C(−1, 2), D(3, −3)	$(x, y) \rightarrow (3x, 3y)$
32.	E(4, 6), F(−6, −2)	$(x, y) \rightarrow (\frac{1}{2}x, \frac{1}{2}y)$
33.	G(−6, 0), H(0, 6)	$(x, y) \rightarrow (\frac{1}{3}x, \frac{1}{3}y)$

34. ΔPQR has vertices P(3, 2), Q(3, −1), and R(2, −2). Find the image of ΔPQR under the mapping $(x, y) \rightarrow (2x, 2y)$.

35. Quadrilateral ABCD has vertices A(−6, 2), B(−4, −6), C(2, −8), and D(6, 10). Find the image of quadrilateral ABCD under the mapping $(x, y) \rightarrow \left(\frac{1}{2}x, \frac{1}{2}y\right)$.

The pairs of triangles are similar. Find the unknown sides.

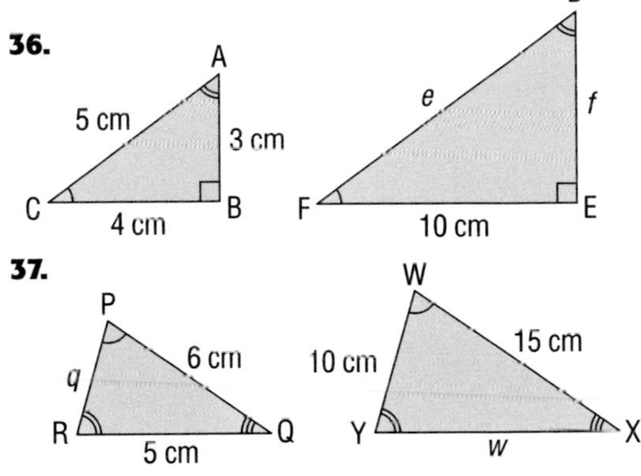

How many lines of symmetry does each figure have?

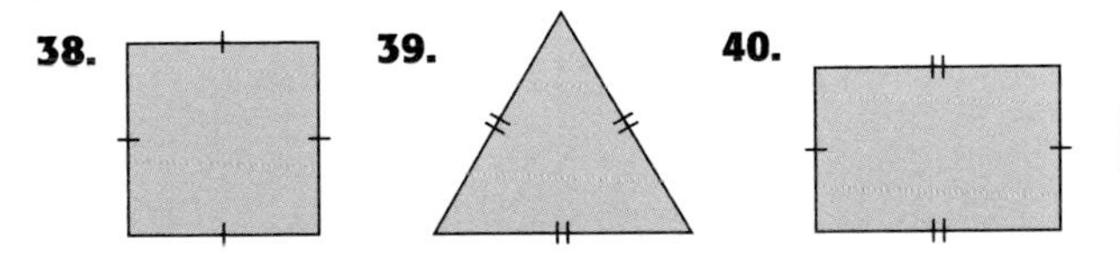

What is the order of the turn symmetry for each figure?

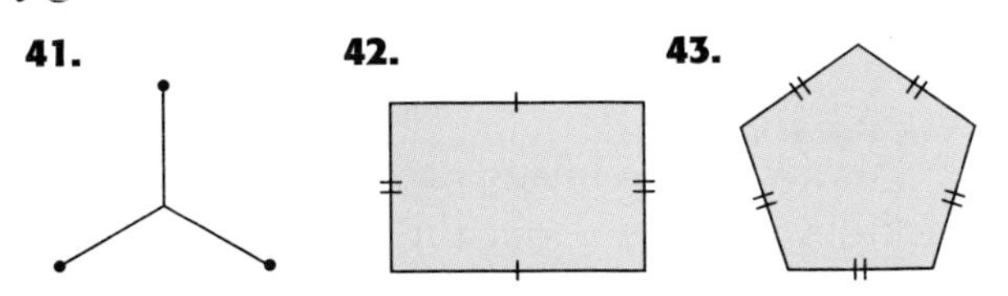

Group Decision Making
Math in the Workplace

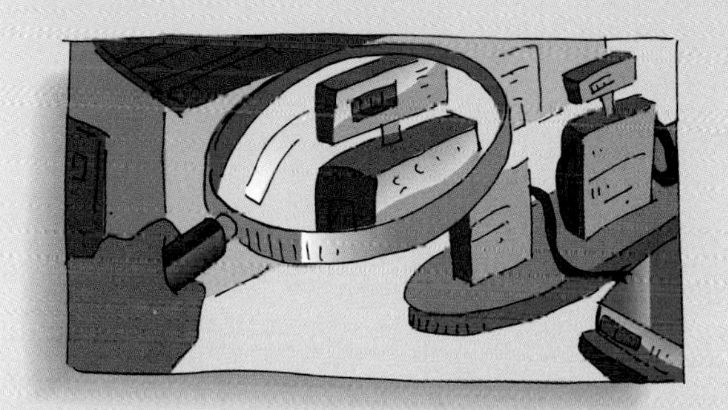

1. Decide as a class what local businesses, trades, or professions you would like to investigate to see how math is used. Your choices might include a jewellery store, clothing store, gas station, dentist's office, doctor's office, and a professional sports team. List your choices on the chalkboard.

2. Form pairs. Each pair chooses which workplace to research.

3. In your pair, research how math is used in your chosen workplace. You may have to visit the workplace, interview people, and use the library.

4. Prepare a visual presentation for the class to show how math is used. Presentations should include a description of how you conducted the research.

5. Write 2 problems based on your research. Be creative.

6. Make a presentation to the class. Include the math problems.

7. As a class, evaluate the visual presentations and the math problems. Check that the math problems can be solved.

Chapter Check

1. ΔABC has vertices A(−2, 0), B(0, −3), and C(3, −2). Find the coordinates of the image of ΔABC under the mapping $(x, y) \rightarrow (x + 5, y + 4)$.

2. ΔPQR has vertices P(2, −1), Q(0, 2), and R(−3, −2). Draw the image of ΔPQR under the mapping $(x, y) \rightarrow (x - 4, y + 3)$.

3. ΔRST has vertices R(−1, 1), S(−3, 4), and T(−6, 3). Draw the image of ΔRST after a reflection in the x-axis.

4. ΔDEF has vertices D(2, 4), E(−4, 1), and F(−1, −3). Draw the image of ΔDEF after a reflection in the y-axis.

5. ΔJKL has vertices J(−4, 0), K(−3, 5), and L(0, −3). Draw the image of ΔJKL after a 180° clockwise rotation about the origin.

6. ΔGHI has vertices G(4, 0), H(4, −3), and I(0, −2). Draw the image of ΔGHI after a 90° counterclockwise rotation about the origin.

7. ΔSTU has vertices S(0, 1), T(−3, 3), and U(4, −3). Draw the image of ΔSTU under the mapping $(x, y) \rightarrow (2x, 2y)$.

The pairs of triangles are similar. Find the unknown side lengths.

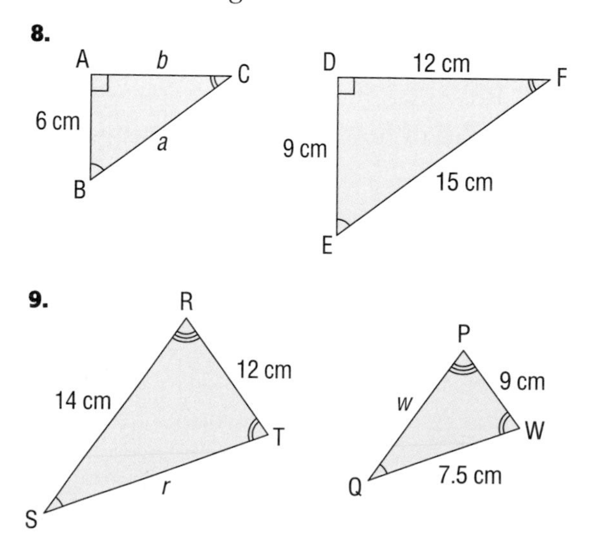

How many lines of symmetry does each figure have?

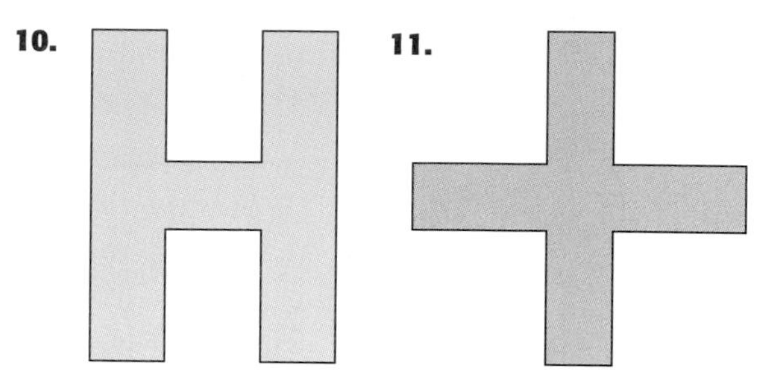

What is the order of turn symmetry for each figure?

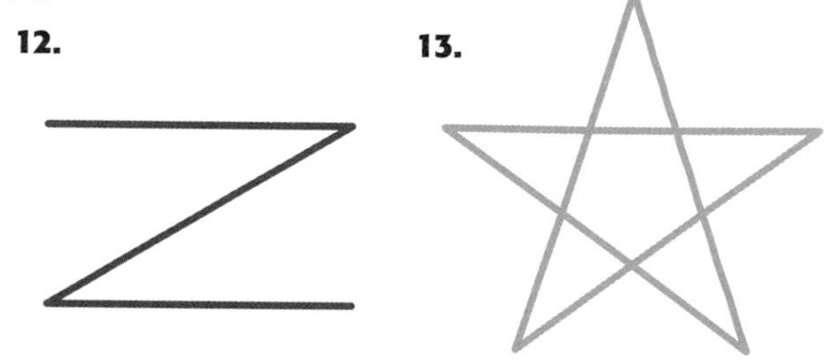

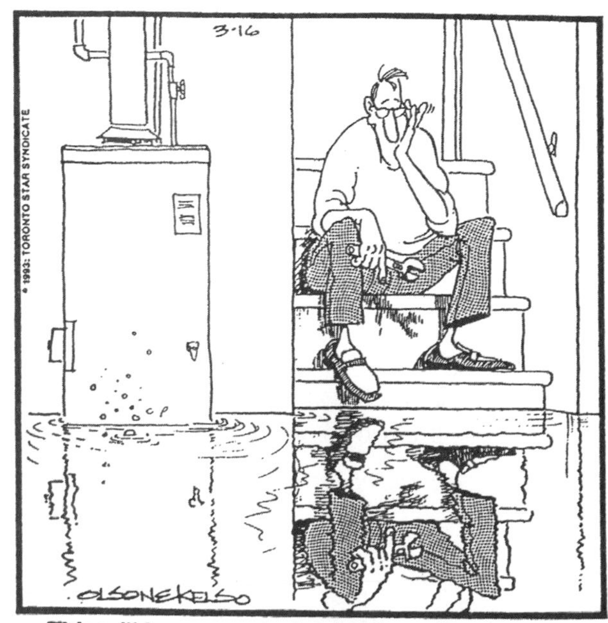

This will be a wonderful day for self reflection.

Using the Strategies

1. How many different sized squares can be constructed on a 5-by-5 geoboard?

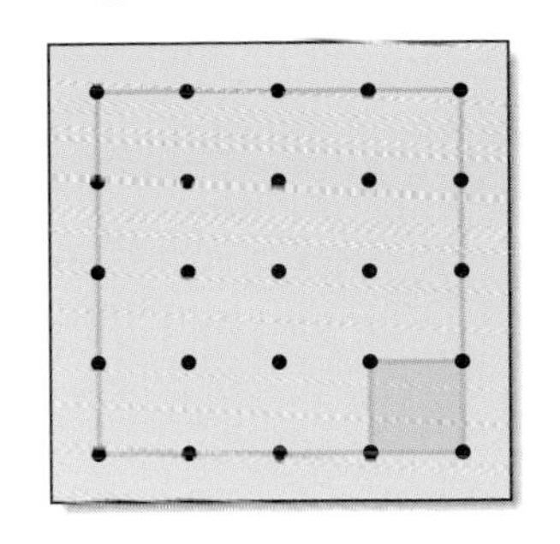

2. Find 2 multiples of 13 that are also multiples of 17.
3. Find 3 consecutive integers whose product is 1716.
4. When you write the whole numbers in words, "one," "two," "three," and so on, what is the first word that has the letters in alphabetical order?
5. The squares are exactly the same size. The total area of the figure is 384 cm^2. What is the perimeter of the figure?

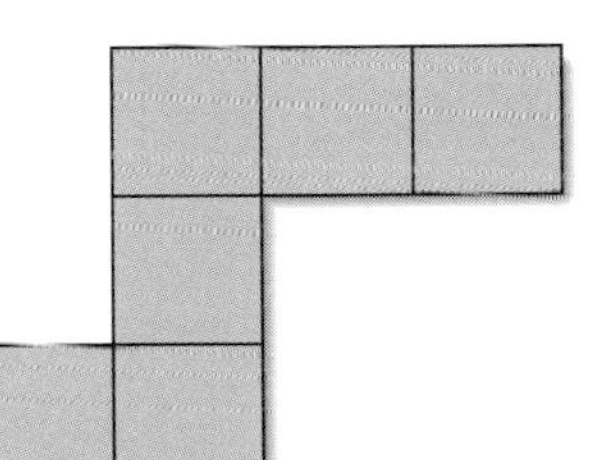

6. On a "prime day," both the month and the day are prime. March 13 is a prime day because March is month 3, which is prime, and so is 13. How many prime days are there this year?
7. The diagram shows one way of covering a 5 cm by 2 cm rectangle with 5 rectangles measuring 2 cm by 1 cm.

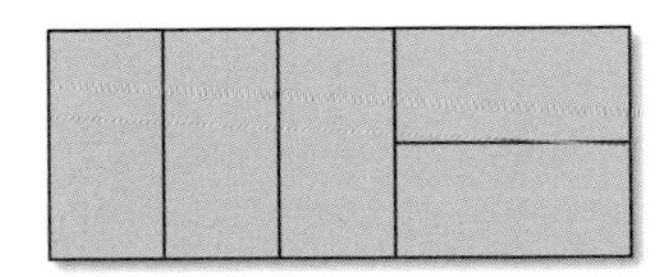

In how many other ways can you cover this rectangle with the smaller rectangles?

8. How many triangles are in the 1st row? the 2nd? the 3rd? the 4th? If the pattern continues, how many triangles are in the 10th row? the 20th row? the 100th row? the nth row?
9. You spend $2.75 in a store and receive $7.25 change from $10.00. Notice that the arrangement of the digits in the amount you spent is a rearrangement of the digits in your change. Find 4 other pairs of amounts spent and change from $10.00 that share this property.
10. A car is racing around this track.

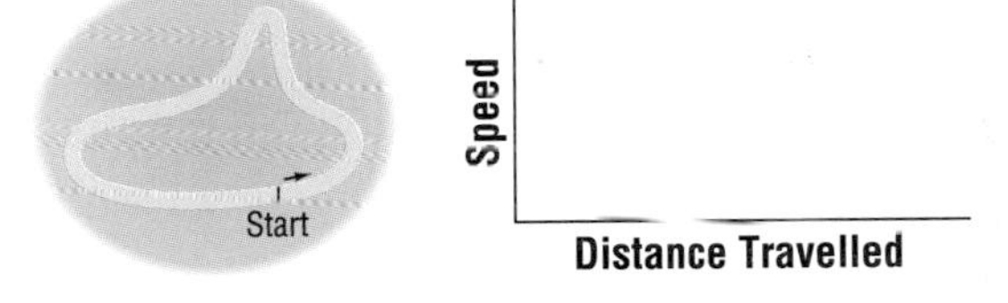

Sketch a graph of speed versus distance travelled for 1 lap of the track.

1. The driving distance, in kilometres, from Moncton to Saint John is given by the expression $0.25d + 26$, where d is the driving distance, in kilometres, from Edmonton to Saskatoon. What is the driving distance from Moncton to Saint John?
2. Use the Data Bank on pages 564 to 569 to write a problem. Have a classmate solve your problem.

TIME ZONES

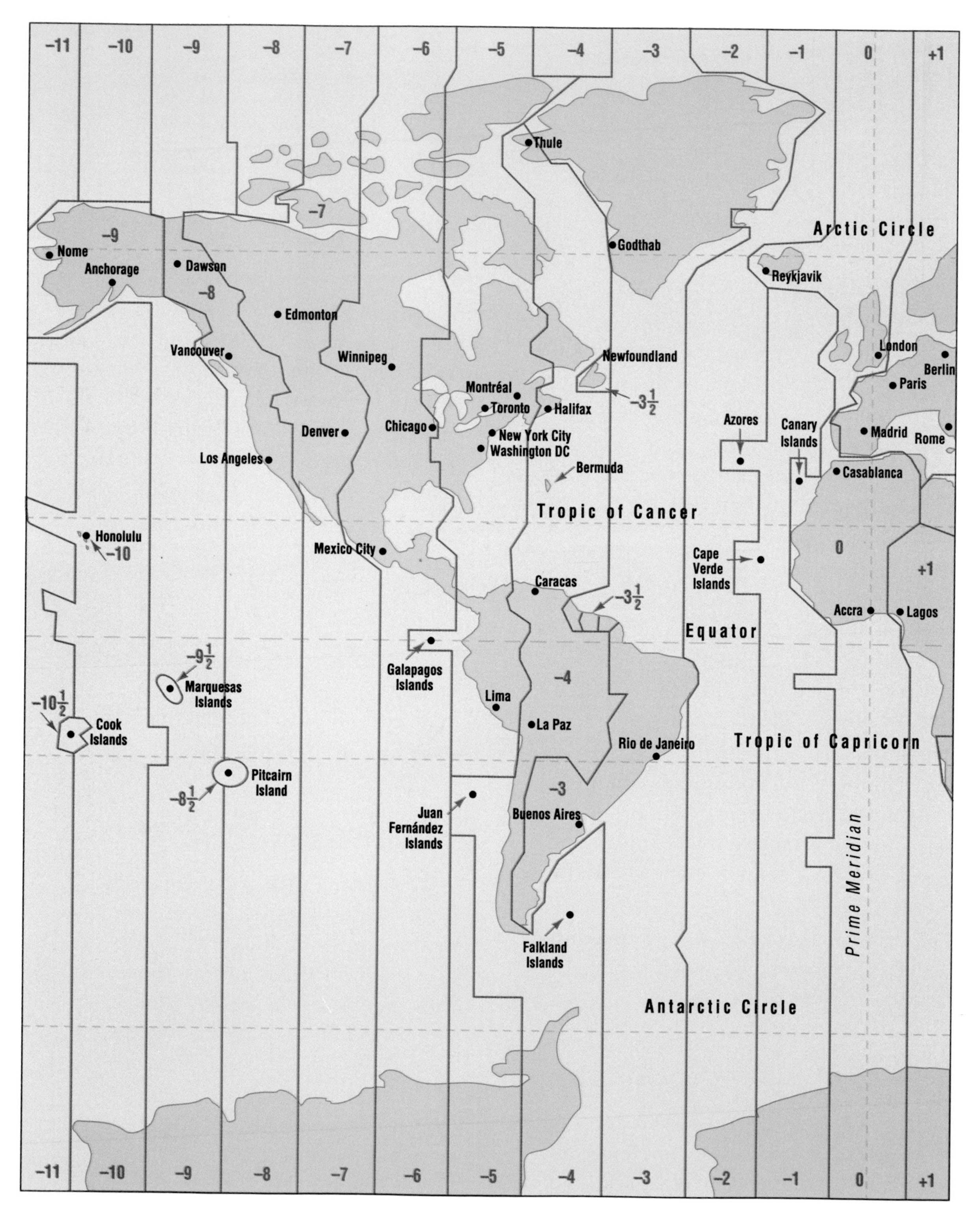
–11
–10
–9
–8
–7
–6
–5
–4
–3
–2
–1
0
+1
Thule
Arctic Circle
Godthab
Nome
Anchorage
Dawson
Reykjavik
Edmonton
London
Berlin
Paris
Newfoundland
Vancouver
Winnipeg
Montréal
Toronto
Halifax
–3½
Azores
Canary Islands
Madrid
Rome
Denver
Chicago
New York City
Washington DC
Los Angeles
Bermuda
Casablanca
Tropic of Cancer
Honolulu
Mexico City
Cape Verde Islands
Caracas
Accra
Lagos
Equator
Galapagos Islands
–9½
Marquesas Islands
–10½
Cook Islands
Lima
La Paz
Rio de Janeiro
Tropic of Capricorn
Pitcairn Island
–8½
Juan Fernández Islands
Buenos Aires
Prime Meridian
Falkland Islands
Antarctic Circle

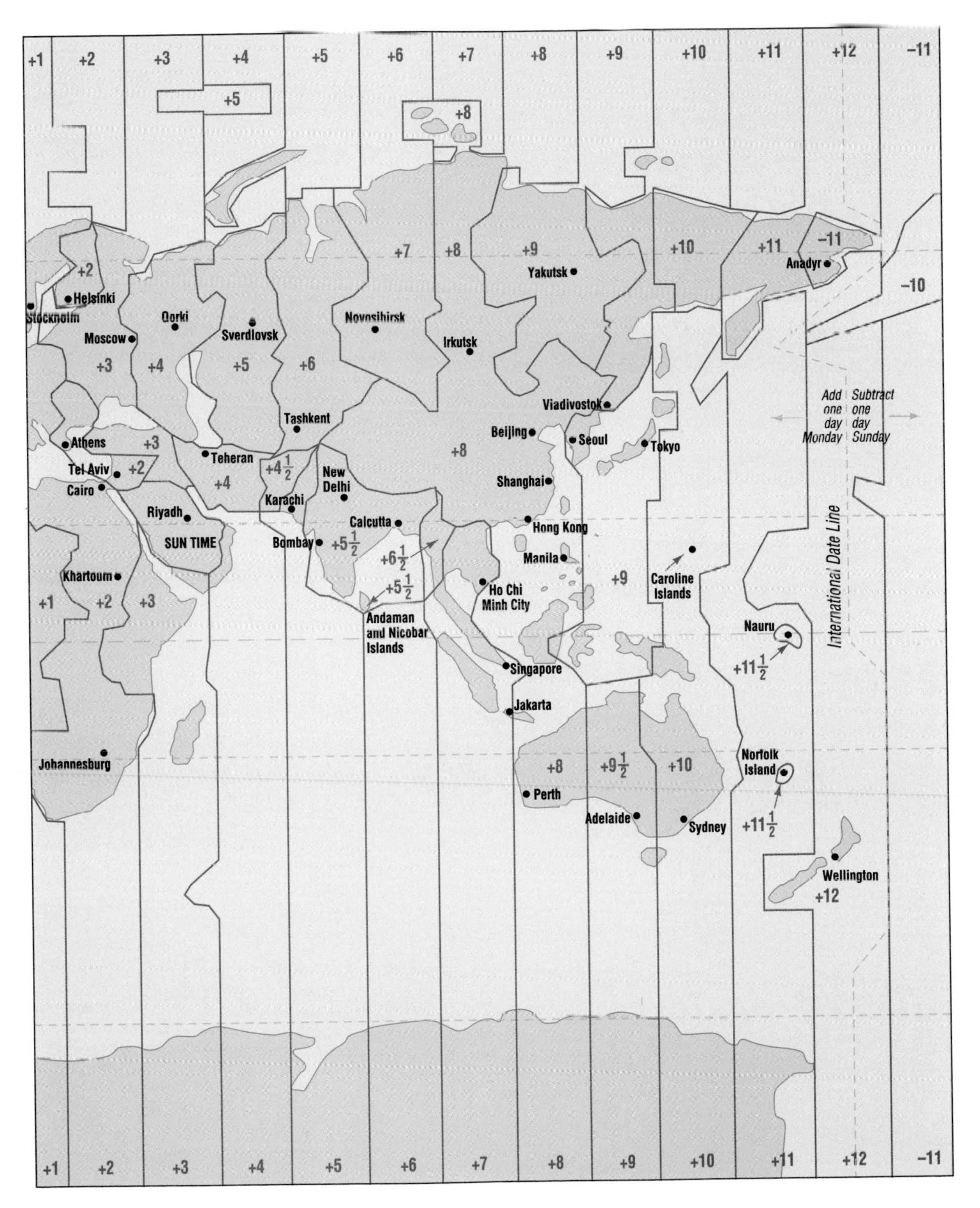
+1
+2
+3
+4
+5
+6
+7
+8
+9
+10
+11
+12
−11
−10
Helsinki
Stockholm
Gorki
Moscow
Sverdlovsk
Novosibirsk
Irkutsk
Yakutsk
Anadyr
Vladivostok
Tashkent
Athens
Tel Aviv
Cairo
Teheran
Riyadh
SUN TIME
Karachi
New Delhi
Calcutta
Bombay
$+4\frac{1}{2}$
$+5\frac{1}{2}$
$+6\frac{1}{2}$
Andaman and Nicobar Islands
Khartoum
Johannesburg
Beijing
Seoul
Tokyo
Shanghai
Hong Kong
Manila
Ho Chi Minh City
Caroline Islands
Singapore
Jakarta
Nauru
$+11\frac{1}{2}$
Norfolk Island
Perth
Adelaide
Sydney
$+9\frac{1}{2}$
Wellington
Add one day Monday
Subtract one day Sunday
International Date Line

FLYING DISTANCES BETWEEN CANADIAN CITIES

From	To	(km)
Calgary	Edmonton	248
	Montreal	3003
	Ottawa	2877
	Regina	661
	Saskatoon	520
	Toronto	2686
	Vancouver	685
	Victoria	725
	Winnipeg	1191
Charlottetown	Ottawa	976
	Toronto	1326
Edmonton	Calgary	248
	Ottawa	2848
	Regina	698
	Saskatoon	484
	Toronto	2687
	Vancouver	826
	Winnipeg	1187
Halifax	Montreal	803
	Ottawa	958
	Saint John	192
	St. John's	880
	Sydney	306
	Toronto	1287
Montreal	Calgary	3003
	Fredericton	562
	Halifax	803
	Moncton	707
	Ottawa	151
	Saint John	614
	St. John's	1618
	Toronto	508
	Vancouver	3679
	Winnipeg	1816
Ottawa	Calgary	2877
	Charlottetown	976
	Edmonton	2848
	Halifax	958
	Montreal	151
	Toronto	363
	Vancouver	3550
	Winnipeg	1687
Regina	Calgary	661
	Edmonton	698
	Saskatoon	239
	Toronto	2026
	Vancouver	1330
	Winnipeg	533
St. John's	Halifax	880
	Montreal	1618
	Toronto	2122

From	To	(km)
Toronto	Calgary	2686
	Charlottetown	1326
	Edmonton	2687
	Halifax	1287
	Montreal	508
	Ottawa	363
	Regina	2026
	St.John's	2122
	Vancouver	3342
	Windsor	314
	Winnipeg	1502
Victoria	Calgary	725
	Vancouver	62
Windsor	Toronto	314
Winnipeg	Calgary	1191
	Edmonton	1187
	Montreal	1816
	Ottawa	1687
	Regina	533
	Saskatoon	707
	Toronto	1502
	Vancouver	1862

DRIVING DISTANCES BETWEEN CANADIAN CITIES

	Edmonton	Halifax	Montreal	Ottawa	Quebec	Regina	Saint John	St. John's	Saskatoon	Toronto	Vancouver	Victoria	Whitehorse	Winnipeg
Calgary	299	4973	3743	3553	4014	764	4664	6334	620	3434	1057	1162	2385	1336
Edmonton		5013	3764	3574	4035	785	4704	6367	528	3455	1244	1349	2086	1357
Halifax			1249	1439	982	4228	309	1503	4485	1788	6050	6154	7099	3656
Montreal				190	270	2979	940	2602	3236	539	4801	4905	5850	2408
Ottawa					460	2789	1130	2792	3046	399	4611	4715	5660	2218
Quebec						3249	673	2363	3507	809	5071	5176	6120	2678
Regina							3919	5581	257	2670	1822	1926	2871	571
Saint John								1727	4176	1479	5741	5845	6790	3347
St. John's									5839	3141	7403	7775	8452	5010
Saskatoon										2927	1677	1782	2614	829
Toronto											4492	4596	5528	2099
Vancouver												105	2697	2232
Victoria													2802	2337
Whitehorse														3524
Winnipeg														

AREAS AND POPULATIONS OF CANADIAN PROVINCES

Province	Area (km²)	Population
Newfoundland and Labrador	405 720	568 474
New Brunswick	73 440	723 900
Nova Scotia	55 491	899 942
Prince Edward Island	5660	129 765
Quebec	1 540 680	6 895 963
Ontario	1 068 582	10 084 885
Manitoba	649 950	1 091 942
Saskatchewan	652 330	988 928
Alberta	661 190	2 545 553
British Columbia	947 800	3 282 061

WORLD BRIDGE RECORDS

Cable Suspension
Akashi Kaikyo, Japan (under construction)
890 m
1780 m
890 m

Cable Suspension
Humber, United Kingdom 1980
1410 m

Stone Arch
Pennsylvania, USA 1901
1161 m

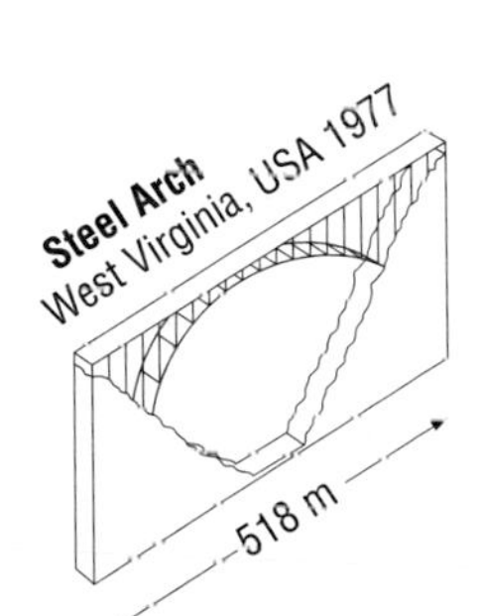

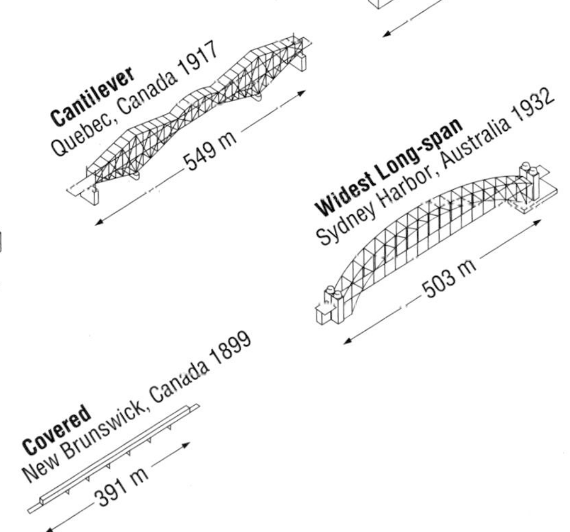

(Bridge widths are exaggerated for the purpose of this diagram.)

PLANETS: DISTANCES, ORBITS, MOONS

Mercury
Distance from the sun: 58 000 000 km
Time to orbit sun: 88 d
Number of moons: 0

Venus
Distance from the sun: 108 000 000 km
Time to orbit sun: 225 d
Number of moons: 0

Earth
Distance from the sun: 150 000 000 km
Time to orbit sun: 1 year
Number of moons: 1

Mars
Distance from the sun: 228 000 000 km
Time to orbit sun: 687 d
Number of moons: 2

Jupiter
Distance from the sun: 779 000 000 km
Time to orbit sun: 12 years
Number of moons: 16

Saturn
Distance from the sun: 1 425 000 000 km
Time to orbit sun: 29.5 years
Number of moons: 18

Uranus
Distance from the sun: 2 870 000 000 km
Time to orbit sun: 84 years
Number of moons: 15

Neptune
Distance from the sun: 4 497 000 000 km
Time to orbit sun: 165 years
Number of moons: 8

Pluto
Distance from the sun: 5 866 000 000 km
Time to orbit sun: 248 years
Number of moons: 1

ELEVATIONS OF THE TEN HIGHEST CITIES

City	Elevation (m)
Bogotá, Colombia	2 639
Addis Ababa, Ethiopia	2 450
Mexico City, Mexico	2 309
Nairobi, Kenya	1 820
Johannesburg, South Africa	1 734
Calgary, Canada	1 045
São Paulo, Brazil	776
Ankara, Turkey	686
Edmonton, Canada	666
Madrid, Spain	655

WORLD'S TOP TEN GOLD PRODUCERS

Country	Annual Production (t)
South Africa	621
United States	205
Australia	152
Canada	128
Brazil	100
Philippines	43
Colombia	33
Papua New Guinea	33
Chile	23
Venezuela	16

WORLD'S LARGEST RESERVOIRS

Location	Capacity (million kL)
Owem Falls, Uganda	2 700 000
Kariba, Zambia	180 600
Bratsk, Russian Federation	169 270
Aswan, Egypt	168 900
Akosombo, Ghana	148 000
Daniel Johnson, Canada	141 852
Guri, Venezuela	138 000
Krasnoyarsk, Russian Federation	73 300
Bennett, WAC, Canada	70 309
Zeya, Russian Federation	68 400

BEAUFORT WIND SCALE

Beaufort Number	Name	Speed (km/h)	Effect on Land
0	Calm	less than 1	Calm; smoke rises vertically.
1	Light air	1-5	Weather vanes inactive; smoke drifts with air.
2	Light breeze	6-11	Weather vanes active; wind felt on face; leaves rustle.
3	Gentle breeze	12-19	Leaves and small twigs move; light flags extend.
4	Moderate breeze	20-28	Small branches sway; dust and loose paper blow about.
5	Fresh breeze	29-38	Small trees sway; waves break on inland waters.
6	Strong breeze	39-49	Large branches sway; umbrellas difficult to use.
7	Moderate gale	50-61	Whole trees sway; difficult to walk against wind.
8	Fresh gale	62-74	Twigs broken off trees; walking against wind very difficult
9	Strong gale	75-88	Slight damage to buildings; shingles blown off roof.
10	Whole gale	89-102	Trees uprooted; considerable damage to buildings.
11	Storm	103-117	Widespread damage; very rare occurrence.
12-17	Hurricane	more than 117	Violent destruction.

WIND CHILL CHART

Wind Speed	Thermometer Reading (degrees Celsius) — Cold, Very Cold, Bitterly Cold, Extremely Cold													
	4	2	–1	–4	–7	–9	–12	–15	–18	–21	–23	–26	–29	–32	–34
Calm	4	2	–1	–4	–7	–9	–12	–15	–18	–21	–23	–26	–29	–32	–34
8 km/h	3	1	–3	–6	–9	–11	–14	–17	–21	–24	–26	–29	–32	–36	–37
16 km/h	–2	–6	–9	–13	–17	–19	–23	–26	–30	–33	–36	–39	–43	–47	–50
24 km/h	–6	–9	–12	–17	–21	–24	–28	–32	–36	–40	–43	–46	–51	–54	–57
32 km/h	–8	–11	–16	–20	–23	–27	–31	–36	–40	–43	–47	–51	–56	–60	–63
40 km/h	–9	–14	–18	–22	–26	–30	–34	–38	–43	–47	–50	–55	–59	–64	–67
48 km/h	–11	–15	–19	–24	–28	–32	–36	–41	–45	–49	–53	–57	–61	–66	–70
56 km/h	–12	–16	–20	–25	–29	–33	–37	–42	–47	–51	–55	–58	–64	–68	–72
64 km/h	–13	–17	–21	–26	–30	–34	–38	–43	–48	–52	–56	–60	–66	–70	–74

Exploring Math

Problem Solving p.xiv

Activity : **1.** 12 **2.** 14 **3.** 8 more, not counting rotations of the square **4.** 8 **6.** 4 **Activity 1:** **1.** 12 km, 12 km, 12 km, 12 km, 12 km **2.** Anywhere on the line segment joining A and B. **Activity 2:** **1.** 16 km, 14 km, 16 km, 12 km, 20 km **2.** At the camp located at B. **Activity 3:** **1.** Anywhere on the line segment joining B and C. **2.** At the camp located at C. **Activity 4:** **1.** Anywhere on the line segment joining the middle two camps. **2.** At the middle camp.

Mathematics as Reasoning p.xvi

Activity 1: **1.** $4 \times 5 + 8 \div 2$ **2.** $(7 - 4) \times (4 + 4)$ **3.** $(7 - 4 \div 4) \times 4$ **Activity 2:** **1.** Sari–blue MATHPOWER; Terri–green COMPUTERPOWER; Dmitri–black SCIENCEPOWER

Algebra p.xviii

Activity 1: **1. a)** 2, 4; 3, 6; 4, 8 **b)** The square must represent a number twice that of the triangle. **2. a)** 1, 3; 2, 6; 3, 9 **b)** 1, 1; 2, 2; 3, 3 **c)** 2, 2; 4, 3; 6, 4 **d)** 3, 2; 6, 4; 9, 6 **e)** 0, 0; 2, 2; no other whole number pairs **f)** 0, 0; no other whole number pairs **Activity 2:** **1.** 7 **2.** 12 **3.** 2 **4.** 6 **5.** 20 **6.** 2 **Activity 3:** **1.** 10 **2.** 3 **3.** 3 **4.** 4 **5.** 0

Functions p.xix

Activity 1: **1. b)** 4, 8, 12, 16, 20, 40, 400 **c)** The perimeter equals 4 times the figure number. **2. a)** 4, 6, 8, 10, 12, 22, 202; the perimeter equals 2 times the figure number, plus 2. **b)** 8, 12, 16, 20, 24, 44, 404; the perimeter equals 4 times the figure number, plus 4. **c)** 8, 10, 12, 14, 16, 26, 206; the perimeter equals 2 times the figure number, plus 6. **Activity 2:** **1.** y equals x less 8; x equals y plus 8 **2.** y equals x divided by 2; x equals 2 times y **3.** y equals 20 less x; x equals 20 less y **4.** y equals 2 times x plus 1; x equals the quantity y less 1, divided by 2

Trigonometry p.xxii

Activity : **2. a)** 1.96 **b)** equal **3.** 11.2 m

Probability p.xxiv

Activity 1: **1. a)** This is the ratio of the number of red marbles to the total number of marbles. **b)** $\frac{2}{10}$, $\frac{5}{10}$ **2.** 50% **3.** 50 **Activity 3:** $\frac{1}{8}$, $\frac{1}{16}$, $\frac{1}{256}$

Mathematics and Counting p.xxv

Activity 1: **1.** 12 **Activity 2:** 8 **Activity 3:** 12 **Activity 4:** **1.** 18, 22, 26 **2.** 63, 127, 255 **3.** 6, 2, $\frac{2}{3}$ **4.** 29, 47, 76

Investigating Limits p.xxvi

Activity 1: **2.** 1 **Activity 2:** **5.** 78.5 cm^2 **7.** Yes; the errors in the approximations would decrease.

Mathematical Structure p.xxvii

Activity 1: **1.** 05:00 **2.** 20 h **3. a)** 8 **b)** 6 **c)** 4 **d)** 1 **e)** 6 **f)** 4 **g)** 10 **h)** 7 **Activity 2:** **1. a)** yes **b)** no **c)** yes **d)** no **2. a)** 2 **b)** 3 **c)** 0 **d)** 2 **e)** 0 **3. a)** yes **b)** no

Chapter 1

Getting Started pp.2-3

Mental Math **1.** 866 **2.** 765 **3.** 868 **4.** 958 **5.** 787 **6.** 777 **7.** 919 **8.** 999 **9.** 434 **10.** 422 **11.** 213 **12.** 121 **13.** 601 **14.** 400 **15.** 111 **16.** 310 **17.** 12 **18.** 12 **19.** 6 **20.** 5 **21.** 23 **22.** 51 **23.** 11 **24.** 22 **25.** 306 **26.** 777 **27.** 8008 **28.** 6060 **29.** 6868 **30.** 4848 **31.** 9009 **32.** 176 **33.** 1111 **34.** 202 **35.** 2020 **36.** 321 **37.** 4321 **38.** 21 **39.** 103 **40.** 41 **41.** 15 **42.** 60 **43.** 4 **44.** 8 **45.** 0 **46.** 15 **47.** 2 **48.** 72

Section 1.1 p.5

Practice **1.** exact **2.** exact **3.** estimate **4.** estimate **5.** exact **6.** estimate **7.** estimate **8.** exact **9.** approx. **10.** exact **11.** exact **12.** Answers may vary **13.** exact **14.** approx. **15.** exact **16.** approx. **17.** exact **18.** approx. **Problems and Applications** **19.** calculator **20.** calculator **21.** lower **23. a)** overestimate **b)** exact **c)** underestimate

Learning Together pp.6-7

Activity 1: **1.** The next term is the sum of the previous two terms **2.** 34 **3.** terms 10 through 14: 55, 89, 144, 233, 377 **Activity 2:** **1.** 1 **2.** 1 **Activity 3:** **1.** 1 **2.** 1 **Activity 4:** **3.** $55 \times 89 = 4\ 895$

Section 1.2 p.9

Practice **1.** 720, 714 **2.** 250, 247 **3.** $93, 92.45 **4.** $270, 270.18 **5.** 200, 212 **6.** $60, $62.00 **7.** $1600, $1602 **8.** 160, 160 **9.** $200, $203.32 **10.** $68, $68.70 **11.** 25, $24.74 **12.** 2150, 2156 **13.** $11, $10.90 **14.** $115, $115.08 **Problems and Applications** **15.** $2.81 **16.** $70.22 **17.** Apr 28: $543.96, Apr 30: $628.96, May 1: $505.51, May 6: $437.91, May 13: $597.91 **18.** yes; $1.00 **19.** 18 524 579

Section 1.3 p.11

Practice **1.** 755.55 **2.** $453.53 **3.** $10.83 **4.** $32.50 **5.** $548.04 **6.** $107.95 **7.** 50.74 **8.** 91.16 **9.** 367.03 (nearest hundredth) **10.** 128.85 (nearest hundredth) **11.** 233.242 **12.** 21.57 (nearest hundredth) **Problems and Applications** **13. a)** 62.5 h **14.** $6.29 **15.** $11.61 **16.** $129.00 **17.** $15.30 **18.** $351 km **19.** 2 h 55 min **20. a)** 2.4 h **b)** 167.1 h **c)** 63 913 h

Section 1.4 p.13

Problems and Applications **1.** 3285 **2.** 18 750 **3.** $372 757.50 **4. a)** 315 km **b)** 280 km **c)** 90 km **5.** $15.50 **6.** $514.80 **7.** 9 397 728 000 000 km **9.** $78.05 **10.** 450 000 000

Section 1.5 p.15

Practice **1.** 7×11 **2.** $3x$ **3.** $5 + 8$ **4.** $9 + y$ **5.** xy m **6.** $+7$ **7.** $y - 8$ **8.** $\frac{x}{y}$ **9.** $4t$ **10. a)** 8 **b)** 12 **c)** 16 **d)** 4 **e)** 0 **f)** 36 **g)** 40 **h)** 140 **11. a)** 7 **b)** 10 **c)** 1 **d)** 31 **e)** 11.5 **f)** 4.3 **g)** 19.6 **h)** 27.4 **12. a)** 3 **b)** 1 **c)** 9 **d)** 19 **e)** 2 **f)** 5.2 **g)** 15.4 **h)** 18 **13. a)** 5 **b)** 17 **c)** 28 **d)** 18 **e)** 16 **f)** 32 **g)** 14 **h)** 7 **i)** 0 **14. a)** 14.6 **b)** 20.6 **c)** 58.5 **d)** 18 **15. a)** 9.6 **b)** 36.6 **c)** 26.5 **d)** 78.23 **e)** 141.88 **f)** 68.852 **Problems and Applications** **16. a)** $11n$ **b)** $451.00; $588.50 **17. a)** $300 + 35n$ **b)** $9050.00 **18. a)** st **b)** 262.5 km **19. a)** $x + 2$ **b)** $x + 1$ **c)** $x - 2$ **d)** $2x + 1$ **e)** $2x + 2$

Section 1.6 p.17

Problems and Applications **1.** $2589.00; assuming $215.75/month **2.** 860; assuming 20 cases per member **3.** 200; assuming equal proportions **4.** 75 km; assuming average speed of 15 km/h **5.** A reduction of 0.1 s per week training **6.** Assuming a representative survey **7.** 26 m **8. a)** 2 **b)** 3 **9.** 2 419 200 s; assuming 28 days **10. a)** 7 h; assuming driving at the speed limit **b)** 23:45 **11. a)** 12.5, 6.25, 3.125 **b)** 32, 39, 47 **c)** 63, 127, 255 **12.** 31.5 s

Section 1.7 pp.19–20

Practice **1.** 5; 3 **2.** 10; 7 **3.** x; 5 **4.** t; 2 **5.** 3^5 **6.** 4^6 **7.** 10^3 **8.** 6^4 **9.** m^5 **10.** r^3 **11.** 5×5 **12.** $1 \times 1 \times 1 \times 1 \times 1 \times 1$ **13.** $2 \times 2 \times 2 \times 2 \times 2$ **14.** $10 \times 10 \times 10 \times 10$ **15.** $0 \times 0 \times 0$ **16.** $y \times y \times y \times y$ **17.** $5 \times x \times x \times x$ **18.** $2m \times 2m \times 2m$ **19.** 8 **20.** 81 **21.** 125 **22.** 100 000 **23.** 2^2 **24.** 4^3 **25.** x^3 **26.** y^2 **27.** 10^2 **28.** 10^3 **29.** 10^5 **30.** 10^6 **31.** 10^8 **32.** 10^7 **33.** 32 **34.** 125 **35.** 256 **36.** 343 **37.** 10 000 000 **38.** 729 **39.** 0.25 **40.** 1.331 **41.** 0.0001 **42.** 2^4 **43.** 2^6 **44.** 2^8 **45.** 5^3 **46.** 5^2 **47.** equal **48.** 2^3 **49.** 58 **50.** 48 **51.** 24 **52.** 8 **53.** 0.1 **54.** 0.005 12 **55.** 1.6 **56.** 0.001 **57.** 76 **58.** 60 **59.** 32 **60.** 4 **61. a)** 64 **b)** 11 **c)** 73 **d)** 64 **62. a)** 13 **b)** 125 **c)** 19 **d)** 14 **e)** 150 **f)** 57 **Problems and Applications** **63. a)** 2000 **b)** 8000 **c)** 256 000 **64. a)** 2^7 **b)** 2^{10} **65. a)** 25 m, 45 m, 40 m **b)** 6 s **66. a)** Each number may be written as a power with the same number base and exponent. **b)** 5^5, 10^{10} **67.** $4a$ **68.** $6a$ **69. a)** 49 **b)** 301 **c)** 0 **70. a)** sometimes true **b)** always true **c)** never true **71. a)** 16, 32, 64 **b)** approx. 1.8×10^{13} km

Section 1.8 p.23

Practice **1.** 5^7 **2.** 2^{10} **3.** 7^{10} **4.** 10^7 **5.** 4^{11} **6.** 3^7 **7.** y^6 **8.** x^9 **9.** a^7 **10.** 4 **11.** 2 **12.** 6 **13.** 1 **14.** 3 **15.** 2 **16.** 3 **17.** 1 **18.** 4^2 **19.** 3^1 or 3 **20.** 9^0 or 1 **21.** 10^1 or 10 **22.** 4^6 **23.** 5^0 or 1 **24.** m^1 or m **25.** x^2 **26.** 4 **27.** 2 **28.** 7 **29.** 10 **30.** 1 **31.** 9 **32.** 4 **33.** 5 **34.** 2^{12} **35.** 3^{10} **36.** 4^{14} **37.** 10^{15} **38.** 5^{16} **39.** x^{20} **40.** y^9 **41.** t^{42} **42.** m^5 **43.** 2 **44.** 3 **45.** 4 **46.** 2 **47.** 3 **48.** 3 **49.** 5 **50.** 1 **Problems and Applications** **51.** 10^5 or 100 000 **52. a)** multiplied bases; added exponents when bases were not equal **b)** 72 **53. a)** divided bases; subtracted exponents when bases were not equal **b)** 54

Section 1.9 p.25

Problems and Applications **1. a)** 10.5 h **b)** No; requires 12.5 h **2.** B3, C2 or D2 **3. a)** 365 050 km **b)** 403 790 km **4. a)** −28.3°C **b)** −5°C **5.** Eiffel Tower: 319.9 m; CN Tower: 553.7 m; Washington Monument: 169.4 m **6. a)** 1.25 m, 0 m **b)** 2.3 m, 4.2 m, 7.8 m, 1.2 m **c)** 5.25 s

Learning Together pp.26–27

Activity 1: **3. a)** 320 **b)** 120 **c)** 346 000 **d)** 20 300 **e)** 60 **f)** 50 **g)** 3280 **h)** 5666 **i)** 70 000 000 **Activity 2:** **3. a)** 8.8 **b)** 0.056 **c)** 8.9 **d)** 0.004 **e)** 0.067 **f)** 0.003 33 **g)** 0.008 **h)** 0.111 **i)** 6 **Activity 3:** **3. a)** 5.6 **b)** 0.062 **c)** 0.97 **d)** 0.2 **e)** 0.06 **f)** 0.55 **g)** 0.328 **h)** 0.066 **i)** 700 **Activity 4:** **3. a)** 340 **b)** 920 **c)** 33 000 **d)** 23 400 **e)** 90 **f)** 4 **g)** 800 **h)** 450 000 **i)** 9900 **Activity 5:** **1. a)** 1200 **b)** 0.12 **c)** 0.7 **d)** 1200 **e)** 880 **f)** 0.88 **g)** 0.4 **h)** 30 **i)** 0.0034 **j)** 0.08 **k)** 8.8 **l)** 960 **m)** 56 **n)** 4 **o)** 9900

Section 1.10 p.29

Practice **1.** $4.5 \times 10\,000$, 4.5×10^4; 85 000, 8.5×10^4; 110 000, 1.1×10^5; 978 000 000, $9.78 \times 100\,000\,000$; 20 300 000, $2.03 \times 10\,000\,000$ **2.** 3 **3.** 6 **4.** 4 **5.** 7 **6.** 5 **7.** 8 **8.** 770 000 **9.** 67 000 **10.** 7.6×10^6 **11.** 9.8×10^8 **12.** 35 **13.** 2.3×10^4 **14.** 6.7×10^3 **15.** 130 000 **16.** 1.7×10^{10} **17.** 4.8×10^{15} **18.** 5.963×10^{14} **19.** 6×10^{12} **20.** 6.3×10^{14} **21.** 1.5×10^3 **22.** 100 **Problems and Applications** **23. a)** 9.1×10^4 **b)** 8×10^{23} **c)** 9.5×10^7 **24. a)** 2.2×10^{16} **b)** 2325.4 years **25.** Since 56 is larger than 10.

Section 1.11 pp.31–32

Problems and Applications **1. a)** 1618 km **b)** 1287 km **c)** 533 km **d)** 1187 km **e)** 3350 km **2.** Regina to Vancouver **3.** 3806 km **4.** 465 km **5.** 70 854 km **6.** via Montreal, by 26 km **7.** 1 h 50 min **8.** 12:55 **9.** No; its range is only 2400 km **10.** L–1011 and B–767 **11.** 1 h 35 min **12.** 10:20 **13.** 16:00 (including take-off and landing) **14.** 21:00 **15.** November 13 at 02:10 **16. b)** 4 h 10 min (not including take-off and landing)

Section 1.12 pp.34–35

Practice **1.** M, A **2.** E, S, M **3.** A, S **4.** M, D **5.** E, D, S **6.** B, M, A **7.** B, E, M **8.** D, A **9.** addition first **10.** subtraction first **11.** addition first **12.** multiplication first **13.** multiplication first **14.** addition first **15.** 13 **16.** 54 **17.** 12 **18.** 12 **19.** 38 **20.** 54 **21.** 3 **22.** 56 **23.** 5 **24.** 2 **25.** 1 **26.** 19 **27.** 3 **28.** 5 **29.** 1 **30.** 29 **31.** 6.4 **32.** 32.3 **33.** 4.4 **34.** 13.9 **35.** 13.39 **36.** 26.068 **37.** 9.84 **38.** 11.9 **39.** 9.8 **40.** 31.5 **41.** 135 **Problems and Applications** **42.** $(7 + 1) \div 2 = 4$

43. $5 \times (8 - 6) = 10$ **44.** $15 \div (3 + 2) + 1 = 4$
45. $(6 + 4) \div (2 \times 2) = 2.5$
46. $4.8 + (1.6 - 1.2) \times 5 = 6.8$
47. $3 \times (1.7 + 5.7) \div 3 = 7.4$
48. $(5 + 2)^2 \div 7 + 12 = 19$
49. a) $2 \times (1.30 + 4.50) + 25 \times (0.65 + 3.00)$ **b)** $102.85
50. $159.60 **51.** $2^2 - 2^2 + 2 = 2$; $2^2 \div 2^2 + 2 = 3$; $2 \div 2 + 2 \div 2 + 2 = 4$; $2^{2-2} + 2^2 = 5$; $2^2 + 2^2 - 2 = 6$; $2^2 \times 2 - 2 \div 2 = 7$; $2 - 2 + 2^2 \times 2 = 8$; $2 \div 2 + 2^2 \times 2 = 9$

Section 1.13 p.37

Problems and Applications **1.** driving from 08:00 until 09:00; at destination from 09:00 until 10:00; driving from 10:00 until 11:00; at destination from 11:00 until 12:00; driving from 12:00 until 12:15; stop for gas at 12:15; driving from 12:15 until 12:30; at destination from 12:30 until 13:30; driving from 13:30 until 14:00; at destination from 14:00 until 16:00; driving from 16:00 until 17:00 **2.** Boat A is not as old as boat B and costs more than boat B. Boat B is faster than boat A and both boats have the same length.
3. Between A and B: car is travelling at 100 km/h Between B and C: car is at rest Between C and D: car is travelling at 50 km/h Between D and E: car is travelling at 25 km/h

Connecting Math and Criminology pp.38–39

Activity 2: **3.** 1024 **Activity 3:** **3.** 25 **Activity 4:** **3.** 25 600

Review pp.40–41

1. a) exact **b)** exact **c)** estimate **d)** estimate **3. a)** $160.00 **b)** $30.00 **c)** $37.00 **d)** $230.00 **4. a)** $120.00 **b)** $30.00 **c)** $60.00 **5. a)** $360.00 **b)** $85.00 **c)** $400.00 **6.** 277 **7.** 1633 **8.** 0.69 **9.** 0.248 **10.** 18 630 **11.** 614 283 **12.** 3.0456 **13.** 10.773 **14.** 23 **15.** 254 **16.** 5.2 **17.** 5.6 **18.** 9352 **19.** 169.45 **20.** 32.1 **21.** 56 **22.** 7.24 **23.** 5.61 **24.** 9 **25.** 4.6 **26.** 3.4 **27.** 0.018 **28.** 420 **29.** 230 **30.** 0.902 **31.** 76 000 **32.** 2^8 **33.** 5^4 **34.** 2^8 **35.** 7^{12} **36.** 3^9 **37.** 6^4 **38.** 29 **39.** 89 **40.** 40 **41.** 15 **42.** 28.9 **43.** 101.4 **44.** 10.21 **45.** 4.3×10^7 **46.** 7.1×10^5 **47.** 8×10^{10} **48.** 3.89×10^{11} **49.** 3500 **50.** 56 000 000 **51.** 99.9 **52.** 12 500 000 000 **53. a)** 49 **b)** 138 **c)** 61 **d)** 95 **e)** 3 **f)** 26 **54. a)** 32.4 **b)** 26.8 **c)** 37.6 **55.** $150.00 **56.** $24.00 **57.** $1600.00 **58.** $49.00 **59.** $393.90 **60.** 13:30, assuming an average speed of 80 km/h **61.** $98.28 **62. a)** 20 m, 30 m, 30 m **b)** 5 s **63.** 3 h 35 min **64.** The present time plus 8 hours. **65. a)** $250 + 35n$ **b)** $5570 **c)** $36.65 **66. b)** 4.5×10^6

Chapter Check p.42

3. 883 **4.** 23.04 **5.** 116 **6.** 35.28 **7.** 2520 **8.** 32.766 **9.** 81 **10.** 0.065 **11.** 23.04 **12.** $80.00 **13.** $20.00 **14.** 3130 **15.** 0.0239 **16.** 5690 **17.** 0.156 **18.** 4.5×10^7 **19.** 2.13×10^5 **20.** 16 **21.** 81 **22.** 125 **23.** 3^9 **24.** 2^4 **25.** 5^8 **26.** 2 **27.** 12 **28.** 21 **29.** 9 **30.** 30.8 **31.** 72.36 **32.** 40 **33.** 85 **34.** 7.5 **35.** 5 **36.** $41.55 **37. a)** 15 m **b)** 20 m **c)** 15 m

Using the Strategies p.43

1. The first number in each pair represents a number from the first set, the second a number from the second set. (1, 8), (2, 2), (3, 13), (4, 12), (5, 11), (6, 10), (7, 9), (8, 1), (9, 7), (10, 6), (11, 5), (12, 4), (13, 3) **2.** $535.00 **4.** Tuesday **5.** 1, 3, 4, 7, 9 **7.** Tuesday **8.** Today's date: January 1 Robert's birthday: December 31 **9.** 15:51 **10.** 4 **11. a)** $8 + 9 + 10 + 11 + 12 + 13 = 63$ **b)** 33, 34, 35, 36 **12.** There are 21 in total, if we include both 1 and 100: 1, 4, 9, 10, 13, 18, 22, 27, 31, 36, 40, 45, 54, 63, 72, 79, 81, 88, 90, 97, 100 **13.** 2.5 min, assuming both can run at their maximum speed for 1 km.

Chapter 2

Getting Started pp.46–47

Activity 1: **1.** Salter, Goldstein, Lemieux, Smith **2.** Olychuk, Delecourt, Donner, Courtner **3.** Carlisle: +2, Benson: −4, Korchinski: +21, Beltz: −9, Kranich: +6, Fragnal: −19, Meechum: 0, Bolchek: −15, Mazza: +2 **Activity 2:** **1.** double bogey: +2, bogey: +1, par: 0, birdie: −1, eagle: −2 **2. a)** over **b)** +2 **c)** 38 **Activity 3:** Third column of table: +16, −1, −5, +3, +7, −3, −13, −6, −22, +4, −8, 0, −1, −12 **Mental Math** **1.** 21 000 **2.** 700 **3.** 1300 **4.** 440 000 **5.** 3610 **6.** 234 000 **7.** 40 000 **8.** 60 000 **9.** 300 000 **10.** 20 000 **11.** 303 **12.** 505 **13.** 808 **14.** 606 **15.** 1010 **16.** 1818 **17.** 2828 **18.** 7272 **19.** 1212 **20.** 6363 **21.** 8008 **22.** 7007 **23.** 8008 **24.** 15 015 **25.** 18 018 **26.** 24 024 **27.** 42 042 **28.** 32 032 **29.** 49 049 **30.** 40 040 **31.** 62 **32.** 750 **33.** 4 **34.** 120 **35.** 760 **36.** 87 **37.** 94 **38.** 465 **39.** 2300 **40.** 141 **41.** 50 **42.** 25 **43.** 200 **44.** 40 **45.** 40 **46.** 110 **47.** 20 **48.** 4 **49.** 60 **50.** 12 **51.** 24 **52.** 36 **53.** 20 **54.** 40 **55.** 30 **56.** 18 **57.** 13 **58.** 4 **59.** 8 **60.** 29

Section 2.1 pp.50–51

Practice **1.** +22 **2.** −9 **3.** +567 **4.** −234 **5.** −13 **6.** +1345 **7.** +3 **8.** −7 **9.** +11 **10.** −3 **11.** +56 **12.** +7 **13.** −8 **14.** +6 **15.** +2 **16.** +3 **17.** −13 **18.** 7 **19.** −3 **20.** 1 **21.** 12 **22.** −6 **23.** 10 **24.** −10 **25.** −8 **26.** 8 **27.** −4 **28.** −1 **29.** −2 **30.** 0 **31.** −6 **32.** 15 **33.** −14 **34.** 7 **35.** 2 **36.** −3 **37.** 8 **38.** 6 **39.** 0 **40.** 6 **41.** −9 **42.** −4 **43.** −5 **44.** 0 **45.** −2 **46.** the integers greater than 2 **47.** the integers greater than −5 **48.** the integers less than 10 **49.** the integers less than −2 **50.** the integers greater than 0 or the positive integers **51.** the integers less than 0 or the negative integers **52.** the integers less than or equal to 0 **53.** $>$ **54.** $>$ **55.** $<$ **56.** $>$ **57.** $>$ **58.** $<$ **59.** $=$ **60.** $>$ **61.** $<$ **62.** $<$ **63.** $+3 > -2$ **64.** $-4 < +7$ **65.** $-2 > -9$ **66.** $-5 < 0$ **67.** 3, 0, −2 **68.** −1, −5, −6, −9 **69.** 5, 4, 0, −3 **70.** 4, 2, 0, −1, −3 **71.** −3, −2, 0, 5 **72.** −4, −3, −2, −1 **73.** −15, −11, 14, 18 **74.** −2, −1, 1, 2 **79.** 2, 3, 4 **80.** −1, 0, 1, 2 **81.** 0, 1, 2, 3 **82.** none **83.** −4, −3, −2, −1 **84.** 4, 5, 6,… **85.** −1, −2, −3,… **86.** 4, 3, 2,… **87.** −2, −1, 0,… **88.** −5, −4, −3,… **89.** −7, −8, −9,… **Problems and Applications** **90. a)** 2 **b)** −3 **c)** −5 **d)** 2 **e)** −10 **f)** 7 **g)** −2 **h)** 3 **i)** −5 **j)** 4 **k)** −6 **l)** 0 **91. a)** 2, 3 **b)** −1, 0, 1, 2, 3, 4, 5 **c)** −2, −1, 0, 1, 2, 3,

4, 5, 6 **d)** −4, −3, −2, −1, 0, 1 **e)** −5, −4, −3, −2, −1, 0 **f)** −8, −7, −6, −5, −4, −3, −2 **92.** yes; called an overdraft **94.** −170, −88, −63, −2, 0, 37, 43, 58, 134

Learning Together pp.52–53

Activity 2: Addition Statements: +7 + (+2) = +9, +5 + 3(−3) = +2, 0 + (−6) = −6, −3 + (+8) = +5, −4 + (−3) = −7, −9 + (+7) = −2 **Activity 3:** Subtraction Statements: +5 − (+2) = +3, +2 − (−3) = +5, 0 − (−2) = +2, −3 − (+3) = −6, −2 − (−3) = +1, −4 − (+5) = −9

Section 2.2 p.55

Problems and Applications **1.** 17, 20 **2.** 6, 4 **3.** 9, 6 **4.** 80, 160 **5.** 16, 8 **6. a)** 16 **b)** 19 **c)** 151 **7. a)** 9 **b)** 36 **c)** 100 **d)** The sum of the first cube is 1^2. To find the sum of the first 2 cubes, add 2 to the base of the previous result: 3^2. To find the sum of the first 3 cubes, add 3 to the base of the previous result: 6^2. etc. **e)** 1 + 2 + 3 + 4 + 5 + 6 + 7 + 8 + 9 = 45, thus the sum of the first 9 cubes is $45^2 = 2025$. **8.** 1, 121, 12321, 1234321, 123454321, 12345654321 **9.** $1 900 000 **10.** 5 **11.** 9 of the digit 0 and 19 of each of the digits 1 to 9 **12.** 65 **13.** 45 **14.** 156 **15.** 28 **16.** 54

Section 2.3 pp.57–59

Practice **1.** 9 **2.** 1 **3.** −6 **4.** 5 **5.** −5 **6.** −1 **7.** −9 **8.** 0 **9.** 5 **10.** −4 **11.** 5 **12.** −4 **13.** −7 **14.** 0 **15.** −5 **16.** 0 **17.** 2 **18.** 4 **19.** −17 **20.** 2 **21.** 6 **22.** −9 **23.** −2 **24.** 1 **25.** −1 **26.** −14 **27.** −7 **28.** 17 **29.** −2 **30.** −19 **31.** 23 **32.** 11 **33.** 2 **34.** 0 **35.** 9 **36.** −1 **37.** 12 **38.** 10 **39.** 1 **40.** −9 **41.** 0 **42.** −13 **43.** −5 **44.** −9 **45.** 15 **46.** 8 **47.** −2 **48.** −8 **49.** 7 **50.** −3 **51.** −3 **52.** −9 **53.** −9 **54.** 0 **55.** 6 **56.** −8 **57.** 45 **58.** −561 **59.** −306 **60.** −143 **61.** −711 **62.** 111 **Problems and Applications** **63.** −1 **64.** −5 **65.** 5 **66.** −8 **67.** −7 **68.** 7 **69.** 7, 6, 5, 4, 3, 2, 1 **70.** 1, 0, −1, −2, −3, −4, −5 **71.** −4, −5, −6, −7, −8 **72.** −2, −3, −4, −5, −6 **73.** −1°C **74.** −9, −6, −8, −6, −12, −14 **75.** $253 **76. a)** −6 **b)** yes **77.** −183°C **78. a)** positive **b)** negative **c)** 0 **d)** positive **e)** negative **79. a)** 2, 3, 4, 5, 6, 7, 8, 9, 10, 11, 12 **b)** −5, −4, −3, −2, −1, 0, 1, 2, 3, 4, 5 **80. a)** −4 + (−5); 5 + 6 + 7 + 8; −7 + (−6) + (−5) + (−4) + (−3) + (−2) + (−1) + 0 + 1 + 2 + 3; 5 + 6 + 7 **b)** −10 + (−11), −8 + (−7) + (−6), −6 + (−5) + (−4) + (−3) + (−2) + (−1); 4 + 5, 2 + 3 + 4, −1 + 0 + 1 + 2 + 3 + 4; −13 + (−14), −8 + (−9) + (−10), −7 + (−6) + (−5) + (−4) + (−3) + (−2) **c)** −2, 2

Section 2.4 pp.61–63

Practice **1.** −4 **2.** 3 **3.** 11 **4.** −7 **5.** 1 **6.** 23 **7.** −21 **8.** 4 **9.** 3 **10.** 13 **11.** −8 **12.** 1 **13.** 15 **14.** 11 **15.** 0 **16.** −1 **17.** 2 **18.** 10 **19.** 1 **20.** −1 **21.** −5 **22.** 2 **23.** −10 **24.** 0 **25.** 5 **26.** 3 **27.** −15 **28.** 10 **29.** 20 **30.** −7 **31.** −2 **32.** 22 **33.** 7 **34.** 29 **35.** −5 **36.** −3 **37.** −9 **38.** −58 **39.** −2 **40.** −70 **41.** 9 **42.** 14 **43.** −8 **44.** −11 **45.** −13 **46.** 6 **47.** 8 **48.** −16 **49.** −1 **50.** −3 **51.** 14 **52.** 2 **53.** −16 **54.** 5 **55.** −6 **56.** −2 **57.** 9 **58.** −1 **59.** −2 **60.** −8 **61.** 2 **62.** 8 **63.** 4 **64.** 3 **65.** −4 **66.** −3 **67.** −2 **68.** 6 **69.** 3 **70.** −2 **Problems and Applications** **71.** +3 **72.** +5 **73.** +5 **74.** −7 **75.** +10 **76.** +9 **77.** 1 **78.** 7 **79.** −2 **80.** 5 **81.** −5 **82.** 6 **83.** −6 **84.** −12 **85.** 1 **86.** −3 **87.** 8 **88.** −8 **89.** 5 **90. a)** 1, 0, −1, −2, −3 **b)** −3, −4, −5, −6, −7 **91. a)** 160 m **b)** 3175 m **c)** 2933 m **d)** 3760 m **e)** 805 m **f)** 3993 m **g)** 987 m **92. a)** 8949 m **b)** Death Valley by 314 m **93.** 15 m **94.** 57°C **95. a)** the non-zero integer **b)** negative or positive integer **96. a)** −2 **b)** −17 **97. a)** always true **b)** never true **c)** sometimes true **98. a)** −1, −5 **b)** −2 **c)** −21

Section 2.5 p.65

Problems and Applications **1.** 4 **2.** 1 **3.** 17 **4.** 16 **5.** 9, assuming each width in times of 15 s and 12 s respectively. **6.** 6, assuming we consider only who sits next to whom is of importance. **7.** 6 **8.** 16 **9.** The farmer must make 7 trips across the river. He begins by taking the goat across and returns. He then takes the wolf across and returns with the goat. He then takes the cabbages across and returns for the goat. **10.** 4

Section 2.6 p.68–69

Practice **1. a)** + **b)** − **c)** − **d)** + **2. a)** + **b)** + **c)** − **d)** − **e)** − **f)** − **3.** 28 **4.** −24 **5.** 18 **6.** −81 **7.** −30 **8.** 33 **9.** 24 **10.** −30 **11.** −35 **12.** 30 **13.** 0 **14.** −56 **15.** −54 **16.** 56 **17.** 0 **18.** −22 **19.** −84 **20.** 20 **21.** −45 **22.** 30 **23.** −56 **24.** 7 **25.** −72 **26.** 77 **27.** 24 **28.** −6 **29.** −40 **30.** −18 **31.** 12 **32.** 50 **33.** 0 **34.** 32 **35.** 60 **36.** 70 **37.** 60 **38.** −48 **39.** 30 **40.** 12 **41.** −48 **42.** −60 **43.** 0 **44.** −12 **45.** 0 **46.** 32 **47.** 49 **48.** 4 **49.** 8 **50.** 8 **51.** 16 **52.** 16 **53.** −1 **54.** 1 **55.** 1 **Problems and Applications** **56.** 3 **57.** −2 **58.** 5 **59.** 4 **60.** 3 **61.** 3 **62.** −4 **63.** −4 **64.** −5 **65.** −5 **66.** 2 **67.** 9 **68.** −2 **69.** 0 **70.** −2 **71.** −3 **72. a)** 6, 3, 0, −3, −6 **b)** −4, −2, 0, 2, 4 **c)** 4, 1, 0, 1, 4 **d)** −1, −3, −5, −7, −9 **73. a)** −15 **b)** 12 **c)** −19 **d)** −5 **e)** −5 **f)** 0 **74. a)** 6 **b)** 18 **c)** −30 **d)** 144 **e)** −15 **f)** 21 **75.** $200 **76. a)** 10°C **b)** −7°C **77. a)** 6400 m **b)** 12 200 m **78. a)** negative **b)** positive **c)** negative **79. a)** never true **b)** always true **c)** sometimes true **d)** always true **e)** sometimes true **80. a)** $(-2)^3$ **b)** no, $(-2)^4 = 16$ **81. a)** 0, 1 **b)** −4, 5 **c)** −1, 3, 4 **d)** −3, −2, −1, 6

Section 2.7 p.71

Practice **1.** positive **2.** negative **3.** positive **4.** negative **5.** −9 **6.** 4 **7.** 2 **8.** −8 **9.** 3 **10.** −9 **11.** 0 **12.** −1 **13.** −1 **14.** 8 **15.** −7 **16.** 5 **17.** −5 **18.** 7 **19.** 4 **20.** −4 **21.** −10 **22.** 6 **23.** −3 **24.** −10 **Problems and Applications** **25.** −2°C **26.** 350 m/min **27. a)** −1 **b)** −1 **28.** equal; the dividend equals 0; opposite **29. a)** sometimes true **b)** sometimes true **c)** never true **30. a)** 2, −8 **b)** −3, −9 **c)** 5, 0

Section 2.8 p.73

Problems and Applications **1.** Multiply the number of listings on a typical page by the number of pages. **2.** Measure the thickness of 100 pages and divide by

100. **3.** Time yourself for 5 pages (say) and multiply by 50. **4.** 100 **5.** 10 000 **6.** 10 100 **7.** 40 **8.** 15 **9.** 162 **10.** 90 **12.** 2^{12} or 4096

Section 2.9 p.75

Practice **1.** 14 **2.** 35 **3.** −2 **4.** 0 **5.** −72 **6.** 320 **7.** −14 **8.** 3 **9.** 10 **10.** 22 **11.** −16 **12.** 13 **13.** −57 **14.** 10 **15.** 13 **16.** 30 **17.** −14 **18.** 3 **19.** 37 **20.** 18 **21.** −30 **22.** −13 **23.** 36 **24.** 11 **25.** −30 **26.** −13 **27.** 1 **28.** 3 **29.** 7 **30.** 21 **31.** −11 **32.** −5 **33.** 5 **34.** 15 **35.** 22 **36.** 26 **37.** −6 **38.** 4 **39.** −1 **40.** −1 **41.** 1 **42.** 4
Problems and Applications **43. a)** $(-3 + 4)^2 \times 5 = 5$ **b)** $(-1 - 3 - 8) \div 4 = -3$ **c)** $3^2 + 4 \times (2 - 5) = -3$ **d)** $(6^2 - 20) \div (2 + 6) = 2$ **44.** For ages ≥ 15; the expression $\frac{1}{2}(n - 9)^3$ is larger.
45. a) $(-3)^2 - 5$; $(-3 - 5)^2$ **b)** 4; 64 **c)** −3 squared decreased by 5; −3 decreased by 5 then squared.

Section 2.10 p.79

Practice **1. a)** 8 **b)** 4 **c)** 0 **d)** −4 **e)** 5 **f)** 8 **2. a)** −6 **b)** 2 **c)** 12 **d)** −4 **e)** 8 **f)** −3 **3. a)** −10 **b)** −24 **c)** 1 **d)** 5 **e)** 18 **f)** −1 **4. a)** 24 **b)** −15 **c)** 25 **d)** −20 **e)** 10 **f)** 2 **g)** −11 **h)** 40 **i)** −38 **5. a)** 5 **b)** −2 **c)** 32 **d)** −3 **e)** −8 **f)** 5 **g)** 81 **h)** 4 **6. a)** −14 **b)** −5 **c)** 17 **d)** −25 **e)** 31 **f)** 125 **g)** 81 **h)** 5 **i)** 27 **j)** 10 **Problems and Applications** **7. a)** 3, 1, −1, −3, −5 **b)** 7, 4, 3, 4, 7 **c)** 3, 2, 1, 0, −1 **d)** 0, −1, −2, −3, −4 **8. a)** a) 91 m, 100 m **b)** 0 m; the rocket has returned to the ground.
9. a) 15 m, 20 m, 15 m **b)** 0 m; the rock is at the same height as the edge of the cliff after 4 s **c)** −25 m; the rock is 25 m below the edge of the cliff. **d)** 6 s
10. a) −3°C **b)** −2°C

Section 2.11 p.81

Problems and Applications **1. a)** 1330 km **b)** 1822 km **c)** 492 km **3.** 13:45 **4.** −47°C **5.** 18:00 on July 1 **6. a)** 6050 km **b)** 67 h **7.** 1.4 **8.** Prince Edward Island, Nova Scotia, New Brunswick, Ontario, Quebec, Alberta, British Columbia, Manitoba, Saskatchewan, Newfoundland and Labrador.

Connecting Math and Archaeology pp.82–83

Activity 1: Percent of C-14 remaining: 100, 50, 25, 12.5, 6.25, 3.13, 1.56, 0.78 **Activity 2:** **1.** about 23 000 years **2.** about 13 200 years **3.** about 12 800 years **4.** about 24 600 years **5.** about 18 900 years

Review pp.84–85

1. a) −6 **b)** 4 **c)** 7 **d)** −9 **2. a)** 5, −1, −4, −6 **b)** 9, 7, 0, −1, −5 **c)** −2, −7, −8, −9 **d)** 2, 1, 0, −1, −2 **3. a)** 5 **b)** −4 **c)** −3 **d)** −9 **5. a)** −1, 0, 1, 2, 3 **b)** −6, −5, −4, −3, −2 **6.** 12 **7.** −2 **8.** −11 **9.** 0 **10.** −1 **11.** −22 **12.** −6 **13.** −9 **14.** −26 **15.** −2 **16.** −4 **17.** +1 **18.** −2 **19.** 2 **20.** −11 **21.** −3 **22.** −7 **23.** 3 **24.** 11 **25.** 3 **26.** −7 **27.** −1 **28.** 10 **29.** 10 **30.** −9 **31.** 5 **32.** 1 **33.** 8 **34.** 10 **35.** 2 **36.** 10 **37.** 1 **38.** 12 **39.** −13 **40.** −2 **41.** 1 **42.** −12 **43.** −3 **44.** −17 **45.** 7 **46.** 12 **47.** 14 **48.** 15 **49.** 110 **50.** −63 **51.** 72 **52.** −32 **53.** −2 **54.** −2 **55.** −3 **56.** 0 **57.** −99 **58.** −36 **59.** 42 **60.** 12 **61.** 24 **62.** 16 **63.** −16 **64.** 12 **65.** 3 **66.** 4 **67.** −9 **68.** −5 **69.** −3 **70.** 3 **71.** 14 **72.** 10 **73.** −10 **74.** −7 **75.** 4 **76.** 3 **77.** 0 **78.** 4 **79.** 0 **80.** −4 **81.** 22 **82.** 4 **83.** −18 **84. a)** 1 **b)** −1 **c)** 28 **d)** 18 **e)** −10 **f)** 10 **85. a)** −12 **b)** 36 **c)** 13 **d)** 174 **e)** −3 **f)** −17 **g)** −1 **h)** 11 **86. a)** 300 m **b)** 400 m **c)** 0 m **87.** 3°C

Chapter Check p.86

1. a) 4, 2, 0, −1, −2, −3 **b)** −3, −5, −6, −7, −9 **2. a)** −1 **b)** 3 **c)** −12 **d)** −3 **3. a)** 2, 1, 0, −1, −2, −3 **b)** −8, −7, −6, −5, −4, −3, −2 **4.** −3 **5.** −11 **6.** 3 **7.** −12 **8.** 15 **9.** 12 **10.** 3 **11.** −3 **12.** −8 **13.** −8 **14.** −12 **15.** −3 **16.** 12 **17.** 18 **18.** 0 **19.** 9 **20.** −16 **21.** 4 **22.** 1 **23.** −3 **24.** −18 **25.** 26 **26.** −4 **27.** −9 **28.** −26 **29. a)** −26 **b)** −11 **c)** 1 **d)** 67 **30.** −5°C **31.** 4°C

Using the Strategies p.87

1. Each entry is the sum of at most the 3 entries immediately above and to the left, that is 1 + 1 = 2, 1 + 2 + 3 = 6, 3 + 6 + 7 = 16 etc. **2.** 2 cm **3.** two 14 years, three 15 years **4.** 15, 21, 28 **5.** 8 **6.** 06:35
7. 163 216, 255 025, 367 236
8. a) 13 + 15 + 17 + 19 = 64
9. b) 21 + 23 + 25 + 27 + 29 = 125
10. c) 31 + 33 + 35 + 37 + 39 + 41 = 216 **d)** It is the middle (or average) value.
e) 43 + 45 + 47 + 49 + 51 + 53 + 55 = 343

Data Bank

1. 1264 m **2.** 12.5%

Chapter 3

Getting Started pp.90–91

Activity 1: **1.** 3, 5; 5, 7; 11, 13; 17, 19; 29, 31; 41, 43; 59, 61; 71, 73 **2.** 3, 5, 7 **3. a)** 24 = 11 + 13 **b)** 30 = 7 + 23 **c)** 42 = 19 + 23 **d)** 100 = 41 + 59
Mental Math **1.** 115 **2.** 144 **3.** 49 **4.** 121 **5.** 330 **6.** 80 **7.** 133 **8.** 412 **9.** 1600 **10.** 6969 **11.** 17 **12.** 12.3 **13.** 31 **14.** 61 **15.** 3342 **16.** 50 **17.** 0.95 **18.** 60 **19.** 30 **20.** 32 **21.** 578 **22.** 695 **23.** 498 **24.** 821 **25.** 210 **26.** 705 **27.** 165 **28.** −82 **29.** 149 **30.** −895 **31.** 32 **32.** −3 **33.** 79 **34.** 19 **35.** 9 **36.** −7 **37.** 20 **38.** 37 **39.** 48 **40.** −26 **41.** 50 **42.** 10 **43.** 80 **44.** 38 **45.** 196 **46.** 34 **47.** 36 **48.** 36 **49.** 150 **50.** −20

Learning Together pp.92–93

Activity 1: **a)** 64 **b)** 100 **c)** 1 000 000 **Activity 2:** **1.** 15, 21, 28 **2.** a square number **Activity 3:** The locker doors that are closed are numbered by perfect squares: 1, 4, 9, 16,…, 900, 961 **Activity 4:**
1. a) $33 = 5^2 + 2^2 + 2^2$ **b)** $42 = 5^2 + 4^2 + 1^2$ **c)** $77 = 6^2 + 5^2 + 4^2$ **d)** $88 = 8^2 + 4^2 + 2^2 + 2^2$ **e)** $153 = 12^2 + 3^2$ **f)** $212 = 14^2 + 4^2$ **g)** $208 = 12^2 + 8^2$ **h)** $903 = 23^2 + 19^2 + 3^2 + 2^2$

Section 3.1 pp.96–97

Practice **1.** $-7, 7$ **2.** $-9, 9$ **3.** $-11, 11$ **4.** $-25, 25$ **5.** $-0.8, 0.8$ **6.** $-0.1, 0.1$ **7.** $-1.4, 1.4$ **8.** $-0.5, 0.5$ **9.** 5 **10.** 10 **11.** 15 **12.** 16 **13.** 13 **14.** 0.6 **15.** 0.2 **16.** 1.1 **17.** 0.9 **18.** 5 **19.** 8 **20.** 10 **21.** 30 **22.** 30 **23.** 60 **24.** 90 **25.** 200 **26.** 900 **27.** 0.9 **28.** 0.9 **29.** 0.2 **30.** 0.03 **31.** 0.09 **32.** 0.05 **33.** 0.03 **34.** 0.02 **35.** 0.001 **36.** 0.002 **37.** 6, 5.6 **38.** 7, 6.6 **39.** 8, 7.9 **40.** 9, 8.9 **41.** 10, 10.0 **42.** 10, Jun **43.** 30, 33.5 **44.** 100, 142.1 **45.** 300, 293.3 **46.** 400,449.6 **47.** 4.4 **48.** 0.9 **49.** 9.9 **50.** 28.8 **51.** 2.4 **52.** 18.2 **53.** 1.7 **54.** 0.2 **Problems and Applications** **55. a)** 4 **b)** 5 **c)** 6 **d)** 7 **e)** -10 **f)** 8 **g)** 20 **h)** -9 **i)** 51.1 **56. a)** 20.8 cm **b)** 14.1 m **57. a)** 16.2 cm^2 **b)** 113.9 cm^2 **c)** 24 m^2 **d)** 13.8 m^2 **58. a)** 6.0 cm **b)** 10.1 m **c)** 28.4 mm **d)** 45.1 cm^2 **59.** 629 m **60. c)** Results are equal since $\sqrt{8} = \sqrt{4 \times 2} = 2\sqrt{2}$ **61.** they are opposites **62.** 6.5 cm - half the side length of the square **63.** an error, negative numbers do not have square roots **64. a)** 17, 24, 28, 33 **b)** The sum of the first n odd numbers is n^2.

Section 3.2 p.99

Practice **1.** 7, 6.8 **2.** 20, 17.4 **3.** $-9, -9.5$ **4.** 40, 40.6 **5.** $-63, -65.7$ **6.** 13, 12.5 **7.** 69, 65.2 **8.** 100, 106.5 **9.** 4.7, 4.5 **10.** 11, 10.8 **11.** 1.4, 1.4 **12.** 2.8, 2.9 **13. a)** 5.0 cm **b)** 20.0 cm **14. a)** 15.0 m **b)** 60.0 m **15. a)** 7.4 m **b)** 29.6 m **16. a)** 9.5 cm **b)** 38 cm **17. a)** 14.1 m **b)** 56.4 m **18. a)** 28.3 cm **b)** 113.2 cm **19. a)** 100 m by 100 m **b)** 50 m by 50 m **20. a)** 7.3 km/h, 10.1 km/h **b)** 18.5 km/h, 15.6 km/h **21. a)** 340.0 m/s **b)** 330.5 m/s **c)** 324.3 m/s **22.** Approximately 11000 km/h **a)** 1, 3, 5, 7, 9 **b)** consecutive odd numbers **c)** $2 \times 245 - 1 = 489$

Section 3.3 p.101

Problems and Applications **1.** Maria–grade 11, Paula–grade 9, Shelly–grade 10 **2.** Al–golf, Bjorn–swimming, Carl–running, Don–bowling **3.** Susan–pilot, Irina–writer, Traci–doctor, Debbie–dentist **4.** 25 m **5.** 3 quarters, 4 dimes, 4 pennies **6.** 110 km **7.** 20 **8.** 24 **9.** 23:59 **10.** Evans–artist, Thompson–plumber, Smith–banker, DiMaggio–teacher

Section 3.4 pp.104–105

Practice **1.** 2 **2.** -5 **3.** 1 **4.** -9 **5.** 5 **6.** 2 **7.** 0 **8.** 1 **9.** $(-4)^3$ **10.** $(-3)^5$ **11.** p^5 **12.** $(-n)^4$ **13.** $3^4 \times (-2)^3$ **14.** $(-2)(-2)(-2)(-2)(-2)$ **15.** $-2 \times 2 \times 2 \times 2 \times 2$ **16.** $(-x)(-x)(-x)$ **17.** 9 **18.** 9 **19.** 1 **20.** -1 **21.** -125 **22.** -125 **23.** -0.125 **24.** 1.4641 **25.** 6.25 **26.** 5^9 **27.** $(-8)^5$ **28.** $(-2)^7$ **29.** 2^6 **30.** $(-2.1)^8$ **31.** $(\ 0.2)^5$ **32.** 5^1 **33.** 6^2 **34.** $(-4)^2$ **35.** $(-9)^5$ **36.** 2^6 **37.** $(-3)^{28}$ **38.** $(-5)^6$ **39.** $(-6)^{15}$ **40.** $(-4)^{42}$ **41.** $(-2.3)^{12}$ **42.** x^6 **43.** y^7 **44.** z^7 **45.** $(-m)^{10}$ **46.** s^0 **47.** $(-r)^6$ **48.** $(-5)^5, -3125$ **49.** 6^7, 279 936 **50.** $(-2)^{10}$, 1024 **51.** $(-1)^{12}$, 1 **52.** $(-3.1)^8$, 8528.9 **53.** $(-3)^2$, 9 **54.** $(-10)^4$, 10 000 **55.** $(-4)^1, -4$ **56.** 16 **57.** 729 **58.** -3125 **59.** 729 **60.** -64 **61.** 256 **62.** 256 **63.** -243 **64.** 64 **65.** 36 **66.** 64 **67.** -384 **68.** 324 **69.** 26 **70.** 781 **71.** -2592 **72.** -10.125 **73.** 6400 **74.** 27.04 **75.** -0.018 **76. a)** 45 **b)** -162 **c)** 1702 **d)** 9 **77. a)** -8 **b)** 405 **c)** 13 **d)** -648 **e)** -125 **f)** 25 **g)** 36 **h)** -432 **i)** -47 **j)** 216 **Problems and Applications** **78. a)** 4 **b)** 9 **c)** 4 **d)** 10 **e)** 4 **f)** -1.3 or 1.3 **g)** 4 **h)** -0.6 **79.** 10 cm, 8 cm; 125 cm^3, 343 cm^3, 74.088 cm^3 **80. a)** 160, 320, 640, 1280, 2560 **b)** 163 840 **c)** 10×2^{40} **81.** positive **82.** negative **83.** one root is negative **84. a)** $\frac{-32}{243}$ **b)** $\frac{14\ 641}{16}$ **c)** $\frac{-27}{8}$ **d)** 1 **e)** $\frac{-1}{32}$ **f)** $\frac{-15}{16}$ **85. a)** yes **b)** no **c)** yes

Learning Together pp.106–107

Activity 1: **1.** The digits 2, 4, 8, 6 repeat in this order. **2.** 6 **3. a)** 6 **b)** 2 **c)** 4 **4.** 8 **Activity 2:** **1.** The digits 3, 9, 7, 1 repeat in this order. **2.** 1 **Activity 3:** 3 **Activity 4:** **1.** 6 **2.** constant **3.** Any base whose ones digit is a 6, for example **Activity 5:** **1. a)** 6 **b)** 4 **2.** The digits 4, 6 repeat in this order **3.** Any base whose ones digit is a 4, for example

Section 3.5 p.109

Problems and Applications **1.** 06:10 **2.** $48.00 **3.** $100.00 **4.** 06:15 **5.** $40.00 **6.** 10:15 **7.** 80 h **8.** 15 years **9.** 264 **10. a)** $32 000 **b)** $243 000 **11.** 3 400 000

Section 3.6 pp.112–113

Practice **1. a)** $\frac{1}{2^3}$ **b)** $\frac{1}{4^1}$ **c)** $\frac{1}{10^7}$ **d)** $\frac{1}{9^8}$ **e)** $\frac{1}{1^4}$ **f)** $\frac{1}{0.5^6}$ **g)** $\frac{1}{-7^6}$ **h)** $\frac{1}{-2^3}$ **i)** 2^3 **j)** 4^2 **k)** 5^4 **l)** $(-3)^5$ **2. a)** 8^{-2} **b)** 7^{-3} **c)** 9^{-4} **3. a)** 2^{-2} **b)** 3^{-3} **c)** 2^{-6} or 4^{-3} **d)** 3^{-5} **4.** 1 **5.** 64 **6.** $\frac{1}{3}$ **7.** $\frac{1}{16}$ **8.** -1 **9.** $\frac{1}{1000}$ **10.** 1 **11.** $\frac{1}{8}$ **12.** $\frac{1}{81}$ **13.** $\frac{-1}{1000}$ **14.** 1000 **15.** 1 **16.** 3 **17.** 16 **18.** -3 **19.** 7^9 **20.** 9^2 **21.** 8^{-8} **22.** 6^4 **23.** 5^{-5} **24.** 4^{-8} **25.** 3^{12} **26.** 9^{-8} **27.** 8^5 **28.** $(-2)^6$ **29.** 2^3 **30.** 3^{-6} **31.** 5^{-2} **32.** 8^1 **33.** $(-2)^{-6}$ **34.** $(-3)^0$ **35.** 81 **36.** 16 **37.** 2 **38.** 27 **39.** $\frac{1}{25}$ **40.** $\frac{1}{36}$ **41.** 7 **42.** 1 **43.** 2000 **44.** $\frac{1}{1728}$ **45.** 10 000 **46.** 100 **47.** $\frac{1}{2}$ **48.** 1 **49.** 28 **50.** 15.5 **51.** 134 **52.** $\frac{5}{4}$ **53.** 9 **54.** $\frac{1}{9}$ **55.** $\frac{1}{2}$ **56.** 7 **57.** $\frac{3}{2}$ **58.** $\frac{-4}{9}$ **59.** 3 **60.** $\frac{1}{25}$ **61.** 1 **62.** 100 **63.** $-\frac{27}{8}$ **64.** $\frac{16}{9}$ **65.** x^7 **66.** x **67.** y^{-4} **68.** t^4 **69.** m^8 **70.** b^2 **71.** m^8 **72.** t^{-8} **73.** y^{10} **74.** m^5 **75.** a^{-8} **76.** t^{-2} **77.** y^{-1} **78.** t^8 **79.** n^2 **Problems and Applications** **80. a)** 0.25 **b)** 0.027 **c)** 2^{-3} **d)** 10^{-2} **81. a)** 8 **b)** 81 **c)** 1 **d)** $\frac{1}{8}$ **e)** $\frac{1}{9}$ **f)** $\frac{1}{16}$ **g)** $\frac{1}{36}$ **h)** $\frac{1}{5}$ **i)** -1 **j)** $\frac{1}{25}$ **82. a)** 3 **b)** 4 **c)** 0 **d)** -2 **e)** -3 **f)** 3 **g)** 2 **h)** 10 **83. a)** 1 **b)** $\frac{27}{8}$ **c)** $\frac{-8}{9}$ **d)** $\frac{7}{16}$ **e)** $\frac{25}{16}$ **f)** $\frac{9}{25}$ **84. a)** $\frac{1}{32}$ **b)** 2^{-5} **c)** $\frac{1}{2^5}$ **d)** 39 900 years **85. a)** sometimes true **b)** sometimes true **c)** always true **86. b)** $\frac{1}{4}$

Section 3.7 p.115

Problems and Applications **1.** 10 **2.** 10 000 000 **3.** 12.25 days, or during the 13th day

Section 3.8 pp.116–117

Practice **1.** 4.5×10^6 **2.** 8.9×10^{-2} **3.** 2.0×10^{-1} **4.** 5.5×10^{-5} **5.** 4.5×10^8 **6.** 3.4×10^5 **7.** 3.3×10^{-7} **8.** 1.0×10^{-8} **9.** 6.0×10^{-9} **10.** 1.0×10^{-12} **11.** 230 000 000 **12.** 0.00 000 47 **13.** 0.000 000 007 **14.** 0.000 001 **15.** 2.3×10^7 **16.** 4.5×10^{-8} **17.** 5×10^{-6} **18.** 10^{-10} **19.** 7.8×10^8 **20.** 6.8×10^{-7} **21.** 8×10^{-11} **22.** 10^{-8} **23.** 1.25×10^{-6} **24.** 9.6×10^{-6} **25.** 1.312×10^{-4} **26.** 2.6×10^5 **27.** 5.0×10^{-2}

Problems and Applications

28. a) $4.35 \times 10^{-3}, 4.3 \times 10^{-3}, 10^{-3}, 8.4 \times 10^{-4}$ **b)** $5.6 \times 10^{-8}, \frac{1}{10^8}, 5.6 \times 10^{-9}, 10^{-9}$ **c)** $10^{-2}, 2.12 \times 10^{-3}, \frac{1}{1000}$ **29.** 4.0×10^{41} **30. a)** 2.0×10^6 J **b)** 8.0×10^{-4} J **c)** 1.5×10^6 J **d)** 8.0×10^{-11} J **31.** 0.23 is not larger than 1 **32.** 5×10^0 **33.** $1 \times 10^{-14}, 1 \times 10^9$

Section 3.9 p.119

Problems and Applications **1.** 6 **2.** 9 **3.** 78, 79, 80 **4.** 46, 48, 50, 52 **5.** 55 **6.** 4 **7.** 15 **8.** 17 m × 13 m **9.** 25 m, 31 m **10.** 16 **11.** 7 **12.** 8 cm by 7 cm × 6 cm **13.** 6 kg cashews, 18 kg peanuts **14.** Companion–15, Officer–46, Member–92

Connecting Math and Logic pp.120–121

Activity 1: **2. a)** TL **b)** TL **c)** TL **d)** TR **e)** LR **f)** TL **3. a)** 5 by 6 **b)** 3 by 3 **4.** 5 by 6 **5.** no **Activity 2:** **2.** odd number **3.** even number **4.** LR **Activity 3:** **2.** even number **3.** odd number **4.** TL **Activity 4:** **2.** odd number **3.** odd number **4.** TR **Activity 5:** **1. a)** The ball will always stop in the lower right corner. **b)** The ball will always stop in the top left corner. **c)** The ball will always stop in the top right corner. **2. a)** TL **b)** TR **c)** LR **d)** TR **e)** TL **f)** LR

Review pp.122–123

1. 5^6 **2.** $(-3)^3$ **3.** $(\frac{1}{2})^4$ **4.** 2^5 **5.** 243 **6.** 144 **7.** −125 **8.** 160 **9.** 1 **10.** 1 **11.** $\frac{1}{2}$ **12.** $\frac{1}{9}$ **13.** $\frac{8}{27}$ **14.** $\frac{25}{9}$ **15.** 9 **16.** 40 **17.** $\frac{-2}{3}$ **18.** $\frac{3}{8}$ **19.** 2^{-2} **20.** n^2 **21.** $(-3)^{-8}$ **22.** $(-x)^{-5}$ **23.** $(-0.4)^{-8}$ **24.** 4^{-2} **25.** 15 625 **26.** 8 **27.** 36 **28.** 4096 **29.** $\frac{1}{2}$ **30.** 16 **31.** $\frac{1}{64}$ **32.** $\frac{1}{9}$ **33. a)** 36 **b)** −32 **c)** 36 **d)** 125 **e)** 172 **f)** 36 **34.** 471.6 **35.** 16 930 **36.** 0.063 **37.** 0.123 94 **38.** 0.0295 **39.** 0.002 695 **40.** 4.639 **41.** 10 **42.** 19.3 **43.** $\frac{1}{100}$ **44.** 2.73×10^7 **45.** 1.93×10^{-8} **46.** 25 300 000 **47.** 0.000 971 **48.** 7.75×10^8 **49.** 2.294×10^{-1} **50.** 3.44×10^{14} **51.** 25 **52.** 49 **53.** 144 **54.** 961 **55.** 0.4 **56.** 9 **57.** 11 **58.** 20 **59.** 36 **60.** 74 **61.** −56 **62.** 11 **63.** −5, 5 **64.** −6, 6 **65.** −8, 8 **66.** −12, 12 **67.** 8, 7.7 **68.** 20, 19.3 **69.** 30, 35.7 **70.** 200, 202.7 **71.** 2, 2.2 **72.** 0.09, 0.1 **73.** −0.4 **74.** 12.4 **75.** 0.3 **76.** 13.6 **77.** 1.9 **78.** 0.6 **79.** 6 m **80.** 8.9 cm **81.** 12.5 cm **82.** 5.3 m **83. a)** 20 cm **b)** 10 000 cm² **c)** 400 cm **84. a)** 341.2 m/s **b)** 348.1 m/s **85. a)** 3 **b)** 6 **86. a)** 5.0×10^{-3} km² **b)** 71 m

Chapter Check p.124

1. 81 **2.** 1 **3.** 1 **4.** $\frac{1}{5}$ **5.** $\frac{1}{81}$ **6.** $\frac{1}{9}$ **7.** −8 **8.** 0.49 **9.** −4 **10.** $\frac{1}{5}$ **11.** $\frac{3}{4}$ **12.** 16 **13.** 4.0×10^{-4} **14.** 2.31×10^{-12} **15.** 0.0046 **16.** 0.000 003 21 **17.** 4.845×10^{-7} **18.** 3.316×10^3 **19.** 6 **20.** 20 **21.** −11 **22.** 10, 9.6 **23.** 18, 17.7 **24.** $(-3)^{-1}$ **25.** s^{-3} **26.** 2^3 **27.** 3^{-2} **28. a)** 48 **b)** −27 **29.** 366 km by 366 km

Using the Strategies p.125

1. 253 × 14 **2.** 7 cm, assuming the cardboard has dimensions 21 cm by 14 cm. **3.** 0, 4 or 8 **4.** 47, 48, 49 **5. a)** 25 **b)** 51 **6. a)** 20 cm² **b)** 5 cm² **7. a)** 12, considering both a.m. and p.m.

Data Bank

1. 1241 km **2.** −7°C with the wind at 32 km/h (−23°C versus −21°C)

Chapter 4

Getting Started pp.128–129

Activity 1: **1.** $\frac{3}{8}$ **2.** $\frac{4}{7}$ **3.** $\frac{7}{8}$ **4.** $\frac{3}{5}$ **Activity 2:** **1.** $\frac{2}{4}, \frac{3}{6}$ **2.** $\frac{4}{10}, \frac{10}{25}$ **3.** $\frac{6}{8}, \frac{12}{16}$ **4.** $\frac{10}{12}, \frac{15}{18}$ **5.** $\frac{2}{6}, \frac{3}{9}$ **6.** $\frac{14}{20}, \frac{21}{30}$ **7.** $\frac{1}{2}$ **8.** $\frac{5}{6}$ **9.** $\frac{4}{5}$ **10.** $\frac{1}{4}$ **11.** $\frac{2}{5}$ **12.** $\frac{2}{5}$ **Activity 4:** **1.** BALL **2.** IN **3.** ANT **4.** NUMB **5.** TANGLE **6.** PEN **7.** TAR **Activity 5:** **1.** A = 4, B = 2, C = 1, D = 3 **2.** $\frac{1}{2}$ **3.** $\frac{1}{4}$ **4.** $\frac{3}{4}$ **5.** $\frac{1}{3}$ **6.** $\frac{2}{3}$ **7.** $\frac{1}{4}, \frac{1}{8}, \frac{1}{16}, \frac{3}{16}$ **Warm Up** **1.** $\frac{3}{4}$ **2.** $\frac{4}{5}$ **3.** $\frac{1}{2}$ **4.** 1 **5.** $\frac{2}{3}$ **6.** $\frac{5}{8}$ **7.** $\frac{7}{10}$ **8.** $\frac{11}{12}$ **9.** $\frac{13}{12}$ **10.** $\frac{9}{10}$ **11.** $\frac{1}{2}$ **12.** $\frac{2}{5}$ **13.** $\frac{1}{4}$ **14.** $\frac{2}{7}$ **15.** $\frac{1}{4}$ **16.** $\frac{5}{8}$ **17.** $\frac{3}{10}$ **18.** $\frac{3}{8}$ **19.** $\frac{1}{6}$ **20.** $\frac{1}{12}$ **21.** $\frac{3}{10}$ **22.** $\frac{1}{12}$ **23.** $\frac{1}{4}$ **24.** $\frac{2}{9}$ **25.** $\frac{3}{8}$ **26.** $\frac{5}{9}$ **27.** $\frac{3}{10}$ **28.** $\frac{3}{16}$ **29.** $\frac{2}{5}$ **30.** $\frac{2}{3}$ **31.** 1 **32.** $\frac{1}{3}$ **33.** $\frac{1}{2}$ **34.** $\frac{3}{2}$ **35.** $\frac{3}{2}$ **36.** $\frac{3}{4}$ **37.** $\frac{2}{3}$ **38.** $\frac{3}{2}$ **39.** $\frac{3}{5}$ **40.** $\frac{4}{5}$ **Mental Math** **1.** 1 **2.** $3\frac{1}{2}$ **3.** $2\frac{1}{2}$ **4.** 2 **5.** 1 **6.** $\frac{5}{7}$ **7.** $\frac{2}{5}$ **8.** 6 **9.** 0 **10.** $\frac{3}{4}$ **11.** 2 **12.** $\frac{1}{4}$ **13.** $\frac{3}{4}$ **14.** $\frac{1}{5}$ **15.** $\frac{1}{6}$ **16.** $\frac{1}{3}$ **17.** $\frac{1}{9}$ **18.** $\frac{1}{8}$ **19.** $\frac{1}{10}$ **20.** $\frac{1}{12}$ **21.** 1 **22.** 1 **23.** 0 **24.** 2 **25.** $1\frac{1}{4}$ **26.** $2\frac{1}{3}$ **27.** $1\frac{5}{8}$ **28.** $2\frac{3}{10}$ **29.** 10 **30.** 20 **31.** 7 **32.** 11

Section 4.1 pp.132–133

Practice **1.** rational **2.** rational **3.** rational **4.** undefined **5.** rational **6.** rational **7.** rational **8.** irrational **9.** irrational **10.** rational **11.** rational **12.** undefined **13.** $\frac{2}{-5}$ **14.** $\frac{7}{1}$ **15.** $\frac{-2}{6}$ **16.** $\frac{-8}{3}$ **17.** $\frac{3}{-4}$ **18.** $\frac{5}{-1}$ **19.** $\frac{-2}{3}$ **20.** $\frac{2}{5}$ **21.** $\frac{1}{2}$ **22.** $\frac{-1}{2}$ **23.** $\frac{-1}{4}$ **24.** -7 **25.** $\frac{-6}{7}$ **26.** $\frac{-4}{5}$ **27.** $\frac{-1}{3}$ **28.** $\frac{4}{5}$ **29.** $\frac{11}{4}$ **30.** $\frac{2}{5}$ **31.** $\frac{-7}{1}$ **32.** $\frac{-6}{5}$ **33.** $\frac{-7}{2}$ **34.** $\frac{9}{4}$ **35.** $\frac{-3}{10}$ **36.** $\frac{-2}{3}$ **37.** $\frac{3}{4}$ **38.** $\frac{8}{3}$ **39.** $\frac{-3}{4}$ **40.** $\frac{-4}{-5}$ **41.** $<$ **42.** $<$ **43.** $=$ **44.** $>$ **45.** $>$ **46.** $<$ **47.** $2, 1\frac{2}{5}, \frac{4}{5}, \frac{-7}{-10}, \frac{1}{2}, 0, \frac{-1}{2}$ **48.** $\frac{9}{4}, 1\frac{7}{8}, \frac{11}{6}, 1\frac{2}{3}, 1\frac{1}{2}, \frac{5}{4}$ **49.** $\frac{11}{9}, \frac{-11}{-9}, \frac{4}{-3}, \frac{-5}{3}$ or $\frac{-11}{-9}, \frac{11}{9}, \frac{4}{-3}, \frac{-5}{3}$ **Problems and Applications** **51. a)** $\frac{1}{6}$ **b)** $\frac{-4}{2}$ **c)** $\frac{1}{-16}$ **52.** $\frac{2}{1}$ **53. a)** $\frac{5}{3}$ **b)** $\frac{4}{3}$ **c)** $\frac{3}{4}$ **d)** $\frac{5}{4}$ **e)** $\frac{1}{1}$ **f)** $\frac{3}{2}$ **g)** $\frac{7}{4}$ **h)** $\frac{2}{1}$ **54. a)** $\frac{2}{1}, \frac{7}{4}, \frac{5}{3}, \frac{3}{2}, \frac{4}{3}, \frac{5}{4}, \frac{1}{1}, \frac{3}{4}$ **b)** Gander **55.** Elephants: $\frac{-2}{9}$, Monkeys: $\frac{2}{1}$, Giraffes: $\frac{5}{14}$, Rhinoceroses: $\frac{-2}{3}$, Hyenas: $\frac{0}{1}$ **56.** $\frac{10}{7}$ **57.** No; both are negative. **59. a)** sometimes true **b)** sometimes true **c)** always true **d)** always true **e)** sometimes true

Learning Together p.134

Activity 1: **1. a)** $\frac{3}{10}$ **b)** $\frac{-3}{5}$ **c)** $1\frac{1}{2}$ **d)** $3\frac{3}{4}$ **e)** $-2\frac{3}{50}$ **f)** $-1\frac{7}{8}$

Activity 2: **1. a)** $\frac{1}{3}$ **b)** $\frac{4}{9}$ **c)** $\frac{23}{99}$ **d)** $\frac{46}{99}$ **e)** $3\frac{7}{9}$ **f)** $5\frac{4}{11}$ **2. a)** $\frac{-2}{9}$ **b)** $\frac{-7}{9}$ **c)** $\frac{-1}{3}$ **d)** $\frac{-52}{99}$ **e)** $-3\frac{1}{9}$ **f)** $-4\frac{2}{11}$ **3. a)** $\frac{4}{15}$ **b)** $\frac{7}{45}$ **c)** $\frac{2}{5}$ **d)** $3\frac{9}{10}$

Section 4.2 pp.136–137

Practice **1.** false **2.** true **3.** true **4.** d **5.** a **6.** c **7.** b **8.** $0.\overline{31}$ **9.** $0.\overline{6}$ **10.** $0.9\overline{5}$ **11.** $5.\overline{61}$ **12.** $32.\overline{4287}$ **13.** $-18\overline{52}$ **14.** $0.\overline{6}$ **15.** -0.875 **16.** 4.5 **17.** $-0.41\overline{6}$ **18.** -2.2 **19.** $0.\overline{36}$ **20.** $0.\overline{428571}$ **21.** $-0.\overline{7}$ **22.** -0.35 **23.** 2, 1 **24.** 63, 2 **25.** 6, 1 **26.** 3, 1 **27.** 3, 1 **28.** 3, 1 **29.** $\frac{-11}{10}$ **30.** $\frac{651}{100}$ **31.** $\frac{7}{100}$ **32.** $\frac{1003}{1000}$ **33.** $\frac{-337}{100}$ **34.** $\frac{-279}{100}$ **35.** $\frac{1}{2}$ **36.** $\frac{-2}{5}$ **37.** $\frac{7}{5}$ **38.** $\frac{4}{5}$ **39.** $\frac{-11}{4}$ **40.** $\frac{5}{4}$ **41.** $\frac{-21}{100}$ **42.** $\frac{13}{4}$ **43.** $\frac{81}{50}$ **44.** $\frac{-37}{5}$ **45.** $\frac{-58}{25}$ **46.** $\frac{31}{10}$ **47.** $<$ **48.** $<$ **49.** $=$ **50.** $>$ **51.** $>$ **52.** $=$ **53.** $>$ **54.** $<$ **55.** $<$ **56.** $=$ **57.** $\frac{1}{3}, \frac{1}{2}, 0.59, \frac{3}{5}$ **58.** $0.71, \frac{711}{1000}, 0.7\overline{1}, 0.\overline{7}$ **59.** $-9.\overline{1}, -9.01, -8.\overline{9}, -8.93$ **60.** $0.\overline{1}, 0.113, 0.\overline{113}, 0.1\overline{13}$ **61.** $\frac{-4}{7}, \frac{3}{8}, 0.3\overline{75}, 0.3\overline{7}$ **Problems and Applications** **62. a)** $-\frac{49}{4}$ **b)** $+\frac{13}{2}$ **c)** $-\frac{176}{5}$ **d)** $+\frac{31}{10}$ **e)** $-\frac{3}{5}$ **63.** $\frac{6}{5}$ **64. a)** bronze, $\frac{7}{18}$; silver, $\frac{5}{18}$; gold, $\frac{1}{3}$ **b)** $0.3\overline{8}, 0.2\overline{7}, 0.\overline{3}$ **65.** $\frac{6}{25}$ **66. a)** $0.\overline{09}, 0.\overline{18}, 0.\overline{27}$ **b)** the periods are multiples of 9 **c)** $0.\overline{63}, 0.\overline{90}$

Learning Together p.138

Activity 1: **4. a)** $\frac{1}{3} \times \frac{1}{2} = \frac{1}{6}$ **b)** $\frac{2}{3} \times \frac{3}{4} = \frac{1}{2}$ **c)** $\frac{1}{2} \times \frac{4}{5} = \frac{2}{5}$ **d)** $\frac{1}{2} \times \frac{2}{3} = \frac{1}{3}$ **5.** The product of two fractions is the product of the numerators divided by the product of the denominators. **Activity 2:** **1.** $\frac{3}{4} \div \frac{1}{4} = 3, \frac{3}{4} \times \frac{4}{1} = 3$ **2.** $\frac{3}{5} \div \frac{1}{10} = 6, \frac{3}{5} \times \frac{10}{1} = 6$ **3. a)** $\frac{5}{8} \div \frac{1}{8} = 5$ **b)** $\frac{5}{6} \div \frac{1}{12} = 10$ **c)** $\frac{1}{2} \div \frac{1}{10} = 5$ **4.** To divide two fractions, invert the divisor and multiply.

Section 4.3 pp.140–141

Practice **1.** $\frac{1}{2} \times \frac{2}{3} = \frac{1}{3}$ **2.** $\frac{1}{2} \times \frac{1}{2} = \frac{1}{4}$ **3.** $\frac{3}{4} \times \frac{1}{2} = \frac{3}{8}$ **4.** 9 **5.** -21 **6.** 12 **7.** -9 **8.** $\frac{-3}{20}$ **9.** 1 **10.** $\frac{-3}{20}$ **11.** $\frac{1}{12}$ **12.** $\frac{-1}{12}$ **13.** $\frac{-3}{8}$ **14.** $\frac{-9}{5}$ **15.** $\frac{1}{2}$ **16.** $\frac{1}{4}$ **17.** $4\frac{7}{8}$ **18.** 0 **19.** $-1\frac{7}{8}$ **20.** $2\frac{1}{2}$ **21.** $4\frac{3}{8}$ **22.** 10 **23.** $\frac{-7}{8}$ **24.** -4 **25.** $1\frac{1}{3}$ **26. a)** $\frac{3}{10}$ **b)** $\frac{9}{20}$ **c)** $\frac{-9}{40}$ **Problems and Applications** **27.** \$150 **28.** $3\frac{7}{13}$ km **29.** $-6\frac{1}{2}$°C **30.** 546 **31.** 15 000 **32. a)** 500 000 **33. a)** never true **b)** always true **c)** sometimes true

Section 4.4 pp.143–144

Practice **1.** $\frac{1}{2} \div 2 = \frac{1}{4}$ **2.** $\frac{2}{3} \div 4 = \frac{1}{6}$ **3.** 2 **4.** -2 **5.** -0.2 **6.** 1.5 **7.** $\frac{4}{1}$ **8.** $\frac{-2}{1}$ **9.** $\frac{-3}{5}$ **10.** $\frac{2}{7}$ **11.** $\frac{1}{3}$ **12.** $\frac{4}{7}$ **13.** none **14.** $\frac{-5}{11}$ **15.** negative **16.** positive **17.** positive **18.** negative **19.** positive **20.** negative **21.** $\frac{3}{4} \div \frac{1}{2} = \frac{3}{2}$ **22.** $1\frac{3}{4} \div \frac{1}{4} = 7$ **23.** $\frac{2}{3} \div \frac{1}{3} = 2$ **24.** 7 **25.** $\frac{7}{2}$ **26.** -42 **27.** -125 **28.** undefined **29.** 1 **30.** $\frac{-14}{3}$ **31.** $\frac{-7}{16}$ **32.** $\frac{9}{4}$ **Problems and Applications** **33.** 5 **34.** 92 km/h **35. a)** \$2.03 **b)** \$506.89 **36.** approximately 124 **37.** 11 **38.** $\frac{1}{16}$ **39.** 8 L **40.** 260 **41.** \$14 200 **42.** -1 **43. a)** $\frac{5}{2}, 2$ **b)** $-\frac{1}{8}, \frac{1}{4}$ **c)** $11\frac{1}{4}, 2\frac{1}{4}$

Section 4.5 pp.146–147

Practice **1.** 10 **2.** 8 **3.** 12 **4.** 4 **5.** 1 **6.** -1 **7.** -3 **8.** -1 **9.** 7 **10.** true **11.** false **12.** false **13.** true **14.** $-\frac{25}{12}$ **15.** 3.75 **16.** $\frac{10}{9}$ **17.** $\frac{3}{5}$ **18.** -0.92 **19.** $\frac{7}{20}$ **20.** $-\frac{19}{10}$ **21.** $\frac{25}{16}$ **22.** $-\frac{68}{9}$ **23.** $-\frac{1}{25}$ **24.** 9.88 **25.** $\frac{23}{100}$ **26.** $\frac{99}{100}$ **27.** -14.56 **28.** -6.45 **29.** -0.66 **30.** $3\frac{3}{4}$ **31.** $-2\frac{1}{6}$ **Problems and Applications** **32.** -1.9°C **33.** $-1\frac{1}{2}$ **34.** $\frac{3}{4}$ **35.** \$315.00 **36.** 1 min 27.66 s **37.** $\frac{1}{5}$ **38. a)** opposites **b)** both equal 0 **39. a)** sometimes true

b) sometimes true **c)** sometimes true **40. a)** $-\frac{7}{10}, \frac{1}{10}$ **b)** $-\frac{1}{4}, -\frac{5}{8}, -\frac{11}{8}$ **c)** $\frac{11}{8}, \frac{3}{4}$

Section 4.6 p.151

Practice **1.** 20 **2.** −29 **3.** 21 **4.** 0 **5.** 7.8 **6.** 49 **7.** −116 **8.** $-\frac{3}{10}$ **9.** −2.75 **10.** 13 **11.** $\frac{2}{3} + 4 \times (\frac{1}{2} + \frac{1}{4}) \div 3 = \frac{5}{3}$ **12.** $(0.5^2 - 0.1) \times 8 \div 2 = 0.6$ **13.** $-2 \times (18.5 - 6.3) \div 4 = -6.1$ **14. a)** 16.5 **b)** 16.5 **c)** 39 **d)** −9 **e)** −36 **f)** 0 **g)** −1 **h)** −0.25 **15. a)** 2.7 **b)** 2.7 **c)** −2 **d)** −6.48 **e)** 2.05 **f)** 52.09 **g)** −51.59 **h)** 12.6 **16.** Across: 1. 19, 3. 25, 5. 12, 6. 50, 7. 84, 8. 81, 9. 95 Down: 1. 113, 2. 92, 3. 254, 4. 50, 7. 81, 8. 85 **17. b)** 2.5 **c)** 11 **19. a)** $\frac{1}{2} + \frac{1}{2} - \frac{1}{2} + \frac{1}{2} = 1, \frac{1}{2} \div \frac{1}{2} + \frac{1}{2} - \frac{1}{2} = 1,$ $\frac{1}{2} \times \frac{1}{2} \div \frac{1}{2} + \frac{1}{2} = 1$ **b)** $-\frac{1}{2} - (-\frac{1}{2}) - (-\frac{1}{2} - \frac{1}{2}) = 1,$ $-\frac{1}{2} \div (-\frac{1}{2}) + (-\frac{1}{2}) - (-\frac{1}{2}) = 1,$ $-\frac{1}{2} \div (-\frac{1}{2}) \div (-\frac{1}{2}) \times (-\frac{1}{2}) = 1$ **c)** $(\frac{1}{2} + \frac{1}{2} + \frac{1}{2}) \div \frac{1}{2} = 3$ **d)** $(-\frac{1}{2} - \frac{1}{2} - \frac{1}{2}) \div (-\frac{1}{2}) = 3$

Section 4.7 p.153

Problems and Applications **1. a)** 25 m **b)** 44.8 m **c)** 60 m **d)** 76.8 m **2.** $v = 5.5$ h **3. a)** $c = 3.5 + 1.4d$, (d in km) **b)** $14.00 **4.** $30.80 **5.** $5.90 **6.** 4.4 s **7. a)** 11.9 years **b)** 0.2 years **c)** 84.1 years

Connecting Math and Physics pp.154–155

Activity 1: **1.** $\frac{1}{3}$ s, 3 vibrations/s **4.** 10 vibrations/s **5.** $\frac{1}{8}$ s **6.** 440 vibrations/s **7.** take the reciprocal **8.** take the reciprocal **9. a)** $C = \frac{1}{f}$ **b)** $f = \frac{1}{C}$

Activity 2: **1.** 900 000 vibrations/s, $\frac{1}{900\ 000}$ s; 101 200 000 vibrations/s, $\frac{1}{101\ 200\ 000}$ s **3.** 1 320 000 vibrations/s, AM, 1320 **4.** 101 300 000 vibrations/s, FM, 101.3

Review pp.156–157

1. rational **2.** rational **3.** rational **4.** rational **5.** rational **6.** irrational **7.** rational **8.** undefined **9.** irrational **10.** rational **11.** rational **12.** rational **13.** $-\frac{7}{10}$ **14.** $\frac{9}{20}$ **15.** $\frac{-3}{1}$ **16.** $\frac{-4}{3}$ **17.** $\frac{2}{3}$ **18.** $\frac{27}{50}$ **19.** $\frac{1}{2}, \frac{3}{4}, \frac{7}{8}$ **20.** $\frac{7}{-5}, \frac{-1}{2}, \frac{-3}{10}$ **21.** $\frac{7}{10}, 0.71, 0.\overline{7}$ **22.** −0.2, 0.02, 0.2 **23.** = **24.** < **25.** > **26.** < **27.** < **28.** > **29.** = **30.** < **31.** −0.6 **32.** $-3.\overline{6}$ **33.** 0.8 **34.** 0.625 **35.** $0.\overline{142\ 857}$ **36.** −0.75 **37.** $\frac{1}{2}$ **38.** $-\frac{6}{25}$ **39.** $\frac{6}{5}$ **40.** $-\frac{3}{50}$ **41.** $\frac{5}{8}$ **42.** $\frac{3}{5}$ **43.** $\frac{7}{2}$ **46.** $\frac{13}{8}$ **48.** $\frac{23}{6}$ **51.** 1.8 **55.** $\frac{15}{8}$ **56.** $\frac{13}{6}$ **57.** 6.58 **58.** 77.5 **59.** 111 **60.** −512 **61.** $\frac{1}{15}$ **62.** −14.4 **63.** 1 **64.** $\frac{1}{3}$ **65. a)** $\frac{1}{3}$ **b)** $-\frac{1}{3}$ **c)** $-\frac{5}{6}$ **d)** $\frac{1}{3}$ **e)** $-\frac{3}{4}$ **f)** $\frac{7}{6}$ **66.** 231 000 km **67.** $70.13 **68. a)** −26.25 cm^2 **b)** 6.76 m^2 **69.** $3.20 **70.** $57\frac{1}{8}$

Chapter Check p.158

1. rational **2.** rational **3.** irrational **4.** rational **5.** rational **6.** irrational **7.** $\frac{4}{5}, \frac{7}{10}, \frac{1}{2}$ **8.** $-1\frac{1}{4}, -1\frac{1}{2}, -2$ **9.** $\frac{8}{7}$, 1.12, 1.1 **10.** 1.7, $\frac{3}{2}, \frac{4}{3}$ **11.** $-\frac{7}{20}$ **12.** $\frac{7}{5}$ **13.** $-\frac{13}{10}$ **14.** $\frac{3}{5}$ **15.** $0.\overline{285\ 714}$ **16.** −2.6 **17.** $3.\overline{3}$ **18.** $-2.\overline{54}$ **19.** $\frac{4}{3}$ **20.** −0.36 **21.** 8 **22.** $\frac{15}{28}$ **23.** 2 **24.** 216 **25.** 9.1 **26.** $-\frac{1}{4}$ **27.** −0.375 **28. a)** $412.50 **b)** $337.50

Using the Strategies p.159

1. Wednesday **2.** 63 **3.** top: 3, 5; middle row: 7, 1, 8, 2; bottom row: 4, 6 **4.** 10 **5.** 212, 213 **6.** 22 or 26 **7.** $5\frac{9}{28}$ **9.** 49 **10.** House numbers

Data Bank

1. 13.5 h **2.** $\frac{9}{4}$, 2.25

Cumulative Review, Chapters 1-4 pp.160–161

Chapter 1 **3.** 1000, 1009 **4.** 24, 23.19 **5.** 5400, 5307 **6.** 360, 408 **7.** 48, 50 **8.** 2, 2.4 **9.** 1, 0.697 **10.** 100, 101 **11.** 46 500 **12.** 0.32 **13.** 0.0023 **14.** 0.63 **15.** 123 000 **16.** 400 **17.** 2.3×10^7 **18.** 8.8×10^6 **19.** 340 000 000 **20.** 4 000 000 **21.** 8880 **22.** 30 100 000 000 **23.** 27 **24.** 16 **25.** 25 **26.** 10 000 **27. a)** 29 **b)** 38 **c)** 49 **d)** 0 **28. a)** 4 **b)** 4.4 **c)** 1.2 **d)** 6.72 **29.** $6.75 **Chapter 2** **1.** −7, −6, −2, −1, 0, 4, 5 **2. a)** −2 **b)** 5 **c)** −2 **d)** −7 **3. a)** 0, 1 **b)** −7, −6, −5, −4 **4.** −4 **5.** −7 **6.** −10 **7.** 18 **8.** −3 **9.** −2 **10.** −3 **11.** −28 **12.** −6 **13.** −13 **14.** 8 **15.** 6 **16.** −9 **17.** 4 **18.** −14 **19.** −4 **20.** 9 **21.** 3 **22.** −8 **23.** 8 **24.** 2 **25.** 6 **26.** 4 **27.** −6 **28.** −12 **29.** 7 **30.** −8 **31.** 17 **32. a)** 1 **b)** −16 **c)** 1 **d)** −1 **e)** −5 **f)** 13 **33.** −9°C **Chapter 3** **1.** 82 **2.** $\frac{1}{2}$ **3.** $\frac{2}{9}$ **4.** 64 **5.** 1 **6.** −1 **7.** −125 **8.** 1 **9.** 42 570 **10.** 0.02 357 **11.** 3.73×10^7 **12.** 1.54×10^{-11} **13.** 9 **14.** 17 **15.** 9 **16.** 7 **17.** 163 **18.** 4 **19. a)** 10.0 cm **b)** 35.7 m **20.** 4.3×10^{-2} cm **Chapter 4** **1.** rational **2.** rational **3.** rational **4.** rational **5.** irrational **6.** $\frac{2}{6}, \frac{1}{6}, -\frac{1}{3}$ **7.** $\sqrt{3}, 1.7, -\frac{3}{2}$ **8.** $3\frac{1}{3}, \frac{10}{3}$, 3.3 or $\frac{10}{3}, 3\frac{1}{3}$, 3.3 **9.** $-1\frac{1}{4}, -1\frac{1}{3}, -1\frac{5}{6}$ **10.** $\frac{31}{50}$ **11.** $\frac{7}{10}$ **12.** $-\frac{29}{4}$ **13.** $-7.\overline{3}$ **14.** $1.\overline{857\ 142}$ **15.** −4.875 **16.** $\frac{1}{5}$ **17.** 15 **18.** $\frac{2}{3}$ **19.** −2.3 **20.** $\frac{13}{6}$ **21.** $-\frac{4}{3}$ **22.** 7.48 **23.** 14.72 **24.** $18.8\overline{3}$ **25.** 52.96 **26.** 3 **27.** 2.01 s, 2.84 s, 3.48 s

Chapter 5

Getting Started pp.164–165

Activity 1: **1.** The third number in the column is the sum of the first two numbers in the column, while the fourth is their product. **2.** The second number is twice the first; the third is twice the second; the fourth is twice the third less the first. **3.** The first number is the third less the fourth; the second is the product of the third and fourth. **4.** The first number is the product of the second and third; the fourth number is the third less the second. **5.** The second number is the product of the first and fourth; the third is the sum of the other three. **6.** The second number is three times the first; the third is the second plus one; the fourth is the sum of the second and third. **Activity 2:** **1.** 6 **2.** 4 **3.** 8 **4.** Count the number of factors **Activity 3:** **1. a)** 8 white, 1 yellow **b)** 13 white, 2 yellow **c)** 21 white, 4 yellow **d)** 18 white, 3 yellow **e)** 29 white, 6 yellow **f)** 40 white, 9 yellow **Mental Math** **1.** 0 **2.** 10 **3.** −6 **4.** 6 **5.** −12 **6.** −30 **7.** −40 **8.** −25 **9.** −20 **10.** 10 **11.** 8 **12.** 9 **13.** −1 **14.** 1 **15.** −25 **16.** −1 **17.** 16 **18.** 4 **19.** −100 **20.** 27 **21.** −1 **22.** −1 **23.** 1 **24.** −27 **25.** −9 **26.** 30 **27.** −20 **28.** −50 **29.** −60 **30.** −24 **31.** 60 **32.** 1000 **33.** 200 **34.** −2 **35.** −3 **36.** −7 **37.** 2 **38.** 4 **39.** 4 **40.** 7 **41.** −7 **42.** 9 **43.** −3 **44.** −11 **45.** 10 **46.** 3^1 **47.** 2^3 **48.** 3^3 **49.** 6^2 **50.** 4^1 **51.** 5^2 **52.** 3^2 **53.** 4^2 **54.** 2^1 **55.** 2^3 **56.** 3^2 **57.** 4^2 **58.** 5^2 **59.** 6^1 **60.** 2^3

Learning Together pp.166–167

Activity 1: **1. a)** $-2x; -4, 6$ **b)** $3x; 6, -9$ **c)** $2x^2 + 2x$; 12, 12 **d)** $-2x^2 + 3x; -2, -27$ **e)** $-3x + 4; -2, 13$ **f)** $x^2 - 2x - 4; -4, 11$ **g)** $-x^2 + 3x + 3; 5, -15$ **h)** $2x^2 - 4x - 2; -2, 28$ **Activity 2:** **1.** $-x + 1, -2, 3$; $-x^2 + 2x, -3, -8$; $x^2 - 2, 7, 2$; $-2x^2 + 4, -14, -4$ **2. a)** 2 long white **b)** 3 square green **c)** 2 long green, 2 white **d)** 3 square green, 1 long green **e)** 1 square white, 3 long green, 4 white **f)** 1 square green, 2 long white, 1 red **g)** 1 long green, 2 white **h)** 4 square white, 2 long green, 5 red

Section 5.1 p.169

Practice **1.** a **2.** c **3.** q **4.** b **5.** k **6.** t **7.** variable: c; constant: 10 **8.** variables: a, b; constant: −6 **9.** variables: x, y; constant: −3 **10.** variables: x, y, z; constant: 0.5 **11.** 2 **12.** 3 **13.** 3 **14.** 2 **15.** 0 **16.** 8 **17. a)** 10 **b)** −25 **c)** 0 **d)** −20 **18. a)** −1 **b)** 4 **c)** 16 **d)** −8 **19. a)** −3 **b)** 3 **c)** −11 **d)** 19 **20. a)** 1 **b)** −4 **c)** −10 **d)** 12 **21. a)** 1 **b)** −24 **c)** 8 **d)** 10 **22. a)** 7.5 **b)** 3 **c)** 0 **d)** 9.8 **e)** 1.9 **f)** 4.5 **g)** 0.25 **h)** −10.25 **Problems and Applications** **23. a)** A **b)** 19 000 **24. a)** $290x + 360y + 210z$ **b)** 1975 kJ

Section 5.2 pp.171–172

Practice **1.** $8x$ **2.** $10t$ **3.** $-7b$ **4.** $-13y$ **5.** $21m$ **6.** $6p$ **7.** $-9s$ **8.** $-23a$ **9.** $9r$ **10.** $4t$ **11.** $2p$ **12.** $4w$ **13.** $-4a$ **14.** $2y$ **15.** $-7q$ **16.** 0 **17.** $t + 5$ **18.** $-3x - 7$ **19.** $a - b$ **20.** $-10x + y$ **21.** $15y - 2z$ **22.** $-p - 10q$ **23.** $4r - 3s$ **24.** $-4j + 7k$ **25.** $3c$ **26.** $8p - 3q$ **27.** $-2j - k$ **28.** $2a - 5b$ **29.** $r - s$ **30.** $4y$ **31.** $2x + 5y + 11$ **32.** $4a - 5b - 5$ **33.** $-15t + 17r + 2$ **34.** $9x - 9z - 19$ **35.** $5p - 11q + 8r - 11$ **36.** $4c - 8x - 13w$ **37.** $2c - 23d + 15j$ **38.** $11n + 11p - 2q$ **39.** $2p + 4q + r + 1$ **40.** $-13x - y + 2z$ **41.** $13q - 15r + 4s$ **42.** $7a - 3c - d - 21$ **Problems and Applications** **43. a)** 20 **b)** 12 **c)** 28 **d)** 48 **e)** 13 **f)** 3 **44. a)** $m + m + n, 2m + n$ **b)** $z + z + z + z, 4z$ **c)** $a + a + 2a + 2a, 6a$ **d)** $f + f + f + q + q + q, 3f + 3q$ **e)** $x + x + y + y + y + y, 2x + 4y$ **f)** $a + b + b + c + c$, $a + 2b + 2c$ **45. a)** $14c$ **b)** $2a + 2d + 4c$ **c)** $4e + 8f$ **d)** $2p + 6r + 2s$ **46.** The sum is meaningless. You are not adding like terms. **47.** $2p + 2q$

Section 5.3 p.175

Practice **1.** monomial **2.** binomial **3.** trinomial **4.** binomial **5.** monomial **6.** trinomial **7.** 1 **8.** 4 **9.** 0 **10.** 5 **11.** 7 **12.** 6 **13.** 4 **14.** 1 **15.** 4 **16.** 8 **17.** 2 **18.** 6 **19.** 7 **20.** $x^5 + x^3 + x^2 + 1$ **21.** $-3x^3 + 2x + 5$ **22.** $-x^2 + 2xy + 5y^2$ **23.** $-4x^4 - 5x^2y + 25xy^2 + 3xy^3$ **24.** $7b^2x^4 - 3x^3 + 4abx^2 + 5ax$ **25.** $-2 + x + 3x^2 - 2x^3 + 5x^5$ **26.** $5 - x + x^2 - 3x^3 + 4x^4$ **27.** $4xy^2 - 2x^2y^2 + 2x^3y - 3x^4$ **28.** $-3 + 2xy^1z + 5x^2yz^2 + 3x^3y^4z^2$ **29.** $z - xy + x^2$ **30.** $16 - 2xy + x^2 - 3x^3$ **31.** $3xy + 2x^3y - x^5$ **32.** $-1 + xy + 3x^3y^2 + x^4y$ **Problems and Applications** **33. a)** monomial **b)** binomial **c)** monomial **34. a)** binomial **b)** binomial **c)** trinomial **d)** monomial **35. a)** 2 of 50 cm², 2 of 100 cm², 2 of 200 cm² **b)** 700 cm² **c)** $2lw + 2lh + 2wh$ **36.** 3822 cm³

Section 5.4 pp.177–178

Practice **1.** $3x^2 + 6x + 3$ **2.** $-2x^2 - 2x - 3$ **3.** $x - 1$ **4.** $-x^2 - 2x + 1$ **5.** $2x, 5x; 3y, -2y; -4xy, 6xy$ **6.** $2a, 5a; -6b, 8b; -2c, 3c$ **7.** $3s^2, 7s^2; 5s, s; -2, -3$ **8.** $7x - 1$ **9.** $4x^2 - 2x - 2$ **10.** $y^2 + 14y - 9$ **11.** $-2y^3 - 8y^2 + 2$ **12.** $6x + 9$ **13.** $7y^2 + 9y + 19$ **14.** $8x - 8y + 15$ **15.** $7x^2 - 8x - 5$ **16.** $9x^2 - x + 2$ **17.** $5y^2 - 6$ **18.** $-z^2 - 2z + 10$ **19.** $7x^2 + 5y^2 - 16$ **20.** $2x^4 + 2x^3 - 5x^2 + 7x - 4$ **21.** $9x^2 - x + 5$ **22.** $5y^2 - 6$ **23.** $m^3 + 9m^2 + 10$ **24.** $3x^2 - x - 6$ **25.** $3x^2 - 2xy + 5y^2$ **26.** $3y^2 - 2y + 1$ **27.** $4x^2y - 2xy + 5y^2$ **Problems and Applications** **28. a)** $12x + 12$ **b)** 60 cm **29. a)** $5x + 2$ **b)** 37 **30. a)** triangle **b)** 786 m **31.** $4x^2 - 2x - 3$ **32.** $8x$

Section 5.5 p.181

Practice **1.** $-x^2 - 4x - 1$ **2.** $-x^2 + 2x + 3$ **3.** $-2x^2 - x + 5$ **4.** $3x^2 + 7x - 2$ **5.** $2x - 7$ **6.** $-2x + 6$ **7.** $2x + 7$ **8.** $2x - 9$ **9.** $3x^2 + 8x - 1$ **10.** $-5x^2 + 2x - 4$ **11.** $-x^2 - 8x + 11$ **12.** −5 **13.** $3x - 2$ **14.** $3x^3 - x^2 - x$ **15.** $-y - 11$ **16.** $7s^2 + 3$ **17.** $3y^2 - 12y - 2$ **18.** $-x^2 - 9x + 6$ **19.** $-7y^2 - 4y + 3$ **20.** $3t^2 - 2t + 7$ **21.** $-3n^2 - 8n - 5$ **22.** $1 - 3x + 6x^2$ **23.** $-4t^2 - 5$ **24.** $2x^2 + 12x - 8$ **25.** $4m^2 + 6m - 7$ **26.** $-2y^2 + 3y - 10$ **Problems and Applications** **27.** $5x + 2$ **28. a)** $4x - 8y - 3$ **b)** $2x^2 + 6x - 10$

c) $3x^2 - y^2$ **d)** $2t^2 + 5t + 8$ **29.** $2x^2 + x - 3$ **30.** 0
31. a) No **b)** Yes; the results are opposite.

Section 5.6 p.183

Practice **1.** $5x + 5$ **2.** $3x - 6$ **3.** $4x + 8$ **4.** $2x - 6$ **5.** $7x - 7$ **6.** $5x + 15$ **7.** $2x + 12$ **8.** $4x - 20$ **9.** $7x + 21$ **10.** $3x - 12$ **11.** $10x + 20$ **12.** $9x - 27$ **13.** $6x + 4$ **14.** $9x + 3$ **15.** $10x + 5$ **16.** $8x + 12$ **17.** $12x - 6$ **18.** $15x - 10$ **19.** $14x + 7$ **20.** $18x + 12$ **21.** $-3x - 6$ **22.** $-8x - 4$ **23.** $-10x + 4$ **24.** $-9x + 6$ **25.** $-10x + 5$ **26.** $-10x + 6$ **27.** $-4x - 20$ **28.** $-2x + 1$ **29.** $-6x + 4y$ **30.** $-15x - 9y$ **31.** $-20x - 8y$ **32.** $-2x - y$ **33.** $5x - 5y$ **34.** $-7x + 21y$ **35.** $-6x - 14$ **36.** $-8x + 4y$ **37.** $-8x - 12y - 4z$ **38.** $15x - 6y + 6$ **39.** $-6x + 18y + 24$ **41.** $6y - 4$ **42.** $7x - 4$ **43.** $x + 26$ **44.** 9 **45.** $12x - 12$ **46.** $-10x + 10$ **47.** $-5x + 12y - 4$ **Problems and Applications** **48.** $5x^2 + 12x + 1$ **49.** $6x^2 + 18x - 18$ **50.** $7y^2 + 7y - 18$ **51.** $8x - 18y + 35$ **52.** $-3x$ **53.** 11 **54.** $14x - 17y + 12$ **55.** $x + 6y - 15$ **56.** $-x^2 + 6y^2 + 7$ **57.** $2x^2 + 8x - 8y - 10$ **58.** $8y^2 + 23y - 12$ **59. a)** $x + 3$, x **b)** $x^2 + 3x$ **60. a)** $2 + y$, y **b)** $2y + y^2$ **61.** $3x - x^2$ **62.** $15x - 3x^2$ **63. a)** $P + Prt$ **b)** $330

Section 5.7 p.185

Practice **1.** $15xy$ **2.** $6mn$ **3.** $35st$ **4.** $24ab$ **5.** $6x^2y$ **6.** $20ab^2$ **7.** $12bc$ **8.** $6ab^2$ **9.** $18st$ **10.** $6xy$ **11.** $12ab$ **12.** $10x^2y^2$ **13.** $15abc$ **14.** $12xy$ **15.** $30xyz$ **16.** $6a^2b^2$ **17.** $9ab$ **18.** $35ab$ **19.** $-15x^2y^2$ **20.** $8at^3$ **21.** $-12abc^2$ **22.** $-24a^2y^2$ **23.** $25xyz$ **24.** $-48x^2y^2$ **25.** $14x^2y^2$ **26.** $10m^3n$ **27.** $-12s^3t^4$ **28.** $6a^3b^5xy$ **29.** $10s^6t^5$ **30.** $-20c^2x^5y^{10}$ **31.** $-15x^2y^2z^2$ **32.** $6cx^2y^2z^3$ **33.** $4x^2y^2z$ **34.** $-70x^2yz$ **35.** $-70xyz$ **36.** $90xyz^2t^2$ **37.** $-8a^2x^6y^2z^3$ **38.** $-6b^2x^2y^5z^3$ **39.** $10a^4b^5$ **40.** $-24a^3b^3c^3$ **41.** $4x^4y^4z^4$ **42.** $-24j^3k^3l^3$ **Problems and Applications** **43. a)** $41x^2$ **b)** $28y^2$ **c)** $68c^2$ **d)** $16x^2$ **e)** $33x^2$ **f)** $20x^2$ **44. a)** the cost of making 24 hats **b)** $36 **45. a)** $27a^3$ **b)** $42y^3$ **46.** $6x$

Section 5.8 p.188

Practice **1.** x^2 **2.** a^3 **3.** p^5 **4.** n^4 **5.** t^6 **6.** $-y^6$ **7.** x^6 **8.** y^6 **9.** m^4 **10.** n^{12} **11.** x^9 **12.** y^6 **13.** z^{12} **14.** m^{20} **15.** p^{36} **16.** s^{20} **17.** $-x^{31}$ **18.** -1 **19.** x^2y^2 **20.** a^3b^3 **21.** x^2y^2 **22.** m^4n^4 **23.** p^3q^3 **24.** $4x^2t^2$ **25.** $16x^2y^2$ **26.** $-8a^3x^3$ **27.** $-27r^3s^3$ **28.** x^6y^6 **29.** x^4y^6 **30.** a^6b^3 **31.** a^2b^6 **32.** m^3n^3 **33.** a^2b^4 **34.** j^6k^8 **35.** x^4y^2 **36.** -1 **37.** $8x^6$ **38.** $9y^6$ **39.** $16x^8$ **40.** $25y^4$ **41.** m^4 **42.** $-n^6$ **43.** $-8n^6$ **44.** $9y^4$ **45.** $9p^2q^2r^2$ **46.** $-27y^3z^3$ **47.** $-64x^6y^9$ **48.** $-9x^2$ **49.** $4x^6y^7$ **50.** $-12x^3y^3$ **51.** $24x^5y^8$ **52.** $-60a^9b^8$ **53.** $-200a^4b^3c^3$ **Problems and Applications** **54. a)** $4x^2y^2$ **b)** $16x^4y^6$ **55. a)** $27x^6y^3$ **b)** $125y^6$ **56.** 157 464 **57. a)** no **b)** yes

Section 5.9 p.191

Practice **1.** $x^2 + 2x$ **2.** $x^2 - 3x$ **3.** $a^2 + a$ **4.** $t^2 - t$ **5.** $y^2 + 4y$ **6.** $m^2 + 5m$ **7.** $x^2 - 5x$ **8.** $y^2 - 7y$ **9.** $a^2 - 10a$ **10.** $3x^2 + 6x$ **11.** $4b^2 - 44b$ **12.** $5t^2 + 15t$ **13.** $6x + 2x^2$ **14.** $7y^2 - 35y$ **15.** $-2x^2 - 8x$ **16.** $-x^2 - 2x$ **17.** $-y^2 + 3y$ **18.** $5x$ **19.** $2y^2 + y$ **20.** $2m^2 - 2m$ **21.** $x^2 + 2$ **22.** $3y^2 - 7y$ **23.** $3a^2$ **24.** $-3x$ **25.** $5x^2 + 16x$ **26.** $x^2 + 2x$ **27.** $9x^2 + 5x$ **28.** $-3y^2 + 5y$ **29.** $5a^2$ **30.** $-x^2 + 2x$ **31.** $22x$ **32.** $x^3 + 2x^2 + 3x$ **33.** $3x^2 + 6x - 15$ **34.** $5x^3 + 10x^2 - 35x$ **35.** $-x^2 + 3x + 1$ **36.** $4m^3 - 20m^2 + 24m$ **37.** $6y^3 - 12y^2 + 9y$ **38.** $-9b^3 + 15b^2 - 3b$ **39.** $-5z^3 + 10z^2 + 25z$ **Problems and Applications** **40.** $2x^2 + 5x - 15$ **41.** $8x^2 + 13x - 35$ **42.** $2x^2 + 5x + 1$ **43.** $3x^3 - 3x^2 + 17x + 12$ **44.** $7m^2 - 17m + 24$ **45.** $y^2 - 7y + 9$ **46.** $x^2 - 8x$ **47. a)** $35y^2 - 20y$ **b)** $14x + 14xy + 9y$ **c)** $38x^2$ **d)** $7x^2 + 6x$

Section 5.10 p.193

Practice **1.** $2x$ **2.** $-3a$ **3.** $3y$ **4.** $4m$ **5.** $-5x$ **6.** $-5y$ **7.** 3 **8.** -6 **9.** 24 **10.** 32 **11.** $-x$ **12.** $4m$ **13.** $3z$ **14.** $-3b$ **15.** 3 **16.** $-9bc$ **17.** $4x$ **18.** $-5s$ **19.** 5 **20.** 2 **21.** 7 **22.** 9 **23.** $12t$ **24.** $3j$ **25.** $4qr$ **26.** $17df$ **27.** $5xy$ **28.** $-3a^2b^3$ **29.** $-2j^3k^7$ **30.** $-5x^3y^{14}$ **31.** $-a^2bc$ **32.** $4xy$ **33.** $2a^2b$ **34.** $3xy$ **35.** $-3m^3$ **36.** $2xyz^2$ **37.** $\frac{4}{3}$ **38.** $-\frac{3}{2}x^2y^2$ **39.** $2x^2y^4$ **40.** $-4a^2b^2$ **41.** $-4m$ **42.** $3x^9y$ **43.** $-\frac{4}{3}xy$ **44.** $9p^3q^{12}r^{-2}$ **Problems and Applications** **45. a)** x **b)** p **c)** $2x^2$ **d)** $3b^4$ **46. a)** 16 cm by 10 cm **b)** 20 cm by 8 cm **47.** $\frac{x}{6}$ **48.** 1

Section 5.11 p.195

Practice **1.** 3 **2.** -6 **3.** $-4n$ **4.** 6 **5.** 11 **6.** y **7.** $4x$ **8.** $-3b^2$ **9.** $-6x$ **10.** $4x - 5y + 8$ **11.** $x^2 + 2x - 3$ **12.** $-y^3 - y^2 + 3$ **13.** $m^2 + 2m - 3$ **14.** $-3x^2 + 8x + 5$ **15.** $-j^4 - 2j^3 - 3j^2$ **16.** $-2x^2 - x + 3$ **17.** $3 - 2m + 3m^2$ **18.** $-2p + 3q - 5p^2q^2$ **19.** $-4a + 3b + 8a^2b^2$ **20.** $2x^2 - 3xy + 4y^2$ **21.** $2ac^3 - 3c^2 + 4bc$ **22.** $3y^2 + 4xy + 6x^2$ **23.** $5n^2 - 6mn - 5m^2$ **24.** $-5abc^3 + 8a^2bc^2 - 7a^4b^2c$ **Problems and Applications** **25. a)** $4x + 1$ **b)** $4x^3 - 5x + 3$ **c)** $3xy - 6 + 7x$ **d)** $2x^2 + 3xy - 6y^3$

Connecting Math and Astronomy pp.196–197

Activity 1: **5.** Mercury, 4847 km; Venus, 12 118 km; Mars, 6761 km; Jupiter, 142 867 km; Saturn, 119 906 km; Uranus, 51 024 km; Neptune, 48 473 km; Pluto, 2551 km **Activity 2:** (in millions of kilometres) **3.** Mercury, 58.3; Venus, 107.7; Mars, 227.4; Jupiter, 777.9; Saturn, 1421.2; Uranus, 2872.3; Neptune, 4503.0; Pluto, 5924.2

Review pp.198–199

1. 7 **2.** 13 **3.** 19 **4.** -2 **5.** -11 **6.** 1 **7.** -17 **8.** 4 **9.** -32 **10.** 2 **11.** -5 **12.** 4 **13.** -6 **14.** -6 **15.** 9 **16.** 9 **17.** 1 **18.** 2 **19.** 9 **20.** 16 **21.** 23 **22.** 15 **23.** -12 **24.** -8 **25.** -1 **26.** -9 **27.** -8.75 **28.** 4.5 **29.** -3.25 **30.** 21.75

31. -5.5 **32.** 23 **33.** -21 **34.** 30.5 **35.** $3x + 7y + 11$ **36.** $3a - 4b - 5$ **37.** $2q - 7r + 1$ **38.** monomial, 3 **39.** monomial, 5 **40.** binomial, 3 **41.** binomial, 2 **42.** trinomial, 4 **43.** binomial, 5 **44.** trinomial, 4 **45.** trinomial, 3 **46.** binomial, 4 **47.** trinomial, 5 **48.** $-5x^3 + x^2 + 3x$ **49.** $y^4 + 2y^2 - 3y + 5$ **50.** $-m^4 + 6m^3 - 3m^2 + 2m + 6$ **51.** $x^4 + x^3 + x^2 + x + 3$ **52.** $-2y^7 + 5y^6 + y^5 - 4y^2 + 3y$ **53.** $3m^2 - 2m + 2$ **54.** $4a^2 + 8a + 2$ **55.** $-b^2 - b + 6$ **56.** $2x^3 - 4x^2 + 8x$ **57.** $2x^2 - 8x + 2$ **58.** $a^2 + 2a - 4$ **59.** $9t^2 - 2t - 14$ **60.** $3a^2 - 6a - 1$ **61.** $-2m^2 + 3m + 2$ **62.** $2x^2 + x - 6$ **63.** $6x^2 - x - 2$ **64.** $-3x^2 + 2x + 2$ **65.** $5x^2 - 5x + 7$ **66.** $5x^2 - 3x - 8$ **67.** $40xy$ **68.** $-150xy^2$ **69.** $12abx^2$ **70.** $10a^2bp$ **71.** $-15x^2y^2$ **72.** $4a^3b$ **73.** $-8x^3y^2z$ **74.** $-3s^5t^5$ **75.** x^6 **76.** y^{12} **77.** x^3y^3 **78.** $9y^6$ **79.** $x^4y^6z^2$ **80.** $-r^9s^{12}t^6$ **81.** $-8x^6y^9$ **82.** $25x^4y^4$ **83.** $-8a^3b^3c^{12}$ **84.** $2x^8y^7z^8$ **85.** $-27a^{12}b^4x^4$ **86.** $8j^{11}k^9l^9$ **87.** $-40a^9b^8$ **88.** $a^6b^7c^6$ **89.** $9x + 14$ **90.** $7a + 14$ **91.** $6y^2 - 19y - 10$ **92.** $-m^2 + m + 2$ **93.** $-2z^2 - 4z + 5$ **94.** $x^2 - 20x$ **95.** $-2y^3 + 6y^2 - 14y$ **96.** $3t^3 + 6t^2 - 3t$ **97.** $4m^3 + 8m^2 - 12m$ **98.** $-6x^3 + 12x^2 - 6x$ **99.** $4xy^2$ **100.** $9xy$ **101.** $5a^2b^2$ **102.** $-2ab^7c^{-2}$ **103.** $3c + 1$ **104.** $2x^3 + 4x - 1$ **105.** $2b^2 - 4ab + a^2$ **106.** $-2abz^3 + 4abz^2$

Chapter Check p.200

1. 10 **2.** $y + 2$ **3.** $6a - b - 10c - 5$ **4.** trinomial, 2 **5.** binomial, 3 **6.** monomial, 3 **7.** $x^5 + x^4 + x^3 + 1$ **8.** $-yx^4 + 5x^3 + 2xy$ **9.** $-3x^4 + x^3y - 2x^2y + 10xy^2$ **10.** $-5b^2x^3 + 5abx^2 + 3ax - 1$ **11.** $6a^2 - 4$ **12.** $b^3 - b^2 + y - 1$ **13.** $5y^2 - 12x + 6$ **14.** $-t^2 - t - 5$ **15.** $-x^2 - 7x + 6$ **16.** $5y^2 + 14y + 1$ **17.** $8x^2 - 9x - 12$ **18.** $-2x^2 + 10x + 1$ **19.** $-50xy$ **20.** $45x^3y^3$ **21.** $-6b^2x^5yz$ **22.** $-15x^4y^2z^3$ **23.** $-27x^{12}y^9$ **24.** $-a^5b^{10}c^{10}$ **25.** x^5y^5z **26.** $9x^8y^8$ **27.** $2x^2 - x + 9$ **28.** $-x^2 + 9x$ **29.** $-3x^3 + 6x^2 - 3x$ **30.** $4m^3 - 12m^2 - 20m$ **31.** $-2a^3b^2$ **32.** $3j^2kl^2$ **33.** $1 - 2y + 3y^2$ **34.** $3a^2 - 2a + 1$

Using the Strategies p.201

1. $\frac{5}{18}$ or approximately 0.3 h **2. a)** 81 **b)** 961 **3.** $36, 37, 38$ **4.** 36 **5.** 84 **6.** 34 **7.** Sue: 16 years, Alex: 12 years **8. a)** more than 20 **b)** less than 20 **c)** 20 **9.** 16 **10.** 5 **11.** 1681 **12. a)** 4 boxes of 6, 1 box of 20 **b)** 1 box of 9, 2 boxes of 6, 1 box of 20 **c)** 4 boxes of 9, 1 box of 6 or 2 boxes of 9, 4 boxes of 6 or 7 boxes of 6 **d)** no

Data Bank

1. Sunday **2.** 484 m

Chapter 6

Getting Started p.205

Warm Up **1.** $7x$ **2.** $3x$ **3.** $9m$ **4.** $6t$ **5.** $-5a$ **6.** $-6n$ **7.** $2s$ **8.** y **9.** $x + 7y$ **10.** $2u$ **11.** $7x - 13$ **12.** $-s - 15t$ **13.** $-3x - 4y + 7$ **14.** $2a - 9b$ **15.** $-15x - 6y + 10$ **16.** $2x + 11$ **17.** $10w - 25$ **18.** $18r - 3$ **19.** $35x - 28$ **20.** $8a - 12b$ **21.** $12p - 6q$ **22.** $-6y + 14$ **23.** $-4x + 9$ **24.** $-10x + 15y + 20$ **25.** $-4x - 5y + 1$ **26.** 1 **27.** 7 **28.** -5 **29.** 5 **30.** 11 **31.** -12 **32.** -24 **33.** 29 **34.** 17 **35.** 15 **36.** -1 **37.** 29 **38.** 4 **39.** -11 **40.** -14 **41.** -2 **42.** -2 **43.** 21 **44.** 37 **45.** 0 **Mental Math** **1.** 3 **2.** 6 **3.** 1 **4.** 3 **5.** -13 **6.** -3 **7.** 13 **8.** 15 **9.** 8 **10.** 12 **11.** 15 **12.** 13 **13.** 15 **14.** -10 **15.** 15 **16.** -4 **17.** -36 **18.** 48 **19.** 1 **20.** 1 **21.** 0 **22.** 7 **23.** 14 **24.** 4 **25.** -5 **26.** 3 **27.** -1 **28.** 6 **29.** 11 **30.** 7 **31.** 7

Section 6.1 p.207

Practice **1.** 4 **2.** -1 **3.** 0 **4.** -2 **5.** 7 **6.** 5 **7.** -7 **8.** -3 **9.** -4 **10.** 4 **11.** -2 **12.** 10 **13.** -6 **14.** -8 **15.** 7 **16.** 22 **17.** -7 **18.** 4 **19.** -3 **20.** -5 **21.** yes **22.** no **23.** yes **24.** no **25.** yes **26.** yes **27.** no **28.** yes **29.** 8 **30.** 4 **31.** 8 **32.** 6 **33.** 12 **34.** 7 **35.** 5 **36.** 2 **37.** 22 **38.** -2 **Problems and Applications** **39.** 1.0 **40.** 6.0 **41.** 4.0 **42.** 2.4 **43.** 32 **44.** 3.1 **45.** 8.7 **46.** 7.4 **47.** 0.8 **48.** -6.5 **49.** 4 cm **50.** 22 km² **51.** 3 **52. a)** $2x = 6$ **b)** $x + 6 = 2$ **c)** $x + 2 = 6$ **d)** $x - 2 = 6$ **e)** $\frac{x}{2} = 6$ **f)** $6x = 2$ **53. a)** $-4, 4$ **b)** $-5, 5$

Section 6.2 p.210

Practice **1.** $x - 2 = 1, 3$ **2.** $x - 4 = 2, 6$ **3.** $x + 2 = 4, 2$ **4.** $x + 4 = 4, 0$ **5.** 3 **6.** 1 **7.** 7 **8.** 4 **9.** 10 **10.** 5 **11.** 6 **12.** 1 **13.** 2 **14.** 5 **15.** 3 **16.** 7 **17.** 1 **18.** 5 **19.** 4 **20.** -16 **21.** -8 **22.** 18 **23.** -6 **24.** 12 **25.** -12 **26.** -10 **Problems and Applications** **27.** 2.0 **28.** 8.0 **29.** 6.0 **30.** 2.1 **31.** -3.0 **32.** 1.0 **33.** 0.8 **34.** 2.7 **35.** 5.8 **36.** 15.7 **37.** -7.7 **38.** -33.8 **39.** c, 5 **40.** a, 123 **41.** d, 2.0 **42.** 10 **43.** 7.1 km **44.** no effect **45. a)** $-3, 3$ **b)** $-8, 8$

Section 6.3 p.213

Practice **1.** $2x = 4, 2$ **2.** $\frac{x}{2} = 3, 6$ **3.** 6 **4.** 2 **5.** 7 **6.** 11 **7.** 6 **8.** 8 **9.** 3 **10.** 2 **11.** 7 **12.** 5 **13.** 4 **14.** 6 **15.** 5 **16.** -6 **17.** 2 **18.** -3 **19.** 4 **20.** 4 **21.** 12 **22.** 5 **23.** 3 **24.** 4 **25.** -8 **26.** 21 **27.** -12 **28.** 32 **29.** -15 **30.** 50 **31.** -24 **32.** 44 **33.** 4.2 **34.** -2.1 **35.** 12 **36.** -3 **37.** 0.7 **38.** 3 **39.** 6 **40.** -42 **Problems and Applications** **41.** a, \$2 **42.** b, 12.6 cm **43.** 360 cm **44.** 70 kg **45.** reduces the equation to $0 = 0$ **46. a)** 0 **b)** no solution **47. a)** $-6, 6$ **b)** $-2, 2$

Learning Together pp.214–215

Activity 1: 3; 1, 7; 4, 5; 6, 0 **Activity 2:** 3; 7, 8; 1, 9; 4, 5 **Activity 3:** 4; 3, 5; 1, 7; 6, 8 **Activity 4:** 6; 5, 4; 8, 9; 3, 1, 0

Section 6.4 pp.218–219

Practice **1.** Left side: Start, $5x + 2$, subtract 2, divide by 5, x, stop; Right side: Start, 22, subtract 2, divide by 5, 4, stop **2.** Left side: Start, $2x + 5$, subtract 5, divide by 2, x, stop; Right side: Start, 25, subtract 5, divide by

2, 10, stop **3.** 4 **4.** 7 **5.** −2 **6.** 4 **7.** −3 **8.** −7 **9.** 5 **10.** −2 **11.** −2 **12.** −3 **13.** −4 **14.** 4 **15.** 3 **16.** −20 **17.** −10 **18.** 16 **19.** 18 **20.** 6 **21.** 6 **22.** −20 **23.** −33 **24.** −4 **25.** −2 **26.** 1 **27.** 2 **28.** −6 **29.** 1 **30.** 2 **31.** −4 **32.** −1 **33.** −4 **34.** −7 **35.** 4 **36.** −2 **37.** 1 **38.** −4 **39.** 7 **40.** −3 **41.** 3.2 **42.** 2.2 **43.** 5 **44.** 5 **45.** −2 **46.** −3.2 **47.** −5 **48.** 7.8 **49.** −1.1 **50.** 2.0 **51.** −1.7 **52.** 1.7 **Problems and Applications** **53.** 17, 15, 8 **54.** 6 years **55. a)** 3 **b)** yes

Section 6.5 p.221

Practice **1.** $3x = 2x + 4, 4$ **2.** $4x + 1 = 2x - 5, -3$ **3.** 7 **4.** 4 **5.** 2 **6.** −2 **7.** −2 **8.** 5 **9.** −1 **10.** 3 **11.** 10 **12.** 5 **13.** 3 **14.** −6 **15.** 0.05 **16.** 0.6 **17.** 1 **18.** 3 **19.** −4 **20.** 2 **21.** 4 **22.** 3 **23.** 2 **24.** −5 **25.** 1 **26.** 3 **27.** −2 **28.** 3 **29.** −5 **30.** −3 **31.** 4 **32.** 7 **33.** −4 **34.** 5 **35.** −3 **36.** −10 **37.** −2 **38.** 1 **39.** 4 **40.** 2 **41.** 6 **42.** −2
Problems and Applications **43.** 20 **44.** 6
45. a) $b + c + y - a$ **b)** $c - y - a + b$ **c)** $(c - y - b)/a$ **d)** $(c - y)/a + b$ **e)** $(a + b + y)/c$ **f)** $a(y - b - c)$
46. 14 km/h **47.** 6

Section 6.6 p.224

Practice **1.** 1 **2.** 4 **3.** 1 **4.** −6 **5.** −5 **6.** −7 **7.** 1 **8.** 3 **9.** 1 **10.** −2 **11.** −9 **12.** −3 **13.** 1 **14.** −1 **15.** 2 **16.** −2 **17.** 2 **18.** 3 **19.** 2 **20.** −30 **21.** 5 **22.** −1 **23.** 4 **24.** 5 **25.** 2 **26.** 2 **27.** 1 **28.** 5 **29.** −15 **30.** −3 **31.** 8
Problems and Applications **32. a)** 3 **b)** 4 **33. a)** 3 **b)** 11 by 7 **34.** 24%

Section 6.7 pp.226–227

Practice **1.** 0.6 **2.** 1.7 **3.** 0.3 **4.** −7.5 **5.** −3 **6.** −6.9 **7.** 1 **8.** −10 **9.** −3 **10.** 4 **11.** 2.4 **12.** −4 **13.** 0.4 **14.** −0.53 **15.** 1.09 **16.** −6 **17.** −11.8 **18.** 1 **19.** 2 **20.** 4 **21.** 4 **22.** −2 **23.** −3 **24.** −2 **25.** −6 **26.** 20 **27.** −1 **28.** 1 **29.** 15 **30.** −4 **31.** −2 **32.** 1 **33.** −9 **34.** 14 **35.** 4 **36.** −9 **37.** −2 **38.** 5 **39.** 2 **40.** 2 **Problems and Applications** **41.** 34 kg **42.** 0.04 or 4%

Technology pp.228–229

Activity 1: 0, 1, 10, 11, 100, 101, 110, 111, 1000, 1001, 1010, 1011, 1100, 1101, 1110, 1111 **Activity 2:** **1.** 1, 10, 100, 1000, 10000, 100000, 1000000, 10000000
2. equal **Activity 3:** **1.** 10100 **2.** 100101 **3.** 111000 **4.** 1001110 **5.** 10010011 **Activity 4:** **1.** 8, 16, 32
2. $7.205\ 7594 \times 10^{16}$ codes, 197 418 066 years
Activity 5: **1.** 21 **2.** 25 **3.** 42 **4.** 51

Section 6.8 pp.232–233

Practice **1. a)** true **b)** true **c)** false **d)** true **e)** false **f)** true **g)** false **h)** true **i)** true **j)** true **k)** false **l)** true **2. a)** 3 **b)** −5 **c)** 1 **d)** 8 **e)** −1 **f)** 1 **g)** 7 **h)** −4 **3.** $x > 2$ **4.** $x < 5$ **5.** $x < 4$ **6.** $x > 7$ **7.** $y > 1$ **8.** $y < 1$ **9.** $z > 3$ **10.** $z < 3$ **11.** $x < 2$ **12.** $y > -4$ **13.** $m < 5$ **14.** $n < -4$ **15.** $s > 0$ **16.** $y \leq 3$ **17.** $p \leq -2$ **18.** $t \leq -3$ **19.** $b \leq 3$ **20.** $m > -3$ **21.** $n < -3$ **22.** $x > 3$ **23.** $y > 3$ **24.** $t \leq 6$ **25.** $y \geq 5$ **26.** $x \geq -3$ **27.** $x \geq -3$ **28.** $x < 5$ **29.** $y > -4$ **30.** $m < -2$ **31.** $n > 3$ **32.** $t \leq -3$ **33.** $s \geq 1$ **34.** $y \leq -2$ **35.** $x \leq 5$ **36.** $x \geq -1$ **37.** $x < 3$ **38.** $x > 6$ **39.** $y \geq 7$ **40.** $y \leq -9$ **41.** $t < -8$ **42.** $a > -2$ **43.** $n < -2$ **44.** $n > -3$ **45.** $x > 13$ **46.** $y < 3$ **47.** $m \geq 1$ **48.** $x > 1$ **49.** $y \leq -1$ **50.** $m < -2$ **51.** $x > -5$ **52.** $m < 5$ **53.** $b \geq -8$ **54.** $w \leq 4$ **55.** $p \geq 6$ **56.** $x \leq 5$ **57.** $x \geq -3$ **58.** $x < 2$ **59.** $y \leq 3$ **60.** $m < 6$ **61.** $x < -2$ **62.** $y \leq 3$ **63.** $t \geq -4$ **64.** $y > -2$ **65.** $x \leq 1$ **66.** $y < 10$ **67.** $t > -6$ **68.** $x > 2$ **69.** $x > -9$ **70.** $x \leq -2$ **71.** $x \leq -2$ **72.** $x \leq 5$ **73.** $x > -24$ **74.** $t \leq 23$ **75.** $m > 2$ **76.** $y \geq 2$ **77.** $x \geq 11$ **Problems and Applications** **78.** 89 **79.** 59 **80.** $90 **81.** 15 **82.** 20 min **83. a)** $s \leq 100$ **b)** 0 km/h; standstill **84. a)** $x \leq 17$ **b)** $x > 3$; otherwise no triangle is possible **85. a)** $x < 4$ **b)** $x > -1$; otherwise there is no rectangle

Connecting Math and Logic p.234

Activity 2: **2.** Two sides; not a Möbius strip.
Activity 3: **3.** One two-sided, one one-sided; only one Möbius strip

Review pp.236–237

1. a) yes **b)** no **c)** no **d)** yes **e)** no **f)** yes **g)** no **h)** no **2.** 4 **3.** 5 **4.** 6 **5.** 4 **6.** −3 **7.** 8 **8.** −7 **9.** −2 **10.** 0.6 **11.** −0.4 **12.** 10 **13.** −10 **14.** −9 **15.** 24 **16.** 6 **17.** 10 **18.** −15 **19.** 12 **20.** 6 **21.** 5 **22.** 2 **23.** −2 **24.** −5 **25.** −9 **26.** 9 **27.** −0.3 **28.** 10 **29.** 9 **30.** −1 **31.** −18 **32.** 24 **33.** 6 **34.** −16 **35.** −6 **36.** −4 **37.** 2 **38.** 3 **39.** −3 **40.** −7 **41.** 2 **42.** −2 **43.** 2 **44.** 3 **45.** −2 **46.** 2 **47.** 4 **48.** 13 **49.** −3 **50.** −2 **51.** −8 **52.** −25 **53.** 3 **54.** 7 **55.** −5 **56.** −3 **57.** −2 **58.** 2.8 **59.** −3.2 **60.** 2 **61.** −0.8 **62.** 3 **63.** 2 **64.** −3 **65.** 1 **66.** −1 **67.** −8 **68.** −4 **69.** 4 **70.** −3 **71.** −1 **72.** −10 **73.** −5 **74.** 16 **75.** $x < 5$ **76.** $y > -5$ **77.** $x > 4$ **78.** $x < 2$ **79.** $x \geq 4$ **80.** $x \geq -1$ **81.** $x > -3$ **82.** $x < 1$ **83.** $x \geq -2$ **84.** $x \leq 3$ **85.** $x \leq 2$ **86.** $t \geq -3$ **87.** $x \leq -2$ **88.** $x < 3$ **89.** $t \geq 7$ **90.** $t > 1$ **91.** $s \leq -9$ **92.** $y \leq 5$ **93.** $x > -10$ **94.** $x \leq -7$ **95.** $x < 8$ **96.** $x < 3$ **97.** 26 **98.** $m \leq 21$

Chapter Check p.238

1. 8 **2.** −3 **3.** 6 **4.** −9 **5.** 8 **6.** 20 **7.** −9 **8.** 11 **9.** 35 **10.** 12 **11.** −3 **12.** −2 **13.** 5 **14.** −2 **15.** −5 **16.** −1 **17.** −0.4 **18.** −2.9 **19.** −9 **20.** −3 **21.** 1 **22.** 3 **23.** −4 **24.** 2 **25.** $x < -7$ **26.** $x > -2$ **27.** $r > -1$ **28.** $x \geq -5$ **29.** $x > -10$ **30.** 0.06 or 6% **31.** $x \leq 7$

Using the Strategies p.239

2. 2, 5 **3.** 79, 80, 81, 82, 83 **4.** A, 4; B, 2; C, 3; D, 1 **5.** 4 in centre; left column: 1, 5, 6; right column: 2, 3, 7 **6.** 3 **7.** 178 (inclusive) **8.** 3 nickels, 2 dimes, 5 guarters

Data Bank

1. 700 km

Chapter 7

Getting Started pp.242–243

Activity 1: **1.** prescription, Yin and Yang, no left trun, poison, copyright, lane ends, male or Mars, explosive, peace, Aquarius, thunderstorm, new paragraph, hurricane, registered trademark, Gemini, female or Venus, hospital, Pisces. **Activity 2:** **1.** −2 **2.** −3 **3.** 5 **4.** 3 **5.** 14 **6.** 0 **7.** 6 **8.** −6 **9.** −10 **10.** 4 **11.** −4 **12.** −6 **13.** 3 **14.** 3 **15.** −5 **16.** 15 **17.** 4 **18.** −6 **19.** −10 **20.** 2 **21.** −4 **22.** 3 **23.** 4 **24.** 1 **25.** 12 **Activity 3:** **1.** 12.4 cm **2.** 20.8 cm **3.** 24.6 cm **4.** 21.5 cm **5.** 45.8 m **6.** 38 cm **7.** 18.7 cm^2 **8.** 22.09 cm^2 **9.** 37 cm^2 **10.** 21 cm^2 **11.** 81 cm^2 **12.** 60 cm^2 **Mental Math** **1.** 20 **2.** 30 **3.** 30 **4.** 50 **5.** 51 **6.** 61 **7.** 60 **8.** 70 **9.** 80 **10.** 30 **11.** 20 **12.** 50 **13.** 30 **14.** 59 **15.** 49 **16.** 69 **17.** 59 **18.** 49 **19.** 100 **20.** 1000 **21.** 20 **22.** 2000 **23.** 40 **24.** 4000 **25.** 36 **26.** 360 **27.** 3600 **28.** 7 **29.** 70 **30.** 5 **31.** 500 **32.** 400 **33.** 40 **34.** 11 **35.** 14 **36.** 40 **37.** 20 **38.** 12 **39.** 18 **40.** 15 **41.** 15 **42.** 22 **43.** 3 **44.** 13 **45.** 10 **46.** 16 **47.** 3 **48.** 20 **49.** 9 **50.** $\frac{5}{8}$ **51.** $\frac{4}{9}$ **52.** $\frac{7}{10}$ **53.** $\frac{2}{5}$ **54.** $\frac{2}{3}$ **55.** $\frac{1}{10}$

Learning Together pp.244–245

Activity 1: 6t **Activity 2:** **1.** $b + 5$ **2.** $n - 1$ **3.** $q \div 9$ **4.** $m - 2$ **5.** $m + 3$ **6.** $5x$ **7.** $6 - t$ **8.** $2 + y$ **9.** $3n$ **10.** $n + 1$ **11.** $n - 5$ **12.** $6n$ **13.** $n \div 5$ **14.** $2n + 3$ **15.** $10 - n$ **16.** $7 \div n$ **17.** $x + 4$ **18.** $7x$ **19.** $5x$ **20.** $3 \div x$ **21.** $x + 7$ **22.** $\frac{1}{2}x$ **23.** $x - 8$ **24.** $2x$ **Activity 3:** **1.** 22, 33, 41, $x - 23$, $y + 23$, $3m - 23$, $4t - 2$, $6r + 18$ **2.** 12, 23, $x - 4$, $m + 4$, $5y - 4$, $4t + 4$, $3w + 4$ **3.** 78, 12, $6x$, $\frac{y}{6}$, $6(m + 7)$, $6(3t + 8)$ **Activity 4:** **1. a)** width **b)** $x + 5$ **d)** $2(2x + 5)$ **2.** $x(x + 4)$ **3.** $4x$ **4.** 13, 18, $x + 3$, $10x + 3$, $4x + 4$ **b)** 46 m, 130 m^2; 66 m, 270 m^2; $2(2x + 3)$, $x(x + 3)$; $2(20x + 3)$, $10x(10x + 3)$; $2(8x + 5)$, $(4x + 1)(4x + 4)$ **5. a)** 17, 17, $x + 7$, $y - 7$, $3t + 7$, $4n - 7$, $5z + 11$ **b)** 54 m, 170 m^2; 82 m, 408 m^2; $2(2x + 7)$, $x(x + 7)$; $2(2y - 7)$, $y(y - 7)$; $2(6t + 7)$, $3t(3t + 7)$; $2(8n - 7)$, $4n(4n - 7)$; $2(10z + 15)$, $(5z + 4)(5z + 11)$

Section 7.1 p.247

Practice **1.** $3x = 18$ **2.** $y - 6 = 4$ **3.** $x + 4 = 18$ **4.** $x - 4 = 10$ **5.** $\frac{m}{4} = 18$ **6.** $4x = 20$ **7.** $\frac{x}{2} = 5$ **8.** $x + 6 = 15$ **9.** $x + 5 = 12$ **10.** $x - 6 = 10$ **11.** $x - 4 = 7$ **12.** $x^2 = 25$ **13.** $10 - x = 2$ **14.** $x + 3 = 18$ **15.** $x - 5 = 9$ **16.** $x + 4 = 21$ **17.** $3x = 9$ **18.** $\frac{x}{5} = 10$ **19.** $x - 6 = -8$ **Problems and Applications** **20.** $2b + 16 = 88$ **21.** $2p + 12 = 84$ **22.** $2c + 7 = 29$ **23.** $3a = 250\ 000\ 000$ **24.** $5.5c = 150\ 000\ 000$ **25. a)** $2(2x + 6) = 36$ **b)** $2(3x + 3) = 36$ **26. a)** $4x + 3 = 39$ **b)** $3x - 3 = 39$ **27.** $\frac{2b + 2}{2} = 58$

Section 7.2 pp.250–251

Practice **1.** x, $35 - x$ **2.** x, $50 - x$ **3.** x, $125 - x$ **4.** x, $36 - x$ **5.** x, $32 - x$ **6.** x, $758 - x$ **7.** x, $468 - x$ **8.** x, $246 - x$ **9.** 10 m, 12 m, 13 m **10.** 4 m, 5 m, 5 m **11.** 3 m, 6 m **12.** 5 m, 7 m **13.** 2 m, 6 m **14.** 5 m, 7 m **Problems and Applications** **15.** 17, 29 **16.** 3058 km, 4241 km **17.** 25 m by 20 m **18.** 34, 35, 36 **19.** 137, 155 **20.** 3 years **21. a)** 3 **b)** 38 **22.** 20 **23.** 10 m, 17 m, 17 m **24.** \$0.75 **25.** 14 **26.** 7 nickels, 5 dimes **27.** 57 nickels, 83 dimes **28. a)** 20 cm **b)** 37 cm by 32 cm **29.** 16, 23 **30.** 8, 3 **31.** 50 **32.** 67 \$2 bills, 78 \$5 bills **33.** 17 small, 21 large **34.** 350 **35.** 3 m **36. a)** 2 cm **b)** 296 cm^2

Section 7.3 p.253

Practice **1. a)** 40 cm^2 **b)** 4 m **c)** 17 m **2. a)** 62.8 cm **b)** 100 cm **3. a)** 24 cm^2 **b)** 20 cm **c)** 6 m **4. a)** 30 m **b)** 14 m **c)** 37 m **5.** $w = \frac{A}{l}$ **6.** $b = \frac{2A}{h}$ **7.** $P = \frac{I}{rt}$ **8.** $r = \frac{C}{2\pi}$ **9.** $m = \frac{E}{c^2}$ **10.** $b = \frac{2A}{h} - a$ **Problems and Applications** **11. a)** 200, 193, 181, 176, 159 **b)** 22 years, 49 years, 37 years **12. a)** 480 000 km, 700 000 km **b)** 3.25 h **13.** 10 250, 12 375, 15 250, 16 500, 20 875 **14.** Paul

Learning Together pp.254–255

Activity 1: **2.** 3, $\frac{1}{2}$; 4, 1; 6, 2; 8, 3; 6, 2; 5, $1\frac{1}{2}$; 7, $2\frac{1}{2}$; 8, 3; 4, 1; 5, $1\frac{1}{2}$; 9, $3\frac{1}{2}$ **3.** equal **4.** 3, 0.5; 4, 1; 5, 1.5; 6, 2; 7, 2.5; 8, 3; 9, 3.5 **5.** 4 square units **6.** $A = \frac{1}{2}(P - 2)$ **8.** 48.5 square units **Activity 2:** **2.** 10, 0, 4; 10, 1, 5; 10, 2, 6; 10, 3, 7; 10, 4, 8; 10, 5, 9; 10, 6, 10 **3.** The area increases by 1 square unit **4.** 11 square units, 12 square units **5.** $A = \frac{1}{2}(P - 2) + I$ **7. a)** 17.5 square units **b)** 29 square units **c)** 23 square units

Section 7.4 p.257

Practice **1.** 15, 18, 21; $b = 3a$ **2.** 10, 12, 14; $m = n - 3$ **3.** 21, 20, 19; $t = \frac{a}{4}$ **4.** 16, $c = 4n$ **5.** 150, $w = 7.5n$ **6.** 120, $p = 0.2s$ **Problems and Applications** **7.** $C = 100 + 3d$ **8.** $C = 3 + 1.5d$ **9. a)** Regions = Roads − Towns + 1 **b)** Towns = Roads − Regions + 1 **c)** Roads = Regions + Towns − 1

Section 7.5 p.259

Practice **1.** 180 km **2.** 170 km **3.** 45 km **4.** 45 km **5.** 0.5 h **6.** 8 h **7.** 0.2 h **8.** 4.5 h **9.** 100 km/h **10.** 80 km/h **11.** 90 km/h **12.** 80 km/h **13.** 4.5 **14.** 195 **15.** 75 **16.** $80x$ **17.** $90(x + 1)$ **18.** $85(x - 1)$ **19.** $\frac{200}{x}$ **20.** $\frac{400}{x}$ **21.** rt **22.** $\frac{D}{r}$ **23.** $\frac{D}{t}$ **Problems and Applications** **24.** 5 h **25. a)** 3.75 h **b)** 20:15 **26. a)** 8 h **b)** 16:00 **c)** uniform motion **27. a)** 1.5 h **b)** 10:15 **28.** 1.25 h

Section 7.6 p.261

Practice **1.** $\frac{1}{2}, \frac{1}{4}$ **2.** $\frac{1}{6}, \frac{1}{8}$ **Problems and Applications** **3.** $2\frac{2}{5}$ min **4.** $1\frac{1}{5}$ h **5.** $\frac{1}{3}$ h **6.** 12 min

7. a) $4\frac{4}{9}$ h **b)** \$80, \$80, \$80 **8.** 4 h **9.** 6 h **10. a)** 20 min **b)** Each is occupied by a separate task.

Review pp.264–265

1. $x + 8 = 20$ **2.** $6x = 72$ **3.** $2c + 8 = 24$ **4.** $2s + 10 = 90$ **5.** $4t = 2400$ **6. a)** $3x + 4 = 64$ **b)** $2(2x + 2) = 64$ **7.** 6, 30 **8.** 8 m by 5 m **9.** 60, 61, 62 **10.** 633 m by 211 m **11.** 41 m, 42 m, 43 m **12.** 12, 48 **13.** 17 g **14. a)** 17 m, 18 m, 20 m **b)** 9 m, 12 m **15.** 38, 13 **16.** 11, 16 **17.** 75 **18.** 29 **19. a)** 95 L, 131 L **b)** 14 min **20. a)** 125 L, 50 L **b)** 13 min **21.** $l = \frac{P - 2w}{2}$ **22. a)** $A = 20 + 25t$ **b)** \$95, \$82.50 **23. a)** $\frac{3}{4}$ h **b)** 09:30 **24.** 5 h **25. a)** 9 h **b)** 19:00 **26.** 6 h **27.** $1\frac{1}{5}$ h **28.** $1\frac{5}{7}$ h

Chapter Check p.266

1. $2a + 4 = 36$ **2.** 81 **3.** 26 m, 11 m **4.** 33 m, 34 m, 35 m **5.** 44 m **6.** $h = \frac{2A}{b}$ **7.** 4 h **8.** $d = 330t$ **9.** 2.5 h **10.** $2\frac{2}{5}$ h

Using the Strategies p.267

1. $\frac{1}{24}$ **2.** 126 m **3.** 15:55 **4.** 1125 m **5.** 343 four cylinder, 142 six cylinder **6.** 34 **7. a)** 14 **b)** 29 years

Data Bank

1. 01:00

Chapter 8

Getting Started pp.270–271

Activity 1: **1.** 1, 2, 3, 4, 6, 12 **2.** 1, 3, 5, 15 **3.** 1, 3, 7, 21 **4.** 1, 2, 4, 8 **5.** 1, 3, 5, 9, 15, 45 **6.** 1, 2, 3, 4, 6, 9, 12, 18, 36 **7.** 1, 2, 4, 13, 26, 52 **8.** 1, 17 **9.** 1, 2, 4, 5, 10, 20, 25, 50, 100 **10.** 1, 2, 3, 4, 6, 8, 12, 24 **11.** 1, 2, 4, 5, 8, 10, 16, 20, 40, 80 **12.** 1, 3, 13, 39 **13.** $2 \times 2 \times 2$ **14.** 3×5 **15.** $2 \times 2 \times 2 \times 3$ **16.** $2 \times 2 \times 5$ **17.** $2 \times 2 \times 3 \times 3$ **18.** $2 \times 3 \times 7$ **19.** $2 \times 2 \times 7$ **20.** 19 **21.** 2×13 **22.** $-5, 6; 2, -35; -7, -18; -4, 20; 3, -3; -1, 9; 4, -12; 4, 5; -3, -4; 2, 10; -5, -6$ **23.** 2 **24.** b **25.** 2 **26.** $-2x$ **27.** $3a$ **28.** $-2x$ **29.** $4y$ **30.** $-4x$ **31.** 2 **32.** $-6xy^2$ **33.** 1 **34.** $-4a$ **Activity 2:** **1.** x^3 **2.** p **3.** a^6 **4.** x^4 **5.** $-5t^3$ **6.** a^2 **7.** $-3x^3$ **8.** $-4a^3$ **9.** $5x^2$ **10.** $6mn$ **11.** $5ab$ **12.** $a + 2$ **13.** $3x - 2y$ **14.** $t - 3$ **15.** $3x - 2$ **16.** $3m - 2n + 1$ **17.** $2x^2 - 5x + 3$ **18.** $3a + 2b$ **19.** $4y^3 - 3x^2$ **Activity 3:** **1.** 5 **2.** 11 **3.** 6 **4.** x **5.** $2a$ **6.** $7x$ **7.** 4 **8.** 25 **9.** 81 **10.** $4x^2$ **11.** $9a^2$ **12.** $25x^6$ **Activity 4:** **1.** $48x^3y^3$ **2.** $9x^2$ **3.** $2x^2 + 3x$ **4.** $5x^2 + 6x$ **5.** $x^2 + 3x + 2$ **6.** $x^2 + 3x$ **7.** $a^2 + 5a + 4$ **8.** $y^2 + 9y + 20$ **9.** $y^2 + 2y$ **10.** $a^2 + 11a + 30$ **11.** $x^2 + 5x + 10$ **12.** $y^2 + 5y + 9$ **Mental Math** **1.** 15 **2.** 14 **3.** 16 **4.** 9 **5.** 34 **6.** 0 **7.** 9 **8.** 90 **9.** 25 **10.** 100 **11.** 25 **12.** 33 **13.** 6 **14.** 3 **15.** 54 **16.** 36 **17.** 26 **18.** 31 **19.** 13 **20.** 41 **21.** 6 **22.** $\frac{1}{6}$ **23.** $\frac{1}{20}$ **24.** 3 **25.** 2 **26.** $\frac{2}{3}$ **27.** $\frac{3}{4}$ **28.** 1 **29.** 2 **30.** $\frac{1}{3}$ **31.** $\frac{1}{2}$ **32.** $\frac{3}{2}$ **33.** $\frac{7}{5}$ **34.** $\frac{1}{12}$ **35.** $\frac{1}{4}$ **36.** 8 **37.** $\frac{3}{4}$ **38.** $\frac{1}{4}$ **39.** $\frac{1}{8}$ **40.** $\frac{2}{15}$ **41.** $\frac{1}{8}$ **42.** $\frac{3}{10}$ **43.** 9 **44.** $\frac{1}{3}$ **45.** $\frac{1}{2}$

Learning Together pp.272–273

Activity 1: **3.** 45 **4.** 45 **5.** equal **6.** Multiply the middle number by 3. **8.** 11 **9.** 11 **10. a)** $x + 8, x + 9; x + 14, x + 15, x + 16$ **b)** $x - 8, x - 7, x - 6; x - 1, x + 1; x + 6, x + 7, x + 8$ **c)** $x - 14, x - 13, x - 12; x - 7, x - 6, x - 5; x + 1, x + 2$
Activity 2: **1.** 5, 7, 9, 11; 7, 9, 11, 13; 9, 11, 13, 15; 11, 13, 15, 17 **3.** 44 **4.** yes **5.** the sum always equals $a + b + c + d + e + f + g + h$

Section 8.1 p.276

Practice **1.** 2, 2, 3 **2.** 2, 2, 2, 2 **3.** 2, 2, 7 **4.** 3, 3, 7 **5.** 2, 2, 2, 2, 3, 3 **6.** 3, 3, 5, 5 **7.** $2 \times 2 \times x \times y \times y$ **8.** $2 \times 3 \times 3 \times a \times a \times b \times b \times b$ **9.** $2 \times 2 \times 3 \times 3 \times x \times x \times y \times z \times z$ **10.** $2 \times 5 \times x \times x \times y$ **11.** $2 \times 3 \times 3 \times 3 \times x \times x \times x \times x \times x$ **12.** $5 \times 5 \times 5 \times a \times a \times a \times a \times b \times b$ **13.** 5 **14.** 8 **15.** 9 **16.** 14 **17.** 24 **18.** 32 **19.** $2a$ **20.** x **21.** $2m^2$ **22.** $3abc$ **23.** 2 **24.** 7 **25.** $5x$ **26.** xy **27.** mn **28.** $2a^2$ **29.** $5bc$ **30.** $3xy$ **31.** 5 **32.** $2x$ **33.** $6a$ **34.** $5xy$ **35.** $7ab$ **36.** $4xy$ **37.** $4ab$ **38.** x^2y^2 **39.** $2x^2y$ **40.** $3xy^2$ **41.** $4ab^3$ **42.** $5s^3t^4$
Problems and Applications **43.** 6 **45. a)** A–Shirley, B–Gustav, C–Bob, D–Collete, E–Karin

Section 8.2 p.278

Practice **1.** 6 **2.** x **3.** a **4.** $5x$ **5.** $4ab$ **6.** 6 **7.** $7a$ **8.** a **9.** x, 3 **10.** $3a$, 2 **11.** $2a^2, 5a, 3$ **12.** $3a^2b, 2a, 1$ **13.** $3x, 2y$ **14.** $3a^2, 4b^3$ **15.** $5(2x + 3)$ **16.** $14(2y - 1)$ **17.** $n(2m - 1)$ **18.** $5x(x + 2)$ **19.** $4x^2(2 + x)$ **20.** $3a^2b(3ab - 2)$ **21.** $2xy^2(2x - 3z^2)$ **22.** $7b^2(2a^2b^2 - 3c^2)$ **23.** $6xy^2z(xy + 2)$ **24.** $3b^4(5a^2b - 3c^5)$ **Problems and Applications** **25.** $3(3a - 2b + 1)$ **26.** $4(a - 2b + 4)$ **27.** $6x(2x^2 - x + 4)$ **28.** $5x(2x^2 - x + 3)$ **29.** $6x^2y(4x^2 - 3x + 2y)$ **30.** $8a(ab + 2b - 3)$ **31.** $5mn(5m^2 - 3mn + n^2)$ **32. a)** $2x + 2y, 2(x + y)$ **b)** $2(x + y)$ **33.** 12 cm by 11 cm **34. a)** $(a + b)(x + y)$ **b)** $(x - 2)(x + 3)$ **c)** $(2x - 3)(x - 5)$ **d)** $(a - b)(2a + b)$ **35. a)** $5t(1 - t)$ **b)** 1.2 m

Section 8.3 p.280

Practice **1.** $(x + 4)(x + 1), x^2 + 5x + 4$ **2.** $(x + 2)(x + 2), x^2 + 4x + 4$ **3.** $(x + 5)(x + 1), x^2 + 6x + 5$ **4.** $(x + 1)(x + 1), x^2 + 2x + 1$ **5.** $3x - 15$ **6.** $2x^2 + 3x$ **7.** $-14x + 42$ **8.** $12x^2 - 4x$ **9.** $15a^3 - 20a^2b$ **10.** $-6x^2 - 10xy$ **11.** $x^2 + 3x + 2$ **12.** $x^2 + 7x + 12$ **13.** $a^2 + 8a + 16$ **14.** $y^2 + 11y + 30$ **15.** $x^2 - 7x + 12$

16. $a^2 - 6a + 8$ **17.** $b^2 - 6b + 5$ **18.** $y^2 - 18y + 81$
19. $x^2 - 3x - 18$ **20.** $c^2 - 6c - 16$ **21.** $t^2 - 100$
22. $q^2 + 3q - 10$ **23.** $c^2 - c - 12$ **24.** $x^2 - 3x - 10$
25. $y^2 + 4y - 12$ **26.** $a^2 + 4a - 45$ **27.** $x^2 - 9$
28. $b^2 + 3b - 70$ **29.** $y^2 - 9y - 36$ **30.** $x^2 - 6x - 7$
31. $2x^2 + 11x + 5$ **32.** $3y^2 + 7y + 2$ **33.** $2x^2 - 3x + 1$
34. $2a^2 - 11a + 15$ **35.** $5y^2 + 8y - 21$
36. $4x^2 - 17x - 15$ **37.** $4x^2 + 8x + 3$
38. $15y^2 - 26y + 8$ **39.** $12x^2 - 17x - 5$
40. $10y^2 - 41y - 18$ **41.** $14y^2 - 55y + 21$
42. $24x^2 - x - 10$ **Problems and Applications**
43. $x^2 + 2.5x + 1$ **44.** $x^2 + 1.8x - 3.6$
45. $x^2 - 12.5x + 25$ **46.** $x^2 - 0.9x - 6.3$
47. $2x^2 + 16x + 30$ **48.** $4x^2 - 16x - 180$
49. $-a^2 - a + 6$ **50.** $10x^2 + 20x - 350$
51. $18x^2 - 21x + 6$ **52.** $2x^3 - 6x^2 - 140x$
53. $0.5x^2 + x - 1.5$ **54.** $1.8x^2 + 3.6x + 1.8$
55. a) 21 cm^3 **b)** 14 cm^2 **c)** 7 cm^2 **56. a)** $21 \neq 13$
b) $x^2 + 8x + 12$ **57. a)** $(x + 10)(x + 5)$ **b)** 750 m^2
58. a) $1000 + 2000r + 1000r^2$ **b)** $1166.40

Section 8.4 p. 283

Practice **1.** $x^2 + 2x + 1, (x + 1)(x + 1)$
2. $x^2 + 4x + 3, (x + 1)(x + 3)$
3. $x^2 + 5x + 6, (x + 2)(x + 3)$
4. $2x^2 + 6x + 4, 2(x + 1)(x + 2)$ **5. a)** 3, 4 **b)** 3, 5 **c)** 1, 12 **d)** 7, 11 **e)** −3, −5 **f)** −5, −5 **g)** −3, −4 **h)** −1, −6
6. a) −4, 3 **b)** −3, 4 **c)** −8, 5 **d)** 10, 15 **e)** −1, 5 **f)** −7, 6 **g)** −12, 5 **d)** −6, 8 **7.** $(x + 2)(x + 5)$ **8.** $(y - 3)(y - 5)$
9. $(w - 8)(w + 7)$ **10.** $(z - 5)(z + 8)$
11. $(x - 6)(x + 5)$ **12.** $(a - 1)(a - 16)$
13. $(x - 10)(x + 1)$ **14.** $(x + 2)(x + 10)$
15. $(x + 5)(x + 5)$ **16.** $(m - 3)(m - 6)$
17. $(a - 3)(a - 3)$ **18.** $(y + 5)(y + 6)$
19. $(x + 1)(x + 9)$ **20.** $(x - 16)(x + 1)$
21. $(a - 2)(a + 8)$ **22.** $(x + 4)(x + 5)$
23. $(a - 1)(a - 24)$ **24.** $(y - 2)(y - 7)$
25. $(y - 9)(y + 2)$ **26.** $(x - 9)(x + 8)$
27. $(s - 10)(s + 8)$ **28.** $(a - 9)(a - 9)$
29. $3(x - 3)(x - 4)$ **30.** $5(x - 2)(x + 1)$
31. $7(x + 2)(x + 3)$ **32.** $4(x - 2)(x - 4)$
33. $6(b - 3)(b + 5)$ **34.** $b(x - 3)(x - 25)$
35. $5j(x - 3)(x - 5)$ **36.** $3t(x + 2)(x + 2)$
37. $t(t - 3)(t + 4)$ **38.** $3k(k - 1)(k + 6)$ **Problems and Applications** **39.** not possible **40.** $(a - 8)(a + 1)$
41. $(b + 6)(b + 8)$ **42.** not possible **43.** $(z - 10)(z - 10)$
44. not possible **45.** not possible **46.** $(y - 2)(y + 10)$
47. a) $(x + 4)(x + 5)$ **b)** $(x + 3)(x + 4)$
48. c) $x^2 + 7x + 12$ **49.** $x^2 + 12x + 20, x^2 + 12x + 36, x^2 + 12x - 28, x^2 + 12x - 13$

Section 8.5 p. 285

Practice **1.** $a + 7$ **2.** $x - 2$ **3.** $3m + 7$ **4.** $9x - 8$
5. $x + y$ **6.** $2a - 3b$ **7.** $x^2 - 1$ **8.** $a^2 - 25$ **9.** $p^2 - 36$
10. $x^2 - 81$ **11.** $y^2 - 64$ **12.** $t^2 - 100$ **13.** $4x^2 - 1$
14. $9y^2 - 1$ **15.** $16a^2 - 9$ **16.** $36t^2 - 25$ **17.** $x^2 - y^2$
18. $100t^2 - 9$ **19.** $p^2 - q^2$ **20.** $9a^2 - b^2$ **21.** $x^2 - 36y^2$
22. $j^2 - 100r^2$ **23.** $x^4 - 9$ **24.** $k^4 - 81$ **Problems and Applications** **25.** $(20 + 4)(20 - 4)$, 384; $(50 - 3)(50 + 3)$, 2491; $(60 + 2)(60 - 2)$, 3596
26. 96 **27.** 216 **28.** 396 **29.** $(10 + 4)(10 - 4)$, 84
30. $(20 - 3)(20 + 3)$, 391 **31.** $(30 + 2)(30 - 2)$, 896
32. $x^4 - 1$ **33.** $a^4 - 16$ **34.** $x^4 - 6561$ **35.** $81x^4 - 16$
36. $x^8 - 1$ **37.** the new field has 1000 m^2 less area.
38. 1 term, 5 terms

Section 8.6 p.287

Practice **1.** $x + 5$ **2.** $w - 10$ **3.** $k + 9$ **4.** $2a + 11$
5. $3 - 4x$ **6.** $x^2 - 6$ **7.** $(t - 2)(t + 2)$
8. $(x - 4)(x + 4)$ **9.** $(b - 3)(b + 3)$
10. $(m - 7)(m + 7)$ **11.** $(p - 5)(p + 5)$
12. $(w - 6)(w + 6)$ **13.** $(a - 9)(a + 9)$
14. $(q - 10)(q + 10)$ **15.** not possible **16.** $(1 - c)(1 + c)$
17. $(8 - x)(8 + x)$ **18.** not possible **19.** $(s - t)(s + t)$
20. $(z - x)(z + x)$ **21.** $(b - g)(b + g)$
22. $(p - q)(p + q)$ **23.** $(2a - 3b)(2a + 3b)$
24. $(4p - 9)(4p + 9)$ **25.** $(5a - 7)(5a + 7)$
26. $(3b - 5)(3b + 5)$ **27.** $(4 - x)(4 + x)$
28. $(5 - 6b)(5 + 6b)$ **29.** $(4 - 7x)(4 + 7x)$
30. $(9 - 2a)(9 + 2a)$ **31.** $(4x - 1)(4x + 1)$
32. $(12 - 11x)(12 + 11x)$ **33.** $(13 - 10t)(13 + 10t)$
34. $(15 - 7w)(15 + 7w)$ **35.** not possible
36. $(4y - 7)(4y + 7)$ **37.** $(3 - 2z)(3 + 2z)$
38. $(5a - 6)(5a + 6)$ **39.** $(xy - 2)(xy + 2)$ **40.** not possible **41.** $((a + b) - (a - b))((a + b) + (a - b))$
42. $(5 - 9pq)(5 + 9pq)$ **43.** $2(m - 5)(m + 5)$
44. $9(x - 2)(x + 2)$ **45.** $5(2r - 3)(2r + 3)$
46. $a(2a - 5)(2a + 5)$ **47.** $10(y - 10)(y + 10)$
48. $x(x - 1)(x + 1)$ **49.** $2(5y - 6)(5y + 6)$
50. $x(x - 3)(x + 3)$ **51.** $8(y - 1)(y + 1)$
52. $4(p - 2)(p + 2)$ **53.** $3(3k - 2)(3k + 2)$
54. $16(t - 1)(t + 1)$ **55.** $(8 - x)(8 + x)$
56. $2(5 - 3x)(5 + 3x)$ **57.** $3(2x - 5y)(2x + 5y)$
58. $2(5x - 7y)(5x + 7y)$ **Problems and Applications** **59. a)** 91 **b)** 13 by 7 **c)** equal **d)** 32, 8 by 4
e) $x^2 - y^2, x + y$ by $x - y$ **60. a)** $(x - 1)(x + 1)(x^2 + 1)$
b) $(x - 1)(x + 1)(x^2 + 1)(x^4 + 1)$
c) $(x - 5)(x + 5)(x^2 + 25)$

Section 8.7 p. 289

Practice **1.** 49 **2.** 81 **3.** 36 **4.** 144 **5.** $4x^2$ **6.** $9a^2$
7. $121y^2$ **8.** x^2 **9.** $16y^2$ **10.** x^2 **11.** a^2 **12.** $4x^2$ **13.** $81t^2$
14. $100b^2$ **15.** $9y^2$ **16.** $49p^2$ **17.** $16j^2$ **18.** $36q^2$ **19.** $-6x$
20. $+16y$ **21.** $+2xy$ **22.** $-2ab$ **23.** $+12x$ **24.** $-40a$
25. $+12xy$ **26.** $-84p$ **27.** $y^2 - 20y + 100$
28. $9a^2 - 6a + 1$ **29.** $25x^2 + 20x + 4$ **30.** $9 - 6x + x^2$
31. $25 - 10y + y^2$ **32.** $25a^2 + 10ab + b^2$
33. $9x^2 + 6xy + y^2$ **34.** $16x^2 - 24xy + 9y^2$
35. $49a^2 - 28ab + 4b^2$ **36.** $16m^2 + 40mn + 25n^2$
Problems and Applications **37.** $(x + 7)^2$ **38.** $(x - 8)^2$
39. $(2a + 3)^2$ **40.** $(3b - 4)^2$ **41.** $(8m - 2)^2$
42. $(9n + 5)^2$ **43.** $x^2 + 12x + 36$ **44.** $a^2 + 8a + 16$
45. $y^2 - 6y + 9$ **46.** $m^2 - 8m + 16$ **47.** $2x^2 - 4x + 2$
48. $9y^2 + 12y + 4$ **49.** $16 \neq 10$
50. $a^2 + b^2 + c^2 + 2ab + 2ac + 2bc$

Section 8.8 p. 291

Practice **1.** $3x^2 + 9x - 15$ **2.** $6a^2 - 10a + 14$ **3.** $2x^2 + 14x + 24$ **4.** $3a^2 + 21a + 30$ **5.** $5y^2 + 20y + 50$ **6.** $x^3 - 10x^2 + 25x$ **7.** $x^3 + 11x^2 + 28x$ **8.** $2a^3 + 8a^2 + 15a$ **9.** $4x^3 - 3x^2 - 4x$ **10.** $6x^3 - 23x^2 + 20x$ **11.** $x^3 + 3x^2 + 5x + 3$ **12.** $x^3 - x^2 - 5x + 1$ **13.** $x^3 + 5x^2 + 3x - 9$ **14.** $x^3 + 3x^2 - 13x - 15$ **15.** $x^3 - 12x^2 + 25x + 28$ **16.** $x^3 - 7x^2 + 16x - 12$ **17.** $x^3 + 5x^2 - 61x + 55$ **18.** $3a^3 - 2a^2 + 4a - 5$ **19.** $5b^3 - 36b^2 + 5b + 14$ **20.** $2x^3 - 19x^2 + 27x - 24$ **21.** $4x^3 - 23x^2 + 8x + 35$ **22.** $3x^3 - 11x^2 + 6x + 8$ **23.** $7a^3 - 8a^2 + 2a - 1$ **24.** $2y^3 - 5y^2 - 23y - 10$ **25.** $2x^3 + 13x^2 + 13x - 10$ **26.** $3a^3 + 4a^2 - 21a - 18$ **27.** $5b^3 - 2b^2 + 3b + 10$ **28.** $7w^3 - 8w^2 + 1$ **29.** $2y^3 - y^2 - 4y - 4$ **30.** $4t^3 - 22t^2 + 9t + 5$ **31.** $x^3 + 2x^2y + x^2 + 2xy + xy^2 + y^2$ **32.** $x^3 - 3x^2y - 2x^2 + 6xy + xy^2 - 2y^2$ **33.** $xb^2 + 3b^2 - 3bxy - 9by - xy^2 - 3y^2$

Problems and Applications
34. $15x^3 - 6x^2y + 10x^2 - 4xy + 3xy^2 + 2y^2$ **35.** $4y^4 - 16y^3 + 15y^2 + 14y - 35$ **36.** $28x^3 + 67x^2 - 3x - 27$ **37.** $6a^3 - 23a^2 + 36a - 35$ **38.** $6y^3 + 11y^2 + 8y + 2$ **39.** $9a^3 - 33a^2 + 9a + 35$ **40. a)** $6a^2 - a - 2$ **b)** $2x^3 + x^2 - 4x + 1$ **c)** $7a^2 + ab - 2b^2$ **41. a)** $x^3 + 3x^2$ **b)** $24x^2 - 12x$ **c)** $6x^2 + 30x + 36$

Section 8.9 p.294

Practice **1.** 8 **2.** 24 **3.** 6 **4.** $21x$ **5.** $15a$ **6.** $24x$ **7.** $\frac{3}{y}, x \neq 0, y \neq 0$ **8.** $\frac{2}{y}, y \neq 0$ **9.** $3x, x \neq 0$ **10.** $2c, a \neq 0, b \neq 0$ **11.** $-2xy^2, x \neq 0, y \neq 0$ **12.** $-4ab, a \neq 0, b \neq 0$ **13.** $\frac{3}{2}$ **14.** $\frac{1}{2}$ **15.** 1 **16.** $\frac{4}{3}$ **17.** $\frac{1}{2}$ **18.** 1 **19.** x^2y **20.** $\frac{5xy^3}{3}$ **21.** $\frac{2pq^2}{r}$ **22.** $7\frac{a^2b^2}{2}$ **23.** xy^2 **24.** x **25.** $\frac{x^3}{4}$ **26.** $\frac{x^2y}{2}$ **27.** $\frac{19}{y}$ **28.** $\frac{x-11}{a}$ **29.** $\frac{1}{4x}$ **30.** $\frac{9xy}{z}$ **31.** $\frac{5-x}{x}$ **32.** $\frac{7}{3x^2}$ **33.** $\frac{x-4}{x^2}$ **34.** $\frac{1-x}{5x^2}$ **35.** $\frac{3x+3}{4}$ **36.** $\frac{x+10}{12}$ **37.** $\frac{5x-2}{6}$ **38.** $\frac{2x-9}{15}$ **39.** $\frac{9-x}{12}$ **40.** $\frac{9x-3}{10}$

Problems and Applications **41. a)** $4x(x + 1)$ **b)** $x(x + 1)$ **c)** $\frac{1}{4}$ **42. a)** $6(x + 1)(x + 2)$ **b)** $(x + 1)(x + 2)$ **c)** $\frac{1}{6}$

Connecting Math and Design pp.296–297

Activity 2: **1.** 314 m, 320.28 m, 326.56 m, 332.84 m, 339.12 m **Activity 3:** **1.** The sides of the track must be 43 m **2.** 400 m, 406.28 m, 412.56 m, 418.84 m, 425.12 m

Review pp.298–299

1. a) 5 **b)** 7 **c)** 17 **d)** 24 **e)** $5a$ **f)** $4x$ **g)** $2ab$ **h)** $5xy$ **2. a)** $3xy$ **b)** 8 **c)** $9xy$ **3.** $5(x - 3)$ **4.** $6x(x - 3)$ **5.** $5a(b + 2c)$ **6.** $7a^2(1 + 5a)$ **7.** $4bc(2a - 3)$ **8.** $3(x^2 + 3y^2)$ **9.** $a(3a - 6b + 1)$ **10.** $2(x + 3y - 5z)$ **11.** $x^2 - x - 6$ **12.** $x^2 + 3x - 28$ **13.** $x^2 + 7x + 10$ **14.** $x^2 - 5x + 6$ **15.** $x^2 + 8x - 20$ **16.** $6x^2 - 19x + 10$ **17.** $10a^2 - 11a - 6$ **18.** $12x^2 - 13x - 14$ **19.** $21x^2 + 6x - 30$ **20.** $10a^2 + 13ab - 3b^2$ **21.** $(x + 1)(x + 7)$ **22.** $(x - 1)(x - 5)$ **23.** $(y + 3)(y + 5)$ **24.** $(a + 2)(a + 6)$ **25.** $(b + 4)(b + 6)$ **26.** $(x - 1)(x - 6)$ **27.** $(x - 4)(x - 7)$ **28.** $(a - 3)(a - 4)$ **29.** $(a - 5)(a + 4)$ **30.** $(x - 6)(x + 5)$ **31.** $(x - 7)(x + 2)$ **32.** $(m - 10)(m + 4)$ **33.** $(x - 3)(x + 7)$ **34.** $(x - 2)(x + 12)$ **35.** $(x - 5)(x + 7)$ **36.** $(x - 5)(x + 3)$ **37.** $2(x + 2)(x + 10)$ **38.** $5(a - 4)(a - 4)$ **39.** $4(w - 6)(w + 5)$ **40.** $3(r - 2)(r - 5)$ **41.** $2(j^2 - 3j + 4)$ **42.** $3(t - 1)(t + 7)$ **43.** $7(y - 4)(y + 5)$ **44.** $3(z - 6)(z - 7)$ **45.** $m^2 - 9$ **46.** $x^2 - 25$ **47.** $4c^2 - 1$ **48.** $4x^2 - 9$ **49.** $16s^2 - 49t^2$ **50.** $9j^2 - 25q^2$ **51.** $(x - 1)(x + 1)$ **52.** $(y - 2)(y + 2)$ **53.** $(2a - 3)(2a + 3)$ **54.** $(a - 2b)(a + 2b)$ **55.** $(2x - y)(2x + y)$ **56.** $(2a - 3b)(2a + 3b)$ **57.** $(3 - x)(3 + x)$ **58.** $(5 - 7x)(5 + 7x)$ **59.** $2(a - 5)(a + 5)$ **60.** $5(x - 2)(x + 2)$ **61.** $(2x - 6)(2x + 6)$ **62.** $4(2a - 3)(2a + 3)$ **63.** yes **64.** no **65.** no **66.** yes **67.** $x^2 + 4x + 4$ **68.** $x^2 - 6x + 9$ **69.** $y^2 + 12y + 36$ **70.** $m^2 - 10m + 25$ **71.** $x^3 - 5x^2 + 8x - 4$ **72.** $x^3 - 8x^2 + 12x + 9$ **73.** $2x^3 - x + 1$ **74.** $3x^3 + 14x^2 - 6x - 5$ **75.** $x^3 + 1$ **76.** $x^3 + x^2 + x - 3$ **77.** $5x^3 - 25x^2 + 17x + 12$ **78.** $4x^3 + 5x^2 - 13x - 14$ **79.** $2x^2$ **80.** $-2b$ **81.** $2x^2y^3$ **82.** $-\frac{x^2y^4}{2}$ **83.** q^3 **84.** $\frac{x^3y^2}{2}$ **85.** $\frac{x^4y^3}{9}$ **86.** x^2y^3 **87.** $\frac{x^2y^3}{4}$ **88.** $\frac{a^3y^2}{3}$ **89.** $2a^2$ **90.** $5a^3b$ **91.** $\frac{5x-11}{6}$ **92.** $\frac{2x-7}{15}$ **93.** $\frac{x}{6}$ **94.** $\frac{19x+1}{42}$ **95.** $\frac{2-x}{x^2}$ **96.** $\frac{2x+3}{7x^2}$ **97.** $\frac{x-17}{6}$ **98.** $\frac{x+17}{12}$ **99.** $\frac{6x+22}{7}$ **100.** $\frac{13x-19}{42}$ **101.** $\frac{2x-9}{35}$ **102.** $\frac{3x-2}{28}$ **103.** $\frac{x+2}{2x}$ **104.** $\frac{5}{x^2}$ **105. a)** $(x + 2)(x + 8)$ **b)** $(x + 4)(x + 10)$ **c)** $x^2 + 14x + 40$

Chapter Check p.300

1. $3x(y - 2z)$ **2.** $12xy(2y + x)$ **3.** $x^2 + 2x - 8$ **4.** $2x^2 + x - 21$ **5.** $8a^2 - 18ab + 9b^2$ **6.** $(x + 2)(x + 5)$ **7.** $(x - 3)(x - 6)$ **8.** $(x - 5)(x + 2)$ **9.** $(x - 5)(x + 7)$ **10.** $(x + 4)^2$ **11.** $(x - 17)(x - 1)$ **12.** $2(x - 2)^2$ **13.** $5(x - 5)(x + 4)$ **14.** $x^2 - 4x + 4$ **15.** $w^2 + 14w + 49$ **16.** $(a - 2)(a + 2)$ **17.** $(2x - 5)(2x + 5)$ **18.** $(9 - x)(9 + x)$ **19.** $(1 - 2b)(1 + 2b)$ **20.** $2(t - 10)(t + 10)$ **21.** $3(x - 2)(x + 2)$ **22.** $(10 - 4t^2)(10 + 4t^2)$ **23.** $2(1 - 3y)(1 + 3y)$ **24.** yes **25.** no **26.** yes **27.** no **28.** $5x^3y^2$ **29.** $5a^2$ **30.** $\frac{5x+19}{6}$ **31.** $\frac{9x-37}{35}$ **32. a)** $(x + 3)(x + 10)$ **b)** $(x + 2)(x + 12)$ **c)** $x^2 + 14x + 24$

Using the Strategies p.301

1. Top Row: $\frac{8}{10}$, $\frac{1}{10}$, $\frac{6}{10}$, Middle Row: $\frac{3}{10}$, $\frac{5}{10}$, $\frac{7}{10}$,

Bottom Row. $\frac{4}{10}, \frac{9}{10}, \frac{2}{10}$ **3.** $a = 3, b = 4$
4. 14, 42, 126 **5.** Saturday **6. a)** 14 **b)** $2(n + 1)$

Data Bank

1. 43%

Cumulative Review, Chapters 5–8 pp.302–303

Chapter 5 **1.** 11 **2.** 15 **3.** 12 **4.** 11 **5.** $3x + 3$ **6.** $2x - 10$ **7.** $8x + 4$ **8.** $15 - 5x$ **9.** $-6x + 6$ **10.** $-12 + 8x$ **11.** $8x - 12y + 20$ **12.** $x - 4$ **13.** $-x + 14$ **14.** $x + 2$ **15.** $4x^2 + 3x - 2$ **16.** $5y^2 - 4y$ **17.** $-x^2 - 2x - 9$ **18.** $-x^2 - 3$ **19.** $6x^2 - 9x$ **20.** $-6x + 2x^2$ **21.** $15xy$ **22.** $-12st$ **23.** $4x^2 - 3x + 3$ **24.** $7x^2 + 5x - 6$ **25.** $5x$ **26.** $9y$ **27.** $5y$ **28.** $-4x$ **29.** $1 - 2x + 3x^2$ **30.** $x^2 - 3x - 2$ **Chapter 6** **1.** 12 **2.** 12 **3.** 6 **4.** 11 **5.** 1 **6.** 4 **7.** 3 **8.** 12 **9.** 3 **10.** 4 **11.** 6 **12.** 1 **13.** 16 **14.** 15 **15.** 50 **16.** 45 **17.** 6 **18.** 1 **19.** 20 **20.** 55 **21.** -4 **22.** 1 **23.** 5 **24.** 9 **25.** 3 **26.** 4 **27.** 9 **28.** 3 **29.** $x \geq 1$ **30.** $x < -2$ **31.** $x < 1$ **32.** $x \geq -3$
Chapter 7 **1.** $n + 4$ **2.** $\frac{n}{2}$ **3.** $10n$ **4.** $n - \frac{1}{2}$ **5.** $3n = 15$ **6.** $n + 5 = 6$ **7.** $2n = n + 10$ **8.** $5(n + 2) = 20$ **9.** 17 cm, 18 cm, 19 cm **10. a)** 20, 30, 40, 50 **b)** $A = 5b$ **c)** 75 **11.** 40 min **12.** 2000 km **13.** 9, 17 **Chapter 8** **1.** 14 **2.** 11 **3.** $2t$ **4.** $7xy$ **5.** $3(x - 7)$ **6.** $5xy(x - 3y)$ **7.** $9x(2x - 3)$ **8.** $2(a - 4b - 5)$ **9.** $x^2 - 2x - 3$ **10.** $x^2 - 9x + 20$ **11.** $6x^2 - 7x + 2$ **12.** $35x^2 + 31x - 18$ **13.** $(x - 8)(x + 1)$ **14.** $(a - 1)(a + 11)$ **15.** $(b - 1)^2$ **16.** $(y - 14)(y + 4)$ **17.** $2(x - 5)^2$ **18.** $3(x - 3)(x + 4)$ **19.** $4(a - 5)(a + 4)$ **20.** $5(a - 2)(a - 3)$ **21.** $2(y - 3)(y - 7)$ **22.** $3(y - 9)(y + 2)$ **23.** $x^2 - 49$ **24.** $9 - t^2$ **25.** $25p^2 - 1$ **26.** $36 - 81y^2$ **27.** $(x - 2)(x + 2)$ **28.** $(x - 6)(x + 6)$ **29.** $(x - 5)(x + 5)$ **30.** $(2p - 7)(2p + 7)$ **31.** $(3q - 8)(3q + 8)$ **32.** $(9 - 5x)(9 + 5x)$ **33.** $8(x - 3)(x + 3)$ **34.** $2(5x - 2)(5x + 2)$ **35.** $4(2 - 3t)(2 + 3t)$ **36.** $3x^2y^4$ **37.** $2x^3$ **38.** $\frac{3x + 20}{10}$ **39.** $\frac{10x - 58}{21}$

Chapter 9

Getting Started pp.306–307

Activity 2: **1.** UVW **2.** POT **3.** L **4.** ERA **5.** I **6.** T
Activity 3: **1.** 13, 16, 19 **2.** 9, 5, 1 **3.** 16, 32, 64 **4.** 22, 29, 37 **5.** 14, 13, 17 **6.** 14, 11, 12 **7.** 32, 16, 8 **Warm Up** **1. b)** $\frac{27}{50}, \frac{54}{100}$ **c)** $\frac{3}{5}, \frac{60}{100}$ **d)** 0.1, $\frac{10}{100}$ **e)** 0.75, $\frac{75}{100}$ **f)** 0.28, $\frac{28}{100}$ **g)** $\frac{1}{2}, \frac{50}{100}$ **h)** $\frac{1}{4}, \frac{25}{100}$ **i)** $\frac{19}{25}, \frac{76}{100}$ **j)** 0.35, $\frac{35}{100}$ **2. a)** $\frac{3}{10}$, 0.3 **b)** $\frac{1}{2}$, 0.5 **c)** $\frac{1}{5}$, 0.2 **d)** $\frac{7}{10}$, 0.7 **e)** $\frac{4}{5}$, 0.8 **f)** $\frac{1}{2}$, 0.5 **g)** $\frac{2}{5}$, 0.4 **h)** $\frac{3}{5}$, 0.6 **Mental Math** **1. a)** 210 **b)** 308 **c)** 489 **d)** 1110 **e)** 750 **f)** 3100 **g)** 1500 **2. a)** 2 **b)** $1\frac{1}{2}$ **c)** $1\frac{7}{10}$ **d)** 8 **e)** 20 **3. a)** 2.5 **b)** 11.8 **c)** 10 **d)** 4.08 **e)** 30

Learning Together pp.308–309

Activity 1: **1.** five to seven **2.** $\frac{3}{2}$ **3.** 7:3 **4.** $\frac{2}{17}$
Activity 2: **1.** 1.28, 1.11, 1.22, 1.14, 1.25, 1.37, 1.14, 1.45 **2.** Henke, Ward, Guzman, Key, Morris, Cone, Stottlemyre, Wells **Activity 3:** **1.** $\frac{177}{571}, \frac{6}{25}, \frac{116}{480}, \frac{164}{622}, \frac{102}{446}, \frac{104}{396}, \frac{130}{458}, \frac{159}{641}, \frac{169}{583}$ **2.** 0.310, 0.240, 0.242, 0.264, 0.229, 0.263, 0.284, 0.248, 0.290 **3.** Alomar, Winfield, Olerud, Carter, Lee, White, Borders, Bell, Gruber

Section 9.1 pp.312–313

Practice **1. a)** $\frac{5}{6}$, 5:6, 5 to 6 **b)** $\frac{5}{4}$, 5:4, 5 to 4 **c)** $\frac{4}{6}$, 4:6, 4 to 6 **d)** $\frac{5}{15}$, 5:15, 5 to 15 **e)** $\frac{15}{6}$, 15:6, 15 to 6 **f)** $\frac{6}{15}$, 6:15, 6 to 15 **g)** $\frac{4}{5}$, 4:5, 4 to 5 **3. a)** $\frac{3}{5}$, 3:5, 3 to 5 **b)** $\frac{3}{1}$, 3:1, 3 to 1 **c)** $\frac{50}{1}$, 50:1, 50 to 1 **b)** $\frac{25}{4}$, 25:4, 25 to 4 **e)** $\frac{2}{3}$, 2:3, 2 to 3 **4. a)** $\frac{2}{5}$ **b)** 1:3 **c)** $\frac{1}{2}$ **d)** 1 to 6 **e)** 3:5 **f)** 2:1 **g)** 6 to 1 **h)** 4:9 **i)** $\frac{12}{5}$ **j)** 9:2 **k)** 2:5 **l)** 2:1 **5. a)** 4 to 3 **b)** 8 to 1 **c)** 1 to 5 **d)** 2 to 1 **e)** 1 to 3 **f)** 4 to 7 **Problems and Applications** **6.** 13:4 **7. a)** 1:1 **b)** 4:5 **c)** 2:1 **d)** 5:2 **e)** 4:3 **f)** 3:4 **g)** 7:4 **h)** 1:1 **i)** 1:1 **j)** 3:2 **8.** 7:6 **9. a)** 5:7 **b)** 12:5 **c)** 7:12 **10. a)** 5 to 3 **b)** 2 to 3 **c)** 4 to 15 **d)** 3 to 2 **11. a)** 4:3 **b)** 1:2 **c)** 3:2 **12.** 3:4 **13.** It is a multiple of the prime number.

Learning Together pp.314–315

Activity 1:
5. AB = 4 cm, AG = $2 + 2\sqrt{5}$ cm or approx. 6.472 cm **6.** 1.618 to 1 **Activity 3:** **1.** 1, 2, 1.5, 1.667, 1.6, 1.625, 1.615, 1.619 **2.** The ratios approach the golden ratio.

Section 9.2 p.319

Practice **1.** 12:10 **2.** 36:3 **3.** 4:20 **4.** 5:8 **5.** $\frac{1}{4}$ **6.** $\frac{9}{5}$ **7.** 3:2 **8.** $\frac{8}{1}$ **9.** $\frac{8}{14}$ **10.** yes **11.** yes **12.** no **13.** no **14.** no **15.** yes **16.** yes **17.** yes **18.** 9 **19.** 15 **20.** 4 **21.** 6 **22.** 4 **23.** 15 **24.** 4 **25.** 27 **26.** 8.75 **27.** 4 **28.** 1.2 **29.** 2.5 **Problems and Applications** **30. a)** 12 **b)** 10 **31.** 1.8 m **32.** $70 **33.** 250 mL powder, 750 mL water **34.** 0.15 L **35.** 10.8 cm **36. a)** 9:4 **b)** 54 cm^2 **37. a)** 7:5 **b)** 12.5 m **38. a)** 8 g tin, 92 g copper **b)** 1150 g **c)** 8.7 g

Section 9.3 pp.321–322

Practice **1.** 3 hotdogs/person **2.** 40 words/min **3.** 80 km/h **4.** $7.60/h **5.** $3.16/can **6.** $21/d **7.** 72 beats/min **8.** 0.15 km/min **9.** 5.5 h/d **10.** $7/barrel **11.** $4/m **12.** 2 km/min **13.** 150 m/min **Problems and Applications** **14.** 2200 beats/s **15.** $9.75, $78.00, $75.00, $14.85 **16.** $7.50/h **17. a)** 35 km/min **b)** 2100 km/h **18. a)** 0.1 h/$ **b)** reciprocal **c)** 1 **19. a)** Elk Island : 25.9 ha/buffalo, Wind Cave : 35.2 ha/buffalo,

Wood Buffalo : 1327.5 ha/buffalo, Yellowstone : 315.2 ha/buffalo **b)** Wood Buffalo, Yellowstone, Wind Cave, Elk Island **20.** 7 L **21.** 150° **22.** 441 kJ **23.** 36 000 years **24. a)** 5.3, 5.8, 5.5, 7.4, 6.6 **b)** D, E, B, C, A **c)** 18.9, 17.3, 18.2, 13.5, 15.2

Section 9.4 pp.325–326

Practice **1.** 34 students/bus **2.** 2 hamburgers/person **3.** 55 km/h **4.** 125 mL/person **5.** 3.5 km/h **6.** $8.75/h **7.** 225 g/cake **8.** $5.10/pizza **9.** $0.14/marker **10.** 12.9 ¢/pen **11.** $1.28/hat **12.** 83.3 ¢/sandwich **13.** 1.8 ¢/g **14.** $16.8/h **Problems and Applications** **15.** Nicole by $0.15/h **16.** Halifax **17.** Nova Scotia **18.** fingernails **19.** sp **20. a)** 19.05 g/cm^3, 18.88 g/cm^3 **b)** uranium **21. a)** $36.00 for 3 people **b)** $340.00 for 35 people **c)** $4.30 for 4 L **d)** 28 g for $0.98 **e)** $72.00 for 5 h **f)** 2.5 L for $2.25 **g)** 2000 sheets for $36.00 **h)** 8 L for $4.08 **22. a)** $0.40/100 g, $0.30/100 g; 250 g for $0.75 **b)** $0.60/100 g, $0.70/100 g; 700 g for $4.20 **c)** $4.50/100 mL, $3.50/100 mL; 0.6 L for $21.00 **d)** $1.38/100 straws, $1.35/100 straws; 110 straws for $1.49 **23. a)** $6.98 **b)** $11.45

Section 9.5 p.329

Practice **1.** 1 cm to 75 cm, 1 cm:75 cm, 1:75 **2.** 1 cm represents 50 cm, 1 cm to 50 cm, 1:50 **3.** 1 unit represents 1000 units, 1 unit to 1000 units, 1 unit:1000 units **4.** 1 cm represents 5 cm, 1 cm:5 cm, 1:5 **5.** 1:25 **6.** 1:100 **7.** 1:300 **8.** 20:1 **9.** 2:1 **10.** 10:1 **11.** 1:2 500 000 **12.** 1:10 000 **13.** 15:1 **14. a)** 1 cm:50 cm **b)** 100 cm:1 cm **15. a)** 36 m **b)** 0.9 cm **16.** 10.5 cm **17.** 5900 m **18.** 5.2 cm **19.** 0.28 mm

Section 9.6 p.331

Practice **1.** 1 cm represents 30 km **2.** 1 cm represents 100 km **3.** 1 cm represents 360 km **4.** 1:300 000 000 **5.** 1:125 000 000 **6.** 1:50 000 000 **7.** 1:2 000 000 **8.** 1 cm represents 78 km **9.** 1 cm represents 6 km **10.** 1 cm represents 4.5 km **11.** 1 cm represents 1 km **12.** 800 km **13.** 420 km **14.** 189 km **15.** 0.4 km **16.** 0.875 km **17.** 16.3 km **18.** 20 cm **19.** 3.5 cm **20.** 4 cm **21.** 4 cm **22.** 30 cm **23.** 10.5 cm **Problems and Applications** **24.** 3024 km **25.** 2856 km **26.** 2982 km **27.** 1932 km **28.** 4872 km **29.** 798 km **30.** 1: 30 000 000, 1 cm represents 30 km, 1 cm:30 km **31.** 82.7 cm

Section 9.7 p.335

Practice **1.** 36% **2.** 68% **3.** 52% **4.** 46% **5.** 40% **6.** 75% **7.** 30% **8.** 25% **9.** 50% **10.** 60% **17.** 17% **18.** 61% **19.** 90% **20.** 30% **21.** 5% **22.** 8% **23.** 29% **24.** 65% **25.** 58% **26.** 60% **27.** 5% **28.** 2% **Problems and Applications** **29.** 12% **30.** 98% **31.** 40% **32.** 52%

Section 9.8 pp.338–339

Practice **1.** 65% **2.** 83% **3.** 98% **4.** 2% **5.** 8% **6.** 60% **7.** 70% **8.** 75.2% **9.** 86.3% **10.** 0.5% **11.** 15% **12.** 49% **13.** 10% **14.** 60% **15.** 37.5% **16.** 50% **17.** 90% **18.** 35% **19.** 81.25% **20.** 44% **21.** 80% **22.** 33.3% **23.** 45.5% **24.** 77.8% **25.** 29.2% **26.** 50% **27.** 80% **28.** 45% **29.** 12.5% **30.** 56.25% **31.** 8% **32.** 0.19% **33.** 0.35 **34.** 0.9 **35.** 0.01 **36.** 0.666 **37.** 0.375 **38.** 0.045 **39.** 0.008 **40.** 0.0055 **41.** $\frac{3}{10}$ **42.** $\frac{6}{25}$ **43.** $\frac{13}{25}$ **44.** $\frac{3}{4}$ **45.** $\frac{22}{25}$ **46.** $\frac{1}{20}$ **47.** $\frac{1}{50}$ **48.** $\frac{16}{25}$ **49.** 0.7, 70% **50.** $\frac{3}{100}$, 3% **51.** $\frac{17}{100}$, 0.17 **52.** 20% **53.** 45% **54.** 25% **55.** 90% **56.** 25% **Problems and Applications** **57.** 40% **58.** 41.0% **59. a)** 84.2% **b)** 80.8% **c)** 75.1% **60. a)** 7% **b)** 93% **61. a)** 25% **b)** 62.5% **c)** All competitors are not accounted for; only the rivals. **62. a)** $\frac{36}{71}$ **b)** $\frac{43}{67}$ **c)** $\frac{02}{98}$ **d)** $\frac{4}{102}$

Section 9.9 pp.341–342

Practice **1.** 32 **2.** 30 **3.** 136 **4.** 9.9 **5.** $6.00 **6.** $0.48 **7.** $8.00 **8.** 50% **9.** 27% **10.** 10% **11.** 20% **12.** 40% **13.** 40 **14.** 20 **15.** 150 **16.** 700 **17.** 400 **Problems and Applications** **18.** 80% **19.** 48 **20.** 232 **21.** 90% **22.** 50 000 **23.** $46\frac{2}{3}$% **24.** 90 000 **25.** 10 billion **26.** 65.2% **27. a)** 0.83% **b)** 1.67% **28. a)** 4% **b)** 12.5%

Section 9.10 pp.344

Practice **1.** $0.06 **2.** $0.08 **3.** $1.30 **4.** $2.58 **5.** $3.50, $31.50 **6.** $1.50, $8.50 **7.** $11.12, $44.48 **8.** $20.19, $60.57 **9.** $2.93, $2.92 **10.** $5.00, $44.99 **11.** $0.55, $6.05 **12.** $0.90, $6.90 **13.** $2.20, $11.00 **14.** $15.00, $90.00 **15.** $18.00, $78.00 **16.** $3.25, $13.00 **Problems and Applications** **17.** $26.46 **18.** 15.3% **19.** $12.80 **20. a)** $70.00 **b)** $129.99 **21.** 20% **25. a)** $3000 **b)** 20% **26.** $65.50 less 15% **27.** $1344 **28. a)** Subtract the percent discount from 100% to obtain the discount price as a percent of the price. Then multiply the price by this percent. **b)** Add the percent markup to 100% to obtain the marked up price as a percent of the cost price. Then multiply the cost price by this percent.

Section 9.11 p.346

Practice **1.** $\frac{1}{2}$, 0.5 **2.** 25%, 0.25 **3.** $\frac{6}{5}$, 1.2 **4.** 110%, 1.1 **5.** $\frac{3}{1}$, 3 **6.** 150%, 1.5 **7.** 160%, $\frac{8}{5}$ **8.** $\frac{21}{20}$, 1.05 **9.** 200%, $\frac{2}{1}$ **10.** 2 **11.** 4 **12.** 1.5 **13.** 5 **14.** 1.1 **15.** 1.15 **16.** 1.25 **17.** 1.01 **18.** 2.5 **19.** 120% **20.** 175% **21.** 250% **22.** 300% **23.** 150% **24.** 260% **25.** 40 **26.** 800 **27.** 180 **28.** 2100 **29.** 4240 **30.** 300 **Problems and Applications** **31. a)** 20% **b)** 120% **32. a)** 35% **b)** 135% **33.** 3000% **34.** 2000% **35.** 420% **36. a)** 15% **b)** 90 000% **37.** 4240 km **38. a)** 200% represents an increase; thus the price has tripled **b)** A dollar is worth 400% of a quarter; not 400% more than a quarter. **c)** The old price would be 80% of the new price.

Section 9.12 p.348

Practice **1.** 0.09 **2.** 0.12 **3.** 0.06 **4.** 0.14 **5.** 0.085 **6.** 0.1125 **7.** 0.1475 **8.** 0.205 **9.** 0.069 **10.** $\frac{1}{5}$ **11.** $\frac{2}{5}$ **12.** $\frac{3}{5}$ **13.** $\frac{4}{5}$ **14.** $\frac{6}{73}$ **15.** $\frac{7}{73}$ **16.** $\frac{12}{73}$ **17.** $\frac{18}{73}$ **Problems and Applications** **18. a)** $15 **b)** $265 **19. a)** $2.88 **b)** $137.88 **20.** $141.37 **21.** $137.13 **22.** 1.5 years **23.** $1250

Learning Together p.349

Activity 1: Wednesday: 12 100, Thursday: 13 310, Friday: 14 641 **Activity 2:** 4410, 4630.50, 4862.03, 5105.13 **Activity 3:** **1.** $2000 **2.** $2166.53 **3.** 166.53 **4.** Interest is earned on previous year's interest when compounded.

Connecting Math and the Environment pp. 350–351

Activity 1: **1.** 290 000; using known estimates. **2.** 22 200 **3.** 8% **4.** 22%, 24%, 17% **5.** 63%

Review pp.352–353

1. 2:3 **2.** 5:4 **3.** 3:4 **4.** 4:1 **5.** 200:3 **6.** 3 **7.** 12 **8.** 40 **9.** 3 **10.** 80 km/h **11.** $17.65/CD **12.** 0.79 **13.** 0.2 **14.** 0.08 **15.** $\frac{1}{2}$ **16.** $\frac{2}{5}$ **17.** $\frac{1}{20}$ **18.** 80% **19.** 250% **20.** 43% **21.** 25% **22.** 640% **23.** 0.8% **24.** 0.39 **25.** 8 **26.** 220 **27.** $8.75 **28. a)** 10:3 **b)** 2:1 **c)** 1:1 **29.** 15 m, 25 m **30.** 3.6 L **31.** $1600 **32.** 10.7 m/s **33.** 75 min **34.** Maria **35.** 350 mL for $4.59 **36. a)** 4.5 cm:6400 m **b)** 1 cm:1 mm **37. a)** 1 cm represents 45 km, 1 cm:45 km **b)** 1:10 000 000, 1 cm:100 km **c)** 1 cm represents 150 km, 1:15 000 000 **38.** 300 **39. a)** 1386 km **b)** 3528 km **40. a)** 60% **b)** 50% **c)** 20% **d)** 0.25% **41. a)** 25% **b)** 25% **c)** 65% **42. a)** $13.50 **43.** $10, 50% **44.** 60% **45.** 504 585 **46.** 61.6 m **47.** $1511.25 **48.** $6.44

Chapter Check p.354

1. 2:3 **2.** 1 g raisins:3 g peanuts **3.** $0.33/can **4.** $1\frac{1}{3}$ km/min **5.** 15 **6.** 6 **7.** 20 **8.** 2.5 **9.** 0.85, $\frac{17}{20}$ **10.** 0.06, $\frac{3}{50}$ **11.** 0.005, $\frac{1}{200}$ **12.** 175 **13.** 55 **14.** 63 mg **15.** 75 g **16.** 50 **17.** 2.5 km/h **18.** 24 355-mL cans **19.** 30 m **20.** 465 km **21.** 109% **22.** 16 **23.** $243.75 **24.** $11.40 **25.** $99 **26.** $600

Using the Strategies p.355

1. 419 **2.** $2.80 **4. a)** A; 1; 2; 27; B; 45, 4; D, 5; Home, 60, 6 **5.** $9 \times 9 + 9 + 9 + 9 \div 9$

Data Bank

1. 7:4 **2.** 9.5%

Chapter 10

Getting Started pp.358–359

Activity 1: **1.** Vancouver, Halifax and St John's; Toronto and Montreal **2.** Regina **3.** Edmonton **4.** Toronto **5.** Halifax and St. John's **6. a)** 2 **b)** 46 **c)** 28 **Warm Up** **1.** $\frac{2}{3}$ **2.** $\frac{4}{5}$ **3.** $\frac{2}{3}$ **4.** $\frac{5}{6}$ **5.** $\frac{3}{4}$ **6.** $\frac{1}{3}$ **7.** 60% **8.** 25% **9.** $33\frac{1}{3}$% **10.** 50% **11.** 25% **12.** 61.25% **13.** 18° **14.** 43.2° **15.** 90° **16.** 108° **17.** 216° **18.** 60.12° **23.** 144° **24.** 216° **25.** 72° **26.** 3.6° **27.** 324° **28.** 79.2° **29.** 8–30%, 9–37.5°, 10–22.5°, 11–10% **Mental Math** **1.** 80 **2.** 700 **3.** 70 **4.** 1200 **5.** 600 **6.** 180 **7.** $21.00 **8.** $7.75 **9.** $7.25 **10.** $11.00 **11.** $14.50 **12.** $15.00 **13.** 90 **14.** 95 **15.** 98 **16.** $5.00 **17.** $7.00 **18.** $14.00 **19.** $10.00 **20.** 24 **21.** 39 **22.** 400 **23.** 1500 **24.** 7 **25.** 25 **26.** 8 **27.** 16

Section 10.1 p.363

Practice **1.** census **2.** sample **3.** sample **4.** census **5.** sample **6.** secondary **7.** secondary **8.** secondary **9.** primary **10.** primary **11.** primary **12.** secondary **13.** primary **14.** secondary **Problems and Applications** **16. a)** 4; 40% of the members are boys **b)** by age, for example **17. b)** 30 **c)** 40% **d)** No **18. a)** yes **b)** yes **c)** no

Section 10.2 p.365

Problems and Applications **1. a)** pump gas **b)** 2 **2. a)** Quebec **b)** Alberta **c)** Ontario **4. b)** Asia, Europe **c)** Asia, Europe **d)** Antarctica

Section 10.3 p.367

Problems and Applications **1. a)** 40°C **b)** March and April **c)** October and November **2. a)** 1984, 1988 **b)** 3 m **c)** part of the graph has been deleted **3. b)** Scotia **4. b)** average age is rising

Section 10.4 p.369

Problems and Applications **1. a)** dry and Chemical problems **b)** 55 ha **2. a)** Canada, South America, U.S.A., Australia, France, Other **b)** 118.8° **c)** due to rounding **5. a)** bar graph **b)** circle graph **c)** broken-line graph

Section 10.5 p.370

Problems and Applications **1. a)** Brazil, India and Russia **b)** United States, India

Section 10.6 p.373

Practice **1.** mode **2.** median **3.** mode **4.** mean **5.** median **6.** mean **7.** 24.8, 25, 25, 6 **8.** 23, 23, no mode, 19 **9.** 18.3, 16, 8 and 16, 33 **10.** 51.3, 51, 80, 69 **11.** 2522, 2175, no mode, 3380 **Problems and Applications** **12. a)** 21.8, 18, 20.5 **b)** Each shot total is weighted according to the number of times it occurred. **13. a)** increases by 2 **b)** decreases by 2 **c)** doubles

14. a) no; too low **b)** yes; more representative **15. b)** \$47 500, \$40 000, \$40 000 **c)** median or mode **d)** \$40 000, \$40 000, \$40 000 **e)** \$50 000, \$40 000, \$40 000 **f)** mean **16. a)** always true **b)** sometimes true **c)** sometimes true **d)** sometimes true **17. a)** 11, 11, 12, 15, 26; other lists are possible **b)** 10, 10, 13, 17; only possibility **c)** 3, 19, 20, 25, 26, 27; other lists are possible **d)** 8, 16, 16, 16, 24; other lists are possible

Section 10.7 p.375

Problems and Applications **1. a)** 12 **b)** 31 **c)** No **d)** 4, 16, 22 **2. b)** 87 **c)** 133.5 km/h **3. a)** 3 **b)** 58 h **4. b)** Home : 64, Away : 57

Section 10.8 p.377

Problems and Applications **1. a)** $\frac{1}{2}$ **b)** 20 h–45 h **c)** $\frac{3}{4}$ **2. b)** 37th percentile **3. a)** 50% of the data lies below the median **b)** 75th percentile, 25th percentile **4. a)** 60 and 70 **b)** Carlos **c)** 25% **5.** Not necessarily; everyone else may have scored higher

Learning Together p.378

Activity 1: **1.** Bars are used to represent the data. **2.** The histogram measures frequencies **3.** The horizontal axis is continuous **4.** No; only the total for 1985 and 1986 **5.** 52

Learning Together p.379

Activity 1: **1. a)** Same horizontal axis **b)** Different vertical axis **c)** rapidly escalating profits

Section 10.9 p.383

Practice **1.** 1, 2, 3, 4, 5, 6 **2.** head, tail **3. a)** 2 **b)** 4
Problems and Applications **4. a)** 12 **b)** H1, H2, H3, H4, H5, H6, T1, T2, T3, T4, T5, T6 **5. a)** 1, 2, 3, 4, 5, 6, 7, 8; yellow, blue, red **c)** 24 **d)** No **7. b)** 2, 3, 4, 5, 6, 7, 8, 9, 10, 11, 12 **c)** 6 **8.** HHH, HHT, HTH, THH, HTT, THT, TTH, TTT

Section 10.10 p.385

Practice **1. a)** $\frac{1}{5}$ **b)** $\frac{3}{5}$ **c)** $\frac{4}{5}$ **d)** 0 **2. a)** none **b)** all **c)** none **3. a)** $16\frac{2}{3}$% **b)** 50% **c)** 50% **d)** 100% **4. a)** $\frac{1}{2}$ **b)** $\frac{1}{4}$ **c)** 2 **5. a)** $\frac{1}{10\ 000}$ **b)** $\frac{1}{5000}$ **6. a)** $\frac{1}{2}$ **b)** $\frac{1}{6}$ **c)** 0 **d)** 1 **7. a)** 0.2 **b)** 0 **c)** 0.2 **d)** 0.1 **8. a)** $\frac{1}{52}$ **b)** $\frac{1}{2}$ **c)** $\frac{1}{4}$ **d)** $\frac{1}{26}$ **9.** 1; one of the possible outcomes must occur **10.** 0.8 **11. a)** 20 **b)** 0.80

Section 10.11 p.387

Problems and Applications **1. a)** $\frac{1}{4}$ **b)** $\frac{1}{6}$ **c)** $\frac{1}{24}$ **d)** $\frac{1}{8}$ **2. a)** $\frac{1}{36}$ **b)** $\frac{1}{36}$ **c)** $\frac{1}{4}$ **3. a)** $\frac{6}{25}$ **b)** $\frac{9}{25}$ **4.** same chance of winning either game **5. a)** $\frac{1}{2}$ **b)** $\frac{1}{4}$ **c** $\frac{1}{8}$ **6. a)** $\frac{1}{5}$ **b)** $\frac{1}{10}$ **c)** $\frac{1}{20}$ **d)** $\frac{2}{15}$ **e)** $\frac{1}{30}$ **f)** $\frac{1}{60}$ **7.** No; the events are independent.

Section 10.12 p.389

Problems and Applications **1. a)** $\frac{1}{2}$ **b)** $\frac{2}{5}$ **c** $\frac{1}{5}$ **2. a)** $\frac{9}{25}$ **b)** $\frac{3}{10}$ **c)** $\frac{4}{25}$ **d)** $\frac{1}{10}$ **e)** $\frac{6}{25}$ **f)** $\frac{3}{10}$ **g)** $\frac{6}{25}$ **h)** $\frac{3}{10}$ **3. a)** $\frac{1}{4}, \frac{1}{9}, \frac{1}{4}$ **b)** $\frac{1}{5}, \frac{1}{15}, \frac{1}{5}$ **4. a)** $\frac{1}{15}$ **b)** $\frac{2}{15}$ **c)** $\frac{1}{3}$ **d)** $\frac{4}{15}$ **e)** $\frac{1}{5}$ **f)** $\frac{1}{30}$ **5. a)** $\frac{1}{6}$ **b)** $\frac{1}{15}$ **c)** $\frac{2}{9}$ **d)** $\frac{1}{45}$ **6. a)** $\frac{1}{15}$ **b)** $\frac{1}{15}$ **c)** $\frac{1}{30}$ **d)** $\frac{2}{15}$ **e)** $\frac{2}{9}$ **f)** $\frac{2}{45}$ **7. a)** $\frac{1}{4}$ **b)** $\frac{25}{102}$ **c)** $\frac{1}{16}$ **d)** $\frac{1}{17}$ **e)** $\frac{1}{169}$ **f)** $\frac{1}{221}$ **8.** with replacement; otherwise there are fewer red marbles for the second selection

Review pp.394–395

1. b) Mount Royal **4. b)** 51 km/h **5. a)** New Brunswick and Nova Scotia **b)** 900 000 **6.** grade 9: 3, grade 10: 4, grade 11: 2, grade 12: 1 **7. b)** Alice Springs; their summer occurs at a different time of the year than in Montreal. **8. a)** 57.2, 41, 37 **b)** Answers may vary **10.** 58 th percentile **11.** 10 **12. a)** $\frac{1}{6}$ **b)** $\frac{1}{3}$ **c)** $\frac{1}{3}$ **d)** $\frac{1}{6}$ **e)** 0 **f)** $\frac{1}{2}$ **g)** $\frac{1}{2}$ **h)** 1 **13. a)** $\frac{4}{20}$ **b)** $\frac{2}{15}$

Chapter Check p.396

1. a) carbon **b)** oxygen **3.** 75.7, 76, 72, and 80 **4.** 41st percentile **6.** 30 **7. a)** $\frac{1}{6}$ **b)** $\frac{1}{2}$ **c)** $\frac{1}{3}$ **d)** 0 **e)** 1 **f)** $\frac{1}{3}$ **8. a)** $\frac{9}{25}$ **b)** $\frac{1}{3}$

Using the Strategies p.397

1. Beginning at the 5 and moving clockwise, place the numbers in the following order: 5, 6, 3, 10, 1, 9, 4, 8, 2, 7 **2. a)** 180 **b)** 8 **c)** 44 **d)** 48 **3.** Move two sheep into the top right pen, one from each of its neightbouring pens; move two sheep into the bottom left pen, one from each of its neighbouring pens. **4.** 14 **5.** 1, 4, 9, 16, 25, 36, 49 **6.** 6 **7. a)** remove any corner square **b)** remove any two corner squares **c)** remove any side square **d)** remove any corner square and nonadjacent side square **e)** remove two side squares **9.** February and March

Data Bank

1. 20:18 **2.** gain

Chapter 11

Getting Started pp.400–401

Activity 4: Area: A = 2, B = 2, C = 1, D = 2, E = 1, F = 4, G = 4 **Mental Math** **1.** 31.4 **2.** 314 **3.** 3140 **4.** 31 400 **5.** 0.314 **6.** 0.0 314 **7.** 0.003 14 **8.** 0.000 314 **9.** 0.125 **10.** 96 **11.** 98 **12.** 15 **13.** 120 **14.** 630 **15.** 52 **16.** 3.1 **17.** 3.768 **18.** 5.652 **19.** 18.212 **20.** 2.826

21. 5.5, 5.5 **22.** 4, 4 **23.** 6, 6 **24.** 10, 10 **25.** 12 **26.** 27 **27.** 75 **28.** 75 **29.** 18 **30.** 30 **31.** 90 **32.** 210 **33.** 24 **34.** 192 **35.** 3000 **36.** 375

Section 11.1 p.403

Practice **1.** 16 **2.** 18 **3.** 22 **4.** 18 **5.** 75.2 cm **6.** 23.4 m **7.** 54.2 m **8.** 59 cm **9.** 9 m **10.** 14.2 cm **11.** 27.6 cm **12.** 6.1, 6.1 **Problems and Applications** **13. a)** no **b)** yes **c)** yes **14. a)** 16.8 m **b)** Approximately \$185.00 **15.** 281 m **16.** 20 cm by 5 cm **17.** Smallest perimeter is 10 times the side length.

Section 11.2 p.405

Practice **1.** 5 cm **2.** 7.5 m **3.** 35 cm **4.** 3.5 m **5.** 4 cm **6.** 13.6 m **7.** 20.4 cm **8.** 2 m **9.** 30 cm, 31 cm **10.** 42 cm, 44 cm **11.** 9 cm, 8 cm **12.** 36 cm, 38 cm **13.** 42 cm, 44 cm **14.** 3 m, 3 m **15.** 69.1 cm **16.** 37.4 m **17.** 31.4 cm **18.** 75.4 m **19.** 1 cm **20.** 3 m **21.** 9 cm **22.** 2 m **Problems and Applications** **23. a)** semicircular **b)** $P = \frac{1}{2}\pi d + d$ **c)** 308.4 cm **24.** 522.6 cm **25. a)** 4.7 cm, 1.6 cm, 3.1 cm **b)** 3 cm **c)** 9.4 cm **26. a)** 40 000 km **27. b)** 28

Section 11.3 p.407

Practice **1.** 10 **2.** 9 **3.** 8 **4.** 10 **5.** 100 cm^2 **6.** 39.69 cm^2 **7.** 105 cm^2 **8.** 36.12 cm^2 **Problems and Applications** **9.** 44 m^2 **10.** 125.12 m^2 **11.** 78 m^2 **12.** 22.94 **14. a)** 76 m^2 **b)** \$54.45 **15. a)** 10 m **b)** 15 m **16.** 95.4 m **17.** 708 m^2 **18.** square **19.** 50 cm by 50 cm

Section 11.4 p.410

Practice **1.** 54 cm^2 **2.** 64 cm^2 **3.** 42 cm^2 **4.** 137.5 m^2 **5.** 92.96 m^2 **6.** 191.99 cm^2 **7.** 39.22 cm^2 **8.** 3.125 cm^2 **Problems and Applications** **9. a)** 60 cm^2 **b)** 40 cm^2 **10.** 140 cm **11.** Front or Back: 1120 cm^2, Side: 1504 cm^2 **12.** 4.5 cm **13.** equal **14.** 5 cm

Section 11.5 p.413

Practice **1.** 113 cm^2 **2.** 50 cm^2 **3.** 163 cm^2 **4.** 88 cm^2 **5.** 2 cm, 12.6 cm^2 **6.** 1.5 cm, 7.1 cm^2 **Problems and Applications** **7. a)** 14 m^2 **b)** 15 cm^2 **c)** 136 cm^2 **d)** 6 m^2 **8. a)** 1.5 m^2 **b)** \$29.99 **9.** 62 m^2 **10. a)** 157 m^2 **b)** 43 m^2 **11.** 42.14 m^2

Learning Together pp.414–415

Activity 1: **1.** rectangles **2.** rectangular prism **Activity 2:** **1.** squares **2.** square prism **3.** 96 cm^2 **3.** 8 **Activity 3:** **1.** triangular prism **2.** triangles and rectangles **Activity 4:** **1.** 3 **2.** 5 rectangles, 2 pentagons

Section 11.6 pp.418–419

Practice **1.** rectangular prism **2.** triangular prism **3.** square prism **4.** triangular prism **5.** square prism **6.** triangular prism **7.** 346 m^2 **8.** 43.6 cm^2 **9.** 150 cm^2 **10.** 958 m^2 **11.** 24 m^3 **12.** 180 cm^3 **13.** 1716 cm^3 **14.** 108 m^3 **15.** 278 m^2, 198 m^3 **16.** 685 m^2, 850 m^3 **17.** 12 000 cm^2, 60 000 cm^3 **18.** 152 cm^2, 112 cm^3 **19.** 132 200 m^2, 3 036 000 m^3 **20.** 1637 m^2, 3300 m^3 **Problems and Applications** **21. a)** 9 m^2 **b)** 1.8 m^3 **22. a)** 94 m^3 **b)** 5 **23. a)** 4.6 m^2 **b)** \$45.91 **c)** 27 227 **24.** 6 cm by 6 cm by 6 cm **25. a)** 232 cm^2 **c)** rectangular prism **26.** 5.44 m^3 **27.** 661.9 m^2, 840.65 m^3 **28. a)** 115 m^3 **b)** Approximately 6 h 23 min

Technology pp.420–421

Activity 1: **1. a)** 288 cm^3 **b)** 192 cm^3 **c)** 92.288 m^3 **d)** 11 200 m^3 **Activity 2:** **1.** 3200 cm^3, 4800 cm^3, 6400 cm^3 **2.** The volume is also doubled, tripled and quadrupled. **3.** 6400 cm^3, 14 400 cm^3, 25 600 cm^3 **4.** The volume is multipled by factors of $2^2, 3^2, 4^2$. **5.** The volume is multipled by factors of $2^3, 3^3, 4^3$.

Section 11.7 p.423

Practice **1.** 5000 cm^3 **2.** 32 000 cm^3 **3.** 400 cm^3 **4.** 10 000 000 cm^3 **5.** 5 m^3 **6.** 0.045 m^3 **7.** 0.000 001 m^3 **8.** 25.2 m^3 **9.** 0.275 L **10.** 0.015 L **11.** 1500 L **12.** 100 L **13.** 7000 mL **14.** 250 mL **15.** 1 000 000 mL **16.** 45 mL **17.** 15 000 g **18.** 4.5 g **19.** 0.325 **20.** 350 g **21.** 28 kg **22.** 15 000 kg **23.** 0.54 kg **24.** 0.0001 kg **25.** 10 cm^3 **26.** 6100 cm^3 **27.** 0.0033 cm^3 **28.** 0.1 cm^3 **Problems and Applications** **29.** 50 kL **31.** 7 **32.** 1.42 L **33. a)** 4.5 m^3 **b)** 6 m^3 **c)** 7.5 min **34.** different inner dimensions **36. a)** 1 L **b)** no

Section 11.8 p.428

Practice **1.** 641 cm^2 **2.** 144 cm^2 **3.** 283 cm^2 **4.** 703 cm^2 **5.** 157 m^3 **6.** 7598.8 cm^3 **7.** 1570 cm^3 **8.** 37.68 m^3 **9.** 100.48 m^2, 75.36 m^3 **10.** 75.36 cm^2, 37.68 cm^3 **Problems and Applications** **11. a)** 6.7 cm **b)** 91 cm^2 **12.** 16.4 m^2 **13.** 2 L **14.** 8 m^3

Learning Together p.429

Activity 1: **1.** triangle, square, pentagon, rectangle **2.** triangular-based pyramid, square-based pyramid, pentagonal-based pyramid, rectangular-based pyramid **4.** 1

Section 11.9 p.432

Practice **1.** 161 cm^2 **2.** 72 m^2 **3.** 50.24 cm^2 **4.** 113.04 m^2 **5.** 50 cm^3 **6.** 15 m^3 **7.** 268 m^3 **8.** 268 cm^3 **Problems and Applications** **9.** 360 cm^3 **10.** 5024 mm^2 **11.** 2 574 467 m^3 **12. a)** 510 926 783 km^2 **b)** spherical **13.** $\frac{1}{50}$ **14.** 76 000 m^2 including the base **15.** The orange of diameter 9 cm has more than 3 times the volume for only 2 times the price of the orange of diameter 6 cm. Assuming the skins are the same thickness, the orange of diameter 9 cm is the better value.

Technology p.433

Activity 2: **2. a)** 120 cm^2, 120 cm^3 **b)** 50 cm^2, 32 cm^3 **c)** 400 m^2, 800 m^3 **d)** 120 m^2, 1200 m^3 **5.** New 10 PRINT "VOLUME OF A" 20 PRINT "SQUARE-BASED PYRAMID" 30 INPUT "SIDE LENGTH IS"; L 40 INPUT "HEIGHT IS"; H 50 V = L^2 * H/3 60 PRINT "THE VOLUME IS"; V 70 END

Review pp.436–437

1. 25.2 cm **2.** 23 m **3.** 26.7 cm **4.** 131.88 m **5.** 12 m **6.** 8.8 cm **7.** 2401 mm^2 **8.** 152 cm^2 **9.** 66.03 m^2 **10.** 4.5 cm^2 **11.** 153.86 cm^2 **12.** 3.52 cm^2 **13.** rectangular prism **14.** triangular prism **15.** 1734 m^2 **16.** 315.34 cm^2 **17.** 648 cm^3 **18.** 29 335.5 cm^3 **19.** 7234.56 cm^2 **20.** 3014.4 m^2 **21.** 99 m^2 **22.** 1055.04 cm^2 **23.** 864 cm^3 **24.** 192 m^3 **25.** 707 m^3 **26.** 33 493 m^3 **27.** hexagonal prism; 2 hexagons, 6 rectangles **28. a)** 1.8 dm^3 **b)** 1.8 kg **c)** 1.8 L **29.** 434 m^2 **30.** 334 cm^2

Chapter Check p.438

1. 120 cm, 900 cm^2 **2.** 53 cm, 100 cm^2 **3.** 21.9 cm, 27 cm^2 **4.** 62.8 m, 314 m^2 **5.** 220 cm^2 **6.** 2512 cm^2 **7.** 3.84 m^2 **8.** 628 cm^2 **9.** 117 m^3 **10.** 1099 m^3 **11.** 4187 cm^3 **12.** 6 510 000 cm^3 **13.** square prism; 2 squares, 4 rectangles **14.** triangular-based pyramid; 1 equilateral triangle, 3 isosceles triangles **15.** 16 **16. a)** 216 cm^2 **b)** 180 mL

Using the Strategies p.439

1. 6 **2.** 35, 12, 23 **3.** 15:00 **4.** 2 × $5, 1 × $2, 1 × 25¢, 1 × 10¢, 1 × 5¢, 3 × 1¢ **6.** 78 **7.** The 35 cm pizza costs less per square centimeter. **8.** Take the middle full glass and empty it into the middle empty glass. **9.** 4

Data Bank

1. Approximately 19 000 000 000 000 **2.** Manitoba and Saskatchewan; Saskatchewan is larger by 2380 km^2.

Chapter 12

Getting Started pp.442–443

Mental Math **1.** 40 m **2.** 24 m **3.** 44 m **4.** 38 m **5.** 150 m^2 **6.** 100 m^2 **7.** 96 m^2 **8.** 150 m^2 **9.** 90 **10.** 50 **11.** 80 **12.** 70 **13.** 30 **14.** 40 **15.** 40

Section 12.1 p.447

Practice **1.** line CD **2.** ray ST **3.** angle DEF **4.** line segment RT **5.** angle M **6.** angle CAR **7.** A, C, E, G, H; $\overleftrightarrow{AB}$, $\overleftrightarrow{DF}$, $\overleftrightarrow{AG}$, $\overleftrightarrow{HK}$, $\overleftrightarrow{BH}$ **8.** $\overline{AB}$, $\overline{AC}$, $\overline{AD}$, $\overline{AE}$, $\overline{CD}$ **9.** ∠ABC, ∠CBD, ∠BDE, ∠EDG, ∠DGF, ∠FGK, ∠GKH, ∠HKM **10.** R, S, T **11.** $\overleftrightarrow{RS}$, $\overleftrightarrow{ST}$, $\overleftrightarrow{TR}$ **12.** ∠MRN, ∠TRS, ∠QSW, ∠RST, ∠XTY, ∠RTS **13.** $\overrightarrow{RW}$, $\overrightarrow{SM}$, $\overrightarrow{RX}$ **14.** $\overline{RS}$, $\overline{ST}$, $\overline{TR}$ **Problems and Applications** **21.** 7, $\overline{AB}$, $\overline{BC}$, $\overline{AF}$, $\overline{BE}$, $\overline{CD}$, $\overline{FE}$, $\overline{ED}$ **22. a)** 12 **b)** There are 3 angles associated with each vertex; for example ∠BAD, ∠BAE, ∠DAE; for a total of 24.

Section 12.2 pp.450–451

Practice **1.** 30° **2.** 65° **3.** 86° **4.** 90° **5.** 38° **6.** 71° **7.** 7° **8.** 90° **9.** 115° **10.** 142° **11.** 173° **12.** 180° **13.** 1-acute, 2-acute, 3-acute, 4-right, 5-acute, 6-acute, 7-acute, 8-right, 9-obtuse, 10-obtuse, 11-obtuse, 12-straight **22.** 90°-right **23.** 180°-straight **24.** 45°-acute **25.** 60°-acute **26.** 80°-acute **27.** 250°-reflex **28.** 130°-obtuse **29.** 60°-acute **38.** 60° **39.** 48° **40.** 15° **41.** 81° **42.** 1° **43.** 24° **44.** 75° **45.** 89° **46.** 140° **47.** 110° **48.** 95° **49.** 30° **50.** 55° **51.** 172° **52.** 70° **53.** 1° **54.** $a = 131°$, $b = 49°$, $c = 131°$ **55.** $x = 14°$, $y = 166°$, $z = 14°$ **56.** $x = 40°$, $y = 80°$, $z = 40°$, $t = 60°$ **57.** $a = 54°$, $b = 39°$, $c = 87°$, $d = 54°$ **58.** $x = 15°$ **59.** $y = 45°$ **60.** $a = 107°$ **61.** $m = 63°$ **62.** $x = 49°$, $y = 49°$, $m = 131°$ **63.** $m = 152°$, $n = 118°$ **Problems and Applications** **68.** opposite angles are equal, adjacent angles sum to 180° **69.** 180° − 108° = 72° **70.** 271° **71. a)** 90°, 180°, 270° **72. a)** ∠BED = 142°, ∠BEC = 38°, ∠AEC = 142° **b)** opposite angles are equal **73. a)** 270° **b)** 150°

Section 12.3 pp.454–455

Practice **1.** yes **2.** no **3.** yes **4.** no **5.** acute **6.** obtuse **7.** acute **8.** acute **9.** scalene **10.** isosceles **11.** equilateral **12.** isosceles **13.** 41° **14.** 26° **15.** $x = 5$ cm, $y = 70°$, $z = 40°$ **16.** $x = 9$ cm, $y = 60°$, $z = 60°$ **Problems and Applications** **17.** The measure of an exterior angle equals the sum of the measures of its 2 interior and opposite angles. **18.** 70°and 70°or 40°and 100° **19.** 20°and 20° **20.** $x = 128°$, $y = 43°$, $z = 137°$ **21.** $x = 65°$, $y = 137°$, $m = 43°$ **22.** $a = 73°$, $b = 61°$, $c = 46°$ **23.** $r = 54°$, $t = 91°$, $s = 35°$, $w = 89°$ **24.** $a = 45°$, $b = 83°$, $c = 83°$, $d = 97°$ **25.** $x = 64°$, $y = 64°$, $t = 78°$ **26.** $r = 40°$, $s = 92°$, $p = 14°$ **27.** $x = 120°$ **28.** $x = 50°$, $y = 25°$ **29.** $x = 53°$, $y = 37°$ **30.** ∠C = ∠A, ∠B = 180° − 2 × ∠A **31.** ∠A = 2 × ∠D **32. a)** never **b)** sometimes **c)** always **d)** sometimes **33. a)** 3 **b)** 1 **c)** 0 **34.** Yes, it could be isosceles. **35.** 11.9 m **36. a)** Triangles F and I do not exist.

Section 12.4 p.459

Practice **1.** $x = 75°$ (alternate angles are equal), $y = 105°$ (co-interior angles are supplementary) **2.** $m = 98°$ (co-interior angles are supplementary), $n = 82°$ (alternate angles are equal) **3.** $n = 110°$ (corresponding angles are equal), $m = 70°$ (co-interior angles are supplementary), $t = 110°$ (opposite angles are equal) **4.** $s = 92°$ (opposite angles are equal), $m = 92°$ (alternate angles are equal), $n = 88°$ (co-interior angles are supplementary) **5.** $t = 70°$, $x = 35°$, $y = 70°$ **6.** $m = 100°$, $x = 100°$, $y = 80°$ **7.** $t = 85°$, $m = 95°$, $x = 40°$ **8.** $x = 51°$, $y = 57°$ **9.** $t = 106°$, $x = 30°$, $y = 44°$ **10.** $a = 75°$, $b = 25°$, $c = 120°$, $d = 35°$,

$e = 35°, f = 45°, g = 45°$
Problems and Applications **11.** ∠ABC = 55°, ∠ACB = 55°, ∠APQ = 55°, ∠AQP = 55°, ∠DQC = 70°, ∠QDC = 55°, ∠QDB = 125°, ∠BPQ = 125°, ∠PQD = 55° **12.** Using the fact that a straight angle equals 180° and that alternate angles are equal.

Section 12.5 p.462

Practice **1.** not composed entirely of line segments **2.** not closed **3.** not joined at endpoints **4.** 52° **5.** 57° **6.** 44° **7.** 69° **8.** 128.6° **9.** 144° **10.** 140° **11.** 150°
Problems and Applications **12.** 1620° **15.** 120° **16. a)** 135° **b)** 150° **17. a)** 360° **b)** 360° **c)** 360° **d)** 360° **e)** 360° **f)** 360° **18.** The sum of the exterior angles of a polygon is 360° **19. a)** 120° **b)** 90° **c)** 72° **d)** 60° **e)** 45° **f)** 36°

Learning Together pp.463

Activity 1: **1.** a, l; c, o; e, m; f, j; d, i; h, k

Learning Together pp.464–465

Activity 1: **2.** No **Activity 2:** **2.** Yes, ∠G and ∠J could differ. **Activity 3:** **2.** Yes, the side lengths could differ. **Activity 4:** **2.** No **Activity 5:** **2.** No **Activity 6:** **2.** Yes, the other side lengths could differ.

Section 12.6 pp.468–469

Practice **1.** BI = CA, BG = CT, IG = AT, ∠B = ∠C, ∠I = ∠A, ∠G = ∠T **2.** HO = CA, HT = CR, OT = AR, ∠H = ∠C, ∠O = ∠A, ∠T = ∠R **3.** ASA **4.** No **5.** No **6.** SSS **7.** No **8.** No **9.** SAS **10.** JC = EK or ∠A = ∠I **11.** RK = PC **12.** ASA; ∠Q = ∠Y, ∠P = ∠X, ∠R = ∠Z, XY = PQ, XZ = PR, RQ = ZY **13.** No **14.** SSS; ∠A = ∠F, ∠B = ∠E, ∠C = ∠D, AB = FE, AC = FD, BC = ED **15.** No **16.** SAS; LU = LY, LJ = LJ, JU = JY, ∠U = ∠Y, ∠UJL = ∠YJL, ∠ULJ = ∠YLJ **17.** ASA; ∠S = ∠Q, ∠SPR = ∠QRP, ∠SRP = ∠QPR, PS = RQ, PR = RP, SR = QP **18.** No **19.** No **20.** SAS; ∠P = ∠B, ∠PAT = ∠BTA, ∠PTA = ∠BAT, PT = BA, PA = BT, AT = TA
Problems and Applications **22.** No, different sizes. **23. a)** otherwise SSS quarantees congruence **b)** may change just the angles **24. a)** sometimes congruent **b)** sometimes congruent **c)** always congruent **d)** sometimes congruent

Learning Together pp.470–471

Activity 1: **1.** 18 **2.** 8 **3.** 32 **4.** 10 **5.** 13 **6.** 20 **7.** 25 **8.** 34 **Activity 2:** **1.** 29 **2.** 52 **3.** 17 **4.** 26 **Activity 3:** **1.** 25, 16, 9 **2.** 34, 9, 25 **3.** 20, 4, 16 **4.** 13, 9, 4 **5.** 8, 4, 4 **Activity 4:** **1.** 29, 4, 25 **2.** 52, 36, 16 **3.** 41, 16, 25 **4.** 32, 16, 16 **5.** 45, 36, 9 **Activity 5:** The sum of the area of the squares on the legs equals the area of the square on the hypotenuse.

Section 12.7 p.473

1. $3^2 + 4^2 = 5^2$ **2.** $6^2 + 7^2 = x^2$ **3.** $9^2 + y^2 = 11^2$ **4.** $m^2 + y^2 = 12^2$ **5.** 26 m **6.** 24 m **7.** 15 m **8.** 12 m **9.** 5.4 m **10.** 4.2 m **11.** 11.7 m^2 **12.** 4.1 m **Problems and Applications** **13.** 5.7 m **14.** 106.3 m **15.** 72.1 m **16.** 165.1 m **17. a)** 7.4 m **b)** 17.0 m **18. a)** 100 cm **b)** 102 cm

Section 12.8 pp.476–477

Practice **1.** ∠AXB; ∠AOB **2.** ∠ARB, ∠ATB; ∠AOB **3.** ∠APB, ∠AQB; ∠AOB **4.** ∠APB, ∠AQB, ∠ASB; ∠AOB **5.** 50° **6.** 38° **7.** 72° **8.** 40° **9.** 90° **10.** 105° **11.** 61° **12.** $x = 42°, y = 37°$ **13.** $a = 15°, b = 90°$
14. $m = 43°, n = 44°$ **15.** $x = 44°, y = 44°$
16. $m = 31°, n = 62°$ **17.** $x = 51°, y = 51°$
18. $r = 90°, t = 90°$ **19.** $m = 30°, t = 77°, x = 73°$
20. $r = 49°, s = 87°, x = 49°, y = 44°$
21. $a = 52°, b = 99°, c = 81°, d = 29°$
22. $r = 70°, x = 48°, y = 48°, z = 22°$
23. $a = 55°, b = 33°, y = 55°, x = 92°$
24. $a = 46°, b = 42°, y = 46°, x = 92°$
25. $x = 42°, y = 42°, t = 42°$
26. $m = 76°, r = 38°, x = 38°, y = 38°$
Problems and Applications **27.** ∠XOZ **28.** 150°, 240° **29. a)** 45° **b)** 3 **30. a)** AB = 12, CD = 16 **31.** 10 cm

Section 12.9 pp.481–482

Practice **1.** $x = 89°, y = 85°$ **2.** $a = 78°, b = 92°$
3. $x = 77°, y = 90°$ **4.** $a = 99°, b = 98°$
5. $x = 110°, y = 105°$ **6.** $m = 119°, n = 82°$
7. $x = 60°, y = 120°$ **8.** $x = 98°, y = 84°, z = 82°$
9. $a = 76°, b = 96°, c = 104°, d = 84°$
10. $m = 69°, x = 114°, y = 66°$
11. $m = 90°, x = 90°, y = 90°$
12. $x = 40°, y = 40°, g = 80°$
13. $m = 94°, r = 79°, t = 32°$
14. $a = 70°, b = 110°, x = 65°, y = 115°$
Problems and Applications
15. $m = 87°, n = 89°, r = 93°, t = 91°, x = 91°, y = 93°$
16. $m = 95°, w = 85°, z = 85°, t = 95°, y = 95°, u = 85°$
17. $m = 71°, n = 109°, x = 66°, y = 43°$
18. $m = 46°, t = 50°, x = 44°$
19. $a = 38°, m = 48°, r = 55°, t = 48°, x = 94°$
20. $m = 65°, r = 29°, t = 65°, n = 94°, y = 34°$
21. $y = 180° - x$ **22.** ∠C = x, ∠B = y
23. supplementary; their sum is 180° **24. a)** 360° **b)** yes; true for all quadrilaterals **25. a)** 90° **b)** square
26. $x = 180° - 2 \times 35°, y = 180° - x$ **27. a)** equal **c)** The size of an exterior angle of a quadrilateral inscribed in a circle is equal to the opposite of its corresponding supplementary angle. **28. a)** $x = 110°, y = 110°, z = 89°$ **b)** Find x by considering a straight angle.

Review pp.490–491

1. A, B, S, R, T **2.** $\overleftrightarrow{AB}$, $\overleftrightarrow{SR}$, $\overleftrightarrow{TM}$, $\overleftrightarrow{AT}$, $\overleftrightarrow{BR}$ **3.** $\overrightarrow{AD}$, $\overrightarrow{BF}$, $\overrightarrow{RH}$, $\overrightarrow{BC}$, $\overrightarrow{TK}$ **4.** ∠EAB, ∠EAF, ∠FAS, ∠SAB, ∠SRM, ∠MRH, ∠KTL, ∠NMP **5.** $\overline{AB}$, SR, $\overline{TM}$, $\overline{AS}$, $\overline{BM}$
6. acute **7.** right **8.** acute **9.** straight **10.** obtuse **11.** reflex **12.** $a = 143°, b = 37°, c = 143°$ **13.** $a = 47°, b = 89°, c = 44°, d = 47°$ **14.** 43° **15.** 62° **16.** 60°

17. 292° **18.** 43° **19.** $x = 42°, y = 42°, z = 138°, t = 138°$ **20.** $x = 85°, y = 85°, m = 66°$ **21.** $x = 62°, y = 79°, r = 141°$ **22.** isosceles **23.** scalene **24.** $a = 115°, b = 65°, c = 115°, d = 115°, e = 65°, x = 65°, y = 115°$ **25.** $a = 100°, b = 80°, c = 100°, d = 80°, e = 100°$ **26.** $a = 47°, b = 62°, c = 71°, d = 109°, e = 109°, f = 118°, g = 118°$ **27.** $a = 34°, b = 87°, c = 34°, d = 59°$ **28.** $x = 46°, y = 51°, z = 83°, t = 83°$ **29.** $a = 65°, b = 105°, c = 115°, d = 115°, e = 105°, f = 75°, g = 65°, x = 40°, y = 75°$ **30.** $180° \times (n - 2)$ **31.** 540° **32.** 900° **33.** 1800° **34.** 2340° **35.** 3240° **36.** 5040° **37.** congruent, SSS **38.** not congruent **39.** congruent, SAS **40.** congruent, ASA **41.** 13 **42.** 8 **43.** 15.8 m **44.** 8.6 m **45.** 25° **46.** $x = 47°, y = 38°$ **47.** $x = 93°, y = 99°$ **48.** $m = 100°, n = 30°, x = 50°, y = 50°$

Chapter Check p.492

1. $a = 115°, b = 115°, c = 65°$ **2.** $a = 49°, b = 42°, c = 49°, d = 89°$ **3.** $a = 108°, b = 72°, c = 108°, d = 108°, e = 72°, f = 108°, g = 72°$ **4.** $a = 40°, t = 79°, x = 79°, y = 61°$ **5.** octagon **6.** quadrilateral **7.** hexagon **8.** pentagon **9.** congruent, ASA **10.** not congruent **11.** congruent, SSS **12.** congruent, SAS **13.** $m = 90°, x = 49°, y = 41°, z = 90°$ **14.** $x = 106°, y = 90°$ **15.** $x = 79°, y = 92°$ **16.** $a = 132°, m = 24°, t = 53°, x = 24°, y = 24°$ **17.** 10.6 m **18.** 11.2 cm

Using the Strategies p.493

1. a) 9 **b)** 16 **c)** 25 **d)** 36 **e)** 625 **f)** 11 025 **2.** February, March **3.** 1 cm^2 **4.** 47 **5.** 1024 **6.** 27 **7. a)** 6 **b)** 4 **c)** 2 **8.** 4 quarters; 3 quarters, 5 nickels; 2 quarters, 10 nickels; 1 quarter, 15 nickels; 20 nickels **9.** 6 **10.** 32 cm **11.** 05:55

Data Bank

1. approximately 50 times

Cumulative Review, Chapters 9–12 pp.494–495

Chapter 9 **3.** \$0.95 **4.** 400 km **5.** 15 m **6. a)** 60 m in 10 s **b)** \$7.08 for 12 containers **7.** green $\frac{2}{35}$, 0.06, 6%; yellow $\frac{1}{14}$, 0.07, 7%; purple $\frac{8}{35}$, 0.23, 23%; orange $\frac{4.5}{70}$, 0.06, 6%; pink $\frac{1}{10}$, 0.1, 10%; white $\frac{33.5}{70}$, 0.48, 48% **8. a)** \$75.54 **b)** \$494 **Chapter 10** **4.** mean, 83.6; median, 81; mode, 79 **5. a)** $\frac{5}{11}$ **b)** $\frac{2}{11}$ **c)** $\frac{4}{11}$ **d)** 0 **e)** $\frac{9}{11}$ **Chapter 11** **1.** $P = 24.8$ cm, $A = 38.4$ cm^2 **2.** $P = 53.5$ cm, $A = 97.8$ cm^2 **3.** $A = 248$ m^2, $V = 240$ m^3 **4.** $A = 562.7$ cm^2, $V = 803.8$ cm^3 **5.** $A = 1626.52$ m^2, $V = 4615.8$ m^3 **6.** 24 416.6 cm^3 **7.** 240 boxes **Chapter 12** **1.** $x = 77°, y = 41°, w = 62°, z = 77°$ **2.** $x = 60°, y = 120°$ **3.** $a = 109°, b = 109°, c = 71°, d = 71°, e = 109°$ **4.** $a = 50°, b = 130°, c = 50°, d = 130°$ **5.** $\Delta ABC \cong \Delta ADC$, SSS **6.** $\Delta PQR \cong \Delta SRT$, SAS **7.** $\Delta DFE \cong \Delta DFG$, SAS **8.** $\Delta ABC \cong \Delta DEC$, ASA **9.** $x = 10.6$ m **10.** $x = 8.1$ m **11.** $x = 99°, y = 75°$ **12.** $x = 30°, y = 29°$

Chapter 13

Getting Started p.499

Warm Up **1.** 20, 12, 5, 2 **2.** 1, −2, −7, −9 **3.** −0.4, −0.2, 0.2, 0.4 **4.** 4, 2, −6, −10 **5.** 2 **6.** 75 **7.** 1 **8.** −5 **9.** −1 **10.** 12 **11.** −1, 2, 5, 8, 11 **12.** $-1, -\frac{1}{2}, 0, \frac{1}{2}, 1$ **13.** 4, 2, 0, −2, −4 **14.** 9, 11, 16, 19, 22 **15.** 1, −1, 6, 3, 0 **16.** −5, −6, −8, −12, −14 **17.** 24, 18, −2, −6, −18 **Mental Math** **1.** 3 **2.** 24 **3.** −12 **4.** −3 **5.** −9 **6.** 24 **7.** $-8\frac{1}{2}$ **8.** 15 **9.** $y = x$ **10.** $y = 2x$ **11.** $y = -3x$ **12.** $y = x + 1$ **13.** $y = x + 5$ **14.** $y = x - 3$ **15.** $y = \frac{x}{2}$ **16.** $y = -x$ **17.** $y = x^2$ **18.** $y = -x^3$

Section 13.1 pp.502–503

Practice **1. a)** The later in the day the longer the tree's shadow. **2.** (1, 0), (2, 0), (3, 0), (1, 1), (2, 1), (3, 1), (1, 2), (2, 2), (3, 2). There is no relationship between the 2 numbers. **3. a)** The sum of the x- and y-values equals four. **b)** 2, 3, 4, 5, 6 **c)** (2, 2), (1, 3), (0, 4), (−1, 5), (−2, 6) **4. a)** The x-value less the y-value equals one. **b)** 2, 3, 4, 5, 6 **c)** (3, 2), (4, 3), (5, 4), (6, 5), (7, 6) **5. a)** The sum of the m- and n-values equals negative one. **b)** −3, −2, −1, 0, 1 **c)** (2, −3), (1, −2), (0, −1), (−1, 0), (−2, 1) **6. a)** The s-value less the t-value equals three. **b)** 1, 0, −1, −2, −3 **c)** (4, 1), (3, 0), (2, −1), (1, −2), (0, −3) **7. a)** 7 **b)** 2 **c)** 13 **d)** 6 **e)** 10 **f)** 13 **8. a)** 3 **b)** 6 **c)** −5 **d)** 9 **e)** 2 **f)** −1 **9. a)** The value of y is four times the value of x, plus two. **b)** 10, 6, 2, −2, −6 **c)** (2, 10), (1, 6), (0, 2), (−1, −2), (−2, −6) **10. a)** The value of y is three times the value of x, less four. **b)** 2, −1, −4, −7, −10 **c)** (2, 2), (1, −1), (0, −4), (−1, −7), (−2, −10) **11. a)** The value of y is negative two times the value of x, plus seven. **b)** 3, 5, 7, 9, 11 **c)** (2, 3), (1, 5), (0, 7), (−1, 9), (−2, 11) **12. a)** The value of y is negative one times the value of x, less two. **b)** 2, 1, 0, −1, −2 **c)** (−4, 2), (−3, 1), (−2, 0), (−1, −1), (0, −2) **13.** (0, 11), (11, 0), (1, 10), (10, 1), (−2, 13) **14.** (3, 0), (0, −3), (1, −2), (2, −1), (−1, −4) **15.** (0, 5), (1, 9), (2, 13), (3, 17) **16.** (0, −7), (1, −5), (2, −3), (3, −1) **17.** $x + y = 9$ **18.** $y = 3x$ **19.** $y = x - 2$ **20.** $y = -2x$ **Problems and Applications** **21. a)** 80, 160, 240, 320 **b)** 400 km, 480 km **c)** Successive entries differ by 80. **d)** (1, 80), (2, 160), (3, 240), (4, 320) **e)** The distance travelled, in kilometres, equals eighty times the time travelled, in hours. **f)** $D = 80t$ **g)** 1040 km, 1680 km

22. a) 8, 16, 32, 64 **b)** Each successive entry is double the previous entry **c)** 128, 256
d) (1, 2), (2, 4), (3, 8), (4, 16), (5, 32), (6, 64) **e)** The number of regions equals two raised to a power equal to the number of folds. **f)** $R = 2^n$ **g)** 1024, 4096
23. a) 5, 9, 14, 20, 27, 35, 44 **b)** The number of diagonals increases by successive numbers 3, 4, 5, ...
c) 54, 65 **d)** (4, 2), (5, 5), (6, 9), (7, 14), (8, 20), (9, 27), (10, 35), (11, 44) **e)** number of diagonals equals the number of sides times the number of sides minus 3 all divided by 2 **f)** $D = \frac{s \times (s - 3)}{2}$ **g)** 170, 1175, 1850

Section 13.2 p.506

Practice **1.** J **2.** G **3.** N **4.** H **5.** P **6.** F **7.** M **8.** K **9.** L **10.** A **11.** D **12.** C **13.** B **14.** E **15.** I
16. A(−6, 5), B(−5, 3), C(−2, 1), D(−4, −2), E(−1, −1), F(−6, −6), G(−2, −5), H(5, 6), I(0, 5), J(2, 2), K(4, 1), L(6, 0), M(5, −1), N(3, −2), P(5, −6), Q(4, −4) **Problems and Applications**
32. a) triangle, 10 **b)** rectangle, 18 **c)** parallelogram, 15 **d)** rectangle, 12 **e)** hexagon, 64 **33.** (5, 3), (−2, −1)
34. Some of the infinitely many possibilities are (1, −1), (7, −3), (9, −1) and (3, 5).
35. b) (4, 5), (6, −5), (−12, −3) **c)** 3 **36. a)** sometimes true **b)** always true **c)** always true

Section 13.3 p.509

Practice **1.** (0, 0), (1, 1), (2, 2), (3, 3), (4, 4), (5, 5) **2.** (0, 25), (1, 20), (2, 15), (3, 10), (4, 5), (5, 0) **3.** 270, 360, 450 **4.** 18.00, 24.00, 30.00 **5.** 90, 120, 150 **6.** 180, 230, 280 **Problems and Applications** **7. b)** 3 mL **c)** 882 mL, 906 mL **d)** 819 mL **8. a)** (0, 35), (0.5, 50), (1, 65), (1.5, 80), (2, 95), (2.5, 110), (3, 125) **c)** No, since there is a fixed charge of $35.00 **9. b)** $36.00 **c)** $48.00

Learning Together pp.510–511

Activity 1: **2.** 7 **3. a)** 18 **b)** (3, 3), (3, 4), (3, 5), (3, 6), (4, 2), (4, 3), (4, 4), (4, 5), (4, 6), (5, 2), (5, 3), (5, 4), (5, 5), (5, 6), (6, 2), (6, 3), (6, 4), (6, 5) **4.** No **5.** The sum of the differences between the co-ordinates of U and L equals the shortest total distance. **6.** yes, at the points with coordinates (3, 1), (4, 1), (5, 1), (6, 1), (2, 3), (2, 4), (2, 5), (2, 6), (3, 7), (4, 7), (5, 7), (6, 7), (7, 2), (7, 3), (7, 4), (7, 5) **Activity 2:** **3. a)** 1 **b)** (5, 2) **4.** 4 **Activity 3:** **2.** (3, 3), (4, 3), (5, 3), (3, 4), (4, 4), (5, 4), (6, 4), (3, 5), (4, 5), (5, 5), (6, 5), (3, 6), (4, 6), (5, 6), (6, 6), (3, 7), (4, 7), (5, 7), (6, 7) **3.** 19

Section 13.4 p.514

Practice **1.** −3, −2, −1, 0, 1 **2.** 0, 1, 2, 3, 4 **3.** 6, 5, 4, 3, 2 **4.** −4, −3, −2, −1, 0 **5. a)** $y = x + 2$ **b)** $y = -x + 5$ **14.** (1, 0), (0, 1) **15.** (5, 0), (0, −5) **16.** (−2, 0), (0, 2) **17.** (6, 0), (0, 4) **18.** (4, 0), (0, −3) **19.** (−5, 0), (0, −3) **Problems and Applications**
20. a) (3, 0), (4, 1), (5, 2), (6, 3), (7, 4), (8, 5), (9, 6), (10, 7), (11, 8), (12, 9) **c)** $y = x - 3$
21. a) (0, 2.20), (5, 8.70), (10, 15.20), (15, 21.70), (20, 28.20), (25, 34.70) **c)** initial charge or fixed cost
22. a) (1, 6), (2, 11), (3, 16), (4, 21), (5, 26) **c)** $y = 5x + 1$ **23. a)** 6, 12, 18, 24 **c)** $A = 3h$

Learning Together pp.516–517

Activity 1: **1.** No, the data does not describe a straight line. **2.** The later the date the greater the height.
Activity 2: **1.** The data does not describe a straight line but, rather, a curve. **4.** 50%, 20% **5.** 15 m **6.** 12% **7.** the answers to questions 5 and 6

Section 13.5 p.519

Practice **1.** (0, 1) **2.** (−3, −2) **3.** (0, 0) **4.** (−1, −2) **5.** $y = x, y = -x; y = x - 1, y = 2x$ **6.** (3, 6) **7.** (2, 1) **8.** (0, 3) **9.** (2, 6) **10.** (−2, 6) **11.** (5, 7) **12.** $(\frac{1}{2}, -1)$
13. b) triangle **14. a)** $C = 60 + 0.20n$, $C = 0.5n$ **d)** (200, 100); both plans cost $100 for 200 km **15. c)** 2.5 h **d)** 20 km **16.** Concurrent means that they all intersect at a common point.

Section 13.6 p.522

Practice **1.** AB−positive, CD−negative, EF−positive, GH−positive, IJ−negative, KL−negative **2.** MN, vertical, $x = -4$; PQ, slope = 0, $y = 3$; RS, vertical, $x = -1$; TV, slope = 0, $y = 1$; WX, vertical, $x = 5$; YZ, slope = 0, $y = -3$ **3.** AB, −1; CD, 0; EF, 1; GH, −1; IJ, vertical; KL, $\frac{2}{3}$; MN, $\frac{1}{4}$; PQ, −2; RS, 3; TV, −4; WX, $\frac{1}{4}$; YZ, −3 **4.** $-\frac{1}{2}$ **5.** 0 **6.** $\frac{2}{9}$ **7.** vertical **8.** $-\frac{3}{2}$ **9.** $\frac{6}{5}$ **10.** positive **11.** negative **12.** positive **13.** negative
Problems and Applications **14. a)** 27.5 km **b)** 1.5 h **c)** Danzel's speed **15. b)** $\frac{2}{3}$ for both **c)** equal slopes **16. c)** PR, $-\frac{1}{2}$; QS, 2; negative reciprocals of each other **d)** yes **17. a)** 698 m **b)** The 760 m slope is steeper since the vertical change over the same distance is greater.

Learning Together pp.523–524

Activity 1: **1.** No **2.** 10 s **3.** 21 m, 9 m **4.** 4 s, 6 s **5.** at times equidistant from 5 s; 3 s and 7 s for example **6.** 25 m **7.** 10 s **Activity 2:** **1. a)** 0, 4, 8, 12, 16, 20, 24 **b)** 0, 1, 4, 9, 16, 25, 36 **2.** $y = 4x$ **3.** Yes; the highest power of any term is 1. **4.** $y = x^2$ **5.** No, it is a second degree relation. **Activity 3:** **2.** 6, 24, 54, 96 **3. a)** 0, 1, 8, 27, 64 **b)** 0, 6, 24, 54, 96 **4.** $y = 6x^2$ **5.** No, it is a second degree relation. **6.** $y = x^3$ **7.** No, it is a third degree relation.

Learning Together p.525

1. AB = 6, CD = 4, EF = 6, GH = 4 **2.** the difference in the unequal coordinates of the endpoints **3.** no; lengths are nonnegative. **4.** AB, (3, 7); CD, (2, 3); EF, (8, 4); GH, (12, 4) **5.** The x-coordinate of the

midpoint is the average of the x-coordinates of the endpoints and the y-coordinate of the midpoint is the average of the y-coordinates of the endpoints.
Activity 2: **3.** A(2, 2), B(4, 6) and (4, 2) or (2, 6); C(5, 2), D(9, 4) and (9, 2) or (5, 4) **4.** 2, 4; 4, 2 **6.** 4.5 for both **7.** AB, (3, 4); CD, (7, 3) **8.** See activity 1, question 5. **Activity 3:** **2.** RS = 4.2, MN = 4.2, WX = 3.6, JK = 6.3

Review pp.528–529

1. 1, 2, 3 **2.** 9, 8, 7 **3.** −1, 0, 1 **4.** 3, 5, 7 **5.** (−1, −5), (0, −1), (1, 3) **6.** (−1, 3), (0, 1), (1, −1) **7.** (−1, −1), (0, 0), (1, 1) **8.** (−1, −2), (0, 0), (1, 2) **9.** (−1, −6), (0, −1), (1, 4) **10.** (−1, 1), (0, 0), (1, −1) **11.** A(−6, 5), B(−4, 2), C(−2, 0), D(−1, 3), E(0, 1), F(1, 4), G(3, 1), H(4, 3), I(5, 2), J(6, 0), K(1, −1), L(4, −2), M(2, −3), N(3, −4), P(−1, −2), Q(−2, −3), R(−4, −4), S(−5, −2), T(−3, −1), W(0, −4) **12. b)** D(2, −2) **c)** 8 by 5 **13. a)** triangle **b)** parallelogram **c)** square **14. b)** (2, 0), (5, −3) **c)** 2 **15. a)** (0, 2), (−2, 0), (1, 3), (−1, 1), (2, 4) **b)** (0, −2), (2, 0), (1, −1), (−1, −3), (3, 1) **c)** (0, 6), (6, 0), (1, 5), (2, 4), (−1, 7) **d)** (0, −4), (4, 0), (5, 1), (1, −3), (3, −1) **17. a)** $y = x + 2$ **b)** $y = 3x$ **18. a)** (0, 3), (3, 0) **b)** (0, 1), (−1, 0) **c)** (0, 2), (3, 0) **d)** (0, 2), (−5, 0) **19. a)** (0, 0) **b)** (0, −2) **20. a)** (−2, 7), (−1, 6), (0, 5), (1, 4), (2, 3); (−2, 5), (−1, 6), (0, 7), (1, 8), (2, 9) **c)** (−1, 6) **21. a)** 100 km **b)** 0.25 h **22. a)** 2.5 m **b)** 17.5 m **23. a)** 1 **b)** −1 **c)** 0 **d)** vertical **e)** −2 **f)** $\frac{3}{2}$ **24.** Both have slope 1. **25. b)** No; it is a second degree relation.

Chapter Check p.530

1. a) 4, 3, 0, 1 **b)** 1, 2, 3, 4 **c)** 0, 1, 2, 3 **d)** 7, 4, 1, −2 **2. a)** (−1, −2), (0, 1), (1, 4) **b)** (−1, 2), (0, 1), (1, 0) **c)** (−1, 7), (0, 0), (1, −7) **d)** (−1, −2), (0, −1), (1, 0) **e)** (−1, −3), (0, 1), (1, 5) **f)** (−1, 2), (0, 3), (1, 4) **3.** A(−6, 3), B(−5, 1), C(−3, 2), D(−2, 0), E(0, 2), F(1, 0), G(3, 2), H(5, 1), I(4, −1), J(3, −2), K(0, −1), L(−3, −2), M(−5, −4), N(−6, −1), P(−4, −1) **4. a)** rectangle **b)** 8 by 2 **c)** AB and DC have a slope of 0; AD and BC are vertical and so the slope is undefined. **5.** (0, −1), (1, 2), (2, 5), (−1, −4), (−2, −7) **6. a)** $y = x + 3$ **b)** $y = -5x$ **7.** (0, 5), (−3, 0) **8. a)** 10.5 m **b)** 1.25 s **9.** (4, 1) **10. a)** $\frac{6}{5}$ **b)** $(0, \frac{17}{5})$, $(1, \frac{23}{5})$ **11. a)** for (−2, 3) and (−1, 0): slope = −3 for (−1, 0) and (0, −1): slope = −1 **b)** No, the points do not fall on the same line, since different pairs give different slopes.

Using the Strategies p.531

1. a) 4 × 5 − 12 **b)** 14 + 3 − 7 **c)** 6 + 7 + 8 ÷ 2 **d)** 11 + 6 + 5 − 3 **2. a)** 120 **b)** 24 **c)** 6 **d)** 2 **3.** 7 **5.** 17 **6.** Yes; begin by guessing 512; if it is too high, guess 512 ÷ 2 = 256; if it is too low, guess 512 + 512 ÷ 2 = 768. Depending on the results of your second guess, the third guess will be either 768 + 256 ÷ 2 = 896, 768 − 256 ÷ 2 = 640, 256 + 256 ÷ 2 = 384 or 256 ÷ 2 = 128. Continue in this fashion. **7.** $51.25 **8.** 69 **9.** Tuesday

Data Bank

1. a) 6.656 m^3 **b)** 1.9 m

Chapter 14

Getting Started pp.534–535

Activity 2: **1. a)** flip **b)** turn **c)** slide **Activity 3:** **1.** $\frac{1}{4}$ turn clockwise **2.** vertical flip **3.** horizontal flip **4.** $\frac{1}{4}$ turn clockwise, horizontal flip **5.** $\frac{1}{2}$ turn clockwise **6.** slide **7.** $\frac{1}{4}$ turn clockwise, vertical flip **8.** $\frac{1}{4}$ turn counterclockwise **Activity 4:** **1. a)** flip **b)** turn **c)** slide **d)** turn **2. a)** turn **b)** turn **c)** flip **d)** slide **3. a)** slide **b)** turn **c)** turn **d)** flip **4. a)** flip **b)** slide **c)** turn **d)** turn **5. a)** turn **b)** flip **c)** turn **d)** slide **Mental Math** **1.** 22 **2.** 20 **3.** 3 **4.** 4.5 **5.** 0.02 **6.** 60 000 **7.** 3 **8.** 1 **9.** $8.98 **10.** $7.75 **11.** $8.96 **12.** $10.94 **13.** $8.85 **14.** $3.44 **15.** $4.90 **16.** $8.20 **17.** −4 **18.** 2 **19.** 20 **20.** −12 **21.** −11 **22.** −4 **23.** −6 **24.** 10 **25.** $5.98 **26.** $11.92 **27.** $4.95 **28.** $59.94 **29.** $15.80 **30.** $29.88 **31.** $8.85 **32.** $9.90 **33.** $\frac{1}{6}$ **34.** $\frac{7}{8}$ **35.** $\frac{3}{5}$ **36.** $\frac{3}{10}$ **37.** 24 **38.** 21 **39.** 15 **40.** 0 **41.** 97 **42.** 56 **43.** 77 **44.** 29 **45.** 222 **46.** 69

Section 14.1 pp.537–538

Practice **1.** A, 6 units right; B, 2 units right and 5 units down; D, 6 units right and 4 units down; F, 10 units right and 7 units down. **3. a)** [2, 0] **b)** [4, 4] **c)** [1, −3] **12.** [4, −5] **13.** [−4, −3] **14.** 3 units to the right and 2 units up **15.** 1 unit to the left and 4 units up **16.** 2 units to the left and 3 units down **17.** 5 units to the right and 1 unit down **18.** 6 units up **19.** 3 units to the left **20.** 2 units to the right and 3 units down **21.** 3 units to the left and 5 units up **22.** 4 units to the left and 2 units up **23.** 1 unit to the right and 6 units down **24.** (5, 7) **25.** (−1, −3) **26.** (4, 3) **27.** (−7, 1) **28.** [4, 4] **29.** [1, −1] **30.** [−1, −1] **31.** [4, −5] **32.** [−5, 8] **33.** [−1, −6] **34.** A′(−1, 4), B′(−5, 6), C′(−7, −3) **35.** D′(1, −3), E′(7, −4), F′(5, −8) **36.** R′(2, 3), S′(−2, 3), T′(−1, −2) **Problems and Applications** **37.** A(1, 1), B(−2, 4), C(−4, −4) **38. b)** R′(7, 2), S′(3, 7), T′(1, 3) **c)** R″(8, −3), S″(4, 2), T″(2, −2) **d)** [5, −3]

Section 14.2 pp.540–541

Practice **1.** A, vertical reflection; B, horizontal reflection **8.** A′(2, −4), B′(1, −1), C′(6, −2) **9.** D′(0, −3), E′(5, −4), F′(2, 0) **10.** R′(−2, 1), S′(1, 4), T′(3, 2) **11.** P′(1, −2), Q′(−3, 2), R′(3, 1) **12.** A′(−1, 3), B′(−2, 1), C′(−6, 3) **13.** D′(−1, 2), E′(0, −2), F′(−3, 1) **14.** P′(2, 1), Q′(3, −3), R′(−1, −2) **15.** R′(−2, 2), S′(3, −1), T′(−2, −2)

16. a) (1, −4), (−1, 4) **b)** (2, −3), (−2, 3) **c)** (−1, 2), (1, −2) **d)** (−3, 2), (3, −2) **e)** (−3, −2), (3, 2) **f)** (4, 0), (−4, 0) **17.** A′(−1, 1), B′(−5, 2), C′(−3, 6) **18.** R′(2, −5), S′(−2, −4), T′(−1, 2) **Problems and Applications** **19. a)** A, H, I, O, T, V, W, X, Y **b)** B, C, D, E, H, I, O, X **c)** H, I, O, X **d)** 0, 3, 8 **20.** A′(−7, 4), B′(−5, 1), C′(−8, 1) **21.** R′(2, −2), S′(−2, 0), T′(4, 1) **22.** P′(0, 3), Q′(4, −1), R′(−1, −2) **23.** D′(−1, −5), E′(−3, −1), F′(3, −1) **24.** D′(1, −7), E′(−2, 0), F′(5, −3) **25. b)** P′(5, 6), Q′(7, 2), R′(1, 1) **c)** P″(5, −6), Q″(7, −2), R″(1, −1) **26. a)** x-coordinate **b)** y-coordinate

Section 14.3 pp.543–544

Practice **1.** 90° ccw **2.** 180° cw **3.** 270° ccw **4.** 90° cw, 270° ccw **5.** 180° cw, 180° ccw **6.** 90° cw, 270° ccw **19.** D′(0, 0), E′(−2, −4), F′(−6, 0) **20.** A′(0, −1), B′(−4, −3), C′(0, −3) **24.** (−6, 0), (0, −6), (6, 0) **25.** (0, −4), (4, 0), (0, 4) **26.** (−3, 0), (0, 3), (3, 0) **27.** (−4, 3), (−3, −4), (4, −3) **28.** (−3, −4), (4, −3), (3, 4) **Problems and Applications** **29.** A′(−1, 0), B′(−1, −2), C′(−6, −2), D′(−6, 0) **30.** A′(0, 1), B′(2, 2), C′(2, 6), D′(0, 5) **31. a)** P′(1, 0), Q′(3, 0), R′(3, −5), S′(1, −5) **b)** P′(0, −1), Q′(2, −1), R′(2, −6), S′(0, −6) **32.** P′(6, 5), Q′(9, 3), R′(9, −2), S′(6, 0); parallelogram **33.** a, b and d **34.** 90° **35. a)** parallel **b)** equal **c)** The line $y = x$ passes through the turn centre. **36. a)** 450° cw **b)** 270° ccw **37.** the North Star

Section 14.4 pp.546–547

Practice **1.** 2 **2.** $\frac{1}{3}$ **5.** A′(6, 4), B′(2, 8) **6.** C′(3, 2), D′(−1, 1) **7.** E′(−3, −3), F′(3, 6) **8.** G′(3, 1), H′(−2, 0) **9.** R′(6, 9), S′(−3, 12), T′(−9, −6) **10.** D′(3, 2), E′(−1, 3), F′(−2, −2), G′(2, −3) **Problems and Applications** **13. b)** 8 **c)** P′(−6, 6), Q′(−6, −6), R′(6, −6) **d)** 72 **e)** P″(−1, 1), Q″(−1, −1), R″(1, −1) **f)** 2 **g)** 72:8 or 9:1; 2:8 or 1:4 **h)** the square of the scale factor **15.** scale factor: 2 **16.** scale factor: 2 **17.** scale factor: 3 **18.** identical **19. b)** A′(4, 6), B′(−5, −6), C′(16, −9)

Technology pp.548–549

Activity 1: **1.** square **2.** equilateral triangle **3.** four lines emanating from a common point, consecutive lines are at an angle of 30° **Activity 2:** **1.** square **2.** four lines of length 60 units enclosing a square of side length 40 units **3.** rectangle **Activity 3:** three consecutive rectangles, joined along a side **Activity 4:** three consecutive squares, separated by 30 units

Section 14.5 pp.552–553

Practice **5.** $d = 14$ cm, $f = 8$ cm **6.** $r = 30$ cm, $s = 6$ cm **7.** $b = 7.5$ cm, $w = 6$ cm **8.** $p = 6.75$ cm, $r = 7.5$ cm **9.** $d = 12.5$ cm, $e = 15$ cm **10.** 8 cm **11.** 9 cm **Problems and Applications** **12.** 37.5 m **13.** 260 m **14.** 233 m **15.** $x = 15$ cm, $y = 20$ cm, $z = 4.8$ cm **16. a)** yes **b)** no; ratio of heights may not equal ratio of widths

Section 14.6 p.555

Practice **1.** 2 **2.** 1 **3.** 4 **4.** 5 **5.** 3 **6.** 4 **7.** 2 **8.** 5 **Problems and Applications** **9. a)** A, H, I, O, T, V, W, X, Y **b)** B, C, D, E, H, I, O, X **c)** H, I, O, X **d)** H, I **10. a)** 2 **b)** 0 **c)** 5 **d)** 7

Learning Together p.556

Activity 1: **1. a)** horizontal stretch **b)** vertical stretch **2.** always a kite **3.** no; may be rectangular **Activity 2:** **a)** vertical stretch and a reflection in the y-axis **b)** horizontal stretch and a reflection in the x-axis **Activity 3:** **a)** horizontal contraction and vertical stretch **b)** horizontal stretch and vertical contraction

Review pp.560–561

1. [4, 2] **2.** [−4, 1] **3.** [−4, −2] **4.** [2, −4] **5.** 2 units to the right and 3 units up **6.** 3 units to the left and 1 unit down **7.** 4 units to the left and 5 units up **8.** 6 units up **9.** [1, −2] **10.** [−4, 6] **11.** [2, −3] **12.** A′(7, 5), B′(8, 2), C′(2, 4) **13.** R′(−5, 1), I′(0, −2), K′(−1, −5) **14.** (4, −5), (−4, 5) **15.** (5, 2), (−5, −2) **16.** (−3, −6), (3, 6) **17.** (−3, 2), (3, −2) **22.** D′(2, −3), E′(5, −1), F′(4, −6) **23.** R′(−3, −1), S′(3, 2), T′(4, −3) **28.** D′(−3, 0), E′(−3, 4), F′(0, 4) **29.** A′(0, −3), B′(5, −3), C′(0, −6) **30.** A′(4, 6), B′(10, 0) **31.** C′(−3, 6), D′(9, −9) **32.** E′(2, 3), F′(−3, −1) **33.** G′(−2, 0), H′(0, 2) **34.** P′(6, 4), Q′(6, −2), R′(4, −4) **35.** A′(−3, 1), B′(−2, −3), C′(1, −4), D′(3, 5) **36.** $e = 12.5$ cm, $f = 7.5$ cm **37.** $q = 4$ cm, $w = 12.5$ cm **38.** 4 **39.** 3 **40.** 2 **41.** 3 **42.** 2 **43.** 5

Chapter Check p.561

1. A′(3, 4), B′(5, 1), C′(8, 2) **2.** P′(−2, 2), Q′(−4, 5), R′(−7, 1) **3.** R′(−1, −1), S′(−3, −4), T′(−6, −3) **4.** D′(−2, 4), E′(4, 1), F′(1, −3) **5.** J′(4, 0), K′(3, −5), L′(0, 3) **6.** G′(0, 4), H′(3, 4), I′(2, 0) **7.** S′(0, 2), T′(−6, 6), U′(8, −6) **8.** $a = 10$ cm, $b = 8$ cm **9.** $w = 10.5$ cm, $r = 10$ cm **10.** 2 **11.** 4 **12.** 2 **13.** 5

Using the Strategies p.563

1. 4 **2.** 221, 442 **3.** 11, 12, 13 **4.** forty **5.** 112 cm **7.** 7 others **8.** 1, 3, 5, 7, 19, 39, 199, $2n - 1$ **9.** $0.95, $9.05; $1.85, $8.15; $3.65, $6.35; $4.55, $5.45

Data Bank

1. 158 km

Glossary

A

Acute Angle An angle whose measure is less than 90°.

Acute Triangle A triangle with all 3 angles less than 90°.

Alternate Angles Two angles formed by two lines and a transversal. The angles are on opposite sides of the transversal in a Z or Ƨ pattern.

Angle The figure formed by 2 rays or 2 line segments with a common endpoint, called the vertex.

Angle Bisector A line that divides an angle into 2 equal parts.

Arc A part of the circumference of a circle.

Average The mean of a set of numbers, found by dividing the sum of the numbers by the number of numbers.

Axes The intersecting number lines on a graph.

B

Bar Graph A graph that uses bars to represent data visually.

Base (of a power) The number used as a factor for repeated multiplication. In 6^3, the base is 6.

Binary System The number system that consists of the two digits 0 and 1.

Binomial A polynomial with 2 terms.

Broken-line Graph A graph that represents data, using line segments joined end-to-end.

C

Capacity The greatest volume that a container can hold, usually measured in litres or millilitres.

Census A survey in which data are collected from every member of a population.

Central Angle An angle subtended by an arc of a circle and with its vertex at the centre.

Chord A line segment that joins 2 points on the circumference of a circle.

Circle Graph A graph that uses sectors of a circle to show how data are divided into parts.

Circumference The perimeter of a circle.

Common Denominator A number that is a common multiple of the denominators of a set of fractions. The common denominator of $\frac{1}{2}$ and $\frac{1}{3}$ is 6.

Common Factor A number that is a factor of two or more numbers. The CF of 4 and 6 is 2.

Common Multiple A number that is a multiple of two or more numbers. CMs of 6 and 10 are 30, 60, 90, ...

Complementary Angles Two angles whose sum is 90°.

Co-interior Angles Two angles formed by two lines and a tranversal. Co-interior angles are on the same side of the transversal in a ⊏ or ⊐ pattern.

Composite Number A number that has more than 2 different factors. $6 = 1 \times 2 \times 3$

Congruent Figures Figures with the same size and shape.

Coordinates An ordered pair, (x, y), that locates a point on a coordinate plane.

Corresponding Angles Angles that have the same relative positions in geometric figures.

Corresponding Angles (from 2 lines and a transversal) Two angles that form an F, ꟻ, Ⅎ, or ⅃ pattern.

Corresponding Sides Sides that have the same relative positions in geometric figures.

D

Database An organized and sorted list of information.

Degree (measure of an angle) The unit for measuring angles. $1° = \frac{1}{360}$ of a complete turn.

Degree of a Monomial The sum of the exponents of the variables.

Degree of a Polynomial The largest sum of the exponents of the variables in any one term of the polynomial.

Dependent Events Events whose outcomes influence each other.

Diagonal A line segment joining 2 nonadjacent vertices in a polygon.

Diameter A chord that passes through the centre of a circle.

Dilatation A transformation that changes the size of an object.

E

Edge The straight line formed where 2 faces of a polyhedron meet.

Equation A number sentence that contains the symbol =.

Equilateral Triangle A triangle with all sides equal.

Equivalent Fractions Fractions such as $\frac{1}{3}$, $\frac{2}{6}$, and $\frac{3}{9}$ that represent the same part of a whole or group.

Equivalent Ratios Ratios such as 1:3, 2:6, and 3:9 that represent the same fractional number or amount.

Event Any possible outcome of an experiment in probability.

Expanded Form The way in which numbers are written to show the total value of each digit. $235 = 2 \times 100 + 3 \times 10 + 5 \times 1$

Exponent The raised number used in a power to show the number of repeated multiplications of the base. In 4^2, the exponent is 2.

Exponential Form A shorthand method for writing numbers expressed as repeated multiplications. $81 = 3 \times 3 \times 3 \times 3 = 3^4$

Expression A mathematical phrase made up of numbers and variables, connected by operators.

Exterior Angle of a Polygon An angle formed by one extended side of a polygon and the other side at the same vertex.

F

Face A plane surface of a polyhedron.

Factors The numbers multiplied to give a specific product.

Factor Tree A diagram used to factor a number into its prime factors.

Flow Chart An organized diagram that displays the steps in a problem's solution.

Frequency The number of times an item or event occurs.

Frequency Table A table that uses tallies to count data.

G

Greatest Common Factor (GCF) The largest factor that two or more numbers have in common. The GCF of 8, 12, and 24 is 4.

H

Histogram A type of bar graph used to summarize and display a large set of data.

Hypotenuse The side opposite the right angle in a right triangle.

I

Image The figure produced by a transformation.

Included Angle An angle whose rays contain 2 sides of a triangle.

Independent Events Events whose outcomes do not influence each other.

Inequality The statement that one expression is greater than, less than, or not equal to another expression.

Inscribed Angle An angle subtended by an arc of a circle and with its vertex on the circumference.

Inscribed Polygon A polygon with its vertices on a circle.

Integers Numbers in the sequence ..., −3, −2, −1, 0, 1, 2, 3, ...

Intersecting Lines Two lines that cross each other at 1 point.

Irrational Number A real number that cannot be expressed in the form $\frac{a}{b}$ where a and b are integers and $b \neq 0$.

Isosceles Triangle A triangle with 2 equal sides.

L

Like Terms Terms, such as x and $4x$, with the same variable raised to the same exponent.

Line of Symmetry A mirror line that reflects an object onto itself.

Line Segment A part of a line. A line segment has 2 endpoints.

Lowest Common Denominator (LCD) The lowest multiple shared by two or more denominators. The LCD of $\frac{1}{8}$ and $\frac{1}{6}$ is 24.

Lowest Common Multiple (LCM) The lowest multiple shared by two or more numbers. The LCM of 5, 10, and 15 is 30.

Lowest Terms A way of writing a fraction so that the numerator and denominator have no common factors other than 1.

M

Mapping A correspondence of points between an object and its image.

Mean The sum of the numbers divided by the number of numbers in a set.

Median The middle number in a set of numbers arranged in order. If there is an even number of numbers, the median is the average of the 2 middle numbers.

Midpoint The point that divides a line segment into 2 equal parts.

Mode The number that occurs most frequently in a set of data. In 1, 2, 2, 6, 6, 6, the mode is 6.

Monomial A number, a variable, or a product of numbers and variables.

N

Net A pattern used to construct a polyhedron.

Non-terminating Decimal A decimal that continues without end, such as 0.33333....

O

Obtuse Angle An angle whose measure is more than 90° but less than 180°.

Obtuse Triangle A triangle with 1 obtuse angle.

Opposite Angles The equal angles formed by 2 intersecting lines.

Order of Operations The rules to be followed when simplifying expressions: **B**rackets, **E**xponents, **D**ivision and **M**ultiplication, **A**ddition and **S**ubtraction.

Ordered Pair A pair of numbers, (x, y), indicating the x and y-coordinates of a point on a graph.

Origin The intersection of the horizontal and vertical axes on a graph. The origin has the coordinates $(0, 0)$.

Outcome The result of an experiment.

P

Parallel Lines Lines in the same plane that never meet.

Parallelogram A quadrilateral with opposite sides parallel and equal in length.

Percent A fraction or ratio in which the denominator is 100.

Periods Groups of 3 digits separated by spaces in numbers with more than 4 digits. There are 3 periods in 123 456 789.

Perpendicular Bisector The line that cuts a line segment into 2 equal parts at right angles.

Perpendicular Lines Two lines that intersect at a 90° angle.

Perspective The different views of an object—top, bottom, side, front.

Pi (π) The quotient that results when the circumference of a circle is divided by its diameter.

Pictograph A graph that uses pictures or symbols to represent similar data.

Polygon A closed figure formed by 3 or more line segments.

Polyhedron A 3-dimensional figure with polygons as faces.

Polynomial A monomial or the sum of monomials.

Population The entire set of items from which data can be taken.

Power A number, such as 3^4, written in exponential form.

Prime Number A number with exactly 2 different factors, 1 and itself. $3 = 1 \times 3$

Principal Square Root The positive square root of a number.

Prism A polyhedron with 2 parallel and congruent bases in the shape of polygons. The other faces are parallelograms.

Probability The ratio of the number of ways an outcome can occur to the total number of possible outcomes.

Proportion An equation, such as $\frac{3}{4} = \frac{6}{8}$, which states that 2 ratios are equal.

Pyramid A polyhedron with 1 base and the same number of triangular faces as there are sides on the base.

Pythagorean Theorem The area of the square drawn on the hypotenuse of a right triangle is equal to the sum of the areas of the squares drawn on the other 2 sides.

Q

Quadrant One of the 4 regions formed by the intersection of the x-axis and y-axis.

R

Random Sample A sample in which each member of the population has an equal chance of being selected.

Range The difference between the highest and lowest in a set of numbers.

Rate A comparison of 2 measurements with different units, such as $\frac{9 \text{ m}}{2 \text{ s}}$.

Ratio A comparison of 2 numbers, such as 4:5 or $\frac{4}{5}$.

Rational Expression An expression that can be written as the quotient of 2 polynomials.

Rational Number A number that can be expressed as the ratio of 2 integers.

Ray A part of a line. A ray contains 1 endpoint.

Real Numbers All the rational and irrational numbers.

Reciprocals Two numbers whose product is 1.

Reflection A flip transformation of an object in a mirror line or reflection line.

Reflex Angle An angle whose measure is more than 180° but less than 360°.

Regular Polygon A closed figure with all sides equal and all angles equal.

Regular Polyhedron A 3-dimensional figure with identical regular polygons as faces.

Repeating Decimal A decimal in which a digit or digits repeat without end.
$\frac{9}{11} = 0.818181\ldots$ or $0.\overline{81}$

Relation A set of ordered pairs.

Rhombus A quadrilateral with opposite sides parallel and all 4 sides equal.

Right Angle An angle whose measure is 90°.

Right Bisector A line that divides another line into 2 congruent segments at a 90° angle.

Right Triangle A triangle with 1 right angle.

Rise The difference in the values of the y-coordinates of a pair of points.

Rotation A turn transformation of an object about a fixed point or turn centre.

Rotational Symmetry A figure has rotational symmetry if it maps onto itself more than once in a complete turn.

Run The difference in the values of the x-coordinates of a pair of points.

S

Sample A selection from a population.

Sample Space The possible outcomes from an experiment.

Scale Drawing An accurate drawing that is either an enlargement or a reduction of an actual object.

Scalene Triangle A triangle with no sides equal.

Similar Figures Figures that have the same shape but not always the same size.

Simplest Form The form of a fraction whose numerator and denominator have no common factors other than 1.

Slope of a Line The quotient $\frac{\text{rise}}{\text{run}}$ for any 2 points on the line.

Square Root of a Number The number that multiplies itself to give the number.

Stem-and-Leaf Plot Information tabulated so that the last digit is the leaf, and the digit or digits in front of it are the stem.

Straight Angle An angle whose measure is 180°.

Supplementary Angles Two angles whose sum is 180°.

Surface Area The sum of the areas of the faces of a 3-dimensional figure.

Survey A sampling of information.

T

Terminating Decimal A decimal, such as 3.154, whose digits terminate.

Tessellation A repeated pattern of geometric figures that will completely cover a surface.

Tetrahedron A polyhedron with 4 triangular faces.

Transformation A mapping of the points of a plane onto the points of the same plane.

Translation A slide transformation of an object.

Transversal A line that intersects 2 lines in the same plane in 2 distinct points.

Trapezoid A quadrilateral with exactly 2 parallel sides.

Tree Diagram A diagram that shows the possible outcomes of consecutive events.

Trinomial A polynomial with 3 terms.

U

Uniform Motion Movement at a constant speed.

Unit Rate A comparison of 2 measurements in which the second term is 1, such as $\frac{3\text{ m}}{1\text{ s}}$ or 3 m/s.

Unlike Terms Terms, such as $2x$, $3y$, and $2z$, with different variables or with the same variable raised to different exponents.

V

Variable A letter or symbol used to represent a number.

Vertex The common endpoint of 2 rays or line segments.

W

Whole Numbers Numbers in the sequence 0, 1, 2, 3, 4, 5, ...

X

***x*-axis** The horizontal number line in the Cartesian coordinate system.

Y

***y*-axis** The vertical number line in the Cartesian coordinate system.

Index

D

E

Q

R

S

Text Credits

42 Reprinted by permission of United Feature Syndicate, **82** *Newsweek*, Special Issue, Fall/Winter 1991, Hamilton, **86** BENT OFFERINGS by Don Addis. By permission of Don Addis and Creators Syndicate, **124** Reprinted by permission of United Feature Syndicate, **158** Reprinted by permission: Tribune Media Services, **200** Reprinted by permission–The Toronto Star Syndicate: Tribune Media Services., **266** THE FAR SIDE copyright 1987 FarWorks, Inc. Dist. by UNIVERSAL PRESS SYNDICATE. Reprinted with permission. All rights reserved., **300** BENT OFFERINGS by Don Addis. By permission of Don Addis and Creators Syndicate, **332** Reprinted by permission of the cartoonist, Warren Clements, **333 top left** Gallup Canada, Toronto, Ontario, **333 middle left** Gallup Canada, Toronto, Ontario, **351** Ontario Ministry of Natural Resources, **358** Canadian Weather Trivia Calendar, 1988. Reprinted with the permission of the Minister of Supply and Services Canada, 1993., **360** Gallup Canada, Inc., May 3, 1993, **392** From *Literary Lapses* by Stephen Leacock. Used by permission of the Canadian Publishers, McClelland & Stewart, Toronto., **396** Reprinted by permission: Tribune Media Services, **562** Reprinted with permission–The Toronto Star Syndicate

Photo Credits

xiii, xiv Ian Crysler, **xv** J.A. Kraulis/Masterfile, **xvii top** Damir Frkovic/Masterfile, **xvii bottom** Mike Dobel/Masterfile, **xix** Animals Animals/Donald Specker, **xx** J.A. Kraulis/Masterfile, **xxii** Dan Paul, **xxiii** Joe Lepiano, **xxvi** Telegraph Colour Library/Masterfile, **4** Dan Paul, TI 108 Instructional Calculator is reprinted by permission of Texas Instruments, Richmond Hill, Ontario, **7** Joe Lepiano, **8** Dan Paul, **12** Royal Canadian Mint, **14** Canadian Sport Images, **16** Ministry of Tourism, British Columbia, **18, 22** Dan Paul, **26, 27** Ian Crysler, **28** Mike Barrett photo/Courtesy The Liberal (Richmond Hill), **30, 32** Air Canada, **33** Dan Paul, **36** Ian Crysler, **48 bottom, 60** Dan Paul, **67** Ian Crysler, **83** Photography by Dick Fish and Donald Chrisman., Cat Scan courtesy of the Cooley-Dickinson Hospital. Drawing by Lynn Thompson. Photography by Dick Fish., **92** Dan Paul, **93 bottom** The Bettmann Archive, **94** Alexandre Lesiège, **98 top** Joe Lepiano, **100 top** Greg Franklin, **100 bottom** Lynn Holden, **101** IMS Creative Communications, **102** © Bill Ivy, **106–107** Ian Crysler, **110 bottom** Dan Paul, **111** Ian Crysler, **114** ATS Aerospace, **120, 121** Dan Paul, **126–127** Tom Ives/Masterfile, **130, 135** Toronto Stock Exchange, **142–143** Dan Paul, **147** Canadian Sport Images/Mike Ridewood, **154** Dan Paul, **155** Joe Lepiano, **159** Colin Erricson, **168** The Canadian Academy of Recording Arts and Sciences (CARAS), **170 top** Dan Paul, **173** Canadian Sport Images/Greg Kinch, **176–177, 180** Dan Paul, **182, 186–187** Ian Crysler, **184** Marineland, Niagara Falls, Canada, **192** Canapress Photo Service/Hans Deryk, **196** NASA, **197 bottom** Dan Paul, **204** Colin Erricson, **206** RCMP–GRC/89–798, **208** Canapress Photo Service/Dave Buston, **209** Colin Erricson, **211** Canadian Amateur Hockey Association Archives, **214–215** Ian Crysler, **220** Young Su Choung, **229** Joe Lepiano, **230** NASA, **240–241** Richard Steedman/Masterfile, **244–245** Dale Wilson/First Light, **246** Canadian Coast Guard, **249** Joe Lepiano, **254–255** Ian Crysler, **260** Dan Paul, **262–263** Ian Crysler, **268–269** NASA, **272–273** Ian Crysler, **274** John de Visser/Masterfile, **275** Emmanuel Carcano, **277** Jack Chow Insurance, **279** Dan Paul, **284** Ian Crysler, **290 top** Dan Paul, **292** Toronto Blue Jays, **304–305** Daryl Benson/Masterfile, **308** Dan Paul, **309** Toronto Blue Jays, **310 top, centre and bottom** Nova Scotia Tourism & Culture, **311** Public Information Office, House of Commons/Service d'information publique, Chambre des communes, **315** Ian Crysler, **316** Dan Paul, **320** Canadian Sport Images/Ted Grant, **324, 325** Dan Paul, **326** Canadian Sport Images/Ted Grant, **332 top left** Paul Joe, Environment Canada, **332 top right** Joe Lepiano, **332 bottom, 332–333 centre** Dan Paul, **333 top right** Ministry of Tourism, British Columbia, **333 bottom** David Madison/Tony Stone Images, **334** Canapress Photo Service, **336** Dofasco Inc., **337 top** Dan Paul, **338** Animals Animals/Jack Wilburn, **340** Animals Animals/Alfred B. Thomas, **341** Dan Paul, **343** Reproduction rights courtesy of Boshkung, Inc. and Mill Pond Press, Inc., **345** Joe Lepiano, **347** Toronto Dominion Bank Archives, **349** Warren Galloway, **350–351** Wayne Lynch/Masterfile, **356–357** Derek Caron/Masterfile, **358** Paul Joe, Environment Canada, **360** Ian Crysler, **372** Colin Erricson, **374** Canapress Photo Service, **376** Canapress Photo Service/Tom Hanson, **378** NASA, **379** Joe Lepiano, **384, 388, 390–391** Colin Erricson, **393** Courtesy of Stephen Leacock Museum, Orillia, Ontario, Canada, **398–399** Paul Chesley/Tony Stone Images, **400** Dan Paul, **402, 404** Joe Lepiano, **408** Doug MacLellan/Hockey Hall of Fame, **412** Courtney Milne, **416** Dan Paul, **421** Ian Crysler, **422** Courtesy of Hydro–Quebec, **424–425** Dan Paul, **426** Ministry of Tourism, British Columbia, **430** Royal Ontario Museum, **431** Courtesy of Ontario Place, **432** Royal Ontario Museum, **434** Giovanni Paolo Panini, Interior Saint Peter's, Rome, Ailsa Mellon Bruce Fund, ©1993 National Gallery of Art, Washington, **440–441** Joe Lepiano, **441 bottom left and right** Dan Paul, **441 top** Environment Canada - Parks Service/Alexander Graham Bell National Historic Site, **442** Dan Paul, **444 left** Nova Scotia Tourism, Culture & Recreation, **444 bottom right** Steve Drexler/Image Bank, **444 top right** Anne Rippy/Image Bank, **445** Joe Lepiano, **448** Ministry of Tourism, British Columbia, **452** Royal Canadian Mint, **453** Ian Crysler, **456** Joe Lepiano, **462** Courtesy of Science North, Sudbury, **464–465** Dan Paul, **466** L'institut de tourisme et d'hôtellerie du Québec, **474** Denis Drever NCC/CCN, **478** Daryl Hepting, Simon Fraser University, **480** Dan Paul, **485** Ian Crysler,

486 Dan Paul, **488 top** Canadian Sport Images/F. Scott Grant, **488 bottom, 489, 498** Ian Crysler, **500** Canadian Sport Images/F. Scott Grant, **504** National Research Council Canada, **512 top and bottom** Photograph by Malak, Ottawa, **515** Dan Paul, Photo courtesy of Casio Canada Ltd., Scarborough, Ontario., **516 top** Canadian Olympic Association, **516 bottom** Canadian Sport Images/F. Scott Grant, **517** The Ontario Ministry of Culture, Tourism and Recreation © 1993, **518** Peter Tittenberger, **520** Telegraph Colour Library/Masterfile, **539** Ian Crysler, **542** Dan Paul, **550** Air Canada, **557** Ian Crysler

Illustration Credits

xvi, xxi, xxiv, xxv Jun Park, **1** Doug Martin, **6** Pronk&Associates, **10** Bernadette Lau, **13** Margo Davies Leclair/Visual Sense Illustration, **21** Kevin Ghiglione, **24** Margo Davies Leclair/Visual Sense Illustration, **44–45** Ted Nasmith, **48 top** Rose Zgodzinski, **52** Jun Park, **54** Clarence Porter, **56** Jun Park, **64** Dave McKay, **66** Mary Jane Gerber, **70** Bernadette Lau, **72** Rose Zgodzinski, **76–77 birds** Bernadette Lau, **76–77** Kevin Ghiglione, **78** Jun Park, **80** Mike Herman, **82 top** Graham Bardell, **88–89** Wesley Bates, **93 top** Jun Park, **98 bottom** Stephen MacEachern, **108** Paul Rivoche, **110 top** Margo Davies Leclair/Visual Sense Illustration, **114** Mike Herman, **116, 118** Jun Park, **123** Stephen Harris, **131, 134** Mike Herman, **148–149** Kevin Ghiglione, **150** Mike Bishop, **152** Jun Park, **162–163** Paul Rivoche, **170 bottom** Mike Herman, **174** Jun Park, **179** Kevin Ghiglione, **189** Jun Park, **190** Mike Herman, **197 top** Sarah Jane English, **202–203** Margo Stahl, **212** Jun Park, **216** Don Gauthier, **222–223** Kevin Ghiglione, **225** Stephen MacEachern, **228** Jun Park, **237** Stephen MacEachern, **248** Jun Park, **252, 256** Sarah Jane English, **285** Margo Davies Leclair/Visual Sense Illustration, **288** Bernadette Lau, **295** Kevin Ghiglione, **296** Andrew Plewes, **297, 305 bottom** Margo Davies Leclair/Visual Sense Illustration, **314, 317** Bernadette Lau, **323** Kevin Ghiglione, **327** Bernadette Lau, **328 right** Jun Park, **329** Margo Davies Leclair/Visual Sense Illustration, **330** James Laish, **333 top left, centre left** Margo Davies Leclair/Visual Sense Illustration, **337 bottom** Jun Park, **371** Andrew Plewes, **380** Margo Davies Leclair/Visual Sense Illustration, **386, 406 top, 414, 446** Jun Park, **460** Bernadette Lau, **472** Jun Park, **496–497** Tomio Nitto, **508** Stephen Harris, **510–511** Anne Stanley, **523** Margo Davies Leclair/Visual Sense Illustration, **524** Jun Park, **526** Kevin Ghiglione, **527** Jun Park, **532–533** Doug Martin, **536** Jun Park, **545, 546** Wesley Bates, **548–549** Jun Park, **553 bottom right** Margo Davies Leclair/Visual Sense Illustration, **554** Andrew Plewes, **561** Stephen MacEachern, **564–565, 567** Margo Davies Leclair/Visual Sense Illustration

Technical Art by Pronk&Associates